Universitext

Universitext

Universitext is a series of textbooks that presents material from a wide variety of mathematical disciplines at master's level and beyond. The books, often well class-tested by their author, may have an informal, personal, even experimental approach to their subject matter. Some of the most successful and established books in the series have evolved through several editions, always following the evolution of teaching curricula, into very polished texts.

Thus as research topics trickle down into graduate-level teaching, first textbooks written for new, cutting-edge courses may make their way into *Universitext*.

For further volumes:
www.springer.com/series/223

Olivier Bordellès

Arithmetic Tales

Translated by Véronique Bordellès

 Springer

Olivier Bordellès
allée de la Combe 2
Aiguilhe, France

Translator
Véronique Bordellès
allée de la Combe
Aiguilhe, France

Translation from the French language edition:
Thèmes d'Arithmétiques
by Olivier Bordellès
Copyright © 2006 Edition Marketing S.A
www.editions-ellipses.fr/
All Rights Reserved

ISSN 0172-5939   ISSN 2191-6675 (electronic)
Universitext
ISBN 978-1-4471-4095-5   ISBN 978-1-4471-4096-2 (eBook)
DOI 10.1007/978-1-4471-4096-2
Springer London Heidelberg New York Dordrecht

Library of Congress Control Number: 2012940391

Mathematics Subject Classification: 11-01, 11A, 11L07, 11M06, 11N05, 11N13, 11N25, 11N37, 11R04, 11R09, 11R11, 11R16, 11R18, 11R21, 11R27, 11R29, 11R42

Printed on acid-free paper

Springer is part of Springer Science+Business Media (www.springer.com)

Preface

"Mathematical science is the queen of sciences, and arithmetic is the queen of mathematics", Gauss said. Indeed, number theory is the study of whole numbers, also called positive integers, the first ones we learn at school. Thus, the theory of numbers deals with problems that are often both easy to understand and very hard to solve. For instance, one of the most famous number theory problems is *Fermat's last theorem*, abbreviated as FLT, stating that the *Fermat equation* $x^n + y^n = z^n$, where x, y, z are positive integers and $n \geqslant 3$ is an integer, has no solution. This proof came from Andrew Wiles in 1995, after more than 350 years of efforts from many mathematicians, such as Ernst Kummer, Sophie Germain, André Weil, Jean-Pierre Serre, Gerd Faltings, Kenneth Ribet and Yves Hellegouarch.

The author's initial aim was simply to have his book entitled *Thèmes d'Arithmétique*, published in 2006 by Ellipses eds, translated into English. But things turned out differently as what you are holding here is an extended, more complete version of the French edition. Not only have the chapters doubled in size but many exercises, all of them with complete solutions, have been added and, more importantly, the sections called *Further Developments* included in each chapter have been significantly enlarged.

Each chapter is divided into three parts. The course itself is suitable for undergraduates. As for the exercises, they either illustrate the course or are designed as springboards for approaching other related topics. Finally, the section Further Developments introduces trickier notions and even occasionally topics that researchers are familiar with. Many results are proved and whenever the proof goes beyond the scope of the book, the reader is cross-referred to the standard sources and references in the subject area. The book includes among other things an almost exhaustive exposition of the recent *discrete Hardy–Littlewood method* developed by Enrico Bombieri, Martin Huxley, Henryk Iwaniec, Charles Mozzochi and Nigel Watt, applications of Vaughan's famous identity, a historico-mathematical introduction to the class field theory together with a detailed illustration of the contribution of analytic tools to the tricky problems of algebraic number theory, such as obtaining upper bounds for class numbers or lower bounds for discriminants and regulators of algebraic number fields.

The first two chapters are intended to supply the main basic tools an undergraduate student should have a good grasp of to acquire the necessary grounding for subsequent work. The emphasis is on summation formulae such as Abel and Euler–MacLaurin summations that are unavoidable in modern number theory.

Chapter 3 is devoted to the study of prime numbers, from the beginning with Euclid's work to modern analysis relating the distribution of primes to the non-trivial zeros of the Riemann zeta-function. A fairly complete account of Chebyshev's benchmarking method is given, along with totally explicit estimates for the usual prime number functions.

Chapter 4 extends the analysis of the previous chapter by dealing with multiplicative functions. A large number of these are given, their average order most of the time being studied in detail through the Möbius inversion formula and through some basic results in summation methods. A complete study of the Dirichlet series from an arithmetic viewpoint is supplied. Furthermore, some estimates for other types of summation are investigated, such as multiplicative functions over short segments or additive functions. Finally, a brief account of Selberg's sieve and the large sieve is also given.

The study of the local law of a certain class of multiplicative functions requires counting the number of points with integer coordinates very near smooth plane curves. The aim of Chap. 5 is to provide some nice results of the theory in a very intricate, but elementary[1] way. The methods of Martin Huxley and Patrick Sargos and Michael Filaseta and Ognian Trifonov are completely investigated to show how some clever combinatorial ideas, introduced in the 1950s essentially by Heini Halberstam, Klaus Roth and Hans-Egon Richert, and in the 1970s by Sir Henry Peter Francis Swinnerton-Dyer, may lead to very good results which appear to be well beyond the scope of any current exponential-sums method.

As can be seen with the famous Dirichlet divisor problem, many questions in analytic number theory reduce to estimate certain exponential sums. Chapter 6 is devoted to the theory of such sums, following the lines of van der Corput's method, eventually leading to its A- and B-processes and, after some rearrangements by Eric Phillips, to the exponent-pairs method, systematically used nowadays. Historically, three methods were developed independently in the 1920s: among other things, Hermann Weyl treated exponential sums with polynomials, Johannes Gualtherus van der Corput extended Weyl's ideas to quasi-monomial functions combining the Poisson summation formula and the stationary phase method, and Ivan Matveevich Vinogradov's work dealt with counting the number of solutions of certain tricky Diophantine systems. This chapter could be viewed as an analytic equivalent to Chap. 5.

Finally, Chap. 7 is an introduction to algebraic number theory, which arose from both a generalization of the arithmetic in $\mathbb{Z}$ and the necessity to solve certain Diophantine equations. Although the idea of using a larger field than $\mathbb{Q}$ was already known at that time, the theory really took off in the 19th century, and among the

---

[1] Note that the word "elementary" means only that the complex analysis is not used.

founding fathers the names of Ernst Kummer, Richard Dedekind, David Hilbert, Leopold Kronecker and Hermann Minkowski may be mentioned. The chapter is aimed at showing that the *ideal numbers* were the right tool to restore unique factorization. Furthermore, the reader is invited to compare the major results, such as the fundamental theorem of ideal theory or the zero-free region of the Dedekind zeta-function, with the corresponding ones from Chap. 3.

Aiguilhe, France   Olivier Bordellès

Acknowledgements

When writing this book, I was helped in many crucial ways.

Thanks must first go to my long-suffering wife, who among other things provided invaluable help in translating the French edition into English.

It is also my delight to acknowledge the help of the following colleagues whose careful reading of the manuscript was expertly done and who suggested numerous improvements on preliminary drafts of this book: Guy Bénat, Jean-Jacques Galzin, Roger Mansuy, Bruno Martin, Landry Salle and Patrick Sargos.

Last but not least, I am gratefully indebted to the Springer staff for trustfully accepting this project. Among them I would like to thank especially Lauren Stoney for her very efficient support.

Olivier Bordellès

Notation

General Notation

$\mathbb{Z}$, $\mathbb{Q}$, $\mathbb{R}$, $\mathbb{C}$ are respectively the sets of integers, rational numbers, real numbers and complex numbers. If $a \in \mathbb{Z}$, we may also adopt the notation $\mathbb{Z}_{\geqslant a}$ of all integers $n \geqslant a$. When $a = 1$, the corresponding set is usually denoted[1] by $\mathbb{N}$. Finally, if p is a prime number, $\mathbb{F}_p$ is the finite field with p elements.

Letters n, m and a, b, c, d, k, l, r, $\dots$ refer to integers, whereas p indicates a prime number.

$a \mid b$ means a divides b, i.e. there exists $k \in \mathbb{Z}$ such that $b = ka$. Similarly, $a \nmid b$ means a does not divide b.

$a \mid b^\infty$ means $p \mid a \Longrightarrow p \mid b$.

$p^k \parallel n$ means $p^k \mid n$ and $p^{k+1} \nmid n$.

$P^+(n)$ is the greatest prime factor of $n \in \mathbb{Z}_{\geqslant 2}$, with the convention $P^+(1) = 1$. This symbol is sometimes abbreviated as $P(n)$.

(a, b) and $[a, b]$ are respectively the greatest common divisor and the least common multiple of a and b. We set the gcd and lcm of three positive integers in the same way, as for instance

$$(a, b, c) = ((a, b), c)$$

and extend this definition by induction to a finite number of positive integers.

For any positive integer n, the number $n! = 1 \times 2 \times \cdots \times n$ is the factorial of n, with the convention that $0! = 1$.

For any $x \in \mathbb{R}$, $[x]$ is the integer part[2] of x, the unique integer verifying

$$x - 1 < [x] \leqslant x$$

---

[1] This is a difference between Anglo-Saxon countries and France where the symbol $\mathbb{N}$ denotes the set of *non-negative* integers, sometimes denoted by $\mathbb{N}_0$ in the UK and the US. The aim of the notation $\mathbb{Z}_{\geqslant a}$ is then to avoid any risk of confusion, so that in this book $\mathbb{N} = \mathbb{Z}_{\geqslant 1}$ and $\mathbb{Z}_{\geqslant 0}$ is the set of non-negative integers.

[2] Some authors also use the *floor* and the *ceiling* functions, denoted respectively by $\lfloor x \rfloor$ and $\lceil x \rceil$, but there will be no need to make such a distinction in this book.

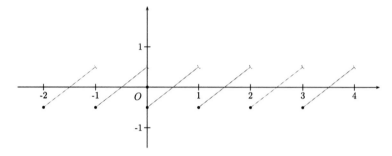

Fig. 1 Function ψ

and $\lfloor x \rceil$ is the nearest integer to x. The notation $\{x\}$ means the fractional part of x defined by $\{x\} = x - [x]$. Hence, for any $x \in \mathbb{R}$, we have

$$0 \leqslant \{x\} < 1.$$

We will also make use of the functions

$$\psi(x) = \{x\} - \frac{1}{2} \quad \text{and} \quad \psi_2(x) = \int_0^x \psi(t)\,dt = \frac{\psi(x)^2}{2} - \frac{1}{8}.$$

The function ψ, see Fig. 1, called the *first Bernoulli function*, is an odd, 1-periodic function and then admits a Fourier series development. Since $\psi_2(0) = \psi_2(1) = 0$, the function ψ_2 is also 1-periodic and then bounded. Furthermore, it is not difficult to check that

$$\left|\psi(x)\right| \leqslant \frac{1}{2} \quad \text{and} \quad -\frac{1}{8} \leqslant \psi_2(x) \leqslant 0.$$

The distance of a real number x to its nearest integer is written $\|x\|$. Hence we have

$$\|x\| = \min\left(\frac{1}{2} + \psi(x), \frac{1}{2} - \psi(x)\right).$$

$\log x$ is the natural logarithm and e^x or $\exp x$ is the exponential function. It is also convenient to define the functions $e(x) = e^{2\pi i x}$ and $e_a(x) = e(x/a) = e^{2\pi i x/a}$.

If $\mathcal{E}$ is a finite set of integers, $|\mathcal{E}|$ is the number of elements belonging to $\mathcal{E}$.

Sums and Products

If $N \geqslant 1$ is any integer, we set

$$\sum_{n=1}^N f(n) = f(1) + f(2) + \cdots + f(N)$$

whilst

$$\sum_{p \leqslant N} f(p) = f(2) + f(3) + f(5) + \cdots$$

where the latest summation runs through prime numbers $p \leqslant N$. The two sums are related thanks to the following characteristic function of primes

$$\mathbf{1}_{\mathbb{P}}(n) = 1 + \left[\frac{2 - \tau(n)}{n} \right] = \begin{cases} 1, & \text{if } n \text{ is prime} \\ 0, & \text{otherwise} \end{cases}$$

where $\tau(n)$ counts the number of positive divisors of n (see Chap. 4), so that

$$\sum_{p \leqslant N} f(p) = \sum_{n=2}^{N} \mathbf{1}_{\mathbb{P}}(n) f(n).$$

If $x \geqslant 1$ is a real number, then by convention

$$\sum_{n \leqslant x} f(n) = \sum_{n=1}^{[x]} f(n) \quad \text{and} \quad \sum_{p \leqslant x} f(p) = \sum_{p \leqslant [x]} f(p).$$

Certain sums run through some special subsets of $\mathbb{Z}$. For instance

$$\sum_{d \mid N} f(d)$$

means that the sum is taken over the positive divisors of N, e.g.

$$\sum_{d \mid 15} f(d) = f(1) + f(3) + f(5) + f(15) \quad \text{and} \quad \sum_{p \mid 15} f(p) = f(3) + f(5).$$

These examples are also valid for the products. For instance

$$\prod_{p \leqslant x} f(p) = f(2) \times f(3) \times f(5) \times \cdots$$

where the product runs through all prime numbers $p \leqslant [x]$.

It is important to note that, in all cases, the index p means a sum or a product running through prime numbers exclusively.

Functions

Let $a < b$ be real numbers and k be a non-negative integer. The notation $f \in C^k[a, b]$ means that f is a real-valued function k-times differentiable on $[a, b]$ and

$f^{(k)}$ is also continuous on $[a, b]$. By convention, $f^{(0)} = f$, $f^{(1)} = f'$, $f^{(2)} = f''$ and $f^{(3)} = f'''$.

Let $x \longmapsto f(x)$ and $x \longmapsto g(x)$ be functions defined for all sufficiently large x and $a, b > 0$.

▷ *Landau.* $f(x) = O(g(x))$, also sometimes written as $g = O(f)$, means that $g > 0$ and that there exist a real number x_0 and a constant $c_0 > 0$ such that, for all $x \geqslant x_0$, we have

$$|f(x)| \leqslant c_0 g(x).$$

For instance, $f(x) = O(1)$ means that f is bounded for all $x \in [x_0, +\infty[$.

▷ *Vinogradov.* $f(x) \ll g(x)$ is equivalent to $f(x) = O(g(x))$.

▷ *Titchmarsh.* $a \asymp b$ means that there exist $c_2 \geqslant c_1 > 0$ such that $c_1 b \leqslant a \leqslant c_2 b$. If a, b represent two functions f and g, then $f \asymp g$ is equivalent to $f \ll g$ and $g \ll f$.

▷ *Landau.* $f(x) = o(g(x))$ for $x \longrightarrow x_0$ means that $g \neq 0$ and

$$\lim_{x \to x_0} \frac{f(x)}{g(x)} = 0.$$

▷ *Landau.* $f(x) \sim g(x)$ for $x \longrightarrow x_0$ means that $g \neq 0$ and

$$\lim_{x \to x_0} \frac{f(x)}{g(x)} = 1.$$

An *asymptotic estimate* for the function f is a relation of the shape $f(x) \sim g(x)$. An *asymptotic formula* for f means a relation of the form $f(x) = g(x) + O(r(x))$, or equivalently $f(x) - g(x) = O(r(x))$, where $g(x)$ is called the *main term* and $O(r(x))$ is an *error term*. Obviously, such a relation is only meaningful if the error term $r(x)$ is of smaller order than $g(x)$. Otherwise, this relation is equivalent to $f(x) = O(r(x))$, so that the estimate is only an *upper bound*.

It is important to understand the difference between $f \asymp g$ and $f \sim g$. The first relation is less precise than the second one but can be used in a larger range. For instance, the Chebyshev estimates from Corollary 3.45 assert that

$$\pi(x) \asymp \frac{x}{\log x}$$

for all $x \geqslant 5$ while the Prime Number Theorem (Theorem 3.85), which was proved some forty years later, implies that

$$\pi(x) \sim \frac{x}{\log x}$$

as soon as $x \longrightarrow \infty$.

Finally, it should be mentioned that the constants implied in some error terms of the form $f(x) \ll g(x)$ depend sometimes on extra parameters. For instance, it is proven in Exercise 2 in Chap. 3 that

$$\tau(n) \ll n^\varepsilon$$

where the implied constant depends on $\varepsilon > 0$. This means that, for all $\varepsilon > 0$, there exists a constant $c(\varepsilon) > 0$ depending on $\varepsilon > 0$ such that, for all $n \geqslant 1$, we have $\tau(n) \leqslant c(\varepsilon)n^{\varepsilon}$. Such a situation[3] is sometimes denoted by

$$\tau(n) \ll_{\varepsilon} n^{\varepsilon}.$$

---

[3] It can be shown that $c(\varepsilon) = \exp(2^{1/\varepsilon} / \log 2^{\varepsilon})$ is admissible.

Contents

1 Basic Tools 1
  1.1 Euclidean Division 1
  1.2 Binomial Coefficients 3
  1.3 Integer and Fractional Parts 4
  1.4 Rolle, Mean Values and Divided Differences 6
  1.5 Partial Summation 8
  1.6 Harmonic Numbers 11
  1.7 Further Developments 14
    1.7.1 The Riemann–Stieltjes Integral 14
    1.7.2 The Euler–MacLaurin Summation Formula 19
  1.8 Exercises 22
  References 25

2 Bézout and Gauss 27
  2.1 Bachet–Bézout's Theorem 27
  2.2 The Euclidean Algorithm 29
  2.3 Gauss's Theorem 31
  2.4 Linear Diophantine Equations 34
  2.5 Congruences 35
  2.6 Further Developments 41
    2.6.1 The Ring $(\mathbb{Z}/n\mathbb{Z}, +, \times)$ 41
    2.6.2 Denumerants 43
    2.6.3 Generating Functions 47
  2.7 Exercises 51
  References 55

3 Prime Numbers 57
  3.1 The Fundamental Theorem of Arithmetic 58
  3.2 Euclid's Theorem 64
  3.3 Fermat, Lagrange and Wilson 67
    3.3.1 Important Tools for Primes 67
    3.3.2 Multiplicative Order 70

	3.3.3	Primitive Roots and Artin's Conjecture	72
	3.3.4	Some Applications of Primitive Roots	76
3.4	Elementary Prime Numbers Estimates		78
	3.4.1	Chebyshev's Functions of Primes	78
	3.4.2	Chebyshev's Estimates	81
	3.4.3	An Alternative Approach	86
	3.4.4	Mertens' Theorems	87
3.5	The Riemann Zeta-Function		92
	3.5.1	Euler, Dirichlet and Riemann	92
	3.5.2	The Gamma and Theta Functions	94
	3.5.3	Functional Equation	96
	3.5.4	Estimates for $|\zeta(s)|$	99
	3.5.5	A Zero-Free Region	102
3.6	Prime Numbers in Arithmetic Progressions		105
	3.6.1	Euclid vs Euler	105
	3.6.2	Dirichlet Characters	108
	3.6.3	Dirichlet L-Functions	111
	3.6.4	The Convergence of the Series $\sum_p \chi(p)p^{-1}$	112
	3.6.5	The Non-vanishing of $L(1, \chi)$	114
3.7	Further Developments		118
	3.7.1	Sieves	118
	3.7.2	Other Approximate Functional Equations for $\zeta(s)$	124
	3.7.3	The Prime Number Theorem	126
	3.7.4	The Riemann–von Mangoldt Formula and the Density Hypothesis	132
	3.7.5	Explicit Formula	133
	3.7.6	The Prime Number Theorem for Arithmetic Progressions	138
	3.7.7	Explicit Estimates	141
	3.7.8	The Piatetski-Shapiro Prime Number Theorem	145
	3.7.9	The Riemann Hypothesis	146
	3.7.10	Some Consequences of the Riemann Hypothesis	149
	3.7.11	The Mean Square of the Riemann Zeta-Function	153
	3.7.12	Additive Characters and Gauss Sums	155
	3.7.13	Incomplete Character Sums	156
3.8	Exercises		158
References			161
4	**Arithmetic Functions**		165
4.1	Definition and Fundamental Examples		165
4.2	Additive and Multiplicative Functions		167
4.3	The Dirichlet Convolution Product		171
4.4	The Möbius Inversion Formula		176
4.5	Summation Methods		180
4.6	Tools for Average Orders		186
	4.6.1	Introduction	186

	4.6.2	Auxiliary Lemmas	190
	4.6.3	The Proof of Theorem 4.22	191
	4.6.4	A Second Theorem	193
4.7	Further Developments		195
	4.7.1	The Ring of Arithmetic Functions	195
	4.7.2	Dirichlet Series—The Formal Viewpoint	196
	4.7.3	Dirichlet Series—Absolute Convergence	198
	4.7.4	Dirichlet Series—Conditional Convergence	201
	4.7.5	Dirichlet Series—Analytic Properties	203
	4.7.6	Dirichlet Series—Multiplicative Aspects	209
	4.7.7	The Von Mangoldt Function of an Arithmetic Function	212
	4.7.8	Twisted Sums with the Möbius Function	214
	4.7.9	Mean Values of Multiplicative Functions	215
	4.7.10	Lower Bounds	220
	4.7.11	Short Sums of Multiplicative Functions	222
	4.7.12	Sums of Sub-multiplicative Functions	225
	4.7.13	Sums of Additive Functions	226
	4.7.14	The Selberg's Sieve	227
	4.7.15	The Large Sieve	234
4.8	Exercises		240
References			247

5 Integer Points Close to Smooth Curves ... 249

5.1	Introduction		249
	5.1.1	Squarefree Numbers in Short Intervals	249
	5.1.2	Definitions and Notation	252
	5.1.3	Basic Lemma in the Squarefree Number Problem	253
	5.1.4	Srinivasan's Optimization Lemma	256
5.2	Criteria for Integer Points		256
	5.2.1	The First Derivative Test	256
	5.2.2	The Second Derivative Test	259
	5.2.3	The kth Derivative Test	262
5.3	The Theorem of Huxley and Sargos		264
	5.3.1	Preparatory Lemmas	265
	5.3.2	Major Arcs	267
	5.3.3	The Proof of Theorem 5.12	273
	5.3.4	Application	274
	5.3.5	Refinements	275
5.4	Further Developments		281
	5.4.1	The Method of Filaseta and Trifonov—Introduction	281
	5.4.2	The Method of Filaseta and Trifonov—The Basic Result	282
	5.4.3	The Method of Filaseta and Trifonov—Higher Divided Differences	283
	5.4.4	The Method of Filaseta and Trifonov—Epilog	286
	5.4.5	The Method of Filaseta and Trifonov—Application	288
	5.4.6	The Method of Filaseta and Trifonov—Generalization	289

      5.4.7   Counting Integer Points on Smooth Curves 290
   5.5 Exercises 291
   References 295

6 Exponential Sums 297
   6.1 Introduction 297
   6.2 Kusmin–Landau's Inequality 301
   6.3 Van der Corput's Inequality 304
   6.4 The Third Derivative Theorem 308
      6.4.1   Weyl's Shift 308
      6.4.2   Van der Corput's A-Process 310
      6.4.3   Main Results 311
   6.5 Applications 315
   6.6 Further Developments 316
      6.6.1   The mth Derivative Theorem 316
      6.6.2   Van der Corput's B-Process 317
      6.6.3   Exponent Pairs 321
      6.6.4   An Improved Third Derivative Theorem 325
      6.6.5   Double Exponential Sums 327
      6.6.6   The Discrete Hardy–Littlewood Method 328
      6.6.7   Vinogradov's Method 334
      6.6.8   Vaughan's Identity and Twisted Exponential Sums 340
      6.6.9   Explicit Estimates for $\Delta(x)$ 349
   6.7 Exercises 350
   References 351

7 Algebraic Number Fields 355
   7.1 Introduction 355
   7.2 Algebraic Numbers 356
      7.2.1   Rings and Fields 356
      7.2.2   Modules 363
      7.2.3   Field Extensions 367
      7.2.4   Tools for Polynomials 368
      7.2.5   Algebraic Numbers 374
      7.2.6   The Ring $\mathcal{O}_{\mathbb{K}}$ 379
      7.2.7   Integral Bases 382
      7.2.8   Tools for $\mathcal{O}_{\mathbb{K}}$ 386
      7.2.9   Examples of Integral Bases 390
      7.2.10 Units and Regulators 399
   7.3 Ideal Theory 403
      7.3.1   Arithmetic Properties of Ideals 403
      7.3.2   Fractional Ideals 405
      7.3.3   The Fundamental Theorem of Ideal Theory 408
      7.3.4   Consequences of the Fundamental Theorem 410
      7.3.5   Norm of an Ideal 411
      7.3.6   Factorization of (p) 415

	7.3.7	Prime Ideal Decomposition in Quadratic Fields	420
	7.3.8	The Class Group	422
	7.3.9	The PARI/GP System	427
7.4		Multiplicative Aspects of the Ideal Theory	428
	7.4.1	The Function $\nu_{\mathbb{K}}$	428
	7.4.2	The Dedekind Zeta-Function	431
	7.4.3	Application to the Class Number	434
	7.4.4	Lower Bounds for $|d_{\mathbb{K}}|$	436
	7.4.5	The Dedekind Zeta-Function of a Quadratic Field	439
7.5		Further Developments	440
	7.5.1	Euler Polynomials and Gauss Class Number Problems	440
	7.5.2	The Brauer–Siegel Theorem	443
	7.5.3	Computations of Galois Groups	446
	7.5.4	The Prime Ideal Theorem and the Ideal Theorem	451
	7.5.5	Abelian Extensions and the Kronecker–Weber Theorem	453
	7.5.6	Class Field Theory over $\mathbb{Q}$	455
	7.5.7	The Class Number Formula for Abelian Extensions	460
	7.5.8	Primes of the Form $x^2 + ny^2$—Particular Cases	462
	7.5.9	Primes of the Form $x^2 + ny^2$—General Case	467
	7.5.10	Analytic Methods for Ideal Classes	470
	7.5.11	Lower Bounds for the Regulator	473
7.6		Exercises	477
		References	479
Appendix		**Hints and Answers to Exercises**	483
A.1		Chapter 1	483
A.2		Chapter 2	486
A.3		Chapter 3	494
A.4		Chapter 4	503
A.5		Chapter 5	522
A.6		Chapter 6	531
A.7		Chapter 7	537
		References	546
Index			549

Chapter 1
Basic Tools

1.1 Euclidean Division

Some results depend on the following axiom.

Axiom 1.1 *Any non-empty subset S of $\mathbb{Z}_{\geqslant 0}$ contains a smallest element. Furthermore, if S is upper bounded, then it contains also a greatest element.*

This result will enable us to get the Euclidean division between two non-negative integers, and thus to study the arithmetic properties of integers.

Theorem 1.2 *Given non-negative integers a and b with $b \geqslant 1$, there exists a unique couple (q, r) of natural numbers such that*

$$a = bq + r \quad and \quad 0 \leqslant r < b.$$

q is the quotient and r is the remainder obtained when b is divided into a.

Proof Let S be the set defined by

$$S = \{a - nb : n \in \mathbb{Z} \text{ and } a - nb \geqslant 0\}.$$

The set S is clearly a subset of $\mathbb{Z}_{\geqslant 0}$ and $S \neq \varnothing$ since $a \in S$. Using Axiom 1.1 we infer that S contains a smallest element denoted by r. Thus, r is a non-negative integer and we call q the integer satisfying $r = a - bq$.

Let us show that $r < b$. If $r = a - bq \geqslant b$, then $a - b(q + 1) \geqslant 0$ so that $a - b(q+1) \in S$, and therefore $r = a - bq \leqslant a - b(q+1)$ since r is the smallest element in S. The latest inequality easily gives $1 \leqslant 0$ which is obviously impossible. We thus proved that $r < b$.

To show the uniqueness, suppose there exists another pair (q', r') of integers such that $q \neq q'$, $r \neq r'$ and $a = bq' + r'$ with $0 \leqslant r' < b$. Since $a = bq + r$, we deduce that $b(q' - q) = r - r'$, and then $b|q' - q| = |r - r'|$, and thus $b \leqslant |r - r'|$ since $q \neq q'$ implies that $|q' - q| \geqslant 1$. But the inequalities $0 \leqslant r < b$ and $0 \leqslant r' < b$ imply that $|r - r'| < b$ giving a contradiction. The proof is complete. □

O. Bordellès, *Arithmetic Tales*, Universitext,
DOI 10.1007/978-1-4471-4096-2_1, © Springer-Verlag London 2012

Remark 1.3 One can compute q and r. In $\mathbb{Q}$, we have

$$\frac{a}{b} = q + \frac{r}{b}$$

and the inequalities $0 \leqslant \frac{r}{b} < 1$ imply $q \leqslant \frac{a}{b} < q + 1$ so that

$$q = \left[\frac{a}{b}\right] \quad \text{and} \quad r = a - bq = a - b\left[\frac{a}{b}\right].$$

Remark 1.4 There exists a version of the Euclidean division in $\mathbb{Z}$. The result is similar to that of Theorem 1.2, except that the condition $0 \leqslant r < b$ must be replaced by $0 \leqslant r < |b|$. We leave the details to the reader.

Remark 1.5 The particular case $r = 0$ is interesting in itself. We will say that b divides a denoted by $b \mid a$. Thus, $b \mid a$ is equivalent to the existence of an integer q such that $a = bq$. Recall that one of the most important properties is the following

$$\begin{cases} a \mid b \\ a \mid c \end{cases} \implies a \mid (ax + by) \quad (x, y \in \mathbb{Z}).$$

Lemma 1.6 *Let $a, b \in \mathbb{C}$ and n be a positive integer. Then*

$$a^n - b^n = (a - b)\sum_{k=0}^{n-1} a^k b^{n-k-1}.$$

In particular, if $a, b \in \mathbb{Z}$, then $(a - b) \mid (a^n - b^n)$.

Proof Indeed, the right-hand side is equal to

$$\sum_{k=0}^{n-1} a^{k+1} b^{n-k-1} - \sum_{k=0}^{n-1} a^k b^{n-k} = a^n + \sum_{k=1}^{n-1} a^k b^{n-k} - \sum_{k=1}^{n-1} a^k b^{n-k} - b^n = a^n - b^n$$

as required.   □

Proposition 1.7 *For all $|x| < 1$ we have*

$$\sum_{k=0}^{\infty} x^k = \frac{1}{1 - x}. \tag{1.1}$$

Proof This is well known and left to the reader.   □

1.2 Binomial Coefficients

This subject is well known and proofs and examples can easily be found in any book of combinatorial theory. We only recall here the main properties required.

Definition 1.8 Let $n \in \mathbb{N}$ and $k \in \{0, \ldots, n\}$. The binomial coefficient $\binom{n}{k}$ is defined by the formula

$$\binom{n}{k} = \frac{n!}{k!(n-k)!}.$$

Together with well-known results such as *Newton's formula*

$$(a+b)^n = \sum_{k=0}^{n} \binom{n}{k} a^{n-k} b^k$$

valid for all $a, b \in \mathbb{C}$, it may be useful to have at our disposal some basic estimates for binomial coefficients.

Proposition 1.9 *Let $n \geqslant 2$ be an integer.*

(i) *For all integers $k \in \{1, \ldots, n\}$, we have*

$$\frac{n^k}{k^k} \leqslant \binom{n}{k} \leqslant \frac{n^k}{k!}.$$

(ii) *For all integers $k \in \{1, \ldots, n-1\}$, we have*

$$e^{-1/8} \sqrt{\frac{2}{\pi n}} \times \frac{n^n}{k^k (n-k)^{n-k}} \leqslant \binom{n}{k} \leqslant \frac{1}{\sqrt{\pi}} \times \frac{n^n}{k^k (n-k)^{n-k}}.$$

(iii) *We have*

$$e^{-1/8} \frac{4^n}{\sqrt{\pi n}} \leqslant \binom{2n}{n} \leqslant \frac{4^n}{\sqrt{\pi n}}.$$

Proof

▷ The first bounds follow easily from

$$\binom{n}{k} = \frac{1}{k!} \prod_{j=0}^{k-1} (n-j) \leqslant \frac{n^k}{k!} \quad \text{and} \quad \binom{n}{k} = \prod_{j=0}^{k-1} \left(\frac{n-j}{k-j}\right) \geqslant \left(\frac{n}{k}\right)^k.$$

▷ The remaining inequalities are immediate consequences of the following Stirling type estimates [Bee69]

$$\sqrt{2\pi n} \left(\frac{n}{e}\right)^n e^{\frac{1}{12n+1}} \leqslant n! \leqslant \sqrt{2\pi n} \left(\frac{n}{e}\right)^n e^{\frac{1}{12n}} \tag{1.2}$$

valid for all $n \in \mathbb{N}$. □

Remark 1.10 With a little more work, it can be proved that

$$e^{-1/8}\sqrt{\frac{n}{2\pi k(n-k)}} \times \frac{n^n}{k^k(n-k)^{n-k}} \leqslant \binom{n}{k} \leqslant \sqrt{\frac{n}{2\pi k(n-k)}} \times \frac{n^n}{k^k(n-k)^{n-k}}$$

for all integers $k \in \{1, \ldots, n-1\}$, which is indeed stronger than Proposition 1.9(ii) since

$$\sqrt{\frac{2}{\pi n}} \leqslant \sqrt{\frac{n}{2\pi k(n-k)}} \leqslant \frac{1}{\sqrt{\pi}}.$$

1.3 Integer and Fractional Parts

Proposition 1.11 *Let $x, y \in \mathbb{R}$. The following assertions hold.*

(i) $[x] = x + O(1)$. *More precisely, one can write $x = [x] + \theta$ with $\theta \in [0, 1[$.*
(ii) *Let $n \in \mathbb{Z}$. Then $[x + n] = [x] + n$ and $\{x + n\} = \{x\}$.*
(iii) $[x] + [y] \leqslant [x + y] \leqslant [x] + [y] + 1$.
(iv) *Suppose $x \geqslant 0$. Then*

$$\sum_{n \leqslant x} 1 = [x].$$

(v) *Let $d \in \mathbb{N}$ and suppose $x \geqslant 0$. Then $[\frac{x}{d}]$ is the number of multiples of d which are not greater than x.*
(vi) *Let $a < b$ be real numbers. Then the number of integers in the interval $[a, b]$ is*

$$[b - a] \quad or \quad [b - a] + 1.$$

(vii) *Let $0 \leqslant \delta < \frac{1}{2}$ be any small real number. Then*

$$\begin{cases} [x + \delta] - [x] = 1 & \Longleftrightarrow \quad \{x\} \geqslant 1 - \delta, \\ [x] - [x - \delta] = 1 & \Longleftrightarrow \quad \{x\} < \delta. \end{cases}$$

We deduce that

$$[x + \delta] - [x - \delta] = \begin{cases} 1, & \text{if } \|x\| < \delta, \\ 0, & \text{if } \|x\| > \delta, \\ 0 \text{ or } 1, & \text{if } \|x\| = \delta. \end{cases}$$

Proof

(i) It is sufficient to note that $x = [x] + \{x\}$ with $0 \leqslant \{x\} < 1$.
(ii) Since $[x + n] = x + n + \theta_1$ and $[x] = x + \theta_2$ with $-1 < \theta_i \leqslant 0$, we have

$$[x + n] - ([x] + n) = \theta_1 - \theta_2$$

so that $|[x + n] - ([x] + n)| < 1$ and we conclude by noting that $[x + n] - ([x] + n) \in \mathbb{Z}$. For the second equality, we have

$$\{x + n\} = x + n - [x + n] = x - [x] = \{x\}.$$

(iii) Using (ii) we have on the one hand

$$[x] + [y] = [[x] + y] \leqslant [x + y].$$

On the other hand, if $x = [x] + \theta_1$ and $y = [y] + \theta_2$ with $0 \leqslant \theta_i < 1$, then we have

$$[x + y] = [[x] + [y] + \theta_1 + \theta_2] = [x] + [y] + [\theta_1 + \theta_2] \leqslant [x] + [y] + 1$$

since $0 \leqslant \theta_1 + \theta_2 < 2$ implies $[\theta_1 + \theta_2] = 0$ or 1.

(iv) $\sum_{n \leqslant x} 1 = [x]$ if $x \geqslant 1$ by convention. If $0 \leqslant x < 1$, then $\sum_{n \leqslant x} 1 = 0$.

(v) If $0 \leqslant x < 1$, then there is no multiple of d which is $\leqslant x$ and $[x/d] = 0$ in this case. Now suppose $x \geqslant 1$. An integer $m \leqslant x$ is a multiple of d if and only if there exists a positive integer k such that $m = kd \leqslant x$. Therefore we must count all integers k between 1 and x/d. Using (iv) we conclude that this number is

$$\sum_{n \leqslant x/d} 1 = \left[\frac{x}{d}\right].$$

(vi) The number of integers in $[a, b]$ is

$$\sum_{a \leqslant n \leqslant b} 1 = \sum_{n \leqslant b} 1 - \sum_{n < a} 1 = \begin{cases} [b] - [a], & \text{if } a \notin \mathbb{Z}, \\ [b] - a + 1, & \text{if } a \in \mathbb{Z}. \end{cases}$$

Now using (ii), if $a \in \mathbb{Z}$ then $[b] - a + 1 = [b - a] + 1$. If $a \notin \mathbb{Z}$, let us set $a = [a] + \theta_1, b = [b] + \theta_2$ and $b - a = [b - a] + \theta_3$ with $0 \leqslant \theta_1 < 1$ and $\theta_1 \neq 0$. We thus get

$$[b - a] - ([b] - [a]) = \theta_2 - \theta_1 - \theta_3 \in]-2, 1[$$

and since $[b - a] - ([b] - [a]) \in \mathbb{Z}$ we conclude that $[b - a] - ([b] - [a]) \in \{-1, 0\}$.

(vii) Using (iii) we get

$$0 \leqslant [x + \delta] - [x] \leqslant \delta + 1 < \frac{3}{2}$$

and then $[x + \delta] - [x] \in \{0, 1\}$. Furthermore, $[x + \delta] - [x] = 1$ if and only if there exists an integer n such that $x < n \leqslant x + \delta$, and in this case we have $n = [x + \delta]$, which is equivalent to

$$\{x\} = x - [x] = x - (n - 1) \geqslant x - (x + \delta - 1) = 1 - \delta.$$

The computations for $[x] - [x - \delta]$ are similar. Finally, if $\|x\| < \delta$, then either $\{x\} < \delta$ or $\{x\} > 1 - \delta$ and thus

$$[x + \delta] - [x - \delta] = [x + \delta] - [x] + [x] - [x - \delta] = 1.$$

In the same way, if $\|x\| > \delta$, then $\delta < \{x\} < 1 - \delta$ so that $[x + \delta] - [x - \delta] = 0$. If $\|x\| = \delta$, then either $\{x\} = \delta$ or $\{x\} = 1 - \delta$ and thus $[x + \delta] - [x - \delta] = 0$ in the first case and $[x + \delta] - [x - \delta] = 1$ in the second case.

The proof is complete.   □

1.4 Rolle, Mean Values and Divided Differences

The following result will be used in Chap. 5.

Theorem 1.12

(i) (Rolle 1) *Let $x_0 < x_1$ and $F \in C^1[x_0, x_1]$ such that $F(x_0) = F(x_1)$. Then there exists a real number $t \in]x_0, x_1[$ such that $F'(t) = 0$.*

(ii) (Mean-value theorem) *Let $x_0 < x_1$ and $f \in C^1[x_0, x_1]$. Then there exists a real number $t \in]x_0, x_1[$ such that*

$$f(x_1) - f(x_0) = f'(t)(x_1 - x_0).$$

(iii) (Rolle 2) *Let k be a positive integer, $x_0 < x_1 < \cdots < x_k$ be real numbers and $F \in C^k[x_0, x_k]$ such that $F(x_0) = F(x_1) = \cdots = F(x_k)$. Then there exists a real number $t \in]x_0, x_k[$ such that $F^{(k)}(t) = 0$.*

(iv) (Divided differences) *Let k be a positive integer, $x_0 < x_1 < \cdots < x_k$ be real numbers and $f \in C^k[x_0, x_k]$. Set $\mathcal{P}(x) = b_k x^k + \cdots + b_0$ the unique polynomial of degree $\leq k$ such that $\mathcal{P}(x_i) = f(x_i)$ for $i = 0, \ldots, k$. Then there exists a real number $t \in]x_0, x_k[$ such that*

$$b_k = \frac{f^{(k)}(t)}{k!}.$$

Proof

(i) Since F is continuous on the interval $[x_0, x_1]$, we have $F([x_0, x_1]) = [m, M]$ for some real numbers m and M. If $m = M$, then F is a constant function on $[x_0, x_1]$ and then $F'(x) = 0$ for all $x \in [x_0, x_1]$. If $m < M$, and since $F(x_0) = F(x_1)$, at least one of the two extrema m or M is attained at a point $t \in]x_0, x_1[$. The function F being differentiable at this point, we then know that $F'(t) = 0$.

(ii) We use (i) with the function

$$x \longmapsto F(x) = f(x) - \frac{(x - x_0)f(x_1) - (x - x_1)f(x_0)}{x_1 - x_0}$$

which satisfies all the hypotheses with

$$F'(x) = f'(x) - \frac{f(x_1) - f(x_0)}{x_1 - x_0}$$

and $F(x_0) = F(x_1) = 0$.

(iii) We use induction on $k \geqslant 1$, the case $k = 1$ being the proposition (i). Suppose the result is true with k replaced by $k - 1$. We use (i) applied to each interval $[x_i, x_{i+1}]$ which implies that F' possesses k zeros y_i such that $x_i < y_i < x_{i+1}$ for $i = 0, \ldots, k - 1$. The induction hypothesis applied to F' implies that there exists a point $t \in]y_0, y_{k-1}[$ such that $(F')^{(k-1)}(t) = 0$. Since $]y_0, y_{k-1}[\subset]x_0, x_k[$, we then find a point $t \in]x_0, x_k[$ such that $F^{(k)}(t) = 0$.

(iv) We use (iii) applied to the function $x \longmapsto F(x) = f(x) - \mathcal{P}(x)$ by noting that $\mathcal{P}^{(k)}(x) = k! \, b_k$.

This proves the theorem. $\square$

Remark 1.13 The polynomial $\mathcal{P}$ is called a *Lagrange polynomial* and straightforward computations give

$$\mathcal{P}(x) = \sum_{j=0}^{k} \left(\prod_{\substack{i=0 \\ i \neq j}}^{k} \frac{x - x_i}{x_j - x_i} \right) f(x_j).$$

Its leading coefficient b_k is called the *divided difference* of f at the points $x_0, x_1, \ldots, x_k$ and is usually denoted by $f[x_0, x_1, \ldots, x_k]$. One can show that

$$b_k = \sum_{j=0}^{k} \frac{f(x_j)}{\prod_{0 \leqslant i \leqslant k, i \neq j}(x_j - x_i)} = \frac{A}{\prod_{0 \leqslant i < j \leqslant k}(x_j - x_i)} \tag{1.3}$$

with

$$A = \begin{vmatrix} 1 & 1 & \cdots & 1 \\ x_0 & x_1 & \cdots & x_k \\ \vdots & \vdots & \ddots & \vdots \\ x_0^{k-1} & x_1^{k-1} & \cdots & x_k^{k-1} \\ f(x_0) & f(x_1) & \cdots & f(x_k) \end{vmatrix}$$

and therefore it is important to note that $A \in \mathbb{Z}$ as soon as $x_j \in \mathbb{Z}$ and $f(x_j) \in \mathbb{Z}$.

Thus proposition (iv) shows that there exists a real number $t \in]x_0, x_k[$ such that

$$\sum_{j=0}^{k} \frac{f(x_j)}{\prod_{0 \leqslant i \leqslant k, i \neq j}(x_j - x_i)} = \frac{f^{(k)}(t)}{k!}. \tag{1.4}$$

This generalizes the mean-value theorem seen in (ii) since the left-hand side is equal to

$$\frac{f(x_1) - f(x_0)}{x_1 - x_0}$$

when $k = 1$. The case $k = 2$ can be written as

$$\frac{f(x_2)(x_1 - x_0) - f(x_1)(x_2 - x_0) + f(x_0)(x_2 - x_1)}{(x_1 - x_0)(x_2 - x_0)(x_2 - x_1)} = \frac{f''(t)}{2}.$$

1.5 Partial Summation

This is the famous integration by parts for sums.

Theorem 1.14 *Let $x \geqslant 0$ be any real number, $a \in \mathbb{Z}_{\geqslant 0}$ and let $f : [a, x] \longrightarrow \mathbb{C}$ and $g \in C^1[a, x]$.*

(i) *We have*

$$\sum_{a \leqslant n \leqslant x} f(n)g(n) = g(x) \sum_{a \leqslant n \leqslant x} f(n) - \int_a^x g'(t) \left(\sum_{a \leqslant n \leqslant t} f(n) \right) dt.$$

(ii) *If $x \geqslant 2$, then*

$$\sum_{p \leqslant x} f(p)g(p) = g(x) \sum_{p \leqslant x} f(p) - \int_2^x g'(t) \left(\sum_{p \leqslant t} f(p) \right) dt.$$

Proof

(i) For any integer $n \in [a, x]$ and any real number $t \in [a, x]$ we set

$$1_n(t) = \begin{cases} 1, & \text{if } n \leqslant t, \\ 0, & \text{otherwise.} \end{cases}$$

Then we have

$$\int_a^x 1_n(t)g'(t)\, dt = \int_n^x g'(t)\, dt = g(x) - g(n)$$

and therefore multiplying by $f(n)$ we get

$$f(n)g(n) = f(n)g(x) - \int_a^x 1_n(t) f(n)g'(t)\, dt$$

and summing over $n \in \{a, \ldots, [x]\}$ we obtain

$$\sum_{a \leqslant n \leqslant x} f(n)g(n) = g(x) \sum_{a \leqslant n \leqslant x} f(n) - \int_a^x g'(t) \left(\sum_{a \leqslant n \leqslant x} \mathbf{1}_n(t) f(n) \right) dt$$

and we easily see that

$$\sum_{a \leqslant n \leqslant x} \mathbf{1}_n(t) f(n) = \sum_{a \leqslant n \leqslant t} f(n).$$

(ii) This readily follows from (i) by using the following function

$$\mathbf{1}_{\mathbb{P}}(n) = \begin{cases} 1, & \text{if } n \text{ is a prime number,} \\ 0, & \text{otherwise.} \end{cases}$$

The formula (i) with $a = 2$ gives

$$\sum_{p \leqslant x} f(p)g(p) = g(x) \sum_{2 \leqslant n \leqslant x} \mathbf{1}_{\mathbb{P}}(n) f(n) - \int_2^x g'(t) \left(\sum_{2 \leqslant n \leqslant t} \mathbf{1}_{\mathbb{P}}(n) f(n) \right) dt$$

$$= g(x) \sum_{p \leqslant x} f(p) - \int_2^x g'(t) \left(\sum_{p \leqslant t} f(p) \right) dt$$

as asserted.

The proof is complete. $\qquad\qquad\qquad\qquad\qquad\qquad\qquad\qquad\qquad\qquad\square$

Remark 1.15 The partial summation formula is also called the *Abel summation formula* since a proof of Theorem 1.14 can be achieved by using the following discrete version of (i) discovered by Abel. If $m < n$ are any non-negative integers and (a_k), (b_k) are any sequences of complex numbers, then we have

$$\sum_{k=m+1}^n a_k b_k = b_n \sum_{k=m}^n a_k - b_{m+1} a_m - \sum_{k=m+1}^{n-1} (b_{k+1} - b_k) \sum_{h=m}^k a_h$$

which can be proved by using the obvious identity $a_k = s_k - s_{k-1}$ where $s_j = \sum_{h=m}^j a_h$ for all $m \leqslant j \leqslant n$, so that

$$\sum_{k=m+1}^n a_k b_k = \sum_{k=m+1}^n b_k (s_k - s_{k-1}) = \sum_{k=m+1}^n b_k s_k - \sum_{k=m}^{n-1} b_{k+1} s_k$$

implying the asserted result. One can deduce from this equality the following useful bound

$$\left| \sum_{k=m+1}^n a_k b_k \right| \leqslant \left\{ 2 \max \left(|b_{m+1}|, |b_n| \right) + V_{m,n} \right\} \max_{m \leqslant k \leqslant n} \left| \sum_{h=m}^k a_h \right|$$

where

$$V_{m,n} = \sum_{k=m+1}^{n-1} |b_{k+1} - b_k|$$

and if (b_k) is a monotone sequence of positive real numbers, then

$$\left| \sum_{k=m+1}^{n} a_k b_k \right| \leqslant 2 \max(b_{m+1}, b_n) \max_{m \leqslant k \leqslant n} \left| \sum_{h=m}^{k} a_h \right|.$$

Remark 1.16 Within the context of the Riemann–Stieltjes integral (see Sect. 1.7), the partial summation is nothing but the formula of integration by parts. Indeed, using (1.5), we can write

$$\sum_{a \leqslant n \leqslant x} f(n)g(n) = f(a)g(a) + \int_a^x g(t) \, d\left(\sum_{a < n \leqslant t} f(n) \right)$$

$$= f(a)g(a) + \left[g(t) \sum_{a < n \leqslant t} f(n) \right]_a^x - \int_a^x g'(t) \left(\sum_{a < n \leqslant t} f(n) \right) dt$$

$$= g(x) \sum_{a \leqslant n \leqslant x} f(n) - \int_a^x g'(t) \left(\sum_{a \leqslant n \leqslant t} f(n) \right) dt$$

$$+ f(a)\big(g(a) - g(x)\big) + f(a)\big(g(x) - g(a)\big)$$

$$= g(x) \sum_{a \leqslant n \leqslant x} f(n) - \int_a^x g'(t) \left(\sum_{a \leqslant n \leqslant t} f(n) \right) dt$$

which is the result of Theorem 1.14.

As an example, here is a simplified version of the well-known *Stirling formula*.

Corollary 1.17 *Let $n \geqslant 2$ be an integer. Then*

$$\log(n!) = n \log n - n + 1 + R_1(n)$$

with $0 \leqslant R_1(n) < \log n$.

Proof Since

$$\log(n!) = \sum_{k=1}^{n} \log k$$

we get, using Theorem 1.14 with $f(t) = 1$, $g(t) = \log t$, $a = 1$ and $x = n$

$$\log(n!) = \log n \sum_{k=1}^{n} 1 - \int_{1}^{n} \frac{1}{t} \left(\sum_{k \leqslant t} 1 \right) dt$$

$$= n \log n - \int_{1}^{n} \frac{[t]}{t} dt$$

$$= n \log n - \int_{1}^{n} \frac{t - \{t\}}{t} dt$$

$$= n \log n - n + 1 + \int_{1}^{n} \frac{\{t\}}{t} dt$$

and using $0 \leqslant \{t\} < 1$ we obtain

$$0 \leqslant \int_{1}^{n} \frac{\{t\}}{t} dt < \int_{1}^{n} \frac{dt}{t} = \log n$$

which is the desired result. $\qquad\square$

1.6 Harmonic Numbers

Definition 1.18 Let x_0 be a real number and f be a bounded integrable function on the interval $[x_0, +\infty[$. The integral of f is said to *converge* on $[x_0, +\infty[$ if the function F defined on $[x_0, +\infty[$ by

$$F(x) = \int_{x_0}^{x} f(t) \, dt$$

has a *finite* limit as x tends to $+\infty$. In this case, the limit is denoted by

$$\int_{x_0}^{+\infty} f(t) \, dt.$$

Examples 1.19

$$\int_{1}^{+\infty} \frac{dt}{t^{\alpha}} = \frac{1}{\alpha - 1} \qquad (\alpha > 1),$$

$$\int_{e}^{+\infty} \frac{dt}{t (\log t)^{\alpha}} = \frac{1}{\alpha - 1} \qquad (\alpha > 1),$$

$$\int_{0}^{+\infty} e^{-\lambda t} \, dt = \frac{1}{\lambda} \qquad (\lambda > 0),$$

$$\int_{0}^{+\infty} e^{-(\lambda t)^2} \, dt = \frac{\sqrt{\pi}}{2\lambda} \qquad (\lambda > 0).$$

There is a complete theory of the improper integral, but this book is not meant to deal with this theory. We leave the details to the reader who may refer to any undergraduate textbook in analysis. We only mention the following rule that could be taken as an axiom.

Rule 1.20 (Riemann–Bertrand rule) *Let $x_0 \geqslant 1$ be any real number and f be a bounded integrable function on $[x_0, +\infty[$. If there exist two real numbers $C_1 > 0$ and $\alpha > 1$ such that the following inequality holds*

$$\left| f(t) \right| \leqslant \frac{C_1}{t^{\alpha}} \quad (t \geqslant x_0)$$

then the integral of f converges on $[x_0, +\infty[$. The same conclusion holds if there exist two real numbers $C_2 > 0$ and $\beta > 1$ such that

$$\left| f(t) \right| \leqslant \frac{C_2}{t (\log t)^{\beta}} \quad (t \geqslant x_0 \geqslant 2).$$

Since

$$\sum_{k=1}^{n} \frac{1}{k} \geqslant \int_{1}^{n} \frac{dt}{t} = \log n$$

the sequence (u_n) defined by

$$u_n = \sum_{k=1}^{n} \frac{1}{k} - \log n$$

is non-negative, and since

$$u_{n+1} - u_n = \frac{1}{n+1} - \log\left(1 + \frac{1}{n}\right) \leqslant 0$$

we infer that (u_n) is non-increasing, and then converges.

Definition 1.21 We call the *Euler–Mascheroni constant*, or more simply the Euler constant, the real number γ defined by

$$\gamma = \lim_{n \to +\infty} \left(\sum_{k=1}^{n} \frac{1}{k} - \log n \right).$$

This number was introduced for the first time by Euler in 1734 who used the letter C. The symbol γ, now unanimously adopted, might have been first used by Lorenzo Mascheroni in 1790. An approximation of γ is

$$\gamma \approx 0.577\,215\,664\,901\,532\ldots.$$

It is almost paradoxical to notice that although many formulae involving γ are known, we do not know much about γ itself. In particular, the problem of the irrationality of γ still remains open. However, by using a representation as a continued fraction, it has been proved that if γ is a rational number p/q, then the integer q must possess at least $242\,080$ digits. The number γ is perhaps slightly less famous than π or e, but plays an important part in number theory.

Theorem 1.22 *Let n be a positive integer. Then*

$$\sum_{k=1}^{n} \frac{1}{k} = \log n + \gamma + R_2(n)$$

with $0 \leqslant R_2(n) < \frac{1}{n}$.

Proof We use partial summation with $f(t) = 1$ and $g(t) = \frac{1}{t}$ which gives

$$\sum_{k=1}^{n} \frac{1}{k} = \frac{1}{n} \sum_{k=1}^{n} 1 + \int_1^n \frac{1}{t^2} \left(\sum_{k \leqslant t} 1 \right) dt$$

$$= 1 + \int_1^n \frac{t - \{t\}}{t^2} dt$$

$$= \log n + 1 - \int_1^n \frac{\{t\}}{t^2} dt.$$

Now since $0 \leqslant \{t\} < 1$, we have $\{t\}/t^2 \leqslant 1/t^2$ and the Riemann–Bertrand Rule 1.20 implies that the last integral converges on $[1, +\infty[$. We deduce that

$$\sum_{k=1}^{n} \frac{1}{k} - \log n = 1 - \int_1^{+\infty} \frac{\{t\}}{t^2} dt + \int_n^{+\infty} \frac{\{t\}}{t^2} dt$$

and by making $n \longrightarrow +\infty$ we get

$$\gamma = 1 - \int_1^{+\infty} \frac{\{t\}}{t^2} dt$$

which gives

$$\sum_{k=1}^{n} \frac{1}{k} - \log n = \gamma + \int_n^{+\infty} \frac{\{t\}}{t^2} dt$$

and the inequalities $0 \leqslant \{t\} < 1$ give

$$0 \leqslant \int_n^{+\infty} \frac{\{t\}}{t^2} dt < \int_n^{+\infty} \frac{dt}{t^2} = \frac{1}{n}$$

which concludes the proof.   □

The number

$$H_n = \sum_{k=1}^{n} \frac{1}{k}$$

is called *nth harmonic number* and can be encountered in many branches of mathematics. A lot of properties have been discovered for H_n, but at the present time the question of a closed formula remains open. However, we shall see in Chap. 3 that H_n is allied to the famous Riemann hypothesis through a very elegant inequality due to Lagarias.

1.7 Further Developments

1.7.1 The Riemann–Stieltjes Integral

In what follows, $a < b$ are two real numbers.

The usual Riemann integral can be generalized in the following way. Let f and g be two real-valued functions defined on the interval $[a, b]$. We denote by Δ a subdivision of $[a, b]$ by the points $a = x_0 < x_1 < \cdots < x_n = b$ and the *norm* of Δ is defined by

$$N(\Delta) = \max_{0 \leqslant k \leqslant n-1} (x_{k+1} - x_k).$$

A *tagged subdivision* of $[a, b]$ is a pair (Δ, ξ) where $\Delta = \{x_0, \ldots, x_n\}$ is a subdivision of $[a, b]$ and $\xi = \{\xi_1, \ldots, \xi_{n-1}\}$ with $\xi_k \in [x_k, x_{k+1}]$.

We call Riemann–Stieltjes sum of f with respect to g for the tagged subdivision (Δ, ξ) of $[a, b]$ the sum $S_\Delta(f, g)$ given by

$$S_\Delta(f, g) = \sum_{k=0}^{n-1} f(\xi_k)\big(g(x_{k+1}) - g(x_k)\big).$$

Definition 1.23 If the limit $\lim_{N(\Delta) \to 0} S_\Delta(f, g)$ exists independently of the manner of subdivision and of the choice of the number ξ_k, then this limit is called the *Riemann–Stieltjes integral* of f with respect to g from a to b and is denoted by

$$\int_a^b f(x) \, dg(x).$$

Note that this integral reduces to the Riemann integral if $g(x) = x$. This definition easily extends to complex-valued functions by setting

$$\int_a^b f(x) \, dg(x) = \int_a^b \operatorname{Re} f(x) \, d\big(\operatorname{Re} g(x)\big) - \int_a^b \operatorname{Im} f(x) \, d\big(\operatorname{Im} g(x)\big)$$
$$+ i\left(\int_a^b \operatorname{Re} f(x) \, d\big(\operatorname{Im} g(x)\big) + \int_a^b \operatorname{Im} f(x) \, d\big(\operatorname{Re} g(x)\big) \right).$$

We do not precisely determine the pairs (f, g) for which the Riemann–Stieltjes integral exists, and the reader interested in this subject should refer to [Wid46]. However, the following class of functions plays an important role in number theory.

Definition 1.24 A function g defined on $[a, b]$ is of *bounded variation* on $[a, b]$ if there exists $M > 0$ such that, for all subdivisions $a = x_0 < x_1 < \cdots < x_n = b$ we have

$$\sum_{k=0}^{n-1} |g(x_{k+1}) - g(x_k)| \leqslant M.$$

The smallest number M satisfying this inequality is called the *total variation* of g in $[a, b]$ and is denoted by $V_{[a,b]}(g)$ or $V_a^b(g)$.

It can be shown that a real-valued function of bounded variation on $[a, b]$ is the difference of two non-decreasing bounded functions. Furthermore, if f and g are of bounded variation on $[a, b]$, then so are $f + g$ and fg and we have

$$V_a^b(f + g) \leqslant V_a^b(f) + V_a^b(g),$$

$$V_a^b(fg) \leqslant \sup_{x \in [a,b]} |f(x)| \times V_a^b(g) + \sup_{x \in [a,b]} |g(x)| \times V_a^b(f).$$

Finally, if $g \in C^1[a, b]$, then we have

$$V_a^b(g) = \int_a^b |g'(x)|\, \mathrm{d}x.$$

A sufficient condition of the existence of the Riemann–Stieltjes integral is given by the following result.

Proposition 1.25 *If f is continuous on $[a, b]$ and if g is of bounded variation on $[a, b]$, then the integral $\int_a^b f(x)\, \mathrm{d}g(x)$ exists.*

The usual properties of the Riemann–Stieltjes integral are similar to those of the Riemann integral. We list some of them in the following proposition.

Proposition 1.26 *Let f, f_1, f_2 be continuous functions on $[a, b]$ and g, g_1, g_2 be functions of bounded variation on $[a, b]$.*

(i) *For all constants α_1, α_2, β_1, β_2, we have*

$$\int_a^b \{\alpha_1 f_1(x) + \alpha_2 f_2(x)\}\, \mathrm{d}\{\beta_1 g_1(x) + \beta_2 g_2(x)\} = \sum_{i,j=1}^{2} \alpha_i \beta_j \int_a^b f_i(x)\, \mathrm{d}g_j(x).$$

(ii) *For any $a, b, c \in \mathbb{R}$ we have*

$$\int_a^b f(x)\, \mathrm{d}g(x) = \int_a^c f(x)\, \mathrm{d}g(x) + \int_c^b f(x)\, \mathrm{d}g(x).$$

(iii) *We have*

$$\left|\int_a^b f(x)\,dg(x)\right| \leqslant \int_a^b |f(x)|\,dV_a^x(g) \leqslant \sup_{x\in[a,b]} |f(x)| \times V_a^b(g).$$

(iv) *Suppose f_1 is of bounded variation on $[a, b]$, f_2 and g bounded on $[a, b]$. Then*

$$\left|\int_a^b f_1(x)f_2(x)\,dg(x)\right| \leqslant \left(|f_1(b)| + V_a^b(f_1)\right) \sup_{t\in[a,b]} \left|\int_a^t f_2(x)\,dg(x)\right|$$

provided that the integrals exist.

The differential $dV_a^x(g)$ is sometimes abbreviated $|dg(x)|$, so that (iii) can be rewritten in the form

$$\left|\int_a^b f(x)\,dg(x)\right| \leqslant \int_a^b |f(x)|\,|dg(x)|.$$

One of the main weaknesses of the Riemann–Stieltjes integral is that the integral

$$\int_a^b f(x)\,dg(x)$$

does not exist if f and g have a common discontinuity in $]a, b[$. Nevertheless, if f is continuous, this integral is often used in number theory to express various sums in terms of integrals. In particular, the Riemann–Stieltjes integral provides a natural context for Abel's summation seen in Sect. 1.5. More precisely, suppose $a < b$ are real numbers and f is continuous on $[a, b]$. Then we have

$$\sum_{a<n\leqslant b} f(n)g(n) = \int_a^b f(x)\,d\left(\sum_{a<n\leqslant x} g(n)\right). \tag{1.5}$$

Note that there is some freedom in the interval of integration, since the left endpoint can be any number in $[[a], [a]+1[$ and the right endpoint can be chosen in $[[b], [b]+1[$, without changing the value of the integral. However, one must be careful in choosing these endpoints of integration. For instance, we have

$$\sum_{k=1}^n \frac{1}{k} = 1 + \sum_{k=2}^n \frac{1}{k} = 1 + \int_1^n \frac{1}{x}\,d\left(\sum_{k\leqslant x} 1\right) = 1 + \int_1^n \frac{d[x]}{x}.$$

In view of (1.5), the Abel summation formula is nothing but the integration by parts for the Riemann–Stieltjes integral.

Proposition 1.27 (Integration by parts) *If f is of bounded variation and g is continuous on $]a, b[$, then the integral $\int_a^b f(x)\,dg(x)$ exists and we have*

$$\int_a^b f(x)\,dg(x) = f(b)g(b) - f(a)g(a) - \int_a^b g(x)\,df(x).$$

Proof We choose an arbitrary subdivision Δ of $[a, b]$ and numbers $\xi_k \in [x_k, x_{k+1}]$. For convenience we set $\xi_{-1} = a$ and $\xi_n = b$, so that we have a subdivision Ξ of $[a, b]$ such that

$$\xi_{-1} = a \leqslant \xi_0 \leqslant x_1 \leqslant \xi_1 \leqslant x_2 \leqslant \xi_2 \leqslant \cdots \leqslant x_{n-1} \leqslant \xi_{n-1} \leqslant x_n = \xi_n = b.$$

Since

$$\xi_{k+1} - \xi_k \leqslant \begin{cases} x_1 - a, & \text{if } k = -1, \\ x_{k+2} - x_k, & \text{if } 0 \leqslant k \leqslant n - 2, \\ b - x_{n-1}, & \text{if } k = n - 1, \end{cases}$$

we have $N(\Xi) \leqslant 2N(\Delta)$. Using Abel's summation as in Remark 1.15, we get

$$S_\Delta(f, g) = f(\xi_0)\big(g(x_1) - g(a)\big) + \sum_{k=1}^{n-1} f(\xi_k)\big(g(x_{k+1}) - g(x_k)\big)$$

$$= f(\xi_0)\big(g(x_1) - g(a)\big) + f(\xi_{n-1}) \sum_{k=1}^{n-1} \big(g(x_{k+1}) - g(x_k)\big)$$

$$- \sum_{k=1}^{n-2} \big(f(\xi_{k+1}) - f(\xi_k)\big) \sum_{j=1}^{k} \big(g(x_{j+1}) - g(x_j)\big)$$

$$= f(\xi_0)\big(g(x_1) - g(a)\big) + f(\xi_{n-1})\big(g(b) - g(x_1)\big)$$

$$- \sum_{k=1}^{n-2} \big(f(\xi_{k+1}) - f(\xi_k)\big)\big(g(x_{k+1}) - g(x_1)\big)$$

$$= f(\xi_0)\big(g(x_1) - g(a)\big) + f(\xi_{n-1})\big(g(b) - g(x_1)\big)$$

$$- \sum_{k=2}^{n-1} g(x_k)\big(f(\xi_k) - f(\xi_{k-1})\big) + g(x_1)\big(f(\xi_{n-1}) - f(\xi_1)\big)$$

$$= f(b)g(b) - f(a)g(a) - \sum_{k=0}^{n} g(x_k)\big(f(\xi_k) - f(\xi_{k-1})\big)$$

$$= f(b)g(b) - f(a)g(a) - S_\Xi(g, f).$$

Since $N(\Xi) \leqslant 2N(\Delta)$, the sum on the right-hand side tends to $\int_a^b g(x) \, df(x)$ as $N(\Delta)$ tends to 0, which concludes the proof. $\qquad\square$

The proof shows that Abel's summation and integration by parts are closely related, and more precisely that the latter depends on the former. Conversely, one can prove that Abel's summation can be recovered from Proposition 1.27 (see [MV07]). The following example is very useful.

Examples 1.28 (Harmonic numbers) For all $x \geqslant 1$, we have

$$\sum_{n \leqslant x} \frac{1}{n} = \log x + \gamma - \frac{\psi(x)}{x} + O\left(\frac{1}{x^2}\right).$$

Proof Using (1.5) and Proposition 1.27, we get

$$\sum_{n \leqslant x} \frac{1}{n} = 1 + \int_1^x \frac{d[t]}{t}$$

$$= 1 + \int_1^x \frac{dt}{t} - \int_1^x \frac{d\psi(t)}{t}$$

$$= 1 + \log x - \frac{\psi(x)}{x} + \psi(1) - \int_1^x \frac{\psi(t)}{t^2} dt$$

$$= \frac{1}{2} + \log x - \frac{\psi(x)}{x} - \int_1^x \frac{d(\psi_2(t))}{t^2}$$

$$= \frac{1}{2} + \log x - \frac{\psi(x)}{x} - \frac{\psi_2(x)}{x^2} - 2 \int_1^x \frac{\psi_2(t)}{t^3} dt.$$

Now since $|\psi_2(t)| \leqslant 1/8$, the last integral converges by Rule 1.20, so that we get

$$\sum_{n \leqslant x} \frac{1}{n} = \log x + \frac{1}{2} - 2 \int_1^\infty \frac{\psi_2(t)}{t^3} dt - \frac{\psi(x)}{x} - \frac{\psi_2(x)}{x^2} + 2 \int_x^\infty \frac{\psi_2(t)}{t^3} dt.$$

Since $\gamma = \lim_{x \to \infty} (\sum_{n \leqslant x} \frac{1}{n} - \log x)$, by letting $x \longrightarrow \infty$ we get

$$\gamma = \frac{1}{2} - 2 \int_1^\infty \frac{\psi_2(t)}{t^3} dt$$

and therefore

$$\sum_{n \leqslant x} \frac{1}{n} = \log x + \gamma - \frac{\psi(x)}{x} - \frac{\psi_2(x)}{x^2} + 2 \int_x^\infty \frac{\psi_2(t)}{t^3} dt$$

and we conclude by the estimate

$$\left| -\frac{\psi_2(x)}{x^2} + 2 \int_x^\infty \frac{\psi_2(t)}{t^3} dt \right| \leqslant \frac{1}{8x^2} + \frac{1}{4} \int_x^\infty \frac{dt}{t^3} = \frac{1}{4x^2}$$

as required.   □

1.7.2 The Euler–MacLaurin Summation Formula

Let f be a function of bounded variation on $[a, b]$. Then using (1.5) and Proposition 1.27 we get

$$\sum_{a < n \leqslant b} f(n) = \int_a^b f(x) \, d[x] = \int_a^b f(x) \, dx - \int_a^b f(x) \, d\psi(x)$$

$$= \int_a^b f(x) \, dx + f(a)\psi(a) - f(b)\psi(b) + \int_a^b \psi(x) \, df(x)$$

so that we have by Proposition 1.26 (iii)

$$\sum_{a < n \leqslant b} f(n) = \int_a^b f(x) \, dx + f(a)\psi(a) - f(b)\psi(b) + O\left(V_a^b(f)\right) \qquad (1.6)$$

and if $f \in C^1[a, b]$ then we get the following more accurate identity

$$\sum_{a < n \leqslant b} f(n) = \int_a^b f(x) \, dx + f(a)\psi(a) - f(b)\psi(b) + \int_a^b f'(x)\psi(x) \, dx \qquad (1.7)$$

which is usually called the *Euler summation formula*. If f has derivatives of higher orders, this can be generalized by integrating by parts repeatedly. The process makes some polynomials appear, often denoted by $B_k(x)$ and called the *Bernoulli polynomials*. These polynomials can be defined by induction by first setting $B_0(x) = 1$ and, for all $k \in \mathbb{N}$, by putting

$$\frac{d}{dx} B_k(x) = k B_{k-1}(x).$$

Thus, apart from the constant term, $B_k(x)$ is determined by this differential equation. The constant term, still denoted by B_k and called the kth *Bernoulli number*, is given by the additional following condition

$$\int_0^1 B_k(x) \, dx = 0 \quad (k \geqslant 1)$$

which is equivalent to $B_{k+1}(1) - B_{k+1}(0) = 0$ for all positive integers k, so that we have

$$B_k(1) = B_k(0) = B_k \quad (k \geqslant 2).$$

One can prove by induction that, for all $k \geqslant 0$, we have

$$B_k(x) = \sum_{j=0}^k \binom{k}{j} B_j x^{k-j}.$$

Table 1.1 Bernoulli numbers

k	0	1	2	4	6	8	10
B_k	1	$-\frac{1}{2}$	$\frac{1}{6}$	$-\frac{1}{30}$	$\frac{1}{42}$	$-\frac{1}{30}$	$\frac{5}{66}$

Either one of these formulae can be used to discover inductively some properties of the Bernoulli numbers and polynomials. For instance, if k is odd, then we have $B_k = 0$ for all $k \geqslant 3$ and $B_k(x) = -B_k(1-x)$ for all $k \geqslant 1$. If k is even, then $B_k(x) = B_k(1-x)$ for all $k \geqslant 0$. Table 1.1 gives the first values of B_k.

The repeated integration by parts of the Euler summation formula gives the functions $B_k(\{x\})$, usually called the *Bernoulli functions*. These functions are periodic of period 1 and, for all $k \geqslant 2$, they are continuous since $B_k(0) = B_k(1)$. Also note that

$$B_1(\{x\}) = \{x\} - \frac{1}{2} = \psi(x) \quad \text{and} \quad B_2(\{x\}) = 2\psi_2(x) + \frac{1}{6}.$$

The periodicity of the functions $B_k(\{x\})$ enables us to consider their expansions in Fourier series. When $k \geqslant 2$, the series is absolutely convergent and $B_k(\{x\})$ is continuous, so that the series converges uniformly to $B_k(\{x\})$. For $k = 1$, the function $\psi(x)$ has a jump discontinuity at the integers, but is also of bounded variation in $[0,1]$ so that the partial sums of its Fourier series converges to $\psi(x)$ when $x \notin \mathbb{Z}$. The usual computations from Fourier analysis provide the following result.

Proposition 1.29 (Fourier series expansions) *If $x \notin \mathbb{Z}$, we have*

$$\psi(x) = -\sum_{h=1}^{\infty} \frac{\sin(2\pi h x)}{\pi h}.$$

When $x \in \mathbb{Z}$, the series converges to 0. Furthermore, for all $k \in \mathbb{N}$, we have uniformly in x

$$B_{2k+1}(\{x\}) = 2(-1)^{k+1}(2k+1)! \sum_{h=1}^{\infty} \frac{\sin(2\pi h x)}{(2\pi h)^{2k+1}},$$

$$B_{2k}(\{x\}) = 2(-1)^{k+1}(2k)! \sum_{h=1}^{\infty} \frac{\cos(2\pi h x)}{(2\pi h)^{2k}}.$$

These formulae are of great use. For instance, taking $x = 0$ in the second identity above provides the following result discovered by Euler.

Proposition 1.30 (Zeta-function at even integers) *For all integers $k \geqslant 1$, we have*

$$\zeta(2k) = (-1)^{k+1} 2^{2k-1} \pi^{2k} \frac{B_{2k}}{(2k)!}$$

where

$$\zeta(\sigma) = \sum_{n=1}^{\infty} \frac{1}{n^{\sigma}} \quad (\sigma > 1).$$

Hence $\zeta(2k)$ is a rational multiple of π^{2k}. For example, $\zeta(2) = \pi^2/6$ and $\zeta(4) = \pi^4/90$. It is interesting to note that no similar formula is known for $\zeta(2k+1)$, although it has been proved that it is not a rational multiple of π^{2k+1}. One can also notice that, since $1 < \zeta(2k) < 1 + 4^{1-k}$, we have for all $k \geqslant 1$

$$\frac{(2k)!}{2^{2k-1}\pi^{2k}} < |B_{2k}| < \left(1 + 4^{1-k}\right)\frac{(2k)!}{2^{2k-1}\pi^{2k}}.$$

Starting from (1.7) and using induction and the properties of the Bernoulli polynomials seen above, we get the following important result.

Theorem 1.31 (Euler–MacLaurin summation formula) *Let $a < b$ be real numbers, $k \in \mathbb{N}$ and $f \in C^k[a, b]$. Then we have*

$$\sum_{a < n \leqslant b} f(n) = \int_a^b f(x)\,\mathrm{d}x + \sum_{j=1}^{k} \frac{(-1)^j}{j!}\left(B_j(\{b\})f^{(j-1)}(b) - B_j(\{a\})f^{(j-1)}(a)\right)$$
$$- \mathcal{R}_f(x)$$

where

$$\mathcal{R}_f(x) = \frac{(-1)^k}{k!}\int_a^b B_k(\{x\})f^{(k)}(x)\,\mathrm{d}x.$$

In practice, it can be interesting to have at our disposal some simpler versions. Along with identity (1.7), we will prove the following corollary.

Corollary 1.32 (Euler–MacLaurin summation formula of order 2) *Let $x \geqslant 1$ be a real number and $f \in C^2[1, +\infty[$. Then we have*

$$\sum_{n \leqslant x} f(n) = \int_1^x f(t)\,\mathrm{d}t + \frac{f(1)}{2} - \psi(x)f(x) + \psi_2(x)f'(x) - \int_1^x \psi_2(t)f''(t)\,\mathrm{d}t.$$

Furthermore, if $f''(t)$ has a constant sign on $[1, +\infty[$ and if $\lim_{t \to \infty} f'(t) = 0$, then there exists a constant γ_f such that

$$\sum_{n \leqslant x} f(n) = \int_1^x f(t)\,\mathrm{d}t + \gamma_f - \psi(x)f(x) + \frac{1}{2}\left\{\psi(x)^2 f'(x) + \int_x^{\infty} \psi(t)^2 f''(t)\,\mathrm{d}t\right\}.$$

Proof The first identity follows from (1.7) and an integration by parts. Indeed, by (1.7), we have

$$\sum_{n \leqslant x} f(n) = f(1) + \sum_{1 < n \leqslant x} f(n)$$

$$= f(1) + \int_1^x f(t)\,dt + \psi(1)f(1) - \psi(x)f(x) + \int_1^x f'(t)\,d\psi_2(t)$$

$$= \int_1^x f(t)\,dt + \frac{f(1)}{2} - \psi(x)f(x) + \psi_2(x)f'(x) - \psi_2(1)f'(1)$$

$$- \int_1^x \psi_2(t)f''(t)\,dt$$

which is the asserted result since $\psi_2(1) = 0$. For the second identity, suppose that $f'' \geqslant 0$ and fix a real number $\varepsilon > 0$. Since $\lim_{t \to \infty} f'(t) = 0$, there exists a real number $A = A(\varepsilon) > 0$ such that, for all $t \geqslant A$, we have $|f'(t)| \leqslant \varepsilon$. Now let $z > y \geqslant A$ be two real numbers. Since $f'' \geqslant 0$ and is continuous, we have

$$\left| \int_y^z \psi_2(t)f''(t)\,dt \right| \leqslant \frac{1}{8} \int_y^z f''(t)\,dt = \frac{f'(z) - f'(y)}{8} \leqslant \frac{\varepsilon}{4}$$

so that the integral

$$\int_1^\infty \psi_2(t)f''(t)\,dt$$

converges by Cauchy's theorem. Setting

$$\gamma_f = \frac{f(1)}{2} - \int_1^\infty \psi_2(t)f''(t)\,dt$$

and using the first identity above, we therefore get

$$\sum_{n \leqslant x} f(n) = \int_1^x f(t)\,dt + \gamma_f - \psi(x)f(x) + \psi_2(x)f'(x) + \int_x^\infty \psi_2(t)f''(t)\,dt$$

and we conclude the proof with $\psi_2(t) = \frac{\psi(t)^2}{2} - \frac{1}{8}$. The case $f'' \leqslant 0$ is similar. $\square$

1.8 Exercises

1 Let a, b be positive integers. In the Euclidean division of a by b, the quotient q and the remainder r satisfy $r \geqslant q$. Show that, in the Euclidean division of a by $b + 1$, we get the same quotient.

2 Let a, q be positive integers. We denote by $\mathcal{S}_q$ the set of positive integers b such that q is the quotient of the Euclidean division of a by b. Show that

$$\sum_{q=1}^{a} |\mathcal{S}_q| = a.$$

3 Let $m, n \in \mathbb{Z} \setminus \{0\}$. Show that

(i) $[\frac{n}{m}] \geqslant \frac{n+1}{m} - 1$.

(ii) If $m \nmid n$, then $[\frac{n}{m}] \leqslant \frac{n-1}{m}$.

(iii) Suppose $1 \leqslant m \leqslant n$. Then

$$\left[\frac{n}{m}\right] - \left[\frac{n-1}{m}\right] = \begin{cases} 0, & \text{if } m \nmid n, \\ 1, & \text{if } m \mid n. \end{cases}$$

4 Let $x \geqslant 1$ be a real number and f be any complex-valued function. Let $g \in C^1[1, x]$ be a complex-valued function. Suppose that

$$g(x) \sum_{n \leqslant x} f(n) \longrightarrow 0$$

as x tends to $+\infty$. Using partial summation, show that

$$\sum_{n=1}^{\infty} f(n)g(n) = -\int_1^{\infty} g'(t) \left(\sum_{n \leqslant t} f(n) \right) dt$$

in the sense that if either side converges, then so does the other one, to the same value. Deduce that we also have

$$\sum_{n > x} f(n)g(n) = -g(x) \sum_{n \leqslant x} f(n) - \int_x^{\infty} g'(t) \left(\sum_{n \leqslant t} f(n) \right) dt.$$

5 Let (a_n) be a sequence of complex numbers such that $|a_n| \leqslant 1$ and suppose there exists a positive real number M such that

$$\left| \sum_{k=1}^{n} a_k \right| \leqslant M \quad (n \geqslant 1).$$

Prove that

$$\left| \sum_{n=1}^{\infty} \frac{a_n}{n} \right| \leqslant \log M + 1.$$

6 If $f \in C^1[a, b]$ where $a < b \in \mathbb{R}$, then the following inequality due to Sobolev

$$\left| f(x) \right| \leqslant \frac{1}{b-a} \int_a^b \left| f(t) \right| dt + \int_a^b \left| f'(t) \right| dt$$

valid for any real number $x \in [a, b]$, plays an important part in analytic number theory where it was used by Gallagher in the theory of the so-called *large sieve* (see Lemma 4.76 and Theorem 4.77). This exercise proposes a discrete version of this inequality.

Let N be a positive integer and $a_1, \ldots, a_N$ be any complex numbers. Show that for any $n \in \{1, \ldots, N\}$ the following inequality holds

$$\left| a_n \right| \leqslant \frac{1}{N} \left| \sum_{k=1}^{N} a_k \right| + \sum_{k=1}^{N-1} \left| a_{k+1} - a_k \right|.$$

7 Let $x \geqslant 2$ be any real number, $f \in C^1[2, +\infty[$ and

$$\theta(x) = \sum_{p \leqslant x} \log p$$

be the (first) Chebyshev prime number function (see Chap. 3). Prove that

$$\sum_{p \leqslant x} f(p) = \frac{f(x)\theta(x)}{\log x} - \int_2^x \theta(t) \frac{d}{dt} \left(\frac{f(t)}{\log t} \right) dt.$$

8 In what follows, $x \geqslant 2$ is a sufficiently large real number. We define the *logarithmic integral* by

$$\mathrm{Li}(x) = \int_2^x \frac{dt}{\log t}.$$

(a) Prove that

$$\frac{x}{\log x} + \int_2^x \frac{dt}{(\log t)^2} = \mathrm{Li}(x) + \frac{2}{\log 2}.$$

(b) We set

$$\pi(x) = \sum_{p \leqslant x} 1$$

the prime number counting function (see Chap. 3). Suppose that we have at our disposal the following upper bound

$$\left| \theta(x) - x \right| \leqslant R(x)$$

where R is an integrable, non-decreasing function satisfying $\sqrt{x}\log x \leqslant R(x) <$ x for all $x \geqslant 2$. Using the result of Exercise 7, show that

$$\left|\pi(x) - \mathrm{Li}(x)\right| < \frac{5R(x)}{\log x}.$$

References

[Bee69] Beesack PR (1969) Improvements of Stirling's formula by elementary methods. Pub Elek Fak Univ Beograda 277:17–21

[MV07] Montgomery HL, Vaughan RC (2007) Multiplicative number theory Vol. I. Classical theory. Cambridge studies in advanced mathematics

[Wid46] Widder DV (1946) The Laplace transform. Princeton University Press, Princeton

Chapter 2
Bézout and Gauss

2.1 Bachet–Bézout's Theorem

Let a be a positive integer. In this section, we denote by $\mathcal{D}(a)$ (resp. $\mathcal{M}(a)$) the set of positive divisors (resp. multiples) of a.

Theorem 2.1 *Let a, b be any positive integers.*

(i) *The set $\mathcal{D}(a, b) = \mathcal{D}(a) \cap \mathcal{D}(b)$ has a greatest element d called the greatest common divisor of a and b denoted by $d = (a, b)$.*

(ii) *The set $\mathcal{M}(a, b) = \mathcal{M}(a) \cap \mathcal{M}(b)$ has a smallest element m called the least common multiple of a and b denoted by $m = [a, b]$.*

Proof We only prove (i). The set $\mathcal{D}(a, b)$ is a non-empty subset of $\mathbb{Z}_{\geqslant 0}$ since it contains 1 and is upper bounded by $\min(a, b)$. We conclude the proof by appealing to Axiom 1.1. □

Definition 2.2 Two positive integers a and b are said to be *coprime*, or *relatively prime*, if and only if $(a, b) = 1$.

Examples 2.3 $(6, 15) = 3$ and $[6, 15] = 30$. The integers 25 and 14 are coprime and $[25, 14] = 350$.

Remark 2.4 Theorem 2.1 shows at once that $(b, a) = (a, b)$ and $[b, a] = [a, b]$. We obviously have $(a, b) \mid [a, b]$. We also extend Theorem 2.1 to all integers by setting

$$(a, b) = (|a|, |b|) \quad \text{and} \quad [a, b] = [|a|, |b|]$$

so that, for any integers $a, b \in \mathbb{Z} \setminus \{0\}$, we will always have

$$(a, b) \geqslant 1 \quad \text{and} \quad [a, b] \geqslant 1.$$

Also, we set $(a, 0) = |a|$ and $[a, 0] = 0$ so that $(a, 1) = 1$, $[a, 1] = |a|$ and, if $a \mid b$ then $(a, b) = |a|$ and $[a, b] = |b|$ for all $a, b \in \mathbb{Z}$.

O. Bordellès, *Arithmetic Tales*, Universitext,
DOI 10.1007/978-1-4471-4096-2_2, © Springer-Verlag London 2012

The following result gives essential information about the gcd.

Theorem 2.5 *Let a, b be any integers and d be a positive integer. Then $d = (a, b)$ if and only if $d \mid a$, $d \mid b$ and there exist integers u, v such that $d = au + bv$.*

Proof Suppose first that $d = (a, b)$ so that obviously $d \mid a$ and $d \mid b$. Now define the set E by

$$E = \{ax + by : (x, y) \in \mathbb{Z}^2 \quad \text{and} \quad ax + by \geqslant 1\}.$$

Clearly $E \subset \mathbb{Z}_{\geqslant 0}$ and $E \neq \varnothing$ since $|a|$ or $|b|$ is an element of E. By Axiom 1.1, we infer that E contains a smallest element $\delta \geqslant 1$ and we set $(u, v) \in \mathbb{Z}^2$ such that $\delta = au + bv$. Let us prove that $\delta = d$. If we had $\delta \nmid a$, the Euclidean division of δ into a would give

$$a = \delta q + r \quad \text{and} \quad 0 < r < \delta$$

and then we would have

$$r = a - \delta q = a - q(au + bv) = a(1 - qu) + b(-qv)$$

which implies that $r \in E$, and then $r \geqslant \delta$, which contradicts the inequality $r < d$. Hence we have $\delta \mid a$, and we prove $\delta \mid b$ in a similar way. Now let c be another positive integer dividing a and b. We then have $c \mid (au + bv) = \delta$, and thus $c \leqslant \delta$, which proves that $d = \delta = au + bv$. Conversely, since $d \mid a$ and $d \mid b$, we infer that $d \leqslant (a, b)$. On the other hand, since $(a, b) \mid a$ and $(a, b) \mid b$, we have $(a, b) \mid (au + bv) = d$ which implies that $(a, b) \leqslant d$, so that $d = (a, b)$, and the proof is complete. $\square$

As a corollary we get at once the important Bachet–Bézout theorem.

Corollary 2.6 (Bachet–Bézout) *Let a, b be any integers. Then a and b are coprime integers if and only if there exist integers u, v such that $au + bv = 1$.*

Remark 2.7 The integers u and v, called *Bézout's coefficients* of a and b, are not unique since we can write

$$au + bv = a(u + kb) + b(v - ka) \quad (k \in \mathbb{Z}).$$

Theorem 2.5 provides a useful criterion for the gcd.

Corollary 2.8 *Let a, b be integers and d be a positive integer. Then*

$$d = (a, b) \quad \Longleftrightarrow \quad \begin{cases} (i) & d \mid a, \, d \mid b \\ (ii) & c \mid a, \, c \mid b \implies c \mid d. \end{cases}$$

Proof The sufficient condition is obvious. Conversely, if $d = (a, b)$, then d satisfies (i) and Theorem 2.5 provides integers u, v such that $d = au + bv$, so that if $c \mid a$ and $c \mid b$, then $c \mid (au + bv) = d$ as required. $\square$

2.2 The Euclidean Algorithm

The computation of the gcd can be achieved by using the following observation discovered by Euclid.

Theorem 2.9 *Given* $a, b, k \in \mathbb{Z} \setminus \{0\}$ *we have*

$$(a, b) = (a, b + ka).$$

Proof If $d_1 = (a, b)$ and $d_2 = (a, b + ka)$, then it is easy to see that $d_1 \mid a$ and $d_1 \mid b$ imply that $d_1 \mid a$ and $d_1 \mid (b + ka)$ so that $d_1 \mid d_2$. Conversely, since $d_2 \mid a$ and $d_2 \mid (b + ka)$, we infer that $d_2 \mid a$ and $d_2 \mid (b + ka - ka) = b$ so that $d_2 \mid d_1$. This is the desired result since $d_1, d_2 \geqslant 1$. □

Now let $a \geqslant b$ be positive integers for which we want to compute their gcd. By making use of the former result, we could proceed in the following way.

(i) If $b \mid a$ then $(a, b) = b$. Otherwise, the Euclidean division of $r_0 = a$ by $r_1 = b$ provides two unique integers q_1 and r_2 such that $a = bq_1 + r_2$ and $0 < r_2 < b$, and Theorem 2.9 with $k = -q_1$ gives

$$(a, b) = (b, a - bq_1) = (b, r_2).$$

(ii) If $r_2 \mid b$ then $(a, b) = (b, r_2) = r_2$. Otherwise, the proof proceeds as before so that we get two unique integers q_2 and r_3 such that

$$(b, r_2) = (r_2, r_3).$$

(iii) And so on ...

This algorithm, derived by Euclid, gradually constructs a *strictly decreasing* sequence of natural integers (r_k). Hence there exists a non-negative integer n such that $r_n \neq 0$ and $r_{n+1} = 0$ and we get

$$(a, b) = (b, r_2) = (r_2, r_3) = \cdots = (r_{n-1}, r_n) = (r_n, 0) = r_n.$$

With a little more work, one can even estimate the number of divisions required in this algorithm to obtain the gcd. The following result is a version of a theorem proved by Lamé.

Proposition 2.10 (Lamé) *Let* $1 \leqslant b \leqslant a$ *be integers. Then the number of divisions necessary to compute* (a, b) *in the Euclidean algorithm does not exceed*

$$\left[\frac{\log b}{\log \Phi} \right] + 1$$

where $\Phi = \frac{1 + \sqrt{5}}{2}$ *is the* golden ratio.

Proof The proof uses the Fibonacci numbers F_n and their connection with the golden ratio. Recall that Φ is the positive solution of the equation $x^2 = x + 1$ and that Fibonacci numbers are the terms of the *Fibonacci sequence* (F_n) defined by

$$\begin{cases} F_0 = 0, \ F_1 = 1 \\ F_{n+2} = F_{n+1} + F_n \quad (n \geqslant 0). \end{cases}$$

An easy induction shows that the following lower bound

$$F_n \geqslant \Phi^{n-2} \tag{2.1}$$

holds for any positive integer n. Now consider the Euclidean algorithm with $n \geqslant 2$ steps written as

$$\begin{array}{ll} r_0 = r_1 q_1 + r_2 & 0 < r_2 < r_1 \\ r_1 = r_2 q_2 + r_3 & 0 < r_3 < r_2 \\ \vdots & \vdots \\ r_{n-2} = r_{n-1} q_{n-1} + r_n & 0 < r_n < r_{n-1} \\ r_{n-1} = r_n q_n \end{array}$$

(with $a = r_0$ and $b = r_1$). Let us prove by induction that

$$r_{n-k} \geqslant F_{k+2} \quad (k = 1, \ldots, n). \tag{2.2}$$

Note first that $q_n \geqslant 2$ since $r_{n-1} = q_n r_n$ and $r_n < r_{n-1}$ and therefore

$$r_{n-1} = r_n q_n \geqslant q_n \geqslant 2 = F_3$$

so (2.2) is true for $k = 1$. It is also true for $k = 2$ since

$$r_{n-2} = r_{n-1} q_{n-1} + r_n \geqslant r_{n-1} + r_n \geqslant 2 + 1 = 3 = F_4.$$

Now suppose that (2.2) is true for some k. We have

$$r_{n-k-1} = r_{n-k} q_{n-k} + r_{n-k+1} \geqslant r_{n-k} + r_{n-k+1}$$

and by induction hypothesis we get

$$r_{n-k-1} \geqslant F_{k+2} + F_{k+1} = F_{k+3}$$

which proves (2.2). We now use (2.2) with $k = n - 1$ to get

$$b = r_1 \geqslant F_{n+1}$$

and (2.1) implies then that

$$n \leqslant \frac{\log b}{\log \Phi} + 1$$

which completes the proof.   □

Example 2.11 Let us show how the Euclidean algorithm could be used to get a gcd and Bézout's coefficients associated to it. Let $d = (18\,459, 3\,809)$. We have

	4	1	5	2
18 459	3 809	3 223	586	293
3 223	586	293	0	

where the quotients are written in the first line and the remainders are in the second and third lines. We then obtain $d = 293$. We also have

$$\begin{cases} 18\,549 = 3\,809\,(4) + 3\,223 \\ 3\,809 = 3\,223\,(1) + 586 \\ 3\,223 = 586\,(5) + 293 \end{cases}$$

which we can rewrite as

$$\begin{cases} 18\,459 + 3\,809\,(-4) + 3\,223\,(-1) = 0 \ \big| \times 6 \\ 3\,809(1) + 3\,223\,(-1) + 586\,(-1) = 0 \ \big| \times(-5) \\ 3\,223(1) + 586\,(-5) = 293 \ \big| \times 1 \end{cases}$$

which immediately gives

$$18\,459 \times (6) + 3\,809 \times (-29) = 293.$$

2.3 Gauss's Theorem

The following result is fundamental.

Theorem 2.12 (Gauss) *Let a, b, c be any integers. Then*

$$\left. \begin{array}{c} a \mid bc \\ (a, b) = 1 \end{array} \right\} \implies a \mid c.$$

Proof Since $(a, b) = 1$, using Bachet–Bézout's theorem we have integers u, v such that $au + bv = 1$. Multiplying both sides of this identity by c and using the fact that there exists $k \in \mathbb{Z}$ such that $bc = ka$, we get

$$c = acu + bcv = acu + kav = a(cu + kv)$$

which is the desired result since $cu + kv \in \mathbb{Z}$. □

This result has, along with Bachet–Bézout's theorem, a lot of consequences. The following proposition summarizes some of them.

Proposition 2.13 *Let n, r be positive integers and $a, b, c, a_1, a_2, \ldots, a_r, k$ be any integers.*

(i) $(ka, kb) = |k|(a, b)$ and $[ka, kb] = |k|[a, b]$.

(ii) If $d = (a, b)$ and $a = da'$ and $b = db'$, then $(a', b') = 1$.

(iii) $(a, b) \times [a, b] = |ab|$.

(iv) $(a, b) = 1 \Rightarrow (a, bc) = (a, c)$.

(v) $(a^n, b^n) = (a, b)^n$ and $[a^n, b^n] = [a, b]^n$.

(vi)

$$\left. \begin{array}{c} a \mid c \\ b \mid c \\ (a, b) = 1 \end{array} \right\} \quad \Longrightarrow \quad ab \mid c.$$

More generally, we have

$$\left. \begin{array}{c} a_1 \mid b, \ldots, a_r \mid b \\ i \neq j \quad \Longrightarrow \quad (a_i, a_j) = 1 \end{array} \right\} \quad \Longrightarrow \quad a_1 \cdots a_r \mid b.$$

(vii) *Suppose $(a, b) = 1$ and $k \geqslant 1$. If $ab = c^k$ for some positive integer c, then there exist positive integers r, s such that $a = r^k$ and $b = s^k$.*

(viii) *Suppose $1 \leqslant k < n$. Then*

$$\frac{n}{(n, k)} \, \Big| \, \binom{n}{k}.$$

Proof Without loss of generality, we prove the theorem by considering that a, b, c and $a_1, a_2, \ldots, a_r, k$ are positive integers.

(i) We set $d = (a, b)$. Clearly we have $kd \mid (ka, kb)$ since $kd \mid ka$ and $kd \mid kb$. By Theorem 2.5, there exist integers u, v such that $d = au + bv$ so that $kd = u(ka) + v(kb)$ and therefore, using Theorem 2.5 again, we infer that $kd = (ka, kb)$.

Next we set $m = [a, b]$ and $M = [ka, kb]$. km is a common multiple of ka and kb so that $km \geqslant M$. Conversely, since $ka \mid M$ and $kb \mid M$, we deduce that the integer M/k is a common multiple of a and b, and thus $M/k \leqslant m$, which gives $M \leqslant km$, and therefore $M = km$.

(ii) We have $d = (a, b) = (da', db') = d(a', b')$ so that $(a', b') = 1$.

(iii) Suppose first that $(a, b) = 1$. We obviously have $[a, b] \leqslant ab$ since these two numbers are common multiples of a and b. On the other hand, since $a \mid [a, b]$, there exists a positive integer q such that $[a, b] = qa$. Since $b \mid [a, b] = qa$ and $(a, b) = 1$, Gauss's theorem implies that $b \mid q$, so that $b \leqslant q$ and thus $ab \leqslant qa = [a, b]$. Hence we proved that $[a, b] = ab$ as soon as $(a, b) = 1$. Now if $(a, b) = d > 1$, we set $a = da'$ and $b = db'$ with $(a', b') = 1$. We then have

$$[a, b] = [da', db'] = d[a', b'] = da'b' = \frac{ab}{d} = \frac{ab}{(a, b)}$$

giving the asserted result.

(iv) We set $d = (a, c)$ and $D = (a, bc)$. Clearly, $d \mid D$. Conversely, since $D \mid a$ and $D \mid bc$, we have $D \mid ac$ and $D \mid bc$ so that $D \mid (ac, bc) = c \times (a, b) = c$, and thus $D \mid d$.

(v) We first note that (iii) implies by induction that $(a, b) = 1 \Longrightarrow (a, b^n) = 1$ and similarly $(b^n, a) = 1 \Longrightarrow (b^n, a^n) = 1$, so that we have

$$(a, b) = 1 \quad \Longrightarrow \quad (a^n, b^n) = 1.$$

Now set $d = (a, b)$ and $a = da'$, $b = db'$ with $(a', b') = 1$. We get

$$(a^n, b^n) = (d^n a'^n, d^n b'^n) = d^n (a'^n, b'^n) = d^n = (a, b)^n.$$

Next we use (iii)

$$[a^n, b^n] = \frac{a^n b^n}{(a^n, b^n)} = \left(\frac{ab}{(a, b)}\right)^n = [a, b]^n.$$

(vi) There exists a positive integer q such that $c = qa$ and then we get

$$\left.\begin{array}{c} b \mid qa \\ (a, b) = 1 \end{array}\right\} \quad \Longrightarrow \quad b \mid q$$

by Gauss's theorem. Thus, $q = hb$ for some positive integer h and then $c = qa = hba$ so that $ab \mid c$. The second assertion follows by an easy induction.

(vii) Let $d = (a, c)$. First, we have $d^k \mid c^k = ab$ and since $(d, b) = ((a, c), b) = (c, (a, b)) = (c, 1) = 1$, we have $(d^k, b) = 1$ and Gauss's theorem implies that $d^k \mid a$. On the other hand, if we set $a = da'$ and $c = dc'$ so that $(a', c') = 1$, then the equality $ab = c^k$ is equivalent to $a'b = d^{k-1}c'^k$, so that $a' \mid d^{k-1}c'^k$ and since $(a', c'^k) = 1$, we have $a' \mid d^{k-1}$ by Gauss's theorem again, so that $a \mid d^k$. Hence we get $a = d^k$ and $b = (c/d)^k$.

(viii) We set $d = (n, k)$ and $n = dn'$, $k = dk'$ with $(n', k') = 1$. We have

$$k'\binom{n}{k} = n'\binom{n-1}{k-1}$$

and the binomial coefficient of the right-hand side is an integer since $n, k \geqslant 1$, therefore we have

$$n' \mid k'\binom{n}{k}.$$

Since $(n', k') = 1$, Gauss's theorem implies that

$$n' \mid \binom{n}{k}$$

which is the desired result.

The proof is complete.   □

2.4 Linear Diophantine Equations

In this section, a, b, n are any integers satisfying $ab \neq 0$ and we set $d = (a, b)$. Any equation of the form $ax + by = n$ with $(x, y) \in \mathbb{Z}^2$ is called a *linear Diophantine equation*. The aim of this section is the resolution of such equations. We begin with an important observation.

Lemma 2.14 *Let $a, b, n \in \mathbb{Z}$ such that $ab \neq 0$ and set $d = (a, b)$. Then the Diophantine equation $ax + by = n$ has a solution in $\mathbb{Z}^2$ if and only if $d \mid n$.*

Proof If $n = 0$ then $d \mid n$ and $(x, y) = (0, 0)$ is a solution. Now suppose $n \neq 0$. We also define $(u, v) \in \mathbb{Z}^2$ such that $d = au + bv$. Suppose first that the equation has at least one solution $(x, y) \in \mathbb{Z}^2$. Since $d \mid a$ and $d \mid b$, we get $d \mid (ax + by) = n$. Conversely, suppose $d \mid n$. Set $x_0 = nu/d$ and $y_0 = nv/d$. Note that $(x_0, y_0) \in \mathbb{Z}^2$ and we have

$$ax_0 + by_0 = \frac{anu + bnv}{d} = \frac{n}{d}(au + bv) = n.$$

Thus, the pair $(x_0, y_0) = (\frac{nu}{d}, \frac{nv}{d})$ is solution of the equation. □

The solutions are completely determined by the following result.

Proposition 2.15 *Let $a, b, n \in \mathbb{Z}$ such that $ab \neq 0$ and set $d = (a, b)$. Suppose that $d \mid n$ and let (x_0, y_0) be a particular solution of the Diophantine equation $ax + by = n$. Then the solutions are given by the following formulae*

$$\begin{cases} x = x_0 + \dfrac{kb}{d} \\ y = y_0 - \dfrac{ka}{d} \end{cases} \quad (k \in \mathbb{Z}).$$

Proof Define $a = da'$, $b = db'$ and $n = dn'$ so that $(a', b') = 1$. The equation is equivalent to $a'x + b'y = n'$. Let (x, y) be a solution distinct from (x_0, y_0). From

$$a'x + b'y = a'x_0 + b'y_0$$

we get

$$a'(x - x_0) = b'(y_0 - y) \tag{2.3}$$

with $y - y_0 \in \mathbb{Z}$, so that $b' \mid a'(x - x_0)$. Since $(a', b') = 1$, Gauss's theorem implies that

$$b' \mid (x - x_0)$$

so that there exists $k \in \mathbb{Z}$ such that $x = x_0 + kb'$. Replacing $x - x_0$ by kb' in (2.3) gives $y = y_0 - ka'$. Conversely, we check that the pairs $(x, y) = (x_0 + kb', y_0 - ka')$ with $k \in \mathbb{Z}$ are solutions of the equation. □

Example 2.16 Solve in $\mathbb{Z}^2$ the equation

$$18\,459x + 3\,809y = 879.$$

Answer Set $d = (18\,459, 3\,809)$. Using Example 2.11 we know that $d = 293$ and since $879 = 3 \times 293$, Lemma 2.14 implies that this equation has at least one solution in $\mathbb{Z}^2$. It is equivalent to $63x + 13y = 3$ and we have seen in Example 2.11 that $63(6) + 13(-29) = 1$, so that the pair $(x_0, y_0) = (18, -87)$ is a particular solution of the equation. Let $(x, y) \in \mathbb{Z}^2$ be a solution. We have

$$63x + 13y = 63(18) + 13(-87) \quad \Longleftrightarrow \quad 63(x - 18) = 13(-y - 87)$$

so that $13 \mid 63(x - 18)$. Since $(63, 13) = 1$, Gauss's theorem implies that $13 \mid (x - 18)$, and hence there exists $k \in \mathbb{Z}$ such that $x = 18 + 13k$. This gives $y = -87 - 63k$. Conversely, we check that the pairs $(18 + 13k, -87 - 63k)$ are solutions of the equation. Thus, the set of solutions is

$$S = \{(18 + 13k, -87 - 63k) : k \in \mathbb{Z}\}.$$

Example 2.17 (Money changing problem) In how many ways can we obtain a sum of \$34 with only \$2 coins and \$ 5 bills?

Answer We have to count the number of solutions in $(\mathbb{Z}_{\geqslant 0})^2$ of the equation $2x + 5y = 34$. Using Proposition 2.15 we get

$$x = 2 + 5k \quad \text{and} \quad y = 6 - 2k \quad (k \in \mathbb{Z})$$

as solutions in $\mathbb{Z}^2$. The condition $(x, y) \in (\mathbb{Z}_{\geqslant 0})^2$ holds if and only if $-2/5 \leqslant k \leqslant 3$, which gives $k \in \{0, 1, 2, 3\}$, and hence the equation has four solutions in $(\mathbb{Z}_{\geqslant 0})^2$, and thus there are four ways to get \$34 with only \$2 coins and \$5 bills.

2.5 Congruences

Introduced by Gauss, congruences are a very efficient tool in number theory. In what follows, n is a positive integer.

Definition 2.18 Let a, b be any integers and let n be a positive integer. We say that a is *congruent to b modulo n*, and we write $a \equiv b \pmod{n}$, if and only if $n \mid (a - b)$. If $n \nmid (a - b)$, we say that a and b are *incongruent modulo n* and write $a \not\equiv b \pmod{n}$.

Examples 2.19

1. $42 \equiv 0 \pmod 2$, $42 \equiv 10 \pmod 2$, $42 \not\equiv 15 \pmod 2$, $42 \equiv 120 \pmod 2$.
2. $10 \equiv 1 \pmod 9$ and $10 \equiv -1 \pmod{11}$.
3. n is even (resp. odd) if and only if $n \equiv 0 \pmod 2$ (resp. $n \equiv 1 \pmod 2$).

4. If the decimal representation of a positive integer n is $n = a_r a_{r-1} \cdots a_0$ with $0 \leqslant a_i \leqslant 9$ (for $i = 0, \ldots, r$) and $a_r \neq 0$, then, for any integer $k \in \{1, \ldots, r+1\}$ we have

$$n \equiv a_{k-1} \cdots a_0 \ (\mathrm{mod}\ 10^k).$$

5. For any integers a, b, we have $a \equiv b \ (\mathrm{mod}\ 1)$.

Remark 2.20 Given a fixed integer a, there is no unicity of the integer b such that $a \equiv b \ (\mathrm{mod}\ n)$ since this relation means nothing but that *there exists $k \in \mathbb{Z}$ such that $a = b + kn$*. Thus, we can choose b depending on the result aimed at. In particular, it is of great interest to take b as the remainder of the Euclidean division of a by n. Indeed, we have in this case $0 \leqslant b < n$, i.e. a *reduction phenomenon*. Let us also mention that the relation $a \equiv 0 \ (\mathrm{mod}\ n)$ means $n \mid a$, and it is not hard to see that, if $a \equiv b \ (\mathrm{mod}\ n)$, then for any positive divisor d of n, we also have $a \equiv b \ (\mathrm{mod}\ d)$. The following theorem summarizes the main properties of congruences.

Proposition 2.21 *Let $a, b, c, d, x, y \in \mathbb{Z}$ and n, d be positive integers.*

(i)

$$\begin{cases} a \equiv a \ (\mathrm{mod}\ n) & \text{(reflexivity)} \\[2mm] a \equiv b \ (\mathrm{mod}\ n) \iff b \equiv a \ (\mathrm{mod}\ n) & \text{(symmetry)} \\[2mm] \left. \begin{array}{l} a \equiv b \ (\mathrm{mod}\ n) \\ b \equiv c \ (\mathrm{mod}\ n) \end{array} \right\} \implies a \equiv c \ (\mathrm{mod}\ n) & \text{(transitivity).} \end{cases}$$

In other words, congruence is an equivalence relation.

(ii) *If $a \equiv b \ (\mathrm{mod}\ n)$ and $c \equiv d \ (\mathrm{mod}\ n)$, then we have*

$$ax + cy \equiv bx + dy \ (\mathrm{mod}\ n) \quad and \quad ac \equiv bd \ (\mathrm{mod}\ n).$$

(iii) *If m is a positive integer, then we have*

$$a \equiv b \ (\mathrm{mod}\ n) \implies a^m \equiv b^m \ (\mathrm{mod}\ n).$$

(iv) (Gauss's theorem) *Here is a congruence version of Theorem 2.12.*

$$\left. \begin{array}{l} ac \equiv bc \ (\mathrm{mod}\ n) \\ (c, n) = 1 \end{array} \right\} \implies a \equiv b \ (\mathrm{mod}\ n).$$

(v) (Generalization of Gauss's theorem) *We set $d = (c, n)$. Then we have*

$$ac \equiv bc \ (\mathrm{mod}\ n) \iff a \equiv b \ (\mathrm{mod}\ n/d).$$

Proof We leave the proofs of (i) and (ii) to the reader as an exercise. They directly derive from the definition.

(iii) Using Lemma 1.6 we know that $(a - b)$ divides $(a^m - b^m)$ so if $n \mid (a - b)$, then $n \mid (a^m - b^m)$.

(iv) The system can be written as

$$\begin{cases} n \mid (a - b)c \\ (c, n) = 1 \end{cases}$$

so that Gauss's theorem immediately yields $n \mid (a - b)$.

(v) We set $d = (c, n)$ and $c = dc'$, $n = dn'$ so that $(c', n') = 1$. Suppose first that $ac \equiv bc \pmod{n}$. Then there exists $k \in \mathbb{Z}$ such that $ac - bc = kn$ and so $ac' - bc' = kn'$. Thus we have

$$\begin{cases} n' \mid (a - b)c' \\ (c', n') = 1 \end{cases}$$

and Gauss's theorem then implies that $n' \mid (a - b)$. Conversely, if $(n/d) \mid (a - b)$, then

$$(nc/d) = nc' \mid c(a - b)$$

and since $n \mid nc'$, we obtain $n \mid c(a - b)$.

The proof is complete.   □

Example 2.22 Find the remainder of the Euclidean division by 7 of

$$A = \sum_{k=1}^{10} 10^{10^k}.$$

Answer $10^2 = 2 + 7 \times 14$ and hence $10^2 \equiv 2 \pmod{7}$, and thus[1]

$$10^6 = \left(10^2\right)^3 \equiv 2^3 \equiv 8 \equiv 1 \pmod{7}.$$

On the other hand, for every integer $k \geqslant 2$, we have

$$3 \mid \left(4^{k-1} - 1\right)$$

by using Lemma 1.6, and hence 6 divides $4(4^{k-1} - 1)$, and so $4^k \equiv 4 \pmod{6}$. Since $10 \equiv 4 \pmod{6}$, we deduce that $10^k \equiv 4 \pmod{6}$ for $k \geqslant 2$. The congruence being

---

[1] We could also use Fermat's little theorem. See Theorem 3.15.

also true for $k = 1$, we have

$$A = \sum_{k=1}^{10} 10^{10^k} = \sum_{k=1}^{10} 10^{4+6h}$$

$$\equiv \sum_{k=1}^{10} 10^4 \equiv \sum_{k=1}^{10} 4$$

$$\equiv 10 \times 4 \equiv 3 \times 4 \equiv 5 \ (\mathrm{mod}\,7).$$

Example 2.23 Let $a, b \in \mathbb{Z}$ and n be a positive integer such that $a \equiv b \ (\mathrm{mod}\,n)$. Prove that

$$a^n \equiv b^n \ (\mathrm{mod}\,n^2).$$

Answer Using Lemma 1.6 we get

$$a^n - b^n = (a - b) \sum_{k=0}^{n-1} a^k b^{n-k-1}$$

and we know that $a - b \equiv 0 \ (\mathrm{mod}\,n)$ and

$$\sum_{k=0}^{n-1} a^k b^{n-k-1} \equiv \sum_{k=0}^{n-1} a^{n-1} \equiv n a^{n-1} \equiv 0 \ (\mathrm{mod}\,n).$$

Example 2.24 Prove that $13 \mid (2^{70} + 3^{70})$.

Answer We have $2^6 = 5 \times 13 - 1$ and $3^3 = 2 \times 13 + 1$, hence $2^6 \equiv -1 \ (\mathrm{mod}\,13)$ and $3^3 \equiv 1 \ (\mathrm{mod}\,13)$. Writing $70 = 11 \times 6 + 4$ and $70 = 3 \times 23 + 1$ we obtain

$$2^{70} + 3^{70} = 2^{11 \times 6+4} + 3^{3 \times 23+1} \equiv -2^4 + 3 \equiv -13 \equiv 0 \ (\mathrm{mod}\,13).$$

Example 2.25 Solve the Diophantine equation $2^m - 3^n = n$ where m, n are non-negative integers such that $n \not\equiv 5 \ (\mathrm{mod}\,8)$.

Answer $(0, 0)$ is the only solution when $m = 0$ and there is no solution when $m = 1$. Furthermore, if $m \geqslant 1$, then $n = 2^m - 3^n \equiv 1 \ (\mathrm{mod}\,2)$ so that one may suppose that $m \geqslant 2$ and $n \geqslant 1$ is an odd integer satisfying $n \not\equiv 5 \ (\mathrm{mod}\,8)$ by assumption. Writing $n = 2k + 1$ for some $k \in \mathbb{Z}_{\geqslant 0}$, we infer that $k \not\equiv \pm 2 \ (\mathrm{mod}\,8)$ and therefore

$$3^n + n = 3 \times 9^k + 2k + 1 \equiv 4 + 2k \not\equiv 0 \ (\mathrm{mod}\,8).$$

Now let m and n be two solutions with $m \geqslant 2$ and $n \geqslant 1$ is odd such that $n \not\equiv 5 \ (\mathrm{mod}\,8)$. By the argument above we get $8 \nmid 2^m$, so that $m \leqslant 2$, and then $m = 2$, giving $n = 1$. Therefore, under the above hypotheses, the equation has two solutions, namely the pairs $(0, 0)$ and $(2, 1)$.

Our next goal is to count the number of solutions of the equation

$$ax \equiv b \ (\mathrm{mod}\, n).$$

Here we mean the number of integer solutions x such that

$$ax \equiv b \ (\mathrm{mod}\, n) \quad \text{and} \quad 0 \leqslant x \leqslant n - 1.$$

Lemma 2.14 and Proposition 2.15 then immediately give

Proposition 2.26 *Let $a, b \in \mathbb{Z}$ and n be a positive integer. We set $d = (a, n)$. Then the congruence*

$$ax \equiv b \ (\mathrm{mod}\, n)$$

has a solution if and only if $d \mid b$. Furthermore, if x_0 is a particular solution, then there are exactly d solutions given by

$$x_0 + \frac{kn}{d} \quad with \quad 0 \leqslant k \leqslant d - 1.$$

In particular, if $(a, n) = 1$, then the congruence $ax \equiv b \ (\mathrm{mod}\, n)$ has exactly one solution.

An important application of this result is the so-called *Chinese remainder theorem* which treats several congruences simultaneously. The name of this result comes from the fact that a 2-congruence version was discovered by the Chinese mathematician Sun Tse (1st century AD). Suppose we have a system of congruences

$$\begin{cases} x \equiv a_1 \ (\mathrm{mod}\, n_1) \\ \vdots \qquad \vdots \\ x \equiv a_k \ (\mathrm{mod}\, n_k) \end{cases}$$

where $a_1, \ldots, a_k \in \mathbb{Z}$ and $n_1, \ldots, n_k$ are positive integers. The following result is very useful.

Theorem 2.27 (Chinese remainder theorem) *Suppose that the positive integers $n_1, \ldots, n_k$ are pairwise coprime, i.e. for every pair $(i, j) \in \{1, \ldots, k\}^2$ such that $i \neq j$, we have $(n_i, n_j) = 1$. Then the system*

$$\begin{cases} x \equiv a_1 \ (\mathrm{mod}\, n_1) \\ \vdots \qquad \vdots \\ x \equiv a_k \ (\mathrm{mod}\, n_k) \end{cases}$$

has a unique solution modulo $n_1 \cdots n_k$.

Proof The proof is constructive, meaning that we construct a solution and then show that it is unique. We set $n = n_1 \cdots n_k$ and, for every integer $i = 1, \ldots, k$, we define $n'_i = n/n_i$. Obviously we have $n'_i \in \mathbb{N}$ and $(n_i, n'_i) = 1$. Proposition 2.26 implies that there exist integers n''_i such that $n'_i n''_i \equiv 1 \pmod{n_i}$. We then set

$$x = a_1 n'_1 n''_1 + \cdots + a_k n'_k n''_k.$$

We notice that if $i \neq j$, then $n'_j \equiv 0 \pmod{n_i}$ and thus

$$x \equiv a_i n'_i n''_i \equiv a_i \pmod{n_i} \quad (i = 1, \ldots, k).$$

Suppose now that the system has two solutions x and y. Then $x - y \equiv 0 \pmod{n_i}$ for every $i = 1, \ldots, k$. Since the integers n_i are pairwise coprime, Proposition 2.13 (vi) implies that $x - y \equiv 0 \pmod{n}$, which completes the proof. □

Example 2.28 Solve the system

$$\begin{cases} x \equiv 1 \pmod{2} \\ x \equiv 2 \pmod{3} \\ x \equiv 3 \pmod{5} \\ x \equiv 4 \pmod{7} \\ x \equiv 5 \pmod{11} \end{cases}$$

Answer Note that the modules are pairwise distinct prime numbers, and then are trivially pairwise coprime. We have $n = 2310$ and the computations are summarized in Table 2.1 so that

$$x = 1155 + 3080 + 4158 + 1320 + 1050 \equiv 1523 \pmod{2310}.$$

Remark 2.29

1. Let a, n be coprime integers. It will be shown in Theorem 4.12 that

$$a^{\varphi(n)} \equiv 1 \pmod{n}$$

where φ is Euler's totient function. It follows that $x = a^{\varphi(n)-1}$ is a solution of the equation $ax \equiv 1 \pmod{n}$ and thus one may take $n''_i = (n/n_i)^{\varphi(n_i)-1}$. Therefore

Table 2.1 Calculations of the n'_i	n'_1	n'_2	n'_3	n'_4	n'_5
	1155	770	462	330	210
	n''_1	n''_2	n''_3	n''_4	n''_5
	1	2	3	1	1

an explicit formula for the system of Theorem 2.27 in terms of n_i and a_i can be derived as follows

$$x \equiv \sum_{i=1}^{k} \left(\frac{n}{n_i} \right)^{\varphi(n_i)} a_i \ (\mathrm{mod}\, n_1 \cdots n_k).$$

2. An important particular case, often used in some recreational mathematics, is when we have $a_1 = \cdots = a_k = a$, where the whole computational machinery is not needed. The trivial solution $x \equiv a \ (\mathrm{mod}\, n_1 \cdots n_k)$ is guaranteed to be the only one by Theorem 2.27.
3. When the moduli are not pairwise coprime, some potential sets of remainders may be ruled out. For instance, the system

$$\begin{cases} x \equiv 2 \ (\mathrm{mod}\, 6) \\ x \equiv 3 \ (\mathrm{mod}\, 4) \end{cases}$$

has no solution in $\mathbb{Z}$, otherwise there would exist $(h, k) \in \mathbb{Z}^2$ such that $6h = 1 + 4k$, and hence $2h \equiv 1 \ (\mathrm{mod}\, 4)$, contradicting Proposition 2.26. A necessary and sufficient condition for a solution to exist is

$$a_i \equiv a_j \ \big(\mathrm{mod}(n_i, n_j)\big)$$

for all $i, j \in \{1, \ldots, k\}^2$. In this case, all solutions are congruent modulo $[n_1, \ldots, n_k]$.

2.6 Further Developments

2.6.1 The Ring $(\mathbb{Z}/n\mathbb{Z}, +, \times)$

A *ring* $(R, +, \times)$ is a non-empty set endowed with two binary operations, usually denoted by $+$ and $\times$, such that $(R, +)$ is an abelian group with identity element denoted by 0, the operation $\times$ being left- and right-distributive over $+$. If the operation $\times$ also has an identity element, often denoted by 1, then the ring is called a *unitary ring*. An element $a \in R$, where R is unitary, is a *unit* of R if there exists $b \in R$ such that $ab = ba = 1$. Such a b is then unique and is called the *inverse* of a. Finally, an element $a \in R$ is called a *zero divisor* if there exists $b \in R \setminus \{0\}$ *not* necessarily unique such that $ab = 0$, and if R is unitary, neither a nor b are units.

Now let n be a positive integer. The set of residue classes modulo n, often denoted by $\overline{0}, \overline{1}, \ldots, \overline{n-1}$, is a unitary abelian ring, denoted by $\mathbb{Z}/n\mathbb{Z}$, endowed with the binary operations $+$ and $\times$ defined by $\overline{x + y} = \overline{x} + \overline{y}$ and $\overline{x \times y} = \overline{x} \times \overline{y}$ (one can check that they are indeed well-defined). The basic result is then the following theorem.

Theorem 2.30 *The ring $\mathbb{Z}/n\mathbb{Z}$ is a disjoint union $\mathbb{Z}/n\mathbb{Z} = E \cup F$ where E, resp. F, is the set of units, resp. zero divisors, of $\mathbb{Z}/n\mathbb{Z}$. Furthermore, $\overline{x} \in \mathbb{Z}/n\mathbb{Z}$ is a unit if and only if $(x, n) = 1$.*

Proof Let $d = (x, n)$. Suppose first that $\overline{x}$ is a unit and define $\overline{y} = (\overline{x})^{-1}$. We then have $xy \equiv 1 \pmod{n}$ and then there exists $k \in \mathbb{Z}$ such that $xy - 1 = kn$. Thus we have $d \mid (xy - kn) = 1$ and then $d = 1$. Conversely, suppose that $d = 1$. By Bachet–Bézout's theorem there exist $u, v \in \mathbb{Z}$ such that $xu + vn = 1$. Therefore $xu \equiv 1 \pmod{n}$ and then $\overline{x}$ is a unit in $\mathbb{Z}/n\mathbb{Z}$. If $d > 1$, we set $x = dx'$, $n = dn'$ with $(x', n') = 1$. Then we have

$$n'x = dn'x' = nx' \equiv 0 \pmod{n}$$

so that $\overline{x}$ is a zero divisor in $\mathbb{Z}/n\mathbb{Z}$. Conversely, a zero divisor cannot be a unit, and then $d > 1$.   □

Euler's totient function is the function φ which counts the number of units in $\mathbb{Z}/n\mathbb{Z}$, with the convention that $\varphi(1) = 1$. Theorem 2.30 has then the following immediate application.

Theorem 2.31 *Let n be a positive integer. Then $\varphi(1) = 1$ and, for every integer $n \geqslant 2$, $\varphi(n)$ is the number of positive integers $m \leqslant n$ such that $(m, n) = 1$. In other words, we have*

$$\varphi(n) = \sum_{\substack{m \leqslant n \\ (m,n)=1}} 1.$$

We shall find this important function again in Chap. 4.

Finally, note that Lemma 3.4 tells us that if $n = p$ is prime, then $\mathbb{Z}/p\mathbb{Z}$ does not have any zero divisor, and then the set of units of $\mathbb{Z}/p\mathbb{Z}$ is $(\mathbb{Z}/p\mathbb{Z})^* = \{\overline{1}, \ldots, \overline{p-1}\}$. Thus, for every prime number p, we have $\varphi(p) = p - 1$. More generally, group theory tells us that, if $n = p_1^{\alpha_1} \cdots p_r^{\alpha_r}$ is the factorization of n into prime powers, then we have the following isomorphism of cyclic groups

$$\mathbb{Z}/n\mathbb{Z} \simeq \mathbb{Z}/p_1^{\alpha_1}\mathbb{Z} \oplus \mathbb{Z}/p_2^{\alpha_2}\mathbb{Z} \oplus \cdots \oplus \mathbb{Z}/p_r^{\alpha_r}\mathbb{Z}$$

which induces the group isomorphism

$$(\mathbb{Z}/n\mathbb{Z})^* \simeq (\mathbb{Z}/p_1^{\alpha_1}\mathbb{Z})^* \oplus (\mathbb{Z}/p_2^{\alpha_2}\mathbb{Z})^* \oplus \cdots \oplus (\mathbb{Z}/p_r^{\alpha_r}\mathbb{Z})^*$$

whence we deduce that

$$\varphi(n) = \varphi(p_1^{\alpha_1})\varphi(p_2^{\alpha_2}) \cdots \varphi(p_r^{\alpha_r})$$

characterizing the multiplicativity of Euler's totient function. We shall prove this again in Chap. 4 by using purely arithmetic arguments.

2.6.2 Denumerants

Let k, n be positive integers and $A_k = \{a_1, \ldots, a_k\}$ be a finite set of pairwise relatively prime positive integers.[2] The aim of this section is to study the number $D_k(n)$
of solutions of the Diophantine equation

$$a_1 x_1 + \cdots + a_k x_k = n$$

where the unknowns are $(x_1, \ldots, x_k) \in (\mathbb{Z}_{\geqslant 0})^k$. The number $D_k(n)$ is called the
denumerant of n with respect to k and to the set A_k. When $a_1 = \cdots = a_k = 1$, this
denumerant will be denoted by $D_{(1,\ldots,1)}(n)$ where the vector $(1, \ldots, 1)$ is supposed
to have k components.

Exercise 10 deals with the case $k = 2$ where it is shown that

$$D_2(n) = \left[\frac{n}{ab} \right] + r$$

where $r = 0$ or $r = 1$ and for which it is convenient to rename the coefficients a,
b instead of a_1, a_2. Thus, if $n < ab$, there is at most one solution. The *Frobenius
problem* deals with the largest number n for which this equation is not soluble, and
it can be proved that this number is given by

$$n = ab - a - b.$$

By using the theory of generating functions, several authors derived some formulae
for the exact value of $D_2(n)$. The following lines borrow the ideas essentially from
[Tri00]. Expanding the generating function into partial fractions and comparing the
coefficients of z^n in each side gives[3]

$$D_2(n) = \frac{n}{ab} + \frac{1}{2}\left(\frac{1}{a} + \frac{1}{b}\right) + \frac{1}{a}\sum_{k=1}^{a-1} \frac{e_a(nk)}{1 - e_a(-bk)} + \frac{1}{b}\sum_{k=1}^{b-1} \frac{e_b(nk)}{1 - e_b(-ak)} \quad (2.4)$$

where $e_a(x) = e^{2\pi i x/a}$. Note that the sums are periodic in n, the first one with period
a and the second one with period b, and since $(a, b) = 1$, the sum of these two sums
is periodic of period ab, and then the expression of $D_2(n)$ is essentially determined
modulo ab. Now define a' and b' to be integers satisfying

$$\begin{cases} 1 \leqslant a' \leqslant b \\ 1 \leqslant b' \leqslant a \\ aa' \equiv -n \pmod{b} \\ bb' \equiv -n \pmod{a}. \end{cases} \quad (2.5)$$

---

[2]When $k = 1$, this obviously means $A_1 = \{1\}$.
[3]See Exercise 11 for another proof of this identity.

Note that

$$\{1 - e_a(-bk)\} \sum_{j=0}^{b'-1} e_a(-bjk) = 1 - e_a\left(-b'bk\right) = 1 - e_a(nk)$$

hence we get

$$\frac{e_a(nk)}{1 - e_a(-bk)} = \frac{1}{1 - e_a(-bk)} - \sum_{j=0}^{b'-1} e_a(-bjk)$$

and using the formulae

$$\sum_{k=1}^{a-1} e_a(-mk) = \begin{cases} a - 1, & \text{if } a \mid m \\ -1, & \text{otherwise} \end{cases}$$

valid for positive integers a, m, and

$$\sum_{k=1}^{a-1} \frac{1}{1 - e_a(-mk)} = \frac{a - 1}{2}$$

valid for positive integers a, m such that $(a, m) = 1$, we then obtain

$$\sum_{k=1}^{a-1} \frac{e_a(nk)}{1 - e_a(-bk)} = \sum_{k=1}^{a-1} \frac{1}{1 - e_a(-bk)} - \sum_{j=0}^{b'-1} \sum_{k=1}^{a-1} e_a(-bjk)$$

$$= \frac{a - 1}{2} - (a - 1) + \sum_{j=1}^{b'-1} 1$$

$$= b' - \frac{a + 1}{2}.$$

and the same is true for the second sum, so that we finally get

$$D_2(n) = \frac{n}{ab} - 1 + \frac{a'}{b} + \frac{b'}{a}.$$

Note that by (2.5) both a and b divide $n + aa' + bb'$, so that $n + aa' + bb'$ is divisible by ab by Proposition 2.13 (vi). Furthermore, since $n + aa' + bb' \geqslant n + a + b$, we infer that the number on the right-hand side above is indeed a non-negative integer. We can then state the following result.

Theorem 2.32 *Let $a, b \geqslant 1$ be coprime integers and define positive integers a', b' as in (2.5). Then the number $D_2(n)$ of solutions in non-negative integer pairs (x, y)*

of the equation $ax + by = n$ is given by

$$D_2(n) = \frac{n}{ab} - 1 + \frac{a'}{b} + \frac{b'}{a}.$$

Note that the following four properties characterize uniquely $D_2(n)$.

$$D_2(n + kab) = D_2(n) + k \quad (k \geqslant 0).$$

$$D_2(n) = 1 \quad \text{if } ab - a - b < n < ab.$$

$$D_2(n) + D_2(m) = 1 \quad \text{if } m + n = ab - a - b \quad (m, n \geqslant 0).$$

$$D_2(n) = 1 \quad \text{if } n = ax_0 + by_0 < ab - a - b \quad (x_0, y_0 \geqslant 0).$$

See [BR04, BZ04, Shi06, Wil90] for more details.

Another formula, discovered by Popoviciu [Pop53] in 1953, is given in Exercise 11.

In the general case $k \geqslant 3$, a similar method involving the generating function can be used, leading to the asymptotic formula

$$D_k(n) = \frac{n^{k-1}}{(k-1)! \, a_1 \cdots a_k} + O\left(n^{k-2}\right)$$

as $n \longrightarrow \infty$, but computations become harder as soon as the number of unknowns increases. On the other hand, a strong upper bound can be derived in a more elementary way by taking into account the following observation. Since

$$a_1 x_1 + \cdots + a_k x_k = n \quad \Longleftrightarrow \quad a_1 x_1 + \cdots + a_{k-1} x_{k-1} = n - a_k x_k$$

we obtain $x_k \leqslant [n/a_k]$ and thus

$$D_k(n) = \sum_{j=0}^{n/a_k} D_{k-1}(n - ja_k). \tag{2.6}$$

The aim is then to prove the following result.

Theorem 2.33 *For every positive integer k we set*

$$s_k = \begin{cases} 0, & \text{if } k = 1 \\ a_2 + \dfrac{a_3 + \cdots + a_k}{2}, & \text{if } k \geqslant 2 \end{cases}$$

and

$$r_k = \begin{cases} 1, & \text{if } k = 1 \\ a_2 \cdots a_k, & \text{if } k \geqslant 2 \end{cases}$$

with $s_2 = a_2$. Then

$$\mathcal{D}_k(n) \leqslant \frac{(n + s_k)^{k-1}}{(k-1)! \, r_k}.$$

Furthermore, if $a_1 = \cdots = a_k = 1$, then

$$\mathcal{D}_{(1,\ldots,1)}(n) = \binom{k+n-1}{n}.$$

The proof uses induction and the following lemma.

Lemma 2.34 *For every real number $x \geqslant 0$ and every integer $k \geqslant 2$ we have*

$$\sum_{0 \leqslant j \leqslant x} (x - j)^{k-1} \leqslant \frac{1}{k}\left(x + \frac{1}{2}\right)^k.$$

Proof We shall in fact prove the slightly stronger following inequality

$$\sum_{0 \leqslant j \leqslant x} (x - j)^{k-1} \leqslant \frac{x^k}{k} + \frac{x^{k-1}}{2} + \frac{(k-1)x^{k-2}}{8}$$

since we easily see using Newton's formula that

$$\frac{1}{k}\left(x + \frac{1}{2}\right)^k = \frac{x^k}{k} + \frac{x^{k-1}}{2} + \frac{(k-1)x^{k-2}}{8} + \text{positive terms}$$

$$\geqslant \frac{x^k}{k} + \frac{x^{k-1}}{2} + \frac{(k-1)x^{k-2}}{8}.$$

By partial summation with $a = 0$, $f(t) = 1$ and $g(t) = (x - t)^{k-1}$ we get

$$\sum_{0 \leqslant j \leqslant x} (x - j)^{k-1} = (k - 1) \int_0^x (x - t)^{k-2}([t] + 1) \, dt$$

and since $[t] = t - \psi(t) - 1/2$, we obtain

$$\sum_{0 \leqslant j \leqslant x} (x - j)^{k-1} = (k - 1)\left(\int_0^x (x - t)^{k-2}\left(t + \frac{1}{2}\right) dt - \int_0^x (x - t)^{k-2} \psi(t) \, dt \right)$$

and integrating by parts gives

$$\sum_{0 \leqslant j \leqslant x} (x - j)^{k-1} = \frac{x^k}{k} + \frac{x^{k-1}}{2} - (k - 1)\left(\left[\psi_2(t)(x - t)^{k-2}\right]_0^x \right.$$

$$\left. + (k - 2) \int_0^x (x - t)^{k-3} \psi_2(t) \, dt \right)$$

$$= \frac{x^k}{k} + \frac{x^{k-1}}{2} - (k-1)(k-2) \int_0^x (x-t)^{k-3} \psi_2(t)\, dt$$

$$\leqslant \frac{x^k}{k} + \frac{x^{k-1}}{2} + \frac{(k-1)(k-2)}{8} \int_0^x (x-t)^{k-3}\, dt$$

$$= \frac{x^k}{k} + \frac{x^{k-1}}{2} + \frac{(k-1)x^{k-2}}{8}$$

which completes the proof of the lemma.   □

Proof of Theorem 2.33 By induction on k, the case $k = 1$ being immediate since

$$\mathcal{D}_1(n) \leqslant 1 = \frac{n^{1-1}}{(1-1)!} = \frac{(n+s_1)^{1-1}}{(1-1)!r_1}.$$

Now suppose $k \geqslant 2$. We let the case $k = 2$ in Exercise 10, and suppose the inequality is true for some $k \geqslant 2$. Using (2.6), induction hypothesis and Lemma 2.34 we get

$$\mathcal{D}_{k+1}(n) = \sum_{j=0}^{n/a_{k+1}} \mathcal{D}_k(n - ja_{k+1})$$

$$\leqslant \frac{1}{(k-1)!\, r_k} \sum_{j=0}^{n/a_{k+1}} (n - ja_{k+1} + s_k)^{k-1}$$

$$\leqslant \frac{a_{k+1}^{k-1}}{(k-1)!\, r_k} \sum_{j=0}^{(n+s_k)/a_{k+1}} \left(\frac{n+s_k}{a_{k+1}} - j \right)^{k-1}$$

$$\leqslant \frac{a_{k+1}^{k-1}}{k!\, r_k} \left(\frac{n+s_k}{a_{k+1}} + \frac{1}{2} \right)^k = \frac{(n+s_{k+1})^k}{k!\, r_{k+1}}$$

which is the desired result. The last part of the theorem may be treated similarly, using the well-known identity (see [Gou72, identity 1.49])

$$\sum_{j=0}^{n} \binom{k+j-1}{j} = \binom{k+n}{n}$$

in the induction argument.   □

2.6.3 Generating Functions

We saw in the last section that it was essential to study the generating function of a sequence to get some information about the sequence itself. Let (u_n) be a sequence.

The *generating function* of (u_n) is the function $F(z)$ formally defined by

$$F(z) = \sum_{n \geq 0} u_n z^n.$$

If $F(z)$ can be written as an elementary function of z, then the comparison of the coefficients of z^n can give a closed formula for u_n. Nevertheless, we must notice that such a series could either be studied formally or must be taken as a function of the complex variable z. In the latter case, convergence problems must be studied. Generating functions are the main tools in combinatorial theory, where results from complex analysis are frequently used (see [Odl95] or [Wil90]). For example, if $F(z)$ is analytic in the open disc $|z| < R$, then, for any r with $0 < r < R$ and any non-negative integer n we have

$$|u_n| \leqslant r^{-n} \max_{|z|=r} |F(z)|$$

by Cauchy's theorem.

For a wide class of generating functions, the *saddle point method* enables us to obtain more accurate estimates. To this end, we first pick up the concept of H-function from [Odl95]. A generating function $F(z) = \sum_{n \geq 0} u_n z^n$ is said to be an *H-function*, or *Hayman's function*, if $F(z)$ is analytic in $|z| < R$ for some $0 < R \leqslant \infty$, real when $z \in] - R, R[$ and satisfies

$$\max_{|z|=r} |F(z)| = F(r)$$

for some $R_0 < r < R$. Furthermore, let

$$a(z) = \frac{zF'(z)}{F(z)} \quad \text{and} \quad b(z) = za'(z)$$

and let $r \longmapsto \delta(r)$ be a function defined in the range $]R_0, R[$ and satisfying $0 < \delta(r) < \pi$ in this range. Suppose that

▷ Uniformly for $|\theta| < \delta(r)$, we have

$$F\left(re^{i\theta}\right) \sim F(r)e^{i\theta a(r) - \frac{1}{2}\theta^2 b(r)} \quad \text{and} \quad F\left(re^{i\theta}\right) = o\left(\frac{F(r)}{\sqrt{b(r)}}\right)$$

as $r \longrightarrow R$.

▷ $\lim_{r \to R} b(r) = \infty$.

It can be shown [Odl95] that if $P(z) = a_n z^n + \cdots + a_0 \in \mathbb{R}[z]$ and if $F(z)$ and $G(z)$ are H-admissible in $|z| < R$, then $e^{F(z)}$, $F(z)G(z)$ and $F(z) + P(z)$ are H-admissible in $|z| < R$. Furthermore, if $a_n > 0$, then $P(F(z))$ is H-admissible in $|z| < R$.

Now we may state one of the main results of the saddle point theory.

Theorem 2.35 *Let $F(z) = \sum_{n \geqslant 0} u_n z^n$ be an H-function and set*

$$a(z) = \frac{zF'(z)}{F(z)} \quad and \quad b(z) = za'(z).$$

For all $n \in \mathbb{Z}_{\geqslant 0}$, let z_n be the solution of $a(z_n) = n$ such that $R_0 < z_n < R$. Then, for $n \longrightarrow \infty$, we have

$$u_n \sim \frac{F(z_n)}{z_n^n \sqrt{2\pi b(z_n)}}.$$

For example, with $u_n = 1/n!$ we have $F(z) = e^z$, $R = \infty$ and $a(z) = b(z) = z$ so that $z_n = n$, and we obtain the following version of Stirling's formula

$$n! \sim \left(\frac{n}{e}\right)^n \sqrt{2\pi n}.$$

In addition to (1.1) which is the simplest example of a generating function with $u_n = 1$, it must be interesting to take the following series into account, which generalizes Newton's formula.

For any positive integer k and any real number $|x| < 1$ we have

$$\sum_{j=0}^{\infty} \binom{k+j-1}{j} x^j = \frac{1}{(1-x)^k}. \tag{2.7}$$

As an example, let us try to find a closed formula for the generating function of the denumerant $\mathcal{D}_k(n)$ seen above. Using (1.1) we have formally

$$\frac{1}{(1-z^{a_1})(1-z^{a_2})\cdots(1-z^{a_k})} = \sum_{x_1 \geqslant 0} z^{a_1 x_1} \sum_{x_2 \geqslant 0} z^{a_2 x_2} \cdots \sum_{x_k \geqslant 0} z^{a_k x_k}$$

$$= \sum_{x_1, \ldots, x_k \geqslant 0} z^{a_1 x_1 + \cdots + a_k x_k}$$

$$= \sum_{n \geqslant 0} \left(\sum_{\substack{x_1, \ldots, x_k \geqslant 0 \\ a_1 x_1 + \cdots + a_k x_k = n}} 1 \right) z^n$$

$$= \sum_{n \geqslant 0} \mathcal{D}_k(n) z^n$$

so that the generating function of $\mathcal{D}_k(n)$ is given by

$$F(z) = \prod_{i=1}^{k} \frac{1}{1-z^{a_i}}. \tag{2.8}$$

Now let us show with an example how this series could be used. We take $k = 3$ and $\mathcal{A}_3 = \{1, 2, 3\}$ so that $\mathcal{D}_3(n)$ counts the number of non-negative integer solutions of the equation $x + 2y + 3z = n$. The generating function of $\mathcal{D}_3(n)$ is then

$$F(z) = \frac{1}{(1 - z)(1 - z^2)(1 - z^3)}$$

and partial fractions expansion theory gives

$$\frac{1}{(1 - z)(1 - z^2)(1 - z^3)} = \frac{1}{6(1 - z)^3} + \frac{1}{4(1 - z)^2} + \frac{17}{72(1 - z)}$$
$$+ \frac{1}{8(1 + z)} + \frac{1}{9}\left(\frac{1}{1 - z/\rho} + \frac{1}{1 - z/\rho^2}\right)$$

where $\rho = e_3(1) = e^{2\pi i/3}$. Using (2.7) we obtain

$$\sum_{n \geqslant 0} \mathcal{D}_3(n) z^n = \sum_{n \geqslant 0} z^n \left\{\frac{1}{6}\binom{n + 2}{n} + \frac{1}{4}\binom{n + 1}{n} + \frac{17}{72}\right.$$
$$\left. + \frac{(-1)^n}{8} + \frac{1}{9}\left(\rho^{-n} + \rho^{-2n}\right)\right\}$$

and comparing the coefficients of z^n in each side gives

$$\mathcal{D}_3(n) = \frac{1}{6}\binom{n + 2}{n} + \frac{1}{4}\binom{n + 1}{n} + \frac{17}{72} + \frac{(-1)^n}{8} + \frac{1}{9}\left(\rho^{-n} + \rho^{-2n}\right)$$
$$= \frac{6n^2 + 36n + 47}{72} + \frac{(-1)^n}{8} + \frac{2}{9}\cos\left(\frac{2\pi n}{3}\right).$$

Thus we can check the values of $\mathcal{D}_3(n)$ according to the residue class modulo 6 of n which we summarize in Table 2.2, and then we have $\mathcal{D}_3(n) = \frac{(n+3)^2}{12} + \varepsilon_n$ where $|\varepsilon_n| \leqslant \frac{1}{3}$ so that

$$\mathcal{D}_3(n) = \left\lfloor \frac{(n + 3)^2}{12} \right\rceil$$

where $\lfloor x \rceil$ is the nearest integer to x.

Table 2.2 Values of $\mathcal{D}_3(n)$

$n \bmod 6$	0	± 1	± 2	3
$\mathcal{D}_3(n)$	$\dfrac{n^2 + 6n + 12}{12}$	$\dfrac{(n + 1)(n + 5)}{12}$	$\dfrac{(n + 2)(n + 4)}{12}$	$\dfrac{(n + 3)^2}{12}$

2.7 Exercises

1 Let a, b, n, x, y, z, t be any positive integers.

(a) Prove that if $(a, b) = 1$, then $(a + b, a - b) = 1$ or 2.
(b) Show that if $(a, b) = 1$, then $(a + b, ab) = 1$.
(c) Prove that $a \mid b \Longleftrightarrow a^n \mid b^n$.
(d) Suppose that $|xt - yz| = 1$. Prove that $(a, b) = (ax + by, az + bt)$.

2 (Thue's lemma) The aim of this exercise is to provide a proof of the following result.

Lemma (Thue) *Let $a > 1$ be an integer and $p \geqslant 3$ be a prime number such that $p \nmid a$. Then the equation $au \equiv v \pmod{p}$ has a solution $(u, v) \in \mathbb{Z}^2$ such that*

$$\begin{cases} 1 \leqslant |u| < \sqrt{p}, \\ 1 \leqslant |v| < \sqrt{p}. \end{cases}$$

(a) Let $S = \{0, \ldots, [\sqrt{p}]\}$ and define a map $f : S^2 \longrightarrow \{0, \ldots, p - 1\}$ by

$$f(u, v) \equiv au - v \pmod{p}.$$

Show that f is not injective.
(b) We denote by (u_1, v_1) and (u_2, v_2) two pairs such that $(u_1, v_1) \neq (u_2, v_2)$ and $f(u_1, v_1) = f(u_2, v_2)$, and set $u = u_1 - u_2$ and $v = v_1 - v_2$.
  ▷ Check that $au \equiv v \pmod{p}$ and that $|u| < \sqrt{p}$ and $|v| < \sqrt{p}$.
  ▷ Show that $u \neq 0$, and then $v \neq 0$.

3 (Euclidean algorithm revisited) Let $a > b$ be positive integers and let (r_k) be the sequence of remainders of the divisions in the Euclidean algorithm with $r_0 = a$ and $r_1 = b$

$$\begin{aligned} r_0 &= r_1 q_1 + r_2 & 0 < r_2 < r_1 \\ r_1 &= r_2 q_2 + r_3 & 0 < r_3 < r_2 \\ &\vdots & \vdots \\ r_{k-1} &= r_k q_k + r_{k+1} & 0 < r_{k+1} < r_k \\ &\vdots & \vdots \\ r_{n-1} &= r_n q_n & r_n = (a, b). \end{aligned}$$

We define the sequences (s_k) and (t_k) so that, for $k = 1, \ldots, n$ we have

$$\begin{cases} s_0 = 1 & s_1 = 0 & s_{k+1} = -q_k s_k + s_{k-1} \\ t_0 = 0 & t_1 = 1 & t_{k+1} = -q_k t_k + t_{k-1} \end{cases}$$

(a) Prove that $r_k = s_k a + t_k b \ (k = 0, \ldots, n+1)$.
(b) Prove that $s_k = (-1)^k |s_k|$ and $t_k = (-1)^{k+1} |t_k| \ (k = 0, \ldots, n+1)$. Deduce that

$$|s_{k+1}| = q_k |s_k| + |s_{k-1}|$$
$$|t_{k+1}| = q_k |t_k| + |t_{k-1}|$$

for $k = 1, \ldots, n$.

(c) Deduce that

$$a = |t_k| r_{k-1} + |t_{k-1}| r_k$$

for $k = 1, \ldots, n+1$.

4 (Infinite sequence of coprime integers (Edwards)) Let u_1 be an odd integer and (u_n) be the sequence of integers defined by $u_n = u_{n-1}^2 - 2$ for any $n \geqslant 2$.

(a) Prove that u_n is odd for every integer $n \geqslant 1$.
(b) Deduce that $(u_n, u_{n-1}) = 1$ for every integer $n \geqslant 2$.
(c) Show that, for any integer $n \geqslant 3$, we have

$$u_n - 2 = u_{n-2}^2 u_{n-3}^2 \cdots u_1^2 (u_2 - 2).$$

(d) Deduce that for any integers $n \geqslant 3$ and $r = 2, \ldots, n-1$, we have

$$(u_n, u_{n-r}) = 1.$$

5 Solve the following equations and system:

$$
\begin{aligned}
12\,825x + 9\,450y &= 2\,025 &&(x, y \in \mathbb{Z}_{\geqslant 0}) \\
5(x + y)^2 &= 147[x, y] &&(x, y \in \mathbb{Z}_{\geqslant 0}) \\
\begin{cases} 3x \equiv 1 \pmod 5 \\ 5x \equiv 2 \pmod 7 \end{cases} &&&(x \in \mathbb{Z}).
\end{aligned}
$$

6 Let a, b be positive integers and set $d = (a, b)$. We consider a Cartesian coordinate system with origin O and x- and y-axis. The rectangular coordinates x, y of a point M will be denoted by $M\langle x, y \rangle$. An *integer point* is a point with integer coordinates.

(a) Define $A\langle a, b \rangle$. Prove that the number $\mathcal{N}_1$ of integer points on the segment $]OA]$ is d.

(b) Define $B\langle a, 0\rangle$. Prove that the number $\mathcal{N}_2$ of integer points *inside* the right-angled triangle OAB is[4]

$$\mathcal{N}_2 = \frac{1}{2}(ab - a - b - d) + 1.$$

7 (Oloa) Fermat's last theorem states that the Diophantine equation $x^n + y^n = z^n$ has no non-trivial solutions $(x, y, z) \in \mathbb{N}^3$ as soon as $n \geqslant 3$. This was finally proved by Wiles and Taylor in 1995. Does the equation $x^n + y^n = z^{n+1}$ possess any non-trivial positive integer solutions for every positive integer n?

8 Let n be a non-negative integer.

(a) Prove that $n^2 \equiv 0, 1$ or $4 \pmod 8$. Deduce that an integer congruent to 7 modulo 8 cannot be written as the sum of three squares.
(b) Prove that $n^3 \equiv \pm 1$ or $0 \pmod 9$. Deduce that an integer congruent to 4 modulo 9 cannot be written as the sum of three cubes.
(c) Solve the Diophantine equation $\sqrt{x^3 + y^3 + z^3} = 2005$ where x, y, z are positive integers.

9 Show that $641 \mid (2^{32} + 1)$.

10 Let $a < b$ be positive integers such that $b - a = m^2$ for some positive integer m. The aim of this exercise is to study the difference $x - y$ of the solutions $(x, y) \in (\mathbb{Z}_{\geqslant 0})^2$ of the Diophantine equation

$$ax^2 + x = by^2 + y.$$

(a) Prove that $x > y$ and that

$$(x - y)\{1 + b(x + y)\} = (mx)^2$$
$$(x - y)\{1 + a(x + y)\} = (my)^2.$$

(b) Prove that

$$x - y = (b - a, x - y) \times (x, y)^2.$$

(c) Provide some sufficient conditions so that $x - y$ may be a square.

---

[4]One can make use of *Pick's formula* which states that the number $\mathcal{N}_{\text{int}}$ inside a convex integer polygon $\mathcal{P}$ is given by

$$\mathcal{N}_{\text{int}} = \text{area}(\mathcal{P}) - \frac{\mathcal{N}_{\partial\mathcal{P}}}{2} + 1$$

where $\mathcal{N}_{\partial\mathcal{P}}$ is the number of integer points on the edges of the polygon.

11 Let a, b be positive integers such that $(a, b) = 1$ and let n be a positive integer. Define $\mathcal{D}_2(n)$ to be the number of non-negative integer solutions x, y of the Diophantine equation $ax + by = n$.

(a) Use Bachet–Bézout's theorem to prove that $\mathcal{D}_2(n) = [\frac{n}{ab}] + r$ where $r = 0$ or $r = 1$.

(b) Use Theorem 2.32 with its notation to prove that

$$r = 0 \quad \Longleftrightarrow \quad aa' + bb' \leqslant ab$$
$$r = 1 \quad \Longleftrightarrow \quad aa' + bb' > ab.$$

12 (Beck & Robins) Let a, b be positive integers such that $(a, b) = 1$ and let n be a positive integer. Define $\mathcal{D}_2(n)$ to be the number of non-negative integer solutions x, y of the Diophantine equation $ax + by = n$ and set $\bar{a}, \bar{b}$ to be positive integers such that $a\bar{a} \equiv 1 \pmod{b}$ and $b\bar{b} \equiv 1 \pmod{a}$. In 1953, Popoviciu [Pop53] proved the following very elegant formula

$$\mathcal{D}_2(n) = \frac{n}{ab} + 1 - \left\{ \frac{n\bar{b}}{a} \right\} - \left\{ \frac{n\bar{a}}{b} \right\}$$

where $\{x\}$ is the fractional part of x. The aim of this exercise is to provide an analytic proof of this formula. Recall that $e_a(k) = e^{2\pi i k/a}$ ($k \in \mathbb{Z}_{\geqslant 0}$) and define the complex-valued function f by

$$f(z) = \frac{1}{z^{n+1}(1 - z^a)(1 - z^b)}.$$

(a) Recall that

$$\sum_{j=0}^{a-1} x^j = \prod_{j=1}^{a-1} (x - e_a(j)). \tag{2.9}$$

Prove that, for positive integers a, k, we have

$$\sum_{j=1}^{a-1} \frac{1}{1 - e_a(j)} = \frac{a-1}{2}$$

$$\prod_{\substack{j=1 \\ j \neq k}}^{a} \frac{1}{e_a(k) - e_a(j)} = \frac{e_a(k)}{a}.$$

(b) Use the generating function $F(z)$ of $\mathcal{D}_2(n)$ to prove that

$$\operatorname*{Res}_{z=0} f(z) = \mathcal{D}_2(n)$$

and compute the residues at all non-zero poles of f.

(c) Prove that

$$\lim_{R\to\infty} \frac{1}{2\pi i} \int_{|z|=R} f(z)\,\mathrm{d}z = 0$$

and deduce the identity (2.4).
(d) Check that if $b = 1$ then we have

$$\mathcal{D}_2(n) = \frac{n}{a} - \left\{\frac{n}{a}\right\} + 1.$$

Deduce that

$$\frac{1}{a}\sum_{k=1}^{a-1} \frac{1}{e_a(kn)(1-e_a(k))} = \frac{1}{2}\left(1-\frac{1}{a}\right) - \left\{\frac{n}{a}\right\}.$$

(e) By noticing that

$$\sum_{k=1}^{a-1} \frac{1}{e_a(kn)(1-e_a(kb))} = \sum_{k=1}^{a-1} \frac{1}{e_a(kn\overline{b})(1-e_a(k))}$$

finalize the proof of Popoviciu's identity.

References

[BR04]   Beck M, Robins S (2004) A formula related to the Frobenius problem in 2 dimensions. In: Chudnovsky D, Chudnovsky G, Nathanson M (eds) Number theory. New York seminar 2003. Springer, Berlin, pp 17–23

[BZ04]   Beck M, Zacks S (2004) Refined upper bounds for the linear Diophantine problem of Frobenius. Adv Appl Math 32:454–467

[Gou72] Gould HW (1972) Combinatorial identities. A standardized set of tables listing 500 binomial coefficients summations. Morgantown, West Virginia

[Odl95] Odlyzko AM (1995) Asymptotic enumeration methods. Elsevier, Amsterdam

[Pop53] Popoviciu T (1953) Asupra unei probleme de patitie a numerelor. Stud Cercet Stiint—Acad RP Rom, Fil Cluj 4:7–58

[Shi06] Shiu P (2006) Moment sums associated with linear binary forms. Am Math Mon 113:545–550

[Tri00] Tripathi A (2000) The number of solutions to $ax + by = n$. Fibonacci Q 38:290–293

[Wil90] Wilf HS (1990) Generatingfunctionology. Academic Press, New York

Chapter 3
Prime Numbers

The study of prime numbers is one of the main branches of number theory. The literature, very abundant, goes back to Pythagoras and, above all, to Euclid who was the first to show there are infinitely many prime numbers. One can notice now that Euclid's ideas can be used and generalized to prime numbers belonging to certain arithmetic progressions (see Sect. 3.6).

Every integer $n \geqslant 2$ has at least two *distinct* divisors, namely 1 and n itself. So it would be convenient to study the "minimalist" integers for this relation. We call a *prime number* every positive integer $p \geqslant 2$ for which the number of distinct divisors is *exactly* two. A positive integer which is not prime is said to be *composite*. Let us notice that the least prime number is the only positive integer which is both prime and even

$$2 \quad 3 \quad 5 \quad 7 \quad 11 \quad 13 \quad 17 \quad 19 \quad 23 \quad 29$$
$$31 \quad 37 \quad 41 \quad 43 \quad 47 \quad 53 \quad 59 \quad 61 \quad 67 \quad 71 \quad \ldots$$

This very fruitful idea does not belong solely to the branch of number theory. The reader knowing finite group theory will certainly have noticed the analogy between prime numbers and the so-called *simple groups*.[1] Thus, it is not surprising to notice that these areas of mathematics do really interlace. For example, the decomposition of a positive integer into prime powers (see Theorem 3.3) also exists in finite group theory, under a slightly different form, and is called the *Jordan–Hölder theorem* which states that given a finite group G, there exists a sequence

$$G = G_0 \supset G_1 \supset \cdots \supset G_{r-1} \supset G_r = \{e_G\}$$

of subgroups G_i of G such that G_{i+1} is a normal subgroup of G_i and the group G_i/G_{i+1} is simple for $i = 0, \ldots, r - 1$. This decomposition is unique, apart from the order of the subgroups. On the other hand, if G is *abelian* with $|G| = p_1^{\alpha_1} \cdots p_r^{\alpha_r}$,

---

[1] A non-trivial group G is said to be *simple* if it has no normal subgroup other than $\{e_G\}$ and G itself, where e_G is the identity element of G.

O. Bordellès, *Arithmetic Tales*, Universitext,
DOI 10.1007/978-1-4471-4096-2_3, © Springer-Verlag London 2012

then

$$G \simeq H_1 \oplus \cdots \oplus H_r \tag{3.1}$$

where the H_i are the p_i-Sylow subgroups of G with $|H_i| = p_i^{\alpha_i}$. This shows how algebra and arithmetic can be related.

Euclid noticed very early that these prime numbers had some interesting properties. Indeed, he established the following basic result which is of great importance.

Proposition 3.1 *Every integer $n \geqslant 2$ has a prime factor. Furthermore, if n is composite, then it has a prime factor p satisfying $p \leqslant \sqrt{n}$.*

Proof We suppose that n is composite, otherwise there is nothing to prove. Let $\mathcal{D}^*(n)$ be the set of *proper* divisors of n, i.e. positive integers $d \mid n$ such that $d \neq 1$ and $d \neq n$. Since n is composite, we have $\mathcal{D}^*(n) \neq \varnothing$ so by Axiom 1.1 this set has a smallest element p. This integer p is prime, otherwise it has a proper divisor which is also a proper divisor of n, and hence is smaller than p, giving a contradiction with the fact that p is the smallest element of $\mathcal{D}^*(n)$. Since there exists $k \in \mathbb{N}$ such that $n = kp$ and that $p \in \mathcal{D}^*(n)$, we thus have $k \neq 1$ and $k \neq n$, so that $k \in \mathcal{D}^*(n)$, and hence $k \geqslant p$ and then $n = kp \geqslant p^2$, which concludes the proof. $\square$

What follows is easy to check, so we leave the proof to the reader.

Proposition 3.2 *Let p, q be prime numbers and k, n be positive integers. Then*

(i) $p \mid q \Longleftrightarrow p = q$.
(ii) $p \nmid n \Longleftrightarrow (n, p) = 1$.
(iii) *Suppose $1 \leqslant k < p$. Then p divides $\binom{p}{k}$.*

3.1 The Fundamental Theorem of Arithmetic

Theorem 3.3 *Every integer $n \geqslant 2$ either is prime, or can be written as a product of prime factors in only one way, apart from the order of the factors. More precisely, we have*

$$n = \prod_{i=1}^{r} p_i^{\alpha_i}$$

where p_i are prime numbers and α_i are non-negative integers ($i = 1, \ldots, r$). The exponents α_i are the p_i-adic valuations of n, also denoted by $v_{p_i}(n)$.

The unicity of this result is the trickiest point. The following lemma is required.

Lemma 3.4 *Let $a, b, n, a_1, \ldots, a_n$ be positive integers and p be a prime number. Then*

(i) $p \mid ab \Longrightarrow p \mid a$ or $p \mid b$.
(ii) $p \mid a_1 \cdots a_n \Longrightarrow p \mid a_i$ for some $i \in \{1, \ldots, n\}$.
(iii) $p \mid a^n \Longrightarrow p \mid a$.

Proof (ii) follows from (i) by induction and (iii) follows from (ii) by setting $a_1 = \cdots = a_n = a$. Now to show (i), suppose that $p \nmid a$. Hence $(a, p) = 1$ and since $p \mid ab$, Gauss's theorem implies that $p \mid b$, which is the desired conclusion.   □

Now we are in a position to prove Theorem 3.3.

Proof of Theorem 3.3

▷ Let S be the set of composite integers ≥ 2 which cannot be written as a product of prime factors and suppose that $S \neq \varnothing$. Using Axiom 1.1, we infer that S has a smallest element denoted by m. Since m is not a prime number, we have $m = ab$ with $a > 1$ and $b > 1$. Since $a < m$, $b < m$ and m is the smallest element of S, we have $a, b \notin S$, and then one can write a and b as a product of prime factors, and hence $m = ab$ can also be written as a product of prime factors, giving a contradiction. Therefore we have proved that $S = \varnothing$.
▷ Suppose that

$$n = \prod_{i=1}^{r} p_i^{\alpha_i} = \prod_{j=1}^{s} q_j^{\beta_j}$$

with p_i, q_j prime numbers and r, s, α_i, β_j non-negative integers. Using Lemma 3.4 we infer that every factor p_i is a factor q_j, and hence in particular we have $r = s$, and, without loss of generality, we might suppose that $p_i = q_i$ for $i = 1, \ldots, r$. Finally, if we had $\alpha_i < \beta_i$ for some integer $i \in \{1, \ldots, r\}$, then the number $n / p_i^{\alpha_i}$ would have two decompositions, one of which involves p_i whilst the other does not. This is impossible by the argument above. By symmetry we cannot have $\alpha_i > \beta_i$ either. Thus we have $\alpha_i = \beta_i$ for all i.   □

There are a lot of applications of Theorem 3.3. We must first introduce a useful notion.

Definition 3.5 Given integers $n, k \geq 2$, we say that:

(i) n is *k-free* if, for every prime divisor p of n, we have $v_p(n) < k$. When $k = 2$, we also say that n is *squarefree*.
(ii) n is *k-full* if, for every prime divisor p of n, we have $v_p(n) \geq k$. When $k = 2$, we also say that n is *square-full*.

Examples 3.6

1. Every prime number is *k*-free for any integer $k \geq 2$, and the numbers 6, 10, 15, 21 are squarefree.
2. Every square is square-full, and so are the numbers 108 and 12 500.

3. If n is k-free, then n is l-free for every integer $l \geqslant k$. Similarly, if n is k-full, then n is l-full for every integer $2 \leqslant l \leqslant k$.
4. By convention, 1 is both k-free and k-full for each integer $k \geqslant 2$.

The following result summarizes some applications of Theorem 3.3.

Corollary 3.7 *Let $n \geqslant 2$ be an integer which has the decomposition*

$$n = \prod_{i=1}^{r} p_i^{\alpha_i}$$

where r is a positive integer, p_i are prime numbers and α_i are non-negative integers.

(i) *Let d be a positive integer. Then we have*

$$d \mid n \quad \Longleftrightarrow \quad d = \prod_{i=1}^{r} p_i^{\beta_i} \quad \text{with} \quad 0 \leqslant \beta_i \leqslant \alpha_i.$$

Furthermore, if the α_i are positive integers and if $\tau(n)$ and $\omega(n)$ respectively count the number of positive divisors and the number of distinct prime factors of n (see Chap. 4), then we have

$$\omega(n) = r$$

$$\tau(n) = (\alpha_1 + 1)(\alpha_2 + 1) \cdots (\alpha_r + 1)$$

$$2^{\omega(n)} \leqslant \tau(n).$$

(ii) *Let $m = \prod_{i=1}^{r} p_i^{\beta_i}$ be a positive integer. Then we have*

$$(n, m) = \prod_{i=1}^{r} p_i^{\min(\alpha_i, \beta_i)} \quad \text{and} \quad [n, m] = \prod_{i=1}^{r} p_i^{\max(\alpha_i, \beta_i)}.$$

These formulae can be generalized by induction to $(n_1, \ldots, n_k)$ and $[n_1, \ldots, n_k]$.

(iii) *Suppose $(n, m) = 1$. Then every divisor d of mn can be written as $d = ab$ with $a \mid m$, $b \mid n$ and $(a, b) = 1$.*

(iv) *Let $k \geqslant 2$ be an integer. Then every integer $n \geqslant 2$ can be written in a unique way as*

$$n = q m^k$$

where q is k-free.

(v) *Every square-full number $n \geqslant 2$ can be written in a unique way as*

$$n = a^2 b^3$$

where b is squarefree.

Proof

(i) First, let d be a divisor of n, which we can clearly suppose to be > 1 and written as

$$d = \prod_{i=1}^{r} p_i^{\beta_i}$$

for some non-negative integers β_i. Since $d \mid n$, there exists an integer $k = \prod_{i=1}^{r} p_i^{\gamma_i}$ with $\gamma_i \in \mathbb{N}$ such that $n = kd$. This implies that

$$\prod_{i=1}^{r} p_i^{\alpha_i} = \prod_{i=1}^{r} p_i^{\beta_i + \gamma_i}$$

and Theorem 3.3 implies that $\alpha_i = \beta_i + \gamma_i \geqslant \beta_i$. Conversely, if

$$d = \prod_{i=1}^{r} p_i^{\beta_i}$$

with $0 \leqslant \beta_i \leqslant \alpha_i$, then

$$n = \prod_{i=1}^{r} p_i^{\alpha_i} = d \prod_{i=1}^{r} p_i^{\alpha_i - \beta_i}$$

and since $\alpha_i - \beta_i \geqslant 0$, we have $d \mid n$.

The equality $\omega(n) = r$ needs no explanation. On the other hand, $d = \prod_{i=1}^{r} p_i^{\beta_i}$ divides n if and only if $0 \leqslant \beta_i \leqslant \alpha_i$, so that we have $\alpha_i + 1$ possible choices for $v_{p_i}(n)$, and therefore the number of divisors is $(\alpha_1 + 1) \cdots (\alpha_r + 1)$. Finally, since $\alpha_i \geqslant 1$, we have

$$\tau(n) = (\alpha_1 + 1) \cdots (\alpha_r + 1) \geqslant 2^r = 2^{\omega(n)}.$$

(ii) We set

$$d = \prod_{i=1}^{r} p_i^{\min(\alpha_i, \beta_i)} \quad \text{and} \quad m = \prod_{i=1}^{r} p_i^{\max(\alpha_i, \beta_i)}.$$

Using (i) we have $d \mid n$ and $d \mid m$. Furthermore, if $c = \prod_{i=1}^{r} p_i^{\gamma_i}$ is another common divisor of n and m, then $\gamma_i \leqslant \min(\alpha_i, \beta_i)$ and hence $c \mid n$. Thus we proved that $d = (m, n)$. Next, using the identity $\max(\alpha_i, \beta_i) = \alpha_i + \beta_i - \min(\alpha_i, \beta_i)$, we get the desired formula for $[m, n]$.

(iii) Since $(m, n) = 1$ one can write

$$n = p_1^{\alpha_1} \cdots p_r^{\alpha_r} \quad \text{and} \quad m = p_{r+1}^{\alpha_{r+1}} \cdots p_{r+s}^{\alpha_{r+s}}$$

with r, s positive integers, $\alpha_i \in \mathbb{N}$ and primes p_i being pairwise distinct. Thus we get

$$mn = p_1^{\alpha_1} \cdots p_r^{\alpha_r} p_{r+1}^{\alpha_{r+1}} \cdots p_{r+s}^{\alpha_{r+s}}$$

and using (i) we infer that every divisor d of mn can be written as

$$d = p_1^{\beta_1} \cdots p_r^{\beta_r} p_{r+1}^{\beta_{r+1}} \cdots p_{r+s}^{\beta_{r+s}}$$

with $0 \leqslant \beta_i \leqslant \alpha_i$. We then choose $a = p_{r+1}^{\beta_{r+1}} \cdots p_{r+s}^{\beta_{r+s}}$ and $b = p_1^{\beta_1} \cdots p_r^{\beta_r}$.

(iv) We write

$$n = \prod_{i=1}^{r} p_i^{\alpha_i} = \prod_{\alpha_i \geqslant k} p_i^{\alpha_i} \prod_{\alpha_i < k} p_i^{\alpha_i}.$$

For every integer i such that $\alpha_i \geqslant k$, the Euclidean division of α_i by k gives a unique pair (h_i, β_i) of non-negative integers such that $\alpha_i = \beta_i + h_i k$ with $0 \leqslant \beta_i < k$, and hence

$$n = \left(\prod_{\alpha_i \geqslant k} p_i^{\beta_i} \prod_{\alpha_i < k} p_i^{\alpha_i} \right) \left(\prod_{\alpha_i \geqslant k} p_i^{h_i} \right)^k$$

and therefore we have proved that the decomposition exists. To show its unicity, suppose that we have $n = q_1 m_1^k = q_2 m_2^k$ with q_i being k-free and let p be a prime number. Then

$$v_p(q_1) + k v_p(m_1) = v_p(q_2) + k v_p(m_2)$$

and hence

$$k \mid \left(v_p(q_1) - v_p(q_2) \right)$$

and since the q_i are k-free, we get $v_p(q_1) = v_p(q_2)$. We conclude that $q_1 = q_2$ and $m_1 = m_2$.

(v) Let $n = \prod_{i=1}^{r} p_i^{\alpha_i}$ be a square-full number, so that $\alpha_i \geqslant 2$ for $i = 1, \ldots, r$. Thus, for every α_i, there exists a unique integer β_i such that

$$\begin{cases} \alpha_i = \beta_i + 2 h_i \\ \beta_i \in \{2, 3\} \end{cases} \quad \text{with} \quad h_i = \left[\frac{\alpha_i - 2}{2} \right] \in \mathbb{Z}_{\geqslant 0}$$

and then

$$n = \left(\prod_{i=1}^{r} p_i^{h_i} \right)^2 \prod_{i=1}^{r} p_i^{\beta_i} = \left(\prod_{i=1}^{r} p_i^{h_i} \prod_{\beta_i=2} p_i \right)^2 \prod_{\beta_i=3} p_i^3$$

which proves the existence of the decomposition. To show its unicity, suppose that we have $n = a_1^2 b_1^3 = a_2^2 b_2^3$ with b_i being squarefree and let p be a prime number. Then

$$2\big(v_p(a_1) - v_p(a_2)\big) = 3\big(v_p(b_2) - v_p(b_1)\big)$$

and Gauss's theorem implies that

$$2 \mid \big(v_p(b_2) - v_p(b_1)\big)$$

which implies that $v_p(b_1) = v_p(b_2)$ since the b_i are squarefree, and hence $b_1 = b_2$ and $a_1 = a_2$.

The proof is complete.   □

Remark 3.8 More generally, one can show that every k-full integer $n \geqslant 2$ can be expressed in a unique way as

$$n = a_1^k a_2^{k+1} \cdots a_k^{2k-1}$$

where $a_2 \cdots a_k$ is squarefree and $(a_i, a_j) = 1$ for all integers $2 \leqslant i < j \leqslant k$. For example, the cube-full integer $n = 2^8 \times 3^7 \times 5^4 \times 11^3 \times 19^{32}$ can be written in a unique way as

$$n = \big(2 \times 3 \times 11 \times 19^9\big)^3 \times (3 \times 5)^4 \times (2 \times 19)^5.$$

Example 3.9 Prove that every integer $n \geqslant 1$ can be written in a unique way as $n = 2^e m$ where m is odd.

Answer If $n = 1$, then we have $e = 0$ and $m = 1$. If $n \geqslant 3$ is odd, then $e = 0$ and $m = n$. Now suppose that $n \geqslant 2$ is even. We get

$$n = 2^e p_1^{\alpha_1} \cdots p_r^{\alpha_r}$$

with p_i being odd primes. We deduce that the integer $m = p_1^{\alpha_1} \cdots p_r^{\alpha_r}$ is odd.

Example 3.10 Let n be a positive integer. Show that $\tau(n)$ is odd if and only if n is a square.

Answer We can clearly suppose $n \geqslant 2$ written as $n = \prod_{i=1}^{r} p_i^{\alpha_i}$. If n is not a square, there exists an integer $i \in \{1, \ldots, r\}$ such that α_i is odd and then we get $\alpha_i + 1 \equiv 0 \pmod 2$, and hence

$$\tau(n) = (\alpha_1 + 1) \cdots (\alpha_i + 1) \cdots (\alpha_r + 1) \equiv 0 \pmod 2.$$

Conversely, if n is a square, then *all* valuations α_i are even, and then $\alpha_i + 1 \equiv 1 \pmod 2$ for every integer $i \in \{1, \ldots, r\}$, so that

$$\tau(n) = \prod_{i=1}^{r} (\alpha_i + 1) \equiv 1 \pmod 2.$$

Example 3.11 Prove there exists a polynomial $P \in \mathbb{Z}[X]$ such that, for m values of the variable x, $P(x)$ is a prime number distinct from the former.

Answer It suffices to choose $P = X + (X - p_1)(X - p_2) \cdots (X - p_m)$ where p_k is the kth prime.

Example 3.12 Show that, if $P \in \mathbb{Z}[X]$ is a non-constant polynomial, then there exist infinitely many integers x such that $|P(x)|$ is not a prime number.

Answer We set

$$P = \sum_{j=0}^{n} a_j X^j$$

with $n \in \mathbb{N}$ and $a_j \in \mathbb{Z}$, $a_n \neq 0$. Suppose there exists $x_0 \in \mathbb{N}$ such that $|P(x_0)| = p_0$ with p_0 prime. Since $n \geqslant 1$, we have $\lim_{x \to \infty} |P(x)| = +\infty$. In particular, there exists $x_1 > x_0$ such that $|P(x)| > p_0$ as soon as $x \geqslant x_1$. On the other hand, for every integer m such that $x_0 + mp_0 \geqslant x_1$, we have

$$P(x_0 + mp_0) = \sum_{j=0}^{n} a_j (x_0 + mp_0)^j$$

$$\equiv \sum_{j=0}^{n} a_j x_0^j \ (\mathrm{mod}\ p_0)$$

$$\equiv p_0 \equiv 0 \ (\mathrm{mod}\ p_0)$$

and hence $|P(x_0 + mp_0)| = kp_0$ for some non-negative integer k. The inequality

$$\left| P(x_0 + mp_0) \right| > p_0$$

implies that $k > 1$ and hence $|P(x_0 + mp_0)|$ is composite.

3.2 Euclid's Theorem

We give three proofs of the famous Euclid's theorem which states that there are infinitely many primes. The first one, discovered by Euclid (300 BC), uses a very clever argument but remains elementary. Euler (1707–1783) chose an analytic argument, and his proof gave birth to what we call today *analytic number theory*. Finally, the third proof comes from the work of Erdős (1913–1996) who used some very clever combinatorial ideas. These three mathematicians, considered to be geniuses, made remarkable and extensive work in a lot of domains of mathematics, and are in fact inescapable in number theory. In order to pay tribute to their talent, let us call these proofs *the three E*.

Theorem 3.13 *There are infinitely many prime numbers.*

Proof (Euclid) Suppose that the set of prime numbers is finite, say $\mathcal{P} = \{p_1, p_2, \ldots, p_n\}$. We let $M = p_1 p_2 \cdots p_n$. Without loss of generality, we may suppose that M is composite. Since $M + 1 \geqslant 2$, it has a prime factor q, and since $(M, M + 1) = 1$, we have $q \neq p_i$ for all $i = 1, \ldots, n$. Hence we found a prime number not lying in $\mathcal{P}$, giving a contradiction. $\qquad\square$

Proof (Euler) Euler noticed that Theorem 3.3 implies the following estimate. For every positive integer n, we have

$$\sum_{k=1}^{n} \frac{1}{k} \leqslant \prod_{p \leqslant n} \left(1 - \frac{1}{p}\right)^{-1}. \tag{3.2}$$

Indeed, using (1.1) with $x = 1/p$, we first get

$$\left(1 - \frac{1}{p}\right)^{-1} = \sum_{k=0}^{\infty} \frac{1}{p^k}$$

and expanding the product $\prod_{p \leqslant n}(1 + 1/p + 1/p^2 + 1/p^3 + \cdots)$ gives through Theorem 3.3 the important equality

$$\prod_{p \leqslant n} \left(1 - \frac{1}{p}\right)^{-1} = \sum_{P^+(k) \leqslant n} \frac{1}{k}.$$

Since every integer $k \leqslant n$ obviously satisfies the condition $P^+(k) \leqslant n$, we have

$$\sum_{P^+(k) \leqslant n} \frac{1}{k} \geqslant \sum_{k \leqslant n} \frac{1}{k}$$

giving (3.2). Now using

$$\sum_{k \leqslant n} \frac{1}{k} > \int_1^n \frac{dt}{t} = \log n$$

and the inequality

$$-\log(1 - x) \leqslant x + \frac{x^2}{2(1 - x)}$$

valid for $0 \leqslant x < 1$, we obtain

$$\log \log n < \log \prod_{p \leqslant n} \left(1 - \frac{1}{p}\right)^{-1} = -\sum_{p \leqslant n} \log\left(1 - \frac{1}{p}\right)$$

$$\leqslant \sum_{p \leqslant n} \frac{1}{p} + \frac{1}{2} \sum_{p \leqslant n} \frac{1}{p(p-1)}$$

$$\leqslant \sum_{p \leqslant n} \frac{1}{p} + \frac{1}{2} \sum_{k=2}^{\infty} \frac{1}{k(k-1)} = \sum_{p \leqslant n} \frac{1}{p} + \frac{1}{2}$$

so that

$$\sum_{p \leqslant n} \frac{1}{p} > \log \log n - \frac{1}{2} \tag{3.3}$$

and hence we deduce that the series $\sum_p 1/p$ is divergent, which implies that there are infinitely many primes. $\qquad\square$

Proof (Erdős) N is a fixed positive integer. Using Corollary 3.7 (iv) we infer that every positive integer $n \leqslant N$ can be expressed in a unique way as $n = qm^2$ with q squarefree, i.e.

$$n = p_1^{\alpha_1} \cdots p_r^{\alpha_r} m^2 \quad \text{with} \quad \alpha_i \in \{0, 1\}.$$

Thus we can count such integers in the two following ways. First, there are *exactly* N integers n between 1 and N. On the other hand, there are 2^r possible choices for the valuations α_i, and since

$$m \leqslant n^{1/2} \leqslant N^{1/2}$$

we deduce that there are at most $N^{1/2}$ possible choices for m, and hence there are *at most* $2^r N^{1/2}$ integers n. Therefore we get

$$N \leqslant 2^r N^{1/2}$$

which gives

$$r \geqslant \frac{\log N}{\log 4}$$

and making $N \longrightarrow \infty$ gives the stated result. $\qquad\square$

Remark 3.14 We will see a better estimate than (3.3) in Corollary 3.50. Also, one can ask for a possible generalization of Euclid's theorem. For example, are there infinitely many prime numbers p in arithmetic progressions $p \equiv a \pmod{q}$ where a, q are positive integers such that $(a, q) = 1$? In fact, this is true and the answer is given by *Dirichlet's theorem* (Theorem 3.63). But how to prove it? A first attempt was to generalize Euclid's argument, but it can be proved that this method cannot treat all the cases (see Proposition 3.66). The answer will come from a generalization of Euler's analytic method. Let us notice that the same idea was used by Viggo Brun in 1918 [Bru19] to show that the series of inverses of twin primes, i.e. primes p such that $p + 2$ is also prime, converges (Theorem 3.79). Nobody knows yet if there are infinitely many twin primes.

3.3 Fermat, Lagrange and Wilson

3.3.1 Important Tools for Primes

Theorem 3.15 (Fermat's little theorem) *Let $a \in \mathbb{Z}$ and p be a prime number such that $p \nmid a$. Then we have*

$$a^{p-1} \equiv 1 \pmod{p}.$$

It is equivalent to say that, for all $a \in \mathbb{Z}$ and prime number p, then we have

$$a^p \equiv a \pmod{p}.$$

Proof The numbers $a, 2a, \ldots, (p-1)a$ are $p-1$ multiples of a all distinct modulo p. Indeed, if there exist integers $1 \leqslant m, n < p$ such that $ma \equiv na \pmod{p}$, then Gauss's theorem implies that $m \equiv n \pmod{p}$ since $p \nmid a$. Since $1 \leqslant m, n < p$, we get $m = n$. Thus, if p is prime such that $p \nmid a$ and if $m \neq n$ such that $1 \leqslant m, n < p$, then $ma \not\equiv na \pmod{p}$. Therefore the $p-1$ multiples of a above are all congruent to $1, 2, 3, \ldots, p-1$ modulo p in a certain order, and then

$$a \times 2a \times 3a \times \cdots \times (p-1)a \equiv 1 \times 2 \times 3 \times \cdots \times (p-1) \pmod{p}$$

so that

$$a^{p-1}(p-1)! \equiv (p-1)! \pmod{p}.$$

Note that $((p-1)!, p) = 1$ since $(p-1)! = \prod_{k=1}^{p-1} k$ and $k < p$ implies that $(k, p) = 1$. We then conclude the proof by appealing to Gauss's theorem.

The second statement follows easily. Indeed, if $a^{p-1} \equiv 1 \pmod{p}$, then $a^p \equiv a \pmod{p}$. Conversely, if $p \nmid a$, then Gauss's theorem implies that $a^{p-1} \equiv 1 \pmod{p}$. Thus, there is equivalence if $p \nmid a$. If $p \mid a$, then $a^p \equiv a \equiv 0 \pmod{p}$. $\square$

Remark 3.16 The converse of Fermat's little theorem is untrue in general. For example, $2^{341} \equiv 2 \pmod{341}$, but $341 = 11 \times 31$ is composite. This leads to the following notion. A composite number $n > 1$ is a *pseudoprime* if $2^n \equiv 2 \pmod{n}$. Thus, 341 is a pseudoprime. In the same way, a *Carmichael number* is a composite number $n > 1$ such that $a^n \equiv a \pmod{n}$ for every integer a. The least Carmichael number is $561 = 3 \times 11 \times 17$. In 1994, Alford, Granville and Pomerance showed that there are infinitely many Carmichael numbers (see [AGP94]) by establishing that $C(x) > x^{2/7}$, where $C(x)$ is the counting function of Carmichael numbers.

Theorem 3.17 (Lagrange)

(i) *Let $P \in \mathbb{Z}[X]$ be a non-zero polynomial with degree n and p be a prime number. Then either the congruence*

$$P(x) \equiv 0 \pmod{p}$$

has at most n incongruent zeros modulo p or p divides each coefficient of P.

(ii) *Let $P \in \mathbb{Z}[X]$ be a monic polynomial with degree n and p be a prime number. Suppose that the congruence $P(x) \equiv 0 \pmod{p}$ has at most n incongruent zeros $a_1, \ldots, a_n$ modulo p. Then we have*

$$P \equiv (X - a_1) \cdots (X - a_n) \pmod{p}.$$

Proof

(i) We prove the assertion by induction. The result is true for $n = 0$. Let $N \in \mathbb{N}$ and suppose the proposition is true for every integer $n < N$. Let $P \in \mathbb{Z}[X]$ such that $\deg P = N$. If there is no solution, then the desired conclusion follows for P. So we suppose that the congruence $P(x) \equiv 0 \pmod{p}$ has at least a solution a. Therefore, there exists a polynomial $Q \in \mathbb{Z}[X]$ with degree $N - 1$ such that

$$P(X) = Q(X)(X - a) + kp$$

for some integer k. Since $P(X) \equiv Q(X)(X - a) \pmod{p}$, Lemma 3.4 implies that

$$P(b) \equiv 0 \pmod{p} \quad \Longleftrightarrow \quad Q(b) \equiv 0 \pmod{p} \quad \text{or} \quad a \equiv b \pmod{p}.$$

Since $\deg Q = N - 1$, induction hypothesis implies that either there are at most $N - 1$ integers b such that $Q(b) \equiv 0 \pmod{p}$, or each coefficient of Q is divisible by p. We deduce that either there are at most N integers b such that $P(b) \equiv 0 \pmod{p}$ or p divides each coefficients of P.

(ii) We set $Q = P - (X - a_1) \cdots (X - a_n)$. We have $\deg Q \leqslant n - 1$ since P is monic. Furthermore, the congruence $Q(x) \equiv 0 \pmod{p}$ has the n incongruent solutions $a_1, \ldots, a_n$ modulo p. Using (i) we infer that p divides each coefficient of Q.

The proof is complete. □

Remark 3.18 The condition p prime number is crucial. For instance, the congruence $x^2 - 1 \equiv 0 \pmod{8}$ has four incongruent solutions modulo 8 which are 1, 3, 5 and 7. And yet the reader will have noticed that the leading coefficient and the moduli are coprime. Sierpiński [Sie64] proved that, for a composite moduli m, the case $m = 4$ only works.

If $P \in \mathbb{Z}[X]$ is a non-zero polynomial with degree n such that the leading coefficient is relatively prime to 4, then the congruence $P(X) \equiv 0 \pmod{4}$ has at most n incongruent solutions modulo 4.

Theorem 3.19 (Wilson)

(i) *If p is a prime number then we have $(p - 1)! \equiv -1 \pmod{p}$.*
(ii) *If $n \geqslant 2$ satisfies $(n - 1)! \equiv -1 \pmod{n}$ then n is prime.*

Proof

(i) Let p be a prime number and set $P = X^{p-1} - 1$. Using Fermat's and Lagrange's theorems we get

$$P \equiv (X - 1)(X - 2) \cdots \big(X - (p - 1)\big) \ (\mathrm{mod}\, p).$$

We get the desired result by evaluating $P(0)$.

(ii) Suppose now that $n \mid (n - 1)! + 1$. If n is composite, then there exists an integer d verifying $1 < d < n$ and $d \mid (n - 1)! + 1$. Since $d \leqslant n - 1$, then $d \mid (n - 1)!$ and therefore $d \mid 1$, giving a contradiction.

The proof is complete. $\square$

Remark 3.20 We thus proved the following equivalence.

An integer $n > 1$ is a prime number if and only if $(n - 1)! \equiv -1 \ (\mathrm{mod}\, n)$

which gives a primality test from a theoretical point of view. Unfortunately this test is ineffective if n is too large. Nevertheless one can notice that, since

$$(n - 1)! = (n - 2)!(n - 1) \equiv -(n - 2)! \ (\mathrm{mod}\, n)$$

one can deduce that *an integer $n > 1$ is prime if and only if $(n - 2)! \equiv 1 \ (\mathrm{mod}\, n)$*. Dickson [Sie64] generalized this criterion in the following way.

An integer $n > 1$ is a prime number if and only if there exists an integer $m < n$ such that

$$(m - 1)! \, (n - m)! \equiv (-1)^m \ (\mathrm{mod}\, n).$$

Example 3.21 Prove that $11 \times 31 \times 61 \mid (20^{15} - 1)$.

Answer Since $11, 31$ and 61 are primes, it is sufficient to show that $20^{15} - 1$ is divisible by these three numbers. By Fermat's little theorem, we have $2^{10} \equiv 1 \ (\mathrm{mod}\, 11)$ and since $10 \equiv -1 \ (\mathrm{mod}\, 11)$, we get

$$20^{15} \equiv 2^5 \times 2^{10} \times 10^{15} \equiv -2^5 \equiv 1 \ (\mathrm{mod}\, 11)$$

since $-2^5 = 1 - 3 \times 11$. Another application of Fermat's little theorem implies that $4^{15} = 2^{30} \equiv 1 \ (\mathrm{mod}\, 31)$, and since $5^3 = 1 + 4 \times 31$, we get

$$20^{15} = 4^{15} \times \left(5^3\right)^5 \equiv 1 \ (\mathrm{mod}\, 31).$$

Finally, since $3^4 \equiv 20 \ (\mathrm{mod}\, 61)$ and $3^{60} \equiv 1 \ (\mathrm{mod}\, 61)$ by Fermat's little theorem we get

$$20^{15} = 3^{60} \equiv 1 \ (\mathrm{mod}\, 61).$$

Example 3.22 Let $p \neq q$ be two distinct prime numbers. Show that

$$p^{q-1} + q^{p-1} \equiv 1 \pmod{pq}.$$

Answer Since $p \neq q$ are primes, Fermat's little theorem yields $p^{q-1} + q^{p-1} \equiv 1$ (mod p) and $p^{q-1} + q^{p-1} \equiv 1$ (mod q) and we conclude by using Proposition 2.13 (vi).

Example 3.23 Show that for every positive integer n, the numbers $(2^{2(28n+1)}+1)^2 + 4$ are composite.

Answer Using Fermat's little theorem we get $2^{28} \equiv 1 \pmod{29}$ so that $2^{2(28n+1)} \equiv 4 \pmod{29}$ and thus

$$\left(2^{2(28n+1)} + 1\right)^2 + 4 \equiv 5^2 + 4 \equiv 0 \pmod{29}.$$

3.3.2 Multiplicative Order

Fermat's little theorem implies the following notion.

Theorem 3.24 (Multiplicative order) *Let $a \in \mathbb{Z}$ and p be a prime number such that $(a, p) = 1$. Then the set of positive integers n satisfying $a^n \equiv 1$ (mod p) has a smallest element, denoted by $\operatorname{ord}_p(a)$ and called the* multiplicative order of a *modulo p.*

Proof The set in question is a non-empty subset of $\mathbb{Z}_{\geq 0}$ since $p - 1$ belongs to this set by Fermat's little theorem.   □

The following result gives a characterization of the multiplicative order of a positive integer.

Theorem 3.25 *Let $a \in \mathbb{Z}$ and p be a prime number such that $p \nmid a$. Then $a^n \equiv 1$ (mod p) if and only if $\operatorname{ord}_p(a)$ divides n. In particular, we have*

$$\operatorname{ord}_p(a) \mid (p - 1). \tag{3.4}$$

Furthermore, if $a^n \equiv a^m$ (mod p) then $\operatorname{ord}_p(a)$ divides $n - m$.

Proof Set $\delta = \operatorname{ord}_p(a)$. If $n = k\delta$ then

$$a^n = \left(a^\delta\right)^k \equiv 1 \pmod{p}.$$

Conversely, let $\mathcal{S} = \{n \in \mathbb{N} : a^n \equiv 1 \pmod{p}\}$. By Theorem 3.24, δ is the smallest element in $\mathcal{S}$. The Euclidean division of n by δ supplies a unique pair (q, r) of positive integers such that $n = q\delta + r$ and $0 \leqslant r < \delta$. We then get

$$1 \equiv a^n \equiv a^r \left(a^\delta\right)^q \equiv a^r \pmod{p}.$$

Hence, if $r \geqslant 1$, we infer that $r \in \mathcal{S}$ and thus $r \geqslant \delta$ which contradicts the inequality $r < \delta$. This implies that $r = 0$ and then $\delta \mid n$. (3.4) follows at once from Fermat's little theorem. Finally, suppose for example that $n > m$. Since

$$a^n - a^m = a^m \left(a^{n-m} - 1\right) \equiv 0 \pmod{p}$$

Lemma 3.4 implies that

$$a^m \equiv 0 \pmod{p} \quad \text{or} \quad a^{n-m} - 1 \equiv 0 \pmod{p}.$$

Since $p \nmid a$ the first congruence is never achieved. Hence we obtain $a^{n-m} \equiv 1 \pmod{p}$ and therefore we have δ divides $n - m$ according to the previous statement. $\qquad\square$

Example 3.26 Determine $\operatorname{ord}_{101}(2)$.

Answer Let $\delta = \operatorname{ord}_{101}(2)$. Using (3.4) we get $\delta \mid 100$. Since $2^{10} \equiv 14 \pmod{101}$ and $2^{25} \equiv 10 \pmod{101}$, we obtain $\delta = 100$.

Example 3.27 Let p be an odd prime number. Prove that every prime divisor of $2^p - 1$ is of the shape $1 + 2kp$ for some positive integer k.

Answer Let q be a prime divisor of $2^p - 1$ and $\delta = \operatorname{ord}_q(2)$. Thus we have $2^p \equiv 1 \pmod{q}$ and hence $\delta \mid p$. Since $\delta \neq 1$, otherwise we have $2 \equiv 1 \pmod{q}$ which is impossible, we get $\delta = p$, and then using (3.4) we have $p \mid (q - 1)$. Thus there exists a positive integer h such that $q = 1 + ph$. Furthermore, q is odd since $2^p - 1$ is odd, and hence ph is even, and therefore h is even since p is odd, and then there exists a positive integer k such that $q = 1 + 2kp$.

Example 3.28 Let p be a prime number. Prove that, if $d \mid (p - 1)$, then the congruence $x^d \equiv 1 \pmod{p}$ has exactly d solutions.

Answer Using Lagrange's theorem we know that this congruence has *at most* d solutions. Let us prove that it has *at least* d solutions. Suppose there are $< d$ incongruent solutions modulo p. Since

$$x^{p-1} - 1 = \left(x^d - 1\right) P(x)$$

where $P \in \mathbb{Z}[x]$ has degree $p - 1 - d$, $a \in \mathbb{Z}$ is solution of the equation $x^{p-1} - 1 \equiv 0$ (mod p) if and only if a is solution of $x^d - 1 \equiv 0$ (mod p) or $P(x) \equiv 0$ (mod p). Lagrange's theorem implies that the congruence $P(x) \equiv 0$ (mod p) has at most $p - 1 - d$ solutions. Hence $x^{p-1} - 1 \equiv 0$ (mod p) has $< d + p - 1 - d = p - 1$ solutions, which contradicts Fermat's little theorem.

3.3.3 Primitive Roots and Artin's Conjecture

Let p be a prime number satisfying $p \neq 2$ and $p \neq 5$. The decimal expansion of $1/p$ is then purely periodic. For example

$$\frac{1}{7} = 0.142857\,142857\ldots \quad \text{and} \quad \frac{1}{11} = 0.09\,09\,09\ldots$$

We shall have a look at the length of the period of the decimal expansion of $1/p$. Since $p \nmid 10$, Theorem 3.24 implies that $\delta = \mathrm{ord}_p(10)$ is well-defined. Hence there exists an integer q such that $10^\delta = 1 + qp$. Therefore we have using (1.1)

$$\frac{1}{p} = \frac{q}{10^\delta - 1} = q \sum_{k=1}^{\infty} 10^{-\delta k}$$

so that the length of the period of the decimal expansion of $1/p$ is equal to $\delta = \mathrm{ord}_p(10)$. In particular, this length[2] divides $p - 1$ by (3.4). Therefore one can ask when the period of the decimal expansion of $1/p$ has its maximal length, in other words when $\mathrm{ord}_p(10) = p - 1$. This leads to the following definition.

Definition 3.29 Let p be a prime number. An integer g is a *primitive root* modulo p if $\mathrm{ord}_p(g) = p - 1$.

The examples above show that 10 is a primitive root modulo 7, but is not a primitive root modulo 11.

There are essentially three questions about primitive roots. First of all, for which prime number p does there exist a primitive root modulo p? The next result provides a complete answer to this question.

Proposition 3.30 *Let p be a prime number. Then there exists a primitive root modulo p.*

Proof We shall construct a possible candidate in the following way. Let $\delta_1, \delta_2, \ldots, \delta_r$ be all the possible orders modulo p of the numbers $1, 2, \ldots, p - 1$ and we define

---

[2]This problem has a long history as it goes back at least to J. H. Lambert in 1769 (see [Dic05, Chap. VI]).

$m = [\delta_1, \ldots, \delta_r]$ written as the product

$$m = p_1^{\alpha_1} \cdots p_s^{\alpha_s}.$$

For each prime power $p_i^{\alpha_i}$, there is some δ_j divisible by it, and hence there exists k such that $p_i \nmid k$ and $\delta_j = p_i^{\alpha_i} k$. Since $\delta_1, \ldots, \delta_r$ are the orders modulo p of $1, 2, \ldots, p-1$, there exists an element x_i of order $p_i^{\alpha_i} k$, and then the order of the element $y_i = x_i^k$ is $p_i^{\alpha_i}$. It follows that the order modulo p of the number $y_1 \cdots y_s$ is equal to m, so that $m \mid (p-1)$ by (3.4). On the other hand, the polynomial $x^m - 1$ has $p-1$ roots modulo p since $1, 2, \ldots, p-1$ are roots. Using Example 3.28, we deduce that $m = p - 1$, and the number $y_1 \cdots y_s$ is a primitive root modulo p. $\square$

Another proof will be supplied in Chap. 4 as a consequence of the Möbius inversion formula (see Theorem 4.13).

In view of this result, we may ask the following questions.

1. For a fixed prime p, how many primitive roots are there?
2. For a fixed integer a, for how many primes will a be a primitive root?

For the first question, it can be shown that the number of primitive roots modulo a fixed prime p is equal to $\varphi(p-1)$, where φ is Euler's totient defined in Theorem 2.31.

The second question above is far more difficult. Let us first see this with an example, where we ask whether 2 is a primitive root modulo p. With the help of tables of primitive roots, one can see, among primes $p \leqslant 100$, that

a. 2 is a primitive root modulo 3, 5, 11, 13, 19, 29, 37, 53, 59, 61, 67, 83.
b. 2 is not a primitive root modulo 7, 17, 23, 31, 41, 43, 47, 71, 73, 79, 89, 97.

Since each list contains 12 elements, one can think that the probability that 2 is a primitive root modulo p is 50%. In fact, extending the list, one should see that this probability could approach a number closer to 37%. More precisely, in 1927, Emil Artin formulated the following conjecture.

Let a be a non-zero integer such that $a \neq \pm 1$. If a is not a square, then there are infinitely many prime numbers p such that a is a primitive root modulo p.

This conjecture comes from the following more precise statement also given by Artin.

Let a be a non-zero integer such that $a \neq \pm 1$ and h be the largest integer such that $a = a_0^h$ with $a_0 \in \mathbb{Z}$. If $\mathcal{N}_a(x)$ is the number of primes $p \leqslant x$ such that a is a primitive root modulo p, then

$$\mathcal{N}_a(x) = \mathcal{A}_h \frac{x}{\log x} + o\left(\frac{x}{\log x}\right),$$

where

$$\mathcal{A}_h = \prod_{p \nmid h}\left(1 - \frac{1}{p(p-1)}\right) \prod_{p \mid h}\left(1 - \frac{1}{p-1}\right). \tag{3.5}$$

Note that the product above is equal to zero when h is even. Hence, the condition that a is not a square is *necessary* for the number of prime numbers p, such that a is a primitive root modulo p, to be infinite. When h is odd, then the product is a positive rational multiple of the so-called *Artin constant*

$$A = A_1 = \sum_{n=1}^{\infty} \frac{\mu(n)}{n\varphi(n)} = \prod_p \left(1 - \frac{1}{p(p-1)}\right) \approx 0.373\,955\,813\,619\ldots$$

where μ is the Möbius function.[3] Artin's conjecture is known to be true as soon as a very difficult hypothesis is proved (see below), but the latter hypothesis is likely to be out of reach for several generations of researchers.

Unconditionally, there is no known number a for which the set $S(a)$ of prime numbers p, such that a is a primitive root modulo p, is known to be infinite. Nevertheless, improving on fundamental work by Gupta and Ram Murty [GM84], Heath-Brown [HB86] succeeded in showing that there are at most two "bad" prime numbers a such that $S(a)$ is finite, and also at most three "bad" squarefree numbers a such that $S(a)$ is finite. Furthermore, he proved that the set S of integers for which Artin's conjecture does not hold is rather thin by establishing the following estimate

$$\sum_{\substack{n \leqslant x \\ n \in S}} 1 \ll (\log x)^2.$$

Let us turn our attention for a while to the heuristic arguments that led to Artin's conjecture. A necessary and sufficient condition for a to be a primitive root modulo p is

$$a^{(p-1)/q} \not\equiv 1 \pmod{p} \quad \text{for all primes } q \text{ dividing } p - 1.$$

Indeed, the condition is clearly sufficient by Definition 3.29. It is also necessary by using the following argument. Suppose that $a^{(p-1)/q} \not\equiv 1 \pmod{p}$ for all primes $q \mid (p-1)$ and that a is not a primitive root mod p. Then $a^{(p-1)/k} \equiv 1 \pmod{p}$ for some integer $k > 1$ such that $k \mid (p-1)$, which in turn implies that $a^{(p-1)/r} \equiv 1 \pmod{p}$ for some prime divisor r of k, resulting in a contradiction.

Hence the heuristic idea is that a is a primitive root mod p if the two following events do not occur simultaneously for any prime q

$$E_1 : a^{(p-1)/q} \equiv 1 \pmod{p}.$$

$$E_2 : p \equiv 1 \pmod{q}.$$

Now by Dirichlet's theorem (Theorem 3.61), the probability of E_2 is equal to $1/(q-1)$, while the probability of E_1 is equal to $1/q$ if $q \nmid h$ and to 1 if $q \mid h$, since, in the first case, we note that $a^{(p-1)/q}$ is a solution of the equation $x^q \equiv 1 \pmod{p}$ and we want a solution congruent to $1 \pmod{q}$ amongst the q expected roots, whereas

---

[3] See Example 4.2.

in the second case we always have $a^{(p-1)/q} = a_0^{h(p-1)/q} \equiv 1 \pmod{p}$. Therefore, if we assume that these events are independent, then the probability that both occur is

$$\frac{1}{q(q-1)} \text{ if } q \nmid h \quad \text{and} \quad \frac{1}{q-1} \text{ if } q \mid h$$

leading to the product (3.5).

Artin's conjecture is one of the richest problems in number theory, notably on account of the connexion between algebraic number theory and analytic number theory. It can be shown that the two conditions $a^{(p-1)/q} \equiv 1 \pmod{p}$ and $p \equiv 1 \pmod{q}$ are equivalent to the fact that p splits completely[4] in normal extensions $\mathbb{K}_q = \mathbb{Q}(\zeta_q, a^{1/q})$ of $\mathbb{Q}$, where $\zeta_q = e_q(1)$ is a qth primitive root of unity. In the 1930s, great efforts were made to prove this conjecture. In 1935, Erdős tried in vain to combine infinitely many q. Finally, using a special form of Chebotarëv's density theorem with a sharp error-term only accessible under the extended Riemann hypothesis for the number fields $\mathbb{K}_q$, Hooley [Hoo67] succeeded in proving a somewhat modified conjecture. Indeed, in 1957, Derrick H. Lehmer and his wife Emma numerically tested Artin's conjecture on a computer and found that the naive heuristic approach described above may be wrong in some cases. Artin then introduced a correction factor to explain these discrepancies,[5] and the corrected conjectured density under ERH finally turns out to be

$$\sum_{n=1}^{\infty} \frac{\mu(n)}{[\mathbb{K}_n : \mathbb{Q}]}$$

where $[\mathbb{K}_n : \mathbb{Q}]$ is the degree[6] of the number field $\mathbb{K}_n = \mathbb{Q}(\zeta_n, a^{1/n})$. If the discriminant[7] $d_{\mathbb{K}_2}$ of the quadratic field $\mathbb{K}_2 = \mathbb{Q}(\sqrt{a})$ is even, then the function $n \longmapsto [\mathbb{K}_n : \mathbb{Q}]$ is multiplicative and the expression above reduces to the product $\mathcal{A}_h$ given in (3.5), but when $d_{\mathbb{K}_2}$ is odd, then, for all squarefree n divisible by $d_{\mathbb{K}_2}$, we have

$$[\mathbb{K}_n : \mathbb{Q}] = \frac{1}{2} \prod_{p \mid n} [\mathbb{K}_p : \mathbb{Q}]$$

so that the function $n \longmapsto [\mathbb{K}_n : \mathbb{Q}]$ is no longer multiplicative. Hooley derived the following expression

$$\sum_{n=1}^{\infty} \frac{\mu(n)}{[\mathbb{K}_n : \mathbb{Q}]} = \left\{ 1 - \mu\left(|d_{\mathbb{K}_2}|\right) \prod_{p \mid d_{\mathbb{K}_2}} \left([\mathbb{K}_p : \mathbb{Q}] - 1\right)^{-1} \right\} \mathcal{A}_h$$

---

[4] See Definition 7.101.

[5] "So I was careless but the machine caught up with me". This is how Artin concluded the fourth letter dated January 28, 1958 to Emma Lehmer, explaining that his conjecture may be false if a is a prime number satisfying $a \equiv 1 \pmod{4}$.

[6] See Definition 7.42.

[7] See Definition 7.58.

when $d_{\mathbb{K}_2}$ is odd. The reader interested in Artin's primitive root conjecture should refer to the very fruitful presentation [Ste03] of this subject.

We now end this section with a curious observation concerning decimal expansions of $1/p$ when $p > 5$ is a prime number. In 1836, Midy (see [Dic05]) proved that if this decimal expansion has even period $2d$ and writing $1/p = 0.(A_1 A_2)(A_1 A_2) \ldots$, where the blocks A_i have d digits each, then $A_1 + A_2 = 10^d - 1$. This result has recently been generalized in [GS05] where it is shown that, if the decimal expansion of $1/p$ has period kd with $k > 1$, and writing $1/p = 0.(A_1 \ldots A_k)(A_1 \ldots A_k) \ldots$ where the blocks A_i have d digits each, then there exists an integer $r = r(p, k) \geqslant 1$ such that $A_1 + \cdots + A_k = r(10^d - 1)$. Furthermore, we have $r = 1$ when $k = 2$, $k = 3$ or $p = 2^k - 1$ is a Mersenne prime.

3.3.4 Some Applications of Primitive Roots

We begin with a first result which has its own interest in itself.

Lemma 3.31 *Let p be a prime number and g a primitive root* mod p. *Then $1, g, g^2, \ldots, g^{p-2}$ are incongruent modulo p, and hence are congruent to $1, 2, \ldots, p - 1$ in a certain order.*

Proof It suffices to show that, if $0 \leqslant i < j \leqslant p - 2$, then $p \nmid (g^j - g^i)$. Suppose the contrary. Then we have $p \mid g^i (g^{j-i} - 1)$, and hence $p \mid (g^{j-i} - 1)$ by Gauss's theorem, otherwise p divides g^{p-1} contradicting the equality $g^{p-1} \equiv 1 \pmod{p}$. By using Theorem 3.25, we have $(p - 1) \mid (j - i)$ contradicting the fact that $0 < j - i < p - 1$. $\qquad\square$

Definition 3.32 Let p be a prime number and $n \geqslant 2$ be an integer. An integer a is said to be an *nth power residue* mod p if the congruence $x^n \equiv a \pmod{p}$ has a solution in $\{1, \ldots, p - 1\}$. When $n = 2$, a is a *quadratic residue* mod p.

It is clear that 0 is an nth power residue mod p for every prime p and integer $n \geqslant 2$. On the other hand, if a is an nth power residue mod p, then so is every integer congruent to $a \pmod{p}$. The following result gives a useful criterion and also the number of incongruent nth power residues mod p.

Proposition 3.33 *Let $n \geqslant 2$ be an integer, p be a prime number and set $d = (n, p - 1)$.*

(i) *Let a be an integer satisfying $p \nmid a$. Then a is an nth power residue* mod p *if and only if $a^{(p-1)/d} \equiv 1 \pmod{p}$.*
(ii) *The number of non-zero integers which are nth power residues* mod p *is given by $(p - 1)/d$.*

Proof Let p be a prime and g be a primitive root mod p.

(i) If an integer a is such that $p \nmid a$ and is an nth power residue mod p, then there exists a number x such that $p \nmid x$ and $a \equiv x^n \pmod{p}$, so that we have

$$a^{(p-1)/d} \equiv \left(x^n\right)^{(p-1)/d} \equiv \left(x^{p-1}\right)^{n/d} \equiv 1 \pmod{p}$$

by Fermat's little theorem.

Conversely, suppose that $a^{(p-1)/d} \equiv 1 \pmod{p}$. By Lemma 3.31, there exists an integer $k \in \{0, \ldots, p-2\}$ such that $a \equiv g^k \pmod{p}$, and hence $g^{k(p-1)/d} \equiv 1 \pmod{p}$, which implies that $(p-1) \mid k(p-1)/d$ since g is a primitive root mod p. Therefore we have $d \mid k$, so that $k = dh$ for some $h \in \mathbb{Z}_{\geqslant 0}$. By Theorem 2.5, there exist two non-negative integers u, v such that $d = nu - v(p-1)$, so that we get $k = dh = nhu - vh(p-1)$. By the use of the relation $g^{p-1} \equiv 1 \pmod{p}$, we get

$$a \equiv a\left(g^{p-1}\right)^{vh} \equiv g^k g^{vh(p-1)} \equiv g^{nhu} \equiv \left(g^{hu}\right)^n \pmod{p}$$

and therefore a is an nth power residue mod p.

(ii) If a is an nth power residue mod p, then $a^{(p-1)/d} \equiv 1 \pmod{p}$, hence using Lagrange's theorem, there are at most $(p-1)/d$ non-zero nth power residues mod p.

On the other hand, the numbers $g^n, g^{2n}, \ldots, g^{n(p-1)/d}$ are $(p-1)/d$ non-zero incongruent nth power residues mod p. Indeed, set $n = dh$ and $p - 1 = dk$ with $(h, k) = 1$, and let $1 \leqslant i < j \leqslant k$ be two integers. Since $(p, g) = 1$, if $g^{in} \equiv g^{jn} \pmod{p}$, then $g^{(j-i)n} \equiv 1 \pmod{p}$, so that $(p-1) \mid n(j-i)$ and hence $k \mid (j-i)$ by Gauss's theorem, contradicting the fact that $1 \leqslant i < j \leqslant k$. Furthermore, it is clear that the equation $x^n \equiv g^{jn} \pmod{p}$ has the solution $x = g^j$.

The proof is complete. $\square$

Example 3.34 Let $p = 7$ and $n = 2$, so that $d = 2$. The number $g_7 = 3$ is the smallest primitive root mod 7, so that the three quadratic residues mod 7 are 1, 2 and 4.

Example 3.35 Prove that, if $p \equiv 3 \pmod{4}$, then -1 is not a quadratic residue mod p.

Answer Suppose there exists a number x such that $x^2 \equiv -1 \pmod{p}$. We then have

$$x^{p-1} = \left(x^2\right)^{(p-1)/2} \equiv (-1)^{(p-1)/2} \equiv -1 \pmod{p}$$

which contradicts Fermat's little theorem.[8]

Example 3.36 Use Example 3.35 to show that the Diophantine equation $y^2 = x^3 + 11$ has no solution in $\mathbb{Z}^2$.

---

[8] See also the remark in the solution to Exercise 18.

Answer Suppose there is a solution pair $(x, y) \in \mathbb{Z}^2$. Then we have $y^2 \equiv x^3 - 1$ (mod 4), and since 0 and 1 are the unique squares modulo 4, we get $x^3 \equiv 1$ or 2 (mod 4), and then $x \equiv 1$ (mod 4) since 2 is not a cube modulo 4. Note that, if $y^2 = x^3 + 11$, then

$$y^2 + 16 = x^3 + 27 = (x + 3)(x^2 - 3x + 9)$$

and since $x \equiv 1$ (mod 4), this gives $x^3 - 3x + 9 \equiv 3$ (mod 4). Hence there exists a prime number $p \equiv 3$ (mod 4) dividing $y^2 + 16$. If a is the unique integer such that $4a \equiv 1$ (mod p), then the relation $y^2 + 16 \equiv 0$ (mod p) implies that $(ay)^2 \equiv -1$ (mod p). This contradicts the result of Example 3.35. Thus, the Diophantine equation $y^2 = x^3 + 11$ has no solution in $\mathbb{Z}^2$.

3.4 Elementary Prime Numbers Estimates

Between the years 1850 and 1875, a great number of estimates involving some functions of prime numbers were discovered by a few mathematicians including Chebyshev and Mertens. These results are the cornerstones of the whole research on primes which eventually led to the Prime Number Theorem in 1896. Although these estimates were not sufficient to show the PNT, they were, and still are, often used in many problems in number theory.

3.4.1 Chebyshev's Functions of Primes

We begin with a result which goes back to Legendre and gives a nice application of the integer part of a number.

Theorem 3.37 (Legendre) *Let p be a prime number and n be a positive integer. The p-adic valuation of $n!$ is given by*

$$v_p(n!) = \sum_{k=1}^{\infty} \left[\frac{n}{p^k} \right].$$

Note that the sum is in fact finite, since the integer part vanishes as soon as $k > \log n / \log p$.

First proof It rests on the fact that, if $m \leqslant n$, then every multiple of m not greater than n divides $n!$.

Let p be a prime number. If $p \leqslant n$, then there are $[n/p]$ multiples of p not greater than n by Proposition 1.11. Similarly, if $p^2 \leqslant n$, then there are $[n/p^2]$ multiples of p^2 not greater than n, and so on. We conclude the proof by adding all these multiples. $\square$

Second proof It rests on the fact that the p-adic valuation of a product is the sum of the p-adic valuations. Thus we get

$$v_p(n!) = v_p(1 \times \cdots \times n) = \sum_{m=1}^{n} v_p(m) = \sum_{m=1}^{n} \sum_{k=1}^{v_p(m)} 1 = \sum_{k=1}^{\infty} \sum_{\substack{m=1 \\ v_p(m) \geq k}}^{n} 1$$

and we conclude the proof by noticing that the inner sum counts the number of multiples of p^k not greater than n which is equal to $[n/p^k]$ by Proposition 1.11. $\square$

Example 3.38 Show that the number of zeros at the end of the decimal expansion of $n!$ is given by

$$\sum_{k=1}^{\infty} \left[\frac{n}{5^k} \right].$$

Answer It follows from the fact that we can write $n! = 10^{v_5(n!)} \times m$ with $10 \nmid m$, so that the desired number is equal to $v_5(n!)$. Legendre's theorem then gives the asserted result.

The following functions are of constant use in analytic number theory.

Definition 3.39

1. The *von Mangoldt function* Λ is defined by

$$\Lambda(n) = \begin{cases} \log p, & \text{if } n = p^\alpha \text{ for some prime } p \text{ and } \alpha \in \mathbb{N} \\ 0, & \text{otherwise.} \end{cases}$$

2. The *first Chebyshev function* θ is defined for $x \geqslant 2$ by

$$\theta(x) = \sum_{p \leqslant x} \log p$$

while it is convenient to set $\theta(x) = 0$ for $0 < x < 2$.
3. The *second Chebyshev function* Ψ is defined for $x \geqslant 2$ by

$$\Psi(x) = \sum_{n \leqslant x} \Lambda(n)$$

while it is convenient to set $\Psi(x) = 0$ for $0 < x < 2$.
4. The *prime counting function* π is defined by

$$\pi(x) = \sum_{p \leqslant x} 1$$

while it is convenient to set $\pi(x) = 0$ for $0 < x < 2$.

We will make use of the next lemma.

Lemma 3.40

(i) *For all $x > 0$, we have*

$$\theta(x) \leqslant \Psi(x) \leqslant \theta(x) + \pi(\sqrt{x})\log x.$$

(ii) *For all $x \geqslant 2$ and all $a > 1$, we have*

$$\frac{\theta(x)}{\log x} \leqslant \pi(x) \leqslant \frac{a\theta(x)}{\log x} + \pi(x^{1/a}).$$

Proof

(i) One may suppose $x \geqslant 2$. We first have

$$\Psi(x) - \theta(x) = \sum_{p^\alpha \leqslant x} \log p - \sum_{p \leqslant x} \log p = \sum_{p \leqslant \sqrt{x}} \sum_{\alpha=2}^{[\log x / \log p]} \log p$$

so that $\Psi(x) \geqslant \theta(x)$. On the other hand

$$\Psi(x) - \theta(x) = \sum_{p \leqslant \sqrt{x}} \sum_{\alpha=2}^{[\log x / \log p]} \log p \leqslant \sum_{p \leqslant \sqrt{x}} \log p \left[\frac{\log x}{\log p}\right] \leqslant \sum_{p \leqslant \sqrt{x}} \log x$$

$$= \pi(\sqrt{x})\log x.$$

(ii) We have

$$\pi(x) = \sum_{p \leqslant x} 1 = \sum_{p \leqslant x} \frac{\log p}{\log p} \geqslant \frac{1}{\log x} \sum_{p \leqslant x} \log p = \frac{\theta(x)}{\log x}.$$

For $2 \leqslant T < x$, we also have

$$\pi(x) = \sum_{p \leqslant T} 1 + \sum_{T < p \leqslant x} 1 = \pi(T) + \sum_{T < p \leqslant x} \frac{\log p}{\log p} \leqslant \pi(T) + \frac{\theta(x)}{\log T}$$

and the choice of $T = x^{1/a}$ implies the asserted estimate.

The proof of the lemma is complete.   □

The von Mangoldt function satisfies an important identity which is what we shall call later a *convolution identity* (see Chap. 4).

Lemma 3.41

(i) *For all positive integers n, we have*

$$\sum_{d|n} \Lambda(d) = \log n.$$

(ii) *For all $x \geqslant 1$, we have*

$$\sum_{d \leqslant x} \Lambda(d) \left[\frac{x}{d} \right] = x \log x - x + 1 + R(x)$$

where $0 \leqslant R(x) \leqslant \log x$.

Proof

(i) The identity is clear for $n = 1$. Suppose $n \geqslant 2$ written as

$$n = p_1^{\alpha_1} \cdots p_k^{\alpha_k}$$

with distinct primes p_i and positive integers α_i. By the definition of the von Mangoldt function, it is sufficient to consider in the above sum the divisors d of n written as $d = p_i^{\beta_i}$ with $1 \leqslant \beta_i \leqslant \alpha_i$ for $i = 1, \ldots, k$. Thus we get

$$\sum_{d \mid n} \Lambda(d) = \sum_{i=1}^{k} \sum_{\beta_i = 1}^{\alpha_i} \log p_i = \sum_{i=1}^{k} \log p_i^{\alpha_i} = \log n.$$

(ii) One may suppose $x > 1$. We use (i) to obtain

$$\sum_{n \leqslant x} \log n = \sum_{n \leqslant x} \sum_{d \mid n} \Lambda(d).$$

We will now interchange the summations. This process is of frequent use in analytic number theory (see Proposition 4.17). By setting $n = kd$ for some integer k, we get

$$\sum_{n \leqslant x} \log n = \sum_{d \leqslant x} \Lambda(d) \sum_{k \leqslant x/d} 1 = \sum_{d \leqslant x} \Lambda(d) \left[\frac{x}{d} \right]$$

and the estimate follows by using Corollary 1.17.

The proof is complete. $\qquad\qquad\qquad\qquad\qquad\qquad\qquad\qquad\qquad\qquad\qquad$ □

3.4.2 Chebyshev's Estimates

One of the main goals of Chebyshev was the proof of the Prime Number Theorem which can be stated as

$$\pi(x) \sim \frac{x}{\log x}$$

as $x \longrightarrow \infty$. Although Chebyshev did not reach this estimate, his work in this direction was crucial and Chebyshev's ideas are still often used. The following two results are fundamental.

Lemma 3.42 *For all $x \geqslant 1$, we have*

$$\theta(x) \leqslant x \log 4.$$

Proof We first prove by induction that, for all positive integers n, we have

$$\theta(n) \leqslant n \log 4. \tag{3.6}$$

This inequality is clearly true for $n \in \{1, 2, 3\}$ and we notice that, if $n \geqslant 4$ is even, we have

$$\theta(n) = \theta(n-1) \leqslant 4^{n-1} < 4^n.$$

Hence one may suppose that $n \geqslant 3$ is odd and set $n = 2m + 1$ with m a positive integer. The idea is to use the fact that the product

$$\prod_{m+1 < p \leqslant 2m+1} p$$

divides the binomial coefficient $\binom{2m+1}{m}$. To see this, it suffices to observe that a prime p such that $m + 1 < p \leqslant 2m + 1$ divides $(2m + 1)!$ because of $p \leqslant 2m + 1$, but does not divide $m!\,(m+1)!$ because of $p > m + 1$, so that

$$\prod_{m+1 < p \leqslant 2m+1} p \quad \text{divides } (2m+1)! = m!\,(m+1)! \times \binom{2m+1}{m}$$

and since the product is coprime to $m!(m+1)!$ by the arguments above, Gauss's theorem then gives the desired assertion.

Taking logarithms, we then get

$$\sum_{m+1 < p \leqslant 2m+1} \log p \leqslant \log \binom{2m+1}{m}.$$

Using Proposition 1.9 (iii) we get

$$\theta(2m+1) - \theta(m+1) = \sum_{m+1 < p \leqslant 2m+1} \log p \leqslant \log \binom{2m+1}{m} \leqslant \log(4^m) = m \log 4$$

and the induction hypothesis applied to $\theta(m+1)$ gives

$$\theta(2m+1) \leqslant m \log 4 + (m+1) \log 4 = (2m+1) \log 4$$

which concludes the proof of (3.6). The lemma follows from

$$\theta(x) = \theta([x]) \leqslant [x] \log 4 \leqslant x \log 4.$$

The proof is complete. $\square$

A better upper bound was proved in [Han72] where the author showed that $n \log 4$ could be replaced by $n \log 3$ in (3.6).

Lemma 3.43 *For all $x \geqslant 1537$, we have*

$$\theta(x) > \frac{x}{\log 4}.$$

Proof The proof follows the lines of Chebyshev's work.

We first notice that the function f defined by

$$f(x) = [x] - \left[\frac{x}{2}\right] - \left[\frac{x}{3}\right] - \left[\frac{x}{5}\right] + \left[\frac{x}{30}\right]$$

is periodic of period 30, and that $f(30 - x) = 1 - f(x)$ for $x \notin \mathbb{Z}$. An inspection of its values when $x \in [1, 15[$ allows us to infer that $f(x)$ only takes the values 0 or 1 if $x \notin \mathbb{Z}$. Since f is continuous on the right, we also have $f(x) = 0$ or 1 when $x \in \mathbb{Z}$. By periodicity, we infer that $f(x) = 0$ or 1 for all $x \in \mathbb{R}$. Hence we get

$$\Psi(x) \geqslant \sum_{n \leqslant x} \Lambda(n) f\left(\frac{x}{n}\right)$$

$$= \sum_{n \leqslant x} \Lambda(n) \left(\left[\frac{x}{n}\right] - \left[\frac{x}{2n}\right] - \left[\frac{x}{3n}\right] - \left[\frac{x}{5n}\right] + \left[\frac{x}{30n}\right]\right)$$

$$= \sum_{n \leqslant x} \Lambda(n)\left[\frac{x}{n}\right] - \sum_{n \leqslant x/2} \Lambda(n)\left[\frac{x}{2n}\right] - \sum_{n \leqslant x/3} \Lambda(n)\left[\frac{x}{3n}\right]$$

$$- \sum_{n \leqslant x/5} \Lambda(n)\left[\frac{x}{5n}\right] + \sum_{n \leqslant x/30} \Lambda(n)\left[\frac{x}{30n}\right]$$

where we used the fact that $[x/kn] = 0$ as soon as $n > x/k$ for $k \in \{2, 3, 5, 30\}$, and using the second estimate of Lemma 3.41, we deduce that

$$\Psi(x) \geqslant x \log x - x + 1 - \left(\frac{x}{2} \log \frac{x}{2} - \frac{x}{2} + 1 + \log \frac{x}{2}\right)$$

$$- \left(\frac{x}{3} \log \frac{x}{3} - \frac{x}{3} + 1 + \log \frac{x}{3}\right) - \left(\frac{x}{5} \log \frac{x}{5} - \frac{x}{5} + 1 + \log \frac{x}{5}\right)$$

$$+ \left(\frac{x}{30} \log \frac{x}{30} - \frac{x}{30} + 1\right)$$

$$= x \log\left(2^{7/15} 3^{3/10} 5^{1/6}\right) - 3 \log x + \log 30 - 1.$$

We conclude the proof by using the first estimate of Lemma 3.40. □

Note that $\log(2^{7/15} 3^{3/10} 5^{1/6}) \approx 0.921\,29\ldots$ which is a very good lower bound. It was sufficient to allow Chebyshev to prove Bertrand's famous postulate.

Corollary 3.44 (Bertrand's postulate) *Let n be a positive integer. Then the interval* $]n, 2n]$ *contains a prime number.*

Proof We check numerically the result for $n \in \{1, \ldots, 768\}$ and we suppose $n \geqslant 769$. Using Lemmas 3.42 and 3.43, we get

$$\sum_{n < p \leqslant 2n} \log p = \theta(2n) - \theta(n) > n\left(\frac{2}{\log 4} - \log 4\right) > 0$$

which implies the desired result.   □

Now we are able to get good upper and lower bounds for the function π.

Corollary 3.45 (Chebyshev's estimates) *For all* $x \geqslant 5$, *we have*

$$\frac{1}{\log 4} \frac{x}{\log x} < \pi(x) < \left(2 + \frac{1}{\log 4}\right) \frac{x}{\log x}.$$

Proof We first check the inequalities

$$\frac{1}{\log 4} \frac{n+1}{\log n} < \pi(n) < \left(2 + \frac{1}{\log 4}\right) \frac{n}{\log(n+1)}$$

for all integers $n \in \{5, \ldots, 1537\}$ which imply the validity of the inequalities of the corollary for all $x \in [5, 1357]$, so that we may suppose $x \geqslant 1537$.

The left-hand side follows directly from the lower bound of the second estimate of Lemma 3.40 and from the lower bound of Lemma 3.43.

Using Lemma 3.40 with $a = 3/2$ and Lemma 3.42, we get

$$\pi(x) \leqslant \frac{3x \log 4}{2 \log x} + x^{2/3}$$

and the inequality $\log x < 3e^{-1.55} x^{1/3}$, valid for all $x \geqslant 1537$, implies that

$$\pi(x) < \frac{3x \log 4}{2 \log x} + \frac{3e^{-1.55} x}{\log x} < \left(2 + \frac{1}{\log 4}\right) \frac{x}{\log x}$$

as required.   □

Remark 3.46 The method of Lemma 3.43 also provides an upper bound for $\Psi(x)$, and hence for $\pi(x)$. For instance, it is easy to see that the function f defined in the proof of the lemma satisfies the properties that $f(t) = 1$ for all $1 \leqslant t < 6$, and hence

$$\sum_{n \leqslant x} \Lambda(n) f\left(\frac{x}{n}\right) \geqslant \sum_{x/6 < n \leqslant x} \Lambda(n) = \Psi(x) - \Psi\left(\frac{x}{6}\right)$$

and similar computations allow us to get the following bound

$$\Psi(x) - \Psi\left(\frac{x}{6}\right) \leqslant x \log\left(2^{7/15}3^{3/10}5^{1/6}\right) + 2\log x + 1 - \log 30.$$

Now replacing x successively by $x/6, x/6^2, \ldots, x/6^k$ where $k = [\log x / \log 6]$ and summing all the resulting inequalities gives

$$\Psi(x) \leqslant (x - 1)\left(\frac{6}{5}\log\left(2^{7/15}3^{3/10}5^{1/6}\right)\right) + \frac{(\log x)^2}{\log 6} - \frac{\log 5 - 1}{\log 6}\log x < 1.1186\,x$$

as soon as $x \geqslant 2273$ (see the survey [BV09] for instance). Therefore one may wonder whether there exist optimal linear combinations of quantities $[x/n]$ to get better positive numbers $c_1 < c_2$ such that

$$c_1 x < \Psi(x) < c_2 x \quad (x \geqslant x_0)?$$

Diamond and Erdős [DE80] showed that the answer is *yes*, and even provided some constants arbitrarily close to 1. Thus, could Chebyshev really prove the PNT forty-five years before Hadamard and de la Vallée Poussin, and without using complex analysis? Unfortunately not, since Diamond and Erdős needed the PNT to provide the constants c_1, c_2. But Chebyshev's ideas played a key part in modern analytic number theory.

Corollary 3.45 allows us to slightly improve on the estimates of Lemma 3.40.

Corollary 3.47

(i) *For all $x \geqslant 25$, we have*

$$\Psi(x) < \theta(x) + \left(4 + \frac{1}{\log 2}\right)\sqrt{x}$$

$$\pi(x) < \frac{3}{2\log x}\left\{\theta(x) + \left(2 + \frac{1}{\log 4}\right)x^{2/3}\right\}.$$

(ii) *For all $x \geqslant 1$, we have*

$$\Psi(x) < 2x.$$

(iii) *Let p_n be the nth prime number. Then, for all integers $n \geqslant 3$, we have*

$$\frac{n}{2}\log n < p_n < 2n \log n.$$

Proof

(i) Follows directly from Corollary 3.45 and Lemma 3.40, the second estimate being used with $a = 3/2$.

(ii) The inequality is first numerically checked for all integers $n \in [1, 100]$, and follows from Lemma 3.42 and (i) otherwise.

(iii) We first check the inequalities for $n \in \{3, \ldots, 337\}$ and suppose $n \geqslant 338$ so that $p_n \geqslant 2273$. The proof rests on the fact that $\pi(p_n) = n$.

▷ For the lower bound, we use the second upper bound of (i) above, Lemma 3.42, the inequality $x^{2/3} < x/13$ valid for all $x \geqslant 2273$ and the trivial inequality $p_n > n$, which imply that

$$n = \pi(p_n) < \frac{3p_n}{2\log p_n}\left\{\log 4 + \frac{1}{13}\left(2 + \frac{1}{\log 4}\right)\right\} < \frac{2p_n}{\log p_n} < \frac{2p_n}{\log n}$$

giving the asserted lower bound for p_n.

▷ For the upper bound, we begin by using the inequality

$$\frac{\log x}{x^{1-\log 2}} < \frac{1}{\log 4}$$

valid for all $x \geqslant 2273$. Applied with $x = p_n$ and using the lower bound of Corollary 3.45, we get

$$\frac{\log p_n}{p_n^{1-\log 2}} < \frac{1}{\log 4} < \frac{\pi(p_n)\log p_n}{p_n} = \frac{n\log p_n}{p_n}$$

which implies that $p_n < n^{1/\log 2}$, and therefore $\log p_n < \frac{\log n}{\log 2}$. Thus, we get

$$n = \pi(p_n) > \frac{1}{\log 4}\frac{p_n}{\log p_n} > \frac{p_n}{2\log n}$$

which implies the desired upper bound.

This concludes the proof of the corollary.   □

3.4.3 An Alternative Approach

In [Nai82], M. Nair had the idea to use $d_n = [1, 2, \ldots, n]$ to get lower bounds for the function π. This estimate rests on the following result.

Lemma 3.48 *For all integers $n \geqslant 2$, we have $d_n \geqslant 2^{n-2}$.*

Proof The inequality is clearly true for $n \in \{2, 3\}$ so that we suppose $n \geqslant 4$. Let

$$I_n = \int_0^1 x^n(1-x)^n \, dx.$$

We first notice that $0 < I_n \leqslant 4^{-n}$ since $0 < x(1 - x) \leqslant 1/4$ for all $0 < x < 1$. Now using Newton's formula, we get

$$I_n = \sum_{k=0}^{n} \binom{n}{k} (-1)^k \int_0^1 x^{n+k} \, dx = \sum_{k=0}^{n} \binom{n}{k} \frac{(-1)^k}{n+k+1}$$

so that I_n is a rational number whose denominator is a divisor of d_{2n+1}. We deduce that $d_{2n+1} \times I_n$ is a positive integer, which implies that for all integers $n \geqslant 1$, we have

$$1 \leqslant d_{2n+1} \times I_n \leqslant 4^{-n} d_{2n+1}$$

and hence

$$d_{2n+1} \geqslant 2^{2n} \quad \text{and} \quad d_{2n+2} \geqslant d_{2n+1} \geqslant 2^{2n}$$

which implies the asserted inequality. $\square$

Now using Corollary 3.7 (ii), we have $d_n = p_1^{\alpha_1} \cdots p_k^{\alpha_k}$, where the prime numbers p_i are distinct and satisfy $p_i \leqslant n$ and each exponent α_i is the larger power of $p_i \leqslant n$, so that

$$\alpha_i = \left\lfloor \frac{\log n}{\log p_i} \right\rfloor \leqslant \frac{\log n}{\log p_i}$$

and therefore

$$d_n \leqslant \prod_{p \leqslant n} p^{\log n / \log p} = \prod_{p \leqslant n} n = n^{\pi(n)}$$

so that

$$\pi(n) \geqslant \frac{\log d_n}{\log n}.$$

Using Lemma 3.48 we get

$$\pi(n) \geqslant \frac{(n-2) \log 2}{\log n}.$$

3.4.4 Mertens' Theorems

Around 1875, Mertens proved two fundamental results on asymptotics for some prime number functions.

Theorem 3.49 (First Mertens' theorem) *For all $x \geqslant 2$, we have*

$$\sum_{p \leqslant x} \frac{\log p}{p} = \log x + O(1).$$

First proof Let $n = [x]$. By Theorem 3.37 we have

$$\log(n!) = \sum_{p \leqslant n} \log p \sum_{k=1}^{\infty} \left[\frac{n}{p^k} \right]$$

$$= \sum_{p \leqslant n} \log p \left[\frac{n}{p} \right] + \sum_{p \leqslant n} \log p \sum_{k=2}^{\infty} \left[\frac{n}{p^k} \right]$$

$$= n \sum_{p \leqslant n} \frac{\log p}{p} + O(\theta(n)) + O\left(n \sum_{p \leqslant n} \log p \sum_{k=2}^{\infty} \frac{1}{p^k} \right)$$

and the use of (1.1), Corollary 1.17 and Lemma 3.42 gives

$$\sum_{p \leqslant n} \frac{\log p}{p} = \frac{\log(n!)}{n} + O\left(\frac{\theta(n)}{n} \right) + O\left(\sum_{p \leqslant n} \frac{\log p}{p(p-1)} \right)$$

$$= \frac{1}{n} \left(n \log n + O(n) \right) + O(1) = \log n + O(1)$$

and we conclude the proof with

$$\sum_{p \leqslant x} \frac{\log p}{p} = \sum_{p \leqslant [x]} \frac{\log p}{p} = \log[x] + O(1) = \log x + O(1)$$

as required.   □

Second proof We first notice that

$$\sum_{d \leqslant x} \frac{\Lambda(d)}{d} = \sum_{p \leqslant x} \frac{\log p}{p} + \sum_{\substack{p^\alpha \leqslant x \\ \alpha \geqslant 2}} \frac{\log p}{p^\alpha}$$

and the second sum is

$$\leqslant \sum_{p \leqslant \sqrt{x}} \log p \sum_{\alpha=2}^{\infty} \frac{1}{p^\alpha} = \sum_{p \leqslant \sqrt{x}} \frac{\log p}{p(p-1)} = O(1)$$

so that

$$\sum_{p \leqslant x} \frac{\log p}{p} = \sum_{d \leqslant x} \frac{\Lambda(d)}{d} + O(1).$$

Now appealing to Lemma 3.41 and Corollary 3.47 (ii) we get

$$x \log x + O(x) = \sum_{d \leqslant x} \Lambda(d) \left[\frac{x}{d} \right] = x \sum_{d \leqslant x} \frac{\Lambda(d)}{d} + O(\Psi(x)) = x \sum_{d \leqslant x} \frac{\Lambda(d)}{d} + O(x)$$

and hence

$$\sum_{d \leqslant x} \frac{\Lambda(d)}{d} = \log x + O(1)$$

which concludes the proof.   □

By partial summation, it is fairly easy to prove that Theorem 3.49 implies the following result.

$$\sum_{p \leqslant x} \frac{1}{p} = \log\log x + B + O\left(\frac{1}{\log x}\right) \qquad (3.7)$$

where $B = 1 - \log\log 2 + \int_2^\infty t^{-1}(\log t)^{-2}R(t)\,dt$ and $|R(t)| \ll 1$. This estimate shows that (3.3) is very close to the right order of magnitude of the sum of the reciprocals of prime numbers. Mertens studied more carefully this sum in order to get a more manageable expression of the constant B, called today the *Mertens constant*.

Corollary 3.50 *For all $x \geqslant e$, we have*

$$\sum_{p \leqslant x} \frac{1}{p} = \log\log x + B + O\left(\frac{1}{\log x}\right)$$

where $B = \gamma + \sum_p \{\log(1 - \frac{1}{p}) + \frac{1}{p}\} \approx 0.261\,497\,212\,8\ldots$ is the Mertens constant.

Proof Let $h > 0$.

▷ The starting point is Euler's formula from Proposition 3.52

$$\prod_p \left(1 - \frac{1}{p^{1+h}}\right)^{-1} = \sum_{n=1}^\infty \frac{1}{n^{1+h}}.$$

Comparing the sum to an integral, we infer

$$\sum_{n=1}^\infty \frac{1}{n^{1+h}} = \int_1^\infty \frac{dt}{t^{1+h}} + O(1) = \frac{1}{h} + O(1)$$

so that

$$\prod_p \left(1 - \frac{1}{p^{1+h}}\right)^{-1} = \frac{1}{h} + O(1)$$

and taking logarithms of both sides we get

$$\sum_p \frac{1}{p^{1+h}} + \sum_p \sum_{\alpha=2}^\infty \frac{1}{\alpha p^{\alpha(1+h)}} = \sum_p \log\left(1 - \frac{1}{p^{1+h}}\right)^{-1} = \log\left(\frac{1}{h}\right) + O(h)$$

as $h \longrightarrow 0$, so that

$$\sum_{p} \frac{1}{p^{1+h}} = \log\left(\frac{1}{h}\right) - \sum_{p}\sum_{\alpha=2}^{\infty} \frac{1}{\alpha p^{\alpha(1+h)}} + O(h). \tag{3.8}$$

▷ Let $X > x \geqslant e$ be large real numbers. By (1.5), Proposition 1.27 and (3.7), we have

$$\sum_{x<p\leqslant X} \frac{1}{p^{1+h}} = \int_{x}^{X} \frac{1}{t^h} \mathrm{d}\left(\sum_{x<p\leqslant t} \frac{1}{p}\right)$$

$$= \frac{1}{X^h} \sum_{x<p\leqslant X} \frac{1}{p} + h\int_{x}^{X} \frac{1}{t^{1+h}}\left(\sum_{\log x<n\leqslant \log t} \frac{1}{n} + O\left(\frac{1}{\log x}\right)\right)\mathrm{d}t$$

$$= \frac{1}{X^h} \log\left(\frac{\log X}{\log x}\right) + h\int_{x}^{X}\left(\sum_{\log x<n\leqslant \log t} \frac{1}{n}\right)\frac{\mathrm{d}t}{t^{1+h}} + O\left(\frac{1}{\log x}\right)$$

$$= \frac{1}{X^h} \log\left(\frac{\log X}{\log x}\right) + h\sum_{\log x<n\leqslant \log X} \frac{1}{n}\int_{e^n}^{X} \frac{\mathrm{d}t}{t^{1+h}} + O\left(\frac{1}{\log x}\right)$$

$$= \frac{1}{X^h} \log\left(\frac{\log X}{\log x}\right) + \sum_{\log x<n\leqslant \log X} \frac{e^{-nh} - X^{-h}}{n} + O\left(\frac{1}{\log x}\right)$$

$$= \sum_{\log x<n\leqslant \log X} \frac{e^{-nh}}{n} + O\left(\frac{1}{\log x}\right)$$

so that

$$\sum_{p>x} \frac{1}{p^{1+h}} = \sum_{n>\log x} \frac{e^{-nh}}{n} + O\left(\frac{1}{\log x}\right)$$

$$= -\log\left(1 - e^{-h}\right) - \sum_{n\leqslant\log x} \frac{e^{-nh}}{n} + O\left(\frac{1}{\log x}\right)$$

$$= -\log\left(1 - e^{-h}\right) - \sum_{n\leqslant\log x} \frac{1}{n} + \sum_{n\leqslant\log x} \frac{1 - e^{-nh}}{n} + O\left(\frac{1}{\log x}\right)$$

$$= -\log\left(1 - e^{-h}\right) - \log\log x - \gamma + O(h\log x) + O\left(\frac{1}{\log x}\right).$$

▷ Now combining (3.8) with this estimate we get

$$\sum_{p\leqslant x} \frac{1}{p^{1+h}} = \log\left(\frac{1 - e^{-h}}{h}\right) + \log\log x + \gamma - \sum_{p}\sum_{\alpha=2}^{\infty} \frac{1}{\alpha p^{\alpha(1+h)}} + O(h\log x)$$

$$+ O\left(\frac{1}{\log x}\right)$$

and letting $h \longrightarrow 0$ and noticing that $\sum_p \sum_{\alpha=2}^{\infty} \frac{1}{\alpha p^{\alpha}} = -\sum_p \{\log(1 - \frac{1}{p}) + \frac{1}{p}\}$ we obtain the asserted result.   □

Mertens discovered the identity

$$\sum_p \sum_{\alpha=2}^{\infty} \frac{1}{\alpha p^{\alpha}} = \sum_{n=2}^{\infty} \frac{\mu(n) \log \zeta(n)}{n}$$

which enabled him to compute B very accurately.

From Corollary 3.50 we readily get Mertens' second theorem.

Corollary 3.51 (Mertens' second theorem) *For all $x \geqslant e$, we have*

$$\prod_{p \leqslant x} \left(1 - \frac{1}{p}\right) = \frac{e^{-\gamma}}{\log x} \left\{1 + O\left(\frac{1}{\log x}\right)\right\}.$$

Proof We have

$$\log \prod_{p \leqslant x} \left(1 - \frac{1}{p}\right)^{-1} = \sum_{p \leqslant x} \log\left(1 - \frac{1}{p}\right)^{-1} = \sum_{p \leqslant x} \frac{1}{p} + \sum_{p \leqslant x} \left\{\log\left(1 - \frac{1}{p}\right)^{-1} - \frac{1}{p}\right\}$$

and using the inequalities

$$\log\left(1 - \frac{1}{p}\right)^{-1} - \frac{1}{p} \leqslant \frac{1}{2p(p-1)} \leqslant \frac{1}{p^2}$$

the estimate

$$\sum_{p > x} \frac{1}{p^2} \ll \frac{1}{x \log x}$$

and Corollary 3.50, we infer that

$$\log \prod_{p \leqslant x} \left(1 - \frac{1}{p}\right)^{-1} = \sum_{p \leqslant x} \frac{1}{p} - \sum_p \left\{\log\left(1 - \frac{1}{p}\right) + \frac{1}{p}\right\} + O\left(\frac{1}{x \log x}\right)$$

$$= \log \log x + \gamma + O\left(\frac{1}{\log x}\right)$$

and the use of $e^h = 1 + O(h)$ for all $h \longrightarrow 0$ finally gives

$$\prod_{p \leqslant x} \left(1 - \frac{1}{p}\right)^{-1} = e^{\gamma} \log x \left\{1 + O\left(\frac{1}{\log x}\right)\right\}$$

which is easily seen to be equivalent to the form given in the corollary.   □

3.5 The Riemann Zeta-Function

3.5.1 Euler, Dirichlet and Riemann

Considered as one of the first analytic number theorists, Euler noticed that the function

$$\sigma \longmapsto \sum_{n=1}^{\infty} \frac{1}{n^{\sigma}}$$

well-defined for all $\sigma > 1$, could carry some important arithmetic information, in particular concerning the distribution of primes. Indeed, using Theorem 3.3, Euler showed the following fundamental result.

Proposition 3.52 (Euler) *For all $\sigma > 1$, we have*

$$\prod_p \left(1 - \frac{1}{p^{\sigma}}\right)^{-1} = \sum_{n=1}^{\infty} \frac{1}{n^{\sigma}}.$$

Proof Let N be a positive integer and $\sigma > 1$ be a real number. Expanding the product

$$\prod_{p \leqslant N} \left(1 - \frac{1}{p^{\sigma}}\right)^{-1} = \prod_{p \leqslant N} \left(1 + \frac{1}{p^{\sigma}} + \frac{1}{p^{2\sigma}} + \frac{1}{p^{3\sigma}} + \cdots\right)$$

and using Theorem 3.3, we obtain

$$\prod_{p \leqslant N} \left(1 - \frac{1}{p^{\sigma}}\right)^{-1} = 1 + \frac{1}{n_1^{\sigma}} + \frac{1}{n_2^{\sigma}} + \frac{1}{n_3^{\sigma}} + \cdots$$

where each integer n_i is such that all its prime factors are $\leqslant N$, so that

$$\prod_{p \leqslant N} \left(1 - \frac{1}{p^{\sigma}}\right)^{-1} = \sum_{P^+(n) \leqslant N} \frac{1}{n^{\sigma}}.$$

Since every integer $n \leqslant N$ satisfies this property, we infer that

$$\left| \sum_{n=1}^{\infty} \frac{1}{n^{\sigma}} - \prod_{p \leqslant N} \left(1 - \frac{1}{p^{\sigma}}\right)^{-1} \right| = \left| \sum_{n=1}^{\infty} \frac{1}{n^{\sigma}} - \sum_{P^+(n) \leqslant N} \frac{1}{n^{\sigma}} \right| \leqslant \sum_{n > N} \frac{1}{n^{\sigma}}.$$

We conclude the proof by noting that the latter sum tends to 0 as N tends to ∞. $\square$

Euler used to work with Proposition 3.52 principally as a formal identity and mainly for integer values of σ. Later, Dirichlet based some of his work upon Euler's product formula but used it with $\sigma > 1$ as a real variable, and proved rigorously that this result is true in $]1, \infty[$.

Riemann, who was one of Dirichlet's students, was certainly influenced by his use of Euler's product formula and, as one of the founders of the theory of functions, would naturally consider σ as a complex variable, renamed $s = \sigma + it \in \mathbb{C}$. Since

$$\sum_{p \leqslant x} \frac{1}{|p^s|} = \sum_{p \leqslant x} \frac{1}{p^\sigma}$$

both sides of Proposition 3.52 converge for every complex s such that $\sigma > 1$. This leads to the following definition.

Definition 3.53 (Riemann zeta-function) The Riemann zeta-function $\zeta(s)$ is defined for all complex numbers $s = \sigma + it$ such that $\sigma > 1$ by

$$\zeta(s) = \sum_{n=1}^{\infty} \frac{1}{n^s} = \prod_p \left(1 - \frac{1}{p^s}\right)^{-1}.$$

The product representation, called the *Euler product*, enables us to see that $\zeta(s) \neq 0$ in the half-plane $\sigma > 1$. Furthermore, since

$$\left| \sum_{p \leqslant x} \frac{\log p}{p^s - 1} \right| \leqslant \sum_{p \leqslant x} \frac{\log p}{p^{1+\varepsilon} - 1}$$

for all $s = \sigma + it \in \mathbb{C}$ such that $\sigma \geqslant 1 + \varepsilon$, we infer that the series of the left-hand side is uniformly convergent in this half-plane, and this allows us to take the logarithmic derivative of both sides of Definition 3.53 and justifies the change of the order of summations, which gives

$$-\frac{\zeta'(s)}{\zeta(s)} = \sum_p \frac{\log p}{p^s - 1} = \sum_p \sum_{\alpha=1}^{\infty} \frac{\log p}{p^{\alpha s}} = \sum_{\alpha=1}^{\infty} \sum_p \frac{\log p}{p^{\alpha s}}$$

and hence we get

$$-\frac{\zeta'(s)}{\zeta(s)} = \sum_{n=1}^{\infty} \frac{\Lambda(n)}{n^s}. \tag{3.9}$$

The series in (3.9) converges absolutely and uniformly in the half-plane $\sigma \geqslant 1 + \varepsilon$ for all $\varepsilon > 0$, so that (3.9) holds for $\sigma > 1$ by analytic continuation.

3.5.2 The Gamma and Theta Functions

▶ Euler defined the Gamma function for all real numbers $\sigma > 0$ as

$$\Gamma(\sigma) = \int_0^\infty x^{\sigma-1} e^{-x} \, dx$$

the convergence of the integral being ensured by the estimates $x^{\sigma-1} e^{-x} \sim x^{\sigma-1}$ when $x \longrightarrow 0$ and $x^{\sigma-1} e^{-x} = o(e^{-x/2})$ when $x \longrightarrow \infty$. Clearly, we have $\Gamma(\sigma) > 0$ for all $\sigma > 0$, $\Gamma(1) = 1$ and integrating by parts gives the recursion $\Gamma(\sigma + 1) = \sigma \Gamma(\sigma)$ for all $\sigma > 0$. In particular, we get by induction the identity $\Gamma(n) = (n - 1)!$ for all positive integers n.

Let $s = \sigma + it \in \mathbb{C}$. Since $|x^s| = x^\sigma$, the integral above defines an analytic function Γ in the half-plane $\sigma > 0$, which still satisfies the functional equation $\Gamma(s + 1) = s\Gamma(s)$ for $\sigma > 0$, as can be seen by repeating an integration by parts. For $-1 < \sigma < 0$, we set $\Gamma(s) = s^{-1}\Gamma(s + 1)$ so that the function Γ has a simple pole at $s = 0$. Similarly in the range $-2 < \sigma < -1$, we define $\Gamma(s) = s^{-1}(s+1)^{-1}\Gamma(s+2)$ which gives a simple pole at $s = -1$. Continuing in this way, we see that the functional equation extends Γ to a meromorphic function on $\mathbb{C}$, which is analytic except for simple poles at $s = 0, -1, -2, -3, \ldots$ One may check that the residue at the pole $-n$ is given by $(-1)^n / n!$.

Let us mention the following useful formulae [Tit39, 4.41]. For any $s \in \mathbb{C}$ we have[9]

$$\frac{1}{\Gamma(s)\Gamma(1 - s)} = \frac{\sin \pi s}{\pi}$$

and[10]

$$\Gamma(s) = \pi^{-1/2} 2^{s-1} \Gamma\left(\frac{s}{2}\right) \Gamma\left(\frac{s+1}{2}\right). \tag{3.10}$$

Stirling's formula may be generalized as follows [Tit39, 4.42].

Theorem 3.54 (Complex Stirling's formula) *For any $s \in \mathbb{C}$, we have*

$$\log \Gamma(s) = \left(s - \frac{1}{2}\right) \log s - s + \log \sqrt{2\pi} + O\left(|s|^{-1}\right)$$

as $|s| \longrightarrow \infty$ and uniformly for $|\arg s| \leqslant \pi - \delta$ ($\delta > 0$), where the complex logarithm is chosen by taking the principal value of its argument. This implies in particular that, if $\sigma \in [\sigma_1, \sigma_2]$ is fixed, then

$$\left|\Gamma(\sigma + it)\right| = |t|^{\sigma-1/2} e^{-\pi|t|/2} \sqrt{2\pi} \left(1 + O\left(|t|^{-1}\right)\right)$$

when $|t| \geqslant t_0$, the constant implied in the error-term depending on σ_1 and σ_2.

---

[9]This identity is called the *reflection formula*.

[10]This identity is called the *duplication formula*.

We will see later that most of the Dirichlet series arising in analytic number theory satisfy functional equations of the same shape,[11] namely as in Theorem 4.58 where products of Gamma functions appear. Theorem 3.54, showing that $\Gamma(s)$ tends to zero exponentially fast as $|t| \longrightarrow \infty$ in vertical strips, is then used to shift the contours of integration. This enables us to get important results which are out of reach by other methods, as in Theorem 7.178.

▶ Replacing x by $\pi n^2 x$ in the integral defining Γ gives

$$\pi^{-s/2}\Gamma\left(\frac{s}{2}\right)n^{-s} = \int_0^\infty x^{\sigma/2-1}e^{-n^2\pi x}\,\mathrm{d}x \tag{3.11}$$

for all $\sigma > 0$. The purpose is to sum both sides of this equation. To this end, we define the following two functions. For all $x > 0$, we set

$$\omega(x) = \sum_{n=1}^\infty e^{-n^2\pi x} \quad \text{and} \quad \theta(x) = 2\omega(x) + 1 = \sum_{n\in\mathbb{Z}} e^{-n^2\pi x}.$$

The function $g : t \longmapsto e^{-t^2\pi}$ satisfies $\int_{\mathbb{R}} g(t)\,\mathrm{d}t = 1$. This implies that its Fourier transform is

$$\widehat{g}(u) = e^{-\pi u^2}.$$

Set $f : t \longmapsto e^{-t^2\pi x}$. Using the transposition formula stating that the Fourier transform of $t \longmapsto g(\alpha t)$ is the function $u \longmapsto |\alpha|^{-1}\widehat{g}(u/\alpha)$ for all real numbers $\alpha \neq 0$, we obtain with $\alpha = x^{1/2}$

$$\widehat{f}(u) = x^{-1/2}e^{-u^2\pi/x}.$$

Then Lemma 6.27 implies the following result: for all $x > 0$, we have

$$\theta\left(\frac{1}{x}\right) = x^{1/2}\theta(x). \tag{3.12}$$

Now we may return to (3.11). Summing this equation over n and interchanging the sum and integral, we get for all $\sigma > 1$

$$\pi^{-s/2}\Gamma\left(\frac{s}{2}\right)\zeta(s) = \int_0^\infty x^{\sigma/2-1}\omega(x)\,\mathrm{d}x.$$

The process is justified since the sum and integral converge absolutely in the half-plane $\sigma > 1$. Splitting the integral at $x = 1$ and substituting $1/x$ for x in the first integral yields

$$\pi^{-s/2}\Gamma\left(\frac{s}{2}\right)\zeta(s) = \int_1^\infty x^{\sigma/2-1}\omega(x)\,\mathrm{d}x + \int_1^\infty x^{-\sigma/2-1}\omega\left(\frac{1}{x}\right)\mathrm{d}x$$

---

[11] The set of these functions is now called the *Selberg class*.

and using (3.12) in the form

$$\omega\left(\frac{1}{x}\right) = x^{1/2}\omega(x) + \frac{x^{1/2} - 1}{2}$$

finally gives

$$\pi^{-s/2}\Gamma\left(\frac{s}{2}\right)\zeta(s) = -\frac{1}{s} + \frac{1}{s-1} + \int_1^\infty \omega(x)\left(x^{s/2} + x^{(1-s)/2}\right)\frac{dx}{x} \qquad (3.13)$$

for all $\sigma > 1$.

3.5.3 Functional Equation

The computations made above eventually lead to the functional equation of the Riemann zeta-function. Thus, Riemann's idea to consider s as a complex variable is very fruitful. However, it is interesting to note that Riemann did not think of a piece-by-piece extension of the function represented by $\sum_{n=1}^\infty n^{-s}$ in the way that analytic continuation is usually used today, but rather searched a formula which remains valid for all s (see [Edw74] for instance). The following result, whose real version was conjectured and partially proved by Euler, is fundamental.

Theorem 3.55 (Functional equation) *It is customary to set*

$$\xi(s) = \pi^{-s/2}\Gamma(s/2)\zeta(s).$$

Then the function $\xi(s)$ can be extended analytically in the whole complex plane to a meromorphic function having simple poles at $s = 0$ and $s = 1$, and satisfies the functional equation $\xi(s) = \xi(1 - s)$.

Thus the Riemann zeta-function can be extended analytically in the whole complex plane to a meromorphic function having a simple pole at $s = 1$ with residue 1. Furthermore, for all $s \in \mathbb{C} \setminus \{1\}$, we have

$$\zeta(s) = 2^s \pi^{s-1} \sin\left(\frac{\pi s}{2}\right)\Gamma(1 - s)\zeta(1 - s).$$

Proof

▷ Let $x \geqslant 1$ be a real number and $s = \sigma + it$ with $\sigma > 1$. By (1.7) with $a = 1, b = x$ and $f(x) = x^{-s}$, we get

$$\sum_{n\leqslant x}\frac{1}{n^s} = \frac{1}{2} + \frac{1 - x^{1-s}}{s-1} - \frac{\psi(x)}{x^s} - s\int_1^x \frac{\psi(u)}{u^{s+1}}\,du \qquad (3.14)$$

so that making $x \longrightarrow \infty$ we obtain

$$\zeta(s) = \frac{1}{2} + \frac{1}{s-1} - s \int_1^\infty \frac{\psi(u)}{u^{s+1}} \, du. \tag{3.15}$$

Since $|\psi(x)| \leqslant \frac{1}{2}$, the integral converges for $\sigma > 0$ and is uniformly convergent in any finite region to the right of the line $\sigma = 0$. This implies that it defines an analytic function in the half-plane $\sigma > 0$, and therefore (3.15) extends ζ to a meromorphic function in this half-plane, which is analytic except for a simple pole at $s = 1$ with residue 1. Also notice that (3.15) can be written in the shape

$$\zeta(s) = \frac{s}{s-1} - s \int_1^\infty \frac{\{u\}}{u^{s+1}} \, du$$

and since $\lim_{|s|\to\infty} s/(s-1) = 1$, we deduce that

$$|\zeta(s)| \ll |s|$$

as $|s| \longrightarrow \infty$.

▷ By (3.13), we have for $\sigma > 1$

$$\xi(s) = -\frac{1}{s} + \frac{1}{s-1} + \int_1^\infty \omega(x) \left(x^{s/2} + x^{(1-s)/2} \right) \frac{dx}{x}.$$

Since $\omega(x) \ll e^{-\pi x}$ as $x \longrightarrow \infty$, we infer that the integral is absolutely convergent for all $s \in \mathbb{C}$ whereas the left-hand side is a meromorphic function on $\sigma > 0$. This implies that

a. The identity (3.13) is valid for all $\sigma > 0$.
b. The function $\xi(s)$ can be defined by this identity as a meromorphic function on $\mathbb{C}$ with simple poles at $s = 0$ and $s = 1$.
c. Since the right-hand side of (3.13) is invariant under the substitution $s \longleftrightarrow 1 - s$, we get $\xi(s) = \xi(1-s)$.
d. The function $s \longmapsto s(s-1)\xi(s)$ is entire on $\mathbb{C}$. Indeed, if $\sigma > 0$, the factor $s - 1$ counters the pole at $s = 1$, and the result on all $\mathbb{C}$ follows from the functional equation.

▷ It remains to show that the functional equation can be written in the form

$$\zeta(s) = 2^s \pi^{s-1} \sin\left(\frac{\pi s}{2} \right) \Gamma(1-s)\zeta(1-s).$$

Since $\xi(s) = \xi(1-s)$, we have

$$\Gamma(s/2)\zeta(s) = \pi^{s-1/2} \Gamma\left(\frac{1-s}{2} \right) \zeta(1-s)$$

and multiplying both sides by $\pi^{-1/2}2^{s-1}\Gamma(\frac{1+s}{2})$ and using (3.10) we obtain

$$\Gamma(s)\zeta(s) = (2\pi)^{s-1}\Gamma\left(\frac{1-s}{2}\right)\Gamma\left(\frac{1+s}{2}\right)\zeta(1-s).$$

Now the reflection formula implies that

$$\zeta(s) = (2\pi)^{s-1}\left(\frac{\sin\pi s}{\sin(\pi(1+s)/2)}\right)\Gamma(1-s)\zeta(1-s)$$

and the result follows from the identity $\sin\pi s = 2\sin(\frac{\pi s}{2})\sin(\frac{\pi}{2}(1+s))$. The proof is complete.   □

Remark 3.56 We may deduce the following basic consequences for the Riemann zeta-function.

1. $\zeta(s)$ has simple zeros at $s = -2, -4, -6, -8, \dots$ Indeed, since the integral in (3.13) is absolutely convergent for all $s \in \mathbb{C}$ and since $\omega(x) > 0$ for all x, we have

$$\xi(-2n) = \frac{1}{2n} - \frac{1}{2n+1} + \int_1^\infty \omega(x)\left(x^{-n} + x^{n+1/2}\right)\frac{dx}{x} > 0$$

for all positive integers n. The result follows from the fact that $\Gamma(s/2)$ has simple poles at $s = -2n$.

These zeros are the only ones lying in the region $\sigma < 0$. They are called *trivial zeros* of the Riemann zeta-function.

2. For all $0 < \sigma < 1$, we have $\zeta(\sigma) \neq 0$. Indeed, since for all $\sigma > 0$

$$\zeta(s) = \frac{s}{s-1} - s\int_1^\infty \frac{\{x\}}{x^{s+1}}\,dx$$

we infer that, for all $0 < \sigma < 1$, we get

$$\left|\zeta(\sigma) - \frac{\sigma}{\sigma - 1}\right| < \sigma\int_1^\infty \frac{dx}{x^{\sigma+1}} = 1$$

which implies that $\zeta(\sigma) < 1 + \sigma/(\sigma - 1)$ for all $0 < \sigma < 1$. Hence $\zeta(\sigma) < 0$ for all $\frac{1}{2} \leqslant \sigma < 1$, and the functional equation implies the asserted result.

The functional equation is very important, but also may be insufficient in some applications for it does not express $\zeta(s)$ explicitly. The following tool will be useful to get some estimates of $\zeta(s)$ in the critical strip, especially when σ is close to 1.

Theorem 3.57 (Approximate functional equation) *We have uniformly for $x \geqslant 1$ and $s \in \mathbb{C} \setminus \{1\}$ such that $\sigma > 0$*

$$\zeta(s) = \sum_{n \leqslant x} \frac{1}{n^s} + \frac{x^{1-s}}{s-1} + R_0(s; x)$$

with

$$R_0(s; x) = \frac{\psi(x)}{x^s} - s \int_x^\infty \frac{\psi(u)}{u^{s+1}} \, du$$

and hence

$$|R_0(s; x)| \leqslant \frac{|s|}{\sigma x^\sigma}.$$

Proof Follows by subtracting (3.14) from (3.15). □

3.5.4 Estimates for $|\zeta(s)|$

In this section, we set $\tau = |t| + 3$ for all $t \in \mathbb{R}$.

▶ The next lemma gives inequalities for $|\zeta(s)|$ near the right of the line $\sigma = 1$.

Lemma 3.58 *For all $\sigma > 1$ and $t \in \mathbb{R}$, we have*

$$\frac{1}{\sigma - 1} < \zeta(\sigma) < \frac{\sigma}{\sigma - 1}$$

and

$$\left| -\frac{\zeta'}{\zeta}(\sigma + it) \right| \leqslant -\frac{\zeta'}{\zeta}(\sigma) < \frac{1}{\sigma - 1}.$$

Furthermore, for all $\sigma > 1$ and $t \in \mathbb{R}$, we have

$$\frac{1}{\zeta(\sigma)} \leqslant |\zeta(\sigma + it)| \leqslant \zeta(\sigma) < \frac{1}{\sigma - 1} + 0.64$$

where the last estimate is valid for all $1 < \sigma < 1.12$

Proof We start with

$$\int_1^\infty \frac{du}{u^\sigma} < \zeta(\sigma) < \int_1^\infty \frac{du}{u^\sigma} + 1$$

giving the first line of inequalities and the second line follows from Abel's summation which implies that

$$-\zeta'(\sigma) = \sum_{n=1}^\infty \frac{\log n}{n^\sigma} = \sum_{n=1}^\infty \log\left(1 + \frac{1}{n}\right) \sum_{h=n+1}^\infty \frac{1}{h^\sigma} < \sum_{n=1}^\infty \frac{1}{n} \int_n^\infty \frac{du}{u^\sigma}$$

$$= \sum_{n=1}^\infty \frac{n^{1-\sigma}}{n(\sigma - 1)} = \frac{\zeta(\sigma)}{\sigma - 1}.$$

For the next result, we first have

$$|\zeta(\sigma + it)| \leqslant \sum_{n=1}^{\infty} \frac{1}{n^{\sigma}} = \zeta(\sigma)$$

and

$$\frac{1}{|\zeta(\sigma + it)|} = \left| \prod_{p} \left(1 - \frac{1}{p^{\sigma + it}} \right) \right| \leqslant \prod_{p} \left(1 + \frac{1}{p^{\sigma}} \right) \leqslant \sum_{n=1}^{\infty} \frac{1}{n^{\sigma}} = \zeta(\sigma).$$

Now using (3.15) and integrating by parts we get

$$\zeta(\sigma) = \frac{1}{\sigma - 1} + \frac{1}{2} - \sigma(\sigma + 1) \int_{1}^{\infty} \frac{\psi_2(u)}{u^{\sigma + 2}} \, du$$

and using $0 \leqslant -\psi_2(u) \leqslant \frac{1}{8}$ and $\sigma < 1.12$ we get the asserted result. $\square$

▶ Using Theorem 3.57, we are in a position to estimate $|\zeta(s)|$ near the left of the line $\sigma = 1$.

  ▷ For $\sigma \geqslant 2$, we have trivially

$$|\zeta(s)| \leqslant \zeta(\sigma) \leqslant \zeta(2) = \frac{\pi^2}{6}.$$

  ▷ For $1 - c/\log \tau \leqslant \sigma < 2$ for some $0 \leqslant c < \frac{1}{3}$, we use Theorem 3.57 with $x = \tau$. Note that $\tau^{1-\sigma} \leqslant e^c$ and, for all $n \leqslant \tau$, we also have $n^{-\sigma} \leqslant e^c n^{-1}$, so that[12]

$$|\zeta(s)| \leqslant \sum_{n \leqslant \tau} \frac{1}{n^{\sigma}} + \tau^{-\sigma} \left(1 + \frac{|s|}{\sigma} \right) \leqslant e^c \sum_{n \leqslant \tau} \frac{1}{n} + \frac{e^c}{\tau} \left(2 + \frac{\tau}{\sigma} \right)$$

$$\leqslant e^c (\log \tau + 3) \leqslant 4 e^c \log \tau.$$

In particular, we have $|\zeta(1 + it)| \leqslant 4 \log \tau$ for all $t \in \mathbb{R}$.

  ▷ For $\sigma = 0$ and $t \neq 0$, we have from the functional equation

$$|\zeta(it)| = \pi^{-1} \sinh\left(\frac{\pi t}{2} \right) |\Gamma(1 - it)| |\zeta(1 - it)|$$

and Theorem 3.54 implies that

$$|\Gamma(1 - it)| \ll \tau^{1/2} e^{-\pi \tau / 2}$$

so that by above we get

$$\zeta(it) \ll \tau^{1/2} \log \tau.$$

---

[12]The constraint $c < \frac{1}{3}$ ensures that $\sigma > \frac{1}{2}$ for $\tau \geqslant 3$.

This inequality also holds for $t = 0$ since it can be shown that $\zeta(0) = -\frac{1}{2}$.

▷ For $0 < \sigma < 1$ and $t > 0$, we use the *Phragmén–Lindelöf principle*, which may be stated as follows.

Let $a < b$ be real numbers and suppose that $f : \mathbb{C} \longrightarrow \mathbb{C}$ is a continuous, bounded function in the strip $a \leqslant \sigma \leqslant b$, and analytic in $a < \sigma < b$. If, for all $t \in \mathbb{R}$, we have

$$\left| f(a + it) \right| \leqslant A \quad \text{and} \quad \left| f(b + it) \right| \leqslant B$$

then, for all s such that $a < \sigma < b$, we have

$$\left| f(s) \right| \leqslant \left(A^{b-\sigma} B^{\sigma-a} \right)^{1/(b-a)}.$$

Applied here with the above estimates this result gives

$$\zeta(s) \ll \tau^{(1-\sigma)/2} \log \tau$$

for all $0 < \sigma < 1$.

▷ By differentiation of the equation of Theorem 3.57, for $s \in \mathbb{C}$ satisfying $1 - c/\log \tau \leqslant \sigma < 2$, we get

$$\zeta'(s) = -\sum_{n \leqslant x} \frac{\log n}{n^s} - \frac{x^{1-s} \log x}{s - 1} - \frac{x^{1-s}}{(s - 1)^2} + R_1(s; x)$$

with

$$\left| R_1(s; x) \right| \leqslant \frac{|s|}{\sigma x^\sigma} \left(\log x + \sigma^{-1} \right) < \frac{|s|}{\sigma x^\sigma} (\log x + 2).$$

Proceeding as before and using the easy inequality

$$\sum_{n \leqslant \tau} \frac{\log n}{n} \leqslant \frac{(\log \tau)^2}{2} + 0.11$$

we obtain $|\zeta'(\sigma + it)| \leqslant 9e^c (\log \tau)^2$.

We may summarize these estimates in the following result.

Theorem 3.59 *Let $0 \leqslant c < \frac{1}{3}$, $t \in \mathbb{R}$ and set $\tau = |t| + 3$. Uniformly in σ, we have*

$$\zeta(\sigma + it) \ll \begin{cases} 1, & \text{if } \sigma \geqslant 2 \\ \tau^{(1-\sigma)/2} \log \tau, & \text{if } 0 \leqslant \sigma \leqslant 1. \end{cases}$$

Furthermore, in the region $1 - c/\log \tau \leqslant \sigma < 2$, we have

$$\left| \zeta(\sigma + it) \right| \leqslant 4e^c \log \tau \quad \text{and} \quad \left| \zeta'(\sigma + it) \right| \leqslant 9e^c (\log \tau)^2.$$

More elaborate techniques can be used to improve on this result, as will be seen in Chap. 6. It is customary to denote by $\mu(\sigma)$ the lower bound of numbers α such that, for all $|t| \geqslant 1$, we have

$$\zeta(\sigma + it) \ll |t|^{\alpha}.$$

From the theory of Dirichlet series, it may be proved [Tit39, §9.41] that this function is never negative, non-increasing and convex downwards, and hence continuous. By above, we get $\mu(\sigma) = 0$ for all $\sigma > 1$ and using the functional equation, it may be shown that $\mu(\sigma) = \frac{1}{2} - \sigma$ for all $\sigma < 0$. These equalities also hold by continuity for $\sigma = 1$ and $\sigma = 0$ respectively. However, the exact value of $\mu(\sigma)$ in the critical strip is still unknown. The simplest possible hypothesis is that the graph of this function consists of two straight lines $y = \frac{1}{2} - \sigma$ if $\sigma \leqslant \frac{1}{2}$ and $y = 0$ if $\sigma \geqslant \frac{1}{2}$. This is known as the *Lindelöf hypothesis* and is then equivalent to the statement that

$$\zeta\left(\frac{1}{2} + it\right) \ll t^{\varepsilon}$$

for all $\varepsilon > 0$ and $t \geqslant 3$. The best value to date is due to Huxley [Hux05] who proved that we have[13]

$$\zeta\left(\frac{1}{2} + it\right) \ll t^{32/205+\varepsilon}.$$

3.5.5 A Zero-Free Region

It can be shown [Tit51, §3.7] that a weak version of the Prime Number Theorem, i.e. $\pi(x) \sim x/\log x$ when $x \longrightarrow \infty$, follows from the next result, which goes back to Hadamard and de la Vallée Poussin (1896).

Theorem 3.60 *For all $t \in \mathbb{R}$, we have $\zeta(1 + it) \neq 0$. Furthermore, if $\tau = |t| + 3$, we have*

$$\left|\zeta(1 + it)\right| > \frac{1}{221\,184(\log \tau)^7}.$$

Proof Since 1 is a pole of ζ, we suppose that $1 + it_0$ is a zero or order m of $\zeta(s)$ for some $t_0 \neq 0$, so that there exists $\ell \neq 0$ such that

$$\lim_{\sigma \to 1^+} \frac{\zeta(\sigma + it_0)}{(\sigma - 1)^m} = \ell.$$

---

[13] See Exercise 7 in Chap. 6.

Let $\sigma > 1$ and $t \in \mathbb{R}$. Using (3.9), we have

$$\mathrm{Re}\big(3 \log \zeta(\sigma) + 4 \log \zeta(\sigma + it) + \log \zeta(\sigma + 2it)\big)$$

$$= \sum_{n=1}^{\infty} \frac{\Lambda(n)}{n^{\sigma} \log n} \big(3 + 4\cos(t \log n) + \cos(2t \log n)\big)$$

$$= 2 \sum_{n=1}^{\infty} \frac{\Lambda(n)}{n^{\sigma} \log n} \big(1 + \cos(t \log n)\big)^2 \geqslant 0$$

which implies that

$$\big|\zeta(\sigma)^3 \zeta(\sigma + it)^4 \zeta(\sigma + 2it)\big| \geqslant 1.$$

Therefore we get

$$\big|(\sigma - 1)^3 \zeta(\sigma)^3 (\sigma - 1)^{-4m} \zeta(\sigma + it_0)^4 \zeta(\sigma + 2it_0)\big| \geqslant (\sigma - 1)^{3-4m}.$$

Now making $\sigma \longrightarrow 1^+$ gives a contradiction, since the left-hand side tends to a finite limit, whereas the right-hand side tends to ∞ if $m \geqslant 1$.

For the lower bound of $|\zeta(1 + it)|$, we proceed as follows. The inequality above and the estimates of Lemma 3.58 and Theorem 3.59 give for $1 < \sigma < 2$

$$\frac{1}{|\zeta(\sigma + it)|} \leqslant \zeta(\sigma)^{3/4} \big|\zeta(\sigma + 2it)\big|^{1/4} < (32)^{1/4} (\sigma - 1)^{-3/4} (\log \tau)^{1/4},$$

so that by Theorem 3.59 we get for $1 < \sigma < 2$

$$\big|\zeta(1 + it)\big| \geqslant \big|\zeta(\sigma + it)\big| - 9(\sigma - 1)(\log \tau)^2$$

$$> (32 \log \tau)^{-1/4} (\sigma - 1)^{3/4} - 9(\sigma - 1)(\log \tau)^2$$

and the choice of $\sigma = 1 + 663\,552^{-1}(\log \tau)^{-9}$ gives the stated result. $\square$

The proof given above follows essentially the lines of de La Vallée Poussin's. Hadamard's argument, also exploiting the link between $\zeta(1 + it_0)$ and $\zeta(1 + 2it_0)$, is similar in principle. The trigonometric polynomial used above may be replaced by any trigonometric polynomial $P(\theta) = \sum_{m=0}^{M} a_m \cos(m\theta)$ satisfying $a_m \geqslant 0$ and $P(\theta) \geqslant 0$, which implies that, for $\sigma > 1$, we have

$$\mathrm{Re}\left(\sum_{m=0}^{M} a_m \sum_{n=1}^{\infty} \frac{\Lambda(n)}{n^{\sigma + imt}}\right) \geqslant 0.$$

Since $\zeta(s) \neq 0$ for $\sigma \geqslant 1$, we infer that, apart from the trivial ones, the function $\zeta(s)$ has all its zeros in the so-called *critical strip* $0 < \sigma < 1$. It is customary to call these zeros the *non-trivial zeros* of $\zeta(s)$ and they are denoted by $\rho = \beta + i\gamma$. Since the Riemann zeta-function is real on the real axis, we have $\overline{\zeta(s)} = \zeta(\bar{s})$ by the reflexion principle, so that, if ρ is a zero of $\zeta(s)$, then so is $\overline{\rho}$. By the functional equation, we deduce that $1 - \rho$ and $1 - \overline{\rho}$ are also zeros of $\zeta(s)$.

For a stronger form of the Prime Number Theorem with an estimate of the error-term (see Sect. 3.7), a larger zero-free region to the left of the line $\sigma = 1$ and some fine estimates of $\zeta(s)$ in this region are needed. This is the purpose of the next result.

Theorem 3.61 *Set* $\tau = |t| + 3$. *The Riemann zeta-function has no zero in the region*

$$\sigma \geqslant 1 - \frac{1}{555\,6379(\log \tau)^9} \quad and \quad t \in \mathbb{R}$$

in which we also have the estimates

$$|\zeta(\sigma + it)| > \frac{1}{442\,368(\log \tau)^7} \quad and \quad \left| -\frac{\zeta'}{\zeta}(\sigma + it) \right| < 555\,6379(\log \tau)^9.$$

Proof Let $0 \leqslant c < \frac{1}{3}$ be the constant appearing in Theorem 3.59 and let s be a complex number such that $1 - c/\log \tau \leqslant \sigma < 2$. From

$$\zeta(1 + it) - \zeta(\sigma + it) = \int_\sigma^1 \zeta'(u + it)\,du$$

and Theorems 3.59 and 3.60, we get

$$|\zeta(\sigma + it)| \geqslant |\zeta(1 + it)| - 9e^c(1 - \sigma)(\log \tau)^2$$

$$> \frac{1}{221\,184(\log \tau)^7} - 9e^c(1 - \sigma)(\log \tau)^2$$

so that if $\sigma \geqslant 1 - 3\,981\,312^{-1}e^{-c}(\log \tau)^{-9}$, we infer

$$|\zeta(\sigma + it)| > \frac{1}{442\,368(\log \tau)^7}$$

and by Theorem 3.59 we also obtain

$$\left| -\frac{\zeta'}{\zeta}(\sigma + it) \right| < 9e^{1/3} \times 442\,368 \times (\log \tau)^9 < 555\,6379(\log \tau)^9$$

as required.   □

We will now prove that the function $-\zeta'(s)/\zeta(s)$ has an analytic continuation to the line $\sigma = 1$.

Proposition 3.62 *The function* $-\zeta'(s)/\zeta(s)$ *has an analytic continuation to* $\sigma = 1$ *with only a simple pole at* $s = 1$ *with residue* 1.

Proof By (3.14), we have $(s-1)\zeta(s) = s\varphi(s)$ for all $\sigma > 0$, where

$$\varphi(s) = 1 - (s-1)\int_1^\infty \frac{\{x\}}{x^{s+1}}\,dx$$

so that $\varphi(s)$ is analytic in the half-plane $\sigma > 0$. We deduce that, for all $\sigma > 1$, we have

$$-(s-1)\frac{\zeta'(s)}{\zeta(s)} = 1 - \frac{\varphi(s) + s\varphi'(s)}{\zeta(s)} \xrightarrow[s\to 1^+]{} 1$$

as required.   □

3.6 Prime Numbers in Arithmetic Progressions

Let a and q be two positive integers. The question of the infinity of the prime numbers $p \equiv a \pmod{q}$ arises naturally. First note that we may suppose that $(a, q) = 1$, otherwise the terms of the sequence $(qn + a)$ are all divisible by (a, q). Thus the condition $(a, q) = 1$ is necessary to this problem. Dirichlet's tour de force was precisely to show that this hypothesis is also *sufficient*. The purpose of this section is to provide a proof of Dirichlet's theorem.

Theorem 3.63 (Dirichlet) *Let a, q be positive coprime integers. Then there are infinitely many prime numbers p such that $p \equiv a \pmod{q}$.*

3.6.1 Euclid vs Euler

Let us go back in time. Before Euler, there were many attempts to generalize Euclid's method to the problem of primes in some fixed arithmetic progressions. The following lemma shows that the method works for at least two of them.

Lemma 3.64

(i) *There are infinitely many prime numbers p such that $p \equiv 3 \pmod{4}$.*
(ii) *There are infinitely many prime numbers p such that $p \equiv 1 \pmod{4}$.*

Proof

(i) Suppose the contrary and let $\mathcal{P}_{4,3} = \{p_1, \ldots, p_n\}$ the finite set of all prime numbers p such that $p \equiv 3 \pmod{4}$. Set $M = 4p_1 \cdots p_n - 1$. If M is prime, then we get a prime number of the type $M \equiv 3 \pmod{4}$ such that $M > p_n$. Suppose now that M is composite. Then there exists at least an odd prime divisor p of M of the form $p \equiv 3 \pmod{4}$, otherwise all prime factors of M are $\equiv 1 \pmod{4}$ since M is odd, and we have $M \equiv 1 \pmod{4}$ which is not the case. On the other hand, we have $p \notin \mathcal{P}_{4,3}$, otherwise we have $p \mid M + 1$. Thus we have found a prime number p such that $p \equiv 3 \pmod{4}$ and $p \notin \mathcal{P}_{4,3}$, leading to a contradiction.

(ii) Suppose the contrary and let $\mathcal{P}_{4,1} = \{p_1, \ldots, p_n\}$ the finite set of all prime numbers p such that $p \equiv 1 \pmod 4$. Set $M = 4(p_1 \cdots p_n)^2 + 1$. If M is prime, then we get a prime number of the type $M \equiv 1 \pmod 4$ such that $M > p_n$. Suppose that M is composite and let p be a prime factor of M. Since M is odd, we have $p \neq 2$ and also $p \notin \mathcal{P}_{4,1}$, otherwise it divides $M - 1$. Since $p \mid M$, we have $4(p_1 \cdots p_n)^2 \equiv -1 \pmod p$ and hence -1 is a quadratic residue modulo p. By Example 3.35, we infer that $p \equiv 1 \pmod 4$. Thus we have found a prime number p such that $p \equiv 1 \pmod 4$ and $p \notin \mathcal{P}_{4,1}$, leading to a contradiction.

The proof is complete.   □

This argument may be generalized to certain arithmetic progressions.

Proposition 3.65 *Let q be an odd prime number. Then there are infinitely many prime numbers p such that $p \equiv 1 \pmod q$.*

Proof

▷ Let $a > 1$ be an integer such that $q \mid a$ and set $M = 1 + a + a^2 + \cdots + a^{q-1}$. Let p be a prime divisor of M. Note that $p \nmid a$, otherwise we have $M \equiv 1 \pmod p$ which contradicts $p \mid M$. Hence $p \neq q$. Using Lemma 1.6, we get $p \mid (a^q - 1)$ and hence $\mathrm{ord}_p(a)$ divides q. Since q is prime, we infer that $\mathrm{ord}_p(a) = 1$ or q. If $\mathrm{ord}_p(a) = 1$, then we get $M \equiv q \pmod p$ and hence $q \equiv 0 \pmod p$ which is impossible, since p and q are two distinct prime numbers. Therefore $\mathrm{ord}_p(a) = q$ and then $q \mid (p - 1)$ by Theorem 3.25. In other words, we have proved that

$$p \equiv 1 \pmod q.$$

▷ Suppose that there are finitely many primes $p_1, \ldots, p_n$ such that $p_i \equiv 1 \pmod q$. Consider the integer $a = qp_1 \cdots p_n$ which is a multiple of q. By above, the integer $M = 1 + a + \cdots + a^{q-1}$ has a prime divisor p such that $p \equiv 1 \pmod q$ and also $p \neq p_i$ otherwise we have $p \mid (M - 1)$, giving a contradiction.   □

The proofs above use a Euclidean argument, in the sense that we have at our disposal a polynomial $P \in \mathbb{Z}[X]$ with degree > 0 and whose integer values have prime divisors, almost all lying in some arithmetic progressions. For instance, Theorem 3.13 uses the polynomial $P = X + 1$, Lemma 3.64 uses the polynomials $P = X - 1$ and $P = X^2 + 1$ respectively and Proposition 3.65 uses the *cyclotomic polynomial* $\Phi_q = X^{q-1} + X^{q-2} + \cdots + 1$ (see Chap. 7, Sect. 7.2.9).

One may wonder whether such arguments may be generalized to *all* arithmetic progressions. The following curious result, due to Schur (1912) for the sufficient condition and Ram Murty (1988) for the necessary condition, shows that such a proof cannot exist for all cases.

Proposition 3.66 *Let a, q be positive coprime integers. Then there exists a Euclidean proof for the sequence $a \pmod q$ if and only if $a^2 \equiv 1 \pmod q$.*

For instance, Euclid's argument may be used to show the infinity of the set of primes such that $p \equiv 8 \pmod 9$ but *cannot be used* to prove that there are infinitely many prime numbers p such that $p \equiv 7 \pmod 9$.

Thus, a new idea is needed. Around 1837, Dirichlet succeeded in using a generalization of Euler's proof of Theorem 3.13 and some group-theoretic tools. More precisely, Dirichlet proved the divergence of the series

$$\sum_{p \equiv a \,(\mathrm{mod}\, q)} \frac{1}{p}$$

by discovering a clever expression for the characteristic function

$$\mathbf{1}_{q,a}(n) = \begin{cases} 1, & \text{if } n \equiv a \pmod q \\ 0, & \text{otherwise} \end{cases} \tag{3.16}$$

(see Proposition 3.68) and showing

$$\lim_{\sigma \to 1^+} \sum_p \frac{\mathbf{1}_{q,a}(p)}{p^\sigma} = \infty.$$

An alternative way is to estimate partial sums of the above series. Let us examine this in the following example taking $q = 4$ and $a = 1$. Dirichlet used the function $\mathbf{1}_{4,1}$ defined for all *odd* positive integers n by

$$\mathbf{1}_{4,1}(n) = \frac{1}{2}\left(1 + \sin\left(\frac{n\pi}{2}\right)\right).$$

One may readily check that, for all odd n, we have

$$\mathbf{1}_{4,1}(n) = \begin{cases} 1, & \text{if } n \equiv 1 \pmod 4 \\ 0, & \text{if } n \equiv 3 \pmod 4. \end{cases}$$

Thus, for all $N > 1$, we get

$$\sum_{\substack{p \leqslant N \\ p \equiv 1 \,(\mathrm{mod}\, 4)}} \frac{1}{p} = \sum_{3 \leqslant p \leqslant N} \frac{\mathbf{1}_{4,1}(p)}{p} = \frac{1}{2} \sum_{3 \leqslant p \leqslant N} \frac{1}{p} + \frac{1}{2} \sum_{3 \leqslant p \leqslant N} \frac{\sin(\pi p/2)}{p}.$$

By Corollary 3.50, the first sum tends to ∞ when $N \longrightarrow \infty$. We will prove in Theorem 3.73 that the series

$$\sum_p \frac{\sin(\pi p/2)}{p}$$

converges, which establishes the divergence of the initial series

$$\sum_{p \equiv 1 \,(\mathrm{mod}\, 4)} \frac{1}{p}.$$

3.6.2 Dirichlet Characters

Dirichlet introduced the *characters* which was the main tool in the proof of his theorem. Although these characters may be defined on any finite abelian group, we may restrict ourselves here to the so-called *Dirichlet characters* as extensions over $\mathbb{Z} \setminus \{0\}$ of homomorphisms of the multiplicative group $(\mathbb{Z}/q\mathbb{Z})^*$.

Definition 3.67 Let q be a positive integer. A *Dirichlet character* modulo q is a map $\chi : \mathbb{Z} \setminus \{0\} \longrightarrow \mathbb{C}$ satisfying the following rules for all $a, b \in \mathbb{Z} \setminus \{0\}$.

$$
\begin{aligned}
&\text{(i)} \quad \chi(a) = \chi(a \ (\mathrm{mod}\, q)) \\
&\text{(ii)} \quad \chi(ab) = \chi(a)\chi(b) \\
&\text{(iii)} \quad \chi(a) = 0 \qquad\qquad\quad \text{if } (a, q) > 1.
\end{aligned}
$$

In fact, (i) and (ii) mean that these characters are homomorphisms of the multiplicative group $(\mathbb{Z}/q\mathbb{Z})^*$ and (iii) extends these maps to $\mathbb{Z} \setminus \{0\}$. We shall make frequent use of the q-periodicity of the characters. One can prove that the set of Dirichlet characters modulo q is a group isomorphic to the multiplicative group $(\mathbb{Z}/q\mathbb{Z})^*$ of the units of the ring $\mathbb{Z}/q\mathbb{Z}$. In particular, there are $\varphi(q)$ Dirichlet characters modulo q. The identity element of this group is called the *principal*, or *trivial*, character modulo q and is usually denoted by χ_0. Thus, χ_0 is defined for all $a \in \mathbb{Z}$ by

$$
\chi_0(a) = \begin{cases} 1, & \text{if } (a, q) = 1 \\ 0, & \text{otherwise.} \end{cases}
$$

Let χ be a Dirichlet character modulo q. We define $\overline{\chi}$ by $\overline{\chi}(a) = \overline{\chi(a)}$. Clearly, $\overline{\chi}$ is also a Dirichlet character modulo q called the *conjugate character* of χ. It is also not difficult to see that, if $(a, q) = 1$, then $\chi(a)$ is a $\varphi(q)$th root of unity. Indeed, denoting by a the residue class of the integer a in $(\mathbb{Z}/q\mathbb{Z})^*$, we have

$$
\left(\chi(a)\right)^{\varphi(q)} = \chi\left(a^{\varphi(q)}\right) = \chi(1) = 1.
$$

Dirichlet succeeded in proving that a suitable linear combination of these characters provides the desired characteristic functions $\mathbf{1}_{q,a}$.

Proposition 3.68 *Let a, q be positive coprime integers and define $\mathbf{1}_{q,a}(n)$ as*

$$
\mathbf{1}_{q,a}(n) = \begin{cases} 1, & \text{if } n \equiv a \ (\mathrm{mod}\, q) \\ 0, & \text{otherwise.} \end{cases}
$$

For all positive integers n we have

$$
\mathbf{1}_{q,a}(n) = \frac{1}{\varphi(q)} \sum_{\chi \ (\mathrm{mod}\, q)} \overline{\chi}(a)\chi(n)
$$

where the summation is taken over all Dirichlet characters modulo q.

Proof

▷ Suppose first that $a = 1$. If $n \equiv 1 \pmod{q}$, then we have by Definition 3.67 (i)

$$\sum_{\chi \,(\mathrm{mod}\, q)} \chi(n) = \sum_{\chi \,(\mathrm{mod}\, q)} \chi(1) = \sum_{\chi \,(\mathrm{mod}\, q)} 1 = \varphi(q).$$

Now assume that $n \not\equiv 1 \pmod{q}$ and $(n, q) = 1$. Then there exists a character χ_1 such that $\chi_1(n) \neq 1$. When χ ranges over all Dirichlet characters modulo q, so does the character $\chi_1 \chi$ and hence

$$\sum_{\chi \,(\mathrm{mod}\, q)} \chi(n) = \sum_{\chi \,(\mathrm{mod}\, q)} (\chi_1 \chi)(n) = \chi_1(n) \sum_{\chi \,(\mathrm{mod}\, q)} \chi(n)$$

which implies that $\sum_{\chi \,(\mathrm{mod}\, q)} \chi(n) = 0$ as required.

▷ Suppose that $a \neq 1$ and let a^{-1} be the inverse of a in $(\mathbb{Z}/q\mathbb{Z})^*$. From the relations $\chi(a)\overline{\chi}(a) = 1$ and $\chi(a)\chi(a^{-1}) = \chi(1) = 1$, we infer that $\overline{\chi}(a) = \chi(a^{-1})$. Hence

$$\sum_{\chi \,(\mathrm{mod}\, q)} \overline{\chi}(a)\chi(n) = \sum_{\chi \,(\mathrm{mod}\, q)} \chi(a^{-1}n) = \begin{cases} 1, & \text{if } a^{-1}n \equiv 1 \pmod{q} \\ 0, & \text{otherwise} \end{cases}$$

by above, which completes the proof. $\qquad\qquad\square$

This result thus provides an expression of the partial sums that we wish to estimate.

Corollary 3.69 *Let a, q be positive coprime integers and $N > 1$ be an integer. Then we have*

$$\sum_{\substack{p \leqslant N \\ p \equiv a \,(\mathrm{mod}\, q)}} \frac{1}{p} = \frac{1}{\varphi(q)} \sum_{\substack{p \leqslant N \\ (p,q)=1}} \frac{1}{p} + \frac{1}{\varphi(q)} \sum_{\chi \neq \chi_0} \overline{\chi}(a) \sum_{p \leqslant N} \frac{\chi(p)}{p}.$$

Proof By Proposition 3.68 we get

$$\sum_{\substack{p \leqslant N \\ p \equiv a \,(\mathrm{mod}\, q)}} \frac{1}{p} = \sum_{p \leqslant N} \frac{1_{q,a}(p)}{p} = \frac{1}{\varphi(q)} \sum_{\chi \,(\mathrm{mod}\, q)} \overline{\chi}(a) \sum_{p \leqslant N} \frac{\chi(p)}{p}$$

and we split the first sum according to $\chi = \chi_0$ or $\chi \neq \chi_0$, leading to the asserted result. $\qquad\qquad\square$

It is fairly easy to see the first sum of Corollary 3.69 tends to ∞ as $N \longrightarrow \infty$ since it only differs from the sum $\sum_{p \leqslant N} 1/p$ of a finite number of terms. The crucial point is then to show that the series

$$\sum_p \frac{\chi(p)}{p}$$

converges for all $\chi \neq \chi_0$, implying the divergence of the series

$$\sum_{p \equiv a \,(\mathrm{mod}\,q)} \frac{1}{p}.$$

It should also be mentioned that the result of Corollary 3.69 may be easily generalized to any complex-valued arithmetic function f in the following form.

$$\sum_{\substack{n \leqslant x \\ n \equiv a \,(\mathrm{mod}\,q)}} f(n) = \frac{1}{\varphi(q)} \sum_{\substack{n \leqslant x \\ (n,q)=1}} f(n) + \frac{1}{\varphi(q)} \sum_{\chi \neq \chi_0} \overline{\chi}(a) \sum_{n \leqslant x} \chi(n) f(n). \qquad (3.17)$$

The next result uses the periodicity of the characters to bound partial sums of non-principal Dirichlet characters.

Proposition 3.70 *For all non-principal Dirichlet characters χ modulo q and all non-negative integers $M < N$, we have*

$$\left| \sum_{n=M+1}^{N} \chi(n) \right| \leqslant \varphi(q).$$

Proof Let $K = q[(N - M - 1)/q]$. As in Proposition 3.68, one may check that, for all $\chi \neq \chi_0$, we have

$$\sum_{a \,(\mathrm{mod}\,q)} \chi(a) = 0$$

and hence, by periodicity, we get

$$\sum_{n=M+1}^{M+K} \chi(n) = \sum_{j=1}^{K/q} \sum_{n=M+1+(j-1)q}^{M+jq} \chi(n) = \sum_{j=1}^{K/q} \sum_{n=M+1}^{M+q} \chi(n) = 0.$$

The interval $]M + K, N]$ contains at most q integers $n_1, \ldots, n_r$ with $r \leqslant q$ and denoting by n_i the residue class of the integer n_i in $(\mathbb{Z}/q\mathbb{Z})^*$, we obtain

$$\left| \sum_{n=M+1}^{N} \chi(n) \right| \leqslant \sum_{\substack{i=1 \\ (n_i,q)=1}}^{r} |\chi(n_i)| \leqslant \sum_{\substack{n \leqslant q \\ (n,q)=1}} 1 = \varphi(q)$$

as asserted.   □

Using Abel's summation as in Remark 1.15 and the trivial bound $\varphi(q) \leqslant q$, we may readily deduce the following useful consequence.

Corollary 3.71 *Let $F \in C^1[1, +\infty[$ be a decreasing function such that $F > 0$ and $F(x) \longrightarrow 0$ as $x \longrightarrow \infty$. For all non-principal Dirichlet characters χ modulo q*

and all real numbers $x \geqslant 1$, we have

$$\left| \sum_{n>x} \chi(n) F(n) \right| \leqslant 2q \, F(x).$$

We end this section with the following definitions. A Dirichlet character is called *real* if its values are real, i.e. $\chi(n) \in \{-1, 0, 1\}$. Otherwise a character is *complex*. A character is said to be *quadratic* if it has order 2 in the character group, i.e. $\chi^2 = \chi_0$ and $\chi \neq \chi_0$. Thus a quadratic character is real, and a real character is either principal or quadratic.

3.6.3 Dirichlet L-Functions

It is remarkable that Dirichlet had the idea to introduce generating series of his characters nearly sixty years before the analytic proofs of the Prime Number Theorem and the Prime Number Theorem for Arithmetic Progressions given by Hadamard and de La Vallée Poussin. Not less remarkable is the fact that the infinity of the set of prime numbers in an arithmetic progression is essentially due to the non-vanishing of these generating series at $s = 1$. This very fruitful idea proved to be one of the most crucial points in almost all problems in number theory, whose treatments mimic Dirichlet's work.

Definition 3.72 (*L*-functions) Let χ be a Dirichlet character modulo $q \geqslant 2$. The *L-function*, or *L-series*, attached to χ is the Dirichlet series of χ, i.e. for all $s = \sigma + it \in \mathbb{C}$ such that $\sigma > 1$, we set

$$L(s, \chi) = \sum_{n=1}^{\infty} \frac{\chi(n)}{n^s}.$$

As in Proposition 3.52, it may be shown that, for all $\sigma > 1$ and all $x \geqslant 2$, we have

$$\prod_{p \leqslant x} \left(1 - \frac{\chi(p)}{p^\sigma} \right)^{-1} = \sum_{P^+(n) \leqslant x} \frac{\chi(n)}{n^\sigma}.$$

Let $s = \sigma + it \in \mathbb{C}$. Since

$$\sum_{p \leqslant x} \left| \frac{\chi(p)}{p^s} \right| = \sum_{p \leqslant x} \frac{1}{p^\sigma}$$

both sides of the above identity converge absolutely for all complex s such that $\sigma > 1$ and therefore, as in Definition 3.53, we have

$$L(s, \chi) = \prod_{p} \left(1 - \frac{\chi(p)}{p^s} \right)^{-1} \tag{3.18}$$

for all $s \in \mathbb{C}$ such that $\sigma > 1$. Note also that

$$L(s, \chi_0) = \sum_{\substack{n=1 \\ (n,q)=1}}^{\infty} \frac{1}{n^s} = \zeta(s) \prod_{p \mid q} \left(1 - \frac{1}{p^s}\right) \tag{3.19}$$

and if $\chi \neq \chi_0$, then the series converges for all $\sigma > 0$ by Corollary 3.71. This result is the best possible since the terms in the series do not tend to 0 when $\sigma = 0$.

3.6.4 The Convergence of the Series $\sum_p \chi(p) p^{-1}$

In this section we intend to prove the main tools which enable us to show Dirichlet's theorem. There are many ways to get the desired result. We choose a rather elementary proof which is due to Shapiro. It has the advantage of relating Dirichlet's theorem to quadratic fields, another area in which Dirichlet showed his talent (see Chap. 7).

Theorem 3.73 *If* $\chi \neq \chi_0$ *is a non-principal Dirichlet character modulo* q *satisfying* $L(1, \chi) \neq 0$, *then the series*

$$\sum_p \frac{\chi(p)}{p}$$

converges.

Proof Let $N \geqslant 2$ be an integer. The idea is to estimate the sum $\sum_{n \leqslant N} n^{-1} \chi(n) \log n$ in two different ways.

▷ By Lemma 3.41 (i) we have

$$\sum_{n \leqslant N} \frac{\chi(n) \log n}{n} = \sum_{n \leqslant N} \frac{\chi(n)}{n} \sum_{d \mid n} \Lambda(d).$$

Interchanging the order of summation (see also Chap. 4) and using Definition 3.67 (ii), we get

$$\sum_{n \leqslant N} \frac{\chi(n) \log n}{n} = \sum_{d \leqslant N} \Lambda(d) \sum_{\substack{n \leqslant N \\ d \mid n}} \frac{\chi(n)}{n}$$

$$= \sum_{d \leqslant N} \Lambda(d) \sum_{k \leqslant N/d} \frac{\chi(kd)}{kd}$$

$$= \sum_{d \leqslant N} \frac{\chi(d) \Lambda(d)}{d} \sum_{k \leqslant N/d} \frac{\chi(k)}{k}$$

$$= L(1, \chi) \sum_{d \leqslant N} \frac{\chi(d) \Lambda(d)}{d} - \sum_{d \leqslant N} \frac{\chi(d) \Lambda(d)}{d} \sum_{k > N/d} \frac{\chi(k)}{k}.$$

Since $L(1, \chi) \neq 0$, we infer that

$$\sum_{d \leqslant N} \frac{\chi(d) \Lambda(d)}{d} = \frac{1}{L(1, \chi)} \left\{ \sum_{n \leqslant N} \frac{\chi(n) \log n}{n} + \sum_{d \leqslant N} \frac{\chi(d) \Lambda(d)}{d} \sum_{k > N/d} \frac{\chi(k)}{k} \right\}.$$

$$(3.20)$$

Using Corollary 3.71 we get

$$\left| \sum_{d \leqslant N} \frac{\chi(d) \Lambda(d)}{d} \sum_{k > N/d} \frac{\chi(k)}{k} \right| \leqslant \frac{2q}{N} \sum_{d \leqslant N} \Lambda(d) = \frac{2q \Psi(N)}{N}$$

and Corollary 3.47 (ii) gives

$$\left| \sum_{d \leqslant N} \frac{\chi(d) \Lambda(d)}{d} \sum_{k > N/d} \frac{\chi(k)}{k} \right| < 4q.$$

Inserting this bound in (3.20) provides the estimate

$$\left| \sum_{d \leqslant N} \frac{\chi(d) \Lambda(d)}{d} \right| < \frac{1}{|L(1, \chi)|} \left(\left| \sum_{n \leqslant N} \frac{\chi(n) \log n}{n} \right| + 4q \right). \qquad (3.21)$$

▷ By partial summation we get

$$\sum_{n \leqslant N} \frac{\chi(n) \log n}{n} = \frac{\chi(2) \log 2}{2} + \sum_{3 \leqslant n \leqslant N} \frac{\chi(n) \log n}{n}$$

$$= \frac{\chi(2) \log 2}{2} + \frac{\log N}{N} \sum_{3 \leqslant n \leqslant N} \chi(n)$$

$$+ \int_3^N \frac{\log t - 1}{t^2} \left(\sum_{3 \leqslant n \leqslant t} \chi(n) \right) dt$$

so that using Proposition 3.70 we obtain

$$\left| \sum_{n \leqslant N} \frac{\chi(n) \log n}{n} \right| \leqslant \frac{\log 2}{2} + q \left(\frac{\log N}{N} + \int_3^N \frac{\log t - 1}{t^2} \, dt \right) = \frac{\log 2}{2} + \frac{q \log 3}{3} < q.$$

Inserting this bound in (3.21) gives

$$\left| \sum_{d \leqslant N} \frac{\chi(d) \Lambda(d)}{d} \right| < \frac{5q}{|L(1, \chi)|}. \qquad (3.22)$$

▷ We have

$$\sum_{p \leqslant N} \frac{\chi(p) \log p}{p} = \sum_{d \leqslant N} \frac{\chi(d) \Lambda(d)}{d} - \sum_{p \leqslant N} \log p \sum_{\alpha=2}^{[\log N / \log p]} \frac{\chi(p^\alpha)}{p^\alpha}$$

and the second sum is bounded since

$$\left| \sum_{p \leqslant N} \log p \sum_{\alpha=2}^{[\log N / \log p]} \frac{\chi(p^\alpha)}{p^\alpha} \right| \leqslant \sum_{p \leqslant N} \log p \sum_{\alpha=2}^{[\log N / \log p]} \frac{1}{p^\alpha} \leqslant \sum_{p} \frac{\log p}{p(p-1)} < 1$$

where we used a result from [Rn62] in the last inequality. Now by partial summation we get

$$\sum_{p \leqslant N} \frac{\chi(p)}{p} = \frac{1}{\log N} \sum_{p \leqslant N} \frac{\chi(p) \log p}{p} + \int_2^N \left(\sum_{p \leqslant t} \frac{\chi(p) \log p}{p} \right) \frac{dt}{t (\log t)^2}$$

so that by above we obtain

$$\left| \sum_{p \leqslant N} \frac{\chi(p)}{p} \right| < \frac{1}{\log N} \left(\left| \sum_{d \leqslant N} \frac{\chi(d) \Lambda(d)}{d} \right| + 1 \right)$$

$$+ \int_2^N \left(\left| \sum_{d \leqslant t} \frac{\chi(d) \Lambda(d)}{d} \right| + 1 \right) \frac{dt}{t (\log t)^2}$$

and estimate (3.22) provides

$$\left| \sum_{p \leqslant N} \frac{\chi(p)}{p} \right| < \frac{1}{\log 2} \left(\frac{5q}{|L(1, \chi)|} + 1 \right)$$

which completes the proof.   □

3.6.5 The Non-vanishing of $L(1, \chi)$

By Theorem 3.73, the non-vanishing of $L(1, \chi)$ for all $\chi \neq \chi_0$ is the main point of the proof of Dirichlet's theorem. Once again, there are many ways to show this. For complex Dirichlet characters, one may mimic and adapt the proof of Theorem 3.57 as follows.

Theorem 3.74 *For all* complex *characters* χ *modulo* q*, we have* $L(1 + it, \chi) \neq 0$ *for all* $t \in \mathbb{R}$.

Proof If $\chi \neq \chi_0$, the function $L(s, \chi)$ is analytic for $\sigma > 0$. On the other hand, by (3.19), the function $L(s, \chi_0)$ is analytic in this half-plane except for a simple pole at $s = 1$ with residue $\varphi(q)/q$. Using (3.18), we deduce that in either case a logarithm of $L(s, \chi)$ is given by

$$\sum_{n=2}^{\infty} \frac{\chi(n)\Lambda(n)}{n^s \log n} = \sum_{p} \sum_{\alpha=1}^{\infty} \frac{\chi(p)^{\alpha}}{\alpha p^{\alpha s}}$$

for $\sigma > 1$. Furthermore, for all characters χ, one may write

$$\chi(n) = \chi_0(n)e^{i\omega(n)}$$

where $\omega : \mathbb{N} \longrightarrow [0, 2\pi[$. Now let $\sigma > 1$ and $t \in \mathbb{R}$. Setting $\theta(n, t) = \omega(n) - t \log n$, we have

$$\mathrm{Re}\left(3 \log L(\sigma, \chi_0) + 4 \log L(\sigma + it, \chi) + \log L(\sigma + 2it, \chi^2)\right)$$

$$= \sum_{\substack{n=2 \\ (n,q)=1}}^{\infty} \frac{\Lambda(n)}{n^{\sigma} \log n} \{3 + 4\cos\theta(n, t) + \cos(2\theta(n, t))\}$$

$$= 2 \sum_{\substack{n=2 \\ (n,q)=1}}^{\infty} \frac{\Lambda(n)}{n^{\sigma} \log n} \{1 + \cos\theta(n, t)\}^2 \geqslant 0$$

which implies that

$$\left|L(\sigma, \chi_0)^3 L(\sigma + it, \chi)^4 L(\sigma + 2it, \chi^2)\right| \geqslant 1.$$

Now suppose that $1 + it_0$ is a zero of order $m \geqslant 1$ of $L(s, \chi)$ for some $t_0 \in \mathbb{R}$, so that there exists $\ell \neq 0$ such that

$$\lim_{\sigma \to 1^+} \frac{L(\sigma + it_0, \chi)}{(\sigma - 1)^m} = \ell.$$

By above we infer that

$$\left|(\sigma - 1)^3 L(\sigma, \chi_0)^3 (\sigma - 1)^{-4m} L(\sigma + it_0, \chi)^4 L(\sigma + 2it_0, \chi^2)\right| \geqslant (\sigma - 1)^{3-4m}.$$

Since $\chi^2 \neq \chi_0$, the function $L(s, \chi^2)$ is continuous at all points on the line $\sigma = 1$ and thus does not have a pole at $s = 1$. Therefore letting $\sigma \longrightarrow 1^+$ gives a contradiction as in Theorem 3.57. $\qquad\qquad\square$

It should be mentioned that this proof extends to the case of real characters, proving in this case that $L(1 + it, \chi) \neq 0$ for all $t \in \mathbb{R} \setminus \{0\}$. The difficult point is thus a proof of $L(1, \chi) \neq 0$ for all quadratic Dirichlet characters. Once again, there exist many proofs, almost all related to certain results from algebraic number theory. Before stating the theorem, we shall make use of the following definition.

Definition 3.75 Let χ be a Dirichlet character modulo q. The *conductor* of χ is the smallest positive integer $f \mid q$ such that there exists a Dirichlet character χ^* modulo f satisfying

$$\chi = \chi_0 \chi^*. \tag{3.23}$$

A Dirichlet character χ modulo q is said to be *primitive* if the conductor f of χ is such that $f = q$. In (3.23), the character χ^* is then primitive and uniquely determined by χ. We will say that χ^* *induces* χ or that χ is *induced by* χ^*.

Example 3.76

▷ It is noteworthy that each Dirichlet character modulo q is induced by a unique primitive Dirichlet character modulo a divisor of q. Furthermore, a character χ is imprimitive if and only if there exists $d \mid q$ with $d < q$ such that, for all positive integers a, b satisfying $(a, q) = (b, q) = 1$ and $a \equiv b \pmod{d}$, we have $\chi(a) = \chi(b)$. This implies that the principal character χ_0 modulo $q > 1$ is imprimitive, by taking $d = 1$.

▷ There is only one primitive character modulo 4 defined for all odd positive integers n by

$$\chi_4(n) = (-1)^{(n-1)/2}.$$

▷ There are two primitive characters[14] modulo 8 defined for all odd positive integers n by

$$\chi_8(n) = (-1)^{(n^2-1)/8} \quad \text{and} \quad \chi_4\chi_8(n) = (-1)^{(n-1)/2+(n^2-1)/8}.$$

▷ If $q = p^\alpha$ is a prime power, the only real primitive characters of conductor q are χ_4, χ_8, $\chi_4\chi_8$ and χ_p. Every real primitive character can be obtained as the product of these characters. This implies that the conductor of a real primitive character is of the form 1, m, $4m$ or $8m$ where m is a positive odd squarefree integer.

▷ Lemma 7.107 gives another useful characterization of the real primitive characters.

▷ Let χ be a Dirichlet character modulo q induced by χ^*. Then we have for $\sigma > 1$

$$L(s, \chi) = \prod_p \left(1 - \frac{\chi_0\chi^*(p)}{p^s}\right)^{-1} = \prod_{p \nmid q} \left(1 - \frac{\chi^*(p)}{p^s}\right)^{-1}$$

$$= L(s, \chi^*) \prod_{p \mid q} \left(1 - \frac{\chi^*(p)}{p^s}\right)$$

---

[14]It is noteworthy that the Dirichlet characters χ_4, χ_8 and $\chi_4\chi_8$ are the characters attached to the quadratic fields $\mathbb{Q}(\sqrt{-1})$, $\mathbb{Q}(\sqrt{2})$ and $\mathbb{Q}(\sqrt{-2})$ respectively. See Chap. 7.

and if $\chi \neq \chi_0$, we get

$$L(1, \chi) = L(1, \chi^\star) \prod_{p|q} \left(1 - \frac{\chi^\star(p)}{p}\right). \qquad (3.24)$$

We are now in a position to prove the following result.

Theorem 3.77 *For all* quadratic *characters* χ *modulo* q, *we have* $L(1, \chi) \neq 0$.

First proof By (3.24), we may suppose that χ is primitive since the Euler product of the right-hand side is non-zero. By Lemma 7.107, the real primitive characters modulo q are of the form $(\chi(-1)q/n)$ and (7.26) and the fact that the class number is a positive integer give the asserted result. $\qquad \square$

Second proof As above, we may suppose that χ is real primitive, and hence is the character attached to a quadratic number field $\mathbb{K}$. Following [Mon93], for all $t \in [0, 1[$, let

$$f(t) = \sum_{n=1}^{\infty} \frac{\chi(n)t^n}{1 - t^n}$$

be the *Lambert series* associated to χ. By [PS98, Theorem VIII-65] and Proposition 7.131 (see also Example 4.11), we have

$$f(t) = \sum_{n=1}^{\infty} \left(\sum_{d|n} \chi(d)\right)t^n = \sum_{n=1}^{\infty} v_{\mathbb{K}}(n)t^n$$

where the function $v_{\mathbb{K}}$ is defined in Definition 7.117. This implies that $f(t) \longrightarrow \infty$ as $t \longrightarrow 1^-$. Now suppose that $L(1, \chi) = 0$. Then we have

$$-f(t) = \sum_{n=1}^{\infty} \left(\frac{1}{n(1-t)} - \frac{t^n}{1-t^n}\right)\chi(n) = \sum_{n=1}^{\infty} a_n \chi(n).$$

Observe that we have

$$(1-t)(a_n - a_{n+1}) = \frac{1}{n} - \frac{1}{n+1} - \frac{t^n}{1 + \cdots + t^{n-1}} + \frac{t^{n+1}}{1 + \cdots + t^n}$$

$$= \frac{1}{n(n+1)} - \frac{t^n}{(1 + \cdots + t^{n-1})(1 + \cdots + t^n)}$$

and using the arithmetic-geometric mean inequality we get for all $0 \leqslant t < 1$

$$\sum_{j=0}^{n-1} t^j \geqslant n \prod_{j=0}^{n-1} t^{j/n} = nt^{(n-1)/2} \geqslant nt^{n/2}$$

and

$$\sum_{j=0}^{n} t^j \geqslant (n+1)t^{n/2}$$

so that

$$(1-t)(a_n - a_{n+1}) \geqslant \frac{1}{n(n+1)} - \frac{t^n}{n(n+1)t^n} = 0$$

and hence the sequence (a_n) is non-increasing and tends to 0. By Corollary 3.71 we get

$$\left| -f(t) \right| \leqslant (2q+1)a_1 = 2q+1$$

contradicting the unboundedness of f on $[0, 1[$. □

Now we may conclude this section.

Proof of Theorem 3.63 Theorems 3.73, 3.74 and 3.77, along with Corollary 3.69, give the complete proof of Dirichlet's theorem. □

3.7 Further Developments

3.7.1 Sieves

Let $n \geqslant 2$ be a fixed integer. The well-known sieve of Eratosthenes asserts that an integer $m \in]\sqrt{n}, n]$ which is not divisible by any prime number $p \leqslant \sqrt{n}$ is prime. Let P_n be the set of prime numbers $p \leqslant n$ and S_n be the set of positive integers $m \leqslant n$ which are not divisible by all prime numbers $\leqslant \sqrt{n}$. We then have

$$P_n \subseteq S_n \cup \{1, \ldots, \sqrt{n}\}$$

and hence

$$\pi(n) \leqslant |S_n| + [\sqrt{n}].$$

More generally, let $r \geqslant 2$ be an integer. We define $\pi(n, r)$ to be the number of positive integers $m \leqslant n$ which are not divisible by prime numbers $\leqslant r$ (hence $|S_n| = \pi(n, [\sqrt{n}])$). Similar arguments as above give

$$\pi(n) \leqslant \pi(n, r) + r. \tag{3.25}$$

One may bound $\pi(n, r)$ by appealing to the *inclusion-exclusion principle* which generalizes the well-known formula

$$|A \cup B| = |A| + |B| - |A \cap B|.$$

There exist many statements of this result, but in number theory we often use the following one.

Proposition 3.78 *Consider N objects and r properties denoted by $p_1, \ldots, p_r$. Suppose that N_1 objects satisfy the property p_1, N_2 objects satisfy the property $p_2, \ldots, N_{12}$ objects satisfy the properties p_1 and $p_2, \ldots, N_{123}$ objects satisfy the properties p_1, p_2 and p_3, and so on. Then, the number of objects which satisfy none of those properties is equal to*

$$N - N_1 - N_2 - \cdots - N_r + N_{12} + N_{13} + \cdots + N_{r-1,r} - N_{123} - N_{124} - \cdots$$

For instance, the identity $\max(a,b) = a + b - \min(a,b)$ can be generalized into the following one

$$\max(a_1, \ldots, a_r) = a_1 + \cdots + a_r - \min(a_1, a_2) - \cdots - \min(a_{r-1}, a_r)$$
$$+ \cdots \pm \min(a_1, \ldots, a_r).$$

Applied to $\pi(n, r)$ and using Proposition 1.11 (v), we get

$$\pi(n, r) = n - \sum_{p \leqslant r} \left[\frac{n}{p} \right] + \sum_{p_1 < p_2 \leqslant r} \left[\frac{n}{p_1 p_2} \right] - \sum_{p_1 < p_2 < p_3 \leqslant r} \left[\frac{n}{p_1 p_2 p_3} \right] + \cdots \quad (3.26)$$

Since $x - 1 < [x] \leqslant x$, we obtain

$$\pi(n, r) < n - \sum_{p \leqslant r} \frac{n}{p} + \sum_{p_1 < p_2 \leqslant r} \frac{n}{p_1 p_2} + \cdots + \sum_{p \leqslant r} 1 + \sum_{p_1 < p_2 \leqslant r} 1 + \cdots$$

$$= n - \sum_{p \leqslant r} \frac{n}{p} + \sum_{p_1 < p_2 \leqslant r} \frac{n}{p_1 p_2} + \cdots + \binom{\pi(r)}{1} + \binom{\pi(r)}{2} + \cdots$$

$$= n \prod_{p \leqslant r} \left(1 - \frac{1}{p} \right) + 2^{\pi(r)} - 1.$$

Now inserting this bound in (3.25) implies that

$$\pi(n) < n \prod_{p \leqslant r} \left(1 - \frac{1}{p} \right) + 2^{\pi(r)} + r - 1.$$

Using the inequalities $\log(1 - x) \leqslant -x$ and (3.2), we get

$$\prod_{p \leqslant r} \left(1 - \frac{1}{p} \right) \leqslant \exp\left(-\sum_{p \leqslant r} \frac{1}{p} \right) < \frac{e^{1/2}}{\log r}$$

so that

$$\pi(n) < \frac{ne^{1/2}}{\log r} + 2^r + r - 1.$$

Choosing $r = 1 + [\log n]$ with $n \geqslant 7$ implies that

$$\pi(n) < \frac{3n}{\log \log n}.$$

This is a weaker result than Corollary 3.45, but the ideas developed above are very fruitful and eventually gave birth to an efficient new branch in number theory called *sieves methods*. Nevertheless, it is interesting to note that this inequality is sufficient to assert that the prime numbers *rarefy*, i.e. $\pi(n) = o(n)$ as $n \longrightarrow \infty$.

Let us adopt a more arithmetical point of view. The Möbius function is one of the most important functions in number theory, and we will prove in Chap. 4 the following convolution identity

$$\sum_{d|n} \mu(d) = \begin{cases} 1, & \text{if } n = 1 \\ 0, & \text{otherwise} \end{cases}$$

and hence the characteristic function of the integers n having no prime factor less than a parameter $z \geqslant 2$ and lying in a set $\mathcal{P}$ of primes is given by

$$\sum_{d|(n,P_z)} \mu(d)$$

where $P_z = \prod_{p \in \mathcal{P},\, p \leqslant z} p$. This sum could be identified with the sieve of Eratosthenes.

Now let $\mathcal{A}$ be a finite set of integers. It is customary to denote

$$S(\mathcal{A}, \mathcal{P}; z) = \sum_{\substack{n \in \mathcal{A} \\ (n, P_z) = 1}} 1$$

so that by above we get

$$S(\mathcal{A}, \mathcal{P}; z) = \sum_{d | P_z} \mu(d) A_d \qquad (3.27)$$

where

$$A_d = \sum_{\substack{n \in \mathcal{A} \\ d|n}} 1. \qquad (3.28)$$

Assume that, for all positive integers d, we have

$$A_d = \frac{X\rho(d)}{d} + r_d \qquad (3.29)$$

where $X > 0$, $\rho(d) \geqslant 0$ is multiplicative and the remainder term satisfies $|r_d| \leqslant \rho(d)$. Inserting this estimate into (3.27) implies that

$$S(\mathcal{A}, \mathcal{P}; z) = X \sum_{d \mid P_z} \frac{\mu(d)\rho(d)}{d} + \sum_{d \mid P_z} \mu(d)r_d = X \prod_{p \leqslant z} \left(1 - \frac{\rho(p)}{p}\right) + \sum_{d \mid P_z} \mu(d)r_d.$$

Suppose we want to count the number of prime numbers in an interval $]x, x + y]$ with $x, y \in \mathbb{Z}_{\geqslant 0}$. We take $\mathcal{A} =]x, x+y] \cap \mathbb{Z}$, $\mathcal{P}$ the set of all primes, and since

$$A_d = \sum_{\substack{x < n \leqslant x+y \\ d \mid n}} 1 = \frac{y}{d} + \left\{\frac{x}{d}\right\} - \left\{\frac{x+y}{d}\right\}$$

we have $X = y$, $\rho(d) = 1$ which gives with $z = \sqrt{x}$ and setting $P = P_{\sqrt{x}}$

$$\pi(x+y) - \pi(x) = y \prod_{p \leqslant x^{1/2}} \left(1 - \frac{1}{p}\right) + \sum_{d \mid P} \mu(d)\left(\left\{\frac{x}{d}\right\} - \left\{\frac{x+y}{d}\right\}\right)$$

and using Corollary 3.51 gives

$$\pi(x+y) - \pi(x) = \left(1 + o(1)\right)\frac{2ye^{-\gamma}}{\log x} + \sum_{d \mid P} \mu(d)\left(\left\{\frac{x}{d}\right\} - \left\{\frac{x+y}{d}\right\}\right).$$

A crude estimate of the remainder term then shows that it is larger than the main term, partly because of the fact that as z increases, the factors of P_z become very large and their number too. It seems to be very difficult to take account of some cancellations of the summands. This is the limitation of the sieve of Eratosthenes.

In 1915, Brun [Bru15] came up with a simple, but very efficient idea. Suppose that we have at our disposal a function g such that

$$\sum_{d \mid n} \mu(d) \leqslant \sum_{d \mid n} g(d).$$

Then by repeating the computations above, we get

$$S(\mathcal{A}, \mathcal{P}; z) \leqslant X \prod_{p \mid P_z} \left(1 - \frac{\rho(p)}{p}\right) + O\left(\sum_{d \mid P_z} |g(d)r_d|\right).$$

The problem is then to find a suitable function g which is easier to handle than the Möbius function, and to minimize the right-hand side of this inequality. Moving from the classic world of exactness to the world of inequalities, Brun proved that

$$\sum_{\substack{d \mid (n, P_z) \\ \omega(d) \leqslant 2k+1}} \mu(d) \leqslant \sum_{d \mid (n, P_z)} \mu(d) \leqslant \sum_{\substack{d \mid (n, P_z) \\ \omega(d) \leqslant 2k}} \mu(d)$$

for all integers $n, k \geqslant 0$ and all $z \geqslant 2$, where $\omega(n)$ is the usual additive arithmetic function counting the number of distinct prime factors of n. This implies that

$$\sum_{\substack{d \mid P_z \\ \omega(d) \leqslant 2k+1}} \mu(d) A_d \leqslant S(\mathcal{A}, \mathcal{P}; z) \leqslant \sum_{\substack{d \mid P_z \\ \omega(d) \leqslant 2k}} \mu(d) A_d \qquad (3.30)$$

for all integers $k \geqslant 0$ and all $z \geqslant 2$. These inequalities, named *Brun's pure sieve*, are related to the so-called *Bonferroni's inequalities* stating that, if $A_1, \ldots, A_n$ are subsets of a finite set S, then we have

$$\sum_{j=0}^{2k+1} (-1)^j S_j \leqslant |\overline{A_1} \cap \cdots \cap \overline{A_n}| \leqslant \sum_{j=0}^{2k} (-1)^j S_j$$

for all integers $k \geqslant 0$, where $\overline{A_j}$ is the complement of A_j in S and

$$S_h = \sum_{\{i_1, \ldots, i_h\} \in \{1, \ldots, n\}} |A_{i_1} \cap \cdots \cap A_{i_h}|.$$

Let us try Brun's ideas on the example of *twin primes*. We call a twin prime every prime p such that $p + 2$ is also a prime number. Let $\pi_2(x)$ be the number of twin primes less than x. In this problem, we work with $\mathcal{P}$ the set of all primes, $\mathcal{A} = \{n(n+2) : 1 \leqslant n < x - 2\}$ so that A_d is the number of solutions of the congruence $n(n+2) \equiv 0 \pmod{d}$ with d squarefree. Writing $d = 2^e d'$ with d' odd and $e \in \{0, 1\}$ and applying the Chinese remainder theorem, we deduce that (3.29) holds with $X = x$ and the strongly multiplicative function ρ defined by $\rho(2) = 1$ and $\rho(p) = 2$ for all odd prime numbers p, and we have $r_d \ll \rho(d)$. Now if p is a twin prime less than x, then either $p \leqslant z$ or the number $n(n+2)$ has no prime factor $\leqslant z$, and hence

$$\pi_2(x) \leqslant S(\mathcal{A}, \mathcal{P}; z) + z.$$

Using (3.30) we get

$$\pi_2(x) \leqslant \sum_{\substack{d \mid P_z \\ \omega(d) \leqslant 2k}} \mu(d) A_d + z$$

$$= x \sum_{\substack{d \mid P_z \\ \omega(d) \leqslant 2k}} \frac{\mu(d) \rho(d)}{d} + z + O\left(\sum_{\substack{d \mid P_z \\ \omega(d) \leqslant 2k}} \mu^2(d) \rho(d) \right)$$

$$= x \sum_{d \mid P_z} \frac{\mu(d) \rho(d)}{d} + z + O\left(\sum_{\substack{d \mid P_z \\ \omega(d) \leqslant 2k}} \mu^2(d) \rho(d) + x \sum_{\substack{d \mid P_z \\ \omega(d) > 2k}} \frac{\mu^2(d) \rho(d)}{d} \right)$$

$$= \frac{x}{2} \prod_{3 \leqslant p \leqslant z} \left(1 - \frac{2}{p} \right) + z + O\left(\sum_{\substack{d \mid P_z \\ \omega(d) \leqslant 2k}} \mu^2(d) \rho(d) + x \sum_{\substack{d \mid P_z \\ \omega(d) > 2k}} \frac{\mu^2(d) \rho(d)}{d} \right).$$

Using $\mu^2(d)\rho(d) \leqslant 2^{\omega(d)}$ gives

$$\sum_{\substack{d \mid P_z \\ \omega(d) \leqslant 2k}} \mu^2(d)\rho(d) \leqslant \sum_{d \leqslant z^{2k}} 2^{\omega(d)} \leqslant 2ek\, z^{2k} \log z$$

and

$$\sum_{\substack{d \mid P_z \\ \omega(d) > 2k}} \frac{\mu^2(d)\rho(d)}{d} \leqslant \frac{1}{4^k} \sum_{d \mid P_z} \frac{\mu^2(d)\rho(d)2^{\omega(d)}}{d} \leqslant \frac{1}{4^k} \sum_{d \leqslant z^{2k}} \frac{4^{\omega(d)}}{d} \leqslant \frac{(2e^2 k)^4}{4^k} (\log z)^4$$

for all $z \geqslant 2$, where we used Example 4.29. By Corollary 3.51 we have

$$\frac{x}{2} \prod_{3 \leqslant p \leqslant z} \left(1 - \frac{2}{p}\right) \leqslant 2x \prod_{p \leqslant z} \left(1 - \frac{1}{p}\right)^2 \ll \frac{x}{(\log z)^2}$$

and choosing $k = [\log x / \log(z^3)]$ and $z = \exp(\frac{\log x}{20 \log \log x})$ implies that

$$\pi_2(x) \ll x \left(\frac{\log \log x}{\log x}\right)^2.$$

By partial summation, we obtain Brun's theorem [Bru19].

Theorem 3.79 (Brun) *The sum of reciprocals of twin primes converges.*

Iwaniec [Iwa77] proved that, under certain circumstances, the sieve of Eratosthenes yields an asymptotic formula for $\mathcal{S}(\mathcal{A}, \mathcal{P}; z)$. Let $0 \leqslant \kappa < \frac{1}{2}$ and assume the following hypotheses with $A, B, C, D \geqslant 1$.

(i) For all $2 \leqslant w < z$, we have

$$-A \leqslant \sum_{w \leqslant p < z} \frac{\rho(p) - \kappa}{p} \leqslant B.$$

(ii) $\max_{n \in \mathcal{A}} |n| \leqslant CX$.
(iii) $|r_d| \leqslant D\rho(d)$.
(iv) $0 \leqslant \rho(p)/p \leqslant 1 - B^{-1}$ for all $p \in \mathcal{P}$.

Define the function f by $f(u) = u^{-\kappa}$ for all $0 < u \leqslant 1$ and, for all $u > 1$, by the continuous solution of the differential-difference equation

$$uf'(u) + \kappa f(u) = \kappa f(u - 1).$$

It can be shown that $f(u) = e^{-\kappa\gamma}\Gamma(1 - \kappa) + O(e^{-u})$ as $u \longrightarrow \infty$.

Theorem 3.80 (Iwaniec) *Under the assumptions* (i)–(iv), *we have*

$$S(\mathcal{A}, \mathcal{P}; z) = \frac{e^{\kappa\gamma} X}{\Gamma(1-\kappa)} \prod_{p|P_z}\left(1 - \frac{\rho(p)}{p}\right)$$

$$\times \left\{ f\left(\frac{\log x}{\log z}\right) + O\left(\frac{A(\log x / \log z + 1)}{1 - 2\kappa}(\log z)^{2\kappa - 1}\right)\right\}.$$

3.7.2 Other Approximate Functional Equations for $\zeta(s)$

According to the periodicity of the function ψ, the error-term of Theorem 3.57 is expected to be small. Using the techniques of exponential sums from Chap. 6, we may prove the following result bearing out this conjecture.

Theorem 3.81 (Approximate functional equation) *Let $\sigma_0 > 0$. We have uniformly for $x \geqslant 1, \sigma \geqslant \sigma_0$ and $|t| \leqslant \pi x$*

$$\zeta(s) = \sum_{n \leqslant x} \frac{1}{n^s} + \frac{x^{1-s}}{s-1} + O(x^{-\sigma}).$$

Proof Let $y > x \geqslant 1$ be real numbers and $s = \sigma + it \in \mathbb{C}$ satisfying the hypotheses of the theorem. We have

$$\zeta(s) = \sum_{n \leqslant x} \frac{1}{n^s} + \sum_{x < n \leqslant y} \frac{1}{n^s} + \sum_{n > y} \frac{1}{n^s}.$$

By (1.7), we infer

$$\sum_{n > y} \frac{1}{n^s} = \frac{y^{1-s}}{s-1} - \frac{y^{-s}}{2} - s \int_y^\infty \frac{\psi(u)}{u^{s+1}}\,du = \frac{y^{1-s}}{s-1} - \frac{y^{-s}}{2} + O(|s|y^{-\sigma}).$$

For the second sum, we use partial summation from Theorem 1.14 which gives

$$\sum_{x < n \leqslant y} \frac{1}{n^s} = \sum_{x < n \leqslant y} n^{-\sigma} n^{-it} = y^{-\sigma} \sum_{x < n \leqslant y} n^{-it} + \sigma \int_x^y \left(\sum_{x < n \leqslant u} n^{-it}\right) \frac{du}{u^{\sigma+1}}.$$

Now Lemma 6.28 is applied with $f(v) = -(2\pi)^{-1}t \log v$ to estimate the sums. Since $|t| \leqslant \pi x$, we have $|f'(v)| < \frac{1}{2}$ for all $x < v \leqslant u \leqslant y$, so that in this case Lemma 6.28 may be written in the following simpler form

$$\sum_{x < n \leqslant u} n^{-it} = \sum_{x < n \leqslant u} e(f(n)) = \int_x^u e(f(v))\,dv + O(1) = \frac{u^{1-it} - x^{1-it}}{1-it} + O(1).$$

Hence we get

$$\sum_{x<n\leqslant y}\frac{1}{n^s}=\frac{y^{1-s}-y^{-\sigma}x^{1-it}}{1-it}+O\left(y^{-\sigma}\right)$$

$$+\sigma\int_x^y\left\{\frac{u^{1-it}-x^{1-it}}{u^{\sigma+1}(1-it)}+O\left(\frac{1}{u^{\sigma+1}}\right)\right\}du$$

$$=\frac{y^{1-s}}{1-it}-\frac{x^{1-s}}{1-it}+\frac{\sigma}{1-it}\int_x^y u^{-s}\,du+O\left(x^{-\sigma}\right)$$

$$=\frac{y^{1-s}}{1-s}-\frac{x^{1-s}}{1-s}+O\left(x^{-\sigma}\right).$$

Therefore we obtain

$$\zeta(s)=\sum_{n\leqslant x}\frac{1}{n^s}-\frac{x^{1-s}}{1-s}-\frac{y^{-s}}{2}+O\left(|s|y^{-\sigma}\right)+O\left(x^{-\sigma}\right)$$

and letting $y\longrightarrow\infty$ gives the asserted result. $\qquad\square$

However, it should be noticed that this result has the disadvantage that $\zeta(s)$ is approximated by a sum of length $\gg|t|$ which is difficult to deal with in many applications. Hardy and Littlewood provided another tool that works with some shorter sums.

Theorem 3.82 (Hardy–Littlewood's approximate functional equation) *Let $x,y,t>c>0$ such that $2\pi xy=t$ and set $\Theta(s)=2^s\pi^{s-1}\sin(\pi s/2)\Gamma(1-s)$ so that $\zeta(s)=\Theta(s)\zeta(1-s)$ by the functional equation.*[15] *We have uniformly in $\sigma\in[0,1]$*

$$\zeta(s)=\sum_{n\leqslant x}\frac{1}{n^s}+\Theta(s)\sum_{n\leqslant y}\frac{1}{n^{1-s}}+O\left(x^{-\sigma}+t^{1/2-\sigma}y^{\sigma-1}\right).$$

In [Ivi85], Ivić derived an analogous equation for the function $\zeta(s)^2$ which is the Dirichlet series of the divisor function $\tau(n)$ counting the number of divisors of n (see Chap. 4).

Theorem 3.83 (Approximate functional equation for $\zeta(s)^2$) *Let $x,y,t>c>0$ such that $(2\pi)^2xy=t^2$ and $\Theta(s)$ as in Theorem 3.82. We have uniformly in $\sigma\in\,]0,1[$*

$$\zeta(s)^2=\sum_{n\leqslant x}\frac{\tau(n)}{n^s}+\Theta(s)^2\sum_{n\leqslant y}\frac{\tau(n)}{n^{1-s}}+O\left(x^{1/2-\sigma}\log t\right).$$

---

[15] Some authors use the symbol χ instead of Θ, but we keep this symbol for the Dirichlet characters in this book.

Obviously, these two results are also valid for $t < 0$ with t replaced by $|t|$ in the error-term. Ivić's proof of Theorem 3.83 relies on the use of the important *Voronoï summation formula* [Vor04] which may be stated as follows.

Let $0 < a < b$ be real numbers which are not integers, $f \in C^2[a,b]$ and Y_0, K_0 are Bessel functions of the second and third kind. Then we have

$$\sum_{a \leqslant n \leqslant b} \tau(n) f(n) = \int_a^b f(x)(\log x + 2\gamma) \, dx + \sum_{n=1}^{\infty} \tau(n) \int_a^b f(x)\alpha(nx) \, dx$$

where

$$\alpha(x) = 4K_0(4\pi \sqrt{x}) - 2\pi Y_0(4\pi \sqrt{x}).$$

These functions may have the following integral representations valid for $x > 0$.

$$K_0(x) = \int_0^{\infty} \cos(x \sinh t) \, dt$$

and

$$Y_0(x) = -\frac{2}{\pi} \int_0^{\infty} \cos(x \cosh t) \, dt.$$

3.7.3 The Prime Number Theorem

The proof of the Prime Number Theorem was first given independently by Hadamard and de La Vallée Poussin in 1896. Not only did they provide a proof of the estimate

$$\pi(x) \sim \frac{x}{\log x}$$

as $x \longrightarrow \infty$, but they also gave a quite accurate error-term which has only been slightly improved up to now. The strategy follows the lines of estimating some combinatorial objects, for which one usually computes their generating series giving in return some information by the use of some extraction theorems. The following tool is the basic result of the theory.

Theorem 3.84 (Truncated Perron summation formula) *Let $f(n)$ be any complex numbers with Dirichlet series $F(s) = \sum_{n=1}^{\infty} f(n)n^{-s}$. Assume that $F(s)$ is absolutely convergent in the half-plane $\sigma > \sigma_a$ for some $\sigma_a \in \mathbb{R}$. Then for all real numbers $x, T \geqslant 4$ such that $x \notin \mathbb{Z}$ and all complex numbers s such that $\sigma \leqslant \sigma_a$, we have*

$$\sum_{n \leqslant x} \frac{f(n)}{n^s} = \frac{1}{2\pi i} \int_{\kappa - iT}^{\kappa + iT} \frac{F(s+u)x^u}{u} \, du$$

$$+ O\left\{ x^{-\sigma} \sum_{x/2<n<2x} |f(n)| \min\left(1, \frac{x}{T|n-x|}\right) + \frac{x^\kappa}{T} \sum_{n=1}^{\infty} \frac{|f(n)|}{n^{\sigma+\kappa}} \right\}$$

where $\kappa = \sigma_a - \sigma + 1/\log x$.

Many proofs exist in the literature (see [MV07, Ten95, Tit51] for instance). In practice, the second sum in the error-term is a Dirichlet series often easy to bound, whereas the first term can be handled as follows. Suppose first that, for all positive integers n, we have

$$|f(n)| \leqslant B(n) \tag{3.31}$$

where $B(n)$ is a positive non-decreasing function. Split the sum into three subsums

$$\sum_{x/2<n<2x} |f(n)| \min\left(1, \frac{x}{T|n-x|}\right) = \sum_{x/2<n\leqslant x-1} + \sum_{x-1<n<x+1} + \sum_{x+1\leqslant n<2x}$$

and take 1 for the minimum in the second sum and the other term in the two other sums, and hence, for all $x \geqslant 4$ such that $x \notin \mathbb{Z}$, we get

$$\sum_{x/2<n<2x} |f(n)| \min\left(1, \frac{x}{T|n-x|}\right)$$

$$\ll \left(\frac{x}{T} \sum_{x/2<n\leqslant x-1} \frac{1}{x-n} + \frac{x}{T} \sum_{x+1\leqslant n<2x} \frac{1}{n-x} + 1 \right) B(2x)$$

$$\ll \left(\frac{x\log x}{T} + 1 \right) B(2x).$$

If in addition we have for some $\alpha > 0$

$$\sum_{n=1}^{\infty} \frac{|f(n)|}{n^\sigma} \ll \frac{1}{(\sigma - \sigma_a)^\alpha} \quad (\sigma > \sigma_a) \tag{3.32}$$

then, under the hypotheses of Theorem 3.84 with $4 \leqslant T \leqslant x\log x$, we get

$$\sum_{n\leqslant x} \frac{f(n)}{n^s} = \frac{1}{2\pi i} \int_{\kappa-iT}^{\kappa+iT} \frac{F(s+u)x^u}{u} \, du + O\left\{ \frac{x^{\sigma_a-\sigma}}{T}(\log x)^\alpha + B(2x)\frac{x^{1-\sigma}\log x}{T} \right\}. \tag{3.33}$$

The idea is to apply (3.33) to von Mangoldt's function $\Lambda(n)$. By (3.9), we know that its Dirichlet series is the function $-\zeta'(s)/\zeta(s)$ whose main property is to have an analytic continuation on the line $\sigma = 1$ by Proposition 3.62. We have $\sigma_a = \alpha = 1$, $B(n) = \log n$ and therefore we get with $s = 0$, $4 \leqslant T \leqslant x$, $x \notin \mathbb{Z}$ and $\kappa = 1 + 1/\log x$

$$\Psi(x) = \frac{1}{2\pi i} \int_{\kappa-iT}^{\kappa+iT} -\frac{\zeta'(s)}{\zeta(s)} \frac{x^s}{s} \, ds + O\left(\frac{x(\log x)^2}{T} \right). \tag{3.34}$$

The strategy is then to apply Cauchy's theorem to treat the integral of (3.34) taken over a rectangle surrounding the point 1. Since the residue at this point is known by Proposition 3.62, it remains to evaluate the integral on the three other sides of the rectangle. This strategy is successful only if we have at our disposal estimates of the function in a region to the left of the line $\sigma = 1$. This is done by Theorem 3.61.

Set $\alpha \leqslant 5556379^{-1}$ and $\lambda = 1 - \alpha(\log 2T)^{-9}$. Let $\mathcal{R}$ be the rectangle with vertices $\kappa \pm iT$ and $\lambda \pm iT$. By Theorem 3.61 and Proposition 3.62, the function $-\zeta'(s)/\zeta(s)$ has a simple pole with residue 1 at $s = 1$, and is otherwise analytic within $\mathcal{R}$. We integrate over $\mathcal{R}$ in the anticlockwise direction and by Cauchy's residue theorem, we get

$$\frac{1}{2\pi i} \int_{\mathcal{R}} -\frac{\zeta'(s)}{\zeta(s)} \frac{x^s}{s} \, ds = x.$$

If $\mathcal{H}_1$ and $\mathcal{H}_2$ are the two horizontal sides and $\mathcal{V}$ is the other vertical side of $\mathcal{R}$, we deduce that

$$\Psi(x) = x - \frac{1}{2\pi i} \left(\sum_{j=1}^{2} \int_{\mathcal{H}_j} -\frac{\zeta'(s)}{\zeta(s)} \frac{x^s}{s} \, ds + \int_{\mathcal{V}} -\frac{\zeta'(s)}{\zeta(s)} \frac{x^s}{s} \, ds \right) + O\left(\frac{x(\log x)^2}{T} \right).$$

Now using Theorem 3.61, we get

$$\left| \int_{\mathcal{H}_j} -\frac{\zeta'(s)}{\zeta(s)} \frac{x^s}{s} \, ds \right| \ll (\log T)^9 \int_{\lambda}^{\kappa} \frac{x^\sigma}{|\sigma + iT|} \, d\sigma \ll \frac{(\log T)^9}{T} \int_{\lambda}^{\kappa} x^\sigma \, d\sigma$$

$$\ll \frac{x}{T} (\log x)^8$$

for $j \in \{1, 2\}$, and

$$\left| \int_{\mathcal{V}} -\frac{\zeta'(s)}{\zeta(s)} \frac{x^s}{s} \, ds \right| \ll x^\lambda (\log T)^9 \int_{-T}^{T} \frac{dt}{\sqrt{\lambda^2 + t^2}} \ll x^\lambda (\log x)^{10}$$

so that we get

$$\Psi(x) = x + O\left(x(\log x)^{10} \left(T^{-1} + x^{-\alpha/\log^9(2T)} \right) \right).$$

We choose $T = \frac{1}{2} \exp\{(\alpha \log x)^{1/10}\}$ which implies that the error-term is

$$\ll x(\log x)^{10} \exp\{-(\alpha \log x)^{1/10}\} \ll x \exp(-\alpha(\log x)^{1/10})$$

for x sufficiently large since $0 < \alpha < 1$. By Corollary 3.47, we deduce that

$$\theta(x) = x + O\{x \exp(-\alpha(\log x)^{1/10})\}$$

and using Exercise 8 in Chap. 1 we infer that, for all x sufficiently large, we have

$$\pi(x) = \mathrm{Li}(x) + O\{x \exp(-\alpha(\log x)^{1/10})\}.$$

We may state the above result in the following theorem called the *Prime Number Theorem* (PNT).

Theorem 3.85 (Prime Number Theorem) *There exists an absolute constant $\alpha \in$ $]0, 1[$ such that, for $x \longrightarrow \infty$, we have*

$$\pi(x) = \mathrm{Li}(x) + O\{x \exp(-\alpha(\log x)^{1/10})\}.$$

Note that by repeated integration by parts we get for a fixed positive integer N

$$\mathrm{Li}(x) = \frac{x}{\log x} + x \sum_{k=1}^{N-1} \frac{k!}{(\log x)^{k+1}} + O\left(\frac{x}{(\log x)^{N+1}}\right)$$

which allows us to have at our disposal several more or less precise versions of the PNT. For instance, for $x \longrightarrow \infty$, the estimate

$$\pi(x) = \frac{x}{\log x} + \frac{x}{(\log x)^2} + O\left(\frac{x}{(\log x)^3}\right)$$

is useful in many applications.

With more work and using tools from complex analysis,[16] one may prove that there is an absolute constant $c > 0$ such that $\zeta(s) \neq 0$ in the region $\sigma \geq 1 - c/\log \tau$, which enables us to replace the exponent $\frac{1}{10}$ in Theorem 3.85 by $\frac{1}{2}$.

I.M. Vinogradov's method of exponential sums (see Theorem 6.42) allows us to get estimates of $\zeta(s)$ in the form

$$|\zeta(s)| \leqslant A t^{B(1-\sigma)^{3/2}} (\log t)^{2/3} \tag{3.35}$$

for all $s = \sigma + it \in \mathbb{C}$ such that $\frac{1}{2} \leqslant \sigma \leqslant 1$ and $t \geqslant 3$. This in turn implies the best zero-free region for $\zeta(s)$ up to now which was obtained by Korobov and I.A. Vinogradov, who proved that there exists an absolute constant $c_0 > 0$ such that $\zeta(s)$ has no zero in the region

$$\sigma \geqslant 1 - \frac{c_0}{(\log |t|)^{2/3}(\log\log |t|)^{1/3}} \quad \text{and} \quad |t| \geqslant 3 \tag{3.36}$$

giving the best error-term in the PNT to date, namely

$$\pi(x) = \mathrm{Li}(x) + O\{x \exp(-c_1(\log x)^{3/5}(\log\log x)^{-1/5})\} \tag{3.37}$$

with $0 < c_1 < 1$ being absolute. In fact, any order of magnitude of $\zeta(s)$ in a certain domain implies a zero-free region as may be seen in the next result devised by Landau (see [MV07, Tit51]).

---

[16]Such as Jensen's inequality or Borel–Carathéodory's lemma.

Theorem 3.86 *Let $\theta(t)$ and $\phi(t)$ be positive functions such that $\theta(t)$ is decreasing, $\phi(t)$ is increasing and $e^{-\phi(t)} \leqslant \theta(t) \leqslant \frac{1}{2}$. Assume that $\zeta(s) \ll e^{\phi(t)}$ in the region $\sigma \geqslant 1 - \theta(t)$ and $t \geqslant 2$. Then the following assertions hold.*

(i) *There exists an absolute constant $c_0 > 0$ such that $\zeta(s)$ has no zero in the region*

$$\sigma \geqslant 1 - c_0 \frac{\theta(2t+1)}{\phi(2t+1)}.$$

(ii) *In the region $\sigma \geqslant 1 - (c_0/2)\theta(2t+2)\phi(2t+2)^{-1}$, we have*

$$\frac{1}{\zeta(s)} \ll \frac{\theta(2t+2)}{\varphi(2t+2)} \quad \text{and} \quad \frac{\zeta'}{\zeta}(s) \ll \frac{\theta(2t+2)}{\varphi(2t+2)}.$$

The link between (3.36) and (3.37) is underscored by the following result proved by Ingham. For all $x \geqslant 2$, we set

$$\Delta_1(x) = \pi(x) - \text{Li}(x)$$

$$\Delta_2(x) = \Pi(x) - \text{Li}(x) = \sum_{n=1}^{\infty} \frac{\pi(x^{1/n})}{n} - \text{Li}(x)$$

$$\Delta_3(x) = \theta(x) - x \quad \text{and} \quad \Delta_4(x) = \Psi(x) - x.$$

Theorem 3.87 *Let $\varphi(t)$ be a positive decreasing function of $t \geqslant 0$, having a continuous derivative and satisfying the following hypotheses*

$$0 < \varphi(t) \leqslant \frac{1}{2}, \quad \lim_{t \to \infty} \varphi'(t) = 0 \quad \text{and} \quad \varphi(t) \gg (\log t)^{-1}.$$

Assume that $\zeta(s)$ has no zero in the region $\sigma \geqslant 1 - \varphi(|t|)$. Let $\varepsilon \in {]}0, 1{[}$ and define

$$\omega(x; \varphi) = \min_{t \geqslant 1}(\varphi(t)\log x + \log t).$$

Then, for $i \in \{1, \ldots, 4\}$, we have

$$\Delta_i(x) \ll x \exp\left\{\frac{1}{2}(\varepsilon - 1)\omega(x; \varphi)\right\}.$$

In particular, if $\varphi(t) = c_0(\log(t+3))^{-\beta}$ with $\beta > 0$, then if $\zeta(s) \neq 0$ in the region

$$\sigma \geqslant 1 - \frac{c_0}{(\log(|t|+3))^{\beta}}$$

then we have

$$\Delta_i(x) \ll x \exp\left\{-c_1(\log x)^{1/(\beta+1)}\right\}.$$

The PNT enables us to improve on estimates for some functions of prime numbers by proceeding as follows. Let $x \geqslant 2$ be a large real number and $f \in C^2[2, x]$. Similarly as in Chap. 1, one has

$$\sum_{p \leqslant x} f(p) = \int_{2^-}^x f(u) \, d\pi(u)$$

$$= f(x)\pi(x) - \int_2^x f'(u) \, \mathrm{Li}(u) \, du - \int_2^x f'(u)\big(\pi(u) - \mathrm{Li}(u)\big) \, du$$

$$= \int_2^x \frac{f(u)}{\log u} \, du + f(2) \, \mathrm{Li}(2) + f(x)\big(\pi(x) - \mathrm{Li}(x)\big)$$

$$- \int_2^x f'(u)\big(\pi(u) - \mathrm{Li}(u)\big) \, du$$

and if the integral

$$\int_2^\infty f'(u)\big(\pi(u) - \mathrm{Li}(u)\big) \, du$$

converges, then we get

$$\sum_{p \leqslant x} f(p) = \int_2^x \frac{f(u)}{\log u} \, du + c_f + f(x)\big(\pi(x) - \mathrm{Li}(x)\big) + \int_x^\infty f'(u)\big(\pi(u) - \mathrm{Li}(u)\big) \, du$$

with

$$c_f = f(2) \, \mathrm{Li}(2) - \int_2^\infty f'(u)\big(\pi(u) - \mathrm{Li}(u)\big) \, du.$$

For instance, with the latest version (3.37) of the PNT we get

$$\sum_{p \leqslant x} \frac{1}{p} = \log \log x + B + O\big\{\exp\big(-c_1 (\log x)^{3/5} (\log \log x)^{-1/5}\big)\big\}$$

$$\sum_{p \leqslant x} \frac{\log p}{p} = \log x - E + O\big\{\exp\big(-c_1 (\log x)^{3/5} (\log \log x)^{-1/5}\big)\big\}$$

$$\prod_{p \leqslant x} \left(1 - \frac{1}{p}\right) = \frac{e^{-\gamma}}{\log x} + O\big\{\exp\big(-c_1 (\log x)^{3/5} (\log \log x)^{-1/5}\big)\big\}$$

for some absolute constant $c_1 > 0$, where $B \approx 0.261\,497\,212\ldots$ is the Mertens constant and $E \approx 1.332\,582\,275\ldots$

3.7.4 The Riemann–von Mangoldt Formula and the Density Hypothesis

We have seen that the non-trivial zeros $\rho = \beta + i\gamma$ of the Riemann zeta-function are of crucial importance in the distribution of prime numbers. Let $\sigma \in [0, 1]$ and $T \geqslant 1$ be real numbers. It is customary to define

$$\mathcal{N}(\sigma, T) = \left|\{\rho = \beta + i\gamma : \beta \geqslant \sigma \text{ and } |\gamma| \leqslant T\}\right| = \sum_{\substack{\beta \geqslant \sigma \\ |\gamma| \leqslant T}} 1$$

and

$$\mathcal{N}(T) = \left|\{\rho = \beta + i\gamma : \beta \in {]}0, 1[\text{ and } 0 < \gamma \leqslant T\}\right|$$

If $T = \gamma$, then we set $\mathcal{N}(T) = \frac{1}{2}(\mathcal{N}(T^+) + \mathcal{N}(T^-))$.

The first important result was conjectured by Riemann and proved by von Mangoldt. For a proof, see [Ivi85, MV07, Ten95, Tit51].

Theorem 3.88 (Riemann–von Mangoldt formula) *We have*

$$\mathcal{N}(T) = \frac{T}{2\pi} \log \frac{T}{2\pi} - \frac{T}{2\pi} + \frac{7}{8} + S(T) + O\left(\frac{1}{T}\right)$$

with

$$S(t) = \frac{1}{\pi} \arg \zeta \left(\frac{1}{2} + it\right)$$

where the argument is defined by continuous variation of s in $\zeta(s)$ starting at $s = 2$, then vertically to $s = 2 + it$ and then horizontally to $s = \frac{1}{2} + it$. Furthermore, we have $S(t) = O(\log t)$.

This implies in particular that

$$\mathcal{N}(T + 1) - \mathcal{N}(T) = O(\log T) \tag{3.38}$$

and using partial summation, we also get

$$\sum_{\substack{\rho \\ |\gamma| \leqslant T}} \frac{1}{|\rho|} \ll \int_1^T \frac{\log u}{u} \, du + \log T \ll (\log T)^2. \tag{3.39}$$

Similarly, since $\mathcal{N}(\gamma_n - 1) < n \leqslant \mathcal{N}(\gamma_n + 1)$, we infer that, for $n \longrightarrow \infty$, we have

$$\gamma_n \sim \frac{2\pi n}{\log n}.$$

Table 3.1 Zero density estimates

$(A(\sigma), D)$	$(\frac{3}{2-\sigma}, 5)$	$(\frac{3}{3\sigma-1}, 44)$	$(\frac{12}{5}, 9)$
Range of validity	$\frac{1}{2} \leqslant \sigma \leqslant \frac{3}{4}$	$\frac{3}{4} \leqslant \sigma \leqslant 1$	$\frac{1}{2} \leqslant \sigma \leqslant 1$
Author	Ingham (1940)	Huxley (1972)	Huxley–Ingham–Ivić

By adapting the method to the L-functions attached to primitive Dirichlet characters, one may also show that

$$\mathcal{N}(T, \chi) = \frac{T}{2\pi} \log \frac{qT}{2\pi} - \frac{T}{2\pi} + O(\log qT) \tag{3.40}$$

where χ is a primitive Dirichlet character modulo $q > 1$ and $\mathcal{N}(T, \chi)$ is the number of zeros $\rho = \beta + i\gamma$ of $L(s, \chi)$ in the rectangle $0 < \beta < 1$ and $0 < \gamma \leqslant T$.

Let us now have a look at estimates of the form

$$\mathcal{N}(\sigma, T) \ll T^{A(\sigma)(1-\sigma)} (\log T)^D \tag{3.41}$$

for some $A(\sigma) \geqslant 0$ and $D \geqslant 0$. In view of Theorem 3.88, we must have $A(\sigma) \geqslant 2$. In many applications, such as the gaps between consecutive primes, the results that we may get using the Lindelöf or the Riemann hypothesis can be obtained via a weaker conjecture, namely the *density hypothesis* stating that, for all $\frac{1}{2} \leqslant \sigma \leqslant 1$ and $T \geqslant 3$, we have $A(\sigma) \leqslant 2$ so that

$$\mathcal{N}(\sigma, T) \ll T^{2(1-\sigma)} \log T.$$

A great deal of effort has been made to establish estimates of the form (3.41). We may summarize the main results in Table 3.1.

3.7.5 Explicit Formula

Let us return to (3.34). Instead of integrating over a rectangle containing only 1, suppose we integrate over a contour that proceeds by straight lines from $\kappa - iT$ to $\kappa + iT$ to $-(2K + 1) + iT$ to $-(2K + 1) - iT$ with $K \geqslant 1$ integer. In the interior of this contour, the integrand has poles at $s = 1$, at zeros ρ of $\zeta(s)$ and at the trivial zeros $s = -2k$. Since x^s decays quickly as $\sigma \longrightarrow -\infty$, one may expect that we can pull the contour to the left and thus get a totally explicit formula for $\psi(x)$. In order to show this rigorously, we first need an important result.

Proposition 3.89 *Let $s = \sigma + it$ with $-1 \leqslant \sigma \leqslant 2$ and t not equal to an ordinate of a zero of $\zeta(s)$. Set $\tau = |t| + 3$. Then we have*

$$-\frac{\zeta'}{\zeta}(s) = \frac{1}{s-1} - \sum_{\substack{\rho \\ |t-\gamma| \leqslant 1}} \frac{1}{s-\rho} + O(\log \tau).$$

Proof Set $F(s) = s(s-1)\xi(s) = s(s-1)\pi^{-s/2}\Gamma(s/2)\zeta(s)$. $F(s)$ is an entire function of order 1, so that by the *Hadamard factorization theorem*, there exist suitable constants a, b such that

$$F(s) = e^{a+bs} \prod_{\rho} \left(1 - \frac{s}{\rho}\right) e^{s/\rho}$$

where the product runs through all zeros $\rho = \beta + i\gamma$ of $F(s)$ which are exactly the non-trivial zeros of $\zeta(s)$. The logarithmic differentiation provides

$$\frac{F'}{F}(s) = b + \sum_{\rho} \left(\frac{1}{s-\rho} + \frac{1}{\rho}\right) \tag{3.42}$$

where the sum is absolutely convergent. Taking $s = 0$ gives $b = F'(0)/F(0)$, and using Theorem 3.55 gives $F(s) = F(1-s)$, so that

$$b = \frac{F'(0)}{F(0)} = -\frac{F'(1)}{F(1)} = -b - \sum_{\rho} \left(\frac{1}{\rho} + \frac{1}{1-\rho}\right).$$

Now if ρ is a zero of $F(s)$, so are $\overline{\rho}$ and $1 - \rho$. Using this observation in the equation above we get

$$b = -\frac{1}{2} \sum_{\rho} \left(\frac{1}{\rho} + \frac{1}{\overline{\rho}}\right).$$

Therefore (3.42) becomes

$$\frac{F'}{F}(s) = \frac{1}{2} \sum_{\rho} \left(\frac{1}{s-\rho} + \frac{1}{s-\overline{\rho}}\right).$$

Now using the definition of F gives

$$\frac{F'}{F}(s) = \frac{1}{s} + \frac{1}{s-1} - \frac{\log \pi}{2} + \frac{\zeta'}{\zeta}(s) + \frac{1}{2}\frac{\Gamma'}{\Gamma}\left(\frac{s}{2}\right)$$

and by logarithmically differentiating the function equation $\Gamma(s+1) = s\Gamma(s)$ we obtain

$$\frac{\Gamma'}{\Gamma}(s+1) = \frac{1}{s} + \frac{\Gamma'}{\Gamma}(s)$$

so that

$$\frac{F'}{F}(s) = \frac{1}{s-1} - \frac{\log \pi}{2} + \frac{\zeta'}{\zeta}(s) + \frac{1}{2}\frac{\Gamma'}{\Gamma}\left(\frac{s}{2}+1\right)$$

and hence

$$-\frac{\zeta'}{\zeta}(s) = \frac{1}{s-1} - \frac{\log \pi}{2} + \frac{1}{2}\frac{\Gamma'}{\Gamma}\left(\frac{s}{2}+1\right) - \frac{1}{2}\sum_{\rho}\left(\frac{1}{s-\rho} + \frac{1}{s-\overline{\rho}}\right). \tag{3.43}$$

By logarithmically differentiating the Hadamard product of $\Gamma(s)$ and denoting temporarily $C \approx 0.5772\ldots$ the Euler–Mascheroni constant, we get using Theorem 1.22

$$\frac{\Gamma'}{\Gamma}(s) = -\frac{1}{s} - C - \sum_{n=1}^{\infty}\left(\frac{1}{n+s} - \frac{1}{n}\right) = \log N - \frac{1}{s} - \sum_{n=1}^{N}\frac{1}{n+s} + O\left(|s|N^{-1}\right)$$

for all positive integers $N > |s|$, and by Corollary 1.32 with $f(t) = (t+s)^{-1}$ and letting $N \longrightarrow \infty$ we infer for all s such that $|s| \geqslant \delta$ and $|\arg s| < \pi - \delta$ (with $\delta > 0$)

$$\frac{\Gamma'}{\Gamma}(s) = \log s - \frac{1}{2s} + O\left(|s|^{-2}\right)$$

and therefore the Gamma-term in (3.43) is $\ll \log \tau$. Applying this estimate and (3.43) to s and to $2 + it$ and subtracting gives

$$-\frac{\zeta'}{\zeta}(s) = \frac{1}{s-1} - \frac{1}{2}\sum_{\rho}\left(\frac{1}{s-\rho} - \frac{1}{2+it-\rho} + \frac{1}{s-\overline{\rho}} - \frac{1}{2+it-\overline{\rho}}\right) + O(\log \tau)$$

$$= \frac{1}{s-1} - \sum_{\rho}\left(\frac{1}{s-\rho} - \frac{1}{2+it-\rho}\right) + O(\log \tau).$$

By (3.38) we have

$$\sum_{\substack{\rho \\ |t-\gamma|\leqslant 1}} \frac{1}{2+it-\rho} \ll \sum_{\substack{\rho \\ |t-\gamma|\leqslant 1}} 1 \ll \log \tau.$$

Now let $k \in \mathbb{N}$ and consider the zeros ρ satisfying $k < |\gamma - t| \leqslant k+1$. Since

$$\left|\frac{1}{s-\rho} - \frac{1}{2+it-\rho}\right| = \frac{2-\sigma}{|(s-\rho)(2+it-\rho)|} \leqslant \frac{3}{|\gamma-t|^2} \leqslant \frac{3}{k^2}$$

we infer that such zeros contribute

$$\ll k^{-2}\left(\sum_{\substack{\rho \\ t+k<\gamma\leqslant t+k+1}} 1 + \sum_{\substack{\rho \\ t-k-1\leqslant\gamma<t-k}} 1\right) \ll k^{-2}\log(\tau+k)$$

by (3.38). Summing over k gives the asserted result. $\qquad\square$

In fact, such a result is also a consequence of a more general tool due to Borel and Carathéodory which may be stated as follows.[17]

---

[17]This exposition is due to Ramaré [Ram11].

Let $s_0 \in \mathbb{C}$ and F be an analytic function in the disc $|s - s_0| \leqslant R$ satisfying in this disc the bound $|F(s)| \leqslant M$ and such that $|F(s_0)| \geqslant m$. Then, for all s such that $|s - s_0| \leqslant R/4$, we have

$$\left| \frac{F'(s)}{F(s)} - \sum_{|\rho - s_0| \leqslant R/2} \frac{1}{s - \rho} \right| \leqslant \frac{8}{R} \log\left(\frac{M}{m}\right)$$

where the sum runs through the zeros ρ of F such that $|\rho - s_0| \leqslant R/2$.

Using this result with $F(s) = \zeta(s)$, $s_0 = 1 + it_0$ for some $t_0 \geqslant 4$, $R = 2$, a bound $|\zeta(s)| \ll |s|^{3/2}$ for $-1 \leqslant \sigma \leqslant 2$ and the lower bound of Theorem 3.61, we get for all s such that $|s - 1 - it_0| \leqslant \frac{1}{2}$

$$-\frac{\zeta'}{\zeta}(s) = - \sum_{|\rho - 1 - it_0| \leqslant 1} \frac{1}{s - \rho} + O(\log t_0).$$

Note that the first term in the identity of Proposition 3.89 is significant only for $|t| \leqslant 1$. This result enables us to get a finer estimate of $-\zeta'(s)/\zeta(s)$ in a larger region than that of Theorem 3.61.

Corollary 3.90 *For every real number $T \geqslant 2$, there exists $T' \in [T, T+1]$ such that, uniformly for $-1 \leqslant \sigma \leqslant 2$, we have*

$$-\frac{\zeta'}{\zeta}(\sigma + iT') = O\left(\log^2 T\right).$$

Proof Indeed, by (3.38), the number of zeros ρ such that $\gamma \in [T, T+1]$ is $\ll \log T$. Subdividing the interval into $\ll \log T$ equal parts of length $c/\log T$ for some $c > 0$ chosen so that the number of parts exceeds the number of zeros, we deduce that there is a part that contains no zeros by the Dirichlet pigeon-hole principle. Hence for T' lying in this part, we must have $|T' - \gamma| \gg 1/\log T$. We infer that each summand in Proposition 3.89 is $\ll \log T$ and since there are $\ll \log T$ summands by (3.38), we get the stated estimate. $\square$

It should also be noticed that, using an asymmetric form of the functional equation of $\zeta(s)$, it can be proved that for $\sigma \leqslant -1$, we have

$$-\frac{\zeta'}{\zeta}(s) = O\left(\log(|s| + 1)\right) \tag{3.44}$$

provided that circles of radii $\frac{1}{4}$ around the trivial zeros $s = -2k$ are excluded (see [MV07, Lemma 12.4]).

We are now in a position to prove Landau's explicit formula for $\Psi(x)$.

Theorem 3.91 (Landau) *Let $T_0 \geqslant 2$ be a real number. Uniformly for $T \geqslant T_0$, we have*

$$\Psi(x) = x - \sum_{|\gamma| \leqslant T} \frac{x^\rho}{\rho} - \log 2\pi - \frac{1}{2} \log\left(1 - \frac{1}{x^2}\right) + O\left(\frac{x}{T}(\log xT)^2 + \log x\right).$$

Proof We may suppose that $x \notin \mathbb{Z}$. Let T' be the number supplied by Corollary 3.90, K be a large positive integer and call $\mathfrak{R}$ the rectangle with vertices $\kappa - iT', \kappa + iT'$, $-(2K+1) + iT'$ and $-(2K+1) - iT'$. By (3.43), we see that $-\zeta'(s)/\zeta(s)$ has a simple pole at $s = -2k$ with residue -1. Since the residue at $s = 1$ is equal to 1, we get by Cauchy's residue theorem

$$\frac{1}{2\pi i} \int_{\mathfrak{R}} -\frac{\zeta'(s)}{\zeta(s)} \frac{x^s}{s} \, ds = x - \sum_{|\gamma| \leqslant T'} \frac{x^\rho}{\rho} - \sum_{1 \leqslant k < K+1/2} \frac{x^{-2k}}{-2k} - \frac{\zeta'}{\zeta}(0).$$

It can be shown that $\zeta'(0)/\zeta(0) = \log 2\pi$. By (3.34), we get

$$\Psi(x) = x - \sum_{|\gamma| \leqslant T'} \frac{x^\rho}{\rho} + \sum_{1 \leqslant k < K+1/2} \frac{x^{-2k}}{2k} - \log 2\pi - \sum_{j=1}^{2} I_{\mathcal{H}_j} - I_{\mathcal{V}}$$

$$+ O\left(\frac{x(\log x)^2}{T'} + \log x\right)$$

where $I_{\mathcal{H}_j}$ denotes the integrals taken over the two horizontal sides and $I_{\mathcal{V}}$ is the integral taken over the vertical side. Using Corollary 3.90, (3.44) and the easy estimate $T' \asymp T$, we obtain

$$I_{\mathcal{H}_j} \ll \int_{-2K-1}^{\kappa} \left| -\frac{\zeta'}{\zeta}(\sigma + iT') \right| \frac{x^\sigma}{|\sigma + iT'|} \, d\sigma$$

$$\ll \int_{-2K-1}^{-1} x^\sigma \frac{\log|\sigma + iT'|}{|\sigma + iT'|} \, d\sigma + (\log T)^2 \int_{-1}^{\kappa} \frac{x^\sigma}{|\sigma + iT'|} \, d\sigma$$

$$\ll \frac{(\log T)^2}{T} \int_{-2K-1}^{\kappa} x^\sigma \, d\sigma \ll \frac{x(\log T)^2}{T}$$

for $j \in \{1, 2\}$ and

$$I_{\mathcal{V}} \ll \int_{-T'}^{T'} \left| -\frac{\zeta'}{\zeta}(-2K - 1 + it) \right| \frac{x^{-2K-1}}{|-2K-1+it|} \, dt \ll \frac{x^{-2K-1}T}{2K+1} \log(KT)$$

and letting $K \longrightarrow \infty$ completes the proof. $\qquad\square$

This result has been slightly improved by Goldston [Gol83] where it is proved that the error-term $xT^{-1}(\log xT)^2$ may be replaced by

$$\frac{x}{T}\log x \log\log x.$$

3.7.6 The Prime Number Theorem for Arithmetic Progressions

Let $a, q \geqslant 1$ be integers such that $(a,q) = 1$. It is customary to define the function

$$\pi(x;q,a) = \sum_{\substack{p \leqslant x \\ p \equiv a \pmod q}} 1.$$

By Theorem 3.63, we have $\lim_{x\to\infty} \pi(x;q,a) = \infty$ and hence the question of its order of magnitude arises naturally. We expect the prime numbers to be well distributed in the $\varphi(q)$ reduced residue classes modulo q. Therefore, applying the method of the former section to the function

$$\Psi(x,\chi) = \sum_{n \leqslant x} \Lambda(n)\chi(n)$$

where χ is a non-principal Dirichlet character modulo q and to its Dirichlet series $-\frac{L'}{L}(s,\chi)$, we get

$$\pi(x;q,a) = \frac{\mathrm{Li}(x)}{\varphi(q)} + O_q\left(xe^{-c_0(q)\sqrt{\log x}}\right)$$

for some constant $0 < c_0(q) < 1$ depending on q, the constants implied in the error-term depending also on q. This dependence makes this result useless in practice. A great deal of effort has been made to prove some efficient estimates where the constants do not depend on the modulus. One of the most important results in the theory is called the *Siegel–Walfisz–Page theorem* or *Siegel–Walfisz theorem*.

Theorem 3.92 (Siegel–Walfisz) *Let $a, q \geqslant 1$ be coprime integers.*

(i) *For all $A > 0$, there exists $c_1(A) > 0$ not depending on q such that, for all $q \leqslant (\log x)^A$, we have*

$$\pi(x;q,a) = \frac{\mathrm{Li}(x)}{\varphi(q)} + O_A\left(xe^{-c_1(A)\sqrt{\log x}}\right).$$

(ii) *For all $A > 0$ and all $q \geqslant 1$, we have*

$$\pi(x;q,a) = \frac{\mathrm{Li}(x)}{\varphi(q)} + O_A\left(\frac{x}{(\log x)^A}\right).$$

Obviously, if we define the Chebyshev type functions

$$\theta(x; q, a) = \sum_{\substack{p \leqslant x \\ p \equiv a \,(\mathrm{mod}\, q)}} \log p \quad \text{and} \quad \Psi(x; q, a) = \sum_{\substack{n \leqslant x \\ n \equiv a \,(\mathrm{mod}\, q)}} \Lambda(n)$$

similar estimates hold for these functions, namely

$$\left.\begin{array}{l} \theta(x; q, a) \\ \Psi(x; q, a) \end{array}\right\} = \frac{x}{\varphi(q)} + O_A\left(x e^{-c_1(A)\sqrt{\log x}}\right)$$

in the first case, and

$$\left.\begin{array}{l} \theta(x; q, a) \\ \Psi(x; q, a) \end{array}\right\} = \frac{x}{\varphi(q)} + O_A\left(\frac{x}{(\log x)^A}\right)$$

in the second case.

The proof of Theorem 3.92 rests on an explicit formula for $\Psi(x, \chi)$ similar to that of Theorem 3.91, namely

$$\Psi(x, \chi) = E_0(\chi) x - \sum_{|\gamma| \leqslant T} \frac{x^\rho}{\rho} + O\left(\frac{x}{T}(\log qx)^2 + x^{1/4} \log x\right)$$

where

$$E_0(\chi) = \begin{cases} 1, & \text{if } \chi = \chi_0 \\ 0, & \text{otherwise} \end{cases}$$

which implies using Proposition 3.68

$$\Psi(x; q, a) = \frac{x}{\varphi(q)} + \sum_{\chi \,(\mathrm{mod}\, q)} \sum_{|\gamma| \leqslant T} \frac{x^\rho}{\rho} + O\left(\frac{x}{T}(\log qx)^2 + x^{1/4} \log x\right).$$

The other important tool is the knowledge of a zero-free region for the function $L(s, \chi)$. The arguments generalize those of the function $\zeta(s)$, except that there is an unforeseen difficulty in connection with the possible existence, still unproven, of an exceptional zero $\beta_1 \in \mathbb{R}$ near the point 1 of a function $L(s, \chi)$ attached to a quadratic Dirichlet character. More precisely, we have the following result.

Theorem 3.93 (Zero-free region for L-functions) *Let $q \in \mathbb{N}$ and $\tau = |t| + 3$. There exists an absolute constant $c_0 > 0$ such that if χ is a Dirichlet character modulo q, then the function $L(s, \chi)$ has no zero in the region*

$$\sigma \geqslant 1 - \frac{c_0}{\log(q\tau)}$$

unless χ is a quadratic character, in which case $L(s, \chi)$ has at most one, necessarily real, zero $\beta_1 < 1$ in this region. This zero is called exceptional.

At the present time, we do not know much more about this exceptional zero. Nevertheless, Landau, Page and Siegel provided some very important results, showing in particular that such a zero occurs at most rarely. We summarize their main discoveries in the next theorem.

Theorem 3.94 *Let q be a positive integer and $\tau = |t| + 3$.*

(i) (Landau). *There exists an absolute constant $c_1 > 0$ such that the function $\prod_{\chi \,(\mathrm{mod}\,q)} L(s, \chi)$ has at most one zero in the region*

$$\sigma \geqslant 1 - \frac{c_1}{\log(q\tau)}.$$

If such a zero β_1 exists, then it is necessarily real and associated to a quadratic character χ_1. This character is called the exceptional character.

(ii) (Page). *If χ is a Dirichlet character modulo q, then $L(\sigma, \chi) \neq 0$ in the region*

$$\sigma \geqslant 1 - \frac{c_2}{q^{1/2}(\log(q+1))^2}$$

where $c_2 > 0$ is an effectively computable absolute constant.

(iii) (Siegel). *Let χ be a quadratic Dirichlet character modulo q. For all $\varepsilon > 0$, there exists a* non-effectively computable *constant $c_\varepsilon > 0$ such that*

$$L(1, \chi) > \frac{c_\varepsilon}{q^\varepsilon}.$$

This implies that, if χ is a quadratic character modulo q, then $L(\sigma, \chi) \neq 0$ in the region

$$\sigma \geqslant 1 - \frac{c_\varepsilon}{q^\varepsilon}.$$

The proof relies on the following lemma due to Estermann providing the lower bound of certain Dirichlet series at $s = 1$, which may have its own interest (for a proof, see [MV07, Was82]).

Lemma 3.95 (Estermann) *Let $f(s)$ be an analytic function in the disc $|s - 2| \leqslant \frac{4}{3}$ satisfying the bound $|f(s)| \leqslant M$ in this disc. Assume that, for all $\sigma > 1$, we have*

$$f(s)\zeta(s) = \sum_{n=1}^{\infty} \frac{a_n}{n^s}$$

with $a_1 \geqslant 1$ and $a_n \geqslant 0$ for all n. Finally, suppose that there exists $\alpha \in [\frac{26}{27}, 1[$ such that $f(\alpha) \geqslant 0$. Then we have

$$f(1) \geqslant \frac{1 - \alpha}{4M^{4(1-\alpha)}}.$$

In the proof of Theorem 3.94, one usually considers a primitive quadratic character χ_1 modulo q_1 such that the associated Dirichlet L-function has a real zero $\beta_1 \geqslant 1 - \varepsilon/4$ and one applies Lemma 3.95 to the function $f(s) = L(s, \chi)L(s, \chi_1) \times L(s, \chi\chi_1)$ where $\chi \neq \chi_1$ is a primitive quadratic character. This implies that $f(1) \geqslant c_1(\varepsilon)q^{-\varepsilon}$ and since $f(1) \ll L(1, \chi)(\log qq_1)^2$, the result of Theorem 3.94 (iii) follows for primitive characters. It may be extended to imprimitive Dirichlet characters by using (3.24).

It should be noticed that we have no way of estimating the size of the smallest possible modulus q_1, so that the constant c_ε of Siegel's theorem is ineffective when $\varepsilon < \frac{1}{2}$. All attempts at providing a value to c_ε for a sufficiently small $\varepsilon > 0$ have been unsuccessful.

There are slightly more accurate versions of Theorem 3.92 where the error-term is similar to that of the PNT.

Let a, q be positive integers such that $(a, q) = 1$ and assume that there is an exceptional real Dirichlet character χ_1 modulo q and β_1 is the concomitant zero. Then there exists a constant $c_0 > 0$ such that

$$\pi(x; q, a) = \frac{\mathrm{Li}(x)}{\varphi(q)} - \frac{\chi_1(a)\,\mathrm{Li}(x^{\beta_1})}{\varphi(q)} + O\left(x\,e^{-c_0\sqrt{\log x}}\right). \qquad (3.45)$$

The term containing β_1 can be removed if the exceptional zero does not exist.

In practice, especially when an upper bound is sufficient, the Brun–Titchmarsh inequality (see Theorems 4.73 and 4.80) is often used because of its larger range of validity.

3.7.7 Explicit Estimates

In the last decade, several explicit bounds for (3.35), (3.36) and (3.37) were discovered. The best results until now were obtained by Ford [For02b, For02a] which proved respectively for (3.35), (3.36) and (3.37) the following estimates.

▷ For all $s = \sigma + it \in \mathbb{C}$ such that $\frac{1}{2} \leqslant \sigma \leqslant 1$ and $t \geqslant 3$, we have

$$\left|\zeta(s)\right| \leqslant 76.2\,t^{4.45(1-\sigma)^{3/2}}(\log t)^{2/3}.$$

▷ $\zeta(s)$ has no zero in the region

$$\sigma \geqslant 1 - \frac{1}{57.54\,(\log |t|)^{2/3}(\log\log |t|)^{1/3}} \quad \text{and} \quad |t| \geqslant 3.$$

▷ The PNT can be written as follows.

$$\pi(x) = \mathrm{Li}(x) + O\left\{x\exp\left(-0.2098\,(\log x)^{3/5}(\log\log x)^{-1/5}\right)\right\}.$$

In another direction, one may consider zero-free regions of the type

$$\sigma \geqslant 1 - \frac{1}{R\log(|t|/d)} \quad \text{and} \quad |t| \geqslant t_0 \tag{3.46}$$

trying to get the lowest positive real number R in order to increase the region. Such zero-free regions are generally determined with the help of the real part of the function $-\zeta'(s)/\zeta(s)$. There are two classical ways to deal with this real part.

1. The *global* one. Used by de La Vallée Poussin in 1899, the formula has the form

$$\mathrm{Re}\left(-\frac{\zeta'}{\zeta}(s)\right) = \frac{1}{2}\mathrm{Re}\left(\frac{\Gamma'}{\Gamma}(s-1)\right) - \frac{\log \pi}{2} + \mathrm{Re}\left(\frac{1}{s-1}\right) - \sum_{\rho}\mathrm{Re}\left(\frac{1}{s-\rho}\right)$$

where the sum runs through all non-trivial zeros ρ of the Riemann zeta-function.

2. The *local* one. Used by Landau, the formula takes the form

$$\mathrm{Re}\left(-\frac{\zeta'}{\zeta}(s)\right) = -\sum_{\substack{\rho \\ |s-\rho|\leqslant c/\log|t|}}\mathrm{Re}\left(\frac{1}{s-\rho}\right) + O(\log|t|).$$

In [Kad05], an intermediate way is considered with circles of radius $\asymp 1$. The main tools are then the knowledge of more and more zeros of $\zeta(s)$ lying on the critical line $\sigma = \frac{1}{2}$ and Weil type explicit formulae as follows. Let $r > 0$ and $f \in C^2[0,r]$ with compact support in $[0,r[$ and satisfying $f(r) = f'(0) = f'(r) = f''(r) = 0$. We denote $F(s)$ the Laplace transform of f, i.e.

$$F(s) = \int_0^r e^{-st} f(t)\, dt$$

and set F_2 the Laplace transform of f''. One can prove the following formula.

For all complex numbers s, we have

$$\mathrm{Re}\left(\sum_{n=1}^{\infty} \frac{\Lambda(n)}{n} f(\log n)\right) = f(0)\left\{\mathrm{Re}\left(\frac{1}{2}\frac{\Gamma'}{\Gamma}\left(\frac{s}{2}+1\right)\right) - \frac{\log \pi}{2}\right\}$$

$$+ \mathrm{Re}\, F(s-1) - \sum_{\rho}\mathrm{Re}\, F(s-\rho) + \mathrm{Re}\left(\frac{F_2(s)}{s^2}\right)$$

$$+ \frac{1}{2\pi}\int_{-\infty}^{\infty}\mathrm{Re}\left(\frac{\Gamma'}{\Gamma}\left(\frac{1}{4}+\frac{it}{2}\right)\right)$$

$$\times \mathrm{Re}\left(\frac{F_2(s-1/2-it)}{(s-1/2-it)^2}\right)dt.$$

With an appropriate choice of f, it is proved in [Kad05] that $\zeta(s)$ has no zero in the region

$$\sigma \geqslant 1 - \frac{1}{5.7 \log |t|} \quad \text{and} \quad |t| \geqslant 3.$$

This region supersedes the above region given by Ford as long as $|t| \leqslant e^{9400}$ and the method may be generalized to zero-free regions of L-functions, as it can be seen in [Kad02] where it is proved that the functions $L(s, \chi)$ never vanish in the region

$$\sigma \geqslant 1 - \frac{1}{6.4355 \log(\max(q, q\tau))}$$

except for at most one of them which should be real and vanishes at most once in this region.

In [Rn75, Lemma 8], the authors established the following effective version of Theorem 3.91.

Proposition 3.96 *Let $x > 1$, $0 < \delta < 1 - x^{-1}$ be real numbers and R, d be real numbers satisfying (3.46). Set*

$$E(x) = \Psi(x) - \left\{ x - \log 2\pi - \frac{1}{2} \log\left(1 - \frac{1}{x^2}\right) \right\}.$$

Then we have

$$|E(x)| \leqslant (2 + 2\delta^{-1} + \delta)\left(x \sum_{\gamma > A} F(\gamma) + 0.0463\sqrt{x} \right) + \frac{x\delta}{2}$$

with

$$F(u) = \frac{1}{u^2} \exp\left(-\frac{\log x}{R \log(u/d)} \right)$$

and A is the unique solution of the equation

$$\mathcal{N}(A) = \frac{A}{2\pi} \log\left(\frac{A}{2\pi} \right) - \frac{A}{2\pi} + \frac{7}{8}$$

where $\mathcal{N}(A)$ is the number of zeros $\rho = \beta + i\gamma$ of $\zeta(s)$ such that $0 < \gamma \leqslant A$.

Rosser and Schœnfeld showed that one can take $R \approx 9.6459\ldots$, $d = 17$ and with the knowledge of non-trivial zeros on the critical line at that time, they were able to take $A \approx 1\,894\,438.512\ldots$ The sum is treated by partial summation which enables us to estimate integrals of the form

$$\int_A^\infty F(u) \log\left(\frac{u}{2\pi} \right) du$$

Table 3.2 Rosser & Schoenfeld's explicit bounds

Inequality	Validity
$\theta(x) < 1.000\,081\,x$	$x > 0$
$\theta(x) > 0.75\,x$	$x \geqslant 36$
$\frac{x}{\log x} \leqslant \pi(x) \leqslant \frac{1.25\,506\,x}{\log x}$	lower bound: $x \geqslant 17$ upper bound: $x > 1$
$|\theta(x) - x| < \frac{8.686\,x}{(\log x)^2}$	$x > 1$
$|\psi(x) - x| < \frac{8.686\,x}{(\log x)^2}$	$x > 1$

which in turn may be written as combinations of Bessel functions of the second kind. After using fine estimates of these quantities and choosing δ appropriately, we arrive at the following explicit result.

Theorem 3.97 (Rosser and Schœnfeld) *Define*

$$\epsilon(x) = 0.110\,123\left(1 + \frac{3.001\,5}{\sqrt{\log x}}\right)(\log x)^{3/8} \exp\left(-\sqrt{\frac{\log x}{R}}\right)$$

with $R \approx 9.6459\ldots$ Then, for all $x > 0$, we have

$$\theta(x) - x \leqslant \Psi(x) - x \leqslant x\epsilon(x)$$

and for all $x \geqslant 39.4$, we have

$$\Psi(x) - x \geqslant \theta(x) - x \geqslant -x\epsilon(x).$$

For instance, the next result summarizes some of the estimates the authors obtained with Theorem 3.97.

Corollary 3.98 *The following estimates hold (see Table 3.2) in the specified range of validity.*

The usual functions of prime numbers may be handled as in the previous section (see [Rn62]).

Corollary 3.99 *The following estimates hold (see Table 3.3) in the specified range of validity. p_n is the n-th prime number, $\gamma \approx 0.5772\ldots$ is the Euler–Mascheroni constant, $B \approx 0.261\,497\,212\ldots$ is the Mertens constant and $E \approx 1.332\,582\,275\ldots$*

The lower bound for p_n is due to Dusart [Dus98].

Explicit estimates for the functions $\pi(x; q, a)$, $\theta(x; q, a)$ and $\Psi(x; q, a)$ have been established by McCurley [McC84], Ramaré and Rumely [RR96] and Dusart [Dus01]. In the second paper, the authors showed an analogue of Proposition 3.96 for $\Psi(x; q, a)$. This result is refined in the third paper, where the following estimates are proved.

Table 3.3 Explicit bounds for usual functions of prime numbers

Inequality	Validity
$n(\log n + \log\log n - 1) \leqslant p_n \leqslant n(\log n + \log\log n)$	lower bound: $n \geqslant 2$ upper bound: $n \geqslant 6$
$\log x - E - \frac{1}{2\log x} < \sum_{p\leqslant x} \frac{\log p}{p} < \log x - E + \frac{1}{\log x}$	$x \geqslant 32$
$\log\log x + B - \frac{1}{2(\log x)^2} < \sum_{p\leqslant x} \frac{1}{p} < \log\log x + B + \frac{1}{(\log x)^2}$	$x > 1$
$e^{\gamma} \log x \left(1 - \frac{1}{2(\log x)^2}\right) < \prod_{p\leqslant x}\left(1 - \frac{1}{p}\right)^{-1} < e^{\gamma} \log x \left(1 + \frac{1}{(\log x)^2}\right)$	$x > 1$

Theorem 3.100 (Dusart) *Let q be a positive integer and define*

$$\epsilon(x) = \left(\frac{q^2 \log x}{R\varphi(q)^2}\right)^{1/4} \exp\left(-\sqrt{\frac{\log x}{R}}\right)$$

where $R \approx 9.6459\ldots$ is as in Theorem 3.97. Then, for all $x > x_0(q)$ where $x_0(q)$ is an effectively computable constant, we have

$$\left|\Psi(x; q, a) - \frac{x}{\varphi(q)}\right| < x\epsilon(x).$$

The same inequality holds with $\Psi(x; q, a)$ replaced by $\theta(x; q, a)$.

In fact, Dusart's result is slightly more accurate and makes use of a constant $C_1(q)$ in the function $\epsilon(x)$ which is quite complicated to define. We use here the fact that, under the conditions of the theorem, we always have $C_1(q) \geqslant 9.14$. Dusart then gave some applications of Theorem 3.100. For instance, if $a \in \{1, 2\}$, then we have

$$\pi(x; 3, a) < \frac{0.55 x}{\log x} \quad (x \geqslant 229\,869)$$

and

$$\pi(x; 3, a) > \frac{x}{2 \log x} \quad (x \geqslant 151).$$

Also

$$\left|\theta(x; 3, a) - \frac{x}{2}\right| < \frac{0.262 x}{\log x} \quad (x \geqslant 1531).$$

3.7.8 The Piatetski-Shapiro Prime Number Theorem

A great deal of effort has been made in the search for other types of sequences containing infinitely many prime numbers, and several problems still remain open.

For instance, it is yet not known whether there are infinitely many primes of the form $n^2 + 1$, and we do not know if there are infinitely many Mersenne primes, i.e. of the form $2^n - 1$, or Fermat primes, i.e. of the form $2^{2^n} + 1$. However, in 1953, Piatetski-Shapiro supplied an example of a sequence sparser than the sequence $(qn + a)$ by studying the distribution of prime numbers of the form $[n^c]$, where $1 < c < c_0$ is a fixed real number. This sequence is somewhat the simplest generalization of polynomials with non-integer degrees. Clearly, there are no prime numbers in the sequence n^2, but it is generally believed that 2 is the correct upper bound in this problem.

It is customary to denote by $\pi_c(x)$ the number of positive integers $n \leqslant x$ such that $[n^c]$ is a prime number. Piatetski-Shapiro proved that, if $1 < c < c_0 = \frac{12}{11} \approx 1.0909\ldots$, then we have for $x \longrightarrow \infty$

$$\pi_c(x) \sim \frac{x}{c \log x}.$$

The best result to date in this problem is due to Rivat and Sargos [RS01] who showed that, using latest estimates in exponential sums of type I and II (see Chap. 6), if $1 < c < c_0 = \frac{2817}{2426} \approx 1.16117\ldots$, then we have for $x \longrightarrow \infty$

$$\pi_c(x) \sim \frac{x}{c \log x}.$$

3.7.9 The Riemann Hypothesis

The following quotation is attributed to Hilbert.

> *If I were to awaken after having slept for a thousand years, my first question would be: Has the Riemann hypothesis been proven?*

This shows the crucial importance of what has proved to be one of the most difficult problems in mathematics. In 1859, in his benchmarking Memoir, Bernhard Riemann formulated his conjecture, called today *the Riemann hypothesis*, which makes a very precise connection between two seemingly unrelated objects. There exist many great old unsolved problems in mathematics, but none of them has the stature of the Riemann hypothesis. This is probably due to the large number of ways in which this conjecture may be formulated. This is also certainly due to the personality of Riemann, a true genius ahead of his time and one of the most extraordinary mathematical talents. Finally, the Riemann hypothesis was highlighted at the 1900 International Congress of Mathematicians, in which Hilbert raised 23 problems that he thought would shape the next centuries. In 2000, the Clay Mathematics Institute listed seven hard open problems and promised a one million-dollar prize. Curiously, any disproof of the Riemann hypothesis does not earn the prize.

Riemann's formulation of his conjecture does not make arithmetic statements appear directly. We have seen that the non-trivial zeros of $\zeta(s)$ are all in the strip $0 < \sigma < 1$, and by the functional equation, if there is a zero in $0 < \sigma \leqslant \frac{1}{2}$, there is also a zero in the region $\frac{1}{2} \leqslant \sigma < 1$. Therefore, the following conjecture represents the "best of all possible worlds" for the zeros of the Riemann zeta-function.

Conjecture 3.101 (Riemann, 1859) *All non-trivial zeros of $\zeta(s)$ are on the critical line $\sigma = \frac{1}{2}$.*

Numerical computations have been made since 1859. It was Riemann himself who calculated the first zero of $\zeta(s)$ on the critical line, whose imaginary part is ≈ 14.13 (see [Tit51] for instance). The usual way is the use of the *Hardy function* $Z(t)$, also sometimes called the *Riemann–Siegel function*, defined in the following way. First set for $t \in \mathbb{R}$

$$\vartheta(t) = \arg\left(\pi^{-it/2}\Gamma\left(\frac{1}{4} + \frac{it}{2}\right)\right)$$

where the argument is defined by continuous variation of t starting with the value 0 at $t = 0$, and let

$$Z(t) = e^{i\vartheta(t)}\zeta\left(\frac{1}{2} + it\right).$$

The functional equation of $\zeta(s)$ implies that $\overline{Z(t)} = Z(t)$ so that $Z(t)$ is a real-valued function for $t \in \mathbb{R}$ and we have

$$\left|Z(t)\right| = \left|\zeta\left(\frac{1}{2} + it\right)\right|.$$

Therefore, if $Z(t_1)$ and $Z(t_2)$ have opposite signs, $\zeta(s)$ has a zero on the critical line between $\frac{1}{2} + it_1$ and $\frac{1}{2} + it_2$. Using Theorem 3.82 at $s = \frac{1}{2} + it$ with $x = y = \sqrt{|t|/(2\pi)}$ and multiplying out by $e^{i\vartheta(t)}$, we deduce that for $N = [\sqrt{|t|/(2\pi)}]$ we have

$$Z(t) = e^{i\vartheta(t)}\sum_{n=1}^{N}\frac{1}{n^{1/2+it}} + e^{-i\vartheta(t)}\sum_{n=1}^{N}\frac{1}{n^{1/2-it}} + O\left(t^{-1/4}\right)$$

$$= 2\sum_{n=1}^{N}\frac{\cos(\vartheta(t) - t\log n)}{\sqrt{n}} + O\left(t^{-1/4}\right)$$

which is a concise form of the *Riemann–Siegel formula*. Siegel discovered this identity among Riemann's private papers in 1932. It enables us to get an improvement over Euler–MacLaurin's summation formula in approximating values of $\zeta(s)$. More precise formulae exist (see [Ivi85, Tit51]) that aid computations and add to the empirical evidence for the Riemann hypothesis. The best result up to now is due to Gourdon [Gou04] who found out that the 10^{13} first zeros of $\zeta(s)$ are on the critical line.

The minimal necessary condition for the Riemann hypothesis was proved by Hardy in 1914 [Har14].

Theorem 3.102 (Hardy) *There are infinitely many zeros of $\zeta(s)$ on the critical line.*

There exist several proofs of this result (see [Tit51]), all of which rely on the consideration of moments of the form $\int t^n f(t)\,dt$. A useful tool is then the following lemma due to Fejér.

Let $a > 0$ be a real number and n be a positive integer. The number of sign changes in the interval $]0, a[$ of a continuous function f is at least the number of sign changes of the sequence

$$f(0), \quad \int_0^a f(t)\,dt, \quad \int_0^a tf(t)\,dt, \dots, \int_0^a t^n f(t)\,dt.$$

Proof of Theorem 3.102 Recall the function

$$\omega(x) = \sum_{n=1}^{\infty} e^{-n^2 \pi x}$$

of the functional equation (3.13). One can show [Tit51] that

$$\int_0^\infty \xi\left(\frac{1}{2} + it\right) \cos(xt)\,dt = \pi \left(2e^{-x/2}\omega\left(e^{-2x}\right) - e^{x/2}\right).$$

Putting $x = -iy$ gives

$$\int_0^\infty \xi\left(\frac{1}{2} + it\right) \cosh(yt)\,dt = 2\pi \left\{ e^{iy/2}\left(\omega\left(e^{2iy}\right) + \frac{1}{2}\right) - \cos\frac{y}{2} \right\}.$$

By Theorem 3.59, we have $\zeta(\frac{1}{2} + it) \ll |t|^{1/4}$ so that $\xi(\frac{1}{2} + it) \ll |t|^{1/4}e^{-\pi|t|/4}$ and hence the above integral may be differentiated with respect to y any number of times provided that $y < \pi/4$. We then get

$$\int_0^\infty \xi\left(\frac{1}{2} + it\right) t^{2n} \cosh(yt)\,dt$$

$$= 2\pi \left\{ \frac{d^{2n}}{dy^{2n}}\left(e^{iy/2}\left(\omega\left(e^{2iy}\right) + \frac{1}{2}\right)\right) + (-1)^{n+1}2^{-2n} \cos\frac{y}{2} \right\}.$$

Now using (3.12) one can prove that the first term on the right-hand side tends to 0 as y tends to $\pi/4$ for a fixed integer n, and thus

$$\lim_{y\to\pi/4} \int_0^\infty \xi\left(\frac{1}{2} + it\right) t^{2n} \cosh(yt)\,dt = (-1)^{n+1}2^{1-2n}\pi \cos\frac{\pi}{8}.$$

Let m be a large positive integer. From above we infer that, if $a_m > 0$ is large enough and y_m is close enough to $\pi/4$, the integral

$$\int_0^{a_m} \xi\left(\frac{1}{2} + it\right) t^{2n} \cosh(y_m t)\,dt$$

has the same sign as $(-1)^{n+1}$ for $n = 0, 1, \ldots, m$. By Fejér's theorem, we infer that $\xi(\frac{1}{2} + it)$ has at least m changes of sign in $]0, a_m[$, as required. $\square$

It is customary to define $\mathcal{N}_0(T)$ to be the number of zeros $\rho = \frac{1}{2} + i\gamma$ with $0 < \gamma < T$, so that the Riemann hypothesis may be written as

$$\mathcal{N}(T) = \mathcal{N}_0(T)$$

for all $T > 0$. Hardy's theorem states that $\mathcal{N}_0(T) \longrightarrow \infty$ as $T \longrightarrow \infty$ and his method yields

$$\mathcal{N}_0(T) > CT$$

for some constant $C > 0$.

The second important result was given by Selberg who proved that a positive proportion of zeros lies on the critical line and more precisely that

$$\mathcal{N}_0(T) > CT \log T$$

for T sufficiently large. Selberg's method could be used to yield an effective estimate of C, but this value was not made specific until 1974 when Levinson found an explicit estimate of this constant by proving that at least $1/3$ of the zeros lie on the critical line (see [Ivi85, Chap. 10]). The best result to date is due to Bui, Conrey and Young [BCY11] who proved that more than 41% of the zeros lie on the critical line, slightly improving the previous record due to Conrey who showed in 1989 that at least $2/5$ of the zeros of ζ are on the critical line.

3.7.10 Some Consequences of the Riemann Hypothesis

▶ The first corollary of the Riemann hypothesis is to provide the best error-term in the Prime Number Theorem. Indeed, suppose that all the zeros of $\zeta(s)$ have real parts equal to $\frac{1}{2}$. By (3.39), we then have for all $T \geqslant 2$

$$\left| \sum_{|\gamma| \leqslant T} \frac{x^\rho}{\rho} \right| \ll x^{1/2} (\log T)^2$$

so that by Theorem 3.91 we get

$$\Psi(x) = x + O\left(x^{1/2} (\log T)^2 + \frac{x}{T} (\log xT)^2 + \log x \right)$$

and choosing $T = x^{1/2}$ we obtain

$$\Psi(x) = x + O\left(x^{1/2} (\log x)^2 \right).$$

This implies that

$$\theta(x) = x + O\big(x^{1/2}(\log x)^2\big)$$

and

$$\pi(x) = \mathrm{Li}(x) + O\big(x^{1/2}\log x\big)$$

by Exercise 8 in Chap. 1. In fact, one can prove that the Riemann hypothesis is *equivalent* to the estimate

$$\Psi(x) = x + O\big(x^{1/2+\varepsilon}\big)$$

for all $\varepsilon > 0$ (see [Ten95]).

▶ Perhaps one of the most arithmetic formulations of the Riemann hypothesis can be made with the Möbius function. This function $\mu(n)$ has already been studied above in the sieves section, and will be defined in Chap. 4 where we shall see that the Dirichlet series attached to this function is $\zeta(s)^{-1}$. We need here the summatory function $M(x)$ of $\mu(n)$, called the *Mertens function*, and hence defined by

$$M(x) = \sum_{n \leqslant x} \mu(n).$$

It can be proved that the PNT is equivalent to the estimate

$$M(x) = o(x)$$

for $x \longrightarrow \infty$, and that the Riemann hypothesis is equivalent to

$$M(x) = O\big(x^{1/2+\varepsilon}\big) \tag{3.47}$$

for all $\varepsilon > 0$. Indeed, this estimate implies by partial summation the convergence of the series $\sum_{n=1}^{\infty} \mu(n)n^{-s} = \zeta(s)^{-1}$ in the half-plane $\sigma > \frac{1}{2}$, and hence $\zeta(s)$ has no zero in this half-plane, which is the Riemann hypothesis. Conversely, if the Riemann hypothesis is true, then Littlewood proved that, for all $\varepsilon > 0$, we have

$$\zeta\left(\frac{1}{2} + \varepsilon + it\right)^{-1} \ll |t|^{\varepsilon}$$

and Theorem 3.84 gives (3.47). Several authors improved on the necessary condition. The best result to date is due to Soundararajan [Sou09] who proved that, if the Riemann hypothesis is true, then the estimate

$$M(x) \ll x^{1/2}\exp\big\{(\log x)^{1/2}(\log\log x)^{14}\big\}$$

holds for large x. It has long been believed that the *Mertens conjecture* stating that, for all $n \in \mathbb{N}$, the inequality

$$\big|M(n)\big| < n^{1/2}$$

would be true. In 1885, Stieltjes announced he had found a proof of the weaker bound $M(n) \ll n^{1/2}$ but he died without publishing his result, and no proof was found in his papers posthumously. We know now that Mertens' conjecture is false since Te Riele and Odlyzko [OtR85] showed that

$$\limsup_{n \to \infty} \frac{M(n)}{\sqrt{n}} > 1.06 \quad \text{and} \quad \liminf_{n \to \infty} \frac{M(n)}{\sqrt{n}} < -1.009.$$

These bounds were subsequently improved to 1.218 and -1.229 respectively [KtR06], and it is also known that the smallest number for which the Mertens conjecture is false is $\exp(1.59 \times 10^{40})$, improving on a previous result of Pintz [Pin87].

▶ The Riemann hypothesis also relies on arithmetic functions. For instance, let $\sigma(n)$ be the sum of the positive divisors of n (see Chap. 4). Robin proved that, if the inequality

$$\sigma(n) < e^{\gamma} n \log \log n$$

holds for all $n \geqslant 5041$, then the Riemann hypothesis is true. On the other hand, Lagarias showed that the Riemann hypothesis is equivalent to the elegant inequality

$$\sigma(n) \leqslant e^{H_n} \log H_n + H_n$$

where $H_n = \sum_{k=1}^{n} k^{-1}$ is the nth harmonic number.

▶ The Riemann hypothesis implies the Lindelöf hypothesis, so that we have for all $\varepsilon > 0$ and $t \geqslant t_0 > 0$

$$\zeta\left(\frac{1}{2} + it\right) \ll t^{\varepsilon}$$

if the Riemann hypothesis is true. Furthermore, it also implies the best error-term in the Dirichlet divisor problem (see Chap. 6), namely

$$\sum_{n \leqslant x} \tau(n) = x(\log x + 2\gamma - 1) + O\left(x^{1/4 + \varepsilon}\right)$$

for all $\varepsilon > 0$, where $\tau(n)$ is the number of divisors of n and γ is the Euler–Mascheroni constant.

▶ In 1977, Redheffer [Red77] introduced the matrix $R_n = (r_{ij}) \in \mathcal{M}_n(\{0, 1\})$ defined by

$$r_{ij} = \begin{cases} 1, & \text{if } i \mid j \text{ or } j = 1 \\ 0, & \text{otherwise} \end{cases}$$

and has shown that

$$\det R_n = M(n)$$

where $M(n)$ is the Mertens function.[18] Hence by above the Riemann hypothesis is equivalent to the estimate

$$|\det R_n| = O\left(n^{1/2+\varepsilon}\right)$$

for all $\varepsilon > 0$. This bound remains unproven, but Vaughan [Vau93] showed that 1 is an eigenvalue of R_n with (algebraic) multiplicity $n - [\log n / \log 2] - 1$, that R_n has two "dominant" eigenvalues $\lambda_\pm$ such that $|\lambda_\pm| \asymp n^{1/2}$, and that the other eigenvalues satisfy $\lambda \ll (\log n)^{2/5}$. However, it seems to be very difficult to extract more information from the eigenvalues of R_n by using tools from matrix analysis. For instance, Hadamard's inequality, which states that

$$|\det M|^2 \leqslant \prod_{i=1}^{n} \|L_i\|_2^2$$

for all matrices $M \in \mathcal{M}_n(\mathbb{C})$, where L_i is the ith row of M and $\|\ldots\|_2$ is the Euclidean norm on $\mathbb{C}^n$, gives

$$\left(M(n)\right)^2 \leqslant n \prod_{i=2}^{n} \left(1 + \left[\frac{n}{i}\right]\right) = 2^{n-[n/2]} n \prod_{i=2}^{[n/2]} \left(1 + \left[\frac{n}{i}\right]\right) \leqslant 2^{n-[n/2]} \binom{n + [n/2]}{n}$$

which is very far from the trivial bound $|M(n)| \leqslant n$.

▶ In another direction, the authors in [BC09] investigated the following integer upper triangular matrix. For all integers $i, j \geqslant 1$, set $\mathrm{mod}(j, i)$ to be the remainder in the Euclidean division of j by i. Let $T_n = (t_{ij}) \in \mathcal{M}_n(\mathbb{Z})$ be the upper triangular matrix of size n such that

$$t_{ij} = \begin{cases} \mathrm{mod}(j, 2) - 1, & \text{if } i = 1 \text{ and } 2 \leqslant j \leqslant n \\ \mathrm{mod}(j, i+1) - \mathrm{mod}(j, i), & \text{if } 2 \leqslant i \leqslant n-1 \text{ and } 1 \leqslant j \leqslant n \\ 1, & \text{if } (i, j) \in \{(1, 1), (n, n)\} \\ 0, & \text{otherwise.} \end{cases}$$

For instance with $n = 8$, we have

$$T_8 = \begin{pmatrix} 1 & -1 & 0 & -1 & 0 & -1 & 0 & -1 \\ 0 & 2 & -1 & 1 & 1 & 0 & 0 & 2 \\ 0 & 0 & 3 & -1 & -1 & 2 & 2 & -2 \\ 0 & 0 & 0 & 4 & -1 & -1 & -1 & 3 \\ 0 & 0 & 0 & 0 & 5 & -1 & -1 & -1 \\ 0 & 0 & 0 & 0 & 0 & 6 & -1 & -1 \\ 0 & 0 & 0 & 0 & 0 & 0 & 7 & -1 \\ 0 & 0 & 0 & 0 & 0 & 0 & 0 & 1 \end{pmatrix}.$$

One can prove the following result [BC09, Corollary 2.6].

---

[18] See Exercise 21 in Chap. 4.

Theorem 3.103 *Let σ_n be the smallest singular value of T_n. If the estimate $\sigma_n \gg n^{-1/2-\varepsilon}$ holds for all $\varepsilon > 0$, then the Riemann hypothesis is true.*

This criterion is probably at least as tricky to show as the Riemann hypothesis itself. However, it is interesting to note that this result is in accordance with the heuristic proofs concerning the smallest singular value[19] of a random matrix of size n. For instance, it is shown in [RV08a, RV08b] that this singular value has a high probability of satisfying the estimate $\sigma_n \asymp n^{-1/2}$. More precisely, let $A \in \mathcal{M}_n(\mathbb{R})$ whose entries are i.i.d centered random variables with unit variance and fourth moment bounded by B. Let σ_n be the smallest singular value of A. The authors show that, for all $\varepsilon > 0$, there exist $K_1, K_2 > 0$ and positive integers n_1, n_2, depending polynomially only on B and ε, such that

$$\mathbb{P}\left(\sigma_n > K_1 n^{-1/2}\right) \leqslant \varepsilon \quad (n \geqslant n_1)$$

and

$$\mathbb{P}\left(\sigma_n < K_2 n^{-1/2}\right) \leqslant \varepsilon \quad (n \geqslant n_2)$$

where $\mathbb{P}(E)$ is the probability of the event E.

But we have plenty of examples in number theory showing that, from a heuristic argument to a rigorous proof, the way is very long and hard.

3.7.11 The Mean Square of the Riemann Zeta-Function

Another approach for bounding $\zeta(\frac{1}{2} + it)$ rests on the study of the integral

$$I_1(T) = \int_0^T \left|\zeta\left(\frac{1}{2} + it\right)\right|^2 dt.$$

Indeed, in view of the following inequalities

$$\left|\zeta\left(\frac{1}{2} + it\right)\right|^2 \ll \left\{\int_{t-(\log t)^2}^{t+(\log t)^2} \left|\zeta\left(\frac{1}{2} + iu\right)\right|^2 du + 1\right\} \log t$$

$$\ll \left\{I_1\left(t + \log^2 t\right) - I_1\left(t - \log^2 t\right) + 1\right\} \log t$$

$$\ll \left\{E\left(t + \log^2 t\right) - E\left(t - \log^2 t\right) + (\log t)^3\right\} \log t$$

where $E(t)$ is the error-term in the asymptotic formula

$$I_1(T) = T\left(\log \frac{T}{2\pi} + 2\gamma - 1\right) + E(T)$$

---

[19]The singular values $\sigma_1 \geqslant \cdots \geqslant \sigma_n \geqslant 0$ of a matrix $A \in \mathcal{M}_n(\mathbb{C})$ are the square-roots of the eigenvalues of the hermitian positive semidefinite matrix AA^*, where A^* is the hermitian adjoint of A.

proved in 1918 by Hardy and Littlewood, we see that any estimate of the form $E(t) \ll t^\alpha$ implies the bound

$$\zeta\left(\frac{1}{2} + it\right) \ll t^{\alpha/2}(\log t)^{1/2}.$$

Numerous bounds have been provided by many authors for $E(t)$. For instance, Ingham proved in 1926 that $E(t) \ll t^{1/2}\log t$ whereas Balasubramanian obtained in 1978 the bound $E(t) \ll t^{346/1067+\varepsilon}$. The best result till now is due to Huxley [Hux05] who showed that

$$E(t) \ll t^{131/416+\varepsilon}.$$

More generally, set

$$I_k(T) = \int_0^T \left|\zeta\left(\frac{1}{2} + it\right)\right|^{2k} dt.$$

The long-standing conjecture is that $I_k(T) \sim T P_{k^2}(\log T)$ where P_{k^2} is a polynomial of degree k^2. So far asymptotic formulae have been only proved for the cases $k = 1$ and $k = 2$. Ivić considers kth moments of $E(t)$ in short intervals to deduce bounds for $I_k(T)$ in the following way. Assume that the integral

$$\int_{T/3}^{3T} \left\{\left|E(t + 2G) - E(t - 2G)\right|^k + \left|E(t + G/2) - E(t - G/2)\right|^k\right\} dt$$

is bounded by estimates of the form

$$\ll_\varepsilon T^{\alpha+\varepsilon} G^\beta$$

where $\alpha = \alpha(k) > 0$, $\beta = \beta(k) \leqslant k - 1$ and $T^\varepsilon \leqslant G = G(T) \ll T^{1/3}$. Then we have

$$I_{1+k-\beta}(T) \ll_\varepsilon T^{1+\alpha+\varepsilon}.$$

Another way to get bounds for $I_1(T)$ is the use of the following result, which relies on the techniques of exponential sums we shall see in Chap. 6. It is due to Heath-Brown who used Atkinson's formula (see [Ivi85]). It should be mentioned that Ivić's proof [Ivi85] does not deal with this formula.

Theorem 3.104 Let $T \geqslant 1$, $T^\varepsilon \leqslant G \leqslant T^{1/2-\varepsilon}$ and define

$$f_T(n) = (2\pi)^{-1}\left\{2T \sinh^{-1}\left(\sqrt{\frac{\pi n}{2T}}\right) + \sqrt{\pi^2 n^2 + 2\pi n T}\right\}$$

and

$$S(x) = \sum_{K \leqslant n \leqslant K+x} (-1)^n \tau(n) e\big(f_T(n)\big).$$

Then uniformly in G, we have

$$\frac{1}{G}\int_{T-G}^{T+G}\left|\zeta\left(\frac{1}{2}+it\right)\right|^2 dt$$

$$\ll T^{-1/4}\sum_{\substack{K=2^k\\ T^{1/3}\leqslant K\ll TG^{-2}(\log T)^2}}\frac{e^{-G^2K/T}}{K^{1/4}}\max_{0\leqslant u\leqslant K}\left|S(u)\right|+\log T.$$

3.7.12   Additive Characters and Gauss Sums

The Dirichlet characters modulo q are characters from the multiplicative groups $(\mathbb{Z}/q\mathbb{Z})^*$ to the multiplicative group $\mathbb{C}^*$ of non-zero complex numbers, and are thus called *multiplicative characters*. The dual concept of additive characters is defined in a similar way. Let a, q be positive coprime integers with $1 \leqslant a \leqslant q$ and consider the primitive qth roots of unity $e_q(a)$. From the well-known identity

$$\sum_{k\,(\mathrm{mod}\,q)}\left(e_q(a)\right)^k = 0$$

we get

$$\frac{1}{q}\sum_{k\,(\mathrm{mod}\,q)}e_q(-ka)e_q(kn) = \begin{cases} 1, & \text{if } n \equiv a \pmod{q} \\ 0, & \text{otherwise} \end{cases} \tag{3.48}$$

which is the analogue of Proposition 3.68. Thus the characteristic function of the integers $n \equiv a \pmod{q}$ may be written as a linear combination of the sequence $(e_q(kn))$. These functions are called *additive characters* since they are characters from the additive groups $(\mathbb{Z}/q\mathbb{Z}, +)$ to the multiplicative group $\mathbb{C}^*$.

Let f be an arithmetic function of period q. Multiplying both sides of (3.48) by $f(n)$ and summing over n running through a complete residue system modulo q, we obtain

$$f(a) = \sum_{k\,(\mathrm{mod}\,q)}\frac{e_q(ka)}{q}\sum_{n\,(\mathrm{mod}\,q)}f(n)e_q(-kn).$$

The function

$$\widehat{f}(k) = \frac{1}{q}\sum_{n\,(\mathrm{mod}\,q)}f(n)e_q(-kn)$$

is called the *finite Fourier transform* of f, and we thus have

$$f(n) = \sum_{k\,(\mathrm{mod}\,q)}\widehat{f}(k)e_q(kn).$$

For all q-periodic arithmetic function f, we have

$$\frac{1}{q}\sum_{n=1}^{q}\left|f(n)\right|^2 = \sum_{k=1}^{q}\left|\widehat{f}(k)\right|^2$$

which is the analogue of Plancherel's formula for functions $f \in L^2(\mathbb{R})$ or Parserval's formula for functions $f \in L^2(\mathbb{R}/\mathbb{Z})$.

It may be interesting to deal with a discrete inner product of the multiplicative character χ and the additive character e_q. To this end, we define the sum

$$\tau(\chi) = \sum_{a\,(\mathrm{mod}\,q)} \chi(a)e_q(a).$$

This is called the *Gauss sum* of χ and is an important tool connecting additive number theory and multiplicative number theory. One of its most important properties is to express any primitive Dirichlet character as a linear combination of additive characters.

Theorem 3.105 *Let χ be a primitive Dirichlet character modulo q. For all positive integers n, we have*

$$\chi(n) = \frac{1}{\tau(\overline{\chi})} \sum_{a\,(\mathrm{mod}\,q)} \overline{\chi}(a)e_q(na).$$

It can be proved that, for all primitive characters χ modulo q, we have $|\tau(\chi)| = \sqrt{q}$. This result may be used to estimate twisted exponential sums of the form

$$\sum_{N<n\leqslant 2N} \chi(n)e\big(f(n)\big)$$

arising in many problems in number theory (see [Bor10] for instance).

3.7.13 Incomplete Character Sums

In many applications the bound of Proposition 3.70 may be not sufficient. The next result provides a nearly best possible estimate for these character sums (for a proof, see [MV07, Theorem 9-18], for instance).

Theorem 3.106 (Pólya–Vinogradov inequality) *For all non-principal Dirichlet characters χ modulo q and all integers M and N with $N \geqslant 1$, we have*

$$\left|\sum_{M<n\leqslant M+N} \chi(n)\right| \leqslant \sqrt{3q}\,\log q.$$

If $N \ll q^{1/2}$, the Pólya–Vinogradov inequality is weaker than the trivial bound provided by the triangle inequality. In the early 1960s, Burgess [Bur63, Bur86] discovered another method to estimate character sums whose bound depends on the length of the range of summation.

Theorem 3.107 (Burgess) *For all non-principal* primitive *Dirichlet characters χ modulo q, all real numbers $0 < \varepsilon < 1$ and all integers M and N with $N \geqslant 1$, we have*

$$\sum_{M < n \leqslant M+N} \chi(n) \ll N^{1-1/r} q^{\frac{r+1}{4r^2}+\varepsilon}$$

for $r \in \{1, 2, 3\}$.

The proof uses in a crucial way the Riemann hypothesis for curves over finite fields which was proved by Weil [Wei48a, Wei48b], implying that for prime p and Dirichlet character χ modulo p of order d, then

$$\left| \sum_{x=1}^{p} \chi(P(x)) \right| \leqslant (m-1)\sqrt{p}$$

where $P \in \mathbb{F}_p[X]$ is not a dth power of a polynomial and m is the number of distinct roots of P. Applying Theorem 3.107 with $r = 1$, we recover a slightly weaker version of the Pólya–Vinogradov inequality. Hence this result generalizes Theorem 3.106.

For incomplete Gauss sums, Burgess [Bur89] obtained the following result.

Theorem 3.108 (Burgess) *For all non-principal Dirichlet characters χ modulo q and all integers $a \geqslant 0$, M and N with $1 \leqslant N < q$, we have*

$$\sum_{M < n \leqslant M+N} \chi(n) e_q(an) \ll N^{2/3} q^{1/8} (\log q)^2.$$

If $q = p^\alpha$ is a prime power with $p > 3$, Burgess [Bur92] showed that the term $N^{2/3} q^{1/8}$ may be replaced by $N^{3/4} q^{1/12}$. This improves on Theorem 3.108 as long as $N \leqslant q^{1/2}$.

A *Kloosterman sum* is a character sum of the form

$$\sum_{\substack{n=1 \\ (n,q)=1}}^{q} e_q(an + b\overline{n})$$

where a, b, q are positive integers and $\overline{n}$ is defined by $n \times \overline{n} \equiv 1 \ (\mathrm{mod}\, q)$. Weil's bound for such sums states that

$$\left| \sum_{\substack{n=1 \\ (n,q)=1}}^{q} e_q(an + b\overline{n}) \right| \leqslant \tau(q)\sqrt{dq}$$

where $d = (b, q)$ and $\tau(q)$ is the number of divisors of q. For incomplete Klooster-man sums, we have the following estimate due to Hooley.

Theorem 3.109 (Hooley) *For all real numbers $0 < \varepsilon < 1$, all integers $1 \leqslant a < b$ and $m, q \geqslant 1$ with $d = (m, q)$, we have*

$$\sum_{n=a}^{b} e_q(m \cdot \bar{n}) \ll q^{1/2+\varepsilon} d^{1/2} + dq^{-1}(b-a).$$

3.8 Exercises

1 Is the integer $3^{4^5} + 4^{5^6}$ a prime number?

2 Show that, for each integer $n \geqslant 33$, we have $\pi(n) \leqslant \frac{n}{3}$.

3 Let $k \geqslant 2$ be an integer. Show that the number of prime numbers in the sequence

$$6^2 + 2, \ 7^2 + 2, \ldots, (6k)^2 + 2$$

is $< k$.

4

(a) Show that the series $\sum_p \frac{1}{p \log p}$ is convergent.

(b) Using Exercise 7 in Chap. 1 and Corollary 3.98, prove that[20] $\sum_p \frac{\log p}{p^2} < \frac{1}{2}$.

5 (Putnam, 1977) Let p be a prime number and $a \geqslant b$ be non-negative integers. Show that

$$\binom{pa}{pb} \equiv \binom{a}{b} \pmod{p}.$$

6 Let p be a prime number. Determine the number of positive integer solutions of the Diophantine equation

$$\frac{1}{x} + \frac{1}{y} = \frac{1}{p}.$$

---

[20] One may also use the inequality

$$\sum_{p \leqslant 100} \frac{\log p}{p^2} < 0.484$$

evaluated using the PARI/GP system.

7 Determine the prime numbers p such that $\frac{p^2-1}{8}$ is also a prime number.

8 Solve the Diophantine equation

$$x + y + z + xy + yz + zx + xyz = 2009$$

where $1 < x < y < z$ are integers.

9 Let a be a positive integer such that $(a, 10) = 1$.

(a) Show that $a^8 \equiv 1 \pmod{10}$ and that, for all non-negative integers k, we have $a^{8 \times 10^k} \equiv 1 \pmod{10^{k+1}}$. Deduce that

$$a^{800\,000\,001} \equiv a \pmod{10^9}.$$

(b) Determine a non-negative integer x such that the decimal expansion of x^3 ends with $123\,456\,789$.

10 Prove that, for all integers $a \geqslant 1$, $k \geqslant 0$ and all primes $p \nmid a$, we have

$$a^{p^{k+1}} \equiv a^{p^k} \pmod{p^{k+1}}.$$

11

(a) Let $n \in \mathbb{N}$ and $p \geqslant 5$ be a prime number such that $p \mid (n^2 + n + 1)$. Show that $\operatorname{ord}_p(n) = 3$ and deduce that $p \equiv 1 \pmod 6$.
(b) Show that there are infinitely many prime numbers of the form $p \equiv 1 \pmod 6$.

12 Let $n \in \mathbb{N}$. Show that, for all primes $p \leqslant n$, we have

$$\frac{n-p}{p-1} - \frac{\log n}{\log p} < v_p(n!) \leqslant \frac{n-1}{p-1}$$

and deduce that

$$v_p(n!) = \frac{n}{p-1} + O(\log n).$$

13 Show that $n \in \mathbb{N}$ is a difference of two squares if and only if $n \not\equiv 2 \pmod 4$. Deduce the values of n for which $n!$ is a difference of two squares.

14 Let $P \in \mathbb{Z}[X]$ be a polynomial of degree $n \geqslant 1$. Show that if there exist at least $2n + 1$ distinct integers m such that $|P(m)|$ is prime, then P is irreducible[21] over $\mathbb{Z}$.
   Are the polynomials $P_1 = X^4 - X^3 + 2X - 1$ and $P_2 = X^4 - 4X^2 - X + 1$ irreducible over $\mathbb{Z}$?

---

[21] See Definition 7.21 and Proposition 7.28.

15 Using Dirichlet's theorem (Theorem 3.63) and the sequence $7 + 15n$, show that there exist infinitely many prime numbers p not lying in a pair of twin prime numbers (q, r).

16 Let $n \geqslant 3$ be an integer and $1 = d_1 < d_2 < \cdots < d_k = n$ be all the divisors of n in increasing order. Show that

$$\sum_{j=2}^{k} d_{j-1} d_j < n^2.$$

17 Let $n \in \mathbb{N}$.

(a) Let a, b be two positive coprime integers and $P = c_n X^n + \cdots + c_1 X + c_0 \in \mathbb{Z}[X]$. Show that, if a/b is a root of P, then $a \mid c_0$ and $b \mid c_n$.

(b) Deduce that if P is a monic integer polynomial, then the roots of P are either integer or irrational.

(c) Use the preceding question to show that, if p is a prime number, then $\sqrt{p} \notin \mathbb{Q}$.

18 (Serret's algorithm) This exercise uses the notation and results of Exercises 2 and 3 in Chap. 2.

Part A. The purpose of this section is to show that if p is a prime number, then p is a sum of two squares if and only if either $p = 2$ or $p \equiv 1 \pmod 4$.

(a) Let p be an odd prime number such that $p = a^2 + b^2$ for some $a, b \in \mathbb{Z}$. Show that $p \equiv 1 \pmod 4$.

(b) Let p be a prime number such that $p \equiv 1 \pmod 4$ and define $x < p$ such that

$$x \equiv \left(\frac{p-1}{2} \right)! \pmod p.$$

Prove that $x^2 \equiv -1 \pmod p$.

(c) Using Exercise 2 in Chap. 3, show that p can be expressed as a sum of two squares.

Part B. Let $x > 1$ be an integer, $p > x$ be a prime number such that $p \equiv 1 \pmod 4$ and let (r_k) be the finite sequence of remainders in the Euclidean algorithm associated to p and x.

(a) Explain why there exists a positive integer k such that $r_{k-1} > \sqrt{p} \geqslant r_k$.

(b) Using Exercise 3 in Chap. 2, show that the pair $(u, v) = (t_k, r_k)$ satisfies the conditions of Thue's lemma.

(c) Express $p = 9733$ as a sum of two squares.

19 Let $n \in \mathbb{N}$. Determine an asymptotic formula for $\prod_p p^{\lfloor n/p \rfloor}$.

20 Let $\sigma > 1$ be a real number and N be a positive integer. Show that

$$\sum_{n=N}^{\infty} \frac{1}{n^{\sigma}} \leqslant \frac{\zeta(\sigma)}{N^{\sigma-1}}$$

and if $N \geqslant 3$, then

$$\sum_{n=N}^{\infty} \frac{\log n}{n^{\sigma}} \leqslant \frac{\zeta(\sigma) \log N}{N^{\sigma-1}}.$$

21 (Unitary divisors) Let $n \in \mathbb{N}$. A divisor d of n is called a *unitary divisor* if $(d, n/d) = 1$. The unitary greatest common divisor of m and n is denoted by $(m, n)^*$.

(a) Show that $(m, n)^*$ is a unitary divisor of (m, n).

(b) Let p be a prime number. Show that $(p^e, p^f)^* = \begin{cases} p^e \text{ or } p^f, & \text{if } e=f \\ 1, & \text{otherwise.} \end{cases}$

(c) Set $n = p_1^{e_1} \cdots p_r^{e_r}$ and $d = p_1^{f_1} \cdots p_r^{f_r}$ a divisor of n with $0 \leqslant f_i \leqslant e_i$. Prove that d is a unitary divisor of n if and only if either $f_i = 0$ or $f_i = e_i$. Deduce the number of unitary divisors of n.

Example Provide the unitary divisors of 3024 and 6615.

(d) Set $n = p_1^{e_1} \cdots p_r^{e_r}$ and $m = p_1^{f_1} \cdots p_r^{f_r}$ with $e_i, f_i \in \mathbb{Z}_+$. Prove that

$$(m, n)^* = p_1^{g_1} \cdots p_r^{g_r} \quad \text{with} \quad g_i = \begin{cases} e_i \text{ or } f_i, & \text{if } e_i = f_i \\ 0, & \text{otherwise.} \end{cases}$$

Example Compute $(6615, 3024)^*$.

References

[AGP94] Alford WR, Granville A, Pomerance C (1994) There are infinitely many Carmichael numbers. Ann Math 140:703–722

[BC09] Bordellès O, Cloitre B (2009) A matrix inequality for Möbius functions. JIPAM J Inequal Pure Appl Math 10, Art. 62

[BCY11] Bui HM, Conrey JB, Young MP (2011) More than 41% of the zeros of the zeta function are on the critical line. Acta Arith 150:35–64

[Bor10] Bordellès O (2010) The composition of the gcd and certain arithmetic functions. J Integer Seq 13, Art. 10.7.1

[Bru15] Brun V (1915) Über das Goldbachsche Gesetz und die Anzahl der Primzahlpaare. Arch Math Naturvidensk 34(8)

[Bru19] Brun V (1919) La série $1/5 + 1/7 + 1/11 + 1/13 + 1/17 + 1/19 + 1/29 + 1/31 + 1/41 + 1/43 + 1/59 + 1/61 + \cdots$ où les dénominateurs sont nombres premiers jumeaux est convergente ou finie. Bull Sci Math 43:124–128

[Bur63]   Burgess DA (1963) On character sums and L-series, II. Proc Lond Math Soc 13:524–536

[Bur86]   Burgess DA (1986) The character sum estimate with $r = 3$. J Lond Math Soc (2) 33:219–226

[Bur89]   Burgess DA (1989) Partial Gaussian sums II. Bull Lond Math Soc 21:153–158

[Bur92]   Burgess DA (1992) Partial Gaussian sums III. Glasg Math J 34:253–261

[BV09]   Bordellès O, Verdier N (2009) Variations autour du Postulat de Bertrand. Bull AMQ 49:25–51

[DE80]   Diamond H, Erdős P (1980) On sharp elementary prime number estimates. Enseign Math 26:313–321

[Dic05]   Dickson LE (2005) History of the theory of numbers. Volume 1: Divisibility and primality. Dover, New York

[Dus98]   Dusart P (1998) Autour de la fonction qui compte le nombre de nombres premiers. PhD thesis, Univ. Limoges

[Dus01]   Dusart P (2001) Estimates for $\theta(x; k, l)$ for large values of x. Math Comput 239:1137–1168

[Edw74]   Edwards HM (1974) Riemann's zeta function. Academic Press, San Diego. 2nd edition: Dover, Mineola, New York, 2001

[For02a] Ford K (2002) Vinogradov's integral and bounds for the Riemann zeta function. Proc Lond Math Soc 85:565–633

[For02b] Ford K (2002) Zero-free regions for the Riemann zeta function. In: Bennett MA, et al (eds) Number theory for the Millennium. AK Peters, Boston, pp 25–56

[GM84]   Gupta R, Murty MR (1984) A remark on Artin's conjecture. Invent Math 78:127–130

[Gol83]   Goldston DA (1983) On a result of Littlewood concerning prime numbers II. Acta Arith 43:49–51

[Gou04]   Gourdon X (2004) The 10^{13} first zeros of the Riemann zeta function, and zeros computation at very large height. Available at: http://numbers.computation.free.fr/Constants/Miscellaneous/zetazeros1e13-1e24.pdf

[GS05]   Gupta A, Sury B (2005) Decimal expansion of $1/p$ and subgroup sums. Integers 5, Art. #A19

[Han72]   Hanson D (1972) On the product of primes. Can Math Bull 15:33–37

[Har14]   Hardy GH (1914) Sur les zéros de la fonction $\zeta(s)$ de Riemann. C R Acad Sci Paris 158:1012–1014

[HB86]   Heath-Brown DR (1986) Artin's conjecture for primitive roots. Q J Math Oxford 37:27–38

[Hoo67]   Hooley C (1967) On Artin's conjecture. J Reine Angew Math 226:207–220

[Hux05]   Huxley MN (2005) Exponential sums and the Riemann zeta function V. Proc Lond Math Soc 90:1–41

[Ivi85]   Ivić A (1985) The Riemann zeta-function. Theory and applications. Wiley, New York. 2nd edition: Dover, 2003

[Iwa77]   Iwaniec H (1977) The sieve of Eratosthenes–Legendre. Ann Sc Norm Super Pisa, Cl Sci 2:257–268

[Kad02]   Kadiri H (2002) Une région explicite sans zéro pour les fonctions L de Dirichlet. PhD thesis, Univ. Lille 1

[Kad05]   Kadiri H (2005) An explicit region without zeros for the Riemann ζ function. Acta Arith 117:303–339

[KtR06]   Kotnik T, te Riele HJJ (2006) The Mertens conjecture revisited. In: Algorithmic number theory, 7th international symposium, ANTS-VII, Berlin, Germany, July 23–28, 2006. Lecture notes in computer science, vol 4076. Springer, Berlin, pp 156–167

[McC84]   McCurley KS (1984) Explicit estimates for the error term in the prime number theorem for arithmetic progressions. Math Comput 423:265–285

[Mon93]   Monsky P (1993) Simplifying the proof of Dirichlet's theorem. Am Math Mon 100:861–862

[MV07]   Montgomery HL, Vaughan RC (2007) Multiplicative number theory Vol. I. Classical theory. Cambridge studies in advanced mathematics, vol 97
[Nai82]   Nair M (1982) A new method in elementary prime number theory. J Lond Math Soc 25:385–391
[OtR85]   Odlyzko A, te Riele HJJ (1985) Disproof of the Mertens conjecture. J Reine Angew Math 357:138–160
[Pin87]   Pintz J (1987) An effective disproof of the Mertens conjecture. Astérisque 147/148:325–333
[PS98]   Pólya G, Szegő G (1998) Problems and theorems in analysis II. Springer, Berlin
[Ram11]   Ramaré O (2010/2011) La méthode de Balasubramanian pour une région sans zéro. In: Groupe de Travail d'Analyse Harmonique et de Théorie Analytique des Nombres de Lille
[Red77]   Redheffer RM (1977) Eine explizit lösbare Optimierungsaufgabe. Int Ser Numer Math 36:203–216
[Rn62]   Rosser JB, Schœnfeld L (1962) Approximate formulas for some functions of prime numbers. Ill J Math 6:64–94
[Rn75]   Rosser JB, Schœnfeld L (1975) Sharper bounds for the Chebyshev functions $\theta(x)$ and $\psi(x)$. Math Comput 29:243–269
[RR96]   Ramaré O, Rumely R (1996) Primes in arithmetic progressions. Math Comput 65:397–425
[RS01]   Rivat J, Sargos P (2001) Nombres premiers de la forme $\lfloor n^c \rfloor$. Can J Math 53:190–209
[RV08a]   Rudelson M, Vershynin R (2008) The least singular value of a random square matrice is $O(n^{-1/2})$. C R Math Acad Sci Paris 346:893–896
[RV08b]   Rudelson M, Vershynin R (2008) The Littlewood–Offord problem and invertibility of random matrices. Adv Math 218:600–633
[Sie64]   Sierpiński W (1964) Elementary theory of numbers. Hafner, New York. 2nd edition: North-Holland, Amsterdam, 1987
[Sou09]   Soundararajan K (2009) Partial sums of the Möbius function. J Reine Angew Math 631:141–152
[Ste03]   Stevenhagen P (2003) The correction factor in Artin's primitive root conjecture. J Théor Nr Bordx 15:383–391
[Ten95]   Tenenbaum G (1995) Introduction à la Théorie Analytique et Probabiliste des Nombres. SMF
[Tit39]   Titchmarsh EC (1939) The theory of functions. Oxford University Press, London. 2nd edition: 1979
[Tit51]   Titchmarsh EC (1951) The theory of the Riemann zeta-function. Oxford University Press, London. 2nd edition: 1986, revised by D.R. Heath-Brown
[Vau93]   Vaughan RC (1993) On the eigenvalues of Redheffer's matrix I. In: Number theory with an emphasis on the Markoff spectrum, Provo, Utah, 1991. Lecture notes in pure and applied mathematics, vol 147. Dekker, New York
[Vor04]   Voronoï G (1904) Sur une fonction transcendante et ses applications à la sommation de quelques séries. Ann Sci Éc Norm Super 21:207–268, 459–533
[Was82]   Washington LC (1982) Introduction to cyclotomic fields. GTM, vol 83. Springer, New York. 2nd edition: 1997
[Wei48a]   Weil A (1948) On some exponential sums. Proc Natl Acad Sci USA 34:204–207
[Wei48b]   Weil A (1948) Sur les courbes algébriques et les variétés qui s'en déduisent. Publ Inst Math Univ Strasbourg 7:1–85

Chapter 4
Arithmetic Functions

4.1 Definition and Fundamental Examples

Definition 4.1 An *arithmetic function* is a map $f : \mathbb{N} \longrightarrow \mathbb{C}$, i.e. a sequence of complex numbers, although this viewpoint is not very useful.

Example 4.2 We list the main arithmetic functions the reader may encounter in his studies in analytic number theory. In what follows, $k \geqslant 1$ is a fixed integer.

▷ The function e_1 defined by[1] $e_1(n) = \begin{cases} 1, & \text{if } n=1 \\ 0, & \text{otherwise.} \end{cases}$

▷ The constant function **1** defined by $\mathbf{1}(n) = 1$.

▷ The kth powers $\mathrm{Id}_k(n) = n^k$. We also set $\mathrm{Id}_1 = \mathrm{Id}$.

▷ The function ω defined by $\omega(1) = 0$ and, for all $n \geqslant 2$, $\omega(n)$ counts the number of distinct prime factors of n.

▷ The function Ω defined by $\Omega(1) = 0$ and, for all $n \geqslant 2$, $\Omega(n)$ counts the number of prime factors of n, including multiplicities.

▷ The *Möbius function* μ defined by $\mu(1) = 1$ and, for all $n \geqslant 2$, by

$$\mu(n) = \begin{cases} (-1)^{\omega(n)}, & \text{if } n \text{ is squarefree} \\ 0, & \text{otherwise.} \end{cases}$$

This is one of the most important arithmetic functions of the theory.

▷ The *Liouville function* λ defined by $\lambda(n) = (-1)^{\Omega(n)}$.

▷ The *Dirichlet–Piltz divisor function* τ_k defined by[2] $\tau_1 = \mathbf{1}$, and for $k \geqslant 2$ and $n \geqslant 1$, by

$$\tau_k(n) = \sum_{d \mid n} \tau_{k-1}(d).$$

---

[1] Some authors also use the notation δ or i.

[2] Some authors also use the notation d_k.

O. Bordellès, *Arithmetic Tales*, Universitext,
DOI 10.1007/978-1-4471-4096-2_4, © Springer-Verlag London 2012

It is customary to denote τ_2 by τ and hence

$$\tau(n) = \sum_{d\mid n} 1$$

so that $\tau(n)$ counts the number of divisors of n.

▷ The *Hooley divisor function* Δ_k defined by $\Delta_1 = \mathbf{1}$, and for $k \geqslant 2$ and $n \geqslant 1$, by

$$\Delta_k(n) = \max_{u_1,\dots,u_{k-1}\in\mathbb{R}} \sum_{\substack{d_1 d_2 \cdots d_{k-1}\mid n \\ e^{u_i} < d_i \leqslant e^{u_i+1}}} 1.$$

It is customary to denote Δ_2 by Δ.

▷ The function $\tau_{(k)}$ defined to be the number of k-free divisors of n ($k \geqslant 2$). Note that $\tau_{(2)} = 2^\omega$.

▷ The function γ_k defined to be the greatest k-free divisor of n ($k \geqslant 2$). The function γ_2 is sometimes called the *core*, or *squarefree kernel*, of n.

▷ The *divisor function* σ_k defined by

$$\sigma_k(n) = \sum_{d\mid n} d^k.$$

It is customary to denote σ_1 by σ. Also note that $\sigma_0 = \tau$.

▷ The *Euler totient function* φ defined by

$$\varphi(n) = \sum_{\substack{m \leqslant n \\ (m,n)=1}} 1.$$

▷ The *Jordan totient function* φ_k defined by

$$\varphi_k(n) = \sum_{\substack{(m_1,\dots,m_k)\in\mathbb{Z}_+^k,\, m_i \leqslant n \\ (m_1,\dots,m_k,n)=1}} 1.$$

Note that $\varphi_1 = \varphi$.

▷ The *Dedekind function* Ψ defined by

$$\Psi(n) = n \prod_{p\mid n}\left(1 + \frac{1}{p}\right).$$

▷ The function μ_k defined to be the *characteristic function of the set of k-free numbers* ($k \geqslant 2$). Note that $\mu_2 = \mu^2$.

▷ The function s_k defined to be the *characteristic function of the set of k-full numbers*[3] ($k \geqslant 2$).

---

[3]There is no official notation for this function in the literature. For instance, Ivić [Ivi85] uses the notation f_k.

Table 4.1 Values of the
τ-function

n	1007	1008	1009	1010	1011	1012	1013	1014
$\tau(n)$	4	30	2	8	4	12	2	12

▷ The functions $\tau^{(e)}$ and $\sigma^{(e)}$ are respectively defined to be *the number and the sum of exponential divisors.*

Recall that if $n = p_1^{\alpha_1} \cdots p_r^{\alpha_r}$ and $d = p_1^{\beta_1} \cdots p_r^{\beta_r}$ is a divisor of n, then d is called an *exponential divisor* if $\beta_i \mid \alpha_i$ for all $i \in \{1, \ldots, r\}$.

▷ The function β defined by $\beta(1) = 1$ and, for all $n \geqslant 2$, $\beta(n)$ is *the number of square-full divisors* of n.

▷ The function $a(n)$ defined to be *the number of non-isomorphic abelian groups* of order n.

▷ Let $\mathbb{K}/\mathbb{Q}$ be an algebraic number field. The function $\nu_{\mathbb{K}}$ is defined by $\nu_{\mathbb{K}}(1) = 1$ and, for all $n \geqslant 2$, $\nu_{\mathbb{K}}(n)$ is *the number of non-zero integral ideals of $\mathcal{O}_{\mathbb{K}}$ of norm equal to n* (see Chap. 7).

▷ The *generalized von Mangoldt functions* Λ_k defined by

$$\Lambda_k(n) = \sum_{d \mid n} \mu(d) \left(\log \frac{n}{d} \right)^k.$$

Note that, by Lemma 3.41 and Möbius inversion formula (Theorem 4.13), we have $\Lambda_1 = \Lambda$.

▷ The *Dirichlet characters* $\chi(n)$ modulo q.

Clearly, sums and products of arithmetic functions are still arithmetic functions. It should also be mentioned that the behavior of such functions can be erratic, as we can see in Table 4.1 showing values taken by τ at certain consecutive integers.

Hence the idea is to study these functions *on average*, i.e. to get estimates for the sums

$$\sum_{n \leqslant x} f(n).$$

This is the aim of this chapter. In view of this purpose, we first need some definitions.

4.2 Additive and Multiplicative Functions

Definition 4.3 Let f be an arithmetic function.

▷ f is *multiplicative* if $f(1) \neq 0$ and if, for all positive integers m, n such that $(m, n) = 1$, we have

$$f(mn) = f(m)f(n). \tag{4.1}$$

▷ f is *completely multiplicative* if $f(1) \neq 0$ and if the condition $f(mn) = f(m)f(n)$ holds for *all* positive integers m, n.

▷ f is *strongly multiplicative* if f is multiplicative and if $f(p^\alpha) = f(p)$ for all prime powers p^α.

The condition $f(1) \neq 0$ is a convention to exclude the zero-function of the set of multiplicative functions. Furthermore, it is easily seen that, if f and g are multiplicative, then so are fg and f/g (with $g \neq 0$ for the quotient).

The dual notion of *additivity* is similar.

Definition 4.4 Let f be an arithmetic function.

▷ f is *additive* if, for all positive integers m, n such that $(m, n) = 1$, we have

$$f(mn) = f(m) + f(n). \tag{4.2}$$

▷ f is *completely additive* if the condition $f(mn) = f(m) + f(n)$ holds for *all* positive integers m, n.

▷ f is *strongly additive* if f is additive and if $f(p^\alpha) = f(p)$ for all prime powers p^α.

Similarly, f is said to be *sub-multiplicative* if, for all positive coprime integers m, n, we have

$$f(mn) \leqslant f(m)f(n)$$

and f is said to be *super-multiplicative* if, for all positive coprime integers m, n, we have

$$f(mn) \geqslant f(m)f(n).$$

The notions of sub-additivity and super-additivity are defined in a similar way, as well as the notions of *completely* sub-multiplicative, sub-additive, super-multiplicative and super-additive.

In order to show that a given function is multiplicative or additive, one can use the relations (4.1) or (4.2). However, one often needs to know that a function is multiplicative or additive precisely *in order to* use (4.1) or (4.2). The following result is then a useful criterion.

Lemma 4.5 *Let f be an arithmetic function.*

(i) *f is multiplicative if and only if $f(1) = 1$ and for all $n = p_1^{\alpha_1} \cdots p_r^{\alpha_r}$ where the p_i are distinct primes, we have*

$$f(n) = \prod_{k=1}^{r} f\left(p_k^{\alpha_k}\right). \tag{4.3}$$

(ii) f is additive if and only if $f(1) = 0$ and for all $n = p_1^{\alpha_1} \cdots p_r^{\alpha_r}$ where the p_i are distinct primes, we have

$$f(n) = \sum_{k=1}^{r} f\left(p_k^{\alpha_k}\right). \tag{4.4}$$

Proof The two proofs are similar, so that we only show (i). Assume first that f satisfies $f(1) = 1$ and (4.3). Let $n = p_1^{\alpha_1} \cdots p_r^{\alpha_r}$ and $m = q_1^{\beta_1} \cdots q_r^{\beta_r}$ be two positive coprime integers. Using (4.3) and the fact that $p_i \neq q_j$ we get

$$f(nm) = f\left(p_1^{\alpha_1} \cdots p_r^{\alpha_r} q_1^{\beta_1} \cdots q_r^{\beta_r}\right) = \prod_{k=1}^{r} f\left(p_k^{\alpha_k}\right) \prod_{k=1}^{r} f\left(q_k^{\beta_k}\right) = f(n) f(m)$$

and hence f satisfies (4.1). Since $f(1) = 1 \neq 0$, we infer that f is multiplicative.

Conversely, let f be a multiplicative function. Using (4.1) with $m = n = 1$ gives $f(1) = f(1) f(1)$ so that $f(1) = 1$ since $f(1) \neq 0$. Now let $n_1, \ldots, n_k$ be pairwise coprime integers. By induction using (4.1), we get

$$f(n_1 \cdots n_k) = f(n_1) \cdots f(n_k).$$

Thus, if $n = p_1^{\alpha_1} \cdots p_r^{\alpha_r}$ where the p_i are distinct primes, we infer that f satisfies (4.3) as required. □

Example 4.6 In Example 4.2, the functions ω and Ω are additive, the first one strongly, the second one completely. All the other functions, except the Hooley divisor function Δ_k and the generalized von Mangoldt function Λ_k, are multiplicative.

▷ The functions e_1, $\mathbf{1}$, Id_k, λ and the Dirichlet characters are completely multiplicative.

▷ It is easily seen that, for all positive integers m, n, we have

$$\omega(mn) = \omega(m) + \omega(n) - \omega\left((m, n)\right)$$

since in the sum $\omega(m) + \omega(n)$, the prime factors of (m, n) have been counted twice. This implies easily the additivity of ω.

▷ The functions μ_k and s_k are clearly multiplicative.

▷ Now let us have a look at the Möbius function. Using its definition, we have $\mu(1) = 1$ and, for all prime powers p^α, we also have

$$\mu\left(p^\alpha\right) = \begin{cases} -1, & \text{if } \alpha = 1 \\ 0, & \text{otherwise} \end{cases}$$

so that, if $n = p_1^{\alpha_1} \cdots p_r^{\alpha_r}$ where the p_i are distinct primes, we have

$$\mu\left(p_1^{\alpha_1}\right) \cdots \mu\left(p_r^{\alpha_r}\right) = \begin{cases} (-1)^r, & \text{if } \alpha_1 = \cdots = \alpha_r = 1 \\ 0, & \text{otherwise} \end{cases}$$

and hence $\mu(p_1^{\alpha_1})\cdots\mu(p_r^{\alpha_r}) = \mu(n)$. Therefore, the μ-function is multiplicative by Lemma 4.5.

▷ The multiplicativity of the function $v_{\mathbb{K}}$ will be established in Proposition 7.118.

▷ It has been shown in Corollary 3.7 (i) that the function τ is multiplicative.

▷ In order to prove the multiplicativity of the function $\tau_{(k)}$, let us write uniquely n in the form $n = ab$ where $(a, b) = 1$ and $\mu_k(a) = s_k(b) = 1$ with $a = p_1^{\alpha_1}\cdots p_r^{\alpha_r}$ and $b = p_{r+1}^{\alpha_{r+1}}\cdots p_s^{\alpha_s}$ where the p_i are distinct primes, and consider the product

$$\prod_{i=1}^{r}\left(1 + p_i + \cdots + p_i^{\alpha_i}\right) \times \prod_{i=r+1}^{s}\left(1 + p_i + \cdots + p_i^{k-1}\right).$$

Each term in the expansion of this product is a k-free divisor of n and, conversely, each k-free divisor of n is a term in the expansion of this product. This implies

$$\tau_{(k)}(n) = \tau(a)k^{\omega(b)}$$

and therefore $\tau_{(k)}$ is multiplicative.

▷ Let n be uniquely written in the form $n = ab$ where $(a, b) = 1$ and $\mu_2(a) = s_2(b) = 1$ with $a = p_1^{\alpha_1}\cdots p_r^{\alpha_r}$ and $b = p_{r+1}^{\alpha_{r+1}}\cdots p_s^{\alpha_s}$ where the p_i are distinct primes and $\alpha_i \in \{0, 1\}$ for $i = 1, \ldots, r$ and $\alpha_i \geqslant 2$ for $i = r+1, \ldots, s$. A divisor $d = p_1^{\beta_1}\cdots p_s^{\beta_s}$ of n is square-full if and only if $d = 1$ or $d \mid b$. This is equivalent to $\beta_i = 0$ or $2 \leqslant \beta_i \leqslant \alpha_i$ for $i = r+1, \ldots, s$. Therefore we get

$$\beta(n) = \prod_{i=r+1}^{s} \alpha_i$$

and hence β is multiplicative.

▷ Let n be uniquely written in the form $n = p_1^{\alpha_1}\cdots p_s^{\alpha_s}$. A divisor $d = p_1^{\beta_1}\cdots p_s^{\beta_s}$ is an exponential divisor of n if and only if $\beta_i \mid \alpha_i$. We readily deduce that

$$\tau^{(e)}(n) = \prod_{i=1}^{s} \tau(\alpha_i)$$

and hence $\tau^{(e)}$ is multiplicative.

▷ Let G be an abelian group of order $n = p_1^{\alpha_1}\cdots p_s^{\alpha_s}$. From (3.1), we have the decomposition

$$G \simeq H_1 \oplus \cdots \oplus H_s$$

where H_i is the p_i-Sylow subgroup of G of order $p_i^{\alpha_i}$, which implies that

$$a(n) = \prod_{i=1}^{s} a\left(p_i^{\alpha_i}\right)$$

so that the function a is multiplicative by Lemma 4.5. Furthermore, let G be an abelian group of order p^α. Then G may be factorized in the form

$$G \simeq G_1 \oplus \cdots \oplus G_r$$

for some positive integer r, where the $G_i \neq \{e_G\}$ are cyclic subgroups of G of order p^{β_i} for some $\beta_i \mid \alpha$ by Lagrange's theorem (Theorem 7.1 (i)). We infer that the number of abelian groups of order p^α is equal to the number of decompositions of the form

$$\alpha = \beta_1 + \cdots + \beta_r$$

i.e. the number of unrestricted partitions of α. Denoting by $P(\alpha)$ this number, we finally get

$$a(n) = \prod_{i=1}^{s} P(\alpha_i).$$

▷ The multiplicativity of the other functions is not obvious. We shall see another tool which will prove to be very useful in this problem.

4.3 The Dirichlet Convolution Product

Definition 4.7 Let f and g be two arithmetic functions. The Dirichlet convolution product of f and g is the arithmetic function $f \star g$ defined by

$$(f \star g)(n) = \sum_{d \mid n} f(d) g\left(\frac{n}{d}\right) = \sum_{d \mid n} f\left(\frac{n}{d}\right) g(d).$$

It should be noticed that the second equality above follows from the fact that the map $d \longrightarrow d'$ such that $dd' = n$ is one-to-one.

Example 4.8 We have $\Lambda_k = \mu \star \log^k$, $\tau = 1 \star 1$ and by induction

$$\tau_k = \underbrace{1 \star \cdots \star 1}_{k \text{ times}}.$$

Also $\sigma_k = \mathrm{Id}_k \star 1$. We shall see later that other arithmetic functions may be written as a convolution product of two simpler arithmetic functions. This will be a powerful tool to get estimates for averages of these functions.

The next result shows that this operation behaves well.

Lemma 4.9 *The Dirichlet convolution product is commutative, associative and has an identity element which is the function e_1. Furthermore, if $f(1) \neq 0$, then f is invertible.*

Proof The commutativity follows at once from Definition 4.7. Now let f, g and h be three arithmetic functions and n be a positive integer. We have

$$\big((f \star g) \star h\big)(n) = \sum_{d \mid n}(f \star g)(d)h\left(\frac{n}{d}\right) = \sum_{d \mid n}\sum_{\delta \mid d} f(\delta)g\left(\frac{d}{\delta}\right)h\left(\frac{n}{d}\right)$$

and

$$\big(f \star (g \star h)\big)(n) = \sum_{d \mid n} f(d)(g \star h)\left(\frac{n}{d}\right) = \sum_{d \mid n} f(d) \sum_{\delta \mid (n/d)} g(\delta)h\left(\frac{n}{d\delta}\right).$$

Setting $d' = d\delta$ in the last inner sum gives

$$\big(f \star (g \star h)\big)(n) = \sum_{d' \mid n}\sum_{d \mid d'} f(d)g\left(\frac{d'}{d}\right)h\left(\frac{n}{d'}\right) = \big((f \star g) \star h\big)(n)$$

establishing the associativity. We also have obviously

$$(e_1 \star f)(n) = \sum_{d \mid n} e_1(d)f\left(\frac{n}{d}\right) = f(n).$$

Finally, we prove the invertibility by constructing by induction the inverse g of an arithmetic function f satisfying $f(1) \neq 0$. The function g is the inverse of f if and only if $(f \star g)(1) = 1$ and $(f \star g)(n) = 0$ for all $n > 1$. This is equivalent to

$$\begin{cases} f(1)g(1) = 1 \\ \displaystyle\sum_{d \mid n} g(d)f\left(\frac{n}{d}\right) = 0 \quad (n \geqslant 2). \end{cases}$$

Since $f(1) \neq 0$, we have $g(1) = f(1)^{-1}$ by the first equation. Now let $n > 1$ and assume that we have proved that there exist unique values $g(1), \ldots, g(n-1)$ satisfying the above equations. Since $f(1) \neq 0$, the second equation above is equivalent to

$$g(n) = -\frac{1}{f(1)} \sum_{\substack{d \mid n \\ d \neq n}} g(d)f\left(\frac{n}{d}\right)$$

which determines $g(n)$ in a unique way by the induction hypothesis, and this definition of $g(n)$ shows that the equations above are satisfied, which completes the proof.   □

Therefore the condition $f(1) \neq 0$ is necessary and sufficient to the invertibility of f. By Lemma 4.5, we infer that every multiplicative function is invertible.

The next result is of crucial importance.

Theorem 4.10 *If f and g are multiplicative, then so is $f \star g$.*

Proof Let f and g be two multiplicative functions and let m, n be positive integers. By Corollary 3.7 (iii), each divisor d of mn can be written uniquely in the form $d = ab$ with $a \mid m$, $b \mid n$ and $(a, b) = 1$ so that

$$(f \star g)(mn) = \sum_{d \mid mn} f(d) g\left(\frac{mn}{d}\right) = \sum_{a \mid m} \sum_{b \mid n} f(ab) g\left(\frac{mn}{ab}\right)$$

and since f and g are multiplicative and $(a, b) = (m/a, n/b) = 1$, we infer that

$$(f \star g)(mn) = \sum_{a \mid m} \sum_{b \mid n} f(a) f(b) g\left(\frac{m}{a}\right) g\left(\frac{n}{b}\right) = (f \star g)(m)(f \star g)(n)$$

as required. $\square$

This result enables us:

1. to show that a given arithmetic function is multiplicative by writing it as a product of at least two multiplicative functions;
2. to show by multiplicativity several arithmetic identities.

Example 4.11

1. Since $\tau = 1 \star 1$ and 1 is multiplicative, we deduce that τ is multiplicative. The same conclusion obviously holds for τ_k.
2. We intend to prove the following identity

$$\sum_{d \mid n} \mu(d) = \begin{cases} 1, & \text{if } n = 1 \\ 0, & \text{if } n > 1 \end{cases} \tag{4.5}$$

i.e.

$$\mu \star 1 = e_1. \tag{4.6}$$

Now since μ and 1 are multiplicative, so is the function $\mu \star 1$ by Theorem 4.10 and hence (4.6) is true for $n = 1$. Besides, it is sufficient to prove (4.6) for prime powers by Lemma 4.5, which is easy to check. Indeed, for all prime powers p^α, we have

$$(\mu \star 1)(p^\alpha) = \sum_{j=0}^{\alpha} \mu(p^j) = \mu(1) + \mu(p) = 1 - 1 = 0$$

as asserted.

3. Let us have a look at Euler's totient function. By (4.5), we have

$$\sum_{d \mid (m,n)} \mu(d) = \begin{cases} 1, & \text{if } (m, n) = 1 \\ 0, & \text{otherwise} \end{cases}$$

so that using Proposition 1.11 (v) we have

$$\varphi(n) = \sum_{\substack{m \leqslant n \\ (m,n)=1}} 1 = \sum_{m \leqslant n} \sum_{d \mid (m,n)} \mu(d) = \sum_{d \mid n} \mu(d) \sum_{\substack{m \leqslant n \\ d \mid m}} 1 = \sum_{d \mid n} \mu(d) \left[\frac{n}{d} \right]$$

$$= \sum_{d \mid n} \mu(d) \frac{n}{d}$$

and therefore

$$\varphi = \mu \star \mathrm{Id} . \tag{4.7}$$

We first deduce that φ is multiplicative by Theorem 4.10. Furthermore, if p^α is a prime power, then $\varphi(p^\alpha)$ counts the number of integers $m \leqslant p^\alpha$ such that $p \nmid m$, and hence $\varphi(p^\alpha)$ is equal to p^α minus the number of multiples of p less than p^α, so that by Proposition 1.11 (v) we get

$$\varphi(p^\alpha) = p^\alpha - \left[\frac{p^\alpha}{p} \right] = p^\alpha - p^{\alpha-1} = p^\alpha \left(1 - \frac{1}{p} \right)$$

which gives using Lemma 4.5

$$\varphi(n) = n \prod_{p \mid n} \left(1 - \frac{1}{p} \right). \tag{4.8}$$

4. Let $k \geqslant 2$ be an integer and let us prove the following identity

$$\sum_{d^k \mid n} \mu(d) = \mu_k(n) \tag{4.9}$$

i.e.

$$\mu_k = f_k \star \mathbf{1} \tag{4.10}$$

where

$$f_k(n) = \begin{cases} \mu(d), & \text{if } n = d^k \\ 0, & \text{otherwise.} \end{cases}$$

Since f_k is clearly multiplicative, we deduce that $f_k \star \mathbf{1}$ is also multiplicative. We easily check (4.10) for $n = 1$. For all prime powers p^α, we have

$$(f_k \star \mathbf{1})(p^\alpha) = 1 + \sum_{j=1}^{\alpha} f_k(p^j)$$

$$= 1 + \begin{cases} 0, & \text{if } \alpha < k \\ \mu(p), & \text{if } \alpha \geqslant k \end{cases} = \begin{cases} 1, & \text{if } \alpha < k \\ 0, & \text{otherwise} \end{cases} = \mu_k(p^\alpha)$$

and therefore (4.10) holds by Lemma 4.5.

5. By Example 4.8, we have $\sigma = \mathrm{Id} \star \mathbf{1}$ and hence σ is multiplicative. Now let us prove the following identity

$$\sigma = \tau \star \varphi. \tag{4.11}$$

(4.11) is true for $n = 1$ and if p^α is a prime power, then we have

$$(\tau \star \varphi)(p^\alpha) = \sum_{j=0}^{\alpha} \varphi(p^j) \tau(p^{\alpha-j})$$

$$= \alpha + 1 + \left(1 - \frac{1}{p}\right) \sum_{j=1}^{\alpha} p^j (\alpha - j + 1)$$

$$= \frac{p^{\alpha+1} - 1}{p - 1} = \sigma(p^\alpha)$$

establishing (4.11).

6. Let $d \in \mathbb{Z} \setminus \{0, 1\}$ be squarefree and $\mathbb{K} = \mathbb{Q}(\sqrt{d})$ be a quadratic field with discriminant $d_\mathbb{K}$. By Lemma 7.107, the Dirichlet character χ defined by $\chi(n) = (d_\mathbb{K}/n)$, where (a/b) is the Kronecker symbol, is the quadratic character associated to $\mathbb{K}$. Now Proposition 7.131 implies

$$\nu_\mathbb{K} = \chi \star \mathbf{1}. \tag{4.12}$$

7. Let us prove the convolution identity

$$\tau_{(k)} = \mathbf{1} \star \mu_k. \tag{4.13}$$

(4.13) is easy for $n = 1$. Since the two functions are multiplicative, it suffices to check this identity for prime powers. Now

$$(\mathbf{1} \star \mu_k)(p^\alpha) = \sum_{j=0}^{\alpha} \mu_k(p^j) = \begin{cases} \alpha + 1, & \text{if } \alpha < k \\ k, & \text{if } \alpha \geqslant k. \end{cases} = \tau_{(k)}(p^\alpha)$$

as required.

8. For all prime powers p^α, we have

$$(f \star \mu)(p^\alpha) = f(p^\alpha) - f(p^{\alpha-1}).$$

We end this section with the next result which generalizes Theorem 3.15.

Theorem 4.12 (Euler–Fermat) *Let a, n be positive integers such that $(a, n) = 1$. Then*

$$a^{\varphi(n)} \equiv 1 \pmod{n}.$$

Proof Write $n = p_1^{\alpha_1} \cdots p_r^{\alpha_r}$ where the p_i are distinct primes. By Proposition 2.13 (vi), it suffices to show

$$a^{\varphi(n)} \equiv 1 \pmod{p_i^{\alpha_i}}$$

for $i \in \{1, \dots, r\}$. Now by (4.8) we have

$$a^{\varphi(n)} = a^{\prod_{i=1}^{r} p_i^{\alpha_i - 1}(p_i - 1)}$$

and hence the result follows if

$$b^{p^{\alpha-1}(p-1)} \equiv 1 \pmod{p^{\alpha}}$$

holds for all integers $b, \alpha \geqslant 1$ and all primes $p \nmid b$. Using Exercise 9 in Chap. 3 with $a = b$ and $k = \alpha - 1$ gives

$$b^{p^{\alpha}} \equiv b^{p^{\alpha-1}} \pmod{p^{\alpha}}$$

which is equivalent to the preceding congruence by Proposition 2.21 (v), which concludes the proof.   □

It should be mentioned that one can also prove this result by mimicking the proof of Theorem 3.15. More generally, Euler–Fermat's theorem is a direct consequence of Lagrange's theorem (Theorem 7.1 (i)) which implies that, if G is any finite group of order $|G|$ and identity element e_G, then, for all $a \in G$, we have

$$a^{|G|} = e_G.$$

Theorem 4.12 then follows at once by applying this result with $G = (\mathbb{Z}/n\mathbb{Z})^*$ since, by Theorem 2.30, we have

$$(a, n) = 1 \quad \Longleftrightarrow \quad \bar{a} \in (\mathbb{Z}/n\mathbb{Z})^*$$

and $|(\mathbb{Z}/n\mathbb{Z})^*| = \varphi(n)$.

4.4 The Möbius Inversion Formula

The identity (4.6) may be seen as an arithmetic form of the inclusion-exclusion principle, and then is the starting point of Brun's sieve (3.30). On the other hand, this identity means algebraically that the Möbius function is the inverse of the function **1** for the Dirichlet convolution product. One can exploit this information in the following way. Suppose that f and g are two arithmetic functions such that

$$g = f \star \mathbf{1}.$$

By Lemma 4.9, we infer that

$$g = f \star \mathbf{1} \quad \Longleftrightarrow \quad g \star \mu = f \star (\mathbf{1} \star \mu) = f.$$

This relation is called the *Möbius inversion formula*, and is a key part in the estimates of average orders of certain multiplicative functions. Let us summarize this result in the following theorem.

Theorem 4.13 (Möbius inversion formula) *Let f and g be two arithmetic functions. Then we have*

$$g = f \star \mathbf{1} \quad \Longleftrightarrow \quad f = g \star \mu$$

i.e. for all positive integers n

$$g(n) = \sum_{d \mid n} f(d) \quad \Longleftrightarrow \quad f(n) = \sum_{d \mid n} g(d) \mu\left(\frac{n}{d}\right).$$

Our first application is a second proof of Proposition 3.30 stating that, for each prime number p, there exists a primitive root modulo p.

Proof of Proposition 3.30 One can suppose that $p \geqslant 3$. If $d \mid (p-1)$, let $N(d)$ be the number of elements of a reduced residue system modulo p having an order equal to d. For each divisor δ of d, a solution of the congruence $x^\delta \equiv 1 \pmod{p}$ is also a solution of $x^d \equiv 1 \pmod{p}$. By Example 3.28, this congruence has exactly d solutions, so that

$$\sum_{\delta \mid d} N(\delta) = d$$

or equivalently

$$N \star \mathbf{1} = \mathrm{Id}.$$

By the Möbius inversion formula and (4.7), we get $N = \mu \star \mathrm{Id} = \varphi$ and hence $N(p-1) = \varphi(p-1)$, so that there are $\varphi(p-1) \geqslant 1$ primitive roots modulo p. $\qquad \square$

Our second application concerns the number of monic irreducible polynomials of degree n in $\mathbb{F}_p[X]$.

Proposition 4.14 *Let p be a prime number, n be a positive integer and $N_{n,p}$ be the number of monic irreducible polynomials of degree n in $\mathbb{F}_p[X]$. Then we have*

$$N_{n,p} = \frac{1}{n} \sum_{d \mid n} \mu(d) p^{n/d}.$$

Proof Let $P \in \mathbb{F}_p[X]$ be a monic polynomial. Since P may be factorized uniquely as a product of monic irreducible polynomials in $\mathbb{F}_p[X]$, we have

$$\sum_{\substack{P \in \mathbb{F}_p[X] \\ P \text{ monic}}} T^{\deg P} = \prod_{\substack{Q \in \mathbb{F}_p[X] \\ Q \text{ monic irreducible}}} \left(1 + T^{\deg Q} + T^{2 \deg Q} + \cdots\right)$$

$$= \prod_{\substack{Q \in \mathbb{F}_p[X] \\ Q \text{ monic irreducible}}} \left(1 - T^{\deg Q}\right)^{-1}$$

$$= \prod_{d=1}^{\infty} \left(1 - T^d\right)^{-N_{d,p}}.$$

Since the number of monic polynomials of degree n in $\mathbb{F}_p[X]$ is equal to p^n, we deduce that

$$\sum_{n=1}^{\infty} p^n T^n = \prod_{d=1}^{\infty} \left(1 - T^d\right)^{-N_{d,p}}$$

and taking logarithms of both sides gives

$$\log(1 - pT) = \sum_{d=1}^{\infty} N_{d,p} \log\left(1 - T^d\right) = \sum_{d=1}^{\infty} N_{d,p} \sum_{\delta=1}^{\infty} \delta^{-1} T^{d\delta}$$

and hence

$$\sum_{n=1}^{\infty} n^{-1} p^n T^n = \sum_{d=1}^{\infty} N_{d,p} \sum_{\delta=1}^{\infty} \delta^{-1} T^{d\delta} = \sum_{n=1}^{\infty} n^{-1} T^n \sum_{d|n} d N_{d,p}$$

and comparing the coefficients of T^n implies that

$$p^n = \sum_{d|n} d N_{d,p}.$$

Now Theorem 4.13 gives the asserted result.   □

The Möbius inversion formula may also be used to count the number of *primitive* Dirichlet characters modulo q. We shall actually prove the following slightly more general result.

Proposition 4.15 *Let $q \geqslant 2$ be an integer and $k \in \mathbb{N}$ such that $(k, q) = 1$. Then*

$$\sideset{}{^*}\sum_{\chi \pmod q} \chi(k) = \sum_{d|(q,k-1)} \varphi(d) \mu(q/d)$$

where the star indicates a summation over primitive Dirichlet characters. In particular, the number $\varphi^\star(q)$ of primitive Dirichlet characters modulo q is given by

$$\varphi^\star(q) = (\varphi \star \mu)(q).$$

Hence $\varphi^\star$ is multiplicative and there is no primitive Dirichlet character modulo $2m$ where m is an odd positive integer.

Proof Let $k \in \mathbb{N}$ such that $(k, q) = 1$. Since each Dirichlet character modulo q is induced by a unique primitive Dirichlet character modulo a divisor of q, we have

$$\sum_{d|q} \sum_{\chi \pmod{d}}^{\star} \chi(k) = \sum_{\chi \pmod{q}} \chi(k) = \mathbf{1}_{q,1}(k)\varphi(q) = \begin{cases} \varphi(q), & \text{if } k \equiv 1 \pmod{q} \\ 0, & \text{otherwise} \end{cases}$$

where we used Proposition 3.68 in the second equality. Now by Theorem 4.13 we get

$$\sum_{\chi \pmod{q}}^{\star} \chi(k) = \sum_{d|q} \mathbf{1}_{d,1}(k)\varphi(d)\mu(q/d) = \sum_{d|(q,k-1)} \varphi(d)\mu(q/d)$$

as required. Taking $k = 1$ gives the formula for $\varphi^\star(q)$ and hence $\varphi^\star$ is multiplicative by Theorem 4.10. Furthermore, $\varphi^\star(2) = \varphi(2) - \varphi(1) = 1 - 1 = 0$ and therefore by multiplicativity, for all odd positive integers m, we get $\varphi^\star(2m) = \varphi^\star(2)\varphi^\star(m) = 0$ as asserted. □

Example 4.16

1. By Lemma 3.41 (i), one has $\log = \Lambda \star \mathbf{1}$ and hence $\Lambda = \mu \star \log$ by Theorem 4.13 as asserted in Example 4.2.
2. Let us prove that

$$\sum_{d|n} \varphi(d) = n$$

and

$$\sum_{d|n} \frac{\mu(d)}{d} = \prod_{p|n} \left(1 - \frac{1}{p}\right).$$

Indeed, from (4.7) and Theorem 4.13, we get $\text{Id} = \varphi \star \mathbf{1}$ which is the first identity. For the second one, we have using (4.7)

$$\sum_{d|n} \frac{\mu(d)}{d} = \frac{1}{n} \sum_{d|n} \mu(d)\frac{n}{d} = \frac{(\mu \star \text{Id})(n)}{n} = \frac{\varphi(n)}{n}$$

which implies the asserted result.

4.5 Summation Methods

In this section, we intend to provide some tools to estimate sums of the form

$$S(x) = \sum_{n \leqslant x} (f \star g)(n) = \sum_{n \leqslant x} \sum_{d \mid n} f(d) g(n/d)$$

with $x \geqslant 1$. The first idea which comes to mind is to interchange the order of summation, which amounts to rearranging the terms of the sum favoring the factorization by the terms $f(1)$, $f(2)$, ... We thus get

$$S(x) = \sum_{d \leqslant x} f(d) \sum_{\substack{n \leqslant x \\ d \mid n}} g(n/d).$$

Making the change of variable $n = kd$ in the inner sum and noticing that $n \leqslant x$ and $d \mid n$ is equivalent to $k \leqslant x/d$, we finally obtain

$$S(x) = \sum_{d \leqslant x} f(d) \sum_{k \leqslant x/d} g(k).$$

We may thus state the following result.

Proposition 4.17 *Let $x \geqslant 1$ be a real number and f and g be two arithmetic functions. Then*

$$\sum_{n \leqslant x} (f \star g)(n) = \sum_{d \leqslant x} f(d) \sum_{k \leqslant x/d} g(k).$$

Example 4.18

1. Since $\tau = \mathbf{1} \star \mathbf{1}$, we get using Proposition 4.17 and the easy equality $[x] = x + O(1)$

$$\sum_{n \leqslant x} \tau(n) = \sum_{d \leqslant x} \sum_{k \leqslant x/d} 1 = \sum_{d \leqslant x} \left[\frac{x}{d} \right] = x \sum_{d \leqslant x} \frac{1}{d} + O(x) = x \log x + O(x).$$

   We shall see later how to improve on this result.
2. Since $\sigma = \mathbf{1} \star \mathrm{Id}$, we get using Proposition 4.17

$$\sum_{n \leqslant x} \sigma(n) = \sum_{d \leqslant x} \sum_{k \leqslant x/d} k = \frac{1}{2} \sum_{d \leqslant x} \left[\frac{x}{d} \right] \left(\left[\frac{x}{d} \right] + 1 \right)$$

and using $[x] = x + O(1)$ we obtain

$$\sum_{n \leqslant x} \sigma(n) = \frac{x^2}{2} \sum_{d \leqslant x} \frac{1}{d^2} + O\left(x \sum_{d \leqslant x} \frac{1}{d} \right)$$

and since

$$\sum_{d \leqslant x} \frac{1}{d^2} = \sum_{d=1}^{\infty} \frac{1}{d^2} - \sum_{d > x} \frac{1}{d^2} = \zeta(2) + O\left(\frac{1}{x}\right)$$

where we used Exercise 19 in Chap. 3, we finally get

$$\sum_{n \leqslant x} \sigma(n) = \frac{x^2 \zeta(2)}{2} + O(x \log x). \tag{4.14}$$

3. Since $\mu \star 1 = e_1$, we get using Proposition 4.17

$$1 = \sum_{n \leqslant x} (\mu \star 1)(n) = \sum_{d \leqslant x} \mu(d) \sum_{k \leqslant x/d} 1 = \sum_{d \leqslant x} \mu(d) \left[\frac{x}{d}\right]$$

so that

$$\sum_{n \leqslant x} \mu(n) \left[\frac{x}{n}\right] = 1 \tag{4.15}$$

which is a very important identity. A variant of this identity may be obtained as follows.

$$\sum_{n \leqslant x+y} \mu(n) \left(\left[\frac{x+y}{n}\right] - \left[\frac{x}{n}\right]\right) = \sum_{n \leqslant x+y} \mu(n) \left[\frac{x+y}{n}\right] - \sum_{n \leqslant x+y} \mu(n) \left[\frac{x}{n}\right]$$

$$= 1 - \sum_{n \leqslant x} \mu(n) \left[\frac{x}{n}\right] - \sum_{x < n \leqslant x+y} \mu(n) \left[\frac{x}{n}\right]$$

$$= 1 - 1 = 0$$

since in the last sum we have $[x/n] = 0$. This implies that

$$\sum_{x < n \leqslant x+y} \mu(n) = -\sum_{n \leqslant x} \mu(n) \left(\left[\frac{x+y}{n}\right] - \left[\frac{x}{n}\right]\right).$$

Set $N = [x]$. Using (4.15) we get

$$N \sum_{n \leqslant N} \frac{\mu(n)}{n} = \sum_{n \leqslant N} \mu(n) \left[\frac{N}{n}\right] + \sum_{n \leqslant N} \mu(n) \left\{\frac{N}{n}\right\} = 1 + \sum_{n \leqslant N-1} \mu(n) \left\{\frac{N}{n}\right\}$$

so that

$$\left|\sum_{n \leqslant x} \frac{\mu(n)}{n}\right| \leqslant \frac{1}{N} \left(1 + \sum_{n \leqslant N-1} \mu(n) \left\{\frac{N}{n}\right\}\right) \leqslant \frac{1 + N - 1}{N} = 1.$$

4. Using Proposition 4.35, we shall see that the Dirichlet series of the Möbius function is $\zeta(s)^{-1}$ which is absolutely convergent in the half-plane $\sigma > 1$. This implies in particular that, for all integers $k \geqslant 2$, we have

$$\sum_{n=1}^{\infty} \frac{\mu(n)}{n^k} = \frac{1}{\zeta(k)}. \tag{4.16}$$

Using Proposition 4.17, one may provide another proof of (4.16). It is first easily seen that the series of the left-hand side converges by Rule 1.20, since $|\mu(n)|/n^k \leqslant 1/n^k$. Now applying Proposition 4.17 with $f(n) = \mu(n)/n^k$ and $g(n) = 1/n^k$ we get for all positive integers N

$$1 = \sum_{n \leqslant N} (f \star g)(n) = \sum_{d \leqslant N} \frac{\mu(d)}{d^k} \sum_{m \leqslant N/d} \frac{1}{m^k}$$

$$= \sum_{d \leqslant N} \frac{\mu(d)}{d^k} \sum_{m=1}^{\infty} \frac{1}{m^k} - \sum_{d \leqslant N} \frac{\mu(d)}{d^k} \sum_{m > N/d} \frac{1}{m^k}$$

$$= \zeta(k) \sum_{d \leqslant N} \frac{\mu(d)}{d^k} - R(N)$$

with

$$\left| R(N) \right| \leqslant \sum_{d \leqslant N} \frac{1}{d^k} \sum_{m > N/d} \frac{1}{m^k}.$$

Now using Exercise 19 in Chap. 3, we obtain

$$\sum_{m > N/d} \frac{1}{m^k} \leqslant \zeta(k) \left(\frac{d}{N} \right)^{k-1}$$

and thus

$$\left| R(N) \right| \leqslant \frac{\zeta(k)}{N^{k-1}} \sum_{d \leqslant N} \frac{1}{d} \leqslant \frac{\zeta(k) \log(eN)}{N^{k-1}}$$

and hence

$$\sum_{d \leqslant N} \frac{\mu(d)}{d^k} = \frac{1}{\zeta(k)} + O\left(\frac{\log N}{N^{k-1}} \right)$$

giving (4.16) by letting $N \longrightarrow \infty$.
5. Using (4.9), Proposition 4.17, the estimate $[x] = x + O(1)$, (4.16) and Exercise 19 in Chap. 3, we get

$$\sum_{n \leqslant x} \mu_k(n) = \sum_{n \leqslant x} \sum_{d^k \mid n} \mu(d) = \sum_{d \leqslant x^{1/k}} \mu(d) \sum_{k \leqslant x/d^k} 1$$

$$= \sum_{d \leqslant x^{1/k}} \mu(d) \left[\frac{x}{d^k} \right] = x \sum_{d \leqslant x^{1/k}} \frac{\mu(d)}{d^k} + O\left(x^{1/k}\right)$$

$$= x \sum_{d=1}^{\infty} \frac{\mu(d)}{d^k} - x \sum_{d > x^{1/k}} \frac{\mu(d)}{d^k} + O\left(x^{1/k}\right)$$

$$= \frac{x}{\zeta(k)} + O\left(x \sum_{d > x^{1/k}} \frac{1}{d^k}\right) + O\left(x^{1/k}\right)$$

$$= \frac{x}{\zeta(k)} + O\left(x^{1/k}\right).$$

One may improve on Proposition 4.17 by inserting a parameter which may be optimized. This is indeed a very fruitful idea leading to the next result, called the *Dirichlet hyperbola principle*.

Proposition 4.19 (Dirichlet hyperbola principle) *Let $1 \leqslant T \leqslant x$ be real numbers and f and g be two arithmetic functions. Then*

$$\sum_{n \leqslant x} (f \star g)(n) = \sum_{n \leqslant T} f(n) \sum_{k \leqslant x/n} g(k) + \sum_{k \leqslant x/T} g(k) \sum_{n \leqslant x/k} f(n)$$

$$- \sum_{n \leqslant T} f(n) \sum_{k \leqslant x/T} g(k).$$

Proof Splitting the sum of the right-hand side of Proposition 4.17 gives

$$\sum_{n \leqslant x} (f \star g)(n) = \sum_{d \leqslant T} f(d) \sum_{k \leqslant x/d} g(k) + \sum_{T < d \leqslant x} f(d) \sum_{k \leqslant x/d} g(k)$$

and

$$\sum_{T < d \leqslant x} f(d) \sum_{k \leqslant x/d} g(k) = \sum_{k \leqslant x/T} g(k) \sum_{T < d \leqslant x/k} f(d)$$

$$= \sum_{k \leqslant x/T} g(k) \left(\sum_{d \leqslant x/k} f(d) - \sum_{d \leqslant T} f(d) \right)$$

as required. □

Historically, it was Dirichlet who discovered this principle when he succeeded in improving the error-term in the sum

$$\sum_{n \leqslant x} \tau(n).$$

Fig. 4.1 Dirichlet hyperbola
principle

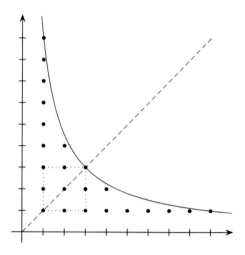

The name comes from the following observation. Since, by above,

$$\sum_{n \leqslant x} \tau(n) = \sum_{d \leqslant x} \left[\frac{x}{d} \right]$$

we are led to count the number of integer points (m, n) with $1 \leqslant m \leqslant x$ lying under
the hyperbola $mn = x$.

Dirichlet used the symmetry of the hyperbola to deduce that the number of integer points is equal to that of the interior of the square $[1, \sqrt{x}]^2$ plus twice the
number of integer points (m, n) such that $\sqrt{x} < m \leqslant x$, see Fig. 4.1. This gives

$$\sum_{n \leqslant x} \tau(n) = [\sqrt{x}]^2 + 2 \sum_{\sqrt{x} < m \leqslant x} \left[\frac{x}{m} \right]$$

$$= [\sqrt{x}]^2 + 2 \sum_{m \leqslant x} \left[\frac{x}{m} \right] - 2 \sum_{m \leqslant \sqrt{x}} \left[\frac{x}{m} \right]$$

$$= [\sqrt{x}]^2 + 2 \sum_{n \leqslant x} \tau(n) - 2 \sum_{m \leqslant \sqrt{x}} \left[\frac{x}{m} \right]$$

and therefore we get

$$\sum_{n \leqslant x} \tau(n) = 2 \sum_{m \leqslant \sqrt{x}} \left[\frac{x}{m} \right] - [\sqrt{x}]^2. \tag{4.17}$$

It is easily seen that using Proposition 4.19 with $f = g = 1$ gives (4.17), so that
this result generalizes the geometric method of Dirichlet. We are now in a position
to show the following estimate which is the first result in the so-called *Dirichlet
divisor problem*.

Corollary 4.20 (Dirichlet) *For x sufficiently large, we have*

$$\sum_{n \leqslant x} \tau(n) = x(\log x + 2\gamma - 1) + O(\sqrt{x}).$$

Proof Using (4.17), the estimate $[x] = x + O(1)$ and Theorem 1.22, we get

$$\sum_{n \leqslant x} \tau(n) = 2x \sum_{m \leqslant \sqrt{x}} \frac{1}{m} + O(\sqrt{x}) - \left(\sqrt{x} + O(1)\right)^2$$

$$= 2x\left(\log \sqrt{x} + \gamma + O\left(x^{-1/2}\right)\right) - x + O(\sqrt{x})$$

$$= x \log x + x(2\gamma - 1) + O(\sqrt{x})$$

as asserted.   □

For a presentation of the Dirichlet divisor problem and related problems, see the introduction in Chap. 6. For more information about this problem, see [Bor09].

The next result provides another application of the Dirichlet hyperbola principle.

Corollary 4.21 *Let χ be a non-principal Dirichlet character modulo q. For each real number $x \geqslant q^{1/2}$ sufficiently large, we have*

$$\sum_{n \leqslant x} (\chi \star \mathrm{Id})(n) = \frac{x^2 L(2, \chi)}{2} - x \sum_{n \leqslant x^{2/3} q^{1/6}} \frac{\chi(n)}{n} \psi\left(\frac{x}{n}\right) + O\left(x^{2/3} q^{1/6} \log q\right).$$

Proof Let T satisfy $1 \leqslant T \leqslant x$. Using Proposition 4.19, we get

$$\sum_{n \leqslant x} (\chi \star \mathrm{Id})(n) = \sum_{n \leqslant T} \chi(n) \sum_{k \leqslant x/n} k + \sum_{n \leqslant x/T} n \sum_{k \leqslant x/n} \chi(k) - \sum_{n \leqslant T} \chi(n) \sum_{n \leqslant x/T} n$$

$$= \frac{1}{2} \sum_{n \leqslant T} \chi(n) \left[\frac{x}{n}\right]\left(\left[\frac{x}{n}\right] + 1\right) + O\left\{\sum_{n \leqslant x/T} n \left|\sum_{k \leqslant x/n} \chi(k)\right|\right\}$$

$$+ O\left\{\left(\sum_{n \leqslant x/T} n\right)\left|\sum_{n \leqslant T} \chi(n)\right|\right\}.$$

Now the use of the Pólya–Vinogradov inequality (Theorem 3.106) and the estimate

$$\left[\frac{x}{n}\right]\left(\left[\frac{x}{n}\right] + 1\right) = \frac{x^2}{n^2} - \frac{2x}{n} \psi\left(\frac{x}{n}\right) + O(1)$$

gives

$$\sum_{n \leqslant x} (\chi \star \mathrm{Id})(n) = \frac{x^2}{2} \sum_{n \leqslant T} \frac{\chi(n)}{n^2} - x \sum_{n \leqslant T} \frac{\chi(n)}{n} \psi\left(\frac{x}{n}\right) + O\left(x^2 T^{-2} q^{1/2} \log q + T\right).$$

Now we have

$$\sum_{n \leqslant T} \frac{\chi(n)}{n^2} = L(2, \chi) - \sum_{n > T} \frac{\chi(n)}{n^2}$$

and, as in Corollary 3.71, we get by Abel's summation and the Pólya–Vinogradov inequality the estimate

$$\left| \sum_{n > T} \frac{\chi(n)}{n^2} \right| \leqslant \frac{2\sqrt{3q} \log q}{T^2}$$

so that the choice of $T = x^{2/3} q^{1/6}$ gives the asserted result. $\square$

When χ is a quadratic character modulo q, and thus is the Dirichlet character attached to the quadratic field $\mathbb{K} = \mathbb{Q}(\sqrt{d})$ where $d = \chi(-1)q$ by Lemma 7.107, it can be proved that

$$(\chi \star \mathrm{Id})(n) = \sum_{i=1}^{n} v_{\mathbb{K}}\big((i, n)\big)$$

(see Definition 7.117 and Exercise 10) so that Corollary 4.21 is used in the problem of the composition of the gcd and the multiplicative function $v_{\mathbb{K}}$ (see [Bor10]).

4.6   Tools for Average Orders

4.6.1   Introduction

Let f be an arithmetic function. By an *average order* of f, we mean finding an asymptotic formula of the form

$$\sum_{n \leqslant x} f(n) = g(x) + O\big(R(x)\big)$$

where the main term $g(x)$ lies in the set of usual functions (polynomials, logarithms, etc) and the error-term $R(x)$ satisfies $R(x) = o(g(x))$ for x sufficiently large. Under these conditions, we shall say that the function $x \longmapsto x^{-1} g(x)$ is an *average order* of f on $[1, x]$ for x sufficiently large. For instance, by Corollary 4.20, the function $x \longmapsto \log x + 2\gamma - 1$ is an average order of the divisor function τ on $[1, x]$ for x sufficiently large.

One of the most important problems in number theory is to find the smallest error-term admissible in the above asymptotic estimate. This problem is sometimes open, as for instance in the PNT, where the best error-term to date (3.37) is presumably far from the conjectured remainder term given by the Riemann hypothesis.

In this section, we consider a certain class of arithmetic functions satisfying the following hypotheses. We shall say that $f \in \mathcal{M}$ if f is a non-negative multiplicative function such that for all $x \geqslant 1$, we have

$$\frac{1}{x} \sum_{p \leqslant x} f(p) \log p \leqslant a \tag{4.18}$$

$$\sum_{p \leqslant x} \sum_{\alpha=2}^{\infty} \frac{f(p^\alpha) \log p^\alpha}{p^\alpha} \leqslant b \tag{4.19}$$

where $a, b > 0$ are independent of x. The purpose of this section is to provide a proof of the following important result.

Theorem 4.22 *Let f be a non-negative multiplicative function satisfying* (4.18) *and* (4.19). *Then, for all $x \geqslant 1$, we have*

$$\sum_{n \leqslant x} f(n) \leqslant e^b (a + b + 1) \frac{x}{\log ex} \exp\left(\sum_{p \leqslant x} \frac{f(p)}{p}\right).$$

In other words, only knowing the values of $f \in \mathcal{M}$ at prime numbers is sufficient to determine an upper bound which in many cases proves to be of the right order of magnitude.

It should be mentioned that $\mathcal{M}$ is not empty. Indeed, the function $\mathbf{1}$ is positive, completely multiplicative and using Lemma 3.42 or Corollary 3.98 and (2.7) with $k = 2$ and $x = 1/p$ we get

$$\frac{1}{x} \sum_{p \leqslant x} \mathbf{1}(p) \log p = \frac{\theta(x)}{x} \leqslant \log 4$$

and

$$\sum_{p \leqslant x} \sum_{\alpha=2}^{\infty} \frac{\mathbf{1}(p^\alpha) \log p^\alpha}{p^\alpha} = \sum_{p \leqslant x} \frac{(2p-1)\log p}{p(p-1)^2} \leqslant 6 \sum_{p \leqslant x} \frac{\log p}{p^2} < 3$$

where we used the inequality

$$\sum_{p \leqslant x} \frac{\log p}{p^2} < \frac{1}{2} \tag{4.20}$$

from Exercise 4.(b) in Chap. 3, so that one may take $(a, b) = (\log 4, 3)$. The reader may also check that the functions 2^ω and τ lie in $\mathcal{M}$ since one can take $(a, b) = (\log 16, 6)$ for the first one and $(a, b) = (\log 16, 14)$ for the second one. The next result provides a useful sufficient condition for a function f to lie in $\mathcal{M}$.

Lemma 4.23 *Let f be a multiplicative function satisfying the* Wirsing *conditions, i.e.*

$$0 \leqslant f(p^\alpha) \leqslant \lambda_1 \lambda_2^{\alpha-1} \tag{4.21}$$

for all prime powers p^α and some real numbers $\lambda_1 > 0$ and $0 \leqslant \lambda_2 < 2$. Then $f \in \mathcal{M}$ with

$$(a, b) = \left(\lambda_1 \log 4, \; \frac{\lambda_1 \lambda_2 (4 - \lambda_2)}{(2 - \lambda_2)^2} \right).$$

Proof We will use the inequality

$$\frac{1}{p - \lambda_2} \leqslant \frac{2}{p(2 - \lambda_2)} \tag{4.22}$$

which readily comes from the fact that $\lambda_2 < 2 \leqslant p$. Note first that f is non-negative by multiplicativity. Furthermore, we have

$$\frac{1}{x} \sum_{p \leqslant x} f(p) \log p \leqslant \frac{\lambda_1 \theta(x)}{x} \leqslant \lambda_1 \log 4$$

by Lemma 3.42. Next we have, using (2.7), (4.22) and (4.20),

$$\sum_{p \leqslant x} \sum_{\alpha=2}^{\infty} \frac{f(p^\alpha) \log p^\alpha}{p^\alpha} \leqslant \lambda_1 \sum_{p \leqslant x} \log p \sum_{\alpha=2}^{\infty} \frac{\alpha}{p} \left(\frac{\lambda_2}{p} \right)^{\alpha - 1} = \lambda_1 \lambda_2 \sum_{p \leqslant x} \frac{(2p - \lambda_2) \log p}{p(p - \lambda_2)^2}$$

$$\leqslant \frac{\lambda_1 \lambda_2 (4 - \lambda_2)}{2 - \lambda_2} \sum_{p \leqslant x} \frac{\log p}{p(p - \lambda_2)} \leqslant \frac{\lambda_1 \lambda_2 (4 - \lambda_2)}{(2 - \lambda_2)^2}$$

as asserted. □

Lemma 4.23 enables us to increase the number of arithmetic functions lying in $\mathcal{M}$.

Lemma 4.24 *The following arithmetic functions lie in $\mathcal{M}$:*

$$e_1, \; \mathbf{1}, \; \mu_k, \; s_k, \; \beta, \; a, \; k^\omega, \; \tau_{(k)}, \; \tau^{(e)} \quad and \quad \tau_k.$$

Proof All these functions are non-negative multiplicative by Example 4.6. Furthermore, if f is such that $0 \leqslant f(n) \leqslant 1$, then the Wirsing conditions (4.21) are obviously satisfied, and this is the case for the first four arithmetic functions. We have

$$\beta(p^\alpha) = \begin{cases} 1, & \text{if } \alpha \in \{0, 1\} \\ \alpha, & \text{if } \alpha \geqslant 2 \end{cases}$$

and hence $\beta(p^\alpha) \leqslant \max(1, \alpha)$, so that the Wirsing conditions are readily satisfied. By [Krä70], we have $P(\alpha) \leqslant 5^{\alpha/4}$ so that

$$a(p^\alpha) \leqslant 5^{1/4} (5^{1/4})^{\alpha - 1}$$

and therefore the function a satisfies (4.21) with $\lambda_1 = \lambda_2 = 5^{1/4}$. Since

$$k^{\omega(p^\alpha)} = k$$

k^ω satisfies (4.21) with $\lambda_1 = k$ and $\lambda_2 = 1$. Similarly, since $\tau^{(e)}(p^\alpha) = \tau(\alpha)$ and

$$\tau_{(k)}(p^\alpha) = \begin{cases} k, & \text{if } \alpha \geqslant k \\ \alpha + 1, & \text{if } \alpha < k \end{cases}$$

we easily see that these two functions satisfy (4.21). Finally, $\tau_k(p^\alpha)$ is the number of solutions of the equation $x_1 \cdots x_k = p^\alpha$, and setting $x_i = p^{\beta_i}$ for some $\beta_i \in \mathbb{Z}_{\geqslant 0}$, we see that we have to count the number of solutions in $(\mathbb{Z}_{\geqslant 0})^k$ of the Diophantine equation

$$\sum_{i=1}^{k} \beta_i = \alpha$$

whose number of solutions is equal to $\mathcal{D}_{(1,\ldots,1)}(\alpha) = \binom{k+\alpha-1}{\alpha}$ by Theorem 2.33, where $\mathcal{D}_{(1,\ldots,1)}$ is the denumerant defined in Chap. 2. Therefore[4]

$$\tau_k(p^\alpha) = \binom{k+\alpha-1}{\alpha} \tag{4.23}$$

and hence

$$\frac{1}{x} \sum_{p \leqslant x} \tau_k(p) \log p = \frac{k\theta(x)}{x} \leqslant k \log 4.$$

Using (2.7) we get

$$\sum_{\alpha=2}^{\infty} \frac{\alpha \tau_k(p^\alpha)}{p^\alpha} = \frac{k}{p} \left\{ \left(1 - \frac{1}{p} \right)^{-k-1} - 1 \right\}$$

and the inequality

$$(1-x)^{-k} \leqslant 1 + kx + k(k+1)2^{k+1}x^2$$

valid for all $0 \leqslant x \leqslant \frac{1}{2}$, implies that

$$\sum_{\alpha=2}^{\infty} \frac{\alpha \tau_k(p^\alpha)}{p^\alpha} \leqslant \frac{k(k+1)2^{k+2}}{p^2}.$$

---

[4]It is noteworthy that $\tau_k(p^\alpha) = \mathcal{D}_{(1,\ldots,1)}(\alpha)$ where the vector $(1,\ldots,1)$ has k components. See also Proposition 7.118.

so that

$$\sum_{p \leqslant x} \log p \sum_{\alpha=2}^{\infty} \frac{\alpha \tau_k(p^\alpha)}{p^\alpha} \leqslant k(k+1)2^{k+2}$$

and therefore $\tau_k \in \mathcal{M}$.   □

4.6.2 Auxiliary Lemmas

Lemma 4.25 *Let $n \in \mathbb{N}$. For all $x \geqslant n$, we have*

$$\log(ex) \leqslant \log n + \frac{x}{n}.$$

Proof Indeed, the function $x \longmapsto \log n + x/n - \log(ex)$ is non-decreasing on $[n, +\infty[$ and vanishes at $x = n$.   □

Lemma 4.26 *Let $x \geqslant 1$ be a real number and f be a positive multiplicative function satisfying (4.19). Then*

$$\sum_{n \leqslant x} \frac{f(n)}{n} \leqslant e^b \exp\left(\sum_{p \leqslant x} \frac{f(p)}{p}\right).$$

Proof Expanding the product

$$\prod_{p \leqslant x} \left(1 + \sum_{\alpha=1}^{\infty} \frac{f(p^\alpha)}{p^\alpha}\right)$$

and using Theorem 3.3 and the multiplicativity, we infer that this product is equal to

$$\sum_{P^+(n) \leqslant x} \frac{f(n)}{n}.$$

Since each positive integer $n \leqslant x$ satisfies the condition $P^+(n) \leqslant x$ and since $f \geqslant 0$, we get

$$\sum_{n \leqslant x} \frac{f(n)}{n} \leqslant \prod_{p \leqslant x} \left(1 + \sum_{\alpha=1}^{\infty} \frac{f(p^\alpha)}{p^\alpha}\right) \leqslant \exp\left(\sum_{p \leqslant x} \sum_{\alpha=1}^{\infty} \frac{f(p^\alpha)}{p^\alpha}\right)$$

$$= \exp\left(\sum_{p \leqslant x} \frac{f(p)}{p} + \sum_{p \leqslant x} \sum_{\alpha=2}^{\infty} \frac{f(p^\alpha)}{p^\alpha}\right) \leqslant e^b \exp\left(\sum_{p \leqslant x} \frac{f(p)}{p}\right)$$

where we used (4.19).   □

4.6.3 The Proof of Theorem 4.22

We are now in a position to prove Theorem 4.22. But instead of dealing with the sum of the theorem, we shall estimate the sum

$$\sum_{n \leqslant x} f(n) g(n)$$

with a suitable choice of the weight $g(n)$ which makes the treatment of the new sum easier. The function g is often chosen among the functions $\log n$ or $(N/n)^{\beta}$ for some real $\beta > 0$. This last choice is called *Rankin's trick* and proves to be a very fruitful idea.

We will choose here the function $g(n) = \log n$ on account of its complete additivity. Hence, if $n = p_1^{\alpha_1} \cdots p_r^{\alpha_r}$, then

$$\log n = \sum_{i=1}^{r} \log p_i^{\alpha_i}$$

and therefore

$$\sum_{n \leqslant x} f(n) \log n = \sum_{p^{\alpha} \leqslant x} \sum_{\substack{k \leqslant x/p^{\alpha} \\ p \nmid k}} f(kp^{\alpha}) \log p^{\alpha} = \sum_{p^{\alpha} \leqslant x} \sum_{\substack{k \leqslant x/p^{\alpha} \\ p \nmid k}} f(k) f(p^{\alpha}) \log p^{\alpha}.$$

Interchanging the summations and neglecting the condition $p \nmid k$ gives

$$\sum_{n \leqslant x} f(n) \log n \leqslant \sum_{k \leqslant x} f(k) \sum_{p^{\alpha} \leqslant x/k} f(p^{\alpha}) \log p^{\alpha}. \tag{4.24}$$

We split the inner sum into two subsums according to either $\alpha = 1$ or $\alpha \geqslant 2$ and, in the second subsum, we use the fact that $p^{\alpha} \leqslant x/k$ is equivalent to $1 \leqslant x/(kp^{\alpha})$ which gives

$$\sum_{p^{\alpha} \leqslant x/k} f(p^{\alpha}) \log p^{\alpha} = \sum_{p \leqslant x/k} f(p) \log p + \sum_{\substack{p^{\alpha} \leqslant x/k \\ \alpha \geqslant 2}} f(p^{\alpha}) \log p^{\alpha}$$

$$\leqslant \frac{ax}{k} + \frac{x}{k} \sum_{\substack{p^{\alpha} \leqslant x/k \\ \alpha \geqslant 2}} \frac{f(p^{\alpha}) \log p^{\alpha}}{p^{\alpha}} \leqslant \frac{(a+b)x}{k}$$

where we used (4.18) and (4.19). Reporting this estimate in (4.24) we get

$$\sum_{n \leqslant x} f(n) \log n \leqslant (a+b)x \sum_{k \leqslant x} \frac{f(k)}{k}. \tag{4.25}$$

By Lemma 4.25, we infer

$$\sum_{n \leqslant x} f(n) \leqslant \frac{1}{\log ex} \sum_{n \leqslant x} f(n) \log n + \frac{x}{\log ex} \sum_{k \leqslant x} \frac{f(k)}{k}$$

and combining this inequality with (4.25) we get

$$\sum_{n \leqslant x} f(n) \leqslant (a+b+1)\frac{x}{\log ex}\sum_{k \leqslant x}\frac{f(k)}{k}$$

and we conclude the proof by using Lemma 4.26.   □

Example 4.27 Theorem 4.22 enables us to get several bounds for average orders of usual arithmetic functions. The sole tool we need is Corollary 3.50 used under the weaker form $\sum_{p \leqslant x} 1/p \ll \log\log x$. We leave the details to the reader.

1. Let $k \geqslant 2$. We have

$$\sum_{n \leqslant x} s_k(n) \ll \frac{x}{\log x}.$$

2. The estimate

$$\sum_{n \leqslant x} f(n) \ll x$$

holds with $f = \beta$, $f = a$, $f = \tau^{(e)}$ and $f = \mu_k$ $(k \geqslant 2)$.
3. Let $k \geqslant 2$. We have

$$\sum_{n \leqslant x} \tau_{(k)}(n) \ll x \log x.$$

4. Let $k \geqslant 1$. The estimate

$$\sum_{n \leqslant x} f(n) \ll x(\log x)^{k-1}$$

holds with $f = k^\omega$ and $f = \tau_k$.

It should be noticed that almost all these bounds are of the right order of magnitude. Indeed, we saw in Example 4.18 that

$$\sum_{n \leqslant x} \mu_k(n) = \frac{x}{\zeta(k)} + O(x^{1/k}).$$

Further, it can be shown that (see [SW00, Wu95, Sur71] respectively)

$$\sum_{n \leqslant x} a(n) = A_1 x + A_2 x^{1/2} + A_3 x^{1/3} + O(x^{55/219}(\log x)^7)$$

$$\sum_{n \leqslant x} \tau^{(e)}(n) = B_1 x + B_2 x^{1/2} + O(x^{1057/4785+\varepsilon})$$

$$\sum_{n \leqslant x} \tau_{(k)}(n) = \frac{x}{\zeta(k)}\left(\log x + 2\gamma - 1 - k\frac{\zeta'}{\zeta}(k)\right) + O(x^{\nu_k})$$

where $A_i = \prod_{j=1, j \neq i}^{\infty} \zeta(\frac{j}{i})$, and

$$B_1 = \prod_p \left(1 + \sum_{\alpha=2}^{\infty} \frac{\tau(\alpha) - \tau(\alpha-1)}{p^{\alpha}}\right) \quad \text{and} \quad B_2 = \prod_p \left(1 + \sum_{\alpha=5}^{\infty} \frac{\tilde{\tau}(\alpha)}{p^{\alpha/2}}\right)$$

with

$$\tilde{\tau}(\alpha) = \tau(\alpha) - \tau(\alpha-1) - \tau(\alpha-2) + \tau(\alpha-3) \tag{4.26}$$

for $\alpha \geqslant 5$, and

$$v_k = \begin{cases} 1/k, & \text{if } k \in \{2, 3\} \\ 131/416 + \varepsilon, & \text{if } k \geqslant 4. \end{cases}$$

Furthermore we shall see in Exercise 13 that

$$\sum_{n \leqslant x} \tau_k(n) = \frac{x(\log x)^{k-1}}{(k-1)!} + O\left(x(\log x)^{k-2}\right)$$

and using Theorem 4.55 with $\kappa = k \in \mathbb{Z}_{\geqslant 2}$ we get

$$\sum_{n \leqslant x} k^{\omega(n)} = \frac{x(\log x)^{k-1}}{(k-1)!} \prod_p \left(1 - \frac{1}{p}\right)^k \left(1 + \frac{k}{p-1}\right) + O\left(x(\log x)^{k-3/2}\right).$$

Only the function s_k is overestimated, since using Corollary 3.7 (v) we shall prove in Exercise 14 that

$$\sum_{n \leqslant x} s_k(n) \ll x^{1/2}.$$

4.6.4 A Second Theorem

The following result provides a much stronger explicit upper bound of the average order subject to restricting the class of multiplicative functions by setting a supplementary condition on f.

Theorem 4.28 *Let f be a multiplicative function satisfying the Wirsing conditions* (4.21) *and*

$$f(p^{\alpha}) \geqslant f(p^{\alpha-1}) \tag{4.27}$$

for all primes p and $\alpha \in \mathbb{N}$. Then we have

$$\sum_{n \leqslant x} f(n) \leqslant x \prod_{p \leqslant x} \left(1 - \frac{1}{p}\right) \left(1 + \sum_{\alpha=1}^{\infty} \frac{f(p^{\alpha})}{p^{\alpha}}\right).$$

Proof By Example 4.11, the condition (4.27) implies that $g = f \star \mu \geqslant 0$. Hence we have by Proposition 4.17

$$\sum_{n \leqslant x} f(n) = \sum_{n \leqslant x} (g \star \mathbf{1})(n) = \sum_{d \leqslant x} g(d) \left[\frac{x}{d} \right] \leqslant x \sum_{d \leqslant x} \frac{g(d)}{d}$$

with

$$\sum_{d \leqslant x} \frac{g(d)}{d} \leqslant \prod_{p \leqslant x} \left(1 + \sum_{\alpha=1}^{\infty} \frac{g(p^{\alpha})}{p^{\alpha}} \right)$$

and we conclude by using $g(p^{\alpha}) = f(p^{\alpha}) - f(p^{\alpha-1})$.   □

Example 4.29

1. If $k \geqslant 1$, the function k^{ω} obviously satisfies (4.27) since $k^{\omega(p^{\alpha})} = k$ for all prime powers p^{α}. Theorem 4.28 gives

$$\sum_{n \leqslant x} k^{\omega(n)} \leqslant x \prod_{p \leqslant x} \left(1 + \frac{k-1}{p} \right) \leqslant x \exp\left((k-1) \sum_{p \leqslant x} \frac{1}{p} \right)$$

and Corollary 3.99 implies that

$$\sum_{n \leqslant x} k^{\omega(n)} \leqslant x \exp\{(k-1)(\log\log x + 1)\} = e^{k-1} x (\log x)^{k-1}$$

for all $x \geqslant 4$.

2. By (4.23), we easily see that τ_k satisfies (4.27), so that using (2.7) we get

$$\sum_{n \leqslant x} \tau_k(n) \leqslant x \prod_{p \leqslant x} \left(1 - \frac{1}{p} \right)^{1-k}.$$

Now the explicit upper bound of the second Mertens theorem provided by Corollary 3.99 implies that

$$\sum_{n \leqslant x} \tau_k(n) < \left(2e^{\gamma} \right)^{k-1} x (\log x)^{k-1}$$

for all $x \geqslant e$.

3. By (4.13), we have $\tau_{(k)} \star \mu = \mu_k \geqslant 0$, so that $\tau_{(k)}$ satisfies (4.27) and thus we get

$$\sum_{n \leqslant x} \tau_{(k)}(n) \leqslant x \prod_{p \leqslant x} \left(1 - \frac{1}{p} \right) \left(1 + \sum_{\alpha=1}^{k-1} \frac{\alpha+1}{p^{\alpha}} + k \sum_{\alpha=k}^{\infty} \frac{1}{p^{\alpha}} \right)$$

$$= x \prod_{p \leqslant x} \left(1 - \frac{1}{p} \right)^{-1} \left(1 - \frac{1}{p^{k}} \right)$$

$$= \frac{x}{\zeta(k)} \prod_{p \leqslant x} \left(1 - \frac{1}{p}\right)^{-1} \prod_{p > x} \left(1 - \frac{1}{p^k}\right)^{-1}$$

$$\leqslant \frac{2e^\gamma}{\zeta(k)} \prod_{p > x} \left(1 - \frac{1}{p^k}\right)^{-1} x \log x$$

as soon as $x \geqslant e$, where we used Corollary 3.99 again.

4. Let $P(\alpha)$ be the number of unrestricted partitions of α. It is proved in [Gup78] that, for all $\alpha \geqslant 1$, we have $P(\alpha) \geqslant P(\alpha - 1)$ so that the function a satisfies (4.27). Theorem 4.28 provides the bound

$$\sum_{n \leqslant x} a(n) \leqslant x \prod_{p \leqslant x} \left(1 - \frac{1}{p}\right) \left(1 + \sum_{\alpha=1}^{\infty} \frac{P(\alpha)}{p^\alpha}\right)$$

and the combinatorial identity

$$1 + \sum_{\alpha=1}^{\infty} P(\alpha) x^\alpha = \prod_{j=1}^{\infty} \frac{1}{1 - x^j}$$

valid for $|x| < 1$, implies

$$\sum_{n \leqslant x} a(n) \leqslant x \prod_{p \leqslant x} \prod_{j=2}^{\infty} \left(1 - \frac{1}{p^j}\right)^{-1} \leqslant x \prod_{j=2}^{\infty} \zeta(j).$$

5. Using Exercise 3, we have $\beta \star \mu = s_2$ and hence β satisfies (4.27). We get

$$\sum_{n \leqslant x} \beta(n) \leqslant x \prod_{p \leqslant x} \left(1 - \frac{1}{p}\right) \left(1 + \frac{1}{p} + \sum_{\alpha=2}^{\infty} \frac{\alpha}{p^\alpha}\right)$$

$$= x \prod_{p \leqslant x} \left(1 - \frac{1}{p^2}\right)^{-1} \left(1 - \frac{1}{p^3}\right)^{-1} \left(1 - \frac{1}{p^6}\right)$$

$$\leqslant x \frac{\zeta(2)\zeta(3)}{\zeta(6)} \prod_{p > x} \left(1 - \frac{1}{p^6}\right)^{-1}.$$

4.7 Further Developments

4.7.1 The Ring of Arithmetic Functions

Lemma 4.9 tells us that the set $\mathcal{A}$ of arithmetic functions with addition and Dirichlet convolution product $\star$ is a unitary commutative ring with identity element e_1. It is also an integral domain (see Chap. 7). To see this we may proceed as follows.

Define a map $N : A \longrightarrow \mathbb{Z}_{\geqslant 0}$ by setting $N(0) = 0$ and, if $f \neq 0$, then $N(f)$ is the smallest non-negative integer n such that $f(n) \neq 0$. Observe that, if $f, g \in A$, then $N(f \star g) = N(f)N(g)$, and thus, if $f \neq 0$ and $g \neq 0$ are such that $N(f) = a$ and $N(g) = b$, then we have $(f \star g)(n) = 0$ for all $n < ab$, for if $n < ab$ and $d \mid n$, then either $d < a$ or $n/d < b$. Furthermore, we also have $(f \star g)(ab) = f(a)g(b) \neq 0$, so that A is an integral domain. One can also prove that this ring is a UFD (see Definition 7.12).

One may lose the integrity property with a slight change in the convolution product. For instance define the *unitary convolution product* by

$$(f \circledast g)(n) = \sum_{\substack{d \mid n \\ (d,n/d)=1}} f(d)g\left(\frac{n}{d}\right).$$

Then it can be shown [Siv89] that the unitary ring $(A, +, \circledast)$ *is not* an integral domain.

One may wonder whether the ring $(A, +, \star)$ is nœtherian (see Definition 7.9). Let $p_1 < p_2 < \dots$ be the increasing sequence of the prime numbers and, for all positive integers k, define the subsets S_k of A as follows

$$S_k = \left\{ f \in A : f(n) = 0 \text{ for all } n \text{ such that } (n, p_1 \cdots p_k) = 1 \right\}.$$

Then the sets S_k are pairwise distinct ideals of A since $(f - g)(n) = f(n) - g(n) = 0$ for all n such that $(n, p_1 \cdots p_k) = 1$ and, for all $f, g \in S_k$ and all $h \in A$, we have

$$(f \star h)(n) = \sum_{d \mid n} f(d)h\left(\frac{n}{d}\right) = 0$$

since if $d \mid n$ and $(n, p_1 \cdots p_k) = 1$, then $(d, p_1 \cdots p_k) = 1$. Now one may check that

$$\{0\} = S_0 \subseteq S_1 \subseteq S_2 \subseteq \cdots \subseteq S_k \subseteq S_{k+1} \subseteq \cdots$$

and

$$\bigcup_{k \in \mathbb{Z}_{\geqslant 0}} S_k = A$$

and hence the ring $(A, +, \star)$ is not nœtherian. One may prove that this ring is also not *artinian* by considering the sequence of pairwise distinct ideals T_k of A defined by

$$T_k = \left\{ f \in A : f(n) = 0 \text{ for all } n \text{ such that } \Omega(n) < k \right\}.$$

4.7.2 Dirichlet Series—The Formal Viewpoint

We have seen in Chap. 2 that the concept of generating function may be fruitful to capture the information of a sequence. In view of the multiplicative properties of certain arithmetic functions, we use *Dirichlet series* rather than power series in analytic number theory.

Definition 4.30 Let f be an arithmetic function. The *formal Dirichlet series* of a variable s associated to f is defined by

$$L(s, f) = \sum_{n=1}^{\infty} \frac{f(n)}{n^s}.$$

As always for formal mathematical objects, we ignore here convergence problems, and $L(s, f)$ is the complex number equal to the sum when it converges.

For instance, $L(s, e_1) = 1$ and $L(s, \mathbf{1}) = \zeta(s)$.

The following proposition reveals the importance of the Dirichlet convolution product.

Proposition 4.31 *Let f, g and h be three arithmetic functions. Then*

$$h = f \star g \quad \Longleftrightarrow \quad L(s, h) = L(s, f)L(s, g).$$

Proof We have

$$L(s, f)L(s, g) = \sum_{k,d=1}^{\infty} \frac{f(k)g(d)}{(kd)^s} = \sum_{n=1}^{\infty} \frac{1}{n^s} \sum_{d|n} f\left(\frac{n}{d}\right) g(d) = \sum_{n=1}^{\infty} \frac{(f \star g)(n)}{n^s}$$

which completes the proof.   □

The next result may be considered as a generalization of Euler's proof of Theorem 3.13.

Proposition 4.32 *Let f be an arithmetic function. Then f is multiplicative if and only if*

$$L(s, f) = \prod_{p} \left(1 + \sum_{\alpha=1}^{\infty} \frac{f(p^\alpha)}{p^{s\alpha}}\right).$$

The above product is called the Euler product *of $L(s, f)$.*

Proof Expanding the product proves that it is equivalent to the conditions $f(1) = 1$ and $f(n) = f(p_1^{\alpha_1}) \cdots f(p_r^{\alpha_r})$ for all $n = p_1^{\alpha_1} \cdots p_r^{\alpha_r}$ and use Lemma 4.5 to complete the proof.   □

4.7.3 Dirichlet Series—Absolute Convergence

If we wish to deal with convergence problems, we need to have at our disposal some tools to determine precisely the region of convergence of a Dirichlet series. Recall that for power series $\sum_{n \geq 0} f(n)s^n$, the domain of convergence is a disc on the boundary of which the behavior of the sum is a priori undetermined. For Dirichlet series, there is an analogous result, except that the domain of absolute convergence is a half-plane.

Proposition 4.33 *For each Dirichlet series $F(s)$, there exists $\sigma_a \in \mathbb{R} \cup \{\pm\infty\}$, called the* abscissa of absolute convergence, *such that $F(s)$ converges absolutely in the half-plane $\sigma > \sigma_a$ and does not converge absolutely in the half-plane $\sigma < \sigma_a$.*

Proof Let S be the set of complex numbers s at which $F(s)$ converges absolutely. If $S = \varnothing$, then put $\sigma_a = +\infty$. Otherwise define

$$\sigma_a = \inf\{\sigma : s = \sigma + it \in S\}.$$

By the definition of σ_a, $F(s)$ does not converge absolutely if $\sigma < \sigma_a$. On the other hand, suppose that $F(s)$ is absolutely convergent for some $s_0 = \sigma_0 + it_0 \in \mathbb{C}$ and let $s = \sigma + it$ such that $\sigma \geqslant \sigma_0$. Since

$$\left| \frac{f(n)}{n^s} \right| = \left| \frac{f(n)}{n^{s_0}} \right| \times \frac{1}{n^{\sigma - \sigma_0}} \leqslant \left| \frac{f(n)}{n^{s_0}} \right|$$

we infer that $F(s)$ converges absolutely at any point s such that $\sigma \geqslant \sigma_0$. Now by the definition of σ_a, there exist points arbitrarily close to σ_a at which $F(s)$ converges absolutely, and therefore by above $F(s)$ converges absolutely at *each* point s such that $\sigma > \sigma_a$.   □

It follows in particular that the series $F(s)$ defines an analytic function in the half-plane $\sigma > \sigma_a$. Note that, by abuse of notation, this function is still denoted by $F(s)$.

Proposition 4.33 implies at once that if $|f(n)| \leqslant \log n$, then the series $F(s)$ is absolutely convergent in the half-plane $\sigma > 1$, and hence $\sigma_a \leqslant 1$.

At $\sigma = \sigma_a$, the series may or may not converge absolutely. For instance, $\zeta(s)$ converges absolutely in the half-plane $\sigma > \sigma_a = 1$ but does not converge on the line $\sigma = 1$. On the other hand, the Dirichlet series associated to the function $f(n) = 1/(\log(en))^2$ has also $\sigma_a = 1$ for the abscissa of absolute convergence, but converges absolutely at $\sigma = 1$.

The partial sums $\sum_{x < n \leqslant y} f(n)$ and the Dirichlet series $F(s)$ are strongly related to each other. The next result shows that if we are able to estimate the order of magnitude of $\sum_{x < n \leqslant y} f(n)$, then a region of absolute convergence of $F(s)$ is known.

Proposition 4.34 *Let $F(s) = \sum_{n=1}^{\infty} f(n)n^{-s}$ be a Dirichlet series. Assume that for all $0 < x < y$, we have*

$$\left| \sum_{x < n \leqslant y} f(n) \right| \leqslant M y^\alpha$$

for some $\alpha \geqslant 0$ and $M > 0$ independent of x and y. Then $F(s)$ converges absolutely in the half-plane $\sigma > \alpha$. Furthermore, we have in this half-plane

$$|F(s)| \leqslant \frac{M|s|}{\sigma - \alpha} \quad and \quad \left| \sum_{x < n \leqslant y} \frac{f(n)}{n^s} \right| \leqslant \frac{M}{x^{\sigma - \alpha}} \left(\frac{|s|}{\sigma - \alpha} + 1 \right).$$

Proof Set $A(x) = \sum_{n \leqslant x} f(n)$ and $S(x, y) = A(y) - A(x)$. By partial summation we have

$$\sum_{x < n \leqslant y} \frac{f(n)}{n^s} = \frac{S(x, y)}{y^s} + s \int_x^y \frac{S(x, u)}{u^{s+1}} \, du$$

and by hypothesis we have $|S(x, y)/y^s| \leqslant M y^{\alpha - \sigma}$, so that $S(x, y)/y^s$ tends to 0 as $y \longrightarrow \infty$ in the half-plane $\sigma > \alpha$. Therefore if one of

$$\sum_{n \geqslant 1} \frac{f(n)}{n^s} \quad or \quad s \int_1^\infty \frac{A(u)}{u^{s+1}} \, du$$

converges absolutely, then so does the other, and the two quantities converge to the same limit. But since

$$\left| \frac{A(u)}{u^{s+1}} \right| \leqslant \frac{M}{u^{\sigma - \alpha + 1}}$$

we infer that the integral converges absolutely for $\sigma > \alpha$ by Rule 1.20, and hence $F(s)$ is absolutely convergent in this half-plane. Therefore for all $\sigma > \alpha$, we get

$$\sum_{n=1}^\infty \frac{f(n)}{n^s} = s \int_1^\infty \frac{A(u)}{u^{s+1}} \, du$$

and hence

$$|F(s)| \leqslant M|s| \int_1^\infty \frac{du}{u^{\sigma - \alpha + 1}} = \frac{M|s|}{\sigma - \alpha}$$

and similarly

$$\left| \sum_{x < n \leqslant y} \frac{f(n)}{n^s} \right| \leqslant \frac{M}{y^{\sigma - \alpha}} + M|s| \int_x^\infty \frac{du}{u^{\sigma - \alpha + 1}} \leqslant \frac{M}{x^{\sigma - \alpha}} \left(\frac{|s|}{\sigma - \alpha} + 1 \right)$$

as required. $\square$

We have seen in Proposition 4.31 that the product of two Dirichlet series is of great importance in the theory of arithmetic functions. The next result gives the domain of absolute convergence of this product.

Proposition 4.35 *Let f and g be two arithmetic functions with associated Dirichlet series $F(s)$ and $G(s)$. Set $h = f \star g$ and let $H(s)$ be the associated Dirichlet series.*

If $F(s)$ and $G(s)$ are absolutely *convergent at a point* s_0, *then* $H(s)$ *converges absolutely at* s_0 *and we have* $H(s_0) = F(s_0)G(s_0)$.

Proof From Proposition 4.31, we have

$$F(s_0)G(s_0) = \sum_{n=1}^{\infty} \frac{h(n)}{n^{s_0}} = H(s_0)$$

where the rearrangement of the terms in the double sums is justified by the absolute convergence of the two series $F(s)$ and $G(s)$ at $s = s_0$. Furthermore, we have

$$\sum_{n=1}^{\infty} \left| \frac{h(n)}{n^{s_0}} \right| \leqslant \left(\sum_{n=1}^{\infty} \left| \frac{f(n)}{n^{s_0}} \right| \right) \left(\sum_{n=1}^{\infty} \left| \frac{g(n)}{n^{s_0}} \right| \right)$$

proving the absolute convergence of $H(s_0)$.   □

In 1885/7, Stieltjes proved the following stronger results.

Proposition 4.36 (Stieltjes) *Let* $F(s)$ *and* $G(s)$ *be two Dirichlet series.*

(i) *If* F *and* G *converge at the point* $\sigma_0 \in \mathbb{R}$ *and if they converge absolutely at* $\sigma_0 + c$ *where* $c \geqslant 0$, *then the series* $F(s)G(s)$ *converges at* $\sigma_0 + \frac{c}{2}$.
(ii) *If* F *and* G *converge at the point* $\sigma_0 \in \mathbb{R}$, *then the series* $F(s)G(s)$ *converges at* $\sigma_0 + \frac{1}{2}$.

Using Proposition 4.35 with $h = e_1$ gives the following result.

Proposition 4.37 *Let* f *be an arithmetic function such that* $f(1) \neq 0$ *and let* $F(s)$ *be its associated Dirichlet series. Let* f^{-1} *be the convolution inverse of* f, *i.e. the arithmetic function such that* $f \star f^{-1} = e_1$ *and let* $G(s)$ *be the associated Dirichlet series of* f^{-1}. *Then*

$$G(s) = \frac{1}{F(s)}$$

at every point s *where* $F(s)$ *and* $G(s)$ *converge absolutely.*

For instance, this proposition enables us to get the Dirichlet series of the Möbius function. Indeed, by (4.6), we have $\mu^{-1} = \mathbf{1}$ and hence, for all $\sigma > 1$, we have

$$\sum_{n=1}^{\infty} \frac{\mu(n)}{n^s} = \frac{1}{\zeta(s)}$$

as stated in Example 4.18.

We end this section with two examples of computation of abscissa of the absolute convergence.

Proposition 4.38 *Let f be an arithmetic function and $F(s)$ be its associated Dirichlet series with abscissa of absolute convergence σ_a.*

(i) *If $|f(n)| \leqslant Mn^\alpha$ for some real numbers $M > 0$ and $\alpha \geqslant 0$, then $\sigma_a \leqslant \alpha + 1$.*

(ii) *We have*

$$\sigma_a \leqslant \limsup_{n \to \infty} \left(1 + \frac{\log |f(n)|}{\log n} \right).$$

Proof

(i) Follows from Proposition 4.34 since we have

$$\left| \sum_{n \leqslant x} f(n) \right| \leqslant M \sum_{n \leqslant x} n^\alpha \leqslant M x^{\alpha+1}.$$

(ii) Set

$$L = \limsup_{n \to \infty} \left(1 + \frac{\log |f(n)|}{\log n} \right)$$

and we may suppose that $L < \infty$. Fix a small real number $\varepsilon > 0$ and let $\sigma > L$. Hence $\sigma \geqslant L + \varepsilon$ and therefore there exists a large positive integer $n_0 = n_0(\varepsilon)$ such that, for all $n \geqslant n_0$, we have

$$1 + \frac{\log |f(n)|}{\log n} < L \leqslant \sigma - \varepsilon$$

and hence, for all $n \geqslant n_0$, we get

$$\left| \frac{f(n)}{n^s} \right| < \frac{n^{\sigma-1-\varepsilon}}{n^\sigma} = \frac{1}{n^{1+\varepsilon}}$$

so that $F(s)$ is absolutely convergent in the half-plane $\sigma > L$.

The proof is complete. □

4.7.4 Dirichlet Series—Conditional Convergence

This is one of the most important differences between power series and Dirichlet series. For power series, the regions of convergence and absolute convergence are identical, except possibly for the boundaries. For Dirichlet series, there may be a non-trivial region, in the form of a vertical strip, in which the series converges but does not converge absolutely. However, we shall see that the width of this strip does not exceed 1.

Proposition 4.39 *For each Dirichlet series $F(s)$, there exists $\sigma_c \in \mathbb{R} \cup \{\pm\infty\}$, called the* abscissa of convergence, *such that $F(s)$ converges in the half-plane $\sigma > \sigma_c$ and does not converge in the half-plane $\sigma < \sigma_c$. Furthermore, we have*

$$\sigma_c \leqslant \sigma_a \leqslant \sigma_c + 1.$$

Proof

▷ Suppose first that $F(s)$ converges at a point $s_0 = \sigma_0 + it_0$ and fix a small real number $\varepsilon > 0$. By Cauchy's theorem, there exists $x_0 = x_0(\varepsilon) \geqslant 1$ such that, for all $y > x \geqslant x_0$, we have

$$\left| \sum_{x < n \leqslant y} \frac{f(n)}{n^{s_0}} \right| \leqslant \varepsilon.$$

Let $s = \sigma + it \in \mathbb{C}$ such that $\sigma > \sigma_0$. Using Proposition 4.34 with s replaced by $s - s_0$ and $\alpha = 0$, we get

$$\left| \sum_{x < n \leqslant y} \frac{f(n)}{n^s} \right| \leqslant \varepsilon \left(\frac{|s - s_0|}{\sigma - \sigma_0} + 1 \right)$$

so that $F(s)$ converges by Cauchy's theorem.

▷ Now we may proceed as in Proposition 4.33. Let S be the set of complex numbers s at which $F(s)$ converges. If $S = \varnothing$, then put $\sigma_c = +\infty$. Otherwise define

$$\sigma_c = \inf\{\sigma : s = \sigma + it \in S\}.$$

By the definition of σ_c, $F(s)$ does not converge if $\sigma < \sigma_c$. On the other hand, there exist points s_0 with σ_0 being arbitrarily close to σ_c at which $F(s)$ converges. By above, $F(s)$ converges at any point s such that $\sigma > \sigma_0$. Since σ_0 may be chosen as close to σ_c as we want, it follows that $F(s)$ converges at any point s such that $\sigma > \sigma_c$.

▷ The inequality $\sigma_c \leqslant \sigma_a \leqslant \sigma_c + 1$ remains to be shown. The left-hand side is obvious. For the right-hand side, it suffices to show that if $F(s_0)$ converges for some s_0, then it converges absolutely for all s such that $\sigma > \sigma_0 + 1$. Now if $F(s)$ converges at some point s_0, then $f(n)/n^{s_0}$ tends to 0 as $n \longrightarrow \infty$. Thus there exists a positive integer n_0 such that, for all $n \geqslant n_0$, we have $|f(n)/n^{s_0}| \leqslant 1$. We deduce that, for any s, we have $|f(n)/n^s| \leqslant n^{\sigma_0 - \sigma}$ so that $F(s)$ is absolutely convergent in the half-plane $\sigma > \sigma_0 + 1$ as required.   □

As in the case of absolute convergence, the behavior of the series on the line $\sigma = \sigma_c$ is a priori undetermined. Furthermore, although there exist series such that $\sigma_c = +\infty$ (take $f(n) = n!$ for instance) or $\sigma_c = -\infty$ (take $f(n) = 1/n!$), these extreme cases do not appear in almost all number-theoretic applications in which we shall have $\sigma_c \in \mathbb{R}$ and $\sigma_a \in \mathbb{R}$. Hence we must deal with half-planes of absolute convergence, of convergence and with a strip in which the series converges conditionally but not absolutely.

The inequality $\sigma_c \leqslant \sigma_a \leqslant \sigma_c + 1$ is sharp. Indeed,

1. If $f(n) \geqslant 0$, then the absolute convergence and the conditional convergence of $F(s)$ are equivalent at any point $s \in \mathbb{R}$. But since the regions of convergence and absolute convergence are half-planes, we infer that in the case of non-negative coefficients, we have $\sigma_c = \sigma_a$.

2. On the other hand, if $f(n) = \chi(n) \neq \chi_0(n)$ is a non-principal Dirichlet character, then we have seen in Chap. 3 that $\sigma_c = 0$, and we clearly have $\sigma_a = 1$.

The next result is an easy consequence of partial summation techniques.

Proposition 4.40 *Let $\sigma_0 > 0$. If the series $\sum_{n=1}^{\infty} f(n) n^{-\sigma_0}$ converges, then as $N \longrightarrow \infty$, we have*

$$\sum_{n=1}^{N} f(n) = o\left(N^{\sigma_0}\right).$$

Proof We first prove the following lemma due to Kronecker.

Let (α_n) be a sequence of complex numbers such that $\sum_{n \geqslant 1} \alpha_n$ converges, and let (β_n) be an increasing sequence of positive real numbers such that $\beta_n \longrightarrow \infty$. Then as $N \longrightarrow \infty$

$$\sum_{n=1}^{N} \alpha_n \beta_n = o(\beta_N).$$

Let $R_n = \sum_{k>n} \alpha_k$. By assumption, we have $R_n = o(1)$ as $n \longrightarrow \infty$, and by Abel's summation as in Remark 1.15 we get

$$\sum_{n=1}^{N} \alpha_n \beta_n = \sum_{n=1}^{N} \beta_n (R_n - R_{n+1})$$

$$= \beta_N \sum_{n=1}^{N} (R_n - R_{n+1}) - \sum_{n=1}^{N-1} (\beta_{n+1} - \beta_n) \sum_{h=1}^{n} (R_h - R_{h+1})$$

$$= R_1 \beta_1 - R_{N+1} \beta_N + \sum_{n=1}^{N-1} R_{n+1} (\beta_{n+1} - \beta_n).$$

Since (β_n) increases to infinity with n, we get $\beta_{n+1} - \beta_n > 0$ and $\sum_{n=1}^{\infty} (\beta_{n+1} - \beta_n) = \infty$, so that as $N \longrightarrow \infty$

$$\sum_{n=1}^{N-1} |R_{n+1}| (\beta_{n+1} - \beta_n) = o\left(\sum_{n=1}^{N-1} (\beta_{n+1} - \beta_n)\right) = o(\beta_N)$$

as required. Now Proposition 4.40 follows at once by applying Kronecker's lemma with $\alpha_n = f(n) n^{-\sigma_0}$ and $\beta_n = n^{\sigma_0}$. □

4.7.5 Dirichlet Series—Analytic Properties

It has been seen above that a Dirichlet series $F(s)$ defines an analytic function in its half-plane of absolute convergence. The next result proves much better: $F(s)$

Fig. 4.2 The disc $\mathcal{D}$

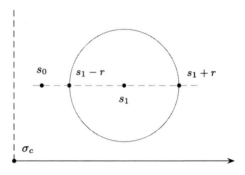

defines in fact an analytic function in its half-plane of *conditional* convergence, which enables us to apply the powerful tools from complex analysis to arithmetic functions. The proof makes use of Weierstrass's double series theorem.[5]

Theorem 4.41 *A Dirichlet series $F(s) = \sum_{n=1}^{\infty} f(n) n^{-s}$ defines an analytic function of s in the half-plane $\sigma > \sigma_c$, in which $F(s)$ can be differentiated termwise so that, for all positive integers k, we have*

$$F^{(k)}(s) = \sum_{n=1}^{\infty} \frac{(-1)^k (\log n)^k f(n)}{n^s}$$

for all $s = \sigma + it$ such that $\sigma > \sigma_c$.

Proof Let $s_1 = \sigma_1 + it_1$ such that $\sigma_1 > \sigma_c$ and we consider the disc $\mathcal{D}$ with s_1 as center and of radius r contained entirely in the half-plane $\sigma > \sigma_c$. We will show that $F(s)$ converges uniformly in $\mathcal{D}$. Fix $\varepsilon > 0$. We must prove that there exists $n_0 = n_0(\varepsilon) \in \mathbb{N}$, independent of $s \in \mathcal{D}$, such that for all $N \geqslant n_0$, we have

$$\left| \sum_{n > N} \frac{f(n)}{n^s} \right| \leqslant \varepsilon$$

for all $s \in \mathcal{D}$. To this end, we choose $s_0 = \sigma_0 + it_0$ such that $t_0 = t_1$ and $\sigma_c < \sigma_0 < \sigma_1 - r$, see Fig. 4.2.

Since $F(s)$ converges at s_0, the partial sums are bounded at this point in absolute value by $M > 0$. Using Proposition 4.34 with s replaced by $s - s_0$ and $\alpha = 0$, we get for all $s \in \mathcal{D}$ and $N_1 > N$

$$\left| \sum_{N < n \leqslant N_1} \frac{f(n)}{n^s} \right| \leqslant \frac{2M}{N^{\sigma - \sigma_0}} \left(\frac{|s - s_0|}{\sigma - \sigma_0} + 1 \right) \leqslant \frac{2M}{N^{\sigma_1 - \sigma_0 - r}} \left(\frac{\sigma_1 - \sigma_0 + r}{\sigma_1 - \sigma_0 - r} + 1 \right).$$

The right-hand side is independent of s and tends to 0 as $N \longrightarrow \infty$, which ensures the existence of n_0. By Weierstrass's double series theorem, we deduce that $F(s)$ is

---

[5]See [Tit39, Theorem 2.8].

analytic in $\mathcal{D}$, and term-by-term differentiation is allowed there. Since s_1 is arbitrary, the asserted result follows. ∎

The next result implies that, if $F(s) = \sum_{n=1}^{\infty} f(n)n^{-s}$ and $G(s) = \sum_{n=1}^{\infty} g(n)n^{-s}$ having the same abscissa of convergence σ_c, then $f(n) = g(n)$ for all $n \in \mathbb{N}$.

Proposition 4.42 *Let $F(s) = \sum_{n=1}^{\infty} f(n)n^{-s}$ be a Dirichlet series with abscissa of convergence σ_c. If $F(s) = 0$ for all s such that $\sigma > \sigma_c$, then $f(n) = 0$ for all $n \in \mathbb{N}$.*

Proof Suppose the contrary and let k be the smallest positive integer such that $f(k) \neq 0$. Let

$$G(s) = k^s \sum_{n=k}^{\infty} \frac{f(n)}{n^s}$$

defined for $\sigma > \sigma_c$. By assumption we have $G(s) = 0$ in this half-plane. On the other hand, if $\sigma > \sigma_c$, we have

$$G(s) = f(k) + \sum_{n=k+1}^{\infty} f(n) \left(\frac{k}{n}\right)^s$$

so that allowing s to become infinite along the real axis, we obtain

$$\lim_{\sigma \to \infty} G(\sigma) = f(k)$$

and thus we get $f(k) = 0$, giving a contradiction. ∎

This result enables us to show arithmetic identities of the form $f(n) = g(n)$ by considering their Dirichlet series $F(s)$ and $G(s)$ and by showing that, for σ sufficiently large, we have $F(s) = G(s)$. For instance, using Proposition 4.32, one may check that the Dirichlet series of the Euler totient function φ is given by

$$\frac{\zeta(s-1)}{\zeta(s)} = \zeta(s-1) \times \frac{1}{\zeta(s)}.$$

Now $\zeta(s-1)$ is the Dirichlet series of Id and $\zeta(s)^{-1}$ is that of μ, and hence $\zeta(s-1)/\zeta(s)$ is the Dirichlet series of $\mu \star$ Id. Since both series converge for $\sigma > 2$, we get $\varphi = \mu \star$ Id by Proposition 4.42 as already proved in (4.7).

Another important problem is the investigation of the behavior of a Dirichlet series $F(s)$ on vertical lines. The following result shows that $F(\sigma + it)$ cannot increase more rapidly than $|t|$ as $|t| \longrightarrow \infty$.

Proposition 4.43 *Let $F(s) = \sum_{n=1}^{\infty} f(n)n^{-s}$ be a Dirichlet series with abscissa of convergence σ_c and let $\sigma_1 > \sigma_c$. Then, uniformly in σ such that $\sigma \geqslant \sigma_1$ and as $|t| \longrightarrow \infty$, we have*

$$F(\sigma + it) = o(|t|).$$

Furthermore, for any $\varepsilon > 0$ and $\sigma_c < \sigma \leqslant \sigma_c + 1$ and as $|t| \longrightarrow \infty$, we have

$$F(\sigma + it) = O\big(|t|^{1-(\sigma-\sigma_c)+\varepsilon}\big).$$

Proof Let $\sigma_0 \in \mathbb{R}$ such that $\sigma_c < \sigma_0 < \sigma_1$. Using Proposition 4.34 with s replaced by $s - \sigma_0$ and $\alpha = 0$, we get for all $x \geqslant 1$ and $\sigma > \sigma_0$

$$\left| \sum_{n>x} \frac{f(n)}{n^s} \right| \leqslant \frac{M}{x^{\sigma-\sigma_0}} \left(\frac{|\sigma - \sigma_0 + it|}{\sigma - \sigma_0} + 1 \right) \leqslant \frac{M}{x^{\sigma-\sigma_0}} \left(\frac{|t|}{\sigma - \sigma_0} + 2 \right)$$

for some $M > 0$, and hence for all $\sigma \geqslant \sigma_1$, we get

$$\left| \frac{F(\sigma + it)}{t} \right| \leqslant \frac{1}{|t|} \left\{ \sum_{n \leqslant x} \frac{|f(n)|}{n^{\sigma}} + \frac{M}{x^{\sigma-\sigma_0}} \left(\frac{|t|}{\sigma - \sigma_0} + 2 \right) \right\}$$

$$\leqslant \frac{1}{|t|} \sum_{n \leqslant x} \frac{|f(n)|}{n^{\sigma_1}} + \frac{M}{x^{\sigma_1-\sigma_0}} \left(\frac{2}{|t|} + \frac{1}{\sigma_1 - \sigma_0} \right)$$

where the right-hand side is independent of $\sigma \in [\sigma_1, +\infty[$. This inequality implies that

$$\limsup_{|t| \to \infty} \left| \frac{F(\sigma + it)}{t} \right| \leqslant \frac{M}{x^{\sigma_1-\sigma_0}(\sigma_1 - \sigma_0)}$$

and since the left-hand side is independent of x, letting $x \longrightarrow \infty$ gives

$$\limsup_{|t| \to \infty} \left| \frac{F(\sigma + it)}{t} \right| = 0$$

uniformly in $\sigma \geqslant \sigma_1$, implying the first asserted result.

For the second one, since $\sigma_c + 1 + \varepsilon > \sigma_a$ by Proposition 4.39, we infer that

$$F(\sigma_c + 1 + \varepsilon) = O(1).$$

By above, we also have $F(\sigma_c + \varepsilon + it) = O(|t|)$ so that the asserted estimate follows from the Phragmén–Lindelöf principle seen in Chap. 3. $\square$

We know that a function defined by a power series has a singularity on the circle of convergence. The situation is rather different for Dirichlet series which need have no singularity on the axis of convergence. For instance, it may be shown that the function $\eta(s) = \sum_{n=1}^{\infty} (-1)^{n+1} n^{-s}$ is entire although we easily see that $\sigma_c = 0$. However, if the coefficients of $F(s)$ are real non-negative, then the next result, due to Landau, shows that the series does have a singularity on the axis of convergence.

Theorem 4.44 (Landau) *Let $F(s) = \sum_{n=1}^{\infty} f(n)n^{-s}$ be a Dirichlet series with abscissa of convergence $\sigma_c \in \mathbb{R}$ and suppose that $f(n) \geqslant 0$ for all n. Then $F(s)$ has a singularity at $s = \sigma_c$.*

Proof Without loss of generality, we may assume that $\sigma_c = 0$ and suppose that the origin is not a singularity of $F(s)$. Thus the Taylor expansion of $F(s)$ about $a > 0$

$$F(s) = \sum_{k=0}^{\infty} \frac{(s-a)^k}{k!} F^{(k)}(a) = \sum_{k=0}^{\infty} \frac{(s-a)^k}{k!} \sum_{n=1}^{\infty} \frac{(-1)^k (\log n)^k f(n)}{n^a}$$

must converge at some point $s = b < 0$ so that the double series

$$\sum_{k=0}^{\infty} \sum_{n=1}^{\infty} \frac{((a-b)\log n)^k f(n)}{n^a k!}$$

must converge, and then may be summed in any order since each term is non-negative. But interchanging the summation implies that this sum is equal to

$$\sum_{n=1}^{\infty} \frac{f(n)}{n^a} \sum_{k=0}^{\infty} \frac{((a-b)\log n)^k}{k!} = \sum_{n=1}^{\infty} \frac{f(n)}{n^b}$$

which does not converge by Proposition 4.39 since $b < 0 = \sigma_c$, giving a contradiction. $\qquad\square$

For instance, the Riemann zeta-function $\zeta(s)$ has a singularity at $\sigma_c = 1$, as we already proved in Theorem 3.55. In fact, this result proves much more, telling us that this singularity is a *pole*.

Landau's theorem is often used in its contrapositive form.

Corollary 4.45 *Let* $F(s) = \sum_{n=1}^{\infty} f(n)n^{-s}$ *be a Dirichlet series with abscissa of convergence* $\sigma_c < \infty$ *and suppose that* $f(n) \geqslant 0$ *for all* n. *If* $F(s)$ *can be analytically continued to the half-plane* $\sigma > \sigma_0$, *then* $F(s)$ *converges for* $\sigma > \sigma_0$.

By the previous results, we infer that a Dirichlet series converges *uniformly* on every compact subset of the half-plane of convergence. The next result, due to Cahen and Jensen, shows that it also converges uniformly in certain regions which extend to infinity.

Proposition 4.46 (Cahen–Jensen) *If the Dirichlet series* $F(s) = \sum_{n=1}^{\infty} f(n)n^{-s}$ *converges at* s_0, *then it converges uniformly on every domain* $C(s_0, \theta)$, *sometimes called the* Stolz domain, *see Fig. 4.3, defined by*

$$C(s_0, \theta) = \left\{ s \in \mathbb{C} : |\arg(s - s_0)| \leqslant \theta < \frac{\pi}{2} \right\}.$$

Proof The map $s \longmapsto s + s_0$ transforms $F(s)$ into a Dirichlet series of the same shape where the coefficient a_n is replaced by the coefficient $a_n n^{-s_0}$, so that we may

Fig. 4.3 Stolz domain

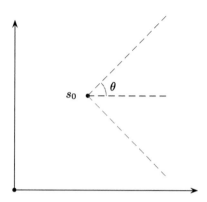

assume that $s_0 = 0$ without loss of generality. Let $\varepsilon > 0$. By Cauchy's theorem, there exists $N_0 = N_0(\varepsilon) \in \mathbb{N}$ such that, for all $N_1 > N > N_0$, we have

$$\left| \sum_{N < n \leqslant N_1} f(n) \right| \leqslant \frac{\varepsilon \cos \theta}{2}$$

since the series converges at 0. Hence using Proposition 4.34 with $\alpha = 0$ and $M = \frac{\varepsilon}{2} \cos \theta$, we get if $\sigma > 0$

$$\left| \sum_{N < n \leqslant N_1} \frac{f(n)}{n^s} \right| \leqslant \frac{\varepsilon \cos \theta}{2N^\sigma} \left(\frac{|s|}{\sigma} + 1 \right) \leqslant \frac{\varepsilon |s| \cos \theta}{\sigma N^\sigma}$$

and we conclude the proof by noticing that, if $s \in C(0, \theta)$, then $\sigma > 0$ and $|s| \cos \theta \leqslant \sigma$. $\qquad \square$

Among other things, this result implies that, if $F(s)$ converges at s_0, then

$$\lim_{s \to s_0} F(s) = F(s_0)$$

provided that s remains inside $C(s_0, \theta)$ as it approaches its limit. When s_0 is off the axis of convergence, this result is clearly a consequence of Theorem 4.41, but when s_0 is on that axis, then the limit above is the analogue of Abel's theorem for power series.

Finally, it should be mentioned that explicit formulae do exist for σ_c and σ_a, but they are not much used in number theory. For instance, Cahen proved that, if the series $\sum_{n \geqslant 1} f(n)$ diverges, then

$$\sigma_c = \limsup_{N \to \infty} \frac{\log |\sum_{n=1}^N f(n)|}{\log N} = \inf \left\{ \sigma_0 : \sum_{n=1}^N f(n) = O\left(N^{\sigma_0}\right) \right\}.$$

If the series $\sum_{n\geqslant 1} f(n)$ converges, then[6]

$$\sigma_c = \limsup_{N\to\infty} \frac{\log|\sum_{n>N} f(n)|}{\log N}.$$

This should be compared to *Hadamard's formula* for the radius R of convergence of the power series $\sum_{n=1}^{\infty} a_n n^s$ given by

$$R^{-1} = \limsup_{n\to\infty} |a_n|^{1/n}.$$

4.7.6 Dirichlet Series—Multiplicative Aspects

In this section, we will now focus on the Dirichlet series $F(s) = \sum_{n=1}^{\infty} f(n) n^{-s}$ with multiplicative coefficients $f(n)$. The first result is of crucial importance.

Theorem 4.47 *Let f be a multiplicative function satisfying*

$$\sum_p \sum_{\alpha=1}^{\infty} |f(p^\alpha)| < \infty. \tag{4.28}$$

Then the series $\sum_{n\geqslant 1} f(n)$ is absolutely convergent and we have

$$\sum_{n=1}^{\infty} f(n) = \prod_p \left(1 + \sum_{\alpha=1}^{\infty} f(p^\alpha)\right).$$

Proof Let us first notice that (4.28) implies the convergence of the product

$$\prod_p \left(1 + \sum_{\alpha=1}^{\infty} |f(p^\alpha)|\right).$$

Now let $x \geqslant 2$ be a real number and set

$$P(x) = \prod_{p\leqslant x} \left(1 + \sum_{\alpha=1}^{\infty} |f(p^\alpha)|\right).$$

The convergence of the series $\sum_{\alpha=1}^{\infty} |f(p^\alpha)|$ enables us to rearrange the terms when we expand $P(x)$ and since $|f|$ is multiplicative, we deduce as in Euler's proof of Theorem 3.13 that

$$P(x) = \sum_{P^+(n)\leqslant x} |f(n)|.$$

---

[6]This result is due to Pincherle.

Since each integer $n \leqslant x$ satisfies the condition $P^+(n) \leqslant x$, we infer that

$$\sum_{n \leqslant x} |f(n)| \leqslant P(x).$$

Since $P(x)$ has a finite limit as $x \longrightarrow \infty$, the above inequality implies that the left-hand side is bounded as $x \longrightarrow \infty$, proving the absolute convergence of the series $\sum_{n \geqslant 1} f(n)$. The second part of the theorem follows from the inequality

$$\left| \sum_{n=1}^{\infty} f(n) - \prod_{p \leqslant x} \left(1 + \sum_{\alpha=1}^{\infty} f(p^\alpha) \right) \right| \leqslant \sum_{n > x} |f(n)|$$

and the fact that the right-hand side tends to 0 as $x \longrightarrow \infty$.   □

Applying this theorem to absolutely convergent Dirichlet series readily implies the next result which is an analytic version of Proposition 4.32.

Corollary 4.48 *Let f be a multiplicative function with Dirichlet series $F(s) = \sum_{n=1}^{\infty} f(n)n^{-s}$ and s_0 be a complex number. Then the three following assertions are equivalent.*

(i)

$$\sum_{p} \sum_{\alpha=1}^{\infty} \frac{|f(p^\alpha)|}{p^{s_0 \alpha}} < \infty. \tag{4.29}$$

(ii) *The series $F(s)$ is absolutely convergent in the half-plane $\sigma > \sigma_0$.*
(iii) *The product*

$$\prod_{p} \left(1 + \sum_{\alpha=1}^{\infty} \frac{f(p^\alpha)}{p^{s\alpha}} \right)$$

is absolutely convergent in the half-plane $\sigma > \sigma_0$.

If one of these conditions holds, then we have for all $\sigma > \sigma_0$

$$F(s) = \prod_{p} \left(1 + \sum_{\alpha=1}^{\infty} \frac{f(p^\alpha)}{p^{s\alpha}} \right). \tag{4.30}$$

In particular, if σ_a is the abscissa of absolute convergence of $F(s)$, then (4.30) holds for all $\sigma > \sigma_a$.

We list in Table 4.2 several arithmetic functions f with their Dirichlet series F and some convolution identities often used. The number Θ is the infimum of the real parts of the non-trivial zeros $\rho = \beta + i\gamma$ of ζ such that $\frac{1}{2} \leqslant \beta < 1$. The Riemann hypothesis is $\Theta = \frac{1}{2}$.

Table 4.2 Table of Dirichlet series of certain multiplicative numbers

f	Convolution	$F(s)$	σ_c	σ_a
e_1	Identity element	1	$-\infty$	$-\infty$
$\mathbf{1}$	$\mathbf{1} \star \mu = e_1$	$\zeta(s)$	1	1
μ	$\mu \star \mathbf{1} = e_1$	$\zeta(s)^{-1}$	$\ominus$	1
λ	$\lambda \star \mu_2 = e_1$	$\dfrac{\zeta(2s)}{\zeta(s)}$	$\ominus$	1
μ_k	$\mu_k \star \mu = f_k$	$\dfrac{\zeta(s)}{\zeta(ks)}$	1	1
s_k	$s_k \star \mu = g_k$	$\prod_{j=k}^{2k-1} \zeta(js) \times \dfrac{G_k(s)}{\zeta((2k+2)s)}$	$\dfrac{1}{k}$	$\dfrac{1}{k}$
β	$\beta \star \mu = s_2$	$\dfrac{\zeta(s)\zeta(2s)\zeta(3s)}{\zeta(6s)}$	1	1
a	$a \star \mu = h$	$\prod_{j=1}^{\infty} \zeta(js)$	1	1
$\tau^{(e)}$	$\tau^{(e)} \star \mu = \kappa$	$\zeta(s)\zeta(2s)P(s)$	1	1
$\tau_{(k)}$	$\tau_{(k)} = \mathbf{1} \star \mu_k$	$\dfrac{\zeta(s)^2}{\zeta(ks)}$	1	1
k^ω	$(k+1)^\omega = \mu_2 k^\omega \star \mathbf{1}$	$\zeta(s)^k H_k(s)$	1	1
τ_k	$\tau_k = \tau_{k-1} \star \mathbf{1}$	$\zeta(s)^k$	1	1
φ	$\mathrm{Id} \times \tau = \sigma \star \varphi$ $\varphi = \mu \star \mathrm{Id}$ $\sigma = \varphi \star \tau$	$\dfrac{\zeta(s-1)}{\zeta(s)}$	2	2
Id_k	$\mathrm{Id}_k = \sigma_k \star \mu$ $\mathrm{Id} = \varphi \star \mathbf{1}$	$\zeta(s-k)$	$k+1$	$k+1$
σ_k	$\sigma_k = \mathrm{Id}_k \star \mathbf{1}$ $\sigma = \tau \star \varphi$	$\zeta(s)\zeta(s-k)$	$k+1$	$k+1$
Λ_k	$\Lambda_k = \mu \star \log^k$	$(-1)^k \dfrac{\zeta^{(k)}(s)}{\zeta(s)}$	1	1
$\chi \neq \chi_0$	$\chi \times \log \times (\mu \star \mathbf{1}) = 0$	$L(s,\chi)$	0	1
$\nu_{\mathbb{K}}$	$\nu_{\mathbb{K}} = \chi \star \mathbf{1}$ if $\mathbb{K}$ is a quadratic field	$\zeta_{\mathbb{K}}(s)$	1	1

In this table, we have

▷ $f_k(n) = \begin{cases} \mu(m), & \text{if } n=m^k \\ 0, & \text{otherwise.} \end{cases}$

▷ $g_k(n) = \begin{cases} \mu(a), & \text{if } n=ab,\ (a,b)=1,\ \mu_2(a)=s_k(b)=1 \\ 0, & \text{otherwise.} \end{cases}$

▷ $G_2(s) = 1$ and, for $k \geqslant 3$, $G_k(s)$ is a Dirichlet series absolutely convergent in the half-plane $\sigma > (2k+3)^{-1}$. For all $k \geqslant 2$, $H_k(s)$ is a Dirichlet series absolutely convergent in the half-plane $\sigma > \frac{1}{2}$.

▷ $P(s) = \prod_p (1 + \sum_{\alpha=5}^{\infty} \frac{\tilde{\tau}(\alpha)}{p^{s\alpha}})$ where $\tilde{\tau}(\alpha)$ is defined in (4.26).

▷ h is the multiplicative function defined by $h(p) = 0$ and $h(p^\alpha) = P(\alpha) - P(\alpha - 1)$ and κ is the multiplicative function defined by $\kappa(p) = 0$ and $\kappa(p^\alpha) = \tau(\alpha) - \tau(\alpha - 1)$.

4.7.7 The Von Mangoldt Function of an Arithmetic Function

The von Mangoldt function Λ may be generalized in the following way.

Definition 4.49 Let f be a complex-valued arithmetic function such that $f(1) \neq 0$. The *von Mangoldt function* Λ_f *attached to* f is implicitly defined by the equation

$$\Lambda_f \star f = f \times \log.$$

We obviously have $\Lambda_1 = \Lambda$ and by Proposition 4.31 we have formally

$$\sum_{n=1}^{\infty} \frac{\Lambda_f(n)}{n^s} = -\frac{F'(s)}{F(s)}$$

where $F(s)$ is the Dirichlet series of f.

Example 4.50 In Table 4.3, we set $v(n) = \begin{cases} \alpha, & \text{if } n = p^\alpha \\ 0, & \text{otherwise.} \end{cases}$

We now summarize the main properties of the function Λ_f. The reader is referred to [LF67] for the proofs.

Proposition 4.51 *Let f, g be complex-valued arithmetic functions such that $f(1) \neq 0$ and $g(1) \neq 0$, p be a prime number, $\alpha \in \mathbb{N}$ and $\lambda_1 \geqslant 1$ and $1 \leqslant \lambda_2 < 2$ be real numbers.*

(i) $\Lambda_{f \star g} = \Lambda_f + \Lambda_g$.
(ii) *f is multiplicative if and only if Λ_f is supported on prime powers.*
(iii) *If f is multiplicative, then $\Lambda_f(p^\alpha) = c_f(p^\alpha) \log p$ with $c_f(p) = p$ and for all $\alpha \geqslant 2$*

$$c_f(p^\alpha) = \alpha f(p^\alpha) - \sum_{j=1}^{\alpha-1} c_f(p^j) f(p^{\alpha-j}).$$

Table 4.3 Von Mangoldt functions of certain arithmetic functions

f	e_1	$\mathbf{1}$	μ	μ_2	$2^\omega \mu$	$(-1)^\omega$	τ_k	φ	σ
Λ_f	0	Λ	$-\Lambda$	$(-1)^{v-1}\Lambda$	$-2^v\Lambda$	$(1-2^v)\Lambda$	$k\Lambda$	$(\mathrm{Id}-1)\Lambda$	$(\mathrm{Id}+1)\Lambda$

This implies that

$$\Lambda_f(p^\alpha) = \log p^\alpha \sum_{j=1}^{\alpha} \frac{(-1)^j}{j} \sum_{k_1+\cdots+k_j=\alpha} f(p^{k_1}) \cdots f(p^{k_j}).$$

Furthermore, if f is multiplicative supported on squarefree numbers, then

$$\Lambda_f(p^\alpha) = (-1)^{\alpha-1} (f(p))^\alpha \log p.$$

(iv) *If f is multiplicative such that $|f(n)| \leqslant 1$, then*

$$|\Lambda_f(p^\alpha)| \leqslant (2^\alpha - 1) \log p$$

and if f is multiplicative such that $0 \leqslant f(p^\alpha) \leqslant \lambda_1 \lambda_2^\alpha$, then

$$|\Lambda_f(p^\alpha)| \leqslant (2\lambda_1\lambda_2)^\alpha \log p.$$

(v) *If $f(n) = g(n)h(n)$ with g multiplicative and h completely multiplicative, then*

$$\Lambda_f(n) = \Lambda_g(n)h(n).$$

One of the main uses of the von Mangoldt function Λ_f associated to a multiplicative function f is to determine the average order of $f(n)/n$ via the following result proved by Iwaniec and Kowalski in [IK04].

Theorem 4.52 *Let f be a complex-valued multiplicative function and $\kappa > -\frac{1}{2}$ such that*

$$\sum_{n \leqslant x} \frac{\Lambda_f(n)}{n} = \kappa \log x + O(1)$$

and

$$\sum_{n \leqslant x} \frac{|f(n)|}{n} \ll (\log x)^{|\kappa|}.$$

Then we have

$$\sum_{n \leqslant x} \frac{f(n)}{n} = \frac{(\log x)^\kappa}{\Gamma(\kappa+1)} \prod_p \left(1 - \frac{1}{p}\right)^\kappa \left(1 + \sum_{\alpha=1}^{\infty} \frac{f(p^\alpha)}{p^\alpha}\right) + O\left((\log x)^{|\kappa|-1}\right).$$

For instance, using the table in Example 4.50, we get for all $k \in \mathbb{N}$

$$\sum_{n \leqslant x} \frac{\tau_k(n)}{n} = \frac{(\log x)^k}{k!} + O\left((\log x)^{k-1}\right).$$

4.7.8 Twisted Sums with the Möbius Function

The Möbius function μ has a particular status in number theory, mostly on account of its link with the PNT and the Riemann hypothesis (Chap. 3). Furthermore, μ is not positive and hence does not satisfy the hypotheses of Theorems 4.22 and 4.28.

The aim of this section is to show how to handle sums of the form

$$\sum_{n \leqslant x} \mu(n) f(n)$$

where f is a multiplicative function taking small values at prime numbers. When $f = 1$, the PNT is equivalent to estimates of the shape

$$\sum_{n \leqslant x} \mu(n) \ll \frac{x}{(\log e^2 x)^A} \tag{4.31}$$

for all $A > 0$. There even exist explicit bounds of this form. For instance, it is shown in [Mar95] that, for all $x > 1$, we have

$$\left| \sum_{n \leqslant x} \mu(n) \right| \leqslant \frac{762.7\, x}{(\log x)^2}$$

and for the logarithmic mean value it is proved that

$$\left| \sum_{n \leqslant x} \frac{\mu(n)}{n} \right| < \frac{726}{(\log x)^2}.$$

When $f = \mu$, we have the following inequalities

$$\frac{\log x}{\zeta(2)} + 0.832\,11 < \sum_{n \leqslant x} \frac{\mu^2(n)}{n} < \frac{\log x}{\zeta(2)} + 1.165\,471.$$

We intend to show the following result.

Theorem 4.53 *Let f be a real-valued multiplicative function such that $0 \leqslant f(p) \leqslant 1$ for all prime numbers p. Then we have for all $x \geqslant e$*

$$\sum_{n \leqslant x} \mu(n) f(n) \ll \frac{x}{\log x} \exp\left(\sum_{p \leqslant x} \frac{1 - f(p)}{p} \right).$$

Proof Let $g = \mu f \star 1$. By Theorem 4.10, g is multiplicative and, for all prime powers p^α, we have

$$g(p^\alpha) = 1 + \sum_{j=1}^{\alpha} \mu(p^j) f(p^j) = 1 - f(p).$$

Hence by assumption on f we infer that $0 \leqslant g(p^\alpha) \leqslant 1$ and, using multiplicativity, we get $0 \leqslant g(n) \leqslant 1$. Using the Möbius inversion formula, Proposition 4.17, the bound (4.31) with $A = 2$ and partial summation, we get

$$
\sum_{n \leqslant x} \mu(n) f(n) = \sum_{n \leqslant x} (g \star \mu)(n) = \sum_{d \leqslant x} g(d) \sum_{k \leqslant x/d} \mu(k)
$$

$$
\ll x \sum_{d \leqslant x} \frac{g(d)}{d \log^2(e^2 x/d)}
$$

$$
\ll \sum_{d \leqslant x} g(d) + x \int_1^x \frac{\log(x/t)}{t^2 (\log e^2 x/t)^3} \left(\sum_{d \leqslant t} g(d) \right) dt
$$

$$
\ll \sum_{d \leqslant x} g(d) + x \left(\int_1^{\sqrt{x}} + \int_{\sqrt{x}}^x \right) \frac{\log(x/t)}{t^2 (\log e^2 x/t)^3} \left(\sum_{d \leqslant t} g(d) \right) dt.
$$

In the first integral, we use the trivial inequality

$$
\sum_{d \leqslant t} g(d) \leqslant t
$$

while in the second integral and the first sum, we use Theorem 4.22 which gives

$$
\sum_{d \leqslant t} g(d) \ll \frac{t}{\log(et)} \exp \left(\sum_{p \leqslant t} \frac{g(p)}{p} \right) \ll \frac{t}{\log(et)} \exp \left(\sum_{p \leqslant x} \frac{1 - f(p)}{p} \right)
$$

for any $1 \leqslant t \leqslant x$, and therefore

$$
\sum_{n \leqslant x} \mu(n) f(n) \ll \frac{x}{\log x} \exp \left(\sum_{p \leqslant x} \frac{1 - f(p)}{p} \right) + \frac{x}{(\log x)^2}
$$

$$
\ll \frac{x}{\log x} \exp \left(\sum_{p \leqslant x} \frac{1 - f(p)}{p} \right)
$$

as required. $\qquad \square$

4.7.9 Mean Values of Multiplicative Functions

In view of the erratic behavior of most multiplicative functions, the question of the existence of a mean value of such functions arises naturally. Since higher prime powers p^α with $\alpha \in \mathbb{Z}_{\geqslant 2}$ are quite rare, one may wonder how the values of f at the prime numbers influence the behavior of f in general. The interest in this problem increased as Erdős and Wintner [Erd57] formulated the following conjecture:

Any multiplicative function assuming only the values -1 and $+1$ has a mean value.

On the other hand, comparing the bounds obtained in Example 4.29 with the asymptotic formulae given afterwards in Example 4.27, one may wonder whether the quantity

$$\prod_{p \leqslant x} \left(1 - \frac{1}{p}\right)\left(1 + \sum_{\alpha=1}^{\infty} \frac{f(p^{\alpha})}{p^{\alpha}}\right)$$

of Theorem 4.28 would be close to the right mean value of multiplicative functions f provided that $f(p)$ is sufficiently close to 1. This product is sometimes called the *heuristical value* for the mean $x^{-1} \sum_{n \leqslant x} f(n)$.

The search for accurate asymptotic formulae for average orders really began in the early 1960s with the following satisfactory result due to Wirsing [Wir61].

Theorem 4.54 (Wirsing) *Let f be a positive multiplicative function satisfying the Wirsing conditions* (4.21) *and*

$$\sum_{p \leqslant x} f(p) \log p = \kappa x + o(x)$$

for some $\kappa > 0$. Then we have as $x \longrightarrow \infty$

$$\sum_{n \leqslant x} f(n) = \left(1 + o(1)\right) \frac{e^{-\gamma \kappa}}{\Gamma(\kappa)} \frac{x}{\log x} \prod_{p \leqslant x} \left(1 + \sum_{\alpha=1}^{\infty} \frac{f(p^{\alpha})}{p^{\alpha}}\right).$$

Wirsing's ideas are similar to those of Theorem 4.22, using equalities instead of inequalities and making also use of the Hardy–Littlewood–Karamata tauberian theorem, although he used elementary arguments in his original paper of 1961. In the same paper, Wirsing also deduced theorems for complex-valued multiplicative functions, but these results neither contained the PNT nor settled the Erdős–Wintner conjecture.

Six years later, Wirsing was able to settle this conjecture, proving in an elementary, but tricky way that if f is a real-valued multiplicative function such that $|f(n)| \leqslant 1$, then f has a mean value. The following year, Halász [Hal68] extended Wirsing's results by establishing the *Wirsing conjecture* stating that, if f is a complex-valued multiplicative function such that $|f(n)| \leqslant 1$, then there exist $\kappa = \kappa_f \in \mathbb{C}$, $a \in \mathbb{R}$ and a slowly varying function[7] L with $|L(u)| = 1$, so that

$$\sum_{n \leqslant x} f(n) = \kappa \, L(\log x) \, x^{1+ia} + o(x)$$

---

[7] A *slowly varying function* is a non-zero Lebesgue measurable function $L : [x_0, +\infty[\longrightarrow \mathbb{C}$ for some $x_0 > 0$ for which

$$\lim_{x \to \infty} \frac{L(cx)}{L(x)} = 1$$

for any $c > 0$.

as $x \longrightarrow \infty$. The proof involves contour integration using an asymptotic formula for the Dirichlet series $\sum_{n=1}^{\infty} f^*(n)n^{-s}$, where f^* is a completely multiplicative function associated with f. Various authors [PT78, Skr74, Tul78] subsequently improved on Halász's theorem by partly removing the unpleasant constraint $|f(n)| \leqslant 1$.

In the 1970s and 1980s, the Russian school, led by Kubilius, proved asymptotic formulae with effective error-terms. For instance, Kubilius [Kub71] showed the following result.

Theorem 4.55 (Kubilius) *Let f be a complex-valued multiplicative function such that there exists a constant $\kappa \in \mathbb{C}$, independent of p, with $|\kappa| \leqslant c_1$ and such that*

$$\sum_{p} \frac{|f(p) - \kappa| \log p}{p} < c_2 \quad and \quad \sum_{p} \sum_{\alpha=2}^{\infty} \frac{|f(p^\alpha)| \log p^\alpha}{p^\alpha} < c_3.$$

Then for any $x \geqslant 20$, we have

$$\sum_{n \leqslant x} f(n) = \frac{x (\log x)^{\kappa-1}}{\Gamma(\kappa)} \prod_{p} \left(1 - \frac{1}{p}\right)^{\kappa} \left(1 + \sum_{\alpha=1}^{\infty} \frac{f(p^\alpha)}{p^\alpha}\right) + O\left(x \, \mathcal{R}_\kappa(x)(\log x)^{|\kappa|/2-1}\right)$$

where

$$\mathcal{R}_\kappa(x) = \begin{cases} (1 - |\kappa|)^{-1/2}, & if \ |\kappa| < 1 - (\log \log x)^{-1} \\ (\log \log x)^{1/2}, & if \ ||\kappa| - 1| \leqslant (\log \log x)^{-1} \\ (|\kappa| - 1)^{-1/2}(\log x)^{(|\kappa|-1)/2}, & if \ |\kappa| > 1 + (\log \log x)^{-1}. \end{cases}$$

The implied constant depends only on c_1, c_2, c_3 and $\Gamma(\kappa)^{-1} = 0$ for $\kappa = 0, -1, -2, \ldots$

Concerning the logarithmic mean value $\sum_{n \leqslant x} f(n)/n$, Martin [Mar02] proved the following precise estimates using some ideas developed by Iwaniec.

Theorem 4.56 (Martin) *Let f be a complex-valued positive multiplicative function satisfying*

$$\sum_{p \leqslant x} \frac{f(p) \log p}{p} = \kappa \log x + O(1)$$

where $\kappa = \sigma_\kappa + it_\kappa \in \mathbb{C}$ is such that $t_\kappa^2 < 2\sigma_\kappa + 1$. Assume also that

$$\sum_{p} \frac{|f(p)| \log p}{p} \sum_{\alpha=1}^{\infty} \frac{|f(p^\alpha)|}{p^\alpha} + \sum_{p} \sum_{\alpha=2}^{\infty} \frac{|f(p^\alpha)| \log p^\alpha}{p^\alpha} < \infty$$

and

$$\prod_{p\leqslant x}\left(1+\frac{|f(p)|}{p}\right)\ll(\log x)^{\lambda}$$

for all $x\geqslant 2$ and some real number $0\leqslant\lambda<\sigma_{\kappa}+1$. Then we have

$$\sum_{n\leqslant x}\frac{f(n)}{n}=\frac{(\log x)^{\kappa}}{\Gamma(\kappa+1)}\prod_{p}\left(1-\frac{1}{p}\right)^{\kappa}\left(1+\sum_{\alpha=1}^{\infty}\frac{f(p^{\alpha})}{p^{\alpha}}\right)+O\left((\log x)^{\lambda-1}\right)$$

and if $q\in\mathbb{N}$, we also have

$$\sum_{\substack{n\leqslant x\\(n,q)=1}}\frac{f(n)}{n}=\frac{(\log x)^{\kappa}}{\Gamma(\kappa+1)}\left(\frac{\varphi(q)}{q}\right)^{\kappa}\prod_{p}\left(1-\frac{1}{p}\right)^{\kappa}\left(1+\sum_{\alpha=1}^{\infty}\frac{f(p^{\alpha})}{p^{\alpha}}\right)$$

$$+O\left\{\left(1+\sum_{p|q}\frac{|f(p)|\log p}{p}\right)(\log x)^{\lambda-1}\right\}.$$

In the case of positive multiplicative functions, the statement is simpler.

Corollary 4.57 *Let f be a positive multiplicative function such that $f(n)\ll n^{\lambda}$ for some $\lambda<\frac{1}{2}$ and satisfying*

$$\sum_{p\leqslant x}\frac{f(p)\log p}{p}=\kappa\log x+O(1)$$

where $\kappa>0$. Then uniformly in $x\geqslant 2$, we have

$$\sum_{n\leqslant x}\frac{f(n)}{n}=\frac{(\log x)^{\kappa}}{\Gamma(\kappa+1)}\prod_{p}\left(1-\frac{1}{p}\right)^{\kappa}\left(1+\sum_{\alpha=1}^{\infty}\frac{f(p^{\alpha})}{p^{\alpha}}\right)+O\left((\log x)^{\kappa-1}\right).$$

Let us apply this result to the function $f(n)=n/S(n)$ where $S(n)=\sum_{i=1}^{n}(i,n)$ is the so-called *Pillai function*. Using Exercise 11, we have $S=\varphi\star\mathrm{Id}$, so that S is multiplicative and since $(i,n)\geqslant 1$, we clearly get $0\leqslant f(n)\leqslant 1$ for all $n\geqslant 1$. Furthermore, we have using Theorem 3.49

$$\sum_{p\leqslant x}\frac{f(p)\log p}{p}=\sum_{p\leqslant x}\frac{\log p}{p}\left(\frac{1}{2}+\frac{1}{4p-2}\right)=\frac{1}{2}\sum_{p\leqslant x}\frac{\log p}{p}+O(1)=\frac{\log x}{2}+O(1)$$

so that the conditions of Corollary 4.57 are satisfied with $\kappa=1/2$, and using $\Gamma(\frac{3}{2})=\frac{\sqrt{\pi}}{2}$, we get the asymptotic formula

$$\sum_{n\leqslant x}\frac{1}{S(n)}=\frac{2}{\sqrt{\pi}}\prod_{p}\left(1-\frac{1}{p}\right)^{1/2}\left(1+\sum_{\alpha=1}^{\infty}\frac{1}{S(p^{\alpha})}\right)\sqrt{\log x}+O\left((\log x)^{-1/2}\right)$$

for all $x \geqslant 2$. Note that, since $\varphi = \mu \star \mathrm{Id}$, we get $S = \mu \star \mathrm{Id} \star \mathrm{Id} = \mu \star (\mathrm{Id} \times \tau)$ (see also Exercise 12) so that

$$S(p^\alpha) = (\alpha + 1)p^\alpha - \alpha p^{\alpha - 1}.$$

In [FI05], the authors have another approach to this problem. They consider a complex-valued multiplicative function f verifying the Ramanujan condition $f(n) \ll n^\varepsilon$ and whose Dirichlet series $F(s)$ satisfies the following properties.

1. $F(s)$ is holomorphic in the half-plane $\sigma > 1$, and has analytic continuation to the whole complex plane where it is holomorphic except possibly for a pole at $s = 1$, not necessarily simple.
2. $F(s)$ satisfies the functional equation in the half-plane $\sigma > 1$

$$F(1 - s) = w\gamma(s)G(s)$$

where w is the so-called *root-number* such that $|w| = 1$, $G(s) = \sum_{n=1}^{\infty} g(n)n^{-s}$ with $g(n) \ll n^\varepsilon$ and $\gamma(s)$ is holomorphic in the half-plane $\sigma > \frac{1}{2}$ having the shape

$$\gamma(s) = \left(\pi^{-m} D\right)^{s - 1/2} \prod_{j=1}^{m} \Gamma\left(\frac{s + \kappa_j}{2}\right) \Gamma\left(\frac{1 - s + \kappa_j}{2}\right)^{-1}$$

where $D \geqslant 1$ is an integer called the *conductor* of $F(s)$, $m \in \mathbb{N}$ is the *degree* of $F(s)$, $\kappa_1, \ldots, \kappa_m$ are the *spectral parameters* of $F(s)$, i.e. complex numbers such that $\operatorname{Re} \kappa_j \geqslant 0$.

We may now state the main result of [FI05].

Theorem 4.58 (Friedlander–Iwaniec) *With the above hypotheses and for any $x \geqslant \sqrt{D}$ and all $\varepsilon > 0$, we have*

$$\sum_{n \leqslant x} f(n) = \operatorname*{Res}_{s=1}\left(F(s)x^s s^{-1}\right) + O_{\varepsilon,\kappa_j}\left(D^{\frac{1}{m+1}} x^{\frac{m-1}{m+1} + \varepsilon}\right)$$

where the implied constant depends only on ε and the spectral parameters $\kappa_1, \ldots, \kappa_m$.

The strength of this result lies in the fact that the error-term depends only on ε and the spectral parameters $\kappa_1, \ldots, \kappa_m$. We may apply it to the multiplicative function $v_{\mathbb{K}}$ studied in Chap. 7, where $\mathbb{K}$ is an algebraic number field of degree $d \geqslant 1$. It is shown in Corollary 7.119 (iii) that

$$v_{\mathbb{K}}(n) \leqslant \tau_d(n) \ll n^\varepsilon.$$

Furthermore, the Dirichlet series of $v_{\mathbb{K}}$ is the Dedekind zeta-function $\zeta_{\mathbb{K}}(s)$ (see Definition 7.121), and using Theorem 7.123 (ii) and the duplication formula (3.10), we infer that $\zeta_{\mathbb{K}}(s)$ satisfies the following functional equation

$$\zeta_{\mathbb{K}}(1-s) = \left(\pi^{-d}|d_{\mathbb{K}}|\right)^{s-1/2}\left(\Gamma\left(\frac{s}{2}\right)\Gamma\left(\frac{1-s}{2}\right)^{-1}\right)^{r_1+r_2}$$

$$\times \left(\Gamma\left(\frac{s+1}{2}\right)\Gamma\left(\frac{2-s}{2}\right)^{-1}\right)^{r_2}\zeta_{\mathbb{K}}(s)$$

where (r_1, r_2) is the signature and $d_{\mathbb{K}}$ is the discriminant of $\mathbb{K}$. Hence the function $v_{\mathbb{K}}$ satisfies the conditions of the above result with $w = 1$, $G(s) = \zeta_{\mathbb{K}}(s)$, $D = |d_{\mathbb{K}}|$, $m = d$ and

$$\kappa_j = \begin{cases} 0, & \text{if } 1 \leqslant j \leqslant r_1 + r_2 \\ 1, & \text{if } 1 \leqslant j \leqslant r_2. \end{cases}$$

Applying Theorem 4.58 we readily get the following important result.

Corollary 4.59 (Ideal Theorem) *Let $\mathbb{K}$ be an algebraic number field of degree d and discriminant $d_{\mathbb{K}}$. For all $x \geqslant |d_{\mathbb{K}}|^{1/2}$ and all $\varepsilon > 0$, we have*

$$\sum_{n \leqslant x} v_{\mathbb{K}}(n) = \kappa_{\mathbb{K}} x + O_\varepsilon\left(|d_{\mathbb{K}}|^{\frac{1}{d+1}} x^{\frac{d-1}{d+1}+\varepsilon}\right)$$

where $\kappa_{\mathbb{K}}$ is given by (7.21) and the implied constant depends only on ε.

This result was first proved by Landau in [Lan27, Satz 210], but with an unspecified constant in the error-term. Both proofs start using the same tools, namely Theorem 3.84 and contour integration methods, but Friedlander and Iwaniec next used the stationary phase (see Lemma 6.29) to estimate some integrals more precisely. They provided a result in a much stronger form than the one given in Theorem 4.58 above, with a supplementary term à la Voronoï as in [Vor04]. The trivial estimate of this term gives Theorem 4.58.

4.7.10 Lower Bounds

Lower bounds for sums $\sum_{n \leqslant x} f(n)$ are not surprisingly harder to obtain than upper bounds. The next result provides a lower bound for the logarithmic mean value on squarefree numbers.

Proposition 4.60 *Let f be a positive multiplicative function satisfying $0 \leqslant f(p) \leqslant \lambda < 1$ for all primes p. Then*

$$\sum_{n \leqslant x} \frac{\mu^2(n) f(n)}{n} \geqslant (1-\lambda) \prod_{p \leqslant x} \left(1 + \frac{f(p)}{p}\right).$$

Note that we have

$$\sum_{n\leqslant x}\frac{\mu^2(n)f(n)}{n} \leqslant \sum_{P^+(n)\leqslant x}\frac{\mu^2(n)f(n)}{n} = \prod_{p\leqslant x}\left(1+\frac{f(p)}{p}\right)$$

so that the result above is of the right order of magnitude.

Proof We have

$$0 \leqslant \prod_{p\leqslant x}\left(1+\frac{f(p)}{p}\right) - \sum_{n\leqslant x}\frac{\mu^2(n)f(n)}{n} = \sum_{\substack{n>x\\P^+(n)\leqslant x}}\frac{\mu^2(n)f(n)}{n}$$

and using $\log n > \log x$ in the right-hand side implies that

$$\prod_{p\leqslant x}\left(1+\frac{f(p)}{p}\right) - \sum_{n\leqslant x}\frac{\mu^2(n)f(n)}{n} < \sum_{\substack{n>x\\P^+(n)\leqslant x}}\frac{\mu^2(n)f(n)}{n}\frac{\log n}{\log x}$$

$$= \frac{1}{\log x}\sum_{\substack{n>x\\P^+(n)\leqslant x}}\frac{\mu^2(n)f(n)}{n}\sum_{p|n}\log p$$

$$\leqslant \frac{1}{\log x}\sum_{p\leqslant x}\frac{f(p)\log p}{p}\sum_{\substack{k\geqslant x/p\\P^+(k)\leqslant x}}\frac{\mu^2(k)f(k)}{k}$$

$$\leqslant \frac{\lambda}{\log x}\sum_{p\leqslant x}\frac{\log p}{p}\sum_{P^+(k)\leqslant x}\frac{\mu^2(k)f(k)}{k}$$

$$= \frac{\lambda}{\log x}\sum_{p\leqslant x}\frac{\log p}{p}\prod_{p\leqslant x}\left(1+\frac{f(p)}{p}\right)$$

and Corollary 3.99 in the form

$$\sum_{p\leqslant x}\frac{\log p}{p} < \log x$$

completes the proof. □

Let us state without proof the next result, which is a consequence of a theorem due to Barban (see [SS94]).

Proposition 4.61 *Let f be a positive multiplicative function such that $f(p) \geqslant \lambda_1 > 0$ for all primes $p \geqslant p_0$ and $0 \leqslant f(p^\alpha) \leqslant \lambda_2$ for all prime powers p^α, where $\lambda_1 > 0$*

and $\lambda_2 > 0$. Then we have

$$\sum_{n \leqslant x} f(n) \gg x \exp\left(\sum_{p \leqslant x} \frac{f(p) - 1}{p}\right).$$

Lemma 4.69 below provides another lower bound using elementary means of a certain class of multiplicative functions.

4.7.11 Short Sums of Multiplicative Functions

By *short sums*, we mean sums having the shape

$$\sum_{x < n \leqslant x+y} f(n)$$

where $x \geqslant 1$, $y > 0$ are real numbers such that $y = o(x)$ as $x \longrightarrow \infty$. Compared to the case of long sums, there are fewer results for such sums in the literature. In particular, there is no known asymptotic formula for the case of positive multiplicative functions satisfying the Wirsing conditions (4.21). The first very important theorem is due to Shiu [Shi80] and provides only an upper bound.

Theorem 4.62 (Shiu) *Let f be a positive multiplicative function, $\delta > 0$, $0 < \varepsilon$, $\theta < \frac{1}{2}$ and $0 < a < q$ be two positive coprime integers. Assume that*

(i) *there exists $\lambda_1 = \lambda_1(\delta) > 0$ such that $f(n) \leqslant \lambda_1 n^\delta$ for all $n \geqslant 1$*
(ii) *there exists $\lambda_2 > 0$ such that $f(p^\alpha) \leqslant \lambda_2^\alpha$ for all prime powers p^α.*

Then, for $x \geqslant 1$ sufficiently large and uniformly in a, q and y such that $q < y^{1-\theta}$ and $x^\varepsilon \leqslant y \leqslant x$, we have

$$\sum_{\substack{x < n \leqslant x+y \\ n \equiv a \,(\mathrm{mod}\, q)}} f(n) \ll \frac{y}{\varphi(q) \log x} \exp\left(\sum_{\substack{p \leqslant x \\ p \nmid q}} \frac{f(p)}{p}\right).$$

The implied constant depends on λ_1, λ_2 and ε.

This result has turned out to be very useful. One may easily check that all the functions of Lemma 4.24 satisfy conditions (i) and (ii) of Shiu's theorem. In order to apply this result to them, the following lemma will be useful.

Lemma 4.63 *Let $q \geqslant 2$ be an integer. For all $x \geqslant \max(e, q)$, we have*

$$\sum_{\substack{p \leqslant x \\ p \nmid q}} \frac{1}{p} < \log\left(2e^\gamma \frac{\varphi(q)}{q} \log x\right).$$

Proof On the one hand, we have

$$\prod_{\substack{p \leqslant x \\ p \nmid q}} \left(1 - \frac{1}{p}\right)^{-1} = \exp\left(\sum_{\substack{p \leqslant x \\ p \nmid q}} \frac{1}{p} + \sum_{\substack{p \leqslant x \\ p \nmid q}} \sum_{\alpha=2}^{\infty} \frac{1}{\alpha p^\alpha}\right) \geqslant \exp\left(\sum_{\substack{p \leqslant x \\ p \nmid q}} \frac{1}{p}\right)$$

and on the other hand, since $q \leqslant x$, we also have

$$\prod_{\substack{p \leqslant x \\ p \nmid q}} \left(1 - \frac{1}{p}\right)^{-1} = \prod_{p \leqslant x} \left(1 - \frac{1}{p}\right)^{-1} \prod_{\substack{p \leqslant x \\ p \mid q}} \left(1 - \frac{1}{p}\right) = \frac{\varphi(q)}{q} \prod_{p \leqslant x} \left(1 - \frac{1}{p}\right)^{-1}$$

and we use Corollary 3.99 to conclude the proof. □

With this lemma at our disposal, Shiu's theorem implies at once the following estimates.

Corollary 4.64 *Let* $x^\varepsilon \leqslant y \leqslant x$, $(a, q) = 1$ *and* $q < y^{1-\theta}$ *as in Theorem 4.62. Then we have for* x *sufficiently large the following estimates.*

(i) *Let* $k \geqslant 2$. *We have*

$$\sum_{\substack{x < n \leqslant x+y \\ n \equiv a \,(\mathrm{mod}\, q)}} s_k(n) \ll \frac{y}{\varphi(q) \log x}.$$

(ii) *The estimate*

$$\sum_{\substack{x < n \leqslant x+y \\ n \equiv a \,(\mathrm{mod}\, q)}} f(n) \ll \frac{y}{q}$$

holds with $f = \beta$, $f = a$, $f = \tau^{(e)}$ *and* $f = \mu_k$ ($k \geqslant 2$).

(iii) *Let* $k \geqslant 2$. *We have*

$$\sum_{\substack{x < n \leqslant x+y \\ n \equiv a \,(\mathrm{mod}\, q)}} \tau_{(k)}(n) \ll \frac{\varphi(q)}{q^2} y \log x.$$

(iv) *Let* $k \geqslant 1$. *The estimate*

$$\sum_{\substack{x < n \leqslant x+y \\ n \equiv a \,(\mathrm{mod}\, q)}} f(n) \ll \frac{y}{q} \left(\frac{\varphi(q)}{q} \log x\right)^{k-1}$$

holds with $f = k^\omega$ *and* $f = \tau_k$.

Once again, only the first bound is overestimated. We shall see in Chap. 5 how to get some improvements on this problem.

The idea of the proof of Shiu's theorem is a combination of sieve methods (see the Selberg sieve below), estimates of particular parts of sums as in Theorem 4.22 and a method used earlier by Wolke, whose ideas go back to Erdős, starting with a decomposition of each integer $n \in \left]x, x+y\right]$ as follows

$$n = ab = \left(p_1^{\alpha_1} \cdots p_r^{\alpha_r}\right)\left(p_{r+1}^{\alpha_{r+1}} \cdots p_s^{\alpha_s}\right)$$

with $a \leqslant y^{\theta/10} < ap_{r+1}^{\alpha_{r+1}}$. The original sum is then split into four classes according to

1. $P^-(b) > y^{\theta/20}$
2. $P^-(b) \leqslant y^{\theta/20}$ and $a \leqslant y^{\theta/20}$
3. $P^-(b) \leqslant \log x \log \log x$ and $a > y^{\theta/20}$
4. $\log x \log \log x < P^-(b) \leqslant y^{\theta/20}$ and $a > y^{\theta/20}$.

Shiu's theorem was later generalized in two directions. First, Nair proved an analogous estimate for sums of the shape

$$\sum_{x<n\leqslant x+y} f(|P(n)|)$$

where $P \in \mathbb{Z}[X]$ is an integer polynomial having non-zero discriminant and no fixed prime divisors, and f is a multiplicative function satisfying the same conditions as in Shiu's theorem. Then, Nair and Tenenbaum [NT98] weakened the property of multiplicativity by showing the next result.

Theorem 4.65 (Nair–Tenenbaum) *Let f be a positive arithmetic function, $0 < \delta \leqslant 1$, $d \in \mathbb{N}$, $0 < \varepsilon < \frac{1}{8d^2}$, $c_0 > 0$ and $P = a_d X^d + \cdots + a_0 \in \mathbb{Z}[X]$ be irreducible. Let $\rho(n)$ be the number of solutions of the congruence $P(x) \equiv 0 \pmod{n}$. Assume that $\rho(p) < p$ for all primes p and that there exist $\lambda_1, \lambda_2 \geqslant 1$ such that, for all positive coprime integers m, n, we have*

$$f(mn) \leqslant \min\left(\lambda_1 m^{\varepsilon\delta/3}, \lambda_2^{\Omega(m)}\right) f(n).$$

Then, for $x \geqslant c_0 \max\{|a_0|, \ldots, |a_d|\}^\delta$ and uniformly in $x^{4d^2\varepsilon} \leqslant y \leqslant x$, we have

$$\sum_{x<n\leqslant x+y} f(|P(n)|) \ll y \prod_{p\leqslant x}\left(1 - \frac{\rho(p)}{p}\right) \sum_{n\leqslant x} \frac{\rho(n)f(n)}{n}.$$

In particular, if $0 < \varepsilon < \frac{1}{2}$ and $x \geqslant c_0$, we have uniformly in $x^\varepsilon \leqslant y \leqslant x$

$$\sum_{x<n\leqslant x+y} f(n) \ll y \prod_{p\leqslant x}\left(1 - \frac{1}{p}\right) \sum_{n\leqslant x} \frac{f(n)}{n}.$$

This result enables us to treat short sums of non-necessarily multiplicative functions. In [NT98], the example of the Hooley's Δ_k-function is given. This function is obviously not multiplicative, but by Exercise 8 we see that it satisfies the condition of Theorem 4.65. Furthermore, the following result is proved in [HT88, Theorem 70].

Proposition 4.66 (Hall–Tenenbaum) *For all integers $k \geqslant 2$ and real numbers $x > e^{e^e}$, we set*

$$\epsilon_k(x) = \sqrt{\frac{\log\log\log x}{\log\log x}} \left\{ k - 1 + \frac{30}{\log\log\log x} \right\}.$$

Then we have

$$\sum_{n \leqslant x} \Delta_k(n) \ll x(\log x)^{\epsilon_k(x)}.$$

A partial summation gives at once

$$\sum_{n \leqslant x} \frac{\Delta_k(n)}{n} \ll (\log x)^{1+\epsilon_k(x)}$$

so that using Theorem 4.65 and the second Mertens theorem (Corollary 3.51), we get

$$\sum_{x < n \leqslant x+y} \Delta_k(n) \ll y(\log x)^{\epsilon_k(x)}$$

for $0 < \varepsilon < \frac{1}{2}$, $x \geqslant c_0$ and uniformly in $x^\varepsilon \leqslant y \leqslant x$.

With $k = 2$ and noticing that, for all $N \in \mathbb{N}$, we have

$$\sum_{N < n \leqslant 2N} \left(\left[\frac{x+y}{n}\right] - \left[\frac{x}{n}\right] \right) = \sum_{N < d \leqslant 2N} \sum_{x < dk \leqslant x+y} 1$$

$$= \sum_{x < n \leqslant x+y} \sum_{\substack{d \mid n \\ N < d \leqslant 2D}} \leqslant \sum_{x < n \leqslant x+y} \Delta(n)$$

we infer the bound

$$\sum_{N < n \leqslant 2N} \left(\left[\frac{x+y}{n}\right] - \left[\frac{x}{n}\right] \right) \ll y(\log x)^{\epsilon_2(x)} = y(\log x)^{o(1)} \tag{4.32}$$

for $0 < \varepsilon < \frac{1}{2}$, $x \geqslant c_0$ and uniformly in $x^\varepsilon \leqslant y \leqslant x$ and $N \in \mathbb{N}$. This is the best result to date for the left-hand side (see also Exercise 9 in Chap. 5).

4.7.12 Sums of Sub-multiplicative Functions

Halberstam and Richert [HR79] proved the next result, which is a generalization of Theorem 4.22 to sub-multiplicative functions and provides an improvement of the constants appearing in this result.

Theorem 4.67 *Let f be a non-negative sub-multiplicative function such that $f(1) = 1$ and satisfying the Wirsing conditions (4.21). Assume that, for all $x \geqslant 2$, we have*

$$\sum_{p \leqslant x} f(p) \log p \leqslant \kappa x + O\left(\frac{x}{(\log x)^2}\right)$$

for some constant $\kappa > 0$. Then

$$\sum_{n \leqslant x} f(n) \leqslant \frac{\kappa x}{\log x}\left(1 + O\left(\frac{1}{\log x}\right)\right) \sum_{n \leqslant x} \frac{f(n)}{n}.$$

The sum of the right-hand side may be estimated by Theorem 4.56 or Corollary 4.57, or by using the usual inequality

$$\sum_{n \leqslant x} \frac{f(n)}{n} \leqslant \prod_{p \leqslant x}\left(1 + \sum_{\alpha=1}^{\infty} \frac{f(p^\alpha)}{p^\alpha}\right).$$

For instance, applying Theorem 4.67 to the function f_q defined by

$$f_q(n) = \begin{cases} 1, & \text{if } (n, q) = 1 \\ 0, & \text{if } (n, q) > 1 \end{cases}$$

where q is a fixed positive integer such that $P^+(q) \leqslant x$, we get using Corollary 3.51

$$\sum_{\substack{n \leqslant x \\ (n,q)=1}} 1 \leqslant e^\gamma \frac{\varphi(q)}{q} x \left(1 + O\left(\frac{1}{\log x}\right)\right).$$

4.7.13 Sums of Additive Functions

The average orders of additive functions are in general easier to estimate than those of multiplicative functions.

Proposition 4.68 *Let f be an additive function and $x \geqslant 2$ be a real number. Then*

$$\sum_{n \leqslant x} f(n) = x \sum_{p \leqslant x} \frac{f(p)}{p} + x \sum_{p^\alpha \leqslant x} \sum_{\alpha=2}^{\infty} \frac{f(p^\alpha) - f(p^{\alpha-1})}{p^\alpha}$$

$$+ O\left(\sum_{p^\alpha \leqslant x} \sum_{\alpha=1}^{\infty} |f(p^\alpha) - f(p^{\alpha-1})|\right).$$

Proof Let $g = f \star \mu$. By Exercise 6, g is supported on prime powers. Furthermore, since f is additive, we have $g(p) = f(p)$. Hence, by Proposition 4.17, we have

$$\sum_{n \leqslant x} f(n) = \sum_{n \leqslant x} (g \star \mathbf{1})(n) = \sum_{d \leqslant x} g(d) \left[\frac{x}{d} \right]$$

$$= \sum_{p \leqslant x} g(p) \left[\frac{x}{p} \right] + \sum_{\substack{p^\alpha \leqslant x \\ \alpha \geqslant 2}} g(p^\alpha) \left[\frac{x}{p^\alpha} \right]$$

$$= x \sum_{p \leqslant x} \frac{f(p)}{p} + O\left(\sum_{p \leqslant x} |f(p)| \right) + x \sum_{\substack{p^\alpha \leqslant x \\ \alpha \geqslant 2}} \frac{g(p^\alpha)}{p^\alpha} + O\left(\sum_{\substack{p^\alpha \leqslant x \\ \alpha \geqslant 2}} |g(p^\alpha)| \right)$$

and we conclude by using $g(p^\alpha) = f(p^\alpha) - f(p^{\alpha-1})$.   □

For instance with $f = \omega$, using a weak form of the PNT, we get

$$\sum_{n \leqslant x} \omega(n) = x \log \log x + Bx + O\left(\frac{x}{\log x} \right)$$

where $B \approx 0.261\,49\ldots$ is the Mertens constant from Corollary 3.50.

4.7.14 The Selberg's Sieve

The sieves of Eratosthenes and Brun have been studied in Chap. 3. The aim of this section is to introduce another powerful tool developed by Atle Selberg in the late 1940s. We first recall the specific notation in sieve methods.

Let $z \geqslant 2$ be a real number, $\mathcal{P}$ be a set of primes, $P_z = \prod_{p \in \mathcal{P}, \, p \leqslant z} p$ and $\mathcal{A}$ be a finite set of integers such that, for all $d \mid P_z$, we have

$$\sum_{\substack{n \in \mathcal{A} \\ d \mid n}} 1 = \frac{X \rho(d)}{d} + r_d$$

with $X > 0$, ρ is a non-negative multiplicative function and r_d is a remainder term satisfying $|r_d| \leqslant \rho(d)$. We set

$$S(\mathcal{A}, \mathcal{P}; z) = \sum_{\substack{n \in \mathcal{A} \\ (n, P_z) = 1}} 1.$$

Instead of using Bonferroni-like inequalities (3.30), Selberg observed that if $\lambda : \mathbb{N} \longrightarrow \mathbb{R}$ is any function such that $\lambda(1) = 1$ then, for all n, we have

$$\sum_{d \mid n} \mu(d) \leqslant \left(\sum_{d \mid n} \lambda(d) \right)^2 \qquad (4.33)$$

and squaring out the inner sum implies that

$$S(\mathcal{A}, \mathcal{P}; z) \leqslant \sum_{n \in \mathcal{A}} \left(\sum_{d \mid (n, P_z)} \lambda(d) \right)^2 = \sum_{d \mid P_z} A_d \sum_{[d_1, d_2] = d} \lambda(d_1)\lambda(d_2).$$

Using (3.29), we get

$$S(\mathcal{A}, \mathcal{P}; z) \leqslant X \sum_{d \mid P_z} \frac{\rho(d)}{d} \sum_{[d_1, d_2] = d} \lambda(d_1)\lambda(d_2) + O\left(\sum_{d \mid P_z} |r_d| \sum_{[d_1, d_2] = d} |\lambda(d_1)\lambda(d_2)| \right).$$

Assume $1 \leqslant \rho(d) < d$ for $d \mid P_z$. Since ρ is multiplicative, we have

$$\rho(d_1)\rho(d_2) = \rho\big((d_1, d_2)\big)\rho\big([d_1, d_2]\big)$$

and setting $\phi(d) = d/\rho(d)$ and $f = \mu \star \phi$ we get

$$\sum_{d \mid P_z} \frac{\rho(d)}{d} \sum_{[d_1, d_2] = d} \lambda(d_1)\lambda(d_2) = \sum_{d_1, d_2 \mid P_z} \frac{\lambda(d_1)\lambda(d_2)}{[d_1, d_2]} \rho\big([d_1, d_2]\big)$$

$$= \sum_{d_1, d_2 \mid P_z} \left(\frac{\lambda(d_1)\lambda(d_2)(d_1, d_2)}{d_1 d_2} \times \frac{\rho(d_1)\rho(d_2)}{\rho((d_1, d_2))} \right)$$

$$= \sum_{d_1, d_2 \mid P_z} \frac{\lambda(d_1)\lambda(d_2)\phi((d_1, d_2))}{\phi(d_1)\phi(d_2)}.$$

Now by the Möbius inversion formula, we have $\phi = f \star 1$ and hence

$$\sum_{d \mid P_z} \frac{\rho(d)}{d} \sum_{[d_1, d_2] = d} \lambda(d_1)\lambda(d_2) = \sum_{d_1, d_2 \mid P_z} \sum_{\delta \mid (d_1, d_2)} \frac{f(\delta)\lambda(d_1)\lambda(d_2)}{\phi(d_1)\phi(d_2)}$$

$$= \sum_{\delta \mid P_z} f(\delta) \left(\sum_{\substack{d \mid P_z \\ \delta \mid d}} \frac{\lambda(d)}{\phi(d)} \right)^2.$$

We have then proved the estimate

$$S(\mathcal{A}, \mathcal{P}; z) \leqslant X \sum_{d \mid P_z} f(d) \left(\sum_{\substack{\delta \mid P_z \\ d \mid \delta}} \frac{\lambda(\delta)}{\phi(\delta)} \right)^2 + O\left(\sum_{d_1, d_2 \mid P_z} |\lambda(d_1)\lambda(d_2)r_{[d_1, d_2]}| \right) \quad (4.34)$$

for all $z \geqslant 2$, if (3.29) holds with $1 \leqslant \rho(d) < d$ for $d \mid P_z$, and where $\lambda : \mathbb{N} \longrightarrow \mathbb{R}$ satisfies $\lambda(1) = 1$. This leads to an optimization problem of a quadratic form which was solved by Selberg. Note that by (4.33) we have the freedom to choose any function λ we like, subject only to the constraint $\lambda(1) = 1$. We begin by setting

$$y_d = \sum_{\substack{\delta \mid P_z \\ d \mid \delta}} \frac{\lambda(\delta)}{\phi(\delta)}. \tag{4.35}$$

For all positive integers $d \mid P_z$, we have by the Möbius inversion formula

$$\sum_{\substack{n \mid P_z \\ d \mid n}} \mu\left(\frac{n}{d}\right) y_n = \frac{\lambda(d)}{\phi(d)}$$

so that

$$\lambda(d) = \phi(d) \sum_{\substack{n \mid P_z \\ d \mid n}} \mu\left(\frac{n}{d}\right) y_n \tag{4.36}$$

which implies in particular that the linear transformation (4.35) is invertible. From (4.35) and (4.36), we have

$$\lambda(d) = 0 \text{ for } d > z \quad \Longleftrightarrow \quad y_d = 0 \text{ for } d > z$$

which we may suppose for convenience. Since $\phi(d) = d/\rho(d)$, we get $\phi(1) = 1$. By (4.36), the condition $\lambda(1) = 1$ may be written as

$$\sum_{\substack{n \mid P_z \\ n \leqslant z}} \mu(n) y_n = 1. \tag{4.37}$$

Hence we set about minimizing the quadratic form

$$\sum_{\substack{d \mid P_z \\ d \leqslant z}} f(d) y_d^2 \tag{4.38}$$

subject to (4.37). Let

$$M_z = \sum_{\substack{d \mid P_z \\ d \leqslant z}} \frac{1}{f(d)}.$$

Using (4.37) and the fact that $d \mid P_z \Longrightarrow \mu^2(d) = 1$, we have

$$\sum_{\substack{d \mid P_z \\ d \leqslant z}} \frac{1}{f(d)} \left(f(d) y_d - \frac{\mu(d)}{M_z} \right)^2 = \sum_{\substack{d \mid P_z \\ d \leqslant z}} f(d) y_d^2 - \frac{2}{M_z} \sum_{\substack{d \mid P_z \\ d \leqslant z}} \mu(d) y_d + \frac{1}{M_z^2} \sum_{\substack{d \mid P_z \\ d \leqslant z}} \frac{\mu^2(d)}{f(d)}$$

$$= \sum_{\substack{d \mid P_z \\ d \leqslant z}} f(d) y_d^2 - \frac{1}{M_z}$$

which implies that (4.38) is minimized when $y_d = M_z^{-1}\mu(d)/f(d)$, the minimum being equal to M_z^{-1}. Note that (4.37) is satisfied with this choice of y_d. By (4.36), we infer that the choice of $\lambda(d)$ ensuring the minimum of (4.38) is given by

$$
\lambda(d) = \frac{\phi(d)}{M_z} \sum_{\substack{n \mid P_z \\ d \mid n \\ n \leqslant z}} \left\{ \mu\left(\frac{n}{d}\right) \times \frac{\mu(n)}{f(n)} \right\} = \frac{\phi(d)}{M_z} \sum_{\substack{m \mid (P_z/d) \\ m \leqslant z/d}} \frac{\mu(m)\mu(md)}{f(md)}
$$

$$
= \frac{\phi(d)}{M_z} \sum_{\substack{m \mid P_z \\ (m,d)=1 \\ m \leqslant z/d}} \frac{\mu(m)\mu(md)}{f(md)} = \frac{\phi(d)\mu(d)}{f(d)M_z} \sum_{\substack{m \mid P_z \\ (m,d)=1 \\ m \leqslant z/d}} \frac{\mu^2(m)}{f(m)}
$$

where we used the equivalence

$$
m \mid (P_z/d) \quad \Longleftrightarrow \quad \begin{cases} m \mid P_z \\ (m,d)=1. \end{cases}
$$

The above sum may be estimated with the following useful tool.

Lemma 4.69 *Let f be a positive multiplicative function, $x \geqslant 2$ be a real number and k be a positive integer. Then*

$$
\prod_{p \mid k}(1+f(p))^{-1} \sum_{n \leqslant x} \mu^2(n) f(n) \leqslant \sum_{\substack{n \leqslant x \\ (n,k)=1}} \mu^2(n) f(n)
$$

$$
\leqslant \prod_{p \mid k}(1+f(p))^{-1} \sum_{n \leqslant kx} \mu^2(n) f(n).
$$

Proof Each positive integer n may be uniquely written as $n = d_1 d_2$ with $d_1 \mid k$ and $(d_2, k) = 1$, and hence $(d_1, d_2) = 1$. Since f is multiplicative, we get for all $y \geqslant 1$

$$
\sum_{n \leqslant y} \mu^2(n) f(n) = \sum_{\substack{d_1 \leqslant y \\ d_1 \mid k}} \mu^2(d_1) f(d_1) \sum_{\substack{d_2 \leqslant y/d_1 \\ (d_2,k)=1}} \mu^2(d_2) f(d_2).
$$

Now since f is positive, using the identity above with $y = kx$ gives

$$
\sum_{n \leqslant kx} \mu^2(n) f(n) = \sum_{d_1 \mid k} \mu^2(d_1) f(d_1) \sum_{\substack{d_2 \leqslant kx/d_1 \\ (d_2,k)=1}} \mu^2(d_2) f(d_2)
$$

$$
\geqslant \sum_{d_1 \mid k} \mu^2(d_1) f(d_1) \sum_{\substack{d_2 \leqslant x \\ (d_2,k)=1}} \mu^2(d_2) f(d_2)
$$

and with $y = x$ we get

$$\sum_{n \leqslant x} \mu^2(n) f(n) \leqslant \sum_{d_1 | k} \mu^2(d_1) f(d_1) \sum_{\substack{d_2 \leqslant x \\ (d_2, k) = 1}} \mu^2(d_2) f(d_2)$$

and the identity

$$\sum_{d_1 | k} \mu^2(d_1) f(d_1) = \prod_{p | k} (1 + f(p))$$

completes the proof. □

Thus, for all $d \mid P_z$ and $d \leqslant z$, we get

$$\sum_{\substack{m \mid P_z \\ (m, d) = 1 \\ m \leqslant z/d}} \frac{\mu^2(m)}{f(m)} \leqslant \prod_{p | d} \left(1 + \frac{1}{f(p)}\right)^{-1} \sum_{\substack{m \mid P_z \\ m \leqslant z}} \frac{\mu^2(m)}{f(m)} = \frac{f(d)}{\phi(d)} \times M_z$$

and hence $|\lambda(d)| \leqslant 1$ for all $d \mid P_z$ such that $d \leqslant z$. Therefore, using $|r_d| \leqslant \rho(d)$ and $\rho(d) \geqslant 1$ for $d \mid P_z$, we infer that the error-term of (4.34) is bounded by

$$\sum_{d_1, d_2 | P_z} |\lambda(d_1)\lambda(d_2)| \frac{\rho(d_1)\rho(d_2)}{\rho((d_1, d_2))} \leqslant \sum_{d_1, d_2 | P_z} \rho(d_1)\rho(d_2) \leqslant z^2 \left(\sum_{\substack{d | P_z \\ d \leqslant z}} \frac{\rho(d)}{d}\right)^2$$

$$\leqslant z^2 \prod_{p | P_z} \left(1 + \frac{\rho(p)}{p}\right)^2 \leqslant z^2 \prod_{p \leqslant z} \left(1 - \frac{\rho(p)}{p}\right)^{-2}.$$

Finally, using $1 \leqslant \rho(d) < d$, we have

$$\sum_{\substack{d | P_z \\ d \leqslant z}} \frac{\mu^2(d)\rho(d)}{\varphi(d)} \leqslant \sum_{\substack{d | P_z \\ d \leqslant z}} \frac{\mu^2(d)\rho(d)}{d} \prod_{p | d} \left(1 - \frac{\rho(p)}{p}\right)^{-1} = \sum_{\substack{d | P_z \\ d \leqslant z}} \frac{\mu^2(d)\rho(d)}{df(d)}$$

$$< \sum_{\substack{d | P_z \\ d \leqslant z}} \frac{\mu^2(d)}{f(d)} = M_z.$$

Putting all together we have proved Selberg's upper bound sieve.

Theorem 4.70 (Selberg) *Let $z \geqslant 2$ be a real number, $\mathcal{P}$ be a set of primes, $P_z = \prod_{p \in \mathcal{P}, \, p \leqslant z} p$ and $\mathcal{A}$ be a finite set of integers such that, for all $d \mid P_z$, we have*

$$\sum_{\substack{n \in \mathcal{A} \\ d | n}} 1 = \frac{X\rho(d)}{d} + r_d$$

*with $X > 0$, ρ is a multiplicative function such that $1 \leqslant \rho(d) < d$ and $|r_d| \leqslant \rho(d)$.
Set*

$$S(\mathcal{A}, \mathcal{P}; z) = \sum_{\substack{n \in \mathcal{A} \\ (n, P_z)=1}} 1.$$

Then we have

$$S(\mathcal{A}, \mathcal{P}; z) \leqslant X \left(\sum_{\substack{d \mid P_z \\ d \leqslant z}} \frac{\mu^2(d)\rho(d)}{\varphi(d)} \right)^{-1} + z^2 \prod_{p \leqslant z} \left(1 - \frac{\rho(p)}{p} \right)^{-2}.$$

In order to have a bound for the sum of the main term, the following lemma will
be useful.

Lemma 4.71 *Let $z \geqslant 2$ be a real number and $k \in \mathbb{N}$. Then*

$$\sum_{\substack{n \leqslant z \\ (n,k)=1}} \frac{\mu^2(n)}{\varphi(n)} > \frac{\varphi(k)}{k} \log z.$$

Proof The lower bound of Lemma 4.69 first gives

$$\sum_{\substack{n \leqslant z \\ (n,k)=1}} \frac{\mu^2(n)}{\varphi(n)} \geqslant \frac{\varphi(k)}{k} \sum_{n \leqslant z} \frac{\mu^2(n)}{\varphi(n)}.$$

Now each positive integer n can be uniquely written as $n = qd$ with q squarefree
and $d \mid q^\infty$. Note also that

$$\sum_{d \mid q^\infty} \frac{1}{d} = \prod_{p \mid q} \left(1 + \frac{1}{p} + \frac{1}{p^2} + \cdots \right) = \prod_{p \mid q} \left(1 - \frac{1}{p} \right)^{-1} = \frac{q}{\varphi(q)}.$$

Therefore we get

$$\log z < \sum_{n \leqslant z} \frac{1}{n} \leqslant \sum_{q \leqslant z} \frac{\mu^2(q)}{q} \sum_{d \mid q^\infty} \frac{1}{d} = \sum_{q \leqslant z} \frac{\mu^2(q)}{\varphi(q)}$$

which concludes the proof. $\square$

When $\rho(d) = 1$, the above results yield a useful estimate.

Corollary 4.72 *With the notation of Theorem 4.70 assuming $\rho(d) = 1$ and $|r_d| \leqslant 1$
for all $d \mid P_z$, we have*

$$S(\mathcal{A}, \mathcal{P}; z) < \frac{X}{\log z} \prod_{\substack{p \leqslant z \\ p \notin P}} \left(1 - \frac{1}{p} \right)^{-1} + z^2.$$

Proof Set

$$Q_z = \prod_{\substack{p \leqslant z \\ p \notin \mathcal{P}}} p.$$

When $\rho(d) = 1$, then $\phi(d) = d$ and hence $f(d) = \varphi(d)$, so that we have by Lemma 4.71

$$M_z = \sum_{\substack{d \leqslant z \\ (Q_z,d)=1}} \frac{\mu^2(d)}{\varphi(d)} > \frac{\varphi(Q_z)}{Q_z} \log z = \prod_{\substack{p \leqslant z \\ p \notin \mathcal{P}}} \left(1 - \frac{1}{p} \right) \log z.$$

Furthermore, since $|\lambda(d)| \leqslant 1$ and $|r_d| \leqslant 1$ for all $d \mid P_z$, we infer that the error-term in (4.34) is bounded by z^2, concluding the proof. □

One of the most famous applications of Selberg's sieve is Brun–Titchmarsh's theorem.

Theorem 4.73 (Brun–Titchmarsh) *Let a, q be positive coprime integers and set*

$$\pi(x; q, a) = \sum_{\substack{p \leqslant x \\ p \equiv a \,(\mathrm{mod}\, q)}} 1.$$

Assume that $x > q$. Then

$$\pi(x; q, a) \leqslant \frac{6x}{\varphi(q) \log(x/q)}.$$

Proof If $1 < x/q \leqslant 20$, then the trivial estimate implies

$$\pi(x; q, a) \leqslant \frac{2x}{q} \leqslant \frac{6x}{\varphi(q) \log(20)} \leqslant \frac{6x}{\varphi(q) \log(x/q)}$$

so we may suppose that $x/q > 20$. Let $z \geqslant 2$ be a real number and we take $\mathcal{P}$ the set of primes $p \nmid q$ and $\mathcal{A}$ the set of numbers $n \leqslant x$ such that $n \equiv a \,(\mathrm{mod}\, q)$. We then have

$$\pi(x; q, a) \leqslant S(\mathcal{A}, \mathcal{P}; z) + z.$$

Now if $(d, q) = 1$, then

$$\sum_{\substack{n \in \mathcal{A} \\ d \mid n}} 1 = \frac{x}{qd} + r_d$$

with $|r_d| \leqslant 1$, so that Selberg's sieve may be used with $X = x/q$ and $\rho(d) = 1$ if $(d, q) = 1$. Therefore by Corollary 4.72 we get

$$\pi(x; q, a) \leqslant \frac{x}{\varphi(q) \log z} + z^2 + z.$$

and choosing $z = (x/q)^{1/2} \log^{-1/2}(x/q)$ gives the asserted result, using $z > (x/q)^{1/4}$ in the first term. $\square$

For an extensive account on sieve methods, the reader is referred to [Bom74, HR11].

4.7.15 The Large Sieve

The large sieve was invented by Ju.V. Linnik in 1941 [Lin41] while investigating the distribution of quadratic non-residues (see Example 4.79). In a series of papers beginning in the late 1940s, Renyi was the first to systematically study the large sieve and to prove that each large even number $2k$ may be expressed in the form $2k = p + P_k$ where $\omega(P_k) \leqslant k$. Subsequently, values of k were given by several authors until Chen proved that $k = 2$ is admissible. The large sieve always played a key part.

The large sieve remained in the area of a few specialists until 1965 when major contributions were made by Roth and then by Bombieri. They paved the way for the recognition that these results rely on an underlying analytic inequality dealing with trigonometric polynomials. The analytic principle of the large sieve was formulated explicitly one year later by Davenport and Halberstam. For an account of the theory of the large sieve, we may refer the reader to [Bom74, FI10, Mon78].

Nowadays, the large sieve is one of the most powerful tools in multiplicative number theory. Roughly speaking, it may be regarded as Fourier analysis of arithmetic progressions, both from the additive and multiplicative points of view. Hence the usual tools from Fourier analysis (quasi-orthogonal systems, Bessel's and Hilbert's inequalities, etc) are the main ingredients of the large sieve.

The analytic additive form of the large sieve can be described as follows. Let $M \in \mathbb{Z}$, N be a positive integer, (a_n) be a sequence of arbitrary complex numbers supported on the interval $]M, M + N]$ and let

$$S(\alpha) = \sum_{M < n \leqslant M+N} a_n e(n\alpha)$$

be a trigonometric polynomial. Let $\delta > 0$ and assume that $\alpha_1, \ldots, \alpha_R$ are δ-well spaced, i.e. for all $r \neq s \in \{1, \ldots, R\}$, we have

$$\|\alpha_r - \alpha_s\| \geqslant \delta. \tag{4.39}$$

The large sieve is an inequality of the shape

$$\sum_{r=1}^{R} |S(\alpha_r)|^2 \leqslant \Delta(N, \delta) \sum_{n=M+1}^{M+N} |a_n|^2. \tag{4.40}$$

Since $|S|$ is 1-periodic, one may assume that $0 < \alpha_1 < \alpha_2 < \cdots < \alpha_R \leqslant 1$ with $\delta \leqslant R^{-1}$, and since

$$\left| \sum_{n=M+K+1}^{M+K+N} a_{n-K} e(n\alpha) \right| = \left| e(K\alpha) S(\alpha) \right| = \left| S(\alpha) \right|$$

we see that the parameter M is irrelevant, and hence we are interested in determining how $\Delta(N, \delta)$ depends on N and δ.

The next lemma is a version of Parseval's well-known identity.

Lemma 4.74 (Parseval's identity) *We have*

$$\int_0^1 \left| S(t) \right|^2 dt = \sum_{M < n \leqslant M+N} \left| a_n \right|^2.$$

Proof Squaring out we get

$$\left| S(t) \right|^2 = \sum_{M < n \leqslant M+N} \left| a_n \right|^2 + \sum_{n_1 \neq n_2} a_{n_1} \overline{a_{n_2}}\, e\big((n_1 - n_2)t\big)$$

and the result follows from the identity

$$\int_0^1 e(kt)\, dt = \begin{cases} 1, & \text{if } k = 0 \\ 0, & \text{otherwise} \end{cases}$$

which concludes the proof. $\qquad\qquad\square$

We first observe that $\Delta(N, \delta)$ cannot be too small. If $a_n = e(-n\alpha_1)$, then

$$\left| S(\alpha_1) \right|^2 = N^2 = N \sum_{M < n \leqslant M+N} \left| a_n \right|^2$$

and hence $\Delta(N, \delta) \geqslant N$. Now assume that the α_r are equally spaced so that $\delta = R^{-1}$. By periodicity and Lemma 4.74, we have

$$\int_0^1 \sum_{r=1}^R \left| S(\alpha_r + t) \right|^2 dt = R \int_0^1 \left| S(t) \right|^2 dt = R \sum_{M < n \leqslant M+N} \left| a_n \right|^2$$

so that for some values of t we have

$$\sum_{r=1}^R \left| S(\alpha_r + t) \right|^2 \geqslant R \sum_{M < n \leqslant M+N} \left| a_n \right|^2$$

and therefore $\Delta(N, \delta) \geqslant [\delta^{-1}] > \delta^{-1} - 1$. The following optimal result, attributed to Selberg, shows that we may take $\Delta(N, \delta)$ to be barely larger than is required by the above observation.

Theorem 4.75 (Large sieve) *Under the hypothesis* (4.39), *the inequality* (4.40) *holds with*

$$\Delta(N, \delta) = N + \delta^{-1} - 1.$$

A possible proof uses generalizations of Bessel's and Hilbert's inequalities. We shall see that one can prove a weaker version of this result, although sufficiently strong to be used in many applications, with the following quicker and elegant ideas due to Gallagher. We start with a useful lemma.

Lemma 4.76 (Gallagher) *Let* $f \in C^1[a, b]$ *with* $a < b$. *Then*

$$\left| f\left(\frac{a+b}{2}\right) \right| \leqslant \frac{1}{b-a} \int_a^b |f(t)| \, dt + \frac{1}{2} \int_a^b |f'(t)| \, dt.$$

Proof An integration by parts provides for all $x \in [a, b]$

$$\int_x^b \frac{t-b}{b-a} f'(t) \, dt + \int_a^x \frac{t-a}{b-a} f'(t) \, dt = f(x) - \frac{1}{b-a} \int_a^b f(t) \, dt$$

and the result follows by noticing that, if $x = \frac{a+b}{2}$, then $|\frac{t-b}{b-a}| \leqslant \frac{1}{2}$ in the first integral and $|\frac{t-a}{b-a}| \leqslant \frac{1}{2}$ in the second one.   □

We are now in a position to prove Gallagher's version of the large sieve [Gal67].

Theorem 4.77 (Gallagher) *Under the hypothesis* (4.39), *the inequality* (4.40) *holds with*

$$\Delta(N, \delta) = \pi N + \delta^{-1}.$$

Proof Lemma 4.76 with $a = \alpha_r - \delta/2$ and $b = \alpha_r + \delta/2$ implies that

$$|f(\alpha_r)| \leqslant \frac{1}{\delta} \int_{\alpha_r - \delta/2}^{\alpha_r + \delta/2} |f(t)| \, dt + \frac{1}{2} \int_{\alpha_r - \delta/2}^{\alpha_r + \delta/2} |f'(t)| \, dt$$

and taking $f(\alpha) = S(\alpha)^2$ we get

$$|S(\alpha_r)|^2 \leqslant \frac{1}{\delta} \int_{\alpha_r - \delta/2}^{\alpha_r + \delta/2} |S(t)|^2 \, dt + \int_{\alpha_r - \delta/2}^{\alpha_r + \delta/2} |S(t)S'(t)| \, dt.$$

By (4.39), the intervals $\mathcal{I}_r = \,]\alpha_r - \delta/2, \alpha_r + \delta/2[$ do not overlap modulo 1, meaning that if $r \neq s$, then no point of $\mathcal{I}_r$ differs by an integer from another point of $\mathcal{I}_s$. Hence summing over r we obtain

$$\sum_{r=1}^R |S(\alpha_r)|^2 \leqslant \frac{1}{\delta} \int_0^1 |S(t)|^2 \, dt + \int_0^1 |S(t)S'(t)| \, dt.$$

Now by Lemma 4.74 we have

$$\int_0^1 |S(t)|^2 \, dt = \sum_{M < n \leqslant M+N} |a_n|^2$$

and using Cauchy–Schwarz's inequality (integral analogue of Lemma 6.17) and Parseval's identity again, we get

$$\int_0^1 |S(t)S'(t)| \, dt \leqslant \left(\int_0^1 |S(t)|^2 \, dt \right)^{1/2} \left(\int_0^1 |S'(t)|^2 \, dt \right)^{1/2}$$

$$= \left(\sum_{M < n \leqslant M+N} |a_n|^2 \right)^{1/2} \left(\sum_{M < n \leqslant M+N} |2\pi n a_n|^2 \right)^{1/2}$$

$$\leqslant 2\pi \left(\max_{M < n \leqslant M+N} |n| \right) \sum_{M < n \leqslant M+N} |a_n|^2.$$

This gives

$$\sum_{r=1}^R |S(\alpha_r)|^2 \leqslant \left\{ \delta^{-1} + 2\pi \left(\max_{M < n \leqslant M+N} |n| \right) \right\} \sum_{M < n \leqslant M+N} |a_n|^2$$

and we conclude the proof by noticing that, since $\Delta(N, \delta)$ is independent of M, we may suppose that $M = -[\frac{1}{2}(N+1)]$ so that $\max_{M < n \leqslant M+N} |n| \leqslant N/2$ as required.   □

The arithmetic version of the large sieve often takes the following form. Consider a finite set $\mathcal{P}$ of prime numbers and, for all $p \in \mathcal{P}$, let Ω_p be a subset of $\mathbb{F}_p$ of residue classes to sieve out. The main purpose of sieve methods is to provide an estimate of

$$\sum_{n \in \mathcal{S}(N, \mathcal{P}, \Omega_p)} a_n$$

where (a_n) is any sequence of complex numbers and $\mathcal{S}(N, \mathcal{P}, \Omega_p)$ is the so-called *sifted set*

$$\mathcal{S}(N, \mathcal{P}, \Omega_p) = \{ n \in \,]M, M+N] \cap \mathbb{Z} : n \pmod{p} \notin \Omega_p \text{ for all } p \in \mathcal{P} \}.$$

We apply Theorem 4.75 or Theorem 4.77 with $\alpha_r = a/q$ for some integers a, q such that $1 \leqslant a \leqslant q \leqslant Q$ and $(a, q) = 1$. (4.39) is then satisfied with $\delta = Q^{-2}$ since, for $r \neq s$, we have

$$\|\alpha_r - \alpha_s\| = \left\| \frac{a}{q} - \frac{a'}{q'} \right\| = \left\| \frac{aq' - a'q}{qq'} \right\| \geqslant \frac{1}{qq'} \geqslant \frac{1}{Q^2}$$

where the inequality $\|n/m\| \geqslant m^{-1}$ with $(m, n) = 1$ comes from Exercise 3 (i) and (ii) in Chap. 1. Using Theorem 4.75 we get

$$\sum_{\substack{q \leqslant Q}} \sum_{\substack{a=1 \\ (a,q)=1}}^{q} \left| S\left(\frac{a}{q}\right) \right|^2 \leqslant (N - 1 + Q^2) \sum_{M < n \leqslant M+N} |a_n|^2.$$

Now [IK04, Lemma 7.15] states that, for all positive integers q, we have

$$\sum_{\substack{a=1 \\ (a,q)=1}}^{q} \left| S\left(\frac{a}{q}\right) \right|^2 \geqslant \mu(q)^2 \prod_{p|q} \left(1 - \frac{\rho(p)}{p}\right)^{-1} \left| \sum_{n \in \mathcal{S}(N, \mathcal{P}, \Omega_p)} a_n \right|^2$$

where, for all primes $p \in \mathcal{P}$, we set

$$\rho(p) = |\Omega_p| \tag{4.41}$$

and hence we obtain the following useful result.

Corollary 4.78 (Arithmetic large sieve) *Let (a_n) be any complex-valued sequence supported on $]M, M + N]$. For all positive integers Q, set*

$$L = L(Q) = \sum_{q \leqslant Q} \mu(q)^2 \prod_{p|q} \left(1 - \frac{\rho(p)}{p}\right)^{-1} \tag{4.42}$$

where $\rho(p)$ is defined in (4.41). If $\rho(p) < p$ for all primes $p \in \mathcal{P}$, then for all $Q \geqslant 1$ we have

$$\left| \sum_{n \in \mathcal{S}(N, \mathcal{P}, \Omega_p)} a_n \right|^2 \leqslant \frac{N - 1 + Q^2}{L} \sum_{M < n \leqslant M+N} |a_n|^2.$$

In particular, we have

$$\sum_{n \in \mathcal{S}(N, \mathcal{P}, \Omega_p)} 1 \leqslant \frac{N - 1 + Q^2}{L}.$$

It is important to keep in mind the following examples given in [Bom74] (with $M = 0$).

Example 4.79

1. **Sieve of Eratosthenes.** $\mathcal{P}$ is the set of primes $p \leqslant \sqrt{N}$ and $\Omega_p = \{\bar{0}\}$.
2. **Twin primes.** $\mathcal{P}$ is the set of primes $p \leqslant \sqrt{N}$ and $\Omega_p = \{\bar{0}, \bar{2}\}$.
3. **Linnik's example on the least quadratic non-residue.**

$$\mathcal{P} = \left\{ p \leqslant \sqrt{N} : \left(\frac{n}{p}\right) = 1 \text{ for all } n \leqslant N^\varepsilon \right\}$$

and $\Omega_p = \{\bar{h} : (h/p) = -1\}$.

A more elaborate large sieve inequality enables the authors in [MV73] to prove the most elegant version of the Brun–Titchmarsh theorem.

Theorem 4.80 (Montgomery and Vaughan) *Let $x, y \geqslant 2$ be real numbers and a, q be positive integers such that $(a, q) = 1$ and $y > q$. Then*

$$\pi(x + y; q, a) - \pi(x; q, a) < \frac{2y}{\varphi(q) \log(y/q)}.$$

This is a powerful version of Brun–Titchmarsh's inequality which may replace Siegel–Walfisz–Page's theorem (Theorem 3.92) in many cases. The factor 2 is of great importance in number theory. More precisely, if the Siegel zero β_1 exists in (3.45), then it is very close to 1 since we may have

$$1 - \frac{c_1}{\log q} \leqslant \beta_1 < 1.$$

On the other hand, it is proved in [Mot79] that if the estimate

$$\pi(x; q, a) \leqslant \frac{(2 - \varepsilon) x}{\varphi(q) \log(x/q)} \tag{4.43}$$

holds for $x \geqslant q^{c_2}$, where $\varepsilon > 0$ is an absolute constant, then

$$\beta_1 \leqslant 1 - \frac{c_3 \varepsilon}{\log q}$$

and hence (4.43) enables us to disprove the existence of this exceptional zero. (4.43) also implies that

$$\pi(x; q, a) = \left\{ 1 + O\left(e^{-c_4 \varepsilon \log x / \log q} \right) \right\} \frac{x}{\varphi(q) \log x}$$

as long as $x \geqslant q^{c_2}$, and thus the constant $2 - \varepsilon$ will be automatically reduced to $1 + \varepsilon$ for all x larger than a sufficiently high power of q.

Vaughan [Vau73] obtained the following useful lower bound for L, in which only a lower estimate of the average values of $\rho(p)$ is sufficient. For all $Q \in \mathbb{N}$, we have

$$L \geqslant \max_{m \in \mathbb{N}} \exp\left(m \log \left(\frac{1}{m} \sum_{p \leqslant Q^{1/m}} \frac{\rho(p)}{p} \right) \right). \tag{4.44}$$

This allows us to get the following effective version of Corollary 4.78.

Corollary 4.81 *Along with the notation of Corollary 4.78, let $\beta > 0$.*

(i) *Suppose that there exists a constant $c_0 > 0$ such that, for all sufficiently large R, we have*

$$\sum_{p \leqslant R} \frac{\rho(p)}{p} > c_0 (\log \log R)^\beta.$$

Then there exists a constant $c_1 > 0$ such that for all sufficiently large N, we have

$$\sum_{n \in S(N,\mathcal{P},\Omega_p)} 1 < \frac{N}{\exp(c_1 (\log \log N)^\beta)}.$$

(ii) Suppose that there exists a constant $c_2 > 0$ such that, for all sufficiently large R, we have

$$\sum_{p \leqslant R} \frac{\rho(p)}{p} > c_2 (\log R)^\beta.$$

Then there exists a constant $c_3 > 0$ such that for all sufficiently large N, we have

$$\sum_{n \in S(N,\mathcal{P},\Omega_p)} 1 < \frac{N}{\exp(c_3 (\log N)^{\beta/(\beta+1)})}.$$

4.8 Exercises

1 Prove that

(a) $\tau(n) \leqslant 2\sqrt{n}$.
(b) $\sigma(n) < n(\log \sqrt{n} + 2 + \log 2)$.
(c) For all *composite* integers $n \in \mathbb{N}$, we have $\varphi(n) \leqslant n - \sqrt{n}$.

2 Let f be a multiplicative function satisfying

$$\lim_{p^\alpha \to \infty} f(p^\alpha) = 0.$$

Show that

$$\lim_{n \to \infty} f(n) = 0.$$

Deduce that, for all $\varepsilon > 0$ and $n \in \mathbb{N}$, we have $\tau(n) \ll n^\varepsilon$.

3 Prove the following identities.

(a) $\sum_{d|n} \tau^3(d) = (\sum_{d|n} \tau(d))^2$.
(b) $\sum_{d|n} \beta(d)\mu(n/d) = s_2(n)$.
(c) $\sum_{d|n} \sigma_k(d)\mu(n/d) = n^k$.
(d) $\sum_{d|n} \mu_2(d)k^{\omega(d)} = (k+1)^{\omega(n)}$.

4 Let k be a positive integer.

(a) Prove that for all $n \in \mathbb{N}$, we have

$$\Lambda_k(n) = \Lambda_{k-1}(n)\log n + (\Lambda_{k-1} \star \Lambda)(n).$$

(b) Prove that for all $m, n \in \mathbb{N}$ such that $(m, n) = 1$, we have

$$\Lambda_k(mn) = \sum_{j=0}^{k} \binom{k}{j} \Lambda_j(m) \Lambda_{k-j}(n).$$

5 In [Gou72, identity 3.93], one may find the following result.

$$\sum_{j=0}^{n} \frac{\binom{2j}{j}\binom{2n-2j}{n-j}}{(2j-1)(2n-2j-1)} = \begin{cases} 1, & \text{if } n = 0 \\ -4, & \text{if } n = 1 \\ 0, & \text{if } n \geqslant 2. \end{cases}$$

With the help of this identity, determine the multiplicative function f such that $\mu = f \star f$.

6 Let f be an additive function and g be a multiplicative function. Prove that, for all $m, n \in \mathbb{N}$ such that $(m, n) = 1$, we have

$$(f \star g)(mn) = (f \star g)(m)(g \star \mathbf{1})(n) + (f \star g)(n)(g \star \mathbf{1})(m).$$

Deduce that, if $n \neq p^\alpha$, then $(f \star \mu)(n) = 0$.

7 Let n be a positive integer.

(a) Prove that

$$\frac{n^{n+1}}{\zeta(n+1)} \leqslant \varphi(n)\sigma\left(n^n\right) \leqslant n^{n+1}.$$

(b) Deduce the nature of the series $\sum_{n \geqslant 1} f(n)$ where

$$f(n) = \frac{n}{\varphi(n)} - \frac{\sigma(n^n)}{n^n}.$$

8 Prove that for all integers m, n not necessarily coprime, we have

$$\Delta_k(mn) \leqslant \tau_k(m)\Delta_k(n).$$

9 (Realizable sequences and Perrin sequences) This exercise first requires the following definitions.

(i) Let E be a non-empty set and $F : E \longrightarrow E$ be any map. Let $n \in \mathbb{Z}_{\geqslant 0}$. An element $x \in E$ is said to be n-periodic by f if

$$F^n(x) = x$$

where $F^0 = \text{Id}$ and $F^n = F \circ \cdots \circ F$ (n times). We define $\text{Per}_n(F)$ to be the set of n-periodic points by F and $\text{Per}_n^*(F)$ to be the set of points having the smallest n-period by F. Finally, we call the *orbit* of $x \in E$ the set

$$\mathcal{O}_x = \left\{ F^k(x) : k \in \mathbb{Z}_{\geqslant 0} \right\}.$$

(ii) A sequence (u_n) of non-negative integers is said to be *realizable* if there exist a set E and a map $F : E \longrightarrow E$ such that, for all $n \in \mathbb{N}$, we have

$$u_n = |\mathrm{Per}_n(F)|.$$

It can be shown that the following sequences are realizable.

(i) The sequence (u_n) defined by $u_n = a^n$ where a is a fixed positive integer.
(ii) The sequence (u_n) defined by $u_n = \mathrm{Tr}(A^n)$ where A is a fixed square matrix with non-negative integer entries.

1. *General properties.*
   (a) Prove that, if $x \in \mathrm{Per}_n(F)$, then the smallest period of x divides n.
   (b) Deduce that, if $\mathrm{Per}_n^*(F)$ is a finite set, then we have

$$|\mathrm{Per}_n(F)| = \sum_{d \mid n} |\mathrm{Per}_d^*(F)|$$

   and

$$|\mathrm{Per}_n^*(F)| = \sum_{d \mid n} |\mathrm{Per}_d^*(F)| \mu(n/d).$$

   (c) Show that, if $x \in \mathrm{Per}_n^*(F)$, then $\mathcal{O}_x = \{x, F(x), F^2(x), \ldots, F^{n-1}(x)\}$ and that $|\mathcal{O}_x| = n$.
   (d) Prove that, if $x \in \mathrm{Per}_n^*(F)$, then $\mathcal{O}_x \subseteq \mathrm{Per}_n^*(F)$.
   (e) For all $x, y \in \mathrm{Per}_n^*(F)$, define

$$x \sim y \quad \Longleftrightarrow \quad x \in \mathcal{O}_y.$$

   Check that this is an equivalence relation and deduce that, if $\mathrm{Per}_n^*(F)$ is a finite set, then we have

$$|\mathrm{Per}_n^*(F)| \equiv 0 \ (\mathrm{mod}\, n).$$

   (f) Deduce that, if $u = (u_n)$ is a realizable sequence, then $u \star \mu \geqslant 0$ and

$$(u \star \mu)(n) \equiv 0 \ (\mathrm{mod}\, n).$$

2. *Applications.*
   (a) Prove that, for all positive integers a, we have

$$\sum_{d \mid n} a^d \mu(n/d) \equiv 0 \ (\mathrm{mod}\, n).$$

   (b) Prove that, for each square matrix A with non-negative integer entries, we have

$$\sum_{d \mid n} \mathrm{Tr}(A^d) \mu(n/d) \equiv 0 \ (\mathrm{mod}\, n).$$

   (c) Deduce *Fermat's little theorem for integer matrices.*

> Let p be a prime number and A be a square matrix with integer
> entries. Then
> $$\mathrm{Tr}(A^p) \equiv \mathrm{Tr}(A) \ (\mathrm{mod}\ p).$$

3. *Perrin sequences.* Let (u_n) be the sequence defined by $u_0 = 3$, $u_1 = 0$, $u_2 = 2$
   and
   $$u_n = u_{n-2} + u_{n-3}$$

   for all $n \geqslant 3$.
   (a) Let A be the matrix
   $$A = \begin{pmatrix} 0 & 0 & 1 \\ 1 & 0 & 1 \\ 0 & 1 & 0 \end{pmatrix}.$$

   Prove that, for all $n \in \mathbb{Z}_{\geqslant 0}$, we have $u_n = \mathrm{Tr}(A^n)$.
   (b) Deduce that, for all primes p, we have $p \mid u_p$.

10 Prove that
$$\sum_{n \leqslant x} \varphi(n) = \frac{x^2}{2\zeta(2)} + O(x \log x).$$

11 (Cesáro, 1885) Let f be an arithmetic function. Show that
$$\sum_{i=1}^{n} f\big((i, n)\big) = (f \star \varphi)(n).$$

12 The *Pillai function* $S(n)$ is defined by
$$S(n) = \sum_{i=1}^{n} (i, n).$$

Let θ be the exponent in the Dirichlet divisor problem,[8] i.e. the smallest positive
real number θ such that the asymptotic estimate
$$\sum_{n \leqslant x} \tau(n) = x(\log x + 2\gamma - 1) + O\big(x^{\theta + \varepsilon}\big)$$

holds for all $\varepsilon > 0$ and $x \geqslant 1$. The aim of this exercise is to prove the following
result.
$$\sum_{n \leqslant x} S(n) = \frac{x^2}{2\zeta(2)} \left(\log x + 2\gamma - \frac{1}{2} - \frac{\zeta'(2)}{\zeta(2)} \right) + O\big(x^{1+\theta+\varepsilon}\big). \tag{4.45}$$

---

[8] The best inequalities to date for θ are $\frac{1}{4} \leqslant \theta \leqslant \frac{131}{416}$. See Chap. 6.

(a) Prove that, for all $\varepsilon > 0$ and $z \geqslant 1$, we have

$$\sum_{n \leqslant z} n\tau(n) = \frac{z^2 \log z}{2} + z^2 \left(\gamma - \frac{1}{4} \right) + O\left(z^{\theta+1+\varepsilon} \right).$$

(b) Using Exercise 11 and (4.7) check that $S = \mu \star (\tau \times \mathrm{Id})$.

(c) Prove (4.45).

13 Let k be a fixed positive integer. For all $x \geqslant 1$, define

$$S_k(x) = \sum_{n \leqslant x} \tau_k(n).$$

(a) Prove that

$$x \int_1^x \frac{S_k(t)}{t^2} \, dt < S_{k+1}(x) \leqslant x \int_1^x \frac{S_k(t)}{t^2} \, dt + S_k(x).$$

(b) Using induction, deduce that

$$x \sum_{j=0}^{k-1} (-1)^{k+j+1} \frac{(\log x)^j}{j!} + (-1)^k < \sum_{n \leqslant x} \tau_k(n) \leqslant x \sum_{j=0}^{k-1} \binom{k-1}{j} \frac{(\log x)^j}{j!}.$$

(c) Deduce that

$$\sum_{n \leqslant x} \tau_k(n) \leqslant \frac{x(\log x + k - 1)^{k-1}}{(k-1)!}.$$

(d) Using induction, prove that for all $k \geqslant 2$

$$\sum_{n \leqslant x} \tau_k(n) = \frac{x(\log x)^{k-1}}{(k-1)!} + O\left(x(\log x)^{k-2} \right).$$

(e) For all $k, n \in \mathbb{N}$, let $\tau_k^*(n)$ be the kth *strict divisor function* of n, i.e. the number of choices of $n_1, \ldots, n_k$ satisfying $n = n_1 \cdots n_k$ with $n_j \geqslant 2$. Prove that, for any $x \geqslant 1$, we have

$$\sum_{n \leqslant x} \tau_k^*(n) \leqslant \frac{x(\log x)^{k-1}}{(k-1)!}.$$

14 Using Corollary 3.7 (v), prove that, for all $x \geqslant 1$, we have

$$\sum_{n \leqslant x} s_2(n) < 3\sqrt{x}$$

and deduce by partial summation that

$$\sum_{n>x} \frac{s_2(n)}{n} < \frac{6}{\sqrt{x}}.$$

15 Let f be a multiplicative function such that $|f(p) - 1| \leqslant p^{-1}$ for all primes p and $|(f \star \mu)(p^\alpha)| \leqslant 1$ for all prime powers p^α with $\alpha \geqslant 2$. Prove that

$$\sum_{n \leqslant x} f(n) = x \prod_p \left(1 - \frac{1}{p}\right)\left(1 + \sum_{\alpha=1}^\infty \frac{f(p^\alpha)}{p^\alpha}\right) + O\left(x^{1/2}\right).$$

16 Prove that

$$\sum_{n \leqslant x} \beta(n) = x \frac{\zeta(2)\zeta(3)}{\zeta(6)} + O\left(x^{1/2}\right).$$

17 Prove that

$$\sum_{n \leqslant x} \frac{\varphi(n)\gamma_2(n)}{n^2} = x \prod_p \left(1 - \frac{1}{p}\right)\left(1 + \frac{1}{p+1}\right) + O\left(x^{1/2}\right).$$

18 (S. Selberg) For all $x \geqslant 1$, define

$$S(x) = \sum_{n \leqslant x} \frac{\mu(n)}{n\tau(n)}.$$

(a) Show that

$$S(x) = \frac{1}{x}\left(\sum_{n \leqslant x} 2^{-\omega(n)} + \sum_{n \leqslant x} \frac{\mu(n)}{\tau(n)}\left\{\frac{x}{n}\right\}\right).$$

(b) Using Theorem 4.22, deduce that there exist $c_0 > 0$ and $x_0 \geqslant e$ such that, for all $x \geqslant x_0$, we have

$$0 < S(x) < c_0(\log x)^{-1/2}.$$

19 (Hall and Tenenbaum) For all $x \geqslant 2$ and $k \in \mathbb{N}$, define

$$N_k(x) = \sum_{\substack{n \leqslant x \\ \Omega(n)=k}} 1.$$

The aim of this exercise is to prove the following bound.

$$N_k(x) \ll k2^{-k}x \log x. \qquad\qquad (4.46)$$

(a) Let $t \in [1, 2[$. Using $f = t^{\Omega} \star \mu$ and Proposition 4.17, prove that

$$\sum_{n \leqslant x} t^{\Omega(n)} \ll \frac{x \log x}{2 - t}.$$

(b) Prove (4.46).

20 (Alladi, Erdős and Vaaler) Let $n, a \geqslant 2$ be integers with n squarefree and $\lambda \geqslant 0$ be a real number. Let f be a multiplicative function such that, for all primes p, we have

$$0 \leqslant f(p) \leqslant \lambda < \frac{1}{a - 1}. \tag{4.47}$$

The aim of this exercise is to show the following inequality.

$$\sum_{d \mid n} f(d) \leqslant \left(\frac{1 + \lambda}{1 + \lambda - a\lambda}\right) \sum_{\substack{d \mid n \\ d \leqslant n^{1/a}}} f(d). \tag{4.48}$$

1. (a) Let p be a prime factor of n. Show that

$$\sum_{d \mid n} f(d) = \sum_{d \mid (n/p)} f(d) + \sum_{k \mid (n/p)} f(kp)$$

and then

$$\sum_{k \mid (n/p)} f(k) = \left(\frac{1}{1 + f(p)}\right) \sum_{d \mid n} f(d). \tag{4.49}$$

(b) Deduce that

$$\sum_{d \mid n} f(d) \log d = \left(\sum_{d \mid n} f(d)\right) \left(\sum_{p \mid n} \frac{f(p) \log p}{1 + f(p)}\right)$$

and then

$$\sum_{d \mid n} f(d) \log d \leqslant \frac{\lambda \log n}{1 + \lambda} \left(\sum_{d \mid n} f(d)\right). \tag{4.50}$$

2. Prove that

$$\sum_{d \mid n} f(d) \frac{\log(n^{1/a}/d)}{\log(n^{1/a})} \leqslant \sum_{\substack{d \mid n \\ d \leqslant n^{1/a}}} f(d)$$

and conclude with (4.50).

21 (Redheffer) Let $R_n = (r_{ij}) \in \mathcal{M}_n(\{0, 1\})$ be the matrix defined by

$$r_{ij} = \begin{cases} 1, & \text{if } i \mid j \text{ or } j = 1 \\ 0, & \text{otherwise.} \end{cases}$$

Show that

$$\det R_n = M(n)$$

where $M(n) = \sum_{k=1}^n \mu(k)$ is the Mertens function.

22 Let $t > 0$.

1. Prove that, for all $n \in \mathbb{N}$, we have

$$\sum_{d \mid n} t^{\omega(d)} \leqslant (1 + t)^{\Omega(n)}.$$

2. Let $k \in \mathbb{N}$. Show that

$$\sum_{\substack{d \mid n \\ \omega(d) \leqslant k}} t^{\omega(d)} \leqslant \sum_{j=0}^k \binom{\Omega(n)}{j} t^j.$$

23 Let $N \in \mathbb{N}$. Prove that

$$\sum_{n \leqslant N} (\tau \star \mu \star \Lambda)(n) = \log(N!).$$

References

[Bom74] Bombieri E (1974) Le Grand Crible dans la Théorie Analytique des Nombres, vol 18. Société Mathématique de France, Paris

[Bor09] Bordellès O (2009) Le problème des diviseurs de Dirichlet. Quadrature 71:21–30

[Bor10] Bordellès O (2010) The composition of the gcd and certain arithmetic functions. J Integer Seq 13, Art. 10.7.1

[Erd57] Erdős P (1957) Some unsolved problems. Mich Math J 4:291–300

[FI05] Friedlander J, Iwaniec H (2005) Summation formulæ for coefficients of L-functions. Can J Math 57:494–505

[FI10] Friedlander J, Iwaniec H (2010) Opera de Cribro. Colloquium publications, vol 57. Am. Math. Soc., Providence

[Gal67] Gallagher PX (1967) The large sieve. Mathematika 14:14–20

[Gou72] Gould HW (1972) Combinatorial identities. A standardized set of tables listing 500 binomial coefficients summations. Morgantown, West Virginia

[Gup78] Gupta H (1978) Finite differences of the partition function. Math Comput 32:1241–1243

[Hal68] Halász G (1968) Über die Mittelwerte multiplikativer zahlentheoretischer Funktionen. Acta Math Acad Sci Hung 19:365–403

[HR79]   Halberstam H, Richert H-E (1979) On a result of R.R. Hall. J Number Theory 11:76–89
[HR11]   Halberstam H, Richert H-E (2011) Sieve methods. Dover, Mineola
[HT88]   Hall RR, Tenenbaum G (1988) Divisors. Cambridge University Press, Cambridge
[Ivi85]   Ivić A (1985) The Riemann zeta-function. Theory and applications. Wiley, New York. 2nd edition: Dover, 2003
[IK04]   Iwaniec H, Kowalski E (2004) Analytic number theory. Colloquium publications, vol 53. Am. Math. Soc., Providence
[Krä70]   Krätzel E (1970) Die maximale Ordnung der Anzahl der wesentlich verschiedenen Abelschen Gruppen n-ter Ordnung. Q J Math (2) Oxford Ser 21:273–275
[Kub71]   Kubilius J (1971) The method of Dirichlet generating series in the theory of distribution of additive arithmetic functions. I. Liet Mat Rink 11:125–134
[Lan27]   Landau E (1927) Einführung in die elementare und analytische Theorie der algebraischen Zahlen und der Ideale. Teubner, Leipzig. 2nd edition: Chelsea, 1949
[LF67]   Levin BV, Fainleb AS (1967) Application of some integral equations to problems of number theory. Russ Math Surv 22:119–204
[Lin41]   Linnik JV (1941) The large sieve. Dokl Akad Nauk SSSR 30:292–294
[Mar95]   Marraki ME (1995) Fonction sommatoire de la fonction de Möbius, 3. Majorations effectives forte. J Théor Nr Bordx 7:407–433
[Mar02]   Martin G (2002) An asymptotic formula for the number of smooth values of a polynomial. J Number Theory 93:108–182
[Mon78]   Montgomery HL (1978) The analytic principle of the large sieve. Bull Am Math Soc 84:547–567
[Mot79]   Motohashi Y (1979) A note on Siegel's zeros. Proc Jpn Acad 55:190–192
[MV73]   Montgomery HL, Vaughan RC (1973) The large sieve. Mathematika 20:119–134
[NT98]   Nair M, Tenenbaum G (1998) Short sums of certain arithmetic functions. Acta Math 180:119–144
[PT78]   Parson A, Tull J (1978) Asymptotic behavior of multiplicative functions. J Number Theory 10:395–420
[Shi80]   Shiu P (1980) A Brun–Titchmarsh theorem for multiplicative functions. J Reine Angew Math 313:161–170
[Siv89]   Sivaramakrishnan R (1989) Classical theory of arithmetical functions. Pure and applied mathematics, vol 126. Dekker, New York
[Skr74]   Skrabutenas R (1974) Asymptotic expansion of sums of multiplicative functions. Liet Mat Rink 14:115–126
[SS94]   Schwarz L, Spilker J (1994) Arithmetical functions. Cambridge University Press, Cambridge
[Sur71]   Suryanarayana D (1971) The number of k-free divisors of an integer. Acta Arith 47:345–354
[SW00]   Sargos P, Wu J (2000) Multiple exponential sums with monomials and their applications in number theory. Acta Math Hung 88:333–354
[Tit39]   Titchmarsh EC (1939) The theory of functions. Oxford University Press, London. 2nd edition: 1979
[Tul78]   Tuljaganova MI (1978) A generalization of a theorem of Halász. Izv Akad Nauk SSSR 4:35–40
[Vau73]   Vaughan RC (1973) Some applications of Montgomery's sieve. J Number Theory 5:641–679
[Vor04]   Voronoï G (1904) Sur une fonction transcendante et ses applications à la sommation de quelques séries. Ann Sci Éc Norm Super 21:207–268
[Wir61]   Wirsing E (1961) Das asymptotische Verhalten vun Summen über multiplikative Funktionen. Math Ann 143:75–102
[Wu95]   Wu J (1995) Problèmes de diviseurs exponentiels et entiers exponentiellement sans facteur carré. J Théor Nr Bordx 7:133–141

Chapter 5
Integer Points Close to Smooth Curves

5.1 Introduction

5.1.1 Squarefree Numbers in Short Intervals

Let x, y be two real numbers satisfying $2 \leqslant y \leqslant x$. We intend to get an asymptotic formula for the sum

$$\sum_{x < n \leqslant x+y} \mu_2(n).$$

First, the trivial bound provides

$$\sum_{x < n \leqslant x+y} \mu_2(n) \leqslant \sum_{x < n \leqslant x+y} 1 = [x+y] - [x] \leqslant y+1$$

and Shiu's theorem (Theorem 4.62) gives no further improvement. It was seen in Example 4.18 that

$$\sum_{n \leqslant x} \mu_2(n) = \frac{x}{\zeta(2)} + O(\sqrt{x})$$

and therefore one may ask for an asymptotic result having the shape

$$\sum_{x < n \leqslant x+y} \mu_2(n) = \frac{y}{\zeta(2)} + O(\sqrt{y}). \tag{5.1}$$

At the present time, (5.1) remains unproven. The long-sum result above immediately yields

$$\sum_{x < n \leqslant x+y} \mu_2(n) = \frac{y}{\zeta(2)} + O(\sqrt{x}). \tag{5.2}$$

O. Bordellès, *Arithmetic Tales*, Universitext,
DOI 10.1007/978-1-4471-4096-2_5, © Springer-Verlag London 2012

so that this estimate is useless if $y \ll \sqrt{x}$. At this point we may formulate our problem more precisely: we ask for the smallest exponent $\theta \in [0, \frac{1}{2}]$ such that the asymptotic formula

$$\sum_{x < n \leqslant x+y} \mu_2(n) = \frac{y}{\zeta(2)} + O\left(x^\theta (\log x)^\beta + y^{1/2}\right) \tag{5.3}$$

holds for some real number $\beta \geqslant 0$. One may weaken this problem by ignoring the logarithmic term and look for estimates of the shape

$$\sum_{x < n \leqslant x+y} \mu_2(n) = \frac{y}{\zeta(2)} + O\left(x^{\theta+\varepsilon} + y^{1/2}\right)$$

valid for all $\varepsilon > 0$ and x sufficiently large. The minimal value $\theta = 0$ would give a slightly weaker version of (5.1). Furthermore, (5.3) has another consequence: there exist constants $c_0 > 0$ and $0 < c_1 < 1$ such that

$$\sum_{x < n \leqslant x+y} \mu_2(n) \geqslant \frac{y}{\zeta(2)} - c_0 x^\theta (\log x)^\beta - c_0 y^{1/2} \geqslant c_1 y - c_0 x^\theta (\log x)^\beta$$

if y is sufficiently large. Therefore, if $y > c_2 x^\theta (\log x)^\beta$ for some constant $c_2 > c_0 c_1^{-1}$, we get

$$\sum_{x < n \leqslant x+y} \mu_2(n) > 0$$

which means that the interval $]x, x + y]$ contains a squarefree integer.

The determination of minimal gaps between squarefree numbers is a long-standing problem in multiplicative number theory. It was introduced by Fogels [Fog41] in 1941 who proved that $\theta = \frac{2}{5}$ is an admissible value in this problem, and then Roth [Rot51], Richert [Ric54], Graham and Kolesnik [GK91] and Filaseta and Trifonov [FT92] successively improved on the value of θ. The best value to date is given in [FT92] in which the authors proved that, *if $y \geqslant c_0 x^{1/5} \log x$, then the interval $]x, x + y]$ contains a squarefree integer*. We shall see their method, based entirely upon elementary arguments, in Sect. 5.4.

Let us have a closer look at the details of the computations. If we use point by point the method of Example 4.18, we get

$$\sum_{x < n \leqslant x+y} \mu_2(n) = \sum_{x < n \leqslant x+y} \sum_{d^2 \mid n} \mu(d)$$

$$= \sum_{d \leqslant \sqrt{x+y}} \mu(d) \sum_{x/d^2 < k \leqslant (x+y)/d^2} 1$$

$$= \sum_{d \leqslant \sqrt{x+y}} \mu(d) \left(\left[\frac{x+y}{d^2} \right] - \left[\frac{x}{d^2} \right] \right).$$

We have

$$\frac{x+y}{d^2} - \frac{x}{d^2} = \frac{y}{d^2}$$

which prompts us to split the above sum into three parts as follows.

$$\sum_{x<n\leqslant x+y} \mu_2(n) = \left(\sum_{d\leqslant 2\sqrt{y}} + \sum_{2\sqrt{y}<d\leqslant\sqrt{x}} + \sum_{\sqrt{x}<d\leqslant\sqrt{x+y}} \right) \mu(d) \left(\left[\frac{x+y}{d^2}\right] - \left[\frac{x}{d^2}\right] \right)$$

$$= S_1 + S_2 + S_3.$$

▷ For S_1, the usual estimate $[t] = t + O(1)$ is sufficient and gives

$$S_1 = y \sum_{d\leqslant 2\sqrt{y}} \frac{\mu(d)}{d^2} + O(\sqrt{y}) = \frac{y}{\zeta(2)} + O(\sqrt{y}).$$

▷ For S_3, the trivial estimate is sufficient and gives

$$|S_3| \leqslant \sum_{\sqrt{x}<d\leqslant\sqrt{x+y}} 1 \leqslant \sqrt{x+y} - \sqrt{x} + 1 \leqslant yx^{-1/2} + 1 \ll \sqrt{y}$$

since $y \leqslant x$. Hence we obtain

$$\sum_{x<n\leqslant x+y} \mu_2(n) = \frac{y}{\zeta(2)} + \sum_{2\sqrt{y}<d\leqslant\sqrt{x}} \mu(d) \left(\left[\frac{x+y}{d^2}\right] - \left[\frac{x}{d^2}\right] \right) + O(\sqrt{y}).$$

It remains to estimate this sum. The triangle inequality gives (5.2). Observe that, since $y/d^2 < 1/4$, the difference

$$\left[\frac{x+y}{d^2}\right] - \left[\frac{x}{d^2}\right]$$

is equal to 1 or 0 depending on the fact that there is either an integer between x/d^2 and $(x+y)/d^2$ or not. Hence, to have a chance to do better than the trivial estimate, we must take these cancellations into account. The idea is then

1. to split the interval $]2\sqrt{y}, \sqrt{x}]$ into $O(\log x)$ dyadic subintervals of the form $]N, 2N]$ which gives

$$|S_2| \ll \max_{2\sqrt{y}<N\leqslant\sqrt{x}} \sum_{N<d\leqslant 2N} \left(\left[\frac{x+y}{d^2}\right] - \left[\frac{x}{d^2}\right] \right) \log x$$

2. to estimate the number of points with integer coordinates, also called *integer points*,[1] lying near the curve of the function $f(u) = x/u^2$ with $N < u \leqslant 2N$.

This chapter is devoted to providing results counting integer points close to sufficiently regular plane curves.

---

[1] Some authors also use the vocable *lattice points*.

5.1.2 Definitions and Notation

▷ In what follows, $N \geq 4$ is a large integer and δ and c_0 are small positive real numbers. We will always suppose that

$$0 < \delta < \frac{1}{4}. \tag{5.4}$$

Note that the constant c_0 may take different values according to the section in which it appears.

▷ The notation $\|t\|$ means the distance from the real number t to its nearest integer denoted by $\lfloor t \rceil$. Some properties of this function are investigated in Exercises 1 and 2.

▷ x, y are large real numbers satisfying $2 \leq y \leq x$, except in the examples where we impose the following more restricted range

$$16 \leq y < \frac{\sqrt{x}}{4}. \tag{5.5}$$

▷ *The set $\mathcal{S}_k$.* Let $k \in \mathbb{N}$. We shall say that $f \in \mathcal{S}_k$ if and only if $f \in C^k[N, 2N]$ such that there exist $\lambda_k > 0$ and $c_k \geq 1$ such that, for all $x \in [N, 2N]$, we have

$$\lambda_k \leq \left| f^{(k)}(x) \right| \leq c_k \lambda_k$$

which also may be denoted using Titchmarsh–Vinogradov's notation by $|f^{(k)}(x)| \asymp \lambda_k$.

For instance, the functions $f : u \longmapsto x/u^2$ and $g = f^{-1} : u \longmapsto (x/u)^{1/2}$ lie in $\mathcal{S}_k$ for all k with

	λ_k	c_k
f	$\left(\frac{(k+1)!}{2^{k+2}}\right)\frac{x}{N^{k+2}}$	2^{k+2}
g	$\left(\frac{(2k-1)!!}{2^{k+1/2}}\right)\frac{x^{1/2}}{N^{k+1/2}}$	$2^{k+1/2}$

where $(2k-1)!! = 1 \times 3 \times 5 \times \cdots \times (2k-1)$.

▷ Finally, we shall need the following integer.

Definition 5.1 Let $f : [N, 2N] \longrightarrow \mathbb{R}$ be any map and δ be a real number satisfying (5.4). We define

$$\mathcal{S}(f, N, \delta) = \left\{ n \in [N, 2N] \cap \mathbb{Z} : \|f(n)\| < \delta \right\}$$

and $\mathcal{R}(f, N, \delta) = |\mathcal{S}(f, N, \delta)|$.

We shall always make use of the following lemma.

Lemma 5.2 *Let $f : [N, 2N] \longrightarrow \mathbb{R}$ be any map, δ be a real number satisfying (5.4) and (δ_n) be a sequence of real numbers supported on $[N, 2N]$ such that $0 \leqslant \delta_n < \delta$. Then we have*

$$\sum_{N \leqslant n \leqslant 2N} \left(\left[f(n) + \delta_n \right] - \left[f(n) - \delta_n \right] \right) \leqslant \mathcal{R}(f, N, \delta)$$

$$\leqslant \sum_{N \leqslant n \leqslant 2N} \left(\left[f(n) + \delta \right] - \left[f(n) - \delta \right] \right).$$

Proof We appeal to Proposition 1.11 (vii) which implies that

$$\sum_{N \leqslant n \leqslant 2N} \left(\left[f(n) + \delta_n \right] - \left[f(n) - \delta_n \right] \right)$$

$$= \sum_{\substack{N \leqslant n \leqslant 2N \\ \|f(n)\| < \delta_n}} 1 + \sum_{\substack{N \leqslant n \leqslant 2N \\ \|f(n)\| = \delta_n}} \left(\left[f(n) + \delta_n \right] - \left[f(n) - \delta_n \right] \right)$$

$$\leqslant \sum_{\substack{N \leqslant n \leqslant 2N \\ \|f(n)\| \leqslant \delta_n}} 1 \leqslant \mathcal{R}(f, N, \delta)$$

$$= \sum_{N \leqslant n \leqslant 2N} \left(\left[f(n) + \delta \right] - \left[f(n) - \delta \right] \right) - \sum_{\substack{N \leqslant n \leqslant 2N \\ \|f(n)\| = \delta}} \left(\left[f(n) + \delta \right] - \left[f(n) - \delta \right] \right)$$

$$\leqslant \sum_{N \leqslant n \leqslant 2N} \left(\left[f(n) + \delta \right] - \left[f(n) - \delta \right] \right)$$

as asserted.   □

It should be noticed that the trivial estimate gives

$$\mathcal{R}(f, N, \delta) \leqslant N + 1. \tag{5.6}$$

This inequality must be kept in mind whenever we obtain a new estimate for $\mathcal{R}(f, N, \delta)$.

5.1.3 Basic Lemma in the Squarefree Number Problem

With the definitions and notation of the previous section, we may state the basic result we shall use to estimate the number of squarefree integers in short intervals.

Lemma 5.3 *Let x, y satisfy (5.5) and $2\sqrt{y} \leqslant A < B \leqslant 2\sqrt{x}$. Then*

$$\sum_{x < n \leqslant x+y} \mu_2(n) = \frac{y}{\zeta(2)} + O\left((R_1 + R_2) \log x + A \right)$$

where $R_1 = R_1(A, B)$ and $R_2 = R_2(B)$ are defined by

$$R_1 = \max_{A < N \leqslant B} \mathcal{R}\left(\frac{x}{n^2}, N, \frac{y}{N^2}\right) \quad and \quad R_2 = \max_{N \leqslant 2x/B^2} \mathcal{R}\left(\sqrt{\frac{x}{n}}, N, \frac{y}{\sqrt{Nx}}\right).$$

Proof By the computations made in the previous sections, we have

$$\sum_{x < n \leqslant x+y} \mu_2(n) = \frac{y}{\zeta(2)} + S_2 + O(\sqrt{y})$$

with

$$S_2 = \sum_{2\sqrt{y} < d \leqslant \sqrt{x}} \mu(d)\left(\left[\frac{x+y}{d^2}\right] - \left[\frac{x}{d^2}\right]\right).$$

Inserting the parameters A and B and estimating the first sum trivially, we get

$$|S_2| \leqslant \left(\sum_{2\sqrt{y} < d \leqslant A} + \sum_{A < d \leqslant B} + \sum_{B < d \leqslant \sqrt{x}}\right)\left(\left[\frac{x+y}{d^2}\right] - \left[\frac{x}{d^2}\right]\right)$$

$$\leqslant A + \sum_{A < d \leqslant B}\left(\left[\frac{x+y}{d^2}\right] - \left[\frac{x}{d^2}\right]\right) + \sum_{B < d \leqslant \sqrt{x}} \sum_{x/d^2 < k \leqslant (x+y)/d^2} 1$$

$$= A + \sum_{A < d \leqslant B}\left(\left[\frac{x+y}{d^2}\right] - \left[\frac{x}{d^2}\right]\right) + \sum_{B < d \leqslant \sqrt{x}} \sum_{\substack{k \\ x < kd^2 \leqslant x+y}} 1$$

and interchanging the summations gives

$$|S_2| \leqslant A + \sum_{A < d \leqslant B}\left(\left[\frac{x+y}{d^2}\right] - \left[\frac{x}{d^2}\right]\right) + \sum_{k \leqslant (x+y)/B^2} \sum_{\sqrt{x/k} < d \leqslant \sqrt{(x+y)/k}} 1$$

$$\leqslant A + \sum_{A < d \leqslant B}\left(\left[\frac{x+y}{d^2}\right] - \left[\frac{x}{d^2}\right]\right) + \sum_{k \leqslant 2x/B^2}\left(\left[\sqrt{\frac{x+y}{k}}\right] - \left[\sqrt{\frac{x}{k}}\right]\right).$$

Now for $k \in [N, 2N]$ and using (5.5), we have

$$\sqrt{\frac{x+y}{k}} - \sqrt{\frac{x}{k}} \leqslant \frac{y}{2\sqrt{Nx}} < \frac{y}{\sqrt{Nx}} < \frac{1}{4}$$

and splitting the sums into $O(\log x)$ subintervals $]N, 2N]$ gives the asserted result. $\qquad\square$

Remark 5.4

1. If we choose $B = x^{1/3}$, then two sums have a range of the same order of magnitude. Also note that the functions $u \longmapsto x/u^2$ and $u \longmapsto \sqrt{x/u}$ are inverse.

Hence one may expect that $R_1(A, x^{1/3})$ and $R_2(x^{1/3})$ have the same order of magnitude. Indeed they have by [Hux96, Lemma 3.1.1], so that Lemma 5.3 enables us to reduce the range of summation.

Choosing $A = 4\sqrt{y}$ and using (5.6) to estimate R_1 and R_2 trivially, we get at once

$$\sum_{x < n \leqslant x+y} \mu_2(n) = \frac{y}{\zeta(2)} + O\left(x^{1/3} \log x\right)$$

if x, y satisfy (5.5). Hence there exists $c_0 > 0$ such that, if $c_0 x^{1/3} \log x \leqslant y < x^{1/2}/4$, the interval $]x, x+y]$ contains a squarefree number.

2. Lemma 5.3 may easily be generalized to the problem of r-free numbers in short intervals, with $r \geqslant 2$. In this case, we have for all $(4y)^{1/r} \leqslant A < B \leqslant x^{1/r}$

$$\sum_{x < n \leqslant x+y} \mu_r(n) = \frac{y}{\zeta(r)} + O\left((R_1 + R_2) \log x + A\right)$$

where

$$R_1 = \max_{A < N \leqslant B} \mathcal{R}\left(\frac{x}{n^r}, N, \frac{y}{N^r}\right) \quad \text{and}$$

$$R_2 = \max_{N \leqslant 2x/B^r} \mathcal{R}\left(\left(\frac{x}{n}\right)^{1/r}, N, \frac{y}{N^{1/r} x^{1-1/r}}\right).$$

The proof is exactly the same as in Lemma 5.3, so we leave the details to the reader.

3. The problem of square-full numbers in short intervals is quite similar in nature to that of the squarefree number problem, except with the following major difference. In [Shi80], the author proved that there exist infinitely many positive integers n such that there is no square-full number between n^2 and $(n+1)^2$. It follows that, if $y < \sqrt{x}$, the interval $]x, x+y]$ may contain no square-full number at all. On the other hand, since there is a square in the interval $]x, x+2\sqrt{x}+1]$ for all $x \geqslant 0$, it follows that there exists a constant $c_0 > 0$ such that, for all $x \geqslant 1$, the interval $]x, x+c_0\sqrt{x}]$ contains a square-full integer. Thus, the maximum size of gaps between square-full numbers is known and one may ask for the distribution of square-full numbers in intervals $]x, x+y]$ with $y > \sqrt{x}$ and the ratio $yx^{-1/2}$ being as small as possible.

In Exercise 6, a basic lemma similar to Lemma 5.3 is established where it is shown that, if $16 x^{1/2} (\log x)^3 \leqslant y \leqslant 4^{-3} x (\log x)^{-1}$, then

$$\sum_{x < n \leqslant x+y} s_2(n) = \frac{\zeta(3/2)}{2\zeta(3)} \frac{y}{\sqrt{x}} + O\left\{(R_1 + R_2) \log x + \frac{y}{\sqrt{x} \log x}\right\}$$

where

$$R_1 = \max_{L < N \leqslant (2x)^{1/5}} \mathcal{R}\left(\sqrt{\frac{x}{n^3}}, N, \frac{y}{\sqrt{x N^3}}\right)$$

and

$$R_2 = \max_{L < N \leqslant (2x)^{1/5}} \mathcal{R}\left(\left(\frac{x}{n^2}\right)^{1/3}, N, \frac{y}{(Nx)^{2/3}}\right)$$

with $L = L(x, y) = y(x \log x)^{-1/2}$. Once again, the number of integer points close to two inverse curves thus appears.

5.1.4 Srinivasan's Optimization Lemma

We end this section with a useful optimization lemma due to Srinivasan [Sri62, Lemma 4], which generalizes the following well-known situation. Suppose we have an estimate of the shape

$$E(H) \ll A H^a + B H^{-b}$$

where $A, B, a, b > 0$ and $H > 0$ is a parameter at our disposal. Choosing H to equalize both terms, we get

$$E(H) \ll \left(A^b B^a\right)^{\frac{1}{a+b}}$$

and this is best possible apart from the value of the implied constant.

Lemma 5.5 (Srinivasan) *Let*

$$E(H) = \sum_{i=1}^{m} A_i H^{a_i} + \sum_{j=1}^{n} B_j H^{-b_j}$$

where $m, n \in \mathbb{N}$ and A_i, B_j, a_i and b_j are positive real numbers. Suppose that $0 \leqslant H_1 \leqslant H_2$. Then

$$\min_{H_1 \leqslant H \leqslant H_2} E(H) \leqslant (m+n) \left\{ \sum_{i=1}^{m} \sum_{j=1}^{n} \left(A_i^{b_j} B_j^{a_i}\right)^{\frac{1}{a_i+b_j}} + \sum_{i=1}^{m} A_i H_1^{a_i} + \sum_{j=1}^{n} B_j H_2^{-b_j} \right\}.$$

This inequality corresponds to the best possible choice of H in the interval $[H_1, H_2]$. Srinivasan pointed out that the case $H_1 = 0$ and $H_2 = \infty$ was already shown by van der Corput in 1922.

5.2 Criteria for Integer Points

5.2.1 The First Derivative Test

The first result we will establish concerns the functions $f \in S_1$ and follows from the classical mean-value theorem. It is only useful when λ_1 is very small, and hence is rather restrictive. However, it will turn out to be the starting point of more elaborate estimates.

Theorem 5.6 (First derivative test) *Let $f \in C^1[N, 2N]$ such that there exist $\lambda_1 > 0$ and $c_1 \geqslant 1$ such that, for all $x \in [N, 2N]$, we have*

$$\lambda_1 \leqslant |f'(x)| \leqslant c_1 \lambda_1. \tag{5.7}$$

Then

$$\mathcal{R}(f, N, \delta) \leqslant 2c_1 N \lambda_1 + 4c_1 N \delta + \frac{2\delta}{\lambda_1} + 1.$$

In practice, the implied constant c_1 is useless, so using Titchmarsh–Vinogradov's notation, this result may be rewritten as follows. If

$$|f'(x)| \asymp \lambda_1$$

then we have

$$\mathcal{R}(f, N, \delta) \ll N\lambda_1 + N\delta + \frac{\delta}{\lambda_1} + 1.$$

Proof If $4c_1\delta \geqslant 1$, then $4c_1 N\delta + 1 \geqslant N + 1 \geqslant \mathcal{R}(f, N, \delta)$ by (5.6). Similarly, if $2c_1\lambda_1 \geqslant 1$, then $2c_1 N\lambda_1 + 1 \geqslant N + 1 \geqslant \mathcal{R}(f, N, \delta)$. Therefore we may suppose that $\max(4c_1\delta, 2c_1\lambda_1) < 1$. Let n and $n + a$ be any integers in $\mathcal{S}(f, N, \delta)$. Using the mean-value theorem, we will prove that either

$$a > \frac{1}{2c_1\lambda_1} = a_1 \tag{5.8}$$

or

$$a < \frac{2\delta}{\lambda_1} = a_2. \tag{5.9}$$

We postpone the proof of these inequalities and assume that either (5.8) or (5.9) holds. Note that the condition $\max(4c_1\delta, 2c_1\lambda_1) < 1$ implies $a_1 > \max(1, a_2)$. Subdividing the interval $[N, 2N]$ into $s = [N/a_1] + 1$ subintervals $\mathcal{I}_1, \ldots, \mathcal{I}_s$ of lengths $\leqslant a_1$, two elements of $\mathcal{S}(f, N, \delta) \cap \mathcal{I}_j$ have a distance $\leqslant a_2$, and hence lie in an interval of length $\leqslant a_2$. Using Proposition 1.11 (v), we infer that

$$|\mathcal{S}(f, N, \delta) \cap \mathcal{I}_j| \leqslant a_2 + 1$$

and thus

$$\mathcal{R}(f, N, \delta) \leqslant \left(\frac{N}{a_1} + 1 \right)(a_2 + 1) = 2c_1 N\lambda_1 + 4c_1 N\delta + \frac{2\delta}{\lambda_1} + 1.$$

The rest of the text is devoted to the proof of the inequalities (5.8) and (5.9).

Since $n, n + a \in \mathcal{S}(f, N, \delta)$, there exist two integers m_1 and m_2 and two real numbers δ_1 and δ_2 such that

$$f(n) = m_1 + \delta_1,$$

$$f(n + a) = m_2 + \delta_2$$

with $|\delta_i| < \delta$ for $i \in \{1, 2\}$. Thus there exist $m_3 \in \mathbb{Z}$ and $\delta_3 \in \mathbb{R}$ such that

$$f(n + a) - f(n) = m_3 + \delta_3$$

with $|\delta_3| < 2\delta$. By the mean-value theorem, there exists $t \in]n, n + a[$ such that

$$f(n + a) - f(n) = af'(t).$$

Since $n + a \in \mathcal{S}(f, N, \delta)$, we have $n + a \leqslant 2N$ and thus $t \in [N, 2N]$. Hence there exist $t \in [N, 2N]$, $m \in \mathbb{Z}$ and δ_3 such that $|\delta_3| < 2\delta < \frac{1}{2}$ satisfying

$$af'(t) = m + \delta_3.$$

Now two cases may occur.

▷ $m \neq 0$. Since $m \in \mathbb{Z}$, we have $|m| \geqslant 1$ and then using (5.7)

$$ac_1\lambda_1 \geqslant a|f'(t)| \geqslant |m| - |\delta_3| > 1 - \frac{1}{2} = \frac{1}{2}$$

which gives (5.8).

▷ $m = 0$. Then we have using (5.7) again

$$a\lambda_1 \leqslant a|f'(t)| = |\delta_3| < 2\delta$$

which gives (5.9) as required.   □

We use Lemma 5.3 with $A = x^{1/3}$ and $B = 2x^{1/2}$. By (5.5), we have $A > 2y^{1/2}$ and hence

$$\sum_{x < n \leqslant x + y} \mu_2(n) = \frac{y}{\zeta(2)} + O\left(R_1(x^{1/3}, 2x^{1/2}) \log x + x^{1/3}\right)$$

and the use of Theorem 5.6 and (5.5) implies that

$$R_1(x^{1/3}, 2x^{1/2}) \ll \max_{x^{1/3} < N \leqslant 2x^{1/2}} \left(\frac{x}{N^2} + \frac{y}{N} + \frac{y}{\sqrt{x}} + 1\right) \ll x^{1/3}$$

and thus

$$\sum_{x < n \leqslant x + y} \mu_2(n) = \frac{y}{\zeta(2)} + O\left(x^{1/3} \log x\right)$$

and hence this result does not improve on the trivial estimate of Remark 5.4 in that problem.

5.2.2 The Second Derivative Test

The next result, coming from a combinatorial argument, will enable us to pass from the first derivative to the second derivative of f. In this section, δ is supposed to verify the inequalities

$$0 < \delta < \frac{1}{8}.$$

Lemma 5.7 (Reduction principle) *Let $f : [N, 2N] \longrightarrow \mathbb{R}$ be any map, A be a real number satisfying $1 \leqslant A \leqslant N$ and, for all integers $a \in [1, A]$, we define on $[N, 2N - a]$ the function $\Delta_a f$ by*

$$\Delta_a f(x) = f(x + a) - f(x).$$

Then

$$\mathcal{R}(f, N, \delta) \leqslant \frac{N}{A} + \sum_{a \leqslant A} \mathcal{R}(\Delta_a f, N, 2\delta) + 1.$$

Proof For all $a \in \mathbb{N}$, define

$$\mathcal{S}(a) = \left\{ n \in [N, 2N] \cap \mathbb{Z} : n \text{ and } n + a \text{ are consecutive in } \mathcal{S}(f, N, \delta) \right\}.$$

▷ We first prove that

$$\mathcal{R}(f, N, \delta) \leqslant \frac{N}{A} + \sum_{a \leqslant A} |\mathcal{S}(a)| + 1. \tag{5.10}$$

Each integer of $\mathcal{S}(f, N, \delta)$, except the largest one, has a successive element and then lies in only one subset $\mathcal{S}(a)$, so that

$$\mathcal{R}(f, N, \delta) = \sum_{a=1}^{\infty} |\mathcal{S}(a)| + 1 = \sum_{a \leqslant A} |\mathcal{S}(a)| + \sum_{a > A} |\mathcal{S}(a)| + 1$$

with $A \in \mathbb{R}$ satisfying $1 \leqslant A \leqslant N$. Now if $\mathcal{S}(f, N, \delta) = \{n_1 \leqslant n_2 \leqslant \cdots \leqslant n_k\}$ and if we set

$$d_1 = n_2 - n_1, \ d_2 = n_3 - n_2, \ldots, d_{k-1} = n_k - n_{k-1}$$

then, for all $a \in \mathbb{N}$, $|\mathcal{S}(a)|$ is the number of indexes $j \in \{1, \ldots, k - 1\}$ such that $d_j = a$ and thus

$$\sum_{a=1}^{\infty} a |\mathcal{S}(a)| = \sum_{j=1}^{k-1} d_j = \sum_{j=1}^{k-1} (n_{j+1} - n_j) = n_k - n_1 \leqslant N$$

since $n_k \leqslant 2N$ and $n_1 \geqslant N$. Therefore

$$N \geqslant \sum_{a=1}^{\infty} a |S(a)| \geqslant \sum_{a>A} a |S(a)| \geqslant A \sum_{a>A} |S(a)|$$

which implies (5.10).

▷ Now let $n \in S(a)$. Then n and $n + a$ are consecutive in $S(f, N, \delta)$, so that

$$\|\Delta_a f(n)\| = \|f(n+a) - f(n)\| \leqslant \|f(n+a)\| + \|f(n)\| < 2\delta$$

and hence $n \in S(\Delta_a f, N, 2\delta)$, which gives

$$|S(a)| \leqslant \mathcal{R}(\Delta_a f, N, 2\delta).$$

Inserting this bound in (5.10) gives the asserted result.   □

Let $f \in S_2$. By the mean-value theorem, we have $|(\Delta_a f)'(x)| \asymp a\lambda_2$. We may apply Theorem 5.6 to $\Delta_a f$ and use the previous lemma to go back to f. These ideas provide the following criterion.

Theorem 5.8 (Second derivative test) *Let $f \in C^2[N, 2N]$ such that there exist $\lambda_2 > 0$ and $c_2 \geqslant 1$ such that, for all $x \in [N, 2N]$, we have*

$$\lambda_2 \leqslant |f''(x)| \leqslant c_2\lambda_2 \tag{5.11}$$

and

$$N\lambda_2 \geqslant c_2^{-1}. \tag{5.12}$$

Then

$$\mathcal{R}(f, N, \delta) \leqslant 6\left\{ (3c_2)^{1/3} N\lambda_2^{1/3} + (12c_2)^{1/2} N\delta^{1/2} + 1 \right\}.$$

In practice, one may use this result in the following way. If

$$|f''(x)| \asymp \lambda_2 \quad \text{and} \quad N\lambda_2 \gg 1$$

then

$$\mathcal{R}(f, N, \delta) \ll N\lambda_2^{1/3} + N\delta^{1/2} + 1.$$

Proof If $\lambda_2 \geqslant (3c_2)^{-1}$, then $(3c_2)^{1/3} N\lambda_2^{1/3} \geqslant N + 1 \geqslant \mathcal{R}(f, N, \delta)$ by (5.6). Similarly, if $\delta \geqslant (12c_2)^{-1}$, then $(12c_2)^{1/2} N\delta^{1/2} + 1 \geqslant N + 1$. Henceforth we assume that

$$0 < \lambda_2 < (3c_2)^{-1} \quad \text{and} \quad 0 < \delta < (12c_2)^{-1}. \tag{5.13}$$

Let $A \in \mathbb{R}$ such that $1 \leqslant A \leqslant N$. For all $x \in [N, 2N]$ and all $a \in [1, A] \cap \mathbb{Z}$ such that $x + a \in [N, 2N]$, the mean-value theorem gives

$$(\Delta_a f)'(x) = f'(x + a) - f'(x) = af''(t)$$

for some $t \in {]}x, x + a[$. Since ${]}x, x + a[\subseteq [N, 2N]$, (5.11) implies that, for all $x \in [N, 2N]$ and all $a \in [1, A] \cap \mathbb{Z}$ such that $x + a \in [N, 2N]$, we have

$$a\lambda_2 \leqslant \left|(\Delta_a f)'(x)\right| \leqslant c_2 a \lambda_2.$$

Therefore, using Lemma 5.7 and Theorem 5.6 we get

$$\mathcal{R}(f, N, \delta) \leqslant \frac{N}{A} + \sum_{a \leqslant A} \left(2c_2 N a \lambda_2 + 8c_2 N \delta + \frac{4\delta}{a\lambda_2} + 1\right) + 1$$

and (5.12) implies that $1 \leqslant c_2 N a \lambda_2$ and

$$4c_2 N \delta \geqslant 4\delta \lambda_2^{-1} \geqslant 4\delta (a\lambda_2)^{-1}$$

for all $a \geqslant 1$, so that

$$\mathcal{R}(f, N, \delta) \leqslant \frac{N}{A} + \sum_{a \leqslant A} (3c_2 N a \lambda_2 + 12 c_2 N \delta) + 1$$

$$\leqslant \frac{N}{A} + 3c_2 A^2 N \lambda_2 + 12 c_2 N A \delta + 1.$$

Now Lemma 5.5 implies that

$$\mathcal{R}(f, N, \delta) \leqslant 3N \left\{(3c_2\lambda_2)^{1/3} + 3c_2\lambda_2 + (12 c_2\delta)^{1/2} + 12 c_2 \delta\right\} + 6$$

and the proof is achieved with the use of (5.13). $\qquad \square$

Example 5.9 We use Lemma 5.3 with $A = 2x^{1/4}$ and $B = x^{1/3}$. By (5.5), we have $A > 2y^{1/2}$ and hence

$$\sum_{x < n \leqslant x + y} \mu_2(n) = \frac{y}{\zeta(2)} + O\left\{R_1\left(2x^{1/4}, x^{1/3}\right) \log x + R_2\left(x^{1/3}\right) \log x + x^{1/4}\right\}.$$

The condition (5.12) is satisfied by the two functions in their respective ranges of summation, and Theorem 5.8 gives

$$R_1\left(2x^{1/4}, x^{1/3}\right) \ll \max_{2x^{1/4} < N \leqslant x^{1/3}} \left(\left(xN^{-1}\right)^{1/3} + y^{1/2}\right) \ll x^{1/4}$$

$$R_2\left(x^{1/3}\right) \ll \max_{N \leqslant 2x^{1/3}} \left((Nx)^{1/6} + y^{1/2}\left(N^3 x^{-1}\right)^{1/4}\right) \ll x^{2/9} + y^{1/2}$$

and thus

$$\sum_{x<n\leqslant x+y} \mu_2(n) = \frac{y}{\zeta(2)} + O\left(x^{1/4} \log x\right).$$

Thus there exists a constant $c_0 > 0$ such that, if $c_0 x^{1/4} \log x \leqslant y < x^{1/2}/4$, the interval $]x, x+y]$ contains a squarefree number.

5.2.3 The kth Derivative Test

We now intend to generalize the results of the previous sections. To this end, we need a generalization of the mean-value theorem. The first idea is to use the Taylor–Lagrange formula which generalizes Theorem 1.12 (ii) by providing more terms if f has higher derivatives. This was done by Konyagin in [Kon98] where the author also used properties of lattices. In what follows, we rather focus on a generalization of the mean-value theorem in the number of points that a function may interpolate. Thus, the divided differences are the main tools in this section.

We first start with a very easy lemma.

Lemma 5.10 *Let* $n = x + y \in \mathbb{Z}$ *with* $x, y \in \mathbb{R}$ *such that* $|y| < |x|$. *Then* $|x| > \frac{1}{2}$.

Proof Since $|y| < |x|$, we have $n \neq 0$ and hence $1 \leqslant |n| \leqslant |x| + |y| < 2|x|$ as asserted. $\qquad\square$

We are now in a position to show the main result of this section.

Theorem 5.11 *Let* $k \geqslant 1$ *be an integer and* $f \in C^k[N, 2N]$ *such that there exist* $\lambda_k > 0$ *and* $c_k \geqslant 1$ *such that, for all* $x \in [N, 2N]$, *we have*

$$\lambda_k \leqslant \left|f^{(k)}(x)\right| \leqslant c_k \lambda_k. \tag{5.14}$$

Assume also that

$$(k+1)! \, \delta < \lambda_k. \tag{5.15}$$

If $\alpha_k = 2k \, (2c_k)^{\frac{2}{k(k+1)}}$, *then*

$$\mathcal{R}(f, N, \delta) \leqslant \alpha_k N \lambda_k^{\frac{2}{k(k+1)}} + 4k.$$

As usual, this result is mostly used in the following form. If

$$\left|f^{(k)}(x)\right| \asymp \lambda_k \quad \text{and} \quad \delta \ll \lambda_k$$

then

$$\mathcal{R}(f, N, \delta) \ll N \lambda_k^{\frac{2}{k(k+1)}} + 1.$$

Proof If $\lambda_k \geqslant \frac{1}{2}$, then

$$\alpha_k N \lambda_k^{\frac{2}{k(k+1)}} + 1 \geqslant N + 1 \geqslant \mathcal{R}(f, N, \delta)$$

so that we may suppose that $\lambda_k < \frac{1}{2}$. We generalize the proof of Theorem 5.6 in the following way. Take $k + 1$ consecutive points $n < n + a_1 < n + a_2 < \cdots < n + a_k$ in $\mathcal{S}(f, N, \delta)$ and we will prove that

$$a_k \geqslant 2k \alpha_k^{-1} \lambda_k^{-\frac{2}{k(k+1)}}. \tag{5.16}$$

Assume first that (5.16) is true. Taking each $(k + 1)$th element of $\mathcal{S}(f, N, \delta)$, we may construct a subset T of $\mathcal{S}(f, N, \delta)$ such that any two distinct elements of T differ by more than $d_k = 2k \alpha_k^{-1} \lambda_k^{-\frac{2}{k(k+1)}}$ and therefore

$$\mathcal{R}(f, N, \delta) \leqslant (k + 1)(|T| + 1) \leqslant 2k \left(\frac{N}{d_k} + 2 \right)$$

giving the asserted result.

The rest of the text is devoted to the proof of (5.16).

By definition, there exist integers $m_0, \ldots, m_k$ and real numbers $\delta_0, \ldots, \delta_k$ such that

$$f(n + a_j) = m_j + \delta_j$$

with $a_0 = 0$ and $|\delta_j| < \delta$ for all $j \in \{0, \ldots, k\}$. Using (1.4), there exists $t \in]n, n + a_k[$ such that

$$\sum_{j=0}^{k} \frac{m_j + \delta_j}{\prod_{0 \leqslant i \leqslant k, i \neq j}(a_j - a_i)} = \frac{f^{(k)}(t)}{k!} \tag{5.17}$$

and by (1.3) and what follows, if $P = b_k X^k + \cdots + b_0$ is the Lagrange polynomial interpolating the points $(n + a_j, m_j)$, then we have

$$b_k = \sum_{j=0}^{k} \frac{m_j}{\prod_{0 \leqslant i \leqslant k, i \neq j}(a_j - a_i)} = \frac{A_k}{D_k} \tag{5.18}$$

for some $A_k \in \mathbb{Z}$ and where $D_k = \prod_{0 \leqslant i < j \leqslant k}(a_j - a_i) > 0$. Hence by (5.17) we get

$$b_k = \frac{f^{(k)}(t)}{k!} - \sum_{j=0}^{k} \frac{\delta_j}{\prod_{0 \leqslant i \leqslant k, i \neq j}(a_j - a_i)}$$

and then

$$k! A_k = k! D_k b_k = D_k f^{(k)}(t) - k! D_k \sum_{j=0}^{k} \frac{\delta_j}{\prod_{0 \leqslant i \leqslant k, i \neq j}(a_j - a_i)} = x + y.$$

Now the condition $|\delta_j| < \delta$ implies

$$|y| < k! \, D_k \, \delta \sum_{j=0}^{k} \frac{1}{\prod_{0 \leqslant i \leqslant k, i \neq j} |a_j - a_i|} \leqslant (k+1)! \, D_k \, \delta$$

where we used the fact that $|a_j - a_i| \geqslant 1$ for all $i \neq j$, and using (5.14) and (5.15) we get

$$|y| < D_k \, \lambda_k \leqslant D_k \big| f^{(k)}(t) \big| = |x|.$$

Lemma 5.10 gives $|x| > \frac{1}{2}$ and the bounds $D_k \leqslant a_k^{\frac{k(k+1)}{2}}$ and (5.14) imply

$$\frac{1}{2} < D_k \big| f^{(k)}(t) \big| \leqslant a_k^{\frac{k(k+1)}{2}} c_k \lambda_k = \frac{\lambda_k}{2} \big((2k)^{-1} a_k \alpha_k \big)^{\frac{k(k+1)}{2}}$$

which implies (5.16), concluding the proof. $\qquad\qquad\qquad\qquad\qquad\qquad\square$

This result does generalize Theorem 5.6, but the proof highlights the following weakness. The lower bound $|a_j - a_i| \geqslant 1$ may probably be improved in many special cases. This would enable us to decrease the upper bound for $|y|$ and then to have a condition less restrictive than (5.15). For instance, in the squarefree number problem, (5.15) requires to estimate R_1 and R_2 of Lemma 5.3 in the range

$$N < \frac{1}{8} \left(\frac{x}{y} \right)^{1/3}$$

which does not allow us to cover all the range of summation.

5.3 The Theorem of Huxley and Sargos

By the remark above, establishing a kth derivative criterion free from condition (5.15) arises naturally. This was done by Huxley and Sargos in [HS95] who proved the important result we shall see in this section. This turns out to be rather difficult, and several tools are needed in the proof. We first state the main result and then provide the proof in several steps. We shall essentially follow the line of [HS95], but our exposition may differ in certain minor points.[2]

Theorem 5.12 (Huxley–Sargos) *Let $k \geqslant 3$ be an integer and $f \in C^k[N, 2N]$ such that there exist $\lambda_k > 0$ and $c_k \geqslant 1$ such that, for all $x \in [N, 2N]$, we have*

$$\lambda_k \leqslant \big| f^{(k)}(x) \big| \leqslant c_k \lambda_k. \tag{5.19}$$

---

[2]For instance, the authors did not make use of Gorny's inequality but proved a Landau–Hadamard–Kolmogorov like result similar to (5.20) using divided differences.

Let δ be a real number satisfying (5.4). Then

$$R(f, N, \delta) \leqslant \alpha_k N \lambda_k^{\frac{2}{k(k+1)}} + \beta_k N \delta^{\frac{2}{k(k-1)}} + 8k^3 \left(\frac{\delta}{\lambda_k}\right)^{1/k} + 2k^2 \left(5e^3 + 1\right)$$

where

$$\alpha_k = 2k^2 c_k^{\frac{2}{k(k+1)}} \quad and \quad \beta_k = 4k^2 \left(5e^3 c_k^{\frac{2}{k(k-1)}} + 1\right).$$

As for the previous results, this inequality is used as

$$R(f, N, \delta) \ll N \lambda_k^{\frac{2}{k(k+1)}} + N \delta^{\frac{2}{k(k-1)}} + \left(\frac{\delta}{\lambda_k}\right)^{1/k} + 1$$

under the sole hypothesis (5.19), where the implied constants depend only on k and c_k.

We shall make use of the following additional notation

$$C_\delta = \left\{(x, y) \in [N, 2N] \times \mathbb{R} : \left|y - f(x)\right| < \delta\right\}.$$

5.3.1 Preparatory Lemmas

The first tool is an easy enumeration principle.

Lemma 5.13 *Let S be a finite set of integers with length $\leqslant N$. If one can recover S by pairwise distinct intervals $\mathcal{I}$ and if $L(\mathcal{I})$ is the length of $\mathcal{I}$, then*

$$|S| \leqslant N \max_{\mathcal{I}} \left(\frac{|S \cap \mathcal{I}|}{L(\mathcal{I})}\right) + 2 \max_{\mathcal{I}} |S \cap \mathcal{I}|.$$

Proof Let $\mathcal{I}_1, \ldots, \mathcal{I}_J$ be such a covering of S where we suppose that $h \neq j \implies \mathcal{I}_h \cap \mathcal{I}_j = \varnothing$ and, if $2 \leqslant j \leqslant J - 1$, then $\mathcal{I}_j \subseteq S$. We have

$$|S| \leqslant \sum_{j=1}^{J} |S \cap \mathcal{I}_j| = \sum_{j=2}^{J-1} \left\{L(\mathcal{I}_j) \times \frac{|S \cap \mathcal{I}_j|}{L(\mathcal{I}_j)}\right\} + |S \cap \mathcal{I}_1| + |S \cap \mathcal{I}_J|$$

$$\leqslant \max_{1 \leqslant j \leqslant J} \left(\frac{|S \cap \mathcal{I}_j|}{L(\mathcal{I}_j)}\right) \sum_{j=2}^{J-1} L(\mathcal{I}_j) + 2 \max_{1 \leqslant j \leqslant J} |S \cap \mathcal{I}_j|$$

$$\leqslant N \max_{1 \leqslant j \leqslant J} \left(\frac{|S \cap \mathcal{I}_j|}{L(\mathcal{I}_j)}\right) + 2 \max_{1 \leqslant j \leqslant J} |S \cap \mathcal{I}_j|$$

where in the last inequality we used the fact that the intervals are pairwise distinct. □

The next tool belongs to a certain class of inequalities, called the *Landau–Hadamard–Kolmogorov inequalities*. In 1913, E. Landau proved that, if I is an interval with length $\geqslant 2$ and $f \in C^2(I)$ satisfies the conditions $|f(x)| \leqslant 1$ and $|f''(x)| \leqslant 1$ on I, then $|f'(x)| \leqslant 2$ and the constant 2 is the best possible. This was generalized by Hadamard who showed that, if $a \in \mathbb{R}$ and $L > 0$, then

$$\sup_{a \leqslant x \leqslant a+L} |f'(x)| \leqslant \frac{2}{L} \sup_{a \leqslant x \leqslant a+L} |f(x)| + \frac{L}{2} \sup_{a \leqslant x \leqslant a+L} |f''(x)|.$$

The generalization to higher orders of derivative was established by several mathematicians. In 1930, L. Neder proved that, if $a \in \mathbb{R}$, $L > 0$ and $f \in C^k[a, a + L]$, then, for all $j \in \{1, \ldots, k - 1\}$, we have

$$\sup_{a \leqslant x \leqslant a+L} |f^{(j)}(x)| \leqslant \frac{(2k)^{2k}}{L^j} \sup_{a \leqslant x \leqslant a+L} |f(x)| + L^{k-j} \sup_{a \leqslant x \leqslant a+L} |f^{(k)}(x)|. \quad (5.20)$$

This result would be sufficient for our proof of Theorem 5.12 but, for the sake of completeness, we mention the following improvement due to Gorny [Gor39].

Lemma 5.14 (Gorny) *Let $k \geqslant 2$ be an integer, $a \in \mathbb{R}$, $L > 0$ and $f \in C^k[a, a + L]$ such that, for all $x \in [a, a + L]$, we have*

$$|f(x)| \leqslant M_0 \quad and \quad |f^{(k)}(x)| \leqslant M_k$$

with $M_0 < \infty$ and $M_k < \infty$. Then, for all $x \in [a, a + L]$ and $j \in \{1, \ldots, k - 1\}$, we have

$$|f^{(j)}(x)| < 4\left(e^2 k/j\right)^j M_0^{1-j/k} \left\{\max\left(M_k, k! M_0 L^{-k}\right)\right\}^{j/k}.$$

In practice, we will use this result in the following form.

$$|f^{(j)}(x)| < 4e(ek/j)^j \left\{k^{j+1} M_0 L^{-j} + e^{j-1} M_0^{1-j/k} M_k^{j/k}\right\}. \quad (5.21)$$

Let us finally mention the following elegant version of Kolmogorov's inequality which can be found in [Man52, Théorème 6.3.III].

Let $f \in C^k(\mathbb{R})$ such that $M_0 = \sup_{x \in \mathbb{R}} |f(x)| < \infty$ and $M_k = \sup_{x \in \mathbb{R}} |f^{(k)}(x)| < \infty$. Then, for all $j \in \{0, \ldots, k\}$ and all $x \in \mathbb{R}$, we have

$$|f^{(j)}(x)| \leqslant 2M_0^{1-j/k} M_k^{j/k}.$$

To end this section, we shall need the next technical tool to calculate the implied constants appearing in Theorem 5.12.

Lemma 5.15 *Let $k \geqslant 3$ be an integer and $a > e(k - 1)$ be a real number. Then*

$$\sum_{j=1}^{k-1} \left(\frac{a}{j}\right)^{2j} \leqslant e^2 k \left(\frac{a}{k}\right)^{2k-2}.$$

Proof Since $a > e(k - 1)$, the function $x \longmapsto (a/x)^{2x}$ is increasing as soon as $1 \leqslant x \leqslant k - 1$ and then

$$\sum_{j=1}^{k-1} \left(\frac{a}{j}\right)^{2j} \leqslant (k - 1)\left(\frac{a}{k - 1}\right)^{2k-2}$$

and we conclude the proof with

$$\frac{k - 1}{k} \times \left(\frac{a}{k - 1} \times \frac{k}{a}\right)^{2k-2} = \exp\left\{(2k - 3)\log\left(1 + \frac{1}{k - 1}\right)\right\}$$

$$\leqslant \exp\left(\frac{2k - 3}{k - 1}\right) \leqslant e^2$$

as asserted. $\qquad\square$

5.3.2 Major Arcs

In this section, we take the notation above. Besides, we recall that $\lfloor x \rceil$ is the nearest integer to x.

Definition 5.16

1. A *major arc* associated to $f^{(k)}$ is a maximal set $\mathcal{A} = \{n_1, \dots, n_J\}$ of consecutive points of $\mathcal{S}(f, N, \delta)$, where $J \geqslant k + 1$ is an integer, such that, for all $j \in \{1, \dots, J\}$, we have

$$\lfloor f(n_j) \rceil = P(n_j)$$

   where $P \in \mathbb{Q}[X]$ is the Lagrange polynomial of degree $< k$ interpolating the points $(n_j, \lfloor f(n_j) \rceil)$. The equation $y = P(x)$ is called the *equation* of $\mathcal{A}$. We set $\mathcal{C}_P$ the curve with equation $y = P(x)$.
2. Let q be the smallest positive integer such that $P \in \frac{1}{q}\mathbb{Z}[X]$. Then q is called the *denominator* of $\mathcal{A}$.

The first result gives a bound for the number of connected components of the set $\mathcal{C}_\delta \cap \mathcal{C}_P$.

Lemma 5.17 *The set $\mathcal{C}_\delta \cap \mathcal{C}_P$ has at most k connected components.*

Proof Let $\widetilde{f} \in C^k(\mathbb{R})$ be the function defined by $\widetilde{f}(x) = f(x)$ for $x \in [N, 2N]$, $\widetilde{f}^{(k)}(x) = f^{(k)}(N)$ if $x \leqslant N$ and $\widetilde{f}^{(k)}(x) = f^{(k)}(2N)$ if $x \geqslant 2N$, so that

$$\left|\widetilde{f}^{(k)}(x)\right| \asymp \lambda_k$$

for all $x \in \mathbb{R}$. Similarly, the set $\widetilde{\mathcal{C}}_\delta$ is the analogue of $\mathcal{C}_\delta$ for the function $\widetilde{f}$.

Since the behaviors of $\tilde{f}$ and P at infinity are different, the set $\tilde{C}_\delta \cap C_P$ is bounded, and hence each connected component of this set, not reduced to a singleton, has two extremities satisfying $P(x) = f(x) \pm \delta$.

Now assume that there exist $k+1$ connected components of $\tilde{C}_\delta \cap C_P$, not reduced to a singleton. Then the equation $P(x) = f(x) + \delta$, say, has $k+1$ solutions $\alpha_0 < \cdots < \alpha_k$. Since the polynomials $P(X)$ and $P(X) - \delta$ have the same leading coefficient, we get by (1.4)

$$0 = \frac{P^{(k)}(t_1)}{k!} = \sum_{j=0}^{k} \frac{P(\alpha_j)}{\prod_{0 \leqslant i \leqslant k, i \neq j}(\alpha_j - \alpha_i)}$$

$$= \sum_{j=0}^{k} \frac{P(\alpha_j) - \delta}{\prod_{0 \leqslant i \leqslant k, i \neq j}(\alpha_j - \alpha_i)}$$

$$= \sum_{j=0}^{k} \frac{f(\alpha_j)}{\prod_{0 \leqslant i \leqslant k, i \neq j}(\alpha_j - \alpha_i)} = \frac{f^{(k)}(t_2)}{k!} \neq 0$$

for some $t_1, t_2 \in \,]\alpha_0, \alpha_k[$, giving a contradiction.   □

This result leads to the following slight refinement of Definition 5.16.

Definition 5.18 Among the connected components of the set $C_\delta \cap C_P$, choose the one having the largest number of points $(n_{h+j}, \lfloor f(n_{h+j}) \rceil)$ for all $j \in \{1, \ldots, l\}$ with $l > k$. Then the set

$$\overline{A} = \{n_{h+1}, \ldots, n_{h+l}\}$$

is called a *proper major arc* extracted from A. The *length* of $\overline{A}$ is the number

$$L = n_{h+l} - n_{h+1}.$$

For convenience, we introduce the following numbers

$$a_k = 36e^{-2}k\left(2e^3\right)^k c_k \quad \text{and} \quad b_k = 20e^3\,k^2 c_k^{\frac{2}{k(k-1)}} \tag{5.22}$$

so that $\beta_k = b_k + 4k^2$.

The next result summarizes the basic properties of the major arcs.

Lemma 5.19 *Let A be a major arc associated to $f^{(k)}$ and $\overline{A}$ be the proper major arc taken from A with denominator q, length L and equation $y = P(x)$.*

(i) *We have*

$$L \leqslant 2k\left(\frac{\delta}{\lambda_k}\right)^{1/k}.$$

(ii) *We have*

$$|\overline{A}| \leqslant 2kLq^{-\frac{2}{k(k-1)}}.$$

(iii) *If* $q \leqslant (a_k\delta)^{-1}$, *then the distance* d *between each point of* $S(f, N, \delta) \setminus \overline{A}$ *and* $\overline{A}$ *satisfies*

$$d > L(a_k q\delta)^{-1/k}$$

where a_k *is defined in* (5.22).

Proof

(i) Set $\overline{A} = \{n_h, \ldots, n_h + L\}$ and define $g(x) = f(x) - P(x)$. Let $\alpha_j = n_h + jL/k$ for $j \in \{0, \ldots, k\}$. Using (1.4) we get

$$\frac{f^{(k)}(t)}{k!} = \frac{g^{(k)}(t)}{k!} = \sum_{j=0}^{k} \frac{g(\alpha_j)}{\prod_{0 \leqslant i \leqslant k, i \neq j}(\alpha_j - \alpha_i)}$$

$$= \left(\frac{k}{L}\right)^k \sum_{j=0}^{k} \frac{g(\alpha_j)}{\prod_{0 \leqslant i \leqslant k, i \neq j}(j - i)}$$

for some $t \in {]\alpha_0, \alpha_k[}$ and taking account of the bound $|g(\alpha_j)| \leqslant \delta$ and (5.19), we obtain

$$\frac{\lambda_k}{k!} \leqslant \delta \left(\frac{k}{L}\right)^k \sum_{j=0}^{k} \frac{1}{(k-j)! j!} = \frac{\delta}{k!}\left(\frac{k}{L}\right)^k \sum_{j=0}^{k} \binom{k}{j} = \frac{2^k \delta}{k!}\left(\frac{k}{L}\right)^k$$

which gives the required bound.

(ii) Let $n_1 < \cdots < n_k$ be k points lying in $\overline{A}$. By the Lagrange interpolation formula of Remark 1.13, we have

$$P(x) = \sum_{j=1}^{k}\left(\prod_{\substack{i=1 \\ i \neq j}}^{k} \frac{x - n_i}{n_j - n_i}\right) P(n_j)$$

and hence q divides $\prod_{1 \leqslant i < j \leqslant k}(n_j - n_i) \leqslant (n_k - n_1)^{\frac{k(k-1)}{2}}$ so that

$$L \geqslant n_k - n_1 \geqslant q^{\frac{2}{k(k-1)}}$$

implying the asserted estimate.

(iii) Take $n \in S(f, N, \delta) \setminus \overline{A}$ and $n_0 \in \overline{A}$. Without loss of generality, one may assume that $n > n_0$ and set $d = n - n_0$ and $m = \lfloor f(n) \rfloor$. Since $m \neq P(n)$, we have

$$|P(n) - m| \geqslant \frac{1}{q} \geqslant \frac{1}{3q} + 2\delta$$

since $q \leqslant (a_k\delta)^{-1} \leqslant (3\delta)^{-1}$ and thus, taking up again the function $g(x) = f(x) - P(x)$ defined in (i), we get

$$
\begin{aligned}
\left|g(n) - g(n_0)\right| &\geqslant \left|g(n)\right| - \left|g(n_0)\right| \geqslant \left|g(n)\right| - \delta \\
&= \left|P(n) - f(n)\right| - \delta \\
&\geqslant \left|P(n) - m\right| - \left|f(n) - m\right| - \delta \\
&\geqslant \frac{1}{3q} + 2\delta - \delta - \delta = \frac{1}{3q}.
\end{aligned}
$$

On the other hand, the Taylor–Lagrange formula yields

$$
g(n) - g(n_0) = \sum_{j=1}^{k-1} g^{(j)}(n_0)\frac{d^j}{j!} + g^{(k)}(t)\frac{d^k}{k!}
$$

for some $t \in \,]n_0, n[$ and hence, using (5.19) and (5.21) applied to the function g with $M_0 = \delta$ and $M_k = c_k\lambda_k$, we get

$$
\left|g(n) - g(n_0)\right| < 4e\delta \sum_{j=1}^{k-1}\left(\frac{ke}{j}\right)^j\frac{d^j}{j!}\left\{k^{j+1}L^{-j} + e^{j-1}\left(\frac{c_k\lambda_k}{\delta}\right)^{j/k}\right\} + \frac{c_k\lambda_k d^k}{k!}
$$

and using (i) in the form

$$
\left(\frac{\lambda_k}{\delta}\right)^{1/k} \leqslant \frac{2k}{L}
$$

along with the easy inequality $j! > e(j/e)^j$ gives

$$
\left|g(n) - g(n_0)\right| < 4e\delta \sum_{j=1}^{k-1}\left(\frac{ke}{j}\right)^j\frac{1}{j!}\left\{k^{j+1} + c_k^{j/k}e^{-1}(2ek)^j\right\}(dL^{-1})^j
$$

$$
+ \frac{(2k)^k c_k \delta d^k}{k! L^k}
$$

$$
\leqslant 4\delta\left(dL^{-1} + (dL^{-1})^{k-1}\right)\sum_{j=1}^{k-1}\left(\frac{ke}{j}\right)^{2j}\left\{k + c_k^{j/k}e^{-1}(2e)^j\right\}
$$

$$
+ (2e)^k e^{-1}c_k\delta(dL^{-1})^k
$$

$$
\leqslant 4\delta\left(dL^{-1} + (dL^{-1})^{k-1}\right)\left(k + c_k 2^{k-1}e^{k-2}\right)\sum_{j=1}^{k-1}\left(\frac{ke}{j}\right)^{2j}
$$

$$
+ (2e)^k e^{-1}c_k\delta(dL^{-1})^k
$$

$$< 2^{k+2} e^{k-2} c_k \, \delta \big(dL^{-1} + (dL^{-1})^{k-1} \big) \sum_{j=1}^{k-1} \left(\frac{ke}{j} \right)^{2j}$$

$$+ (2e)^k e^{-1} c_k \, \delta \big(dL^{-1} \big)^k$$

where we used $c_k 2^{k-1} e^{k-2} \geq 2^{k-1} e^{k-2} > k$. Now Lemma 5.15 implies that

$$|g(n) - g(n_0)| < 4e^{-2} k \, (2e^3)^k c_k \, \delta \big(dL^{-1} + (dL^{-1})^{k-1} \big)$$

$$+ (2e)^k e^{-1} c_k \, \delta \big(dL^{-1} \big)^k$$

$$= 9^{-1} a_k \, \delta \big(dL^{-1} + (dL^{-1})^{k-1} \big) + (2e)^k e^{-1} c_k \, \delta \big(dL^{-1} \big)^k$$

$$\leq 9^{-1} a_k \, \delta \big(dL^{-1} + (dL^{-1})^{k-1} + (dL^{-1})^k \big).$$

Combining with the above inequality, we then get

$$q^{-1} < 3^{-1} a_k \, \delta \big(dL^{-1} + (dL^{-1})^{k-1} + (dL^{-1})^k \big)$$

so that

$$1 < a_k \, q\delta \max \big(dL^{-1}, (dL^{-1})^{k-1}, (dL^{-1})^k \big)$$

and hence

$$d > L \min \big\{ (a_k q\delta)^{-1}, (a_k q\delta)^{-\frac{1}{k-1}}, (a_k q\delta)^{-1/k} \big\}$$

and the inequality $(q\delta)^{-1} \geq a_k$ implies the statement of the lemma.

The proof is complete.   □

We are now in a position to estimate the contribution of the points coming from the major arcs.

Lemma 5.20 *Let R_0 be the contribution of the points coming from the major arcs associated to $f^{(k)}$ to the number $\mathcal{R}(f, N, \delta)$. Then*

$$R_0 \leq b_k N \delta^{\frac{2}{k(k-1)}} + 8k^3 \left(\frac{\delta}{\lambda_k} \right)^{1/k} + 10 e^3 k^2$$

where b_k is defined in (5.22).

Proof Let $\mathcal{M}_0$ be the set of major arcs and $Q_k > 0$ be the real number defined by

$$Q_k = (a_k \delta)^{-1}$$

where a_k is given in (5.22).

Write $\mathcal{M}_0 = \mathcal{M}_1 \cup \mathcal{M}_2$ where $\mathcal{M}_1$ is the set of major arcs with denominator $> Q_k$ and $\mathcal{M}_2 = \mathcal{M}_0 \setminus \mathcal{M}_1$. For $i \in \{1, 2\}$, let

$$S_i = \bigcup_{A \in \mathcal{M}_i} \overline{A} \quad \text{and} \quad R_i = |S_i|.$$

Using Lemma 5.17, we infer that $R_0 \leqslant k(R_1 + R_2)$.

▷ **Estimate of R_1.** Take $2k - 1$ consecutive points of S_1. One may take k consecutive points $n_1, \ldots, n_k$ from the same proper major arc with denominator $q > Q_k$. As in Lemma 5.19 (ii), we deduce that $n_k - n_1 \geqslant q^{\frac{2}{k(k-1)}} > Q_k^{\frac{2}{k(k-1)}}$ so that

$$R_1 \leqslant 2k\big(N Q_k^{-\frac{2}{k(k-1)}} + 1\big) < 10e^3 k\big(N(c_k\delta)^{\frac{2}{k(k-1)}} + 1\big) = b_k(2k)^{-1} N\delta^{\frac{2}{k(k-1)}} + 10e^3 k.$$

▷ **Estimate of R_2.** Without loss of generality, we may assume that $Q_k \geqslant 1$, otherwise $S_2 = \varnothing$. Let $\overline{A}_1, \ldots, \overline{A}_J$ be the ordered sequence of proper major arcs with denominator $q_j \leqslant Q_k$. For each proper major arc $\overline{A}_j$, we set n_j and L_j its first point and length, and define

$$d_j = L_j(a_k q_j \delta)^{-1/k}$$

and $\mathcal{I}_j = [n_j, n_j + d_j]$. We claim that $\mathcal{I}_j$ contains $\overline{A}_j$ and does not contain $\overline{A}_{j+1}$. Indeed, observe that:
▷ $\overline{A}_j$ is lying in an interval of length L_j and since $q_j \leqslant Q_k$, we have

$$d_j \geqslant L_j(a_k Q_k \delta)^{-1/k} = L_j$$

so that $\mathcal{I}_j$ contains $\overline{A}_j$.
▷ Assume that there exists an element of $\overline{A}_{j+1}$ belonging to $\mathcal{I}_j$. Then the distance d between this element and $\overline{A}_j$ satisfies $d \leqslant d_j$, contradicting Lemma 5.19 (iii).
Therefore the intervals $\mathcal{I}_j$ are pairwise distinct, and using Lemma 5.13 with $S = S_2$ we get

$$R_2 \leqslant N \max_j \frac{|\overline{A}_j|}{d_j} + 2 \max_j |\overline{A}_j|.$$

Now by Lemma 5.19 (ii) and the choice of d_j, we have

$$\frac{|\overline{A}_j|}{d_j} \leqslant 2k L_j q_j^{-\frac{2}{k(k-1)}} L_j^{-1} (a_k q_j \delta)^{1/k} = 2k(a_k \delta)^{1/k} q_j^{\frac{k-3}{k(k-1)}}$$

and since $q_j \leqslant Q_k = (a_k\delta)^{-1}$ and $k \geqslant 3$, we obtain

$$\frac{|\overline{A}_j|}{d_j} \leqslant 2k(a_k\delta)^{\frac{2}{k(k-1)}} < 10e^3 k (c_k\delta)^{\frac{2}{k(k-1)}} = b_k(2k)^{-1}\delta^{\frac{2}{k(k-1)}}$$

for $j \in \{1, \dots, J\}$, and therefore

$$R_2 \leqslant b_k(2k)^{-1} N \delta^{\frac{2}{k(k-1)}} + 4k \max_j L_j \leqslant b_k(2k)^{-1} N \delta^{\frac{2}{k(k-1)}} + 8k^2 \left(\frac{\delta}{\lambda_k} \right)^{1/k}$$

by Lemma 5.19 (i) and (ii). The proof is complete.   □

5.3.3 The Proof of Theorem 5.12

We first need to estimate the contribution of the points of $\mathcal{S}(f, N, \delta)$ which do not come from major arcs. The proof of the next result, which provides such an estimate, is similar to that of Theorem 5.11.

Lemma 5.21 *Let* $N \leqslant n_0 < \cdots < n_k \leqslant 2N$ *be* $k + 1$ *points of* $\mathcal{S}(f, N, \delta)$ *which do not lie on the same algebraic curve of degree* $< k$. *Then*

$$n_k - n_0 > \min\left((c_k \lambda_k)^{-\frac{2}{k(k+1)}}, \, 2^{-1} \delta^{-\frac{2}{k(k-1)}} \right).$$

Proof As in the proof of Theorem 5.11, there exist integers $m_0, \dots, m_k$ and real numbers $\delta_0, \dots, \delta_k$ such that

$$f(n_j) = m_j + \delta_j$$

with $|\delta_j| < \delta$ for all $j \in \{0, \dots, k\}$. We take up again the number

$$D_k = \prod_{0 \leqslant h < i \leqslant k} (n_i - n_h) > 0$$

and if $P = b_k X^k + \cdots + b_0$ is the Lagrange polynomial interpolating the points (n_j, m_j), then we have

$$b_k = \sum_{j=0}^{k} \frac{m_j}{\prod_{0 \leqslant i \leqslant k, i \neq j} (n_j - n_i)} = \frac{A_k}{D_k}$$

where $A_k \in \mathbb{Z}$ is analogous to the number in (5.18). Reasoning exactly in the same way as in the proof of Theorem 5.11, we get

$$k! A_k = D_k f^{(k)}(t) - k! D_k \sum_{j=0}^{k} \frac{\delta_j}{\prod_{0 \leqslant i \leqslant k, i \neq j} (n_j - n_i)}.$$

Now since the points (n_j, m_j) do not all lie on the same algebraic curve of degree $< k$, we have $b_k \neq 0$ and hence $|A_k| \geqslant 1$, and using $|\delta_j| < \delta$ and (5.19) we get

$$k! \leqslant k! |A_k| < c_k \lambda_k \, D_k + k! \, \delta \, D_k \sum_{j=0}^{k} \frac{1}{\prod_{0 \leqslant i \leqslant k, i \neq j} |n_j - n_i|}$$

$$= c_k \lambda_k \, D_k + k! \, \delta \sum_{j=0}^{k} \prod_{\substack{0 \leqslant h < i \leqslant k \\ h \neq j, i \neq j}} (n_i - n_h)$$

$$\leqslant c_k \lambda_k \, (n_k - n_0)^{\frac{k(k+1)}{2}} + (k+1)! \, \delta \, (n_k - n_0)^{\frac{k(k-1)}{2}}$$

implying that

$$n_k - n_0 > \min \left(\left(\frac{k!}{2} \right)^{\frac{2}{k(k+1)}} (c_k \lambda_k)^{-\frac{2}{k(k+1)}}, \, (2k+2)^{-\frac{2}{k(k-1)}} \delta^{-\frac{2}{k(k-1)}} \right)$$

which is slightly better than the asserted lower bound.   □

We are now in a position to prove our main result.

Proof of Theorem 5.12 Let S_0 be the set of the points of $S(f, N, \delta)$ coming from the major arcs and $T_0 = S(f, N, \delta) \setminus S_0$. By Lemma 5.20, we have

$$|S_0| = R_0 \leqslant b_k N \delta^{\frac{2}{k(k-1)}} + 8k^3 \left(\frac{\delta}{\lambda_k} \right)^{1/k} + 10e^3 k^2$$

where b_k is given in (5.22). Now let $G = \{n_0, \ldots, n_{k^2}\}$ be a set of $k^2 + 1$ consecutive ordered points of T_0. Since G is not contained in any major arc, one may find an integer $j \in \{k, \ldots, k^2\}$ such that the $j + 1$ points (n_i, m_i) do not lie on the same algebraic curve of degree $< k$. By Lemma 5.21, we have

$$n_{k^2} - n_0 \geqslant n_j - n_0 \geqslant n_k - n_0 > \min \left((c_k \lambda_k)^{-\frac{2}{k(k+1)}}, \, 2^{-1} \delta^{-\frac{2}{k(k-1)}} \right)$$

implying that

$$|T_0| \leqslant 2k^2 \left(N (c_k \lambda_k)^{\frac{2}{k(k+1)}} + 2N \delta^{\frac{2}{k(k-1)}} + 1 \right)$$

and we conclude the proof using $\mathcal{R}(f, N, \delta) \leqslant |T_0| + |S_0|$.   □

5.3.4 Application

We return to the squarefree number problem and intend to use Theorem 5.12 in order to bound the sum S_2. To this end, suppose first that $y \leqslant x^{4/9}$ and we use Lemma 5.3 with $A = 2x^{2/9}$ and $B = x^{1/3}$. We take up again the estimate of $R_2(x^{1/3})$ we obtained in Example 5.9, namely

$$R_2 \left(x^{1/3} \right) \ll x^{2/9} + y x^{-2/9}.$$

Furthermore, we use Huxley–Sargos's result with $k = 3$ for $R_1(2x^{2/9}, x^{1/3})$ which gives

$$R_1\left(2x^{2/9}, x^{1/3}\right) \ll \max_{2x^{2/9} < N \leqslant x^{1/3}} \left\{(Nx)^{1/6} + (Ny)^{1/3} + N\left(yx^{-1}\right)^{1/3}\right\}$$

$$\ll x^{2/9} + x^{1/9}y^{1/3}.$$

Thus, if $y \leqslant x^{1/3}$, we get

$$\sum_{x < n \leqslant x+y} \mu_2(n) = \frac{y}{\zeta(2)} + O\left(x^{2/9}\log x\right)$$

so that there exists a constant $c_0 > 0$ such that, if $c_0 x^{2/9}\log x \leqslant y < x^{1/2}/4$, the interval $]x, x + y]$ contains a squarefree number.

5.3.5 Refinements

The refinements of Theorem 5.12 have been made in several directions.

▶ First, it has been shown that this result also holds for $k = 2$ (see [BS94, HS95]). Furthermore, subject to an additional hypothesis which is satisfied by most of the functions arising in the usual problems, one can prove the following result [BS94].

Theorem 5.22 (Branton–Sargos) *Let $f \in C^2[N, 2N]$ such that there exist $\lambda_1 > 0$ and $\lambda_2 > 0$ such that, for all $x \in [N, 2N]$, we have*

$$\left|f'(x)\right| \asymp \lambda_1 \quad and \quad \left|f''(x)\right| \asymp \lambda_2.$$

Then

$$\mathcal{R}(f, N, \delta) \ll N\lambda_2^{1/3} + N\delta + \lambda_1\left(\frac{\delta}{\lambda_2}\right)^{1/2} + \frac{\delta}{\lambda_1} + 1.$$

▶ In Theorem 5.12, $N\lambda_k^{\frac{2}{k(k+1)}}$ is the main term, also sometimes called the *smoothness term*, and the others are the secondary terms. It is very difficult to improve on the main term, and the quantity $(\delta\lambda_k^{-1})^{1/k}$ is quasi-optimal. Thus, one may wonder whether the term $N\delta^{\frac{2}{k(k-1)}}$ may be improved, since, when δ is small, it increases rapidly as k grows. In [HS06], the authors dealt with this problem. By generalizing the method of [BS94], using a k-dimensional version of the reduction principle (Lemma 5.7) and a new divisibility relation on the divided differences discovered by Filaseta and Trifonov [FT96], they proved the following result.

Theorem 5.23 (Huxley–Sargos) *Let $k \geqslant 3$ be an integer and $f \in C^k[N, 2N]$ such that there exist $\lambda_{k-1} > 0$ and $\lambda_k > 0$ such that, for all $x \in [N, 2N]$, we have*

$$\left| f^{(k-1)}(x) \right| \asymp \lambda_{k-1}, \quad \left| f^{(k)}(x) \right| \asymp \lambda_k \quad \text{and} \quad \lambda_{k-1} = N \lambda_k. \tag{5.23}$$

Then the following upper bounds hold.

(i) *For all $k \geqslant 3$, we have*

$$\mathcal{R}(f, N, \delta) \ll N \lambda_k^{\frac{2}{k(k+1)}} + N \delta^{\frac{2}{(k-1)(k-2)}} + N(\delta \lambda_{k-1})^{\frac{2}{k^2 - k + 2}} + \left(\frac{\delta}{\lambda_{k-1}} \right)^{\frac{1}{k-1}} + 1.$$

(ii) *For $k = 3$, we have*

$$\mathcal{R}(f, N, \delta) \ll N \lambda_3^{1/6} + N \delta^{2/3} + N \left(\delta^3 \lambda_3 \right)^{1/12} + \left(\frac{\delta}{\lambda_2} \right)^{1/2} + 1.$$

(iii) *For all $k \geqslant 4$ and $\varepsilon > 0$, we have*

$$\mathcal{R}(f, N, \delta) \ll \left\{ N \lambda_k^{\frac{2}{k(k+1)}} + N(\delta \lambda_{k-1})^{\frac{2}{k^2 - k + 2}} + N \delta^{\frac{4}{k^2 - 3k + 6}} \right.$$

$$\left. + N \left(\delta^2 N^{-1} \lambda_{k-1}^{-1} \right)^{\frac{2}{k^2 - 3k + 4}} \right\} N^\varepsilon + \left(\frac{\delta}{\lambda_{k-1}} \right)^{\frac{1}{k-1}} + 1.$$

(iv) *For all $k \geqslant 5$, we have*

$$\mathcal{R}(f, N, \delta) \ll N \lambda_k^{\frac{2}{k(k+1)}} + N \delta^{\frac{2}{(k-1)(k-2)}} + \left(\frac{\delta}{\lambda_{k-1}} \right)^{\frac{1}{k-1}} + 1.$$

▶ It can easily be seen that Theorem 5.11 is a simple consequence of Theorem 5.12, since the conditions $\delta \ll \lambda_k \ll 1$ imply that the main term dominates all the others. The purpose of the next result is to provide an estimate analogous to that of Theorem 5.11 but with a hypothesis more flexible than (5.15).

Proposition 5.24 *Let $f \in C^\infty[N, 2N]$ such that there exists $T \geqslant 1$ such that, for all $x \in [N, 2N]$ and all $j \in \mathbb{Z}_{\geqslant 0}$, we have*

$$\left| f^{(j)}(x) \right| \asymp \frac{T}{N^j} \tag{5.24}$$

and

$$N \delta \leqslant T \leqslant \delta^{-1}. \tag{5.25}$$

Then, for all $k \geqslant 1$, we have

$$\mathcal{R}(f, N, \delta) \ll T^{\frac{2}{k(k+1)}} N^{\frac{k-1}{k+1}}.$$

Note that, using (5.24), Huxley–Sargos's result may be stated as

$$\mathcal{R}(f, N, \delta) \ll T^{\frac{2}{k(k+1)}} N^{\frac{k-1}{k+1}} + N\delta^{\frac{2}{k(k-1)}} + N(\delta T^{-1})^{1/k} \tag{5.26}$$

for all $k \geqslant 2$ and note that the term $N(\delta T^{-1})^{1/k}$ is dominated by the term $N\delta^{\frac{2}{k(k-1)}}$ as soon as $k \geqslant 3$. Hence Proposition 5.24 shows that the conditions (5.25) are sufficient to remove this term.

Proof We use induction, the case $k = 1$ being clearly true by Theorem 5.6 in which the conditions (5.25) enable us to eliminate the secondary terms. Now suppose that the result is true for some $k \geqslant 1$. By induction hypothesis and (5.26) used with $k + 1$ instead of k, we get

$$\mathcal{R}(f, N, \delta) \ll \min\left(E, \ T^{\frac{2}{k(k+1)}} N^{\frac{k-1}{k+1}}\right)$$

where

$$E = \max\left(T^{\frac{2}{(k+1)(k+2)}} N^{\frac{k}{k+2}}, \ N\delta^{\frac{2}{k(k+1)}}, \ N(\delta T^{-1})^{\frac{1}{k+1}}\right) = \max(e_1, e_2, e_3)$$

say. The result follows at once if $E = e_1$. The cases $E = e_2$ and $E = e_3$ are treated using the following inequality for means: *if $x, y \geqslant 0$ and $0 \leqslant a \leqslant 1$, then*

$$\min(x, y) \leqslant x^a y^{1-a}.$$

▷ *Case $E = e_2$.* We choose $a = \frac{1}{k+2}$ which gives

$$\min\left(e_2, \ T^{\frac{2}{k(k+1)}} N^{\frac{k-1}{k+1}}\right) \leqslant T^{\frac{2}{(k+1)(k+2)}} N^{\frac{k}{k+2}} (T\delta)^{\frac{2}{k(k+1)(k+2)}} \leqslant T^{\frac{2}{(k+1)(k+2)}} N^{\frac{k}{k+2}}$$

by (5.25).

▷ *Case $E = e_3$.* We choose $a = \frac{2}{k+2}$ which gives

$$\min\left(e_3, \ T^{\frac{2}{k(k+1)}} N^{\frac{k-1}{k+1}}\right) \leqslant T^{\frac{2}{(k+1)(k+2)}} N^{\frac{k}{k+2}} (N\delta T^{-1})^{\frac{2}{(k+1)(k+2)}} \leqslant T^{\frac{2}{(k+1)(k+2)}} N^{\frac{k}{k+2}}$$

by (5.25). The proof is complete. $\qquad \square$

Example 5.25 We return to the squarefree number problem. Taking account of (5.25) we get if $y \leqslant x^{1/3}$

$$\max_{(xy)^{1/4} < N \leqslant x^{1/3}} \mathcal{R}\left(\frac{x}{n^2}, N, \frac{y}{N^2}\right) + \max_{y < N \leqslant 2x^{1/3}} \mathcal{R}\left(\sqrt{\frac{x}{n}}, N, \frac{y}{\sqrt{Nx}}\right) \ll x^{\frac{k^2-k+2}{3k(k+1)}}$$

for all $k \geqslant 3$. With $k = 3$ we get the bound $\ll x^{2/9}$.

▶ The main term in the case $k = 2$ was improved by Huxley [Hux96], Huxley and Trifonov [HT96] and then Trifonov [Tri02], who extended an earlier work by

Swinnerton-Dyer that we will see in Sect. 5.4. The basic idea in this method is that the integer points close to the curve form a convex polygonal line. The Dirichlet pigeon-hole principle is then applied to the determinant formed with the coordinates of consecutive vertices. For an exhaustive exposition of Swinnerton-Dyer's method, the reader may refer to [Hux96]. We provide below one of the many versions of the theorem proved by the author (see [Hux99]).

Theorem 5.26 (Huxley) *Let $f \in C^3[N, 2N]$ such that there exist $C \geqslant 1, 0 < \lambda_2 \leqslant C^{-1}$ and $\lambda_3 > 0$ such that, for all $x \in [N, 2N]$, we have*

$$C^{-1}\lambda_2 \leqslant |f''(x)| \leqslant C\lambda_2, \quad C^{-1}\lambda_3 \leqslant |f'''(x)| \leqslant C\lambda_3 \quad and \quad \lambda_2 = N\lambda_3. \quad (5.27)$$

Then

$$\mathcal{R}(f, N, \delta) \ll \{N^{9/10}\lambda_2^{3/10} + N^{4/5}\lambda_2^{1/5} + N\lambda_2^{3/8}\delta^{1/8} + N^{7/8}\lambda_2^{1/4}\delta^{1/8}$$

$$+ N^{6/7}(\lambda_2\delta)^{1/7} + N\lambda_2^{1/5}\delta^{2/5}\}(\log N)^{2/5} + N\delta + (\delta\lambda_2^{-1})^{1/2} + 1.$$

The implied constant depends only on C.

Example 5.27 Assume $y \leqslant x^{2/5}$.

▷ Theorem 5.26 implies for all $N \geqslant 2x^{1/4}$ that

$$(\log N)^{-2/5}\mathcal{R}\left(\frac{x}{n^2}, N, \frac{y}{N^2}\right) \ll (xN^{-1})^{3/10} + x^{1/5} + (x^3yN^{-6})^{1/8} + (xy)^{1/7}$$

$$+ (xy^2N^{-3})^{1/5} + yN^{-1} + N(yx^{-1})^{1/2}$$

so that

$$\max_{x^{2/7} < N \leqslant x^{1/3}} \mathcal{R}\left(\frac{x}{n^2}, N, \frac{y}{N^2}\right) \ll x^{3/14}(\log x)^{2/5}.$$

Furthermore, using Theorem 5.23 (ii) we get

$$\mathcal{R}\left(\frac{x}{n^2}, N, \frac{y}{N^2}\right) \ll (Nx)^{1/6} + (y^2N^{-1})^{1/3} + (xy^3N)^{1/12} + N(yx^{-1})^{1/2}$$

so that

$$\max_{4\sqrt{y} < N \leqslant x^{2/7}} \mathcal{R}\left(\frac{x}{n^2}, N, \frac{y}{N^2}\right) \ll x^{3/14}.$$

Therefore

$$\max_{4\sqrt{y} < N \leqslant x^{1/3}} \mathcal{R}\left(\frac{x}{n^2}, N, \frac{y}{N^2}\right) \ll x^{3/14}(\log x)^{2/5}. \quad (5.28)$$

▷ Theorem 5.26 implies for all $N \geqslant 3x^{1/5}$ that

$$(\log N)^{-2/5}\mathcal{R}\left(\sqrt{\frac{x}{n}}, N, \frac{y}{\sqrt{Nx}}\right) \ll (Nx)^{3/20} + (N^3x)^{1/10} + y(Nx^{-1})^{1/2}$$
$$+ (xy)^{1/8} + (xy^2N^3)^{1/16} + (N^3y)^{1/7}$$
$$+ (x^{-1}y^4N^3)^{1/10} + N(yx^{-1})^{1/2}$$

so that

$$\max_{3x^{1/5}<N\leqslant 8x^{1/3}} \mathcal{R}\left(\sqrt{\frac{x}{n}}, N, \frac{y}{\sqrt{Nx}}\right) \ll x^{1/5}(\log x)^{2/5}$$

since $y \leqslant x^{2/5}$. Furthermore, using the trivial estimate (5.6) we have

$$\max_{N\leqslant 3x^{1/5}} \mathcal{R}\left(\sqrt{\frac{x}{n}}, N, \frac{y}{\sqrt{Nx}}\right) \ll x^{1/5}.$$

Therefore

$$\max_{N\leqslant 8x^{1/3}} \mathcal{R}\left(\sqrt{\frac{x}{n}}, N, \frac{y}{\sqrt{Nx}}\right) \ll x^{1/5}(\log x)^{2/5}. \tag{5.29}$$

Using Lemma 5.3 with $A = 4\sqrt{y}$ and $B = x^{1/3}$, we then get assuming $y \leqslant x^{2/5}$

$$\sum_{x<n\leqslant x+y} \mu_2(n) = \frac{y}{\zeta(2)} + O\left(x^{3/14}(\log x)^{7/5}\right).$$

We infer that there exists a constant $c_0 > 0$ such that, if $c_0x^{3/14}(\log x)^{7/5} \leqslant y < x^{1/2}/4$, the interval $]x, x+y]$ contains a squarefree number.

A slight improvement of this estimate can be obtained via the following result due to Trifonov [Tri02].

Theorem 5.28 (Trifonov) *Let $f \in C^3[N, 2N]$ such that there exist $C \geqslant 1, 0 < \lambda_2 \leqslant 1$ and $\lambda_3 > 0$ such that, for all $x \in [N, 2N]$, we have*

$$C^{-1}\lambda_2 \leqslant |f''(x)| \leqslant C\lambda_2, \quad C^{-1}\lambda_3 \leqslant |f'''(x)| \leqslant C\lambda_3 \quad \text{and} \quad \lambda_2 = N\lambda_3 \tag{5.30}$$

and

$$N\lambda_2 \geqslant 1 \quad \text{and} \quad N\delta^2 \leqslant C^{-1}. \tag{5.31}$$

Then, for all $\varepsilon > 0$, we have

$$\mathcal{R}(f, N, \delta) \ll \left\{N^{43/54}\lambda_2^{4/27} + N^{4/5}\lambda_2^{4/25} + N^{9/10}\delta^{4/15} + N^{12/13}\delta^{4/13}\right.$$
$$\left. + N^{6/7}\lambda_2^{2/7} + N\lambda_2 + N(\lambda_2\delta)^{1/4}\right\}N^{\varepsilon} + \lambda_2(N\delta)^{5/2}.$$

The implied constant depends only on C and ε.

Example 5.29 We apply this result to the function $f : u \longmapsto x/u^2$ of the squarefree number problem. First, the conditions (5.31) are fulfilled as soon as

$$\max(x^{1/4}, 4y^{2/3}) \leqslant N \leqslant 2^{-1}x^{1/3}.$$

We assume then that $y \leqslant 8^{-1}x^{3/8}$ and use Lemma 5.3 with $A = 4\sqrt{y}$ and $B = 2^{-1}x^{1/3}$. We split the range of R_1 into three parts.

▷ For $4\sqrt{y} < N \leqslant x^{13/48}$, we use Theorem 5.23 (iii) with $k = 4$ giving

$$\mathcal{R}\left(\frac{x}{n^2}, N, \frac{y}{N^2}\right) \ll x^{1/10}N^{2/5} + (xy)^{1/7} + (Ny^2)^{1/5} + N(y^2x^{-1})^{1/4}$$
$$+ N(yx^{-1})^{1/3}$$

and hence

$$\max_{4\sqrt{y}<N\leqslant x^{13/48}} \mathcal{R}\left(\frac{x}{n^2}, N, \frac{y}{N^2}\right) \ll x^{5/24}$$

as long as $y \leqslant 8^{-1}x^{3/8}$.

▷ For $x^{13/48} < N \leqslant x^{41/136}$, we use Theorem 5.28 giving for all $\varepsilon > 0$

$$\mathcal{R}\left(\frac{x}{n^2}, N, \frac{y}{N^2}\right) \ll \{N^{11/54}x^{4/27} + (Nx)^{4/25} + N^{11/30}y^{4/15} + (Ny)^{4/13}$$
$$+ (xN^{-1})^{2/7} + xN^{-3} + (xyN^{-2})^{1/4}\}N^\varepsilon + x(y^5N^{-13})^{1/2}$$

and hence

$$\max_{x^{13/48}<N\leqslant x^{41/136}} \mathcal{R}\left(\frac{x}{n^2}, N, \frac{y}{N^2}\right) \ll x^{57/272+\varepsilon}$$

as long as $y \leqslant x^{101/272}$.

▷ For $x^{41/136} < N \leqslant 2^{-1}x^{1/3}$, we use Theorem 5.26 giving as in Example 5.27

$$\max_{x^{41/136}<N\leqslant x^{1/3}} \mathcal{R}\left(\frac{x}{n^2}, N, \frac{y}{N^2}\right) \ll x^{57/272}(\log x)^{2/5}$$

as long as $y \leqslant x^{127/272}$.

Hence we get for all $\varepsilon > 0$

$$\max_{4\sqrt{y}\leqslant N\leqslant 2^{-1}x^{1/3}} \mathcal{R}\left(\frac{x}{n^2}, N, \frac{y}{N^2}\right) \ll x^{57/272+\varepsilon} \tag{5.32}$$

and taking account of (5.29) and Lemma 5.3, we infer that

$$\sum_{x<n\leqslant x+y} \mu_2(n) = \frac{y}{\zeta(2)} + O(x^{57/272+\varepsilon})$$

if $y \leqslant x^{101/272}$. Hence there exists a constant $c_0 > 0$ such that, if $c_0 x^{57/272+\varepsilon} \leqslant y < x^{1/2}/4$, the interval $]x, x + y]$ contains a squarefree number.

5.4   Further Developments

5.4.1   The Method of Filaseta and Trifonov—Introduction

The exponent $\frac{57}{272} \approx 0.209\,558\ldots$ obtained in Example 5.29 is a very good result which is not attainable by current exponential sum results (see Chap. 6). The bound (5.29) shows that any improvement must come from estimates of integer points close to the curve of the function $f : u \longmapsto x/u^2$. Using divided differences and the polynomial identity (5.34) which takes the particular structure of f into account, we will prove the following theorem, due to Filaseta and Trifonov [FT92, FT96], which supersedes each previous result.

Theorem 5.30 (Filaseta–Trifonov) *Let $x \geqslant 1$ and δ satisfying (5.4) be real numbers and N be an integer such that $4 \leqslant N \leqslant x^{1/2}$. Assume that there exists a small real number $c_0 > 0$ such that*

$$N\delta \leqslant c_0. \tag{5.33}$$

Then for x sufficiently large

$$\mathcal{R}\left(\frac{x}{n^2}, N, \delta\right) \ll x^{1/5} + x^{1/15}\delta N^{5/3}.$$

The authors combined the divided difference techniques with the following identity

$$(X + Y)^2 P(X, Y) - X^2 Q(X, Y) = Y^3 \tag{5.34}$$

where $P, Q \in \mathbb{Z}[X, Y]$ are the homogeneous polynomials in two variables defined by

$$P(X, Y) = -2X + Y \quad \text{and} \quad Q(X, Y) = -2X - 3Y. \tag{5.35}$$

Note that the identity (5.34) has to be used with Y being small compared to X.

In what follows, x, δ and N are as stated in Theorem 5.30 and f is the function supported on $[N, 2N]$ and defined by $f(u) = x/u^2$. The constant c_0 appearing in (5.33) is sufficiently small, say $c_0 < 600^{-1}$. When $\mathcal{I}$ is an interval of $\mathbb{R}$, the notation $|\mathcal{I}|$ always means $|\mathcal{I} \cap \mathbb{Z}|$. Finally, one may assume that (5.33) is always satisfied.

The next tool is similar to Lemma 5.10 and the proof, which is left to the reader, is analogous.

Lemma 5.31 *Let* $n = x + y \in \mathbb{Z}$ *with* $x, y \in \mathbb{R}$. *Then*

(i) *If* $|x| \geqslant \frac{1}{2}$ *and* $|y| < \frac{1}{2}$, *then* $n \neq 0$.
(ii) *If* $|x| \leqslant \frac{1}{2}$ *and* $|y| < \frac{1}{2}$, *then* $n = 0$.

5.4.2 The Method of Filaseta and Trifonov—The Basic Result

The first step in Filaseta–Trifonov's method is the following construction of a subset of $\mathcal{S}(f, N, \delta)$ in which the elements are not too close to each other.

Lemma 5.32 *There exists a subset* T *of* $\mathcal{S}(f, N, \delta)$ *satisfying the following two properties.*

(i) $\mathcal{R}(f, N, \delta) \leqslant 4(|T| + 1)$.
(ii) *Any two consecutive elements of* T *differ by* $> (2x)^{-1/3} N^{4/3}$.

Proof If $N < x^{1/4}$, then one may take $T = \mathcal{S}(f, N, \delta)$ because we have in this case $x^{-1/3} N^{4/3} < 1$. Now suppose that $N \geqslant x^{1/4}$ and let $a, b \in \mathbb{N}$ and n, $n + a$ and $n + a + b$ be three consecutive elements of $\mathcal{S}(f, N, \delta)$ such that

$$1 \leqslant a, b \leqslant (2x)^{-1/3} N^{4/3}.$$

We will show that there are only *two* possibilities for the choice of b. The result will then follow by taking each 4th element of $\mathcal{S}(f, N, \delta)$.

By definition, there exist non-zero integers m_i and real numbers δ_i such that

$$f(n) = m_1 + \delta_1,$$

$$f(n + a) = m_2 + \delta_2,$$

$$f(n + a + b) = m_3 + \delta_3$$

with $|\delta_i| < \delta$ for $i \in \{1, 2, 3\}$. In fact, each integer m_i is positive since, for all $u \in [N, 2N]$, we have $f(u) \geqslant x/(4N^2) \geqslant 1/4$ and $\delta \leqslant c_0 N^{-1}$. Also note that, by (5.34), we get

$$f(n)P(n, a) - f(n + a)Q(n, a) = \frac{xa^3}{n^2(n + a)^2} \leqslant \frac{xa^3}{N^4} \leqslant \frac{1}{2}.$$

On the other hand, we have

$$f(n)P(n, a) - f(n + a)Q(n, a) = m_1 P(n, a) - m_2 Q(n, a) + \varepsilon$$

with

$$|\varepsilon| < \delta \big(|P(n, a)| + |Q(n, a)|\big) \leqslant 4\delta(n + a) \leqslant 8N\delta \leqslant 8c_0 < \frac{1}{2}.$$

Hence by Lemma 5.31, we obtain

$$m_1 P(n,a) - m_2 Q(n,a) = 0. \tag{5.36}$$

Similarly we also have

$$m_2 P(n+a,b) - m_3 Q(n+a,b) = 0,$$
$$m_1 P(n,a+b) - m_3 Q(n,a+b) = 0$$

and eliminating m_3 we get

$$m_2 P(n+a,b)Q(n,a+b) - m_1 P(n,a+b)Q(n+a,b) = 0$$

which gives

$$3b^2(m_1 - m_2) + \kappa_1 b + 2\kappa_2 = 0 \tag{5.37}$$

where

$$\kappa_1 = a(5m_1 + 3m_2) - 4n(m_1 - m_2),$$
$$\kappa_2 = a^2(m_1 + 3m_2) - an(m_1 - 5m_2) - 2n^2(m_1 - m_2).$$

If $m_1 = m_2$, then by (5.36) we have $P(n,a) = Q(n,a)$ and then $-2n + a = -2n - 3a$, so that $a = 0$ which is impossible since $a \geqslant 1$. Therefore $m_1 \neq m_2$ and (5.37) is a quadratic equation in b, concluding the proof. $\qquad\square$

We deduce at once the following result.

Corollary 5.33 *If $N\delta \leqslant c_0$ and $4 \leqslant N \leqslant x^{1/2}$, then*

$$\mathcal{R}\left(\frac{x}{n^2}, N, \delta\right) \leqslant 10\left(\frac{x}{N}\right)^{1/3}.$$

Furthermore, if $x^{2/5} \leqslant N \leqslant x^{1/2}$, then

$$\mathcal{R}\left(\frac{x}{n^2}, N, \delta\right) \leqslant 10x^{1/5}.$$

Note that this result is equivalent to Proposition 5.24 used with $k = 2$ except with the condition $N\delta \ll 1$ instead of (5.25).

5.4.3 The Method of Filaseta and Trifonov—Higher Divided Differences

By Corollary 5.33, it is now sufficient to assume

$$2^{2/3}x^{1/5} \leqslant N < x^{2/5} \tag{5.38}$$

and we set

$$R = \left(2^{-1}N^4x^{-1}\right)^{1/3} \quad \text{and} \quad A = 2^{-1/3}Nx^{-1/5}.$$

By (5.38), we have $R < A \leqslant N^{1/2}$. We take up again the subset T of Lemma 5.32 in which any two consecutive elements differ by $> R$. As in Lemma 5.7, we will use the following subsets of T. For $a \in \,]R, A] \cap \mathbb{Z}$, we define

$$T(a) = \{n \in \mathbb{N} : n \text{ and } n + a \text{ are consecutive in } T\}.$$

By (5.10) we have

$$|T| \ll \frac{N}{A} + \sum_{R < a \leqslant A} |T(a)|. \tag{5.39}$$

The next tool will enable us to get an upper bound for $|T(a)|$.

Lemma 5.34 *Let $a \in \,]R, A] \cap \mathbb{Z}$ and $\mathcal{I}$ be a subinterval of $[N, 2N]$ satisfying*

$$|\mathcal{I}| \leqslant 16^{-1}N^5x^{-1}a^{-3}.$$

Then of each three consecutive elements in $T(a) \cap \mathcal{I}$, there are two consecutive elements that differ by

$$> 6^{-1}(ax)^{-1/3}N^{5/3}.$$

Proof Let n, $n + b$ and $n + b + d$ be three consecutive elements of $T(a) \cap \mathcal{I}$ with $b, d \in \mathbb{N}$. Hence the six integers $n, n + a, n + b, n + a + b, n + b + d, n + a + b + d$ are all lying in $[N, 2N]$. Since $a > R$, observe also that

$$|\mathcal{I}| \leqslant 16^{-1}N^5x^{-1}a^{-3} < 16^{-1}N^5x^{-1}R^{-3} = \frac{N}{8}.$$

Consider the six integers $m_1, \ldots, m_6$ and real numbers $\delta_1, \ldots, \delta_6$ satisfying $|\delta_i| < \delta$ and defined by

$$f(n) = m_1 + \delta_1,$$
$$f(n + a) = m_2 + \delta_2,$$
$$f(n + a + b) = m_3 + \delta_3,$$
$$f(n + b) = m_4 + \delta_4,$$
$$f(n + b + d) = m_5 + \delta_5,$$
$$f(n + a + b + d) = m_6 + \delta_6$$

and we set $F_a(n) = -\Delta_a f(n) = f(n) - f(n + a)$ and

$$D_1 = dF_a(n) - (b + d)F_a(n + b) + bF_a(n + b + d).$$

Observe that, with the notation of Remark 1.13, we have

$$\frac{D_1}{bd(b+d)} = F_a[n, n+b, n+b+d]$$

so that by (1.4) there exists a real number $t \in]n, n+b+d[\subseteq [N, 2N]$ such that

$$D_1 = \frac{bd(b+d)}{2!} F_a''(t) = bd(b+d) \frac{3ax(2t+a)(2t^2+2at+a^2)}{t^4(t+a)^4}.$$

Hence $D_1 > 0$ and, since $t, t+a \in [N, 2N]$, we get

$$D_1 \geqslant bd(b+d) \frac{3ax(2N+a)(2N^2+2aN+a^2)}{(2N)^8} \geqslant bd(b+d) \frac{ax}{25N^5}.$$

Recall that $n+b$ and $n+b+d$ are consecutive in $T(a) \cap \mathcal{I}$ and hence $\min(b, d) \geqslant a > R$ and then

$$D_1 \geqslant (b+d) \frac{R^3 x}{25N^5} = \frac{b+d}{50N}$$

so that

$$b+d \leqslant 50 N D_1. \qquad (5.40)$$

Now set

$$E_1 = d(m_1 - m_2) - (b+d)(m_4 - m_3) + b(m_5 - m_6) \in \mathbb{Z}.$$

Using the definition of the integers $m_1, \ldots, m_6$ and the function F_a, we get

$$\begin{aligned}
E_1 &= d\big(F_a(n) + \delta_1 - \delta_2\big) - (b+d)\big(F_a(n+b) + \delta_4 - \delta_3\big) \\
&\quad + b\big(F_a(n+b+d) + \delta_5 - \delta_6\big) \\
&= d F_a(n) - (b+d) F_a(n+b) + b F_a(n+b+d) + d(\delta_1 - \delta_2) \\
&\quad - (b+d)(\delta_4 - \delta_3) + b(\delta_5 - \delta_6) \\
&= D_1 + R_1
\end{aligned}$$

with $|R_1| < 4\delta(b+d)$. By (5.40) and using (5.33) with $c_0 < 600^{-1}$, we obtain

$$4\delta(b+d) \leqslant 200N\delta D_1 < \frac{D_1}{3}$$

and therefore

$$|E_1 - D_1| < \frac{D_1}{3}$$

which implies that $E_1 \neq 0$. Since $E_1 \in \mathbb{Z}$, we infer that $|E_1| \geqslant 1$ and since

$$|E_1| \leqslant D_1 + |R_1| < \frac{4D_1}{3}$$

we get

$$D_1 > \frac{3}{4}.$$

One may obtain an upper bound for D_1 in a similar way, since $a \leqslant A \leqslant N^{1/2} \leqslant N/2$ and $t \in [N, 2N]$ imply

$$D_1 \leqslant bd(b+d) \frac{3ax(4N + N/2)(8N^2 + 2 \times 2N^2)}{N^8} = 162\, bd(b+d) \frac{ax}{N^5}.$$

Combining both these estimates we get

$$bd(b+d) > \frac{N^5}{216\,ax}$$

so that either b or d is $> 6^{-1}(ax)^{-1/3} N^{5/3}$ as asserted.   □

5.4.4 The Method of Filaseta and Trifonov—Epilog

We are now in a position to prove Theorem 5.30. To this end, we pick up from Lemma 5.34 the interval $\mathcal{I}$ and the integers $m_1, \ldots, m_4$ and define

$$G_a(n) = P(n, a) f(n) - Q(n, a) f(n + a),$$
$$D_2 = -\Delta_b G_a(n) = G_a(n) - G_a(n + b),$$
$$E_2 = P(n, a) m_1 - Q(n, a) m_2 - P(n + b, a) m_4 + Q(n + b, a) m_3 \in \mathbb{Z}.$$

As in the proof of Lemma 5.34, we have $D_2 = E_2 + R_2$ and using (5.33) we have again $|R_2| \leqslant 20N\delta < \frac{1}{2}$. By (1.4), there exists a real number $t \in \,]n, n+b[$ such that

$$D_2 = -b\, G_a'(t) = 2a^3 x b \big(t^{-2}(t + a)^{-3} + t^{-3}(t + a)^{-2} \big).$$

Since $t, t + a \in [N, 2N]$ we obtain as above

$$16^{-1} ba^3 x N^{-5} \leqslant |D_2| \leqslant 4ba^3 x N^{-5} \tag{5.41}$$

and since $b \leqslant |\mathcal{I}| \leqslant 16^{-1} N^5 x^{-1} a^{-3}$, we get

$$|D_2| \leqslant \frac{1}{4}.$$

By Lemma 5.31, we infer that $E_2 = 0$ which implies that $|D_2| = |R_2| \leqslant 20N\delta$ and using (5.41) we get

$$b \ll a^{-3} x^{-1} N^5 |D_2| \ll a^{-3} x^{-1} N^6 \delta.$$

Thus, any two elements of $T(a) \cap \mathcal{I}$ are lying in a sub-interval $\mathcal{J}$ of $\mathcal{I}$ satisfying

$$|\mathcal{J}| \ll a^{-3} x^{-1} N^6 \delta.$$

Now by Lemma 5.34, of each three consecutive elements in $T(a) \cap \mathcal{I}$, there are two consecutive elements that differ by $\gg (ax)^{-1/3} N^{5/3}$ so that

$$|T(a) \cap \mathcal{I}| \ll \frac{|\mathcal{J}|}{(ax)^{-1/3} N^{5/3}} + 1 \ll x^{-2/3} \delta a^{-8/3} N^{13/3} + 1.$$

We saw that $|\mathcal{I}| \leqslant N/8$. Subdividing $[N, 2N]$ into $s = [N/(16^{-1} N^5 x^{-1} a^{-3})] + 1$ distinct sub-intervals $\mathcal{I}_1, \ldots, \mathcal{I}_s$ with lengths $\leqslant 16^{-1} N^5 x^{-1} a^{-3}$, we get

$$|T(a)| \ll \left(\frac{N}{N^5 x^{-1} a^{-3}} + 1 \right) \left(x^{-2/3} \delta a^{-8/3} N^{13/3} + 1 \right)$$

$$\ll \left(x a^3 N^{-4} + 1 \right) \left(x^{-2/3} \delta a^{-8/3} N^{13/3} + 1 \right)$$

$$\ll x a^3 N^{-4} + (Nax)^{1/3} \delta$$

where we used the fact that $a > R = (2x)^{-1/3} N^{4/3}$, implying that $1 < 2x a^3 N^{-4}$, in the last inequality. Using (5.39) we obtain

$$|T| \ll N A^{-1} + \sum_{R < a \leqslant A} \left(x a^3 N^{-4} + (Nax)^{1/3} \delta \right)$$

$$\ll N A^{-1} + x \left(A N^{-1} \right)^4 + \left(N A^4 x \right)^{1/3} \delta$$

and the choice of $A = 2^{-1/3} N x^{-1/5}$ gives

$$|T| \ll x^{1/5} + x^{1/15} \delta N^{5/3}. \tag{5.42}$$

Let us summarize all the results obtained above.

▷ If $4 \leqslant N \leqslant 2^{2/3} x^{1/5}$, we use the trivial bound (5.6) giving

$$\mathcal{R}\left(\frac{x}{n^2}, N, \delta \right) \leqslant N + 1 \ll x^{1/5}.$$

▷ If $2^{2/3} x^{1/5} < N < x^{2/5}$, Lemma 5.32 (i) and (5.42) give

$$\mathcal{R}\left(\frac{x}{n^2}, N, \delta \right) \ll |T| + 1 \ll x^{1/5} + x^{1/15} \delta N^{5/3}.$$

▷ If $x^{2/5} \leqslant N \leqslant x^{1/2}$, Corollary 5.33 provides

$$\mathcal{R}\left(\frac{x}{n^2}, N, \delta \right) \ll x^{1/5}.$$

The proof of Theorem 5.30 is complete. $\square$

5.4.5 The Method of Filaseta and Trifonov—Application

Applying Theorem 5.30 to the gaps between squarefree integers gives the following consequence.

Corollary 5.35 *Let x, y be real numbers satisfying (5.5). Then we have*

$$\sum_{x < n \leqslant x+y} \mu_2(n) = \frac{y}{\zeta(2)} + O\{(x^{1/5} + yx^{-1/60}) \log x + y^{1/2}\}.$$

Proof Let c_0 be the constant in (5.33) and set $c_1 = c_0^{-1}$. We consider two cases.

▷ *Case* $16 \leqslant y \leqslant x^{1/4}$. We may write

$$\sum_{2\sqrt{y} < n \leqslant \sqrt{x}} \left(\left[\frac{x+y}{n^2} \right] - \left[\frac{x}{n^2} \right] \right)$$

$$= \left(\sum_{2\sqrt{y} < n \leqslant c_1 x^{1/4}} + \sum_{c_1 x^{1/4} < n \leqslant \sqrt{x}} \right) \left(\left[\frac{x+y}{n^2} \right] - \left[\frac{x}{n^2} \right] \right)$$

$$= \Sigma_1 + \Sigma_2.$$

We use Theorem 5.23 (i) with $k = 4$ for Σ_1 and Theorem 5.30 for Σ_2 which gives

$$\Sigma_1 \ll \max_{2\sqrt{y} < N \leqslant c_1 x^{1/4}} \left(x^{1/10} N^{2/5} + (Ny)^{1/3} + (xy)^{1/7} + N(yx^{-1})^{1/3} \right) \log x$$

$$\ll \left(x^{1/5} + x^{1/12} y^{1/3} + (xy)^{1/7} \right) \log x \ll x^{1/5} \log x$$

since $y \leqslant x^{1/4}$, and

$$\Sigma_2 \ll \max_{c_1 x^{1/4} < n \leqslant \sqrt{x}} \left(x^{1/5} + x^{1/15} y N^{-1/3} \right) \log x \ll \left(x^{1/5} + yx^{-1/60} \right) \log x.$$

▷ *Case* $x^{1/4} < y < x^{1/2}/4$. We now have

$$\sum_{2\sqrt{y} < n \leqslant \sqrt{x}} \left(\left[\frac{x+y}{n^2} \right] - \left[\frac{x}{n^2} \right] \right) = \left(\sum_{2\sqrt{y} < n \leqslant c_1 y} + \sum_{c_1 y < n \leqslant \sqrt{x}} \right) \left(\left[\frac{x+y}{n^2} \right] - \left[\frac{x}{n^2} \right] \right)$$

and Theorem 5.23 (i) with $k = 4$ applied to the first sum implies that it contributes

$$\ll \left(x^{1/10} y^{2/5} + (xy)^{1/7} + y^{2/3} + y^{4/3} x^{-1/3} \right) \log x \ll yx^{-1/60} \log x$$

where we used the fact that $x^{1/4} < y < x^{1/2}/4$, and the second sum contributes

$$\ll \left(x^{1/5} + x^{1/15} y^{2/3} \right) \log x \ll \left(x^{1/5} + yx^{-1/60} \right) \log x$$

since $y > x^{1/4}$. Applying Lemma 5.3 with $A = 2\sqrt{y}$ and $B = \sqrt{x}$ concludes the proof.   □

5.4.6 The Method of Filaseta and Trifonov—Generalization

The computations made above may be generalized to the case of the function $u \longmapsto x/u^r$ where $r \geqslant 2$ is a fixed integer, which enables us to treat the r-free number problem. To this end, we need to have at our disposal two homogeneous integer polynomials $P_{r-1}(X, Y)$ and $Q_{r-1}(X, Y)$ with total degree $r - 1$ analogous to (5.35) in order to generalize (5.34). It may be proved [FT96] that the polynomials

$$P_{r-1}(X, Y) = r \binom{2r-1}{r} \sum_{j=0}^{r-1} (-1)^{r-1+j} \binom{r-1}{j} \frac{(X+Y)^{r-1-j} Y^j}{2r-j-1},$$

$$Q_{r-1}(X, Y) = r \binom{2r-1}{r} \sum_{j=0}^{r-1} (-1)^{r-1} \binom{r-1}{j} \frac{X^{r-1-j} Y^j}{2r-j-1}$$

with $r \geqslant 1$, have the desired properties and satisfy the identity

$$(X+Y)^r P_{r-1}(X, Y) - X^r Q_{r-1}(X, Y) = Y^{2r-1}. \tag{5.43}$$

Similarly, in order to prove the analogue of Lemma 5.34 using divided differences of the second order, the following polynomials

$$P_{r-2}(X, Y) = (r-1) \binom{2r-2}{r-1} \sum_{j=0}^{r-2} (-1)^{r+j} \binom{r-2}{j} \frac{(X+Y)^{r-2-j} Y^j}{(2r-j-3)(2r-j-2)},$$

$$Q_{r-2}(X, Y) = (r-1) \binom{2r-2}{r-1} \sum_{j=0}^{r-2} (-1)^r \binom{r-2}{j} \frac{X^{r-2-j} Y^j}{(2r-j-3)(2r-j-2)}$$

with $r \geqslant 2$, are homogeneous integer polynomials of total degree $r - 2$ and satisfy the identity

$$(X+Y)^r P_{r-2}(X, Y) - X^r Q_{r-2}(X, Y) = Y^{2r-3}(2X + Y). \tag{5.44}$$

The polynomial identities (5.43) and (5.44) are very difficult to show, and may be replaced by the theory of *Padé approximants* (see [FT96]) which allows us to define r as a rational number. Adapting the proof above to the general case, Filaseta and Trifonov proved the following result.

Theorem 5.36 (Filaseta–Trifonov) *Let $r \geqslant 2$ be an integer, $x \geqslant 1$ and δ satisfying (5.4) be real numbers and N be an integer such that $4 \leqslant N \leqslant x^{1/r}$. Assume that there exists a small real number $c_r > 0$, depending only on r, such that*

$$N^{r-1}\delta \leqslant c_r.$$

Then for x sufficiently large

$$\mathcal{R}\left(\frac{x}{n^r}, N, \delta\right) \ll x^{\frac{1}{2r+1}} + x^{\frac{1}{6r+3}} \delta N^{r-1/3}.$$

We shall see in Exercise 3 that the method may be adapted to the function $u \longmapsto \sqrt{x/u}$ and, more generally, to the function $u \longmapsto (x/u)^{1/r}$ where $r \geqslant 2$. The reader can check that the following dual result holds.

Theorem 5.37 *Let* $r \geqslant 2$ *be an integer,* $x \geqslant 1$ *and* δ *satisfying* (5.4) *be real numbers and* N *be an integer such that* $4 \leqslant N \leqslant x$. *Assume that there exists a small real number* $c_r > 0$, *depending only on* r, *such that*

$$N^{r-1}\delta \leqslant c_r.$$

Then for x sufficiently large

$$\mathcal{R}\left(\left(\frac{x}{n}\right)^{1/r}, N, \delta\right) \ll \left(N^{r^2-1}x\right)^{\frac{1}{r(2r+1)}} + x^{\frac{1}{3r(2r+1)}} \delta N^{r-1/6-\frac{r+2}{6r(2r+1)}}.$$

5.4.7 Counting Integer Points on Smooth Curves

Using Theorem 5.11 and letting $\delta \longrightarrow 0$, we infer that the number of integer points *lying on* the arc of the curve $y = f(x)$ with $N < x \leqslant 2N$ is

$$\ll N\lambda_k^{\frac{2}{k(k+1)}} + 1.$$

Historically, this number was first investigated by Jarnik who proved that a strictly convex arc $y = f(x)$ with length L has at most

$$\leqslant \frac{3}{(2\pi)^{1/3}} L^{2/3} + O\left(L^{1/3}\right)$$

integer points and this is a nearly best possible result under the sole hypothesis of convexity. However, Swinnerton-Dyer and Schmidt proved independently that if $f \in C^3[0, N]$ is such that $|f(x)| \leqslant N$ and $f'''(x) \neq 0$ for all $x \in [0, N]$, then the number of integer points on the arc $y = f(x)$ with $0 \leqslant x \leqslant N$ is $\ll N^{3/5+\varepsilon}$. This result was generalized by Bombieri and Pila who showed the following result.

Proposition 5.38 (Bombieri–Pila) *Let* $N \geqslant 1$, $k \geqslant 4$ *be integers and define* $K = \binom{k+2}{2}$. *Let* $\mathcal{I}$ *be an interval with length* N *and* $f \in C^K(\mathcal{I})$ *satisfying* $|f'(x)| \leqslant 1$, $f''(x) > 0$ *and such that the number of solutions of the equation* $f^{(K)}(x) = 0$ *is* $\leqslant m$. *Then there exists a constant* $c_0 = c_0(k) > 0$ *such that the number of integer points on the arc* $y = f(x)$ *with* $x \in \mathcal{I}$ *is*

$$\leqslant c_0(m + 1)N^{1/2+3/(k+3)}.$$

The ideas of Bombieri and Pila have recently been extended by Huxley [Hux07] to counting the number of integer points which are very close to regular curves. The function is supposed to be C^5 and, along with the usual non-vanishing conditions of the derivatives on $[N, 2N]$, the proof also requires lower bounds of the following determinants

$$D_1(f;x) = \begin{vmatrix} f'''(x) & 3f''(x) \\ f^{(4)}(x) & 4f'''(x) \end{vmatrix},$$

$$D_2(f;x) = \frac{1}{2f''(x)} \begin{vmatrix} f'''(x) & 3f''(x) & 0 \\ f^{(4)}(x) & 4f^{(3)}(x) & 6f''(x)^2 \\ f^{(5)}(x) & 5f^{(4)}(x) & 20f''(x)f'''(x) \end{vmatrix}.$$

Proposition 5.39 (Huxley) *Assume that $f \in C^5[N, 2N]$ such that there exist real numbers $C, T \geqslant 1$ such that*

$$|f^{(j)}(x)| \leqslant C^{j+1} j! \times \frac{T}{N^j} \quad (j = 1, \ldots, 5),$$

$$|f^{(j)}(x)| \geqslant \frac{j!}{C^{j+1}} \times \frac{T}{N^j} \quad (j = 2, 3),$$

$$|D_1(f;x)| \geqslant 144\,C^{-8} \times \frac{T^2}{N^6},$$

$$|D_2(f;x)| \geqslant 4320\,C^{-12} \times \frac{T^3}{N^9}.$$

Let δ be a real number satisfying (5.4). Then we have

$$\mathcal{R}(f, N, \delta) \ll (NT)^{4/15} + N(\delta^{11} T^9)^{1/75}.$$

The implied constant depends only on C.

The main term is a very good result, since Theorem 5.8 only gives $(NT)^{1/3}$ and Theorem 5.26 provides the bound $(NT)^{3/10}$. On the other hand, the secondary term is too large, and thus useless in many applications. In this direction, the author proved that this term may be improved subject to some additional non-vanishing conditions of certain quite complicated determinants.

5.5 Exercises

1 Show that the function $x \longmapsto \|x\|$ is bounded, even, periodic of period 1 and that we have

$$\big|\|x\| - \|y\|\big| \leqslant |x - y| \quad \text{and} \quad \|x + y\| \leqslant \|x\| + \|y\|.$$

2 Prove that, for all $x \in \mathbb{R}$, we have $2\|x\| \leqslant |\sin(\pi x)| \leqslant \pi\|x\|$.

3 This exercise is the analogue of Lemma 5.32 for the function $f : u \longmapsto (x/u)^{1/2}$. The aim is to show the following estimate.

> Let $x \geqslant 1$ and δ satisfying (5.4) be real numbers and N be an integer such that $4 \leqslant N \leqslant x$. Assume that there exists a small real number $c_0 > 0$ such that $N\delta \leqslant c_0$. Then

$$\mathcal{R}\left(\sqrt{\frac{x}{n}}, N, \delta\right) \ll (Nx)^{1/6}.$$

One may use the polynomials

$$P(X, Y) = 4X + Y \quad \text{and} \quad Q(X, Y) = 4X + 3Y.$$

(a) Show that there exists a subset T of $\mathcal{S}(f, N, \delta)$ satisfying the following two properties.
  (i) $\mathcal{R}(f, N, \delta) \leqslant 4(|T| + 1)$.
  (ii) Any two consecutive elements of T differ by $> 2^{2/3}x^{-1/6}N^{5/6}$.
(b) Deduce the desired result.

4 Using Theorem 5.30, show that for $16 \leqslant y < x^{1/2}/4$, the following asymptotic formula holds.

$$\sum_{x < n \leqslant x+y} \mu_2(n) = \frac{y}{\zeta(2)} + O\left(x^{1/15}y^{2/3}\log x\right).$$

5 Let $r \geqslant 2$ be an integer and x, y be real numbers such that $4^r \leqslant y < x^{1/r}$. By adapting the proof of Corollary 5.35 with the use of Theorem 5.36 instead of Theorem 5.30, show that

$$\sum_{x < n \leqslant x+y} \mu_r(n) = \frac{y}{\zeta(r)} + O\left\{\left(x^{\frac{1}{2r+1}} + yx^{-\frac{1}{6r(2r+1)}}\right)\log x + y^{1/r}\right\}.$$

6 This exercise deals with the problem of square-full numbers in short intervals. Let $x \geqslant 10^{35}$ be a large real number and y be a real number satisfying

$$16x^{1/2}(\log x)^3 \leqslant y \leqslant 4^{-3}x(\log x)^{-1}. \tag{5.45}$$

Set $L = L(x, y) = y(x\log x)^{-1/2}$.

(a) Splitting the sum into two subsums, show that

$$\sum_{L < b \leqslant (x+y)^{1/3}} \left(\left[\sqrt{\frac{x+y}{b^3}}\right] - \left[\sqrt{\frac{x}{b^3}}\right]\right) \ll (R_1 + R_2)\log x + L$$

where

$$R_1 = \max_{L < B \leqslant (2x)^{1/5}} \mathcal{R}\left(\sqrt{\frac{x}{b^3}}, B, \frac{y}{\sqrt{xB^3}}\right)$$

and

$$R_2 = \max_{L < A \leqslant (2x)^{1/5}} \mathcal{R}\left(\left(\frac{x}{a^2}\right)^{1/3}, A, \frac{y}{(Ax)^{2/3}}\right).$$

(b) Deduce that we have

$$\sum_{x < n \leqslant x+y} s_2(n) = \frac{\zeta(3/2)}{2\zeta(3)} \frac{y}{x^{1/2}} + O\{(R_1 + R_2)\log x + L\}.$$

(c) Using Theorem 5.22 and 5.23 (i), show that

$$\sum_{x < n \leqslant x+y} s_2(n) = \frac{\zeta(3/2)}{2\zeta(3)} \frac{y}{x^{1/2}} + O\left\{x^{2/15}\log x + \frac{y}{(x\log x)^{1/2}}\right\}.$$

(d) Using Theorem 5.26 to estimate R_1 and R_2 in some specific ranges and using Theorem 5.23 (i) in the complementary ranges, show that, if (5.45) is replaced by

$$x^{37/60}(\log x)^3 \leqslant y \leqslant 4^{-3}x(\log x)^{-1} \tag{5.46}$$

then we have the following improvement

$$\sum_{x < n \leqslant x+y} s_2(n) = \frac{\zeta(3/2)}{2\zeta(3)} \frac{y}{x^{1/2}} + O\left\{x^{1/8}(\log x)^{7/5} + \frac{y}{(x\log x)^{1/2}}\right\}.$$

7 Let $x \geqslant 1$ and δ satisfying (5.4) be real numbers, $r \geqslant 2$ and $N \geqslant 4$ be integers satisfying

$$x^\varepsilon \leqslant 2^{r+1}N^r\delta \leqslant \frac{2x}{3}$$

for all $\varepsilon > 0$. Using Theorem 4.62 or Theorem 4.65, show that

$$\mathcal{R}\left(\frac{x}{n^r}, N, \delta\right) \ll_{r,\varepsilon} N^r\delta.$$

8 This exercise provides a variant of Theorem 5.30, from which we take up all the notation. Furthermore, assume that

$$N\delta \leqslant c_0, \quad N < \left(c_0^{-1}x\right)^{1/3} \quad \text{and} \quad N^2\delta \geqslant 4x^{-1}. \tag{5.47}$$

We set

$$A = 2^{-1/3}N^{2/3}(x\delta)^{-1/6}. \tag{5.48}$$

(a) Check that this choice of A is admissible in the proof of Theorem 5.30, i.e. using (5.47) check that

$$R < A \leqslant \frac{N}{2}.$$

(b) Deduce that, if (5.47) holds, then

$$\mathcal{R}\left(\frac{x}{n^2}, N, \delta\right) \ll x^{1/5} + \left(N^2 \delta x\right)^{1/6} + \left(x \delta^{-2} N^{-4}\right)^{1/3} + \left(x \delta^7 N^{11}\right)^{1/9}.$$

(c) Prove that, if $x^{1/5} \leqslant y \leqslant x^{1/3}$, then

$$\sum_{x < n \leqslant x+y} \mu_2(n) = \frac{y}{\zeta(2)} + O\left(x^{1/9} y^{4/9} (\log x)^{7/5}\right)$$

and check that this error-term is better than that of Exercise 4 apart from the logarithmic power.

9 Let x, y be real numbers satisfying $x^{1/3} \leqslant y \leqslant x$. Using Theorem 5.22 or otherwise, prove that

$$\max_{4y < N \leqslant x} \sum_{N < n \leqslant 2N} \left(\left[\frac{x+y}{n}\right] - \left[\frac{x}{n}\right]\right) \ll y.$$

This is a slight improvement of (4.32) but in the restricted range $x^{1/3} \leqslant y \leqslant x$.

10 Let $a \geqslant 2$ be an integer and $f \in C^\infty[N, 2N]$ such that there exists $T \geqslant 1$ such that, for all $x \in [N, 2N]$ and all $j \in \mathbb{Z}_{\geqslant 0}$, we have

$$\left|f^{(j)}(x)\right| \asymp \frac{T}{N^j} \qquad\qquad (5.49)$$

and

$$N\delta \leqslant T \leqslant N^a. \qquad\qquad (5.50)$$

By mimicking the proof of Proposition 5.24, show that, for all $k \geqslant 2a$, we have

$$\mathcal{R}(f, N, \delta) \ll T^{\frac{2}{k(k+1)}} N^{\frac{k-1}{k+1}} + N\delta^{\frac{1}{a(2a-1)}}.$$

This implies in particular that, if $k \geqslant 4$ and $N\delta \leqslant T \leqslant N^2$, then

$$\mathcal{R}(f, N, \delta) \ll T^{\frac{2}{k(k+1)}} N^{\frac{k-1}{k+1}} + N\delta^{1/6}.$$

References

[BS94]   Branton M, Sargos P (1994) Points entiers au voisinage d'une courbe plane à très faible courbure. Bull Sci Math 118:15–28

[Fog41]   Fogels E (1941) On the average values of arithmetic functions. Proc Camb Philos Soc 37:358–372

[FT92]   Filaseta M, Trifonov O (1992) On gaps between squarefree numbers II. J Lond Math Soc 45:215–221

[FT96]   Filaseta M, Trifonov O (1996) The distribution of fractional parts with application to gap results in number theory. Proc Lond Math Soc 73:241–278

[GK91]   Graham SW, Kolesnik G (1991) Van der Corput's method of exponential sums. London math. soc. lect. note, vol 126. Cambridge University Press, Cambridge

[Gor39]   Gorny A (1939) Contribution à l'étude des fonctions dérivables d'une variable réelle. Acta Math 71:317–358

[HS95]   Huxley MN, Sargos P (1995) Points entiers au voisinage d'une courbe plane de classe C^n. Acta Arith 69:359–366

[HS06]   Huxley MN, Sargos P (2006) Points entiers au voisinage d'une courbe plane de classe C^n, II. Funct Approx Comment Math 35:91–115

[HT96]   Huxley MN, Trifonov O (1996) The square-full numbers in an interval. Math Proc Camb Philos Soc 119:201–208

[Hux96]   Huxley MN (1996) Area, lattice points and exponential sums. Oxford Science Publications, London

[Hux99]   Huxley MN (1999) The integer points close to a curve III. In: Győry et al (eds) Number theory in progress, vol 2. de Gruyter, Berlin, pp 911–940

[Hux07]   Huxley MN (2007) The integer points in a plane curve. Funct Approx Comment Math 37:213–231

[Kon98]   Konyagin S (1998) Estimates for the least prime factor of a binomial coefficient. Mathematika 45:41–55

[Man52]   Mandelbrojt S (1952) Séries Adhérentes, Régularisation des Suites, Applications. Gauthier-Villars, Paris

[Ric54]   Richert HE (1954) On the difference between consecutive squarefree numbers. J Lond Math Soc 29:16–20

[Rot51]   Roth KF (1951) On the gaps between squarefree numbers. J Lond Math Soc 26:263–268

[Shi80]   Shiu P (1980) On the number of square-full integers between successive squares. Mathematika 27:171–178

[Sri62]   Srinivasan B-R (1962) On Van der Corput's and Nieland's results on the Dirichlet's divisor problem and the circle problem. Proc Natl Inst Sci India, Part A 28:732–742

[Tri02]   Trifonov O (2002) Lattice points close to a smooth curve and squarefull numbers in short intervals. J Lond Math Soc 65:309–319

Chapter 6
Exponential Sums

6.1 Introduction

The function $\psi : x \longmapsto \psi(x) = x - [x] - \frac{1}{2}$ appears in most problems in number theory. The reason is quite simple: when we look at asymptotics for average orders of classical arithmetic functions, the summations are often taken over some subsets of the set of divisors of some integer n, and when interchanging the order of the summations, integer parts, and hence the function ψ, arise. For instance, let us consider the number $\tau(n)$ of divisors of n. By (4.17), we first have

$$\sum_{n \leqslant x} \tau(n) = 2 \sum_{d \leqslant \sqrt{x}} \left[\frac{x}{d} \right] - [\sqrt{x}]^2.$$

Now replacing each integer part by ψ and using the estimate

$$\sum_{d \leqslant x} \frac{1}{d} = \log x + \gamma - \frac{\psi(x)}{x} + O\left(\frac{1}{x^2} \right)$$

obtained in Example 1.28, we get

$$\sum_{n \leqslant x} \tau(n) = 2 \sum_{d \leqslant \sqrt{x}} \left(\frac{x}{d} - \psi\left(\frac{x}{d} \right) - \frac{1}{2} \right) - \left(\sqrt{x} - \psi(\sqrt{x}) - \frac{1}{2} \right)^2$$

$$= 2x \sum_{d \leqslant x} \frac{1}{d} - 2 \sum_{d \leqslant \sqrt{x}} \psi\left(\frac{x}{d} \right) - x - \psi(\sqrt{x})^2 + \frac{1}{4} + 2\psi(\sqrt{x})\sqrt{x}$$

$$= x(\log x + 2\gamma - 1) - 2 \sum_{d \leqslant \sqrt{x}} \psi\left(\frac{x}{d} \right) + O(1).$$

O. Bordellès, *Arithmetic Tales*, Universitext,
DOI 10.1007/978-1-4471-4096-2_6, © Springer-Verlag London 2012

The error-term in this problem is usually denoted by $\Delta(x)$, so that

$$\Delta(x) = \sum_{n \leqslant x} \tau(n) - x(\log x + 2\gamma - 1) = -2 \sum_{d \leqslant \sqrt{x}} \psi\left(\frac{x}{d}\right) + O(1).$$

In fact, with a little more work, one can prove [BBR12] that

$$\left|\Delta(x)\right| \leqslant 2 \left| \sum_{d \leqslant \sqrt{x}} \psi\left(\frac{x}{d}\right) \right| + \frac{1}{2}.$$

Using the trivial bound $|\psi(x)| \leqslant 1/2$ enables us to recover Corollary 4.20. The *Dirichlet divisor problem* is about the smallest exponent $\theta \in [0, 1[$ for which the following estimate

$$\sum_{n \leqslant x} \tau(n) = x(\log x + 2\gamma - 1) + O\left(x^\theta (\log x)^\beta\right)$$

holds for some $\beta \geqslant 0$. By Corollary 4.20 or the computations made above, the pair $(\theta, \beta) = (1/2, 0)$ is admissible. Hardy showed in [Har16] that $\Delta(x)$ cannot be a $o(x^{1/4})$, in other words we necessarily have $\theta \geqslant 1/4$ (with some $\beta > 0$).

It is surmised that $\theta = 1/4$ is the right value in the Dirichlet divisor problem, which is supported by the following mean-value result

$$\int_1^T \Delta(x)^2 \, dx = \frac{\zeta(3/2)^4}{36\zeta(2)\zeta(3)} T^{3/2} + O\left(T(\log T)^5\right)$$

due to Tong [Ton56]. However, no one has been in a position to prove it until now, although this conjecture is a consequence of the Lindelöf hypothesis, and hence of the Riemann hypothesis. The best result to date is due to Huxley [Hux03] who showed that the pair

$$(\theta, \beta) = \left(\frac{131}{416}, \frac{26\,947}{8320}\right)$$

is admissible (see Theorem 6.43). Note that $131/416 \approx 0.314\,903\,84\ldots$ so that we are still far from the conjectured value, but this problem gradually gets harder and harder. For example, to reach this value, Huxley used a new method called the *discrete Hardy–Littlewood method* (see Sect. 6.6), which is a tricky mixture of several known methods and new ones.

Similarly, the *Dirichlet–Piltz divisor problem* is about the smallest exponent $\theta_k \in [0, 1[$ for which the following estimate

$$\sum_{n \leqslant x} \tau_k(n) = x P_{k-1}(\log x) + O\left(x^{\theta_k}(\log x)^{\beta_k}\right)$$

holds for some $\beta_k \geqslant 0$, where $P_{k-1}(X)$ is a polynomial of degree $k - 1$ in X and whose coefficients depend on k. By Dirichlet's hyperbola principle (Proposition 4.19), we get

$$\sum_{n \leqslant x} \tau_k(n) = x \sum_{d \leqslant x^{1-1/k}} \frac{\tau_{k-1}(d)}{d} + \sum_{m \leqslant x^{1/k}} \sum_{d \leqslant x/m} \tau_{k-1}(d)$$

$$- x^{1/k} \sum_{d \leqslant x^{1-1/k}} \tau_{k-1}(d) + O\left(\sum_{x^{1-1/k}} \tau_{k-1}(d) \right)$$

and by induction one can prove that the pair

$$(\theta_k, \beta_k) = \left(1 - \frac{1}{k}, 0 \right)$$

is admissible. Using contour integration methods, Hardy and Littlewood showed that $\theta_k = \frac{k-1}{k+2}$ is admissible as soon as $k \geqslant 4$. As for the function τ, one can show that $\theta_k \geqslant \frac{k-1}{2k}$ with some $\beta_k > 0$, and one conjectures this is the right value in this problem. The Lindelöf hypothesis implies that conjecture.[1] For more information on this subject, the reader should refer to [Bor09, Ivi85].

Now let us return to the function ψ. This function is odd and 1-periodic, and hence permits a Fourier series expansion. By Proposition 1.29, the series

$$-\sum_{h=1}^{\infty} \frac{\sin(2\pi h x)}{\pi h} = -\sum_{h \in \mathbb{Z} \setminus \{0\}} \frac{e(hx)}{2\pi i h}$$

converges to $\psi(x)$ if $x \in \mathbb{R} \setminus \mathbb{Z}$ and to 0 if $x \in \mathbb{Z}$. Unfortunately, the convergence is not uniform, so that it would be better to work with partial sums of this series, or more generally with trigonometric polynomials. The following result, due to Vaaler [Vaa85], is a useful answer to this question.

Theorem 6.1 (Vaaler) *For all real numbers $x \geqslant 1$ and all integers $H \geqslant 1$, we have*

$$\psi(x) = - \sum_{0 < |h| \leqslant H} \Phi\left(\frac{h}{H+1} \right) \frac{e(hx)}{2\pi i h} + \mathcal{R}_H(x)$$

where $\Phi(t) = \pi t (1 - |t|) \cot(\pi t) + |t|$ for $0 < |t| < 1$ and

$$|\mathcal{R}_H(x)| \leqslant \frac{1}{2H+2} \sum_{|h| \leqslant H} \left(1 - \frac{|h|}{H+1} \right) e(hx).$$

Note that $0 < \Phi(t) < 1$ for $0 < |t| < 1$ and, using Exercise 3, we have

$$\sum_{|h| \leqslant H} \left(1 - \frac{|h|}{H+1} \right) e(hx) = \frac{1}{H+1} \left| \sum_{h=0}^{H} e(hx) \right|^2$$

---

[1] The Dirichlet–Piltz divisor problem can be generalized to number fields where it is usually called the *General divisor problem* It was originally investigated by Hasse and Suetuna [HS31], and the asymptotics are of the same order of magnitude as in the rational case.

so that the sum in the error-term is a non-negative real number. Using this with $x = f(n)$ and summing over $]N, 2N]$ we get the following inequality.

Corollary 6.2 *Let H, N be positive integers and $f : [N, 2N] \longrightarrow \mathbb{R}$ be any function. Then we have*

$$\left| \sum_{N < n \leqslant 2N} \psi(f(n)) \right| \leqslant \frac{N}{2H + 2} + \left(1 + \frac{1}{\pi}\right) \sum_{h=1}^{H} \frac{1}{h} \left| \sum_{N < n \leqslant 2N} e(hf(n)) \right|.$$

Proof Using Theorem 6.1 we get

$$\sum_{N < n \leqslant 2N} \psi(f(n)) = - \sum_{0 < |h| \leqslant H} \Phi\left(\frac{h}{H + 1}\right) \frac{1}{2\pi i h} \sum_{N < n \leqslant 2N} e(hf(n))$$

$$+ \sum_{N < n \leqslant 2N} \mathcal{R}_H(f(n))$$

where Φ and $\mathcal{R}_H$ are defined in Theorem 6.1, and hence

$$\left| \sum_{N < n \leqslant 2N} \psi(f(n)) \right| \leqslant |\Sigma_1| + |\Sigma_2|$$

with

$$|\Sigma_1| \leqslant \sum_{0 < |h| \leqslant H} \frac{1}{2\pi |h|} \left| \sum_{N < n \leqslant 2N} e(hf(n)) \right| = \frac{1}{\pi} \sum_{h=1}^{H} \frac{1}{h} \left| \sum_{N < n \leqslant 2N} e(hf(n)) \right|$$

and

$$|\Sigma_2| \leqslant \frac{1}{2H + 2} \sum_{|h| \leqslant H} \left(1 - \frac{|h|}{H + 1}\right) \sum_{N < n \leqslant 2N} e(hf(n))$$

$$= \frac{N}{2H + 2} + \frac{1}{H + 1} \sum_{h=1}^{H} \left(1 - \frac{h}{H + 1}\right) \operatorname{Re}\left(\sum_{N < n \leqslant 2N} e(hf(n))\right)$$

$$\leqslant \frac{N}{2H + 2} + \sum_{h=1}^{H} \frac{1}{h} \left| \sum_{N < n \leqslant 2N} e(hf(n)) \right|$$

which implies the asserted result.   □

Remark 6.3 One can also work directly with the partial sums of the Fourier series of the function ψ with the following asymptotic formula. For all positive integers

H, we have (see [MV81] for instance)

$$\psi(x) = - \sum_{0 < |h| \leqslant H} \frac{e(hx)}{2\pi i h} + O\left\{ \min\left(1, \frac{1}{H\|x\|}\right) \right\}$$

where $\|x\|$ is the distance from x to its nearest integer. The two results are equivalent, but Vaaler's theorem is often more useful in applications.

6.2   Kusmin–Landau's Inequality

The first result in exponential sums is the following estimate which was of interest to mathematicians including van der Corput, Kusmin, Landau, Karamata and Tomic (see [Mor58]). We begin by proving the original result, and then provide some more practical versions we shall be able to use later.

We first start with a technical lemma.

Lemma 6.4 *Let $M \in \mathbb{N}$ and $x_1, \ldots, x_M$ be pairwise distinct complex numbers. Then*

$$2 \sum_{n=1}^{M} x_n = x_1 \left(\frac{x_1 + x_2}{x_1 - x_2} + 1 \right) + \sum_{n=2}^{M-1} x_n \left(\frac{x_n + x_{n+1}}{x_n - x_{n+1}} - \frac{x_{n-1} + x_n}{x_{n-1} - x_n} \right)$$

$$+ x_M \left(1 - \frac{x_{M-1} + x_M}{x_{M-1} - x_M} \right).$$

Proof If we set $x_0 = x_{M+1} = 0$ and $T_n = \frac{x_n + x_{n+1}}{x_n - x_{n+1}}$ for $n = 0, \ldots, M$ then the right-hand side above is equal to

$$\sum_{n=1}^{M} x_n (T_n - T_{n-1}).$$

We use Abel's summation as in Remark 1.15 which gives

$$\sum_{n=1}^{M} x_n (T_n - T_{n-1}) = x_M \sum_{n=1}^{M} (T_n - T_{n-1}) - \sum_{n=1}^{M-1} (x_{n+1} - x_n) \sum_{k=1}^{n} (T_k - T_{k-1})$$

$$= x_M (T_M - T_0) - \sum_{n=1}^{M-1} (x_{n+1} - x_n)(T_n - T_0)$$

and using $T_0 = -1$, $T_M = 1$ and $(x_{n+1} - x_n)T_n = -(x_n + x_{n+1})$ we get

$$\sum_{n=1}^{M} x_n (T_n - T_{n-1}) = 2x_M + \sum_{n=1}^{M-1} (x_n + x_{n+1}) - \sum_{n=1}^{M-1} (x_{n+1} - x_n) = 2 \sum_{n=1}^{M} x_n$$

as required. $\square$

We are now in a position to prove the main result of this section.

Theorem 6.5 (Kusmin–Landau's inequality) *Let $M \in \mathbb{N}$ and $f : [1, M] \longrightarrow \mathbb{R}$ be a function such that there exists $\lambda_1 \in \mathbb{R}$ such that*

$$0 < \lambda_1 \leqslant f(2) - f(1) \leqslant \cdots \leqslant f(M) - f(M-1) \leqslant 1 - \lambda_1. \qquad (6.1)$$

Then we have

$$\left| \sum_{n=1}^{M} e(\pm f(n)) \right| \leqslant \frac{2}{\pi \lambda_1}.$$

Proof Let $S_M := \sum_{n=1}^{M} e(f(n))$. We use Lemma 6.4 with $x_n = e(f(n))$ and the fact that

$$\frac{e(x) + e(y)}{e(x) - e(y)} = \frac{1 + e(y - x)}{1 - e(y - x)} = i \cot \pi (y - x)$$

so that

$$\begin{aligned}
2 S_M = {}& e(f(1)) \{ 1 + i \cot \pi (f(2) - f(1)) \} \\
& + e(f(M)) \{ 1 + i \cot \pi (f(M) - f(M-1)) \} \\
& + i \sum_{n=2}^{M-1} e(f(n)) \{ \cot \pi (f(n+1) - f(n)) - \cot \pi (f(n) - f(n-1)) \}.
\end{aligned}$$

Now since the function $x \longmapsto \cot(\pi x)$ is strictly decreasing in $]0, 1[$ and by the use of (6.1), we infer that

$$\begin{aligned}
2 |S_M| \leqslant {}& \frac{1}{\sin \pi (f(2) - f(1))} + \frac{1}{\sin \pi (f(M) - f(M-1))} \\
& + \sum_{n=2}^{M-1} \left| \cot \pi (f(n+1) - f(n)) - \cot \pi (f(n) - f(n-1)) \right| \\
= {}& \frac{1}{\sin \pi (f(2) - f(1))} + \frac{1}{\sin \pi (f(M) - f(M-1))} \\
& + \cot \pi (f(2) - f(1)) - \cot \pi (f(M) - f(M-1)) \\
= {}& \cot\left(\frac{\pi}{2} (f(2) - f(1)) \right) + \tan\left(\frac{\pi}{2} (f(M) - f(M-1)) \right) \\
\leqslant {}& \cot\left(\frac{\pi \lambda_1}{2} \right) + \tan\left(\frac{\pi}{2} (1 - \lambda_1) \right) = 2 \cot\left(\frac{\pi \lambda_1}{2} \right)
\end{aligned}$$

and we conclude the proof by using the inequality $\cot x \leqslant 1/x$ valid for $0 < x \leqslant \pi/2$. $\qquad \square$

Corollary 6.6 *Let $N < N_1 \leqslant 2N$ be positive integers and $f \in C^1[N, N_1]$ such that f' is non-decreasing and there exists a real number $\lambda_1 > 0$ such that, for all $x \in [N, N_1]$, we have*

$$k + \lambda_1 \leqslant f'(x) \leqslant k + 1 - \lambda_1 \quad (k \in \mathbb{Z}). \tag{6.2}$$

Then we have

$$\left| \sum_{N < n \leqslant N_1} e\big(\pm f(n)\big) \right| \leqslant \frac{2}{\pi \lambda_1}.$$

Proof Since $e(f(n)) = e(f(n) - kn)$, we may suppose that $k = 0$. Define the functions g and h by

$$\begin{aligned} g(x) &= f(x + N) \quad &(1 \leqslant x \leqslant N_1 - N), \\ h(x) &= g(x) - g(x - 1) \quad &(2 \leqslant x \leqslant N_1 - N). \end{aligned}$$

Since $g'(x) = f'(x + N)$, the function g' is non-decreasing and we have $\lambda_1 \leqslant g'(x) \leqslant 1 - \lambda_1$. Furthermore, since $h'(x) = g'(x) - g'(x - 1) \geqslant 0$, the function h is non-decreasing. Hence we have

$$\lambda_1 \leqslant g(2) - g(1) \leqslant \cdots \leqslant g(N_1 - N) - g(N_1 - N - 1) \leqslant 1 - \lambda_1$$

so that g satisfies the hypothesis of Theorem 6.5, and we get

$$\left| \sum_{N < n \leqslant N_1} e\big(f(n)\big) \right| = \left| \sum_{n=1}^{N_1 - N} e\big(g(n)\big) \right| \leqslant \frac{2}{\pi \lambda_1}$$

which concludes the proof. □

Here is the version that we shall use in practice.

Corollary 6.7 *Let $N < N_1 \leqslant 2N$ be positive integers and $f \in C^1[N, N_1]$ such that f' is non-decreasing and there exist real numbers $c_1 \geqslant 1$ and $0 < \lambda_1 \leqslant (c_1 + 1)^{-1}$ such that, for all $x \in [N, N_1]$, we have*

$$\lambda_1 \leqslant f'(x) \leqslant c_1 \lambda_1. \tag{6.3}$$

Then the conclusion of Corollary 6.6 still holds.

Proof The result follows directly from Corollary 6.6 since the hypothesis $0 < \lambda_1 \leqslant (c_1 + 1)^{-1}$ implies that $c_1 \lambda_1 \leqslant 1 - \lambda_1$ so that (6.2) is satisfied with $k = 0$. □

Example 6.8 We use Corollary 6.7 on Dirichlet's divisor problem in the following way. We have

$$|\Delta(x)| \leqslant 2 \left| \sum_{n \leqslant \sqrt{x}} \psi\left(\frac{x}{n}\right) \right| + \frac{1}{2} \tag{6.4}$$

and splitting the interval $[1, \sqrt{x}]$ into dyadic subintervals of the type $]N, 2N]$ gives

$$|\Delta(x)| \leqslant \max_{1 \leqslant N \leqslant \sqrt{x}} \left| \sum_{N < n \leqslant 2N} \psi\left(\frac{x}{n}\right) \right| \frac{\log x}{\log 2} + \frac{1}{2}$$

since there are at most $\log x / \log 4$ such subintervals. Now using Corollary 6.2 we get for all integers $H \geqslant 1$

$$|\Delta(x)| \leqslant \max_{1 \leqslant N \leqslant \sqrt{x}} \left\{ \frac{N}{2H+2} + 2 \sum_{h=1}^{H} \frac{1}{h} \left| \sum_{N < n \leqslant 2N} e\left(\frac{hx}{n}\right) \right| \right\} \frac{\log x}{\log 2} + \frac{1}{2}.$$

Now to estimate the exponential sum, Corollary 6.7 could *a priori* be used with $\lambda_1 = hx(4N^2)^{-1}$ and $c_1 = 4$, but the condition $\lambda_1 \leqslant 5^{-1}$ forces to suppose $N \geqslant \sqrt{2Hx}$, which is in contradiction with the range $1 \leqslant N \leqslant \sqrt{x}$ of the sum. Thus, we cannot apply Kusmin–Landau's inequality in this way to the Dirichlet divisor problem.

6.3 Van der Corput's Inequality

Kusmin–Landau's inequality is sharp under the sole hypothesis (6.3) but, as can be shown in Example 6.8, the condition $\lambda_1 \leqslant (c_1 + 1)^{-1}$ is too restrictive to be really used efficiently in usual problems of number theory. In the 1920s, van der Corput established a very useful inequality which can be considered as the starting point of crucial theorems supplying estimates for exponential sums.

Theorem 6.9 (van der Corput's inequality) *Let $N < N_1 \leqslant 2N$ be positive integers and $f \in C^2[N, N_1]$ such that there exist real numbers $c_2 \geqslant 1$ and $\lambda_2 > 0$ such that, for all $x \in [N, N_1]$, we have*

$$\lambda_2 \leqslant f''(x) \leqslant c_2 \lambda_2. \tag{6.5}$$

Then

$$\left| \sum_{N < n \leqslant N_1} e(\pm f(n)) \right| \leqslant \frac{4}{\sqrt{\pi}} \left(c_2 N \lambda_2^{1/2} + 2\lambda_2^{-1/2} \right).$$

Remark 6.10 The condition (6.5) will prove to be much more useful than (6.3). It also should be mentioned that the proof of the theorem will make the following slightly stronger result appear.

Let $N < N_1 \leqslant 2N$ be positive integers and $f \in C^2[N, N_1]$ such that there exists a real number $\lambda_2 > 0$ such that, for all $x \in [N, N_1]$, we have $f''(x) \geqslant \lambda_2$. Then

$$\left| \sum_{N < n \leqslant N_1} e(\pm f(n)) \right| \leqslant \frac{4}{\sqrt{\pi \lambda_2}} \left(f'(N_1) - f'(N) + 2 \right).$$

In practice, we shall make no use of such an improvement.

There are many methods to show Theorem 6.9. The one we will introduce here is the discrete analogue of the proof of the second derivative test for integrals, also proved by van der Corput (see Lemma 6.26). We shall propose other proofs in Sect. 6.6 and Exercise 2. See also [GK91, Mon94].

We begin with an intermediate result.

Lemma 6.11 *Let* $N < N_1 \leqslant 2N$ *be positive integers and* $f \in C^2[N, N_1]$ *such that there exists a real number* $\lambda_2 \in]0, \pi^{-1}[$ *such that, for all* $x \in [N, N_1]$, *we have* $f''(x) \geqslant \lambda_2$. *Assume also that, for all* $x \in]N, N_1[$, *we have* $f'(x) \notin \mathbb{Z}$. *Then*

$$\left| \sum_{N < n \leqslant N_1} e(\pm f(n)) \right| \leqslant \frac{4}{\sqrt{\pi \lambda_2}}.$$

Proof Let $t \in]0, \frac{1}{2}[$ be a parameter at our disposal. Since $f'(x) \notin \mathbb{Z}$ for all $x \in]N, N_1[$, there exist $u, v \in \mathbb{R}$ and non-negative integers M_1, M_2 such that

$$u = f'(N) \quad \text{and} \quad v = f'(N_1)$$
$$f'(M_1) = [u] + t \quad \text{and} \quad f'(M_2) = [u] + 1 - t.$$

We now split the sum into three subsums

$$\sum_{N < n \leqslant N_1} e(f(n)) = \sum_{N < n \leqslant M_1} e(f(n)) + \sum_{M_1 < n \leqslant M_2} e(f(n)) + \sum_{M_2 < n \leqslant N_1} e(f(n))$$

and estimate the first and third sums trivially. The mean-value theorem (Theorem 1.12) implies the existence of a real number $c \in]N, 2N[$ such that

$$\left| \sum_{N < n \leqslant M_1} e(f(n)) \right| \leqslant \max(M_1 - N, 1) = \max\left(\frac{f'(M_1) - f'(N)}{f''(c)}, 1 \right)$$

$$\leqslant \max\left(\frac{[u] + t - u}{\lambda_2}, 1 \right) \leqslant \max\left(\frac{t}{\lambda_2}, 1 \right).$$

Similarly, we have

$$\left| \sum_{M_2 < n \leqslant N_1} e(f(n)) \right| \leqslant \max\left(\frac{v - ([u] + 1) + t}{\lambda_2}, 1 \right) \leqslant \max\left(\frac{t}{\lambda_2}, 1 \right)$$

since $[u] \leqslant u \leqslant v \leqslant [u] + 1$. The second sum is estimated using Corollary 6.6. Since f' is non-decreasing and, in the interval $[M_1, M_2]$, we have $[u] + t \leqslant f'(x) \leqslant [u] + 1 - t$, Corollary 6.6 applies and gives

$$\left| \sum_{M_1 < n \leqslant M_2} e(f(n)) \right| \leqslant \frac{2}{\pi t}.$$

Hence we get

$$\left| \sum_{N < n \leqslant N_1} e(\pm f(n)) \right| \leqslant 2\max\left(\frac{t}{\lambda_2}, 1\right) + \frac{2}{\pi t}$$

and the asserted estimate follows from taking $t = \sqrt{\lambda_2 \pi^{-1}}$ since $\lambda_2 < \pi^{-1}$. $\square$

We are now able to prove van der Corput's inequality.

Proof of Theorem 6.9 If $\lambda_2 \geqslant \pi^{-1}$, then we have

$$\frac{4N\lambda_2^{1/2}}{\sqrt{\pi}} \geqslant \frac{4N}{\pi} > N \geqslant N_1 - N \geqslant \left| \sum_{N < n \leqslant N_1} e(f(n)) \right|$$

so that we may suppose that $\lambda_2 < \pi^{-1}$. We pick up the numbers u and v from the proof of Lemma 6.11 and set

$$[u, v] \cap \mathbb{Z} = \{m + 1, \ldots, m + K\}$$

where $m \in \mathbb{Z}$ and $K \in \mathbb{N}$, and we define for all integers $k \in \{1, \ldots, K + 1\}$ the intervals

$$J_k =]m + k - 1, m + k] \cap [u, v].$$

Lemma 6.11 implies that

$$\left| \sum_{N < n \leqslant N_1} e(f(n)) \right| \leqslant \sum_{k=1}^{K+1} \left| \sum_{n \in J_k \cap \mathbb{Z}} e(f(n)) \right| \leqslant \frac{4(K + 1)}{\sqrt{\pi \lambda_2}}$$

and we have by the mean-value theorem

$$K - 1 \leqslant v - u = f'(N_1) - f'(N) \leqslant c_2(N_1 - N)\lambda_2 \leqslant c_2 N \lambda_2$$

which concludes the proof. $\square$

When the function f is sufficiently smooth, one can improve on the secondary term in van der Corput's inequality with the use of Corollary 6.7. The following result is the practical version of this inequality that we shall use in most of the applications.

Corollary 6.12 *Let $N < N_1 \leqslant 2N$ be positive integers and $f \in C^2[N, N_1]$ such that there exist real numbers $c_1, c_2 \geqslant 1$ and $\lambda_1, \lambda_2, s_1 > 0$ such that, for all $x \in [N, N_1]$, we have*

$$\lambda_1 \leqslant |f'(x)| \leqslant c_1\lambda_1, \quad \lambda_2 \leqslant |f''(x)| \leqslant c_2\lambda_2 \quad and \quad \lambda_1 = s_1(N_1 - N)\lambda_2. \quad (6.6)$$

We set $c = 4\pi^{-1/2}\{c_2 + 2s_1(c_1 + 1)\}$. Then

$$\left| \sum_{N < n \leqslant N_1} e(f(n)) \right| \leqslant c\, N\lambda_2^{1/2} + \frac{2}{\pi\lambda_1}.$$

Proof If $\lambda_1 \leqslant (c_1 + 1)^{-1}$, then we apply Corollary 6.7. Suppose that $\lambda_1 > (c_1 + 1)^{-1}$. Since $\lambda_1 = s_1(N_1 - N)\lambda_2$, we can write

$$2s_1(c_1 + 1)N\lambda_2^{1/2} \geqslant 2s_1(c_1 + 1)(N_1 - N)\lambda_2^{1/2} = 2(c_1 + 1)\lambda_1\lambda_2^{-1/2} > 2\lambda_2^{-1/2}$$

so that by Theorem 6.9 we get

$$\left| \sum_{N < n \leqslant N_1} e(f(n)) \right| \leqslant \frac{4}{\sqrt{\pi}}\left(c_2 N\lambda_2^{1/2} + 2\lambda_2^{-1/2} \right)$$

$$< \frac{4}{\sqrt{\pi}}\left(c_2 N\lambda_2^{1/2} + 2s_1(c_1 + 1)N\lambda_2^{1/2} \right) = c\, N\lambda_2^{1/2}$$

which implies the asserted result. $\square$

Example 6.13 We return to the Dirichlet divisor problem in which we assume that $x \geqslant 416$. We first split the sum (6.4) into the ranges $[1, 6x^{1/3}]$ and $]6x^{1/3}, x^{1/2}]$, then estimate the sum trivially in the first one and use the splitting argument seen in Example 6.8 for the second sum. Since there are at most $\log x / \log 64$ subintervals of the form $]N, 2N]$ in $]6x^{1/3}, x^{1/2}]$, (6.4) becomes for all integers $H \geqslant 1$

$$\left| \Delta(x) \right| \leqslant 6x^{1/3} + \max_{6x^{1/3} < N \leqslant \sqrt{x}} \left\{ \frac{N}{2H + 2} + 2\sum_{h=1}^{H} \frac{1}{h} \left| \sum_{N < n \leqslant 2N} e\left(\frac{hx}{n} \right) \right| \right\} \frac{\log x}{\log 8} + \frac{1}{2}.$$

Corollary 6.12 may be applied with $N_1 = 2N$, $\lambda_1 = hx(4N^2)^{-1}$, $c_1 = 4$, $\lambda_2 = hx(4N^3)^{-1}$, $c_2 = 8$, $s_1 = 1$ so that $c = 72\pi^{-1/2}$ which gives

$$\left| \sum_{N < n \leqslant 2N} e\left(\frac{hx}{n} \right) \right| \leqslant \frac{36}{\sqrt{\pi}}\left(\frac{hx}{N} \right)^{1/2} + \frac{8N^2}{\pi hx}$$

and therefore we get

$$\sum_{h=1}^{H} \frac{1}{h} \left| \sum_{N < n \leqslant 2N} e\left(\frac{hx}{n} \right) \right| \leqslant \frac{36}{\sqrt{\pi}}\left(\frac{x}{N} \right)^{1/2} \sum_{h=1}^{H} \frac{1}{h^{1/2}} + \frac{8N^2}{\pi x} \sum_{h=1}^{H} \frac{1}{h^2}$$

$$\leqslant \frac{72}{\sqrt{\pi}}\left(\frac{Hx}{N} \right)^{1/2} + \frac{8\zeta(2)N^2}{\pi x}.$$

We finally get

$$\left|\Delta(x)\right| \leqslant 6x^{1/3} + \max_{6x^{1/3} < N \leqslant \sqrt{x}} \left\{\frac{N}{2H+2} + \frac{144}{\sqrt{\pi}}\left(\frac{Hx}{N}\right)^{1/2} + \frac{8\pi N^2}{3x}\right\}\frac{\log x}{\log 8} + \frac{1}{2}$$

and the choice of $H = [6^{-1}Nx^{-1/3}]$ gives for all $x \geqslant 416$

$$\left|\Delta(x)\right| < 24\,x^{1/3}\log x.$$

We may state the following result.

Theorem 6.14 (Voronoï, van der Corput) *For all $x \geqslant 416$, we have*

$$\sum_{n \leqslant x} \tau(n) = x(\log x + 2\gamma - 1) + O\left(x^{1/3}\log x\right).$$

Remark 6.15 Theorem 6.14 was rediscovered by van der Corput using his method of exponential sums, but it was first shown by Voronoï [Vor03] in an elementary way, since theorems for exponential sums did not exist at the time. Voronoï improved on Dirichlet's hyperbola principle (Proposition 4.19) by considering triangles instead of rectangles under the hyperbola $mn = x$ in Dirichlet's method. He was then able to prove the following estimate.

Lemma 6.16 (Voronoï) *For all real numbers $x, T \geqslant 1$, we have*

$$\left|\Delta(x)\right| \leqslant \frac{19}{12}\sum_{n \leqslant T}\tau(n) + \left(\frac{\sqrt{x}}{4T} + \frac{\sqrt{T}}{6}\right)\sum_{n \leqslant T}\frac{\tau(n)}{n^{1/2}} + \frac{3x^{1/4}}{4}\sum_{n \leqslant T}\frac{\tau(n)}{n^{3/4}} + \frac{T}{6} + \sqrt{\frac{x}{T}} + \frac{7}{4}.$$

Estimating the sums $\sum_{n \leqslant T}\tau(n)n^{-1/2}$ and $\sum_{n \leqslant T}\tau(n)n^{-3/4}$ by partial summation (Theorem 1.14) and choosing the parameter T optimally, Voronoï deduced that

$$\left|\Delta(x)\right| < 3\,x^{1/3}\log x$$

for all $x \geqslant 308$. Unfortunately, his method did not seem to produce better exponents than $1/3$, whereas van der Corput's was to show great promise.

6.4 The Third Derivative Theorem

6.4.1 Weyl's Shift

Cauchy–Schwarz's (or Cauchy–Bunyakovski–Schwarz's) inequality, which plays an important role in many branches of modern mathematics, is one of the most famous results in inequalities theory. For a survey of this inequality and related results, the reader should refer to [Dra03].

Lemma 6.17 (Cauchy–Schwarz's inequality) *Let M be a positive integer and $a_1, \ldots, a_M$ and $b_1, \ldots, b_M$ be two finite sequences of arbitrary complex numbers. Then we have*

$$\left| \sum_{m=1}^{M} a_m b_m \right|^2 \leqslant \sum_{m=1}^{M} |a_m|^2 \sum_{m=1}^{M} |b_m|^2.$$

Proof A possible proof of this inequality rests on Cauchy–Binet's identity for complex numbers

$$\sum_{m=1}^{M} x_m y_m \sum_{m=1}^{M} z_m t_m - \sum_{m=1}^{M} x_m t_m \sum_{m=1}^{M} z_m y_m = \frac{1}{2} \sum_{m=1}^{M} \sum_{n=1}^{M} (x_m z_n - x_n z_m)(y_m t_n - y_n t_m)$$

which is a generalization of Lagrange's identity, applied with $x_m = \overline{a_m}$, $y_m = a_m$, $z_m = b_m$ and $t_m = \overline{b_m}$ (for $m = 1, \ldots, M$), so that we get at once

$$\sum_{m=1}^{M} |a_m|^2 \sum_{m=1}^{M} |b_m|^2 - \left| \sum_{m=1}^{M} a_m b_m \right|^2 = \frac{1}{2} \sum_{m=1}^{M} \sum_{n=1}^{M} |\overline{a_m} b_n - \overline{a_n} b_m|^2 \geqslant 0$$

as required. $\qquad\square$

In analytic number theory, this inequality is often used to separate certain functions for which better results are obtained when they are summed individually.

The next tool is a device based upon the following observation. Suppose that N is a positive integer and $a_{N+1}, \ldots, a_{2N}$ are N arbitrary complex numbers. If we set for all $n \in \mathbb{Z}$

$$\alpha_n = \begin{cases} a_n, & \text{if } n \in \{N+1, \ldots, 2N\}, \\ 0, & \text{otherwise} \end{cases} \tag{6.7}$$

then we have for all integers h

$$\sum_{N < n \leqslant 2N} a_n = \sum_{n \in \mathbb{Z}} \alpha_n = \sum_{n \in \mathbb{Z}} \alpha_{n+h}. \tag{6.8}$$

Equality (6.8) is called *Weyl's shift*. It allows a great flexibility with the indices of the sums. For instance, for all positive integer H, we can write

$$\sum_{N < n \leqslant 2N} a_n = \frac{1}{H} \sum_{h=1}^{H} \sum_{n \in \mathbb{Z}} \alpha_{n+h}$$

and interchanging the summations we infer that

$$\sum_{N < n \leqslant 2N} a_n = \frac{1}{H} \sum_{n=N-H+1}^{2N-1} \sum_{h=1}^{H} \alpha_{n+h}. \tag{6.9}$$

6.4.2 Van der Corput's A-Process

The following inequality, discovered by van der Corput, is a very clever application of Weyl's shift and Cauchy–Schwarz's inequality. Often called the *A-process*, it is a first result given by van der Corput by which the problem of estimating a given exponential sum with a function $f(n)$ is replaced by estimating another exponential sum with the (upper) finite difference $\Delta_h f(n) = f(n+h) - f(n)$ of f. Since $\Delta_h f(n)$ is of the same order of magnitude as $hf'(n)$, one may expect some improvement in the estimates.

Lemma 6.18 (van der Corput's A-process) *Let N be a positive integer and $a_{N+1}, \ldots, a_{2N}$ be arbitrary complex numbers. For all integers $H \in \{1, \ldots, N\}$, we have*

$$\left| \sum_{N < n \leqslant 2N} a_n \right|^2 \leqslant \frac{2N}{H+1} \sum_{N < n \leqslant 2N} |a_n|^2 + \frac{4N}{H+1} \sum_{h=1}^{H} \left| \sum_{N < n \leqslant 2N-h} \overline{a_n} a_{n+h} \right|.$$

Proof Let S_N be the sum on the left-hand side. By (6.9), we have

$$|S_N| = \frac{1}{H+1} \left| \sum_{n=N-H}^{2N-1} \sum_{h=1}^{H+1} \alpha_{n+h} \right|$$

where the numbers α_n are defined in (6.7), and by Cauchy–Schwarz's inequality, we infer that

$$|S_N|^2 \leqslant \frac{N+H}{(H+1)^2} \sum_{n=N-H}^{2N-1} \left| \sum_{h=1}^{H+1} \alpha_{n+h} \right|^2.$$

Squaring out the modulus we obtain

$$|S_N|^2 \leqslant \frac{N+H}{(H+1)^2} \sum_{n=N-H}^{2N-1} \sum_{i=1}^{H+1} \sum_{j=1}^{H+1} \alpha_{n+i} \overline{\alpha_{n+j}}.$$

We now change the variables in the following way. Set $i = j + h$ with $1 \leqslant i \leqslant H+1$, so that $1 - h \leqslant j \leqslant H + 1 - h$ with $|h| = |i - j| \leqslant H$. This gives

$$|S_N|^2 \leqslant \frac{N+H}{(H+1)^2} \sum_{n=N-H}^{2N-1} \sum_{|h| \leqslant H} \sum_{j=1-h}^{H+1-h} \alpha_{n+h+j} \overline{\alpha_{n+j}}$$

and using Weyl's shift (6.8) we get

$$|S_N|^2 \leqslant \frac{N+H}{(H+1)^2} \sum_{n \in \mathbb{Z}} \sum_{|h| \leqslant H} \sum_{j=1-h}^{H+1-h} \alpha_{n+h} \overline{\alpha_n}$$

$$= \frac{N+H}{(H+1)^2} \sum_{|h|\leqslant H} \sum_{N<n\leqslant 2N-h} (H+1-|h|)\alpha_{n+h}\overline{\alpha_n}$$

$$= \frac{N+H}{H+1} \sum_{|h|\leqslant H} \left(1 - \frac{|h|}{H+1}\right) \sum_{N<n\leqslant 2N-h} \alpha_{n+h}\overline{\alpha_n}.$$

We then separate the case $h=0$ from the other values, which gives

$$|S_N|^2 \leqslant \frac{N+H}{H+1} \sum_{N<n\leqslant 2N} |\alpha_n|^2 + \frac{N+H}{H+1} \sum_{\substack{|h|\leqslant H \\ h\neq 0}} \left(1 - \frac{|h|}{H+1}\right) \sum_{N<n\leqslant 2N-h} \alpha_{n+h}\overline{\alpha_n}$$

$$= \frac{N+H}{H+1} \sum_{N<n\leqslant 2N} |\alpha_n|^2 + \frac{N+H}{H+1} \left\{ \sum_{h=1}^{H} \left(1 - \frac{h}{H+1}\right) \sum_{N<n\leqslant 2N-h} \alpha_{n+h}\overline{\alpha_n} \right.$$

$$\left. + \sum_{h=1}^{H} \left(1 - \frac{h}{H+1}\right) \sum_{N<n\leqslant 2N+h} \alpha_{n-h}\overline{\alpha_n} \right\}$$

and the use of Weyl's shift (6.9) again on the last sum gives

$$|S_N|^2 \leqslant \frac{N+H}{H+1} \sum_{N<n\leqslant 2N} |\alpha_n|^2 + \frac{N+H}{H+1} \left\{ \sum_{h=1}^{H} \left(1 - \frac{h}{H+1}\right) \sum_{N<n\leqslant 2N-h} \alpha_{n+h}\overline{\alpha_n} \right.$$

$$\left. + \sum_{h=1}^{H} \left(1 - \frac{h}{H+1}\right) \sum_{N<n\leqslant 2N-h} \alpha_n\overline{\alpha_{n+h}} \right\}$$

$$= \frac{N+H}{H+1} \sum_{N<n\leqslant 2N} |\alpha_n|^2$$

$$+ \frac{2(N+H)}{H+1} \mathrm{Re}\left\{ \sum_{h=1}^{H} \left(1 - \frac{h}{H+1}\right) \sum_{N<n\leqslant 2N-h} \alpha_{n+h}\overline{\alpha_n} \right\}$$

$$\leqslant \frac{N+H}{H+1} \sum_{N<n\leqslant 2N} |a_n|^2 + \frac{2(N+H)}{H+1} \sum_{h=1}^{H} \left| \sum_{N<n\leqslant 2N-h} a_{n+h}\overline{a_n} \right|$$

and the bound $H\leqslant N$ gives the asserted result.   □

6.4.3 Main Results

The A-process will allow us to work with a function of the same order of magnitude of the derivative of the initial function. Applying van der Corput's inequality from

Theorem 6.9 to this function may then give a criterion depending on the order of magnitude of the third derivative of the initial function. More precisely, we will show the following result.

Theorem 6.19 (Third derivative criterion) *Let* $N \in \mathbb{N}$ *and* $f \in C^3[N, 2N]$ *such that there exist real numbers* $c_2, c_3 \geqslant 1$ *and* $\lambda_2, \lambda_3, s_2 > 0$ *such that, for all* $x \in [N, 2N]$, *we have*

$$\lambda_2 \leqslant |f''(x)| \leqslant c_2\lambda_2, \quad \lambda_3 \leqslant |f'''(x)| \leqslant c_3\lambda_3 \quad and \quad \lambda_2 = s_2N\lambda_3. \tag{6.10}$$

We set $c = 4\pi^{-1/2}\{c_3 + 2s_2(c_2 + 1)\}$ *and suppose that* $(\pi/8)s_2c^{-2} \leqslant 1$. *Then we have*

$$\left| \sum_{N < n \leqslant 2N} e(f(n)) \right| \leqslant 6^{1/2}c^{1/3}N\lambda_3^{1/6} + 8^{1/2}c^{1/3}(s_2\pi)^{-1/2}\lambda_3^{-1/3} \log^{1/2}(eN).$$

Proof We set $S_N = \sum_{N < n \leqslant 2N} e(f(n))$. We first notice that if $\lambda_3 \geqslant c^{-2}$, then $c^{1/3}N\lambda_3^{1/6} \geqslant N$ and if $\lambda_3 \leqslant c^{-2}N^{-3}$, then

$$8^{1/2}c^{1/3}(s_2\pi)^{-1/2}\lambda_3^{-1/3} \geqslant \left(8\pi^{-1}c^2s_2^{-1}\right)^{1/2}N \geqslant N$$

because of $(\pi/8)s_2c^{-2} \leqslant 1$. Therefore we may suppose that

$$c^{-2}N^{-3} < \lambda_3 < c^{-2}. \tag{6.11}$$

By van der Corput's A-process applied with $a_n = e(f(n))$, for all integers $H \in \{1, \ldots, N\}$, we have

$$|S_N|^2 \leqslant \frac{2N^2}{H+1} + \frac{4N}{H+1}\sum_{h=1}^{H}\left| \sum_{N < n \leqslant 2N-h} e(\Delta_h f(n)) \right|$$

where we set $\Delta_h f(n) = f(n+h) - f(n)$. The conditions (6.10) imply that

$$\Lambda_1 \leqslant |(\Delta_h f)'(x)| \leqslant c_2\Lambda_1, \quad \Lambda_2 \leqslant |(\Delta_h f)''(x)| \leqslant c_3\Lambda_2 \quad and \quad \Lambda_1 = s_2N\Lambda_2$$

with $\Lambda_j = h\lambda_{j+1}$ for $j \in \{1, 2\}$. We may apply Corollary 6.12 which gives

$$|S_N|^2 \leqslant \frac{2N^2}{H+1} + \frac{4N}{H+1}\sum_{h=1}^{H}\left(cN(h\lambda_3)^{1/2} + \frac{2}{\pi h\lambda_2}\right)$$

$$\leqslant \frac{2N^2}{H+1} + 4cN^2(H\lambda_3)^{1/2} + \frac{8\log eN}{\pi s_2(H+1)\lambda_3}$$

where we used $\lambda_2 = s_2N\lambda_3$ and the bound $H \leqslant N$. Now the choice of $H = [(c^2\lambda_3)^{-1/3}]$ gives the asserted result. Note that the conditions (6.11) ensure that $1 \leqslant H \leqslant N$. $\square$

As for Corollary 6.12, we may improve on the secondary term in Theorem 6.19 if we suppose that f is sufficiently smooth, a condition that appears frequently in applications.

Corollary 6.20 *Let N be a positive integer and $f \in C^3[N, 2N]$ such that there exist real numbers $c_1, c_2, c_3 \geqslant 1$ and $\lambda_1, \lambda_2, \lambda_3, s_1, s_2 > 0$ such that, for all $x \in [N, 2N]$, we have*

$$\lambda_j \leqslant |f^{(j)}(x)| \leqslant c_j \lambda_j \quad (j = 1, 2, 3) \quad and \quad \lambda_j = s_j N \lambda_{j+1} \quad (j = 1, 2). \quad (6.12)$$

We set

$$c = 4\pi^{-1/2}\{c_3 + 2s_2(c_2 + 1)\},$$

$$\kappa_1 = 4\pi^{-1/2}\{c_2(s_1(c_1 + 1))^{-1/2} + 2(s_1(c_1 + 1))^{1/2}\},$$

$$\kappa_2 = 6^{1/2}c^{1/3} + 8^{1/2}c^{1/3}(s_1\pi^{-1}(c_1 + 1))^{1/2}$$

and $\kappa = \max(\kappa_1, \kappa_2)$. Suppose that $(\pi/8)s_2 c^{-2} \leqslant 1$ and $N \geqslant s_1 s_2(c_1 + 1)\log^3(eN)$. Then we have

$$\left| \sum_{N < n \leqslant 2N} e(f(n)) \right| \leqslant \kappa N\lambda_3^{1/6} + \frac{2}{\pi \lambda_1}.$$

Proof If $\lambda_1 \leqslant (c_1 + 1)^{-1}$, then we apply Corollary 6.7. If

$$(c_1 + 1)^{-1} < \lambda_1 \leqslant (c_1 + 1)^{-1}\log(eN)$$

then we apply Corollary 6.12. In view of $\lambda_2 = s_1^{-1}N^{-1}\lambda_1$, we have

$$(s_1(c_1 + 1))^{-1}N^{-1} < \lambda_2 \leqslant (s_1(c_1 + 1))^{-1}N^{-1}\log(eN)$$

so that

$$\left| \sum_{N < n \leqslant 2N} e(f(n)) \right| \leqslant (4\pi^{-1/2}\{c_2 + 2s_1(c_1 + 1)\})N\lambda_2^{1/2} + \frac{2}{\pi \lambda_1}$$

$$\leqslant \kappa_1(N\log eN)^{1/2} + \frac{2}{\pi \lambda_1}.$$

But we also have $N^3\lambda_3 = (s_1 s_2)^{-1}N\lambda_1 > (s_1 s_2(c_1 + 1))^{-1}N \geqslant \log^3(eN)$ so that

$$(N\log eN)^{1/2} \leqslant N\lambda_3^{1/6}.$$

Finally, if $\lambda_1 > (c_1 + 1)^{-1}\log(eN)$, then the equality $\lambda_1 = s_1 s_2 N^2 \lambda_3$ implies that

$$\lambda_3^{-1/3}\log^{1/2}(eN) \leqslant (s_1 s_2(c_1 + 1))^{1/2}N\lambda_3^{1/6}$$

and we use Theorem 6.19 to conclude the proof. $\square$

Example 6.21 We will apply this result to the Dirichlet divisor problem. As in Example 6.13, we first split the sum (6.4) into the ranges $[1, 2x^{1/4}]$ and $]2x^{1/4}, x^{1/2}]$, estimate the sum trivially in the first one and use the splitting argument seen in Example 6.8 for the second sum. We get for all integers $H \geqslant 1$

$$|\Delta(x)| \leqslant 2x^{1/4} + \max_{2x^{1/4} < N \leqslant \sqrt{x}} \left\{ \frac{N}{2H+2} + 2\sum_{h=1}^{H} \frac{1}{h} \left| \sum_{N < n \leqslant 2N} e\left(\frac{hx}{n}\right) \right| \right\} \frac{\log x}{\log 4} + \frac{1}{2}.$$

Corollary 6.20 may be applied with $\lambda_1 = hx(4N^2)^{-1}$, $c_1 = 4$, $\lambda_2 = hx(4N^3)^{-1}$, $c_2 = 8$, $s_1 = 1$, $\lambda_3 = 3hx(8N^4)^{-1}$, $c_3 = 16$, $s_2 = 2/3$, so that

$$c = 112\pi^{-1/2}, \qquad \kappa_1 = 72(5\pi)^{-1/2} \approx 18.1665\ldots$$

and

$$\kappa_2 = 2^{11/6}(7\pi^{-2})^{1/3}(\sqrt{3\pi} + 2\sqrt{5}) \approx 23.96\ldots$$

which gives for all $N \geqslant 2200$

$$\left| \sum_{N < n \leqslant 2N} e\left(\frac{hx}{n}\right) \right| < 21(hxN^2)^{1/6} + \frac{8N^2}{\pi hx}$$

so that

$$|\Delta(x)| \leqslant 2x^{1/4} + \max_{2x^{1/4} < N \leqslant \sqrt{x}} \left\{ \frac{N}{2H+2} + 252(HxN^2)^{1/6} + \frac{8\pi N^2}{3x} \right\} \frac{\log x}{\log 4} + \frac{1}{2}.$$

The choice of $H = [(N^4 x^{-1})^{1/7}]$ then gives

$$|\Delta(x)| < 3x^{1/4} + \max_{2x^{1/4} < N \leqslant \sqrt{x}} \left\{ 183(xN^3)^{1/7} + \frac{7N^2}{x} \right\} \log x < 184\, x^{5/14} \log x$$

for $x \geqslant 2200^4$, and hence Corollary 6.20 is worse than van der Corput's inequality in this problem if we use it in the whole range $]x^{1/4}, x^{1/2}]$. In fact, the function $N \longmapsto (xN^3)^{1/7}$ gives a better result as long as $x^{1/4} < N < x^{4/9}$. For instance, with the help of the computations above, we infer that, for x sufficiently large, we have

$$\sum_{n \leqslant x^{2/5}} \psi\left(\frac{x}{n}\right) \ll x^{11/35} \log x$$

which is a good result since Corollary 6.12 only gives

$$\sum_{n \leqslant x^{2/5}} \psi\left(\frac{x}{n}\right) \ll x^{1/3} \log x$$

but Corollary 6.20 does not allow us to recover the whole range of summation.

6.5 Applications

We will use the previous results to give explicit upper bounds for sums of the ψ-function. Van der Corput's inequality implies the first following estimate.

Corollary 6.22 *Let N be a positive integer and $f \in C^2[N, 2N]$ such that there exist real numbers $c_1, c_2 \geqslant 1$ and $\lambda_1, \lambda_2, s_1 > 0$ such that, for all $x \in [N, 2N]$, we have*

$$\lambda_j \leqslant |f^{(j)}(x)| \leqslant c_j \lambda_j \quad (j = 1, 2) \quad and \quad \lambda_1 = s_1 N \lambda_2. \qquad (6.13)$$

Set c as in Corollary 6.12 and $\kappa = \frac{3}{2}(2c(1 + \pi^{-1}))^{2/3}$. Then

$$\left| \sum_{N < n \leqslant 2N} \psi(f(n)) \right| < \kappa N \lambda_2^{1/3} + 2\lambda_1^{-1}.$$

Proof If $\lambda_2 \geqslant \kappa^{-3}$, then $\kappa N \lambda_2^{1/3} \geqslant N$, so that we may suppose

$$0 < \lambda_2 < \kappa^{-3}. \qquad (6.14)$$

Using Corollaries 6.2 and 6.12 we get

$$\left| \sum_{N < n \leqslant 2N} \psi(f(n)) \right| \leqslant \frac{N}{2H + 2} + \left(1 + \frac{1}{\pi}\right) \sum_{h=1}^{H} \frac{1}{h} \left| \sum_{N < n \leqslant 2N} e(hf(n)) \right|$$

$$\leqslant \frac{N}{2H + 2} + \left(1 + \frac{1}{\pi}\right) \left(c N \lambda_2^{1/2} \sum_{h=1}^{H} \frac{1}{h^{1/2}} + \frac{2}{\pi \lambda_1} \sum_{h=1}^{H} \frac{1}{h^2} \right)$$

$$< \frac{N}{2H + 2} + \left(\frac{2\kappa}{3}\right)^{3/2} N (H\lambda_2)^{1/2} + \frac{2}{\lambda_1}$$

where we used $\sum_{h=1}^{H} h^{-2} \leqslant \sum_{h=1}^{\infty} h^{-2} = \zeta(2) = \pi^2/6$. The choice of $H = [3(2\kappa)^{-1} \lambda_2^{-1/3}]$ gives the desired result. Note that hypothesis (6.14) implies that $H \geqslant 1$. □

The third derivative theorem implies the following result.

Corollary 6.23 *Let $N \in \mathbb{N}$ and $f \in C^3[N, 2N]$ such that there exist real numbers $c_1, c_2, c_3 \geqslant 1$ and $\lambda_1, \lambda_2, \lambda_3, s_1, s_2 > 0$ such that, for all $x \in [N, 2N]$, we have*

$$\lambda_j \leqslant |f^{(j)}(x)| \leqslant c_j \lambda_j \quad (j = 1, 2, 3) \quad and \quad \lambda_j = s_j N \lambda_{j+1} \quad (j = 1, 2). \quad (6.15)$$

Set c and κ as in Corollary 6.20, $v = \frac{7}{5}(2\kappa(1 + \pi^{-1}))^{6/7}$ and suppose that $(\pi/8)s_2 c^{-2} \leqslant 1$ and $N \geqslant s_1 s_2 (c_1 + 1) \log^3(eN)$. Then

$$\left| \sum_{N < n \leqslant 2N} \psi(f(n)) \right| < v N \lambda_3^{1/7} + 2\lambda_1^{-1}.$$

Proof The proof is exactly the same as in Corollary 6.22, except that we use Corollary 6.20 instead of Corollary 6.12. We leave the details to the reader.   □

6.6 Further Developments

6.6.1 The mth Derivative Theorem

Theorems 6.9 and 6.19 can be generalized to functions having derivatives of higher orders. More precisely, one can prove by induction the following result.

Theorem 6.24 (*m*th derivative test) *Let $m \geqslant 2$ be an integer, $N < N_1 \leqslant 2N$ be positive integers and $f \in C^m[N, N_1]$ such that there exists a real number $\lambda_m > 0$ such that, for all $x \in [N, N_1]$, we have*

$$\left| f^{(m)}(x) \right| \asymp \lambda_m.$$

Then we have

$$\sum_{N < n \leqslant N_1} e\big(f(n)\big) \ll N\lambda_m^{1/(2^m - 2)} + N^{1 - 2^{2-m}} \lambda_m^{-1/(2^m - 2)}.$$

As in Corollaries 6.12 and 6.20, one may improve on the secondary term if f is sufficiently smooth with the help of Kusmin–Landau's inequality.

Corollary 6.25 (Improved *m*th derivative test) *Let $m \geqslant 2$, $N \geqslant 1$ be integers and $f \in C^m[N, 2N]$ such that there exist real numbers $T \geqslant 1$ and $1 \leqslant c_0 \leqslant \cdots \leqslant c_m$ such that $T \geqslant N$ and, for all $x \in [N, 2N]$ and all $j \in \{0, \ldots, m\}$, we have*

$$\frac{T}{N^j} \leqslant \left| f^{(j)}(x) \right| \leqslant c_j \frac{T}{N^j}. \tag{6.16}$$

Then we have

$$\sum_{N < n \leqslant 2N} e\big(f(n)\big) \ll c_m^{4/2^m}\, T^{1/(2^m - 2)} N^{1 - m/(2^m - 2)}$$

where the implied constants are absolute.

For instance, using the same method as in Examples 6.13 and 6.21, we get from Corollary 6.25 used with $m = 4$ the following estimate

$$\sum_{N < n \leqslant 2N} \psi\left(\frac{x}{n}\right) \ll x^{1/15} N^{2/3}$$

for all $N \leqslant \sqrt{x}$, which is worse than Corollary 6.20. Hence we need a new idea to get further improvements. This is the aim of the next section.

6.6.2 Van der Corput's B-Process

The integral analogues of Theorem 6.5 and Lemma 6.11 exist and have also been proved by van der Corput. We put them together for the sake of clarity.

Lemma 6.26 (1st and 2nd derivative tests for integrals) *Let f be a real-valued function defined on an interval $[a, b]$.*

(i) *Suppose $f \in C^1[a, b]$ such that f' is monotone and there exists $\lambda_1 > 0$ such that, for all $x \in [a, b]$, we have $f'(x) \geqslant \lambda_1$. Then*

$$\left| \int_a^b e\big(f(x)\big)\,dx \right| \leqslant \frac{2}{\pi \lambda_1}.$$

(ii) *Suppose $f \in C^2[a, b]$ such that there exists $\lambda_2 > 0$ such that, for all $x \in [a, b]$, we have $f''(x) \geqslant \lambda_2$. Then*

$$\left| \int_a^b e\big(f(x)\big)\,dx \right| \leqslant \frac{4\sqrt{2}}{\sqrt{\pi \lambda_2}}.$$

Proof

(i) We use the following result due to Ostrowski (see [Bul98] for instance).

> Let F be a real-valued monotone integrable function on $[a, b]$ and G be a complex-valued integrable function on $[a, b]$. Then we have
>
> $$\left| \int_a^b F(x)G(x)\,dx \right| \leqslant |F(a)| \max_{a \leqslant t \leqslant b} \left| \int_a^t G(x)\,dx \right| + |F(b)| \max_{a \leqslant t \leqslant b} \left| \int_t^b G(x)\,dx \right|.$$

Applying this with $F(x) = 1/(2\pi f'(x))$ which is monotone and $G(x) = 2\pi f'(x) e(f(x))$, we get

$$\left| \int_a^b e\big(f(x)\big)\,dx \right| \leqslant \frac{1}{2\pi \lambda_1} \left(\max_{a \leqslant t \leqslant b} \left| \big[ie\big(f(x)\big)\big]_a^t \right| + \max_{a \leqslant t \leqslant b} \left| \big[ie\big(f(x)\big)\big]_t^b \right| \right)$$

$$\leqslant \frac{1}{2\pi \lambda_1} \times 4 = \frac{2}{\pi \lambda_1}.$$

(ii) Let $\delta > 0$ be a parameter to be chosen later and write $[a, b] = E \cup F$ where $E = \{x : |f'(x)| < \delta\}$ and $F = \{x : |f'(x)| \geqslant \delta\}$. Since $f'' > 0$, the set E consists of at most one interval and F consists of at most two intervals. If $E = [c, d]$, then $(d - c)\lambda_2 \leqslant |f'(d) - f'(c)| \leqslant 2\delta$ so that

$$\left| \int_E e\big(f(x)\big)\,dx \right| \leqslant d - c \leqslant \frac{2\delta}{\lambda_2}$$

and using (i) we have

$$\left| \int_F e\big(f(x)\big) \, dx \right| \leqslant \frac{4}{\pi \delta}.$$

Therefore

$$\left| \int_a^b e\big(f(x)\big) \, dx \right| \leqslant \frac{2\delta}{\lambda_2} + \frac{4}{\pi \delta}$$

and the choice of $\delta = (2\pi^{-1}\lambda_2)^{1/2}$ gives the desired result.

The proof is complete.   □

Besides proving the A-process (Lemma 6.18), van der Corput provided another tool for estimating exponential sums called the B-process. This result relies on a truncated Poisson summation formula and the method of the stationary phase. The Poisson formula, which is a useful tool in harmonic analysis, allows us to replace an exponential sum by an integral, which may be estimated by the stationary phase. We first state without proof one of the many versions of this theorem (see [MV07] for instance).

Lemma 6.27 (Poisson summation) *Let f be a continuous function on $\mathbb{R}$ such that $\int_{\mathbb{R}} |f(x)| \, dx$ exists and is finite. Then we have*

$$\sum_{k \in \mathbb{Z}} f(k) = \lim_{K \to \infty} \sum_{|k| \leqslant K} \widehat{f}(k)$$

where $\widehat{f}(t) = \int_{\mathbb{R}} f(x) e(-tx) \, dx$.

For our purpose, it will be more useful if we have at our disposal a truncated version of Lemma 6.27. We state here the following two results without proof (see [Mon94]).

Lemma 6.28 (Truncated Poisson summation) *Let $f \in C^1[a,b]$ such that f' is increasing on $[a,b]$ and put $\alpha = f'(a)$ and $\beta = f'(b)$. Then we have*

$$\sum_{a \leqslant n \leqslant b} e\big(f(n)\big) = \sum_{\alpha - 1 \leqslant k \leqslant \beta + 1} \int_a^b e\big(f(x) - kx\big) \, dx + O\big(\log(\beta - \alpha + 2)\big).$$

The range of summation in the sum of the right-hand side may be restricted to $[\alpha - \frac{1}{4}, \beta + \frac{1}{4}]$, say, without changing the error-term (see [Hux96]). Lemmas 6.26 and 6.28 allow us to recover the van der Corput inequality. Indeed, using

Lemma 6.26 (ii) in the integral above, we get with the hypothesis of Theorem 6.9

$$\sum_{N < n \leqslant 2N} e\big(f(n)\big) \ll \big(f'(2N) - f'(N) + 1\big)\lambda_2^{-1/2} + \log\big(f'(2N) - f'(N) + 2\big)$$

$$\ll N\lambda_2^{1/2} + \lambda_2^{-1/2} + \log(N\lambda_2 + 2) \ll N\lambda_2^{1/2} + \lambda_2^{-1/2}$$

since $N\lambda_2^{1/2} + \lambda_2^{-1/2} \geqslant 2N^{1/2} \gg \log(N\lambda_2 + 2)$ if N is sufficiently large.

The next result is an effective version of the method of the stationary phase.

Lemma 6.29 (Stationary phase) *Let $g \in C^4[a, b]$ and suppose that there exists $x_0 \in$ $]a, b[$ such that $g'(x_0) = 0$. We also suppose that there exist $\lambda_2, \lambda_3, \lambda_4 > 0$ such that, for all $x \in [a, b]$, we have*

$$g''(x) \geqslant \lambda_2, \quad |g'''(x)| \leqslant \lambda_3 \quad and \quad |g^{(4)}(x)| \leqslant \lambda_4.$$

Then we have

$$\int_a^b e\big(g(x)\big)\,dx = \frac{e(g(x_0) + 1/8)}{\sqrt{g''(x_0)}} + O\big(E_1(x_0) + E_2\big)$$

where

$$E_1(x_0) = \min\left(\frac{1}{\lambda_2(x_0 - a)}, \lambda_2^{-1/2}\right) + \min\left(\frac{1}{\lambda_2(b - x_0)}, \lambda_2^{-1/2}\right),$$

$$E_2 = (b - a)\big\{\lambda_2^{-2}\lambda_4 + \lambda_2^{-3}\lambda_3^2\big\}.$$

Combining Lemmas 6.28 and 6.29 gives van der Corput's B-process.

Theorem 6.30 (B-process) *Let $N \in \mathbb{N}$, $a \leqslant b \leqslant a + N$ and $f \in C^4[a, b]$ such that there exist $c_2 \geqslant 1$ and $\lambda_2, \lambda_3, \lambda_4 > 0$ such that, for all $x \in [a, b]$, we have*

$$\lambda_2 \leqslant f''(x) \leqslant c_2\lambda_2, \quad |f'''(x)| \leqslant \lambda_3 \quad and \quad |f^{(4)}(x)| \leqslant \lambda_4.$$

We set $\alpha = f'(a)$ and $\beta = f'(b)$ and for each integer $k \in [\alpha, \beta]$, let x_k be defined by $f'(x_k) = k$. Then we have

$$\sum_{a \leqslant n \leqslant b} e\big(f(n)\big) = \sum_{\alpha \leqslant k \leqslant \beta} \frac{e(f(x_k) - kx_k + 1/8)}{\sqrt{f''(x_k)}} + O(R_1 + R_2)$$

where

$$R_1 = \lambda_2^{-1/2} + \log(N\lambda_2 + 2) \quad and \quad R_2 = N^2\big\{\lambda_2^{-1}\lambda_4 + \lambda_2^{-2}\lambda_3^2\big\}.$$

Proof If $\lambda_2 \leqslant N^{-2}$, then $R_1 \gg N$, so that we may suppose that $\lambda_2 > N^{-2}$. By Lemma 6.26 we have as above

$$\int_a^b e(f(x) - kx)\,dx \ll \lambda_2^{-1/2} \tag{6.17}$$

uniformly in k and $\beta - \alpha = f'(b) - f'(a) \leqslant c_2 N \lambda_2$ so that by Lemma 6.28 we get

$$\sum_{a \leqslant n \leqslant b} e(f(n)) = \sum_{\alpha \leqslant k \leqslant \beta} \int_a^b e(f(x) - kx)\,dx + O(R_1).$$

If $\beta - \alpha \leqslant 1$, then the theorem is proved by (6.17). If $\beta - \alpha > 1$, then the sum on the right-hand side is non-empty. We may use Lemma 6.29 to estimate each integral, which gives

$$\sum_{a \leqslant n \leqslant b} e(f(n)) = \sum_{\alpha \leqslant k \leqslant \beta} \frac{e(f(x_k) - kx_k + 1/8)}{\sqrt{f''(x_k)}} + O\left\{ \sum_{\alpha \leqslant k \leqslant \beta} (E_1(x_k) + E_2) \right\}$$
$$+ O(R_1).$$

Now $k - \alpha = f'(x_k) - f'(a) \leqslant c_2 \lambda_2 (x_k - a)$ and similarly $\beta - k \leqslant c_2 \lambda_2 (b - x_k)$ so that

$$E_1(x_k) \leqslant \min\left(\frac{c_2}{k - \alpha}, \lambda_2^{-1/2} \right) + \min\left(\frac{c_2}{\beta - k}, \lambda_2^{-1/2} \right).$$

Furthermore, since $\sum_{\alpha \leqslant k \leqslant \beta} 1 = [\beta - \alpha]$ or $[\beta - \alpha] + 1$ and since $\beta - \alpha > 1$, we have

$$\frac{N\lambda_2}{2} \leqslant \frac{\beta - \alpha}{2} \leqslant \sum_{\alpha \leqslant k \leqslant \beta} 1 \leqslant \beta - \alpha + 1 \leqslant 2c_2 N \lambda_2$$

and therefore

$$\sum_{\alpha \leqslant k \leqslant \beta} (E_1(x_k) + E_2) \ll \lambda_2^{-1/2} + \sum_{\alpha + 1 \leqslant k \leqslant \beta} \frac{1}{k - \alpha} + \sum_{\alpha \leqslant k \leqslant \beta - 1} \frac{1}{\beta - k} + \sum_{\alpha \leqslant k \leqslant \beta} E_2$$
$$\ll \lambda_2^{-1/2} + \log(\beta - \alpha) + N\lambda_2 E_2 \ll R_1 + R_2$$

as required.   □

Note that if we estimate trivially the exponential sum in Theorem 6.30, we recover Theorem 6.9. Thus, van der Corput's inequality follows from the *B*-process and the triangle inequality.

One may state the *B*-process in another form if we introduce the function s_f defined by

$$s_f(y) = f \circ (f')^{-1}(y) - y(f')^{-1}(y) \tag{6.18}$$

for all functions f sufficiently smooth and monotone on $[a, b]$ and all $y \in [\alpha, \beta]$, where $\alpha = f'(a)$ and $\beta = f'(b)$. With this notation, it is easy to see that, for all integers k, we have $s_f(k) = f(x_k) - kx_k$ where x_k is defined in Theorem 6.30, so that van der Corput's B-process may be rewritten in the following form.

Corollary 6.31 (*B-process version 2*) *Let $N \in \mathbb{N}$, $a \leqslant b \leqslant a + N$ and $f \in C^4[a, b]$ satisfying the hypotheses of Theorem 6.30. We set $\alpha = f'(a)$ and $\beta = f'(b)$ and for each integer $k \in [\alpha, \beta]$, let x_k be defined by $f'(x_k) = k$. The function s_f being defined in (6.18), we have*

$$\sum_{a \leqslant n \leqslant b} e\big(f(n)\big) = e\left(\frac{1}{8}\right) \sum_{\alpha \leqslant k \leqslant \beta} \frac{e(s_f(k))}{\sqrt{f'' \circ (f')^{-1}(k)}} + O(R_1 + R_2)$$

where the remainder terms R_i are defined in Theorem 6.30.

It could be interesting to have some properties of the function s_f. For instance, one can see that, if $f \in C^k[a, b]$ for $k \in \{1, 2, 3\}$ and f' is monotone, then $s_f \in C^k[\alpha, \beta]$ and if $f'' \neq 0$ when $k \geqslant 2$, we have

$$(s_f)'(y) = -\big(f'\big)^{-1}(y), \quad (s_f)''(y) = \frac{-1}{f'' \circ (f')^{-1}(y)} \quad \text{and}$$

$$(s_f)'''(y) = \frac{f''' \circ (f')^{-1}(y)}{(f'' \circ (f')^{-1}(y))^3}.$$

6.6.3 Exponent Pairs

The A- and B-processes can be systematized in an algorithmic type procedure which was developed by van der Corput and later simplified by Phillips. To make things more accurate, we start by giving a definition of a class of function on which the process may apply. In practice, this class contains almost all functions that we may encounter in the usual problems of number theory. Roughly speaking, we need to work with functions satisfying hypotheses of the type (6.16) and possibly the logarithm function for the problem of the Riemann-zeta function in the critical strip.

Definition 6.32 Let $N \geqslant 1$ and $r \geqslant 2$ be integers, $T \geqslant 1$, $\sigma > -1$ and $0 < \varepsilon < \frac{1}{2}$ be real numbers. A function f belongs to the class $\mathcal{F}(T, N; r, \sigma, \varepsilon)$ if $f \in C^r(\mathcal{I})$ such that $\mathcal{I} \subseteq [N, 2N]$ and, for all $x \in \mathcal{I}$, we have

$$f(x) = \begin{cases} \pm \dfrac{T N^\sigma}{x^\sigma} + u(x), & \text{if } \sigma \neq 0, \\[2mm] \pm T \log x + u(x), & \text{if } \sigma = 0 \end{cases}$$

where $u \in C^r(\mathcal{I})$ is a *remainder function* satisfying

$$|u^{(j)}(x)| \leqslant \begin{cases} \varepsilon(\sigma)_j\, TN^{-j}, & \text{if } \sigma \neq 0, \\ \varepsilon(j-1)!\, TN^{-j}, & \text{if } \sigma = 0 \end{cases}$$

for all $x \in \mathcal{I}$ and $j \in \{1, \ldots, r\}$, and where the *Pochhammer symbol* $(\sigma)_j$ is defined by

$$(\sigma)_j = \begin{cases} 1, & \text{if } j = 0, \\ \sigma(\sigma+1)\cdots(\sigma+j-1), & \text{if } j \geqslant 1. \end{cases}$$

A function belonging to the class $\mathcal{F}(T, N; r, \sigma, \varepsilon)$ with $\sigma \neq 0$ is called a *monomial function*.

We summarize in the next result the basic properties of this class of functions.

Lemma 6.33 *Let $N \geqslant 1$ and $r \geqslant 2$ be integers, $T \geqslant 1$, $\sigma > -1$ and $\varepsilon > 0$ be real numbers.*

(i) *Let $f \in \mathcal{F}(T, N; r, \sigma, \varepsilon)$, $1 \leqslant h \leqslant \frac{2\varepsilon N}{\sigma+r}$ be an integer and define the function $\Delta_h f(x) = f(x+h) - f(x)$. Then*

$$\Delta_h f \in \mathcal{F}(hTN^{-1}, N; r-1, \sigma+1, 2\varepsilon).$$

(ii) *If $g_\sigma(x) = -T(Nx^{-1})^\sigma$ with $\sigma \neq 0$ or if $g_0(x) = T \log x$, then, for all $\varepsilon > 0$ and all integers $r \geqslant 2$, we have*

$$S_{g_\sigma} \in \mathcal{F}\left(T, TN^{-1}; r, -\frac{\sigma}{\sigma+1}, \varepsilon\right).$$

Proof We will make use of the following result which is a particular case of an inequality by Ostrowski (see [NP04] for instance).

Let $a < b$ be real numbers and $f \in C^1[a, b]$ such that there exists $\lambda_1 > 0$ such that, for all $x \in [a, b]$, we have $|f'(x)| \leqslant \lambda_1$. Then

$$\left| (b-a)f(a) - \int_a^b f(t)\,dt \right| \leqslant \frac{(b-a)^2 \lambda_1}{2}.$$

(i) Let $f \in \mathcal{F}(T, N; r, \sigma, \varepsilon)$ and we set $\mathbb{T} = hTN^{-1}$. We have

$$\Delta_h f(x) = \pm \frac{hTN^\sigma}{x^{\sigma+1}} + \mathcal{R}_{h,\sigma}(x) = \pm \frac{\mathbb{T}N^{\sigma+1}}{x^{\sigma+1}} + \mathcal{R}_{h,\sigma}(x)$$

where

$$\mathcal{R}_{h,\sigma}(x) = \pm TN^\sigma \left(\int_x^{x+h} \frac{dt}{t^{\sigma+1}} - \frac{h}{x^{\sigma+1}} \right) + \Delta_h u(x)$$

so that for all $j \in \{1, \ldots, r-1\}$ we have, using Ostrowski's inequality

$$\left| \mathcal{R}_{h,\sigma}^{(j)}(x) \right| \leqslant (\sigma+1)_j T N^\sigma \left| \int_x^{x+h} \frac{dt}{t^{\sigma+j+1}} - \frac{h}{x^{\sigma+j+1}} \right| + \int_x^{x+h} |u^{(j+1)}(t)| \, dt$$

$$\leqslant \frac{(\sigma+1)_j}{2} h T N^{-j-1} (\sigma+j+1) (hN^{-1}) + \varepsilon (\sigma+1)_j h T N^{-j-1}$$

$$\leqslant \varepsilon (\sigma+1)_j \, \mathbb{T} N^{-j} \left(\frac{\sigma+j+1}{\sigma+r} + 1 \right) \leqslant 2\varepsilon (\sigma+1)_j \, \mathbb{T} N^{-j}.$$

(ii) Easy computations from the definition (6.18) give

$$s_{g_\sigma}(y) = -\left(\sigma^{\frac{1}{\sigma+1}} + \sigma^{-\frac{\sigma}{\sigma+1}} \right) \frac{T (T N^{-1})^{-\frac{\sigma}{\sigma+1}}}{y^{-\frac{\sigma}{\sigma+1}}}$$

and

$$s_{g_0}(y) = -T \log(e y T^{-1}).$$

The proof is complete. □

We are now in a position to define the exponent pairs.

Definition 6.34 (Exponent pairs) Let k and l be real numbers satisfying

$$0 \leqslant k \leqslant \frac{1}{2} \leqslant l \leqslant 1.$$

The pair (k, l) is called an *exponent pair* if, for all $\sigma > -1$, there exist an integer $r \geqslant 2$ and $0 < \varepsilon < \frac{1}{2}$ such that, for all $N \in \mathbb{N}$, all real numbers $T \geqslant 1$ and all $f \in \mathcal{F}(T, N; r, \sigma, \varepsilon)$, the estimate

$$\sum_{a < n \leqslant b} e(f(n)) \ll T^k N^{l-k} + N T^{-1}$$

holds, with $[a, b] \subseteq [N, 2N]$.

Note that the pair $(0, 1)$ is an exponent pair by the triangle inequality and is called the *trivial exponent pair*. By Corollary 6.12, the pair $(\frac{1}{2}, \frac{1}{2})$ is an exponent pair, and by Corollary 6.20 the pair $(\frac{1}{6}, \frac{2}{3})$ is also an exponent pair. Furthermore, there is no need to ever take $l > 1$ in an exponent pair, otherwise we would get an estimate worse than the one provided by the triangle inequality. Similarly, there is no need to take $k > \frac{1}{2}$, for then (k, l) would provide a weaker result than the pair $(\frac{1}{2}, \frac{1}{2})$. More generally, the reader should refer to [GK91] where an interesting explanation is given about the hypotheses $0 \leqslant k \leqslant \frac{1}{2} \leqslant l \leqslant 1$ appearing in this definition.

The term $N T^{-1}$, which comes from Kusmin–Landau's inequality, is here to prevent the case where the order of magnitude of the function in the interval $[N, 2N]$ is very small compared to N. However, we shall mostly have $T \gg N$, so that NT^{-1} is absorbed by the main term $T^k N^{l-k}$. Thus, the exponent pair $(\frac{1}{2}, \frac{1}{2})$ tells us that the exponential sum is bounded by the square-root of the order of magnitude of the function in $[N, 2N]$, which is an interesting result. But it would be more efficient to have the bound $\sum_{a < n \leqslant b} e(f(n)) \ll N^{1/2+\varepsilon}$ instead of $T^{1/2}$. Unfortunately, this conjecture, called the *exponent pair conjecture*, still remains unproven. Note that this conjecture would solve Dirichlet's divisor problem.

Using Corollary 6.2 along with exponent pairs gives the following useful result.

Corollary 6.35 *Let $f \in \mathcal{F}(T, N; r, \sigma, \varepsilon)$ and (k, l) be an exponent pair. Then we have*

$$\sum_{a < n \leqslant b} \psi(f(n)) \ll (T^k N^l)^{\frac{1}{k+1}} + NT^{-1}$$

with $[a, b] \subseteq [N, 2N]$.

Proof If $N \leqslant T^{\frac{k}{1+k-l}}$, then we have

$$(T^k N^l)^{\frac{1}{k+1}} \geqslant N \geqslant b - a \geqslant \left| \sum_{a < n \leqslant b} \psi(f(n)) \right|$$

so that we may suppose that $N > T^{\frac{k}{1+k-l}}$. Using Corollary 6.2 and Definition 6.34, we get for any positive integer H

$$\sum_{a < n \leqslant b} \psi(f(n)) \ll NH^{-1} + (HT)^k N^{l-k} + NT^{-1}$$

and the choice of $H = [(T^{-k} N^{1+k-l})^{\frac{1}{k+1}}]$ gives the asserted result. $\square$

Lemma 6.33 implies that in a certain sense the exponent pairs are compatible with the A- and B-processes. As shown by the next result, this compatibility has the advantage of producing some new exponent pairs from older ones. For the proof, we refer the reader to [GK91, Mon94].

Theorem 6.36 (Exponent pairs by A- and B-processes) *Let (k, l) be an exponent pair. Then*

$$A(k, l) = \left(\frac{k}{2k+2}, \frac{1}{2} + \frac{l}{2k+2} \right) \quad and \quad B(k, l) = \left(l - \frac{1}{2}, k + \frac{1}{2} \right)$$

are exponent pairs.

This theorem shows the great usefulness of the exponent pairs, for one may combine the A- and B-processes successively to produce exponent pairs from the pair

(0, 1). For instance, setting A^j and B^j to symbolize the A- and B-processes applied j times respectively, we have

$$BA^3 (BA^2)^2 B(0, 1) = \left(\frac{97}{251}, \frac{132}{251} \right) \tag{6.19}$$

so that $(\frac{97}{251}, \frac{132}{251})$ is an exponent pair. Note that $B^2(k, l) = (k, l)$, which is not surprising since the Poisson summation formula is one of the ingredients of the B-process.

Using (6.19) in Corollary 6.35 gives the following improvement in the Dirichlet divisor problem.

$$\sum_{N < n \leqslant 2N} \psi \left(\frac{x}{n} \right) \ll x^{97/348} N^{35/348} + N^2 x^{-1}$$

so that, for x sufficiently large, we have

$$\sum_{n \leqslant \sqrt{x}} \psi \left(\frac{x}{n} \right) \ll x^{229/696} \log x.$$

Note that $\frac{229}{696} \approx 0.329\,022\,988 \ldots$ Compared to the conjectured value $1/4$, this estimate represents a saving of a little more than 5% in comparison to the bound provided by van der Corput's inequality (Example 6.13).

The search for the best exponent pair to a given estimate implies a difficult optimization problem which was investigated by Graham and Kolesnik in [GK91]. For the Dirichlet divisor problem, the authors proved that the best exponent accessible by this method is $0.329\,021\,356\,85 \ldots$ which is slightly better than the exponent above.

6.6.4 An Improved Third Derivative Theorem

The question whether the exponents $1/3$ and $1/7$ in Corollaries 6.22 and 6.23 can be increased arises naturally. Concerning the first one, a counter-example, given by Grekos [Gre88], shows that this exponent cannot be improved. In [RS03], the authors investigate the problem of the second exponent. Their idea is to estimate the following exponential sum with a parameter

$$\frac{1}{H} \sum_{H < h \leqslant 2H} \left| \sum_{n \in I_h} e \left(\frac{h}{H} f(n) \right) \right|$$

where H, N are large positive integers and I_h is any subinterval of $]N, 2N]$ which may depends on h for all $h \in [H, 2H]$. The non-trivial treatment of this sum rests on bounds of quadruple exponential sums and the use of the so-called *spacing lemmas*, which, in Huxley's terminology, are tools for estimating certain Diophantine systems. Their main result may be stated as follows.

Theorem 6.37 *Let H, N be large positive integers, I_h be any subinterval of $]N, 2N]$ which may depends on h for all $h \in [H, 2H]$ and $f \in C^3[N, 2N]$ such that there exists a real number $\lambda_3 > 0$ such that, for all $x \in [N, 2N]$ we have*

$$|f'''(x)| \asymp \lambda_3.$$

Then, for all $\varepsilon > 0$, we have

$$\frac{1}{H} \sum_{H < h \leqslant 2H} \left| \sum_{n \in I_h} e\left(\frac{h}{H} f(n)\right) \right| \ll N^{1+\varepsilon} \lambda_3^{1/6} H^{-1/9} + N^{1+\varepsilon} \lambda_3^{1/5} + N^{3/4+\varepsilon} + \lambda_3^{-1/3}.$$

We now use this tool along with Corollary 6.2 in the following way. Suppose first that $N^{-19/3} \leqslant \lambda_3 < 1$. Splitting the interval $[1, H]$ into $O(\log H)$ subintervals of the form $[H_1, 2H_1]$, we get

$$\sum_{N < n \leqslant 2N} \psi(f(n)) \ll \frac{N}{H} + \sum_{h=1}^{H} \frac{1}{h} \left| \sum_{N < n \leqslant 2N} e(hf(n)) \right|$$

$$\ll \frac{N}{H} + \max_{1 \leqslant H_1 \leqslant H} \sum_{H_1 < h \leqslant 2H_1} \frac{1}{h} \left| \sum_{N < n \leqslant 2N} e(hf(n)) \right| \log H$$

$$\ll \frac{N}{H} + \max_{1 \leqslant H_1 \leqslant H} \frac{1}{H_1} \sum_{H_1 < h \leqslant 2H_1} \left| \sum_{N < n \leqslant 2N} e(hf(n)) \right| \log H$$

$$\ll \frac{N}{H} + \max_{1 \leqslant H_1 \leqslant H} \left(N^{1+\varepsilon} (H_1 \lambda_3)^{1/6} H_1^{-1/9} + N^{1+\varepsilon} (H_1 \lambda_3)^{1/5} \right.$$
$$\left. + N^{3/4+\varepsilon} + (H_1 \lambda_3)^{-1/3} \right) \log H$$

$$\ll \frac{N}{H} + \left(N^{1+\varepsilon} (H \lambda_3^3)^{1/18} + N^{1+\varepsilon} (H \lambda_3)^{1/5} \right.$$
$$\left. + N^{3/4+\varepsilon} + \lambda_3^{-1/3} \right) \log H$$

and choosing $H = [\lambda_3^{-3/19}]$ gives

$$\sum_{N < n \leqslant 2N} \psi(f(n)) \ll N^{\varepsilon} \left(N \lambda_3^{3/19} + N^{3/4} + \lambda_3^{-1/3} \right)$$

the second term being absorbed by the first one, and $\log H \leqslant \log N \ll N^{\varepsilon}$ since $\lambda_3 \geqslant N^{-19/3}$. If $\lambda_3 \geqslant 1$, then $N \lambda_3^{3/19} \gg N$ and if $\lambda_3 < N^{-19/3}$, then $\lambda_3^{-1/3} > N^{19/9}$ so that we may state the following result.

Corollary 6.38 *Let $f \in C^3[N, 2N]$ such that there exists a real number $\lambda_3 > 0$ such that, for all $x \in [N, 2N]$ we have $|f'''(x)| \asymp \lambda_3$. Then, for all $\varepsilon > 0$, we have*

$$\sum_{N < n \leqslant 2N} \psi(f(n)) \ll N^\varepsilon \left(N \lambda_3^{3/19} + N^{3/4} + \lambda_3^{-1/3} \right).$$

Applied to the Dirichlet divisor problem, this result implies that

$$\sum_{N < n \leqslant 2N} \psi\left(\frac{x}{n}\right) \ll N^\varepsilon \left(x^{3/19} N^{7/19} + N^{3/4} + N^{4/3} x^{-1/3} \right)$$

so that, for x sufficiently large, we get

$$\sum_{n \leqslant x^{2/5}} \psi\left(\frac{x}{n}\right) \ll x^{29/95 + \varepsilon}$$

which is better than the previous one obtained by the third derivative test, but once again this result does not allow us to recover the whole range of summation.

6.6.5 Double Exponential Sums

In [Vor04], Voronoï derived a very elegant formula for the remainder term $\Delta_k(x)$ of the Dirichlet–Piltz divisor problem. Later, some mathematicians obtained a truncated version of this formula which may be written in the case $k = 2$ as

$$\Delta(x) \ll x^{1/4} \left| \sum_{n \leqslant R} \frac{\tau(n)}{n^{3/4}} e(2\sqrt{nx}) \right| + x^{1/2 + \varepsilon} R^{-1/2} + x^\varepsilon$$

for all $\varepsilon > 0$, where $R \geqslant 1$ is a parameter at our disposal such that $R \ll x^A$ for some $A > 0$. Using the fact that $\tau(n) = \sum_{d \mid n} 1$ we get

$$\Delta(x) \ll x^{1/4} \left| \sum_{d \leqslant R} \sum_{k \leqslant R/d} \frac{e(2\sqrt{kdx})}{(kd)^{3/4}} \right| + x^{1/2 + \varepsilon} R^{-1/2} + x^\varepsilon.$$

Splitting the domain $d \leqslant R$, $kd \leqslant R$ into $O(\log^2 R)$ subdomains of the form $[M, 2M] \times [N, 2N]$ with M, N positive integers such that $MN \asymp R$ gives

$$\Delta(x) \ll x^{1/4} \max_{\substack{M \leqslant R, N \leqslant R \\ MN \asymp R}} \left| \sum_{M \leqslant m \leqslant 2M} \sum_{N \leqslant n \leqslant 2N} \frac{e(2\sqrt{mnx})}{(mn)^{3/4}} \right| \log^2 R + x^{1/2 + \varepsilon} R^{-1/2} + x^\varepsilon$$

and using Abel's summation we get

$$\Delta(x) \ll x^{1/4} R^{-3/4} \max_{\substack{M, N \leqslant R \\ MN \asymp R}} \max_{\substack{M \leqslant M_1 \leqslant 2M \\ N \leqslant N_1 \leqslant 2N}} \left| \sum_{M \leqslant m \leqslant M_1} \sum_{N \leqslant n \leqslant N_1} e(2\sqrt{mnx}) \right| \log^2 R$$
$$+ x^{1/2 + \varepsilon} R^{-1/2} + x^\varepsilon.$$

Hence we have here a double exponential sum to deal with instead of the function ψ. Note that the trivial estimate of this double sum and the choice of $R = x^{1/3}$ enable us to recover Theorem 6.14 with $x^{1/3+\varepsilon}$ instead of $x^{1/3} \log x$. The theory of multiple exponential sums was developed nearly at the same time as the simple exponential sums. In [Kol85], Kolesnik establishes the following result.

Lemma 6.39 *Let $X > 0$ be a real number and $M, N, R \geqslant 1$ be integers such that $M \geqslant N$ and $MN \asymp R$, and let $\mathcal{D} \subseteq [M, 2M] \times [N, 2N]$. Let $\alpha, \beta \in \mathbb{R} \setminus \mathbb{Z}$ such that $\alpha + \beta \neq 3, 4, 13/3$ and $(\alpha - 4)^2 + \beta(\alpha - 3) \neq 0$. Then, for all $\varepsilon > 0$, we have*

$$\sum_{(m,n)\in\mathcal{D}} e\left(X\left(\frac{m}{M}\right)^\alpha \left(\frac{n}{N}\right)^\beta\right) \ll R^{145/173+\varepsilon} X^{11/173} + R^{168/197+\varepsilon} X^{11/197}$$

$$+ R^{263/230+\varepsilon} X^{-22/230} + R^{21/26+\varepsilon} X^{1/13}$$

$$+ R^{255/302+\varepsilon} X^{9/151}.$$

Using this result with $\alpha = \beta = 1/2$ and $X = 2\sqrt{MNx} \asymp \sqrt{Rx}$ we get

$$\Delta(x) \ll x^{195/692} R^{83/692+\varepsilon} + x^{219/788} R^{103/788+\varepsilon} + x^{93/460} R^{159/460+\varepsilon}$$

$$+ x^{15/52} R^{5/52+\varepsilon} + x^{169/604} R^{75/604+\varepsilon} + x^{1/2+\varepsilon} R^{-1/2}$$

and we choose R so that the first and the last terms are equal, which gives $R = x^{151/429}$ and therefore

$$\Delta(x) \ll x^{139/429+\varepsilon}.$$

We have $\frac{139}{429} \approx 0.324\,009\,32\ldots$ This bound is slightly better than the one obtained with exponent pairs. The proof of Lemma 6.39 involves very complicated mathematics, this is the reason why the research then took a new turn.

6.6.6 The Discrete Hardy–Littlewood Method

The idea to adapt Hardy–Littlewood's circle method for exponential integrals to exponential sums first appeared in [BI86] in which the authors obtained an improvement in the problem of bounding $\zeta(\frac{1}{2} + it)$. The method was later used by Iwaniec and Mozzochi [IM88] to treat the Dirichlet divisor problem and the Gauss circle problem, which are similar by nature. This method is one of the trickiest in exponential sum theory. The proof is far beyond the scope of this book,[2] but one can systematize the ideas in the following six steps.

---

[2]For a complete exposition of this subject, the reader should refer to [GK91, Hux96].

In what follows, let $f \in C^r[N/2, 2N]$ with $r \geq 4$ and the derivatives satisfying van der Corput type hypotheses as in (6.16) so that there exist $c_i \geq 1$ and $T \geq 1$ such that, for all $x \in [N/2, 2N]$, we have

$$\frac{T}{N^j} \leq |f^{(j)}(x)| \leq c_j \frac{T}{N^j} \quad (j = 1, \dots, r).$$

We also suppose that $c_3 T \leq N^3$. The purpose is to get an estimate for the exponential sum

$$S = \sum_{N \leq n \leq 2N} e(f(n)).$$

Step 1. Major and Minor Arcs

As in the original circle method (see [IK04] for a complete exposition), the interval $[N, 2N]$ is covered by $[N/M] + 1$ intervals $I_k = [N + (k-1)M, N + kM[$ where M is a positive integer chosen so that $M \leq N/10$, $M^2 T \leq 2c_3 N^3$ and $M^4 T \leq N^4$. Let R be the integer such that

$$(R-1)^2 MT < 2c_3 N^3 \leq R^2 MT$$

so that $R^2 \geq M$ and suppose that $R \leq M$, since this is the range where this method really works. The role of R is to give the right order of magnitude of q when we approximate the real number $f''(x)/2$ by a rational number a/q. With the two parameters defined above, we have ensured that, for all $x \in [N/2, 2N]$,

$$M^{-3} \ll f'''(x) \ll M^{-2} \quad \text{and} \quad f^{(4)}(x) \ll M^{-4}.$$

The problem is then to find a point in I_k such that the derivatives of f at this point are rational numbers, in order to expand $f(x)$ in a Taylor series about this point. Since I_k has length M, as x runs through I_k, $f''(x)/2$ runs through an interval J_k of length $\asymp MTN^{-3} \asymp R^{-2}$. Choose a rational $a/q = a_k/q_k$ in J_k with $(a, q) = 1$ for which q is minimal. If $q \leq R^2/M$, we call I_k or J_k a *major arc*, otherwise a *minor arc*. Having chosen a/q, we pick the integer $n_0 = n_0^{(k)} \in I_k$ for which $f''(n_0)/2$ is the closest to a/q. Since f'' is strictly monotone, we have $|n_0 - x_0| \leq 1$ where x_0 is the unique solution of the equation $f''(x)/2 = a/q$. Hence we have

$$\left| \frac{f''(n_0)}{2} - \frac{a}{q} \right| \leq \frac{c_3 T}{2N^3} \leq \frac{4c_3^2}{M^2}.$$

Next we choose a rational approximation b/s to $qf'(n_0)$ such that

$$\left| \frac{b}{s} - qf'(n_0) \right| \leq qM^{-1} \quad \text{and} \quad s \leq \max(1, Mq^{-1})$$

which is always possible by Dirichlet's approximation theorem, and set

$$g(x) = f(n_0) + \frac{bx}{qs} + \frac{ax^2}{q} + \mu x^3$$

where $\mu = \frac{1}{6} f'''(n_0)$. This polynomial $g(x)$ is the function which is used to approx-imate $f(x + n_0)$. The values of m such that $n_0 + m$ lies in I_k belong to an interval $[M_1, M_2]$ such that $M_1 \geqslant M$, $M_2 \leqslant 3M$ and $M_2 - M_1 \leqslant M$, and writing

$$\sum_{M_1 \leqslant m \leqslant M_2} e\big(f(n_0 + m)\big) = \sum_{M_1 \leqslant m \leqslant M_2} e\big(f(n_0 + m) - g(m)\big) e\big(g(m)\big)$$

and using the fact that

$$f'(n_0 + x) - g'(x) = f'(n_0) - \frac{b}{qs} + x\left(f''(n_0) - \frac{2a}{q}\right) + \frac{x^3}{6} f^{(4)}(n_0 + \theta x)$$

where $0 < \theta < 1$, we get using the estimates above

$$f'(n_0 + x) - g'(x) \ll M^{-1}$$

for all $x \ll M$ such that $n_0 + x \in [N/2, 2N]$, so that by partial summation we get

$$\sum_{M_1 \leqslant m \leqslant M_2} e\big(f(n_0 + m)\big) \ll \max_{M_1 \leqslant M_3 \leqslant M_2} \left| \sum_{M_3 \leqslant m \leqslant M_2} e\big(g(m)\big) \right|.$$

For each k we have a different polynomial $g(m) = g_k(m)$ and different limits $M_3^{(k)}, M_2^{(k)}$. Also note that we have bounded S by the sum of the moduli of $\ll N/M$ shorter sums of length $\asymp M$. Since the best result one can hope for these sums is $\ll M^{1/2}$, the best estimate we could obtain for S would be

$$S \ll \left(\frac{N}{M}\right) M^{1/2} \ll \frac{N}{M^{1/2}}. \tag{6.20}$$

Step 2. Gauss Sums

The sum

$$\sum_{M_3 \leqslant m \leqslant M_2} e\big(g(m)\big) = e\big(f(n_0)\big) \sum_{M_3 \leqslant m \leqslant M_2} e\left(\mu m^3 + \frac{am^2}{q} + \frac{bm}{qs}\right)$$

may be viewed as an incomplete Gauss sum perturbed by the factor $e(\mu m^3)$. Gauss sums are usually defined as sums of additive or multiplicative characters. In our problem, the quadratic Gauss sum

$$G(a, b; q) = \sum_{m \,(\mathrm{mod}\, q)} e_q\big(am^2 + bm\big)$$

plays a key part since

$$\sum_{M_3 \leqslant m \leqslant M_2} e\left(\mu m^3 + \frac{am^2}{q} + \frac{bm}{qs}\right) = \frac{1}{qs} \sum_{k \,(\mathrm{mod}\, qs)} G(as, b+k; qs)$$

$$\times \sum_{M_3 \leqslant m \leqslant M_2} e\left(\mu m^3 - \frac{km}{qs}\right).$$

Gauss sums have some interesting properties. For instance, if $b + k = ds$ for some integer d and if q is odd, then

$$G(as, b+k; qs) = s\, e_q\left(\overline{-4a}\, d^2\right) G(a, 0; q)$$

otherwise the sum is zero. If q is even, there is a slight variation in this identity which does not really affect the result. The notation $\bar{\alpha}$ means the integer β satisfying $\alpha\beta \equiv 1 \pmod{q}$. Since $|G(a, 0; q)| = \sqrt{q}$ if q is odd and $(a, q) = 1$, we get by [Hux96, Lemma 5.4.5]

$$\left|\sum_{M_3 \leqslant m \leqslant M_2} e\big(g(m)\big)\right| = \frac{1}{\sqrt{q}} \left|\sum_{d \,(\mathrm{mod}\, q)} e_q\left(\overline{-4a}\, d^2\right) \sum_{M_3 \leqslant m \leqslant M_2} e\left(\mu m^3 - \frac{m}{q}\left(d - \frac{b}{s}\right)\right)\right|$$

if q is odd, which we shall suppose from now on.

Step 3. Poisson Summation

One may apply the truncated Poisson summation formula of Lemma 6.28 along with a computation about a special integral called the *Airy integral* and the use of Lemma 6.26. This process replaces the sums over d and m above by

$$\left|\sum_{M_3 \leqslant m \leqslant M_2} e\big(g(m)\big)\right| = \frac{1}{\sqrt{q}} \left|\sum_h \left(\frac{q}{12h_1\mu}\right)^{1/4} e_q\left(\overline{-4a}\, h^2\right) e\left(-\frac{2}{\mu^{1/2}}\left(\frac{h_1}{3q}\right)^{3/2}\right)\right|$$

$$+ O\left(\frac{q+R}{\sqrt{q}} \log M\right)$$

for $\mu > 0$, where h is an integer variable, $h_1 = h - b/s$ and the sum is such that

$$3\mu q M_3^2 \leqslant h_1 \leqslant 3\mu q M_2^2.$$

For $\mu < 0$ we use $-g(m)$. Since $R^2 \asymp N^3 (MT)^{-1}$, the size of each summand is

$$\ll \left(\frac{q}{\mu h_1}\right)^{1/4} \ll \left(\frac{q}{\mu^2 M^2 q}\right)^{1/4} \asymp (\mu M)^{-1/2} \asymp R$$

and the length of the range of summation is $\leqslant 18\mu q M^2 < 144 c_3^2 q M R^{-2}$. Hence the trivial estimate on the sum over h is $\ll R q^{-1/2}$ times the number of terms. On major arcs, we have $qMR^{-2} \leqslant 1$ and one can prove [HW88] that the number of major arcs with $Q \leqslant q < 2Q$ is $\ll NQ^2(MR^2)^{-1} + 1$. Thus, dividing the major arcs into ranges $Q \leqslant q < 2Q$ where Q is a power of 2 such that $Q \leqslant R^2/M$, we infer that the major arcs contribute

$$\ll \sum_{Q=2^\alpha \leqslant R^2/M} \frac{R}{Q^{1/2}} \left(\frac{NQ^2}{MR^2} + 1 \right) \log M$$

$$\ll \left(\frac{N}{MR} \cdot \frac{R^3}{M^{3/2}} + R \right) \log M$$

$$\ll \left(\frac{N}{M^{1/2}} + R \right) \log M$$

to the sum S, where we used the inequality $R \leqslant M$. Taking (6.20) into account, this estimate is nearly a best possible result.

Step 4. The Large Sieve on Minor Arcs

We now look for the contribution of the minor arcs for which $q > R^2/M$. The use of the large sieve corresponds to Vinogradov's mean-value theorem in the continuous case (see Lemma 6.41). The sum over h above is first tidied with some analytic tools like partial sums by Fourier transforms, Hölder's inequality, etc. Sums of bilinear forms of the type

$$\sum_{M_3 \leqslant m \leqslant M_2} e(g(m)) \ll Rq^{-1/2} \sum_H \int_{-1/2}^{1/2} \mathcal{K}(t) \left| \sum_{H \leqslant h < 2H} e(-\mathbf{x}^{(h)} \cdot \mathbf{y}(t)) \right| dt$$

$$+ q^{1/2} \log M$$

thus appear, where H runs over all powers of 2 covering the range of h in the summation above, $\mathcal{K}(t)$ is a kernel function satisfying

$$\int_{-1/2}^{1/2} \mathcal{K}(t) \, dt \ll \log M$$

and the vectors $\mathbf{x}^{(h)}$ and $\mathbf{y}(t)$ have respectively integer and rational coordinates in four dimensions, the first one corresponding to h and the second one to the minor arc J_k. After some rearrangements, the sum is treated with the following inequality called *the double large sieve inequality*.

   Let $\mathbf{x}^{(1)}, \ldots, \mathbf{x}^{(K)}$ and $\mathbf{y}^{(1)}, \ldots, \mathbf{y}^{(L)}$ be real vectors in h dimensions satisfying

$$\left| x_i^{(k)} \right| \leqslant \frac{X_i} {2} \quad \text{and} \quad \left| y_i^{(l)} \right| \leqslant \frac{Y_i}{2}$$

for $i = 1, \ldots, h$, $k = 1, \ldots, K$ and $l = 1, \ldots, L$. Let $a_i = X_i/(1 + X_i Y_i)$ and $b_i = 1/X_i$. Then we have

$$\left| \sum_{k=1}^{K} \sum_{l=1}^{L} e\left(\mathbf{x}^{(k)} \cdot \mathbf{y}^{(l)}\right) \right|^2 \leqslant \left(\frac{\pi}{2}\right)^{4h} A(\mathbf{a}) B(\mathbf{b}) \prod_{i=1}^{h} (X_i Y_i + 1)$$

where

$$A(\mathbf{a}) = \sum_k \sum_j \prod_{i=1}^{h} \max\left(0, 1 - \left|\frac{x_i^{(k)} - x_i^{(j)}}{a_i}\right|\right),$$

$$B(\mathbf{b}) = \sum_l \sum_j \prod_{i=1}^{h} \max\left(0, 1 - \left|\frac{y_i^{(l)} - y_i^{(j)}}{b_i}\right|\right).$$

Step 5. Semicubical Powers of Integers

This is the trickiest point of the method. The numbers $A(\mathbf{a})$ and $B(\mathbf{b})$ require to count the number of solutions of a system of four Diophantine inequalities involving semicubical powers of eight integer unknowns. This is done using very complicated combinatorial arguments (see [BI86, GK91]).

Step 6. Final Step

The final step rests on counting the points indexed by the minor arcs. One has to count the number of coincidences of the points $\mathbf{y}$, i.e. to count the number of Diophantine inequalities with rational unknowns. The numbers M and R are then chosen optimally.

This method enables us to get some new exponent pairs which are not attainable by any successive A- and B-processes. In [HW88], Huxley and Watt were able to get the exponent pair $(\frac{9}{56} + \varepsilon, \frac{1}{2} + \frac{9}{56} + \varepsilon)$. They gradually improved their estimates by getting the pairs

$$(k, l) = \begin{cases} \left(\dfrac{2}{13} + \varepsilon, \dfrac{35}{52} + \varepsilon\right) & (1987), \\[2mm] \left(\dfrac{89}{570} + \varepsilon, \dfrac{1}{2} + \dfrac{89}{570} + \varepsilon\right) & (1996), \\[2mm] \left(\dfrac{32}{205} + \varepsilon, \dfrac{1}{2} + \dfrac{32}{205} + \varepsilon\right) & (2005) \end{cases}$$

the last two exponent pairs being obtained by Huxley [Hux96, Hux05].

In [Hux03], Huxley used the ideas exposed above and added some refinements, notably bounding the number of integer points near a specific smooth curve (as in Chap. 5), to estimate the following double exponential sum

$$\sum_{H < h \leqslant 2H} a_h \sum_{N < n \leqslant 2N} b_n e\big(hf(n)\big)$$

where (a_h) and (b_n) are of bounded variation. Using Corollary 6.2, he then deduced the following result.

Theorem 6.40 (Huxley) *Let* $r \geqslant 5$, $N \geqslant 1$ *be integers and* $f \in C^r[N, 2N]$ *such that there exist real numbers* $T \geqslant 1$ *and* $1 \leqslant c_0 \leqslant \cdots \leqslant c_r$ *such that, for all* $x \in [N, 2N]$ *and all* $j \in \{0, \ldots, r\}$, *we have*

$$\frac{T}{N^j} \leqslant \big|f^{(j)}(x)\big| \leqslant c_j \frac{T}{N^j}.$$

Then we have

$$\sum_{N < n \leqslant 2N} \psi\big(f(n)\big) \ll (NT)^{131/416}(\log NT)^{18\,627/8320}.$$

Applied to the Dirichlet divisor problem, this gives the best known result up to now namely

$$\Delta(x) \ll x^{131/416}(\log x)^{26\,947/8320}.$$

Corollary 6.22 with van der Corput's theory gives under the same hypotheses

$$\sum_{N < n \leqslant 2N} \psi\big(f(n)\big) \ll (NT)^{1/3}$$

if $T \geqslant N$. We have $1/3 \approx 0.33\ldots$ and $131/416 \approx 0.314\,903\,84\ldots$ which shows the extreme difficulty in getting some new improvements—this is actually the case with every problem in analytic number theory.

6.6.7 Vinogradov's Method

Let us move on now to another divisor problem and suppose we want to estimate the sum $\sum_{n \leqslant x} \sigma(n)$. As in Chap. 4, we have

$$\sum_{n \leqslant x} \sigma(n) = \frac{1}{2} \sum_{d \leqslant x} \left[\frac{x}{d}\right]\left(\left[\frac{x}{d}\right] + 1\right)$$

$$= \frac{1}{2} \sum_{d \leqslant x} \left(\frac{x}{d} - \psi\left(\frac{x}{d}\right) - \frac{1}{2}\right)\left(\frac{x}{d} - \psi\left(\frac{x}{d}\right) + \frac{1}{2}\right)$$

$$= \frac{1}{2} \sum_{d \leqslant x} \left(\frac{x^2}{d^2} - \frac{2x}{d} \psi\left(\frac{x}{d}\right) + \psi\left(\frac{x}{d}\right)^2 - \frac{1}{4} \right)$$

$$= \frac{x^2}{2} \sum_{d \leqslant x} \frac{1}{d^2} - x \sum_{d \leqslant x} \frac{1}{d} \psi\left(\frac{x}{d}\right) + O(x)$$

$$= \frac{x^2 \zeta(2)}{2} - \frac{x^2}{2} \sum_{d > x} \frac{1}{d^2} - x \sum_{d \leqslant x} \frac{1}{d} \psi\left(\frac{x}{d}\right) + O(x)$$

$$= \frac{x^2 \zeta(2)}{2} - x \sum_{d \leqslant \sqrt{x}} \frac{1}{d} \psi\left(\frac{x}{d}\right) - x \sum_{\sqrt{x} < d \leqslant x} \frac{1}{d} \psi\left(\frac{x}{d}\right) + O(x).$$

We apply (1.6) to the second sum with $f(t) = t^{-1} \psi(x/t)$, $a = \sqrt{x}$ and $b = x$. Since $V_a^b(f) \ll 1 + x^{-1/2}$ we have

$$\sum_{\sqrt{x} < d \leqslant x} \frac{1}{d} \psi\left(\frac{x}{d}\right) = \int_{\sqrt{x}}^x \frac{1}{t} \psi\left(\frac{x}{t}\right) dt + \frac{\psi(x)}{2x} + \frac{\psi(\sqrt{x})^2}{\sqrt{x}} + O(1)$$

$$= \int_1^{\sqrt{x}} \frac{\psi(u)}{u} du + O(1) \ll 1$$

so that

$$\sum_{n \leqslant x} \sigma(n) = \frac{x^2 \zeta(2)}{2} - x \sum_{n \leqslant \sqrt{x}} \frac{1}{n} \psi\left(\frac{x}{n}\right) + O(x).$$

The trivial estimate on the sum gives

$$\left| \sum_{n \leqslant \sqrt{x}} \frac{1}{n} \psi\left(\frac{x}{n}\right) \right| \leqslant \frac{1}{2} \sum_{n \leqslant \sqrt{x}} \frac{1}{n} \ll \log x \tag{6.21}$$

so that we get the asymptotic formula

$$\sum_{n \leqslant x} \sigma(n) = \frac{x^2 \zeta(2)}{2} + O(x \log x).$$

If we intend to improve on the error-term, the use of previous results such as Corollary 6.35 can help to reduce the interval of summation. Indeed, using Abel's summation and Corollary 6.35 with the exponent pair (6.19), we get

$$\sum_{a(x) < n \leqslant \sqrt{x}} \frac{1}{n} \psi\left(\frac{x}{n}\right) \ll \max_{a(x) < N \leqslant \sqrt{x}} \left| \sum_{N < n \leqslant 2N} \frac{1}{n} \psi\left(\frac{x}{n}\right) \right| \log x$$

$$\ll \max_{a(x)<N\leqslant\sqrt{x}} N^{-1} \max_{N\leqslant N_1\leqslant 2N} \left| \sum_{N<n\leqslant N_1} \psi\left(\frac{x}{n}\right) \right| \log x$$

$$\ll \max_{a(x)<N\leqslant\sqrt{x}} N^{-1} \max_{N\leqslant N_1\leqslant 2N} \left(x^{97/348} N^{35/348} + N^2 x^{-1}\right) \log x$$

$$\ll \max_{a(x)<N\leqslant\sqrt{x}} \left(x^{97/348} N^{-313/348} + N x^{-1}\right) \log x \ll 1$$

where $a(x) = x^{97/313} (\log x)^{348/313} \approx x^{0.279} (\log x)^{1.112}$. However, this method does not enable us to cover the whole range of summation.

From 1934 to 1937, Vinogradov created another tool to deal with exponential sums which may supersede Weyl–van der Corput's method when the orders of the derivatives of the function are large. We sketch below the ideas underlying this method, but we omit the proofs. For more information about Vinogradov's work, see [For02, Mon94, Vin54].

Let k be a large integer, say $k \geqslant 5$, and $f \in C^{k+1}[N, 2N]$ such that there exists a real number $\lambda_{k+1} \in [N^{-2}, N^{-1}]$ such that, for all $x \in [N, 2N]$, we have

$$\left| \frac{f^{(k+1)}(x)}{(k+1)!} \right| \asymp \lambda_{k+1}.$$

Let H be a positive integer satisfying $H \leqslant N/2$ and $H^{k+1}\lambda_{k+1} \ll 1$. Using Weyl's shift (6.9) with $a_n = e(f(n))$ and Taylor's series expansion we get

$$\left| \sum_{N<n\leqslant 2N} e(f(n)) \right| \leqslant \frac{1}{H} \sum_{N-H<n<2N} \left| \sum_{h\leqslant H} e(f(n+h)) \right| + H$$

$$= \frac{1}{H} \sum_{N-H<n<2N} \left| \sum_{h\leqslant H} e(Q_n(h) + u_n(h)) \right| + H$$

where

$$Q_n(h) = hf'(n) + \cdots + \frac{h^k}{k!} f^{(k)}(n) \quad \text{and}$$

$$u_n(h) = f(n) + \frac{1}{k!} \int_0^h (h-t)^k f^{(k+1)}(n+t) \, dt.$$

Since $H^{k+1}\lambda_{k+1} \ll 1$, the total variation of $u_n(h)$ is $\ll 1$, so that using Abel's summation we get

$$\sum_{N<n\leqslant 2N} e(f(n)) \ll \frac{1}{H} \sum_{n\sim N} \left| \sum_{h\sim H} e(Q_n(h)) \right| + H$$

and writing $f^{(j)}(n)/j! = [f^{(j)}(n)/j!] + \{f^{(j)}(n)/j!\}$ we have

$$\sum_{N<n\leqslant 2N} e(f(n)) \ll \frac{1}{H} \sum_{n\sim N} \left| \sum_{h\sim H} e(P_n(h)) \right| + H$$

where $P_n(h) = h\{f'(n)\} + h^2\{f''(n)/2\} + \cdots + h^k\{f^{(k)}(n)/k!\}$. The first step in Vinogradov's method is to prove the following estimate

$$\sum_{N<n\leqslant 2N} e(f(n)) \ll \frac{N}{H}\left(H^{k(k+1)/2}\mathcal{B}(H)J_{k,b}(H)N^{-2}\right)^{\frac{1}{2b}} + H \qquad (6.22)$$

where b is a fixed positive integer, $\mathcal{B}(H)$ is the number of pair (n, n') such that

$$\left| \frac{f^{(k)}(n)}{k!} - \frac{f^{(k)}(n')}{k!} \right| \leqslant H^{-k} \quad \text{and} \quad \left| \left\{ \frac{f^{(k-1)}(n)}{(k-1)!} \right\} - \left\{ \frac{f^{(k-1)}(n')}{(k-1)!} \right\} \right| \leqslant H^{1-k}$$

and $J_{k,b}(H)$ is the *Vinogradov integral* defined by

$$J_{k,b}(H) = \int_0^1 \cdots \int_0^1 \left| \sum_{h\leqslant H} e(a_1 h + \cdots + a_k h^k) \right|^{2b} da_1 \cdots da_k.$$

One can get an upper bound for $\mathcal{B}(H)$ in the following way. Set $n' = n + \ell$ with $|\ell| \leqslant L \ll N$ for some $L \in \mathbb{Z}_{\geqslant 0}$. Then the first condition implies that $(k+1)|\ell|\lambda_{k+1} \leqslant H^{-k}$. Taking $L \asymp (\lambda_{k+1} H^k)^{-1}$ and taking the inequality $|\{a\} - \{b\}| \geqslant \|a - b\|$ into account, we get

$$\mathcal{B}(H) \ll \mathrm{Card}\left\{ (n, \ell) : 0 \leqslant |\ell| \leqslant L,\ N/2 < n \leqslant 2N,\ \left\| \frac{\Delta_\ell f^{(k-1)}(n)}{(k-1)!} \right\| \leqslant H^{1-k} \right\}$$

$$\ll N + \sum_{0<|\ell|\leqslant L} \mathcal{R}\left(\frac{\Delta_\ell f^{(k-1)}(n)}{(k-1)!}, N, 2H^{1-k} \right)$$

$$\ll N + \sum_{0<|\ell|\leqslant L} \left(N|\ell|\lambda_{k+1} + NH^{1-k} + \left(|\ell|\lambda_{k+1}H^{k-1}\right)^{-1} + 1 \right)$$

where we used Theorem 5.6. Now since $NL\lambda_{k+1} \ll NH^{1-k}$ and $(|\ell|\lambda_{k+1}H^{k-1})^{-1} \gg H$ we get

$$\mathcal{B}(H) \ll N + \frac{N}{H^{2k-1}\lambda_{k+1}} + \frac{\log N}{\lambda_{k+1}H^{k-1}}$$

and the choice of $H = [\lambda_{k+1}^{-1/(2k-1)}]$ gives

$$\mathcal{B}(H) \ll N \log N \qquad (6.23)$$

where we used the fact that $N^2\lambda_{k+1} \geq 1$. Note that from the hypotheses above we have

$$N^{\frac{1}{2k-1}} \ll H \ll N^{\frac{2}{2k-1}} \quad \text{and} \quad H^{k+1}\lambda_{k+1} \ll N^{-\frac{k-2}{2k-1}}.$$

The next result, which gives a non-trivial upper bound for $J_{k,b}(H)$, lies at the heart of Vinogradov's method.

Lemma 6.41 (Vinogradov's mean-value theorem) *Set $\varepsilon = e^{-b/k^2}$. Then*

$$J_{k,b}(H) \ll H^{2b-(1-\varepsilon)k(k+1)/2}.$$

The proof relies on the observation that $J_{k,b}(H)$ counts the number of integer solutions of the following system of Diophantine equations

$$\begin{cases} h_1 + \cdots + h_b = h'_1 + \cdots + h'_b, \\ h_1^2 + \cdots + h_b^2 = h'^2_1 + \cdots + h'^2_b, \\ \quad \vdots \qquad \quad \vdots \qquad \quad \vdots \\ h_1^k + \cdots + h_b^k = h'^k_1 + \cdots + h'^k_b \end{cases}$$

with $1 \leq h_i, h'_i \leq H$. Indeed, expanding the power in the integrand of $J_{k,b}(H)$ and exchanging the order of summation and integration, we get

$$J_{k,b}(H) = \sum_{\mathbf{h},\mathbf{h}'} \prod_{j=1}^{k} \left(\int_0^1 e\{(h_1^j + \cdots + h_b^j - h'^j_1 - \cdots - h'^j_b)a_j\}\, da_j \right)$$

where $\mathbf{h} = (h_1, \ldots, h_b)$ and $\mathbf{h}' = (h'_1, \ldots, h'_b)$ run independently over $\{1, \ldots, H\}^b$. Hence the integral vanishes unless $h_1^j + \cdots + h_b^j = h'^j_1 + \cdots + h'^j_b$ for all $1 \leq j \leq k$.

Note that Lemma 6.41 is nearly the best possible result since one can show that

$$J_{k,b}(H) \gg H^{2b-k(k+1)/2}.$$

If b is large compared to k^2, say $b \geq ck^2 \log k$ for some $c \geq 1$, then the bound is quasi-optimal.

Now putting (6.23) and Lemma 6.41 into (6.22) we get

$$\sum_{N < n \leq 2N} e(f(n)) \ll N^{1-\frac{1}{2b}} H^{\frac{\varepsilon k(k+1)}{4b}} \log N + H$$

$$\ll N^{1-\frac{1}{2b}} \lambda_{k+1}^{-\frac{\varepsilon k(k+1)}{4b(2k-1)}} \log N + \lambda_{k+1}^{-\frac{1}{2k-1}}$$

$$\ll N^{1-\frac{r(b,k)}{2b}} \log N + N^{\frac{2}{2k-1}}$$

where

$$r(b,k) = 1 - e^{-b/k^2} \frac{k(k+1)}{2k-1}.$$

Taking $b = k^2 \log k$ with $k \geqslant 5$ then gives

$$\sum_{N < n \leqslant 2N} e\big(f(n)\big) \ll N^{1 - \frac{1}{6k^2 \log k}} \log N.$$

We can state the following result.

Theorem 6.42 (Vinogradov) *Let $k \geqslant 5$ be an integer and $f \in C^{k+1}[N, 2N]$ such that there exists a real number $\lambda_{k+1} \in [N^{-2}, N^{-1}]$ such that, for all $x \in [N, 2N]$, we have*

$$\left| \frac{f^{(k+1)}(x)}{(k+1)!} \right| \asymp \lambda_{k+1}.$$

Then

$$\sum_{N < n \leqslant 2N} e\big(f(n)\big) \ll N^{1 - \frac{1}{6k^2 \log k}} \log N.$$

In [Wal63], Walfisz was able to get a non-trivial upper bound for the sum (6.21). For large N, he used a van der Corput type estimate as in Corollary 6.25, and for smaller ones he used a Vinogradov type result. Afterwards, Karacuba [Kar71, Pet98] obtained a refined estimate using similar ideas as in Vinogradov's work which contains both methods. Applying this result to the function $f(n) = T/n$, Pétermann and Wu showed in [PW97] that, if $e^{200} \leqslant N < N_1 \leqslant 2N$ and $T \geqslant N^2$, then there exists $c_0 > 0$ such that we have

$$\sum_{N < n \leqslant N_1} e\left(\frac{T}{n}\right) \ll N \exp\left(-c_0 \frac{(\log N)^3}{(\log T N^{-1})^2}\right).$$

Applying this in Corollary 6.2 we obtain with $e^{200} \leqslant N < N_1 \leqslant 2N$ and $N \leqslant x^{1/2}$

$$\sum_{N < n \leqslant N_1} \psi\left(\frac{x}{n}\right) \ll NH^{-1} + N \exp\left(-c_0 \frac{(\log N)^3}{(\log H x N^{-1})^2}\right) \log H.$$

Set $c = 15(4c_0)^{-1/3}$ and $w(x) = e^{c(\log x)^{2/3}}$. Choosing $H = [\exp\{(\log N)^3/(\log x)^2\}]$ gives for all $w(x) \leqslant N \leqslant x^{1/2}$

$$\sum_{N < n \leqslant N_1} \psi\left(\frac{x}{n}\right) \ll \frac{N(\log N)^3}{(\log x)^2} \exp\left(-c_1 \frac{(\log N)^3}{(\log x)^2}\right)$$

with $c_1 = 64c_0/81$ and where we have used the bounds $N \geqslant w(x)$ and $H \leqslant x^{1/8}$. An application of Abel's summation yields

$$\sum_{N < n \leqslant N_1} \frac{1}{n} \psi\left(\frac{x}{n}\right) \ll \frac{(\log N)^3}{(\log x)^2} \exp\left(-c_1 \frac{(\log N)^3}{(\log x)^2}\right)$$

for $w(x) \leqslant N \leqslant x^{1/2}$ and a similar argument to that used in the proof of [PW97, Lemma 2.3] finally gives

$$\sum_{w(x) < n \leqslant x^{1/2}} \frac{1}{n} \psi\left(\frac{x}{n}\right) \ll (\log x)^{2/3}.$$

Since we have trivially

$$\sum_{n \leqslant w(x)} \frac{1}{n} \psi\left(\frac{x}{n}\right) \ll \sum_{n \leqslant w(x)} \frac{1}{n} \ll (\log x)^{2/3}$$

we may state the following result (compare to (4.14)).

Theorem 6.43 (Walfisz) *For all $x \geqslant 2$ sufficiently large, we have*

$$\sum_{n \leqslant x} \sigma(n) = \frac{x^2 \zeta(2)}{2} + O\left(x(\log x)^{2/3}\right).$$

A similar method was used to get the following improvement for the average order of Euler's totient function (compare to Exercise 10 in Chap. 4).

Theorem 6.44 (Walfisz) *For all $x \geqslant 2$ sufficiently large, we have*

$$\sum_{n \leqslant x} \varphi(n) = \frac{x^2}{2\zeta(2)} + O\left(x(\log x)^{2/3}(\log \log x)^{4/3}\right).$$

6.6.8 Vaughan's Identity and Twisted Exponential Sums

The sums $\sum_{p \leqslant x} e(f(p))$ and $\sum_{n \leqslant x} \mu(n)e(f(n))$ frequently arise in number theory. For instance, if we intend to get an estimate for $\sum_{N < n \leqslant 2N} \mu(n)\psi(f(n))$, we may use the asymptotic formula of Remark 6.3 which implies that, for any positive integer H, we have

$$\sum_{N < n \leqslant 2N} \mu(n)\psi\left(f(n)\right) = -\frac{1}{2\pi i} \sum_{0 < |h| \leqslant H} \frac{1}{h} \sum_{N < n \leqslant 2N} \mu(n)e\left(hf(n)\right)$$

$$+ O\left\{ \sum_{N < n \leqslant 2N} \min\left(1, \frac{1}{H\|f(n)\|}\right)\right\} \qquad (6.24)$$

so that we need to have at our disposal bounds for sums of the type

$$\sum_{N < n \leqslant 2N} \mu(n)e\left(hf(n)\right). \qquad (6.25)$$

For the error-term of (6.24), it can be shown [MV81] that the minimum admits a Fourier series expansion. Nevertheless, it is sometimes simpler to use the following lemma.

Lemma 6.45 *Let $N \geqslant 1$ and $H \geqslant 4$ be integers, and $f : [N, 2N] \longrightarrow \mathbb{R}$ be any function. We set $K = [\log H / \log 2]$. Then we have*

$$\sum_{N \leqslant n \leqslant 2N} \min\left(1, \frac{1}{H \|f(n)\|}\right) < 24NH^{-1} + 2 \sum_{k=0}^{K-2} 2^{-k} \mathcal{R}(f, N, 2^k H^{-1}).$$

Proof We have

$$\sum_{N \leqslant n \leqslant 2N} \min\left(1, \frac{1}{H \|f(n)\|}\right) = \sum_{\substack{N \leqslant n \leqslant 2N \\ \|f(n)\| < H^{-1}}} 1 + \frac{1}{H} \sum_{\substack{N \leqslant n \leqslant 2N \\ \|f(n)\| \geqslant H^{-1}}} \frac{1}{\|f(n)\|}$$

$$= \mathcal{R}(f, N, H^{-1}) + \frac{1}{H} \sum_{\substack{N \leqslant n \leqslant 2N \\ \|f(n)\| \geqslant H^{-1}}} \frac{1}{\|f(n)\|}.$$

Since

$$\left\{ n \in [N, 2N] \cap \mathbb{Z} : \|f(n)\| \geqslant H^{-1} \right\}$$

$$\subseteq \bigcup_{k=1}^{K} \left\{ n \in [N, 2N] \cap \mathbb{Z} : 2^{k-1} H^{-1} \leqslant \|f(n)\| < 2^k H^{-1} \right\}$$

we get

$$\sum_{\substack{N \leqslant n \leqslant 2N \\ \|f(n)\| \geqslant H^{-1}}} \frac{1}{\|f(n)\|} \leqslant \sum_{k=1}^{K} \sum_{\substack{N \leqslant n \leqslant 2N \\ 2^{k-1} H^{-1} \leqslant \|f(n)\| < 2^k H^{-1}}} \frac{1}{\|f(n)\|}$$

$$\leqslant (N+1)\left(2^{1-K} + 2^{2-K}\right) H + \sum_{k=1}^{K-2} \sum_{\substack{N \leqslant n \leqslant 2N \\ 2^{k-1} H^{-1} \leqslant \|f(n)\| < 2^k H^{-1}}} \frac{1}{\|f(n)\|}$$

$$\leqslant 6 \times 2^{-K}(N+1)H + 2H \sum_{k=1}^{K-2} 2^{-k} \sum_{\substack{N \leqslant n \leqslant 2N \\ \|f(n)\| < 2^k H^{-1}}} 1$$

$$< 12(N+1) + 2H \sum_{k=1}^{K-2} 2^{-k} \mathcal{R}(f, N, 2^k H^{-1})$$

since $2^{-K} < 2H^{-1}$. Thus we get

$$\sum_{N \leqslant n \leqslant 2N} \min\left(1, \frac{1}{H \|f(n)\|}\right) < \mathcal{R}(f, N, H^{-1}) + 24NH^{-1}$$

$$+ 2 \sum_{k=1}^{K-2} 2^{-k} \mathcal{R}(f, N, 2^k H^{-1})$$

which implies the desired result. $\square$

In the case of the function $f(n) = x/n$ with $x \geqslant 1$ being a large real number, Shiu's result may be used to get a quasi-optimal estimate.[3]

Lemma 6.46 *Let $x \geqslant 1$ be a real number, $N \geqslant 1$ and $4 \leqslant H \leqslant x$ be integers. Then, for all $\varepsilon > 0$, we have*

$$\sum_{N \leqslant n \leqslant 2N} \min\left(1, \frac{1}{H \|x/n\|}\right) \ll_\varepsilon NH^{-1} x^\varepsilon.$$

Proof Using Lemma 6.45 we get

$$\sum_{N \leqslant n \leqslant 2N} \min\left(1, \frac{1}{H \|x/n\|}\right) \ll NH^{-1} + \sum_{k=0}^{\lfloor \log H / \log 2 \rfloor - 2} 2^{-k} \mathcal{R}\left(\frac{x}{n}, N, \frac{2^k}{H}\right)$$

and interchanging the summations we obtain using Lemma 5.2

$$\mathcal{R}\left(\frac{x}{n}, N, \frac{2^k}{H}\right) \leqslant \sum_{N \leqslant n \leqslant 2N} \left(\left[\frac{x}{n} + \frac{2^k}{H}\right] - \left[\frac{x}{n} - \frac{2^k}{H}\right]\right)$$

$$\leqslant \sum_{x - 2^{k+1}NH^{-1} < m \leqslant x + 2^{k+1}NH^{-1}} \sum_{\substack{d \mid m \\ N \leqslant d \leqslant 2N}} 1$$

$$\leqslant \sum_{x - 2^{k+1}NH^{-1} < m \leqslant x + 2^{k+1}NH^{-1}} \tau(m)$$

$$\ll_\varepsilon 2^k NH^{-1} x^\varepsilon$$

where we used Theorem 4.62 in the last estimate. This implies the asserted result. $\square$

It is obvious that the equality (6.24) still holds for all bounded sequences (a_n) of complex numbers instead of $\mu(n)$. Furthermore, with $f(n) = x/n$, if we replace the

---

[3]One may slightly improve on this result by using Hooley's Δ-function instead of the τ-function, and making use of Theorem 4.65 instead of Shiu's theorem.

Möbius function by any sequence (a_n) of complex numbers satisfying $a_n \ll n^\varepsilon$ in (6.24), then the error-term of Lemma 6.46 is only affected by a factor x^ε if $N \leqslant x$. We may therefore state the following useful result which generalizes Corollary 6.2 in the case of the function $f(n) = x/n$.

Proposition 6.47 *Let $\varepsilon > 0$, $x \geqslant 1$ be real numbers, $N \geqslant 1$, $H \geqslant 4$ be integers satisfying $\max(N, H) \leqslant x$, and (a_n) be a complex-valued sequence supported on $[N, 2N]$ such that, for all $n \in \{N, \ldots, 2N\}$, we have $a_n \ll n^\varepsilon$. Then we have*

$$\sum_{N < n \leqslant 2N} a_n \psi\left(\frac{x}{n}\right) \ll N H^{-1} x^{2\varepsilon} + \sum_{h \leqslant H} \frac{1}{h} \left| \sum_{N < n \leqslant 2N} a_n e\left(\frac{hx}{n}\right) \right|.$$

Another long-standing problem in number theory is to ask whether there is a prime number in the interval $\mathcal{I} =]x, x + x^{1/2}]$ for large x. Even if we assume the Riemann hypothesis, it seems to be extremely difficult to answer this question. As an approximation, Ramachandra [Ram69] suggested the problem of showing that there is $n \in \mathcal{I}$ having a large prime factor p with $p > x^\phi$ and ϕ being as large as possible. We may proceed as follows. Let $P(n)$ be the greatest prime factor of n and, for all positive integers d, set

$$N(d) = \sum_{\substack{n \in \mathcal{I} \\ d \mid n}} 1.$$

The starting point is the following easy estimate. Interchanging the summations and using the convolution identity $\Lambda \star \mathbf{1} = \log$, we get

$$\sum_{d \leqslant x} N(d)\Lambda(d) = \sum_{x < n \leqslant x + x^{1/2}} \sum_{d \mid n} \Lambda(d) = \sum_{x < n \leqslant x + x^{1/2}} \log n = x^{1/2} \log x + O\left(x^{1/2}\right).$$

Now note that $N(d) \leqslant 1$ for all $d > x^{1/2}$ so that, using Chebyshev's estimate of Lemma 3.42, we get for all $\varepsilon > 0$

$$\sum_{\substack{x^{3/5-\varepsilon} < d \leqslant x \\ d \text{ not prime}}} N(d)\Lambda(d) = \sum_{\alpha \geqslant 2} \sum_{x^{3/5-\varepsilon} < p^\alpha \leqslant x} N(p^\alpha) \log p$$

$$\leqslant \sum_{p \leqslant x^{1/2}} \log p + \sum_{\alpha \geqslant 3} \sum_{p^\alpha \leqslant x} \log p$$

$$\leqslant x^{1/2} \log 4 + x^{1/3} \log x \ll x^{1/2}$$

for large x. Similarly, using $N(d) = x^{1/2} d^{-1} + O(1)$, we get

$$\sum_{d \leqslant x^{1/2-\varepsilon}} N(d)\Lambda(d) = x^{1/2} \sum_{d \leqslant x^{1/2-\varepsilon}} \frac{\Lambda(d)}{d} + O\left(\Psi\left(x^{1/2-\varepsilon}\right)\right)$$

$$= \left(\frac{1}{2} - \varepsilon\right) x^{1/2} \log x + O\left(x^{1/2}\right)$$

where we used Chebyshev's estimates of Corollary 3.47 and Lemma 3.42. We then have at this step

$$x^{1/2}\log x + O\left(x^{1/2}\right) = \sum_{d \leqslant x} N(d)\Lambda(d)$$

$$= \sum_{d \leqslant x^{1/2-\varepsilon}} N(d)\Lambda(d) + \sum_{x^{1/2-\varepsilon} < d \leqslant x^{3/5-\varepsilon}} N(d)\Lambda(d)$$

$$+ \sum_{\substack{x^{3/5-\varepsilon} < d \leqslant x \\ d \text{ not prime}}} N(d)\Lambda(d) + \sum_{x^{3/5-\varepsilon} < p \leqslant x^{\phi}} N(p)\log p$$

$$+ \sum_{x^{\phi} < p \leqslant x} N(p)\log p$$

$$= \left(\frac{1}{2} - \varepsilon\right)x^{1/2}\log x + \Sigma_2 + \Sigma_3$$

$$+ \sum_{x^{\phi} < p \leqslant x} N(p)\log p + O\left(x^{1/2}\right)$$

where

$$\Sigma_2 = \sum_{x^{1/2-\varepsilon} < d \leqslant x^{3/5-\varepsilon}} N(d)\Lambda(d) \quad \text{and} \quad \Sigma_3 = \sum_{x^{3/5-\varepsilon} < p \leqslant x^{\phi}} N(p)\log p$$

so that

$$\sum_{x^{\phi} < p \leqslant x} N(p)\log p = \left(\frac{1}{2} + \varepsilon\right)x^{1/2}\log x - \Sigma_2 - \Sigma_3 + O\left(x^{1/2}\right).$$

The next step is to show the following estimate

$$\Sigma_2 = \frac{1}{10}x^{1/2}\log x + O\left(x^{1/2}\right) \tag{6.26}$$

and to find the largest exponent ϕ such that the upper bound

$$\Sigma_3 < \frac{2}{5}x^{1/2}\log x \tag{6.27}$$

holds, so that we shall have

$$\sum_{x^{\phi} < p \leqslant x} N(p)\log p > \frac{1}{2}\varepsilon x^{1/2}\log x$$

for x sufficiently large, which yields the existence of a prime p satisfying $p > x^{\phi}$ such that $N(p) = 1$, so that there exists an integer $n \in \mathcal{I}$ such that $P(n) > x^{\phi}$.

The estimate (6.26) has been proved in [Liu93, Har07] using the following arguments. Since

$$N(d) = \frac{x^{1/2}}{d} - \psi\left(\frac{x+x^{1/2}}{d}\right) + \psi\left(\frac{x}{d}\right)$$

it is sufficient to show that, for all positive integers N such that $x^{1/2} < N \leqslant x^{3/5-\varepsilon}$, we have

$$\sum_{N<d\leqslant 2N} \Lambda(d)f(d) \ll \frac{x^{1/2}}{\log x} \qquad (6.28)$$

where

$$f(d) = \psi\left(\frac{x+x^{1/2}}{d}\right) - \psi\left(\frac{x}{d}\right).$$

The estimate (6.27) is shown in [BH09] with $\phi = 0.7428$, and also in [Har07] with the slightly weaker value $\phi = 0.74$, by using sieve techniques, and more precisely the Rosser–Iwaniec sieve and an alternative sieve.

To treat the sums (6.25) and (6.28), and to get fine estimates for (6.27), one can make use of Vaughan's ingenious identities based upon some decompositions of certain formulae involving the Riemann zeta-function (see [MV81]). Let $U \geqslant 1$ be a real number and set

$$F(s) = \sum_{n\leqslant U} \frac{\Lambda(n)}{n^s} \quad \text{and} \quad G(s) = \sum_{n\leqslant U} \frac{\mu(n)}{n^s}.$$

We have for $\sigma > 1$

$$-\frac{\zeta'(s)}{\zeta(s)} = F(s) - \zeta(s)F(s)G(s) - \zeta'(s)G(s) + \left(-\frac{\zeta'(s)}{\zeta(s)} - F(s)\right)\left(1 - \zeta(s)G(s)\right).$$

For $\sigma > 1$, these functions can be expanded as Dirichlet series and Proposition 4.42 implies that

$$\Lambda(n) = a_1(n) + a_2(n) + a_3(n) + a_4(n)$$

where

$$a_1(n) = \begin{cases} \Lambda(n), & \text{if } n \leqslant U, \\ 0, & \text{otherwise,} \end{cases} \qquad a_2(n) = -\sum_{\substack{mdr=n \\ m,d\leqslant U}} \Lambda(m)\mu(d),$$

$$a_3(n) = \sum_{\substack{kd=n \\ d\leqslant U}} \mu(d)\log k \quad \text{and} \quad a_4(n) = -\sum_{\substack{mk=n \\ m>U, k>U}} \Lambda(m)\left(\sum_{\substack{d|k \\ d\leqslant U}} \mu(d)\right).$$

Multiplying throughout by $f(n)$ and summing we get

$$\sum_{N<n\leqslant 2N} \Lambda(n)f(n) = \sum_{i=1}^{4} \sum_{N<n\leqslant 2N} a_i(n)f(n).$$

Similarly, by using

$$\frac{1}{\zeta(s)} = 2G(s) - G(s)^2\zeta(s) + \left(\frac{1}{\zeta(s)} - G(s)\right)\left(1 - \zeta(s)G(s)\right)$$

we get

$$\mu(n) = b_1(n) + b_2(n) + b_3(n)$$

where

$$b_1(n) = \begin{cases} 2\mu(n), & \text{if } n \leqslant U, \\ 0, & \text{otherwise,} \end{cases} \qquad b_2(n) = -\sum_{\substack{mdr=n \\ m,d\leqslant U}} \mu(m)\mu(d)$$

and

$$b_3(n) = -\sum_{\substack{mk=n \\ m>U, k>U}} \mu(m)\left(\sum_{\substack{d\mid k \\ d\leqslant U}} \mu(d)\right)$$

so that

$$\sum_{N<n\leqslant 2N} \mu(n)f(n) = \sum_{i=1}^{3} \sum_{N<n\leqslant 2N} b_i(n)f(n).$$

Later, Heath-Brown [HB82] generalized Vaughan's identities by providing some formulae which are more flexible. One usually bounds the sums with $a_1(n)$ and $b_1(n)$ trivially. The other sums involve the so-called sums of type I and type II in Vaughan's terminology which may be defined in the following way. We consider integers $M, N, R, R' \geqslant 1$ such that $R < R' \leqslant 2R$. If $f : [R, R'] \longrightarrow \mathbb{C}$ is any function, it is convenient to call *sums of type I* (related to f) the sums

$$S_I = \sum_{M<m\leqslant 2M} \sum_{\substack{N<n\leqslant 2N \\ R<mn\leqslant R'}} a_m f(mn)$$

and *sums of type II* (related to f) the sums

$$S_{II} = \sum_{M<m\leqslant 2M} \sum_{\substack{N<n\leqslant 2N \\ R<mn\leqslant R'}} a_m b_n f(mn)$$

where a_m, b_n are complex numbers supported respectively on $[M, 2M]$ and $[N, 2N]$ and satisfying $a_m \ll_\varepsilon m^\varepsilon$ and $b_n \ll_\varepsilon n^\varepsilon$.

By using Vaughan's or Heath-Brown's identities, one may prove the following useful result (see [Bak07]).

Proposition 6.48 *Let $S > 0$ be a positive real number and suppose that the following estimates*

$$S_I \ll S \quad for \quad N \gg R^{1/2},$$

$$S_{II} \ll S \quad for \quad R^{1/3} \ll N \ll R^{1/2}$$

hold for all sums of type I and type II. Then we have

$$\sum_{R < n \leqslant R'} \mu(n) f(n) \ll S(\log 3R)^5.$$

A similar result holds for the sum (6.28). For instance, using Heath-Brown's identity, the authors in [RS01] proved the following proposition.

Proposition 6.49 *Let $k \geqslant 4$ be an integer, $(2k)^{-1} \leqslant \alpha \leqslant 1/6$ be a real number and $S > 0$ be a positive real number. Suppose that $MN \asymp R$ and that the following estimates*

$$S_I \ll S \quad for \quad N \geqslant R^{(1-\alpha)/2} \quad and \; for \quad R^{2\alpha} < N \leqslant R^{1/3},$$

$$S_{II} \ll S \quad for \quad R^{\alpha} \leqslant N \leqslant R^{2\alpha}$$

hold for all sums of type I and type II. Then we have for all $\varepsilon > 0$

$$\sum_{R < n \leqslant R'} \Lambda(n) f(n) \ll_{k,\varepsilon} S R^{\varepsilon}.$$

One may notice that Vaughan's identity for the function Λ is in a certain sense a rearrangement of the convolution identity $\Lambda = -\mu \log \star 1$. Harman [Har07] points out that the genesis of these identities lies in approximating infinite series by finite sums. Such considerations were certainly around in the 1930s, when Vinogradov adapted the sieve of Eratosthenes–Legendre by replacing a sum over primes by double sums. But it is noteworthy that Vaughan's rearrangement was not used until the 1960s–70s. It also should be mentioned that these identities do not work if f is multiplicative. Indeed, we have in this case, supposing f completely multiplicative and neglecting the multiplicative condition $R < mn \leqslant R'$ for the sake of simplicity,

$$S_{II} = \left(\sum_{M < m \leqslant 2M} a_m f(m) \right) \left(\sum_{N < n \leqslant 2N} b_n f(n) \right)$$

so that the new sums are not easier to deal with than the original sum. In fact, one generally uses these identities with $f(n) = e(g(n))$ or $f(n) = \sum_{h \sim H} e(g(h,n))$ for some real-valued function g.

One can prove that the condition $R < mn \leqslant R'$ can be removed from sums $\mathcal{S}_I$ and $\mathcal{S}_{II}$ at a cost of a factor $\log R$ (see [Bak86, Lemma 15] or [RS01] for instance). Over the last two decades, many authors have provided some non-trivial bounds for sums of type I and type II (see [Bak86, Bak94, Bak07, CZ98, CZ99, CZ00, FI89, GR96, KRW07, Liu94, Liu95, LW99, RS06, SW00, Wu93, Wu02, Zha99] among a lot of references). As an example, we prove an analogue of the third derivative theorem for sums of type II.

Theorem 6.50 *Let $f \in C^3([M, 2M] \times [N, 2N])$ such that there exists $\lambda_3 > 0$ such that, for all $(x, y) \in [M, 2M] \times [N, 2N]$, we have*

$$\left| \frac{\partial}{\partial x} \frac{\partial^2}{\partial y^2} f(x, y) \right| \asymp \lambda_3.$$

Let (a_m) and (b_n) be two complex-valued sequences supported respectively on $[M, 2M]$ and $[N, 2N]$ satisfying $|a_m| \leqslant 1$ and $|b_n| \leqslant 1$. Then we have

$$\sum_{M < m \leqslant 2M} \sum_{N < n \leqslant 2N} a_m b_n e\big(f(m, n)\big) \ll MN\lambda_3^{1/6} + MN^{3/4} + M^{1/2}N$$

$$+ M^{3/4} N^{1/2} \lambda_3^{-1/4}.$$

Proof We may suppose $\lambda_3 \ll 1$ otherwise the estimate is trivial. Let $S_{M,N}$ be the sum on the left-hand side and

$$S_M(n) = \sum_{M < m \leqslant 2M} a_m e\big(f(m, n)\big)$$

so that

$$S_{M,N} = \sum_{N < n \leqslant 2N} b_n S_M(n).$$

Let H be a positive integer such that $H \leqslant M$. By Cauchy–Schwarz's inequality we have

$$|S_{M,N}|^2 \leqslant N \sum_{N < n \leqslant 2N} |b_n S_M(n)|^2 \leqslant N \sum_{N < n \leqslant 2N} |S_M(n)|^2$$

and van der Corput's A-process (Lemma 6.18) gives

$$|S_M(n)|^2 \leqslant \frac{2M^2}{H} + \frac{4M}{H} \operatorname{Re} \left\{ \sum_{h \leqslant H} \left(1 - \frac{h}{H}\right) \sum_{M < m \leqslant 2M - h} a_{m+h} \overline{a_m} e\big(\Delta_h f(m, n)\big) \right\}$$

where

$$\Delta_h f(m, n) = f(m + h, n) - f(m, n).$$

We infer that

$$|S_{M,N}|^2 \ll \frac{(MN)^2}{H} + \frac{MN}{H} \sum_{h \leqslant H} \sum_{M < m \leqslant 2M} \left| \sum_{N < n \leqslant 2N} e\big(\Delta_h f(m,n)\big) \right|$$

and since

$$\left| \frac{\partial^2}{\partial y^2} \Delta_h f(x,y) \right| \asymp h\lambda_3$$

for all $(x,y) \in [M,2M] \times [N,2N]$, using van der Corput's inequality (Theorem 6.9) we get

$$|S_{M,N}|^2 \ll \frac{(MN)^2}{H} + \frac{MN}{H} \sum_{h \leqslant H} \sum_{M < m \leqslant 2M} \left\{ N(h\lambda_3)^{1/2} + (h\lambda_3)^{-1/2} \right\}$$

$$\ll \frac{(MN)^2}{H} + (MN)^2 (H\lambda_3)^{1/2} + M^2 N (H\lambda_3)^{-1/2}$$

so that

$$|S_{M,N}| \ll MNH^{-1/2} + MN(H\lambda_3)^{1/4} + MN^{1/2}(H\lambda_3)^{-1/4}$$

and Lemma 5.5 gives the asserted result plus a secondary term $MN\lambda_3^{1/4}$ which is absorbed by the main term since $\lambda_3 \ll 1$.   □

6.6.9   Explicit Estimates for $\Delta(x)$

Since Voronoï's paper [Vor03], there have been so far few explicit estimates for the remainder term $\Delta(x)$ in the Dirichlet's divisor problem. This was investigated in [BBR12] where a modified version of Lemma 6.16 was used to get the following estimates.

Theorem 6.51 (Berkane, Bordellès and Ramaré) *We have*

$$|\Delta(x)| \leqslant \begin{cases} 0.961\sqrt{x}, & \text{if } x \geqslant 1, \\ 0.482\sqrt{x}, & \text{if } x \geqslant 1981, \\ 0.397\sqrt{x}, & \text{if } x \geqslant 5560, \\ 0.764\, x^{1/3} \log x, & \text{if } x \geqslant 9995. \end{cases}$$

Furthermore, these estimates are sharp in view of

$$|\Delta(1980)| > 0.5\sqrt{x}, \quad |\Delta(5559)| > 0.4\sqrt{x} \quad \text{and} \quad |\Delta(9994)| > 0.8\, x^{1/3} \log x.$$

6.7 Exercises

1 Let $x, y, \alpha, \beta \in \mathbb{R}$ and $M < N$ be integers.

(a) Show that $4\|x - y\| \leqslant |e(x) - e(y)| \leqslant 2\pi \|x - y\|$.

(b) Show that

$$\left| \sum_{n=M+1}^{N} e(\alpha n + \beta) \right| \leqslant \min\left(N - M, \frac{1}{2\|\alpha\|} \right).$$

2 Let $N < N_1 \leqslant 2N$ be large integers and $f : [N, N_1] \longrightarrow \mathbb{R}$ be a function satisfying the hypotheses of Theorem 6.9 with $\lambda_2 \leqslant 10^{-2}$ and let $\delta \in \,]0, 1/10]$ be a small real number. Suppose also that f' is non-decreasing.

(a) Splitting the sum into two subsums, show that

$$\sum_{N < n \leqslant N_1} e(\pm f(n)) \ll \mathcal{R}(f', N, \delta) + (N\lambda_2 + 1)\delta^{-1}.$$

(b) Using Theorem 5.6 and choosing δ optimally, deduce another proof of van der Corput's inequality.

3 Let $a \in \mathbb{R}$ and $H \in \mathbb{N}$. Show that

$$\left| \sum_{h=0}^{H-1} e(ha) \right|^2 = \sum_{|h| \leqslant H-1} (H - |h|) e(ha).$$

4 Let $f : [N, 2N] \longrightarrow \mathbb{R}$ be any function and $\delta \in \,]0, \frac{1}{4}[$. Set $K = [(8\delta)^{-1}] + 1$. The purpose of this exercise is to prove the following result.

For all positive integers $H \leqslant K$, we have

$$\mathcal{R}(f, N, \delta) \leqslant \frac{4N}{H} + \frac{4}{H} \sum_{h=1}^{H-1} \left| \sum_{N \leqslant n \leqslant 2N} e(hf(n)) \right|. \qquad (6.29)$$

One should compare this inequality to Corollary 6.2.

(a) Let $n \in [N, 2N] \cap \mathbb{Z}$ such that $\|f(n)\| < \delta$. Prove that, for all integers h such that $|h| < H$, we have

$$\operatorname{Re}\{e(hf(n))\} \geqslant \frac{\sqrt{2}}{2}.$$

(b) Prove that

$$\mathcal{R}(f, N, \delta) \leqslant \frac{2}{H^2} \sum_{\substack{N \leqslant n \leqslant 2N \\ \|f(n)\| < \delta}} \left| \sum_{h=0}^{H-1} e(hf(n)) \right|^2$$

and show (6.29) by using Exercise 3.

5 Use (6.29) to prove the following result.

Let f be a function satisfying the hypotheses of Definition 6.32 and let (k, l) be an exponent pair. Suppose that $N \leqslant 8T$. Then we have

$$\mathcal{R}(f, N, \delta) \ll N\delta + \left(T^k N^l\right)^{\frac{1}{k+1}} + T^k N^{l-k}. \tag{6.30}$$

6 Apply the previous exercise to the squarefree number and square-full number problems from Chap. 5.

7 Let $s = \sigma + it \in \mathbb{C}$ such that $\frac{1}{2} \leqslant \sigma \leqslant 1$ and $t \geqslant 3$, and $\zeta(s)$ be the Riemann zeta-function.

(a) Show that $\zeta(\sigma + it) \ll |\sum_{n \leqslant t} n^{-\sigma - it}| + t^{1-2\sigma} \log t$.

(b) Let (k, l) be an exponent pair such that $l - k \leqslant \frac{1}{2}$. Prove that

$$\zeta(\sigma + it) \ll t^{\frac{k(1-\sigma)}{1+k-l}} \log t.$$

Deduce that, for all $\varepsilon > 0$, we have $\zeta(\frac{1}{2} + it) \ll t^{32/205 + \varepsilon}$.

References

[Bak86] Baker RC (1986) The greatest prime factor of the integers in an interval. Acta Arith 47:193–231

[Bak94] Baker RC (1994) The square-free divisor problem. Q J Math Oxford 45:269–277

[Bak07] Baker RC (2007) Sums of two relatively prime cubes. Acta Arith 129:103–146

[BBR12] Berkane D, Bordellès O, Ramaré O (2012) Explicit upper bounds for the remainder term in the divisor problem. Math Comput 81:1025–1051

[BH09] Baker RC, Harman G (2009) Numbers with a large prime factor II. In: Chen WWL, Gowers WT, Halberstam H, Schmidt WM, Vaughan RC (eds) Analytic number theory, essays in honour of Klaus Roth. Cambridge University Press, Cambridge

[BI86] Bombieri E, Iwaniec H (1986) On the order of $\zeta(\frac{1}{2} + it)$. Ann Sc Norm Super Pisa, Cl Sci 13:449–472

[Bor09] Bordellès O (2009) Le problème des diviseurs de Dirichlet. Quadrature 71:21–30

[Bul98] Bullen PS (1998) A dictionary of inequalities. Pitman monographs. Addison-Wesley, Reading

[CZ98] Cao X, Zhai W-G (1998) The distribution of square-free numbers of the form $[n^c]$. J Théor Nr Bordx 10:287–299

[CZ99] Cao X, Zhai W-G (1999) On the number of coprime integer pairs within a circle. Acta Arith 89:163–187

[CZ00] Cao X, Zhai W-G (2000) Multiple exponential sums with monomials. Acta Arith 92:195–213

[Dra03] Dragomir SS (2003) A survey on Cauchy–Bunyakovski–Schwarz type discrete inequalities. JIPAM J Inequal Pure Appl Math 4, Article 63

[FI89] Fouvry E, Iwaniec H (1989) Exponential sums with monomials. J Number Theory 33:311–333

[For02] Ford K (2002) Recent progress on the estimation of Weyl sums. In: Modern problems of number theory and its applications; Topical problems Part II, Tula, Russia, 2001, pp 48–66

[GK91]   Graham SW, Kolesnik G (1991) Van der Corput's method of exponential sums. London math. soc. lect. note series, vol 126. Cambridge University Press, Cambridge

[GR96]   Granville A, Ramaré O (1996) Explicit bounds on exponential sums and the scarcity of squarefree binomial coefficients. Mathematika 45:73–107

[Gre88]   Grekos G (1988) Sur le nombre de points entiers d'une courbe convexe. Bull Sci Math 112:235–254

[Har16]   Hardy GH (1916) On Dirichlet's divisor problem. Proc Lond Math Soc 15:1–25

[Har07]   Harman G (2007) Prime-detecting sieves. London math. soc. monographs. Princeton University Press, Princeton

[HB82]   Heath-Brown DR (1982) Prime numbers in short intervals and a generalized Vaughan identity. Can J Math 34:1365–1377

[HS31]   Hasse H, Suetuna Z (1931) Ein allgemeines Teilerproblem der Idealtheorie. J Fac Sci, Univ Tokyo, Sect 1A, Math 2:133–154

[Hux96]   Huxley MN (1996) Area, lattice points and exponential sums. Oxford Science Publications, London

[Hux03]   Huxley MN (2003) Exponential sums and lattice points III. Proc Lond Math Soc 87:591–609

[Hux05]   Huxley MN (2005) Exponential sums and the Riemann zeta function V. Proc Lond Math Soc 90:1–41

[HW88]   Huxley MN, Watt N (1988) Exponential sums and the Riemann zeta function. Proc Lond Math Soc 57:1–24

[IK04]   Iwaniec H, Kowalski E (2004) Analytic number theory. Colloquium publications, vol 53. Am. Math. Soc., Providence

[IM88]   Iwaniec H, Mozzochi CJ (1988) On the divisor and circle problems. J Number Theory 29:60–93

[Ivi85]   Ivić A (1985) The Riemann zeta-function. Theory and applications. Wiley, New York. 2nd edition: Dover, 2003

[Kar71]   Karacuba AA (1971) Estimates for trigonometric sums by Vinogradov's method, and some applications. Proc Steklov Inst Math 112:251–265

[Kol85]   Kolesnik G (1985) On the method of exponent pairs. Acta Arith 45:115–143

[KRW07]   Kowalski E, Robert O, Wu J (2007) Small gaps in coefficients of L-functions and $\mathfrak{B}$-free numbers in short intervals. Rev Mat Iberoam 23:281–326

[Liu93]   Liu H-Q (1993) The greatest prime factor of the integers in an interval. Acta Arith 65:301–328

[Liu94]   Liu H-Q (1994) The distribution of 4-full numbers. Acta Arith 67:165–176

[Liu95]   Liu H-Q (1995) Divisor problems of 4 and 3 dimensions. Acta Arith 73:249–269

[LW99]   Liu H-Q, Wu J (1999) Numbers with a large prime factor. Acta Arith 89:163–187

[Mon94]   Montgomery HL (1994) Ten lectures on the interface between analytic number theory and harmonic analysis. CMBS, vol 84. Amer. Math. Monthly

[Mor58]   Mordell LJ (1958) On the Kusmin–Landau inequality for exponential sums. Acta Arith 4:3–9

[MV81]   Montgomery HL, Vaughan RC (1981) The distribution of squarefree numbers. In: Halberstam H, Hooley C (eds) Recent progress in analytic number theory, vol. I. Academic Press, San Diego

[MV07]   Montgomery HL, Vaughan RC (2007) Multiplicative number theory Vol. I. Classical theory. Cambridge studies in advanced mathematics, vol 97

[NP04]   Niculescu CP, Persson L-E (2003/2004) Old and new on the Hermite–Hadamard inequality. Real Anal Exch 29:663–685

[Pet98]   Pétermann Y-FS (1998) On an estimate of Walfisz and Saltykov for an error term related to the Euler function. J Théor Nr Bordx 10:203–236

[PW97]   Pétermann Y-FS, Wu J (1997) On the sum of exponential divisors of an integer. Acta Math Hung 77:159–175

[Ram69]   Ramachandra K (1969) A note on numbers with a large prime factor. J Lond Math Soc 1:303–306

[RS01]   Rivat J, Sargos P (2001) Nombres premiers de la forme $[n^c]$. Can J Math 53:190–209
[RS03]   Robert O, Sargos P (2003) A third derivative test for mean values of exponential sums with application to lattice point problems. Acta Arith 106:27–39
[RS06]   Robert O, Sargos P (2006) Three-dimensional exponential sums with monomials. J Reine Angew Math 591:1–20
[SW00]   Sargos P, Wu J (2000) Multiple exponential sums with monomials and their applications in number theory. Acta Math Hung 88:333–354
[Ton56]   Tong KC (1956) On divisor problem III. Acta Math Sin 6:515–541
[Vaa85]   Vaaler J (1985) Some extremal functions in Fourier analysis. Bull Am Math Soc 12:183–216
[Vin54]   Vinogradov IM (1954) The method of trigonometric sums in the theory of numbers. Interscience, New York
[Vor03]   Voronoï G (1903) Sur un problème du calcul des fonctions asymptotiques. J Reine Angew Math 126:241–282
[Vor04]   Voronoï G (1904) Sur une fonction transcendante et ses applications à la sommation de quelques séries. Ann Sci Éc Norm Super 21:207–268
[Wal63]   Walfisz A (1963) Weylsche Exponentialsummen in der Neueren Zahlentheorie. VEB, Berlin
[Wu93]   Wu J (1993) Nombres B-libres dans les petits intervalles. Acta Arith 65:97–98
[Wu02]   Wu J (2002) On the primitive circle problem. Monatshefte Math 135:69–81
[Zha99]   Zhai W-G (1999) On sums and differences of two coprime kth powers. Acta Arith 91:233–248

Chapter 7
Algebraic Number Fields

7.1 Introduction

Algebraic number theory came from the necessity to solve certain Diophantine equations for which the classical tools borrowed from arithmetic in $\mathbb{Z}$ were not sufficient enough to provide a satisfying answer. For instance, the Fermat equation $x^n + y^n = z^n$ where x, y, z are positive integers and $n \geqslant 3$ is an integer, can be factored using a primitive nth root of unity $\zeta_n = e_n(1)$ as

$$z^n = \prod_{i=0}^{n-1} (x + \zeta_n^i y).$$

The right-hand side makes numbers appear which are of the form $x + \zeta_n^i y$. These numbers do not belong to $\mathbb{Q}$, but lie in a larger set which may be viewed as an extension of $\mathbb{Q}$, contained in $\mathbb{C}$, and obtained by *adjoining* the number ζ_n to $\mathbb{Q}$. This new set thus obtained is denoted by $\mathbb{Q}(\zeta_n)$ and its elements can be written in the form

$$\sum_{i=0}^{n-1} a_i \zeta_n^i \quad \text{with} \quad a_i \in \mathbb{Q}.$$

One can prove that such a set is a field, called a *cyclotomic field*, which belongs to the sets named *algebraic number fields*. Note that ζ_n is a root of the algebraic polynomial $X^n - 1$ but not a root of $X^d - 1$ for any $d < n$.

Another example is the set $\mathbb{Q}(\sqrt{-5})$ whose elements are of the form $a + b\sqrt{-5}$ with $a, b \in \mathbb{Q}$. This set is also a field, called an *imaginary quadratic field*, and has a subset of *algebraic integers* denoted by $\mathbb{Z}[\sqrt{-5}]$ whose elements are of the form $a + b\sqrt{-5}$ with $a, b \in \mathbb{Z}$. One can prove that $\mathbb{Z}[\sqrt{-5}]$ is a ring and is the analogue of $\mathbb{Z}$ in $\mathbb{Q}$. As shown in Definition 7.6, one can supply this ring with arithmetic tools. For instance, if $\alpha, \beta \in \mathbb{Z}[\sqrt{-5}]$, we say that α *divides* β, written $\alpha \mid \beta$, if there exists $\gamma \in \mathbb{Z}[\sqrt{-5}]$ such that $\beta = \alpha\gamma$. A *unit* is an element which divides 1, an *irreducible* is an element π such that any factorization $\pi = \alpha\beta$ implies that α or β is a unit and

O. Bordellès, *Arithmetic Tales*, Universitext,
DOI 10.1007/978-1-4471-4096-2_7, © Springer-Verlag London 2012

a *prime* is an element π such that if $\pi \mid \alpha\beta$, then $\pi \mid \alpha$ or $\pi \mid \beta$. One can show that $3, 7, 1 \pm 2\sqrt{-5}$ are irreducibles in $\mathbb{Z}[\sqrt{-5}]$ (see Exercise 1), but since

$$21 = 3 \times 7 = (1 + 2\sqrt{-5})(1 - 2\sqrt{-5})$$

we see that in this ring the unique factorization of the elements into irreducibles is false. One says that $\mathbb{Z}[\sqrt{-5}]$ is not a *unique factorization domain* (UFD). In $\mathbb{Z}$, both notions of irreducibles and primes coincide so that $\mathbb{Z}$ is a UFD. Thus, such a generalization of $\mathbb{Z}$ and $\mathbb{Q}$ requires a more careful approach.

This problem lies at the heart of algebraic number theory. Lamé thought that, if x and y are chosen in $\mathbb{Z}$ so that $x + y$ and $x + \zeta_n^i y$ have no common factors in $\mathbb{Z}[\zeta_n]$ for any $0 < i \leqslant n - 1$, then the Fermat equation has a solution only if there are $z_i \in \mathbb{Z}$ such that $x + \zeta_n^i y = z_i^n$ for all $0 \leqslant i \leqslant n - 1$. This was actually his argument when he addressed a meeting of the *Académie des Sciences* on March 1, 1847 and where he announced he had solved Fermat's Last Theorem.[1] Liouville said that the assumption that allows equation $x + \zeta_n^i y = z_i^n$ to follow in the argument is that $\mathbb{Z}[\zeta_n]$ is a UFD for all n. But on April 28, 1847, Kummer proved that this is not the case in general, and later Cauchy showed that the first counterexample occurs for $n = 23$. This led Kummer to invent what he called *ideal numbers* to restore unique factorization and get what we now call the *Fundamental Theorem of Ideal Theory* that we shall see in the next section.[2]

7.2 Algebraic Numbers

7.2.1 Rings and Fields

In this chapter, the reader is supposed to be familiar with the notions of groups and subgroups. For a nice introduction to these subjects, see [GG04]. Let us only quote two of the most important basic results in the theory of finite groups.

Theorem 7.1 *Let G be a finite group of order* $|G|$.

(i) (Lagrange). *Let H be a subgroup of G. Then H is a finite group and we have*

$$|G| = (G : H) \times |H|$$

where $(G : H)$ *is the index of H in G. In particular,* $|H|$ *divides* $|G|$.

(ii) (Cauchy). *Let p be a prime divisor of* $|G|$. *Then G has an element of order p.*

---

[1] The FLT states that the Fermat equation has no solution in positive integers x, y, z as soon as $n \geqslant 3$. This was finally proved by Wiles in 1995.

[2] For an interesting account of the history of the birth of the ideal theory in the late 19th century, the reader is referred to [Mol99, ST02].

Recall[3] that $(R, +, \times)$ is a *ring* if the three following rules occur.

R_1. $(R, +)$ is an abelian group.

R_2. The binary operation $\times$ is associative and has an identity element, usually denoted by 1_R or 1.

R_3. Multiplication is left- and right-distributive over addition.

If multiplication is commutative, then the ring is said to be a *commutative ring*.[4] If a ring has no zero-divisor, then it is called an *integral domain*. An element $a \in R$ is a *unit* if there exists $b \in R$ such that $ab = 1$. Such an element b, which is then unique, is called the *inverse* of a, generally denoted by a^{-1}. A *field* $\mathbb{K}$ is a non-zero ring such that every non-zero element in $\mathbb{K}$ is a unit. The notions of *subrings* and *subfields* can be defined in a similar manner, and one can check that a set $S \subset R$ is a subring of a ring R if $1_R \in S$ and if $a + b, -a, ab \in S$ for all $a, b \in S$. Similarly, a set $\mathbb{k} \subset \mathbb{K}$ is a subfield of a field $\mathbb{K}$ if $1_{\mathbb{K}} \in \mathbb{k}$ and if $a + b, -a, ab, a^{-1} \in \mathbb{k}$ for all $a, b \in \mathbb{k}$ such that $a \neq 0$.

Let R be a ring. The set of the units of R is a multiplicative group denoted by R^*. If $\mathbb{K}$ is a field, we then have $\mathbb{K}^* = \mathbb{K} \setminus \{0\}$ and this group is usually denoted by $\mathbb{K}^\times$.

The *characteristic* of a ring R or a field is the non-negative integer defined by

$$\text{char}(R) = \min\{n \in \mathbb{N} : n 1_R = 0\}.$$

If no such integer exists, then we set $\text{char}(R) = 0$. It may be proved that, if R is a *finite* ring, then $\text{char}(R) > 0$ and divides $|R|$ and if R is an integral domain, then $\text{char}(R) = 0$ or is equal to a prime number. In the latter case, we have the important identity

$$(a + b)^p = a^p + b^p$$

valid for all $a, b \in R$.

Let R, L be two rings. A *homomorphism* f from R to L is a map $f : R \longrightarrow L$ such that, for all $x, y \in R$, we have $f(x + y) = f(x) + f(y)$, $f(xy) = f(x)f(y)$ and $f(1_R) = 1_L$. An *isomorphism* is a bijective homomorphism. The *kernel* and *image* of a homomorphism are defined in the usual way

$$\ker f = \{x \in R : f(x) = 0\} \quad \text{and} \quad \text{Im} f = \{f(x) \in L : x \in R\}.$$

The important concept of ideal will play a key part in this chapter. An *ideal* of a commutative ring R is a non-empty subset $\mathfrak{a}$ of R such that

I_1. $(\mathfrak{a}, +)$ is an additive subgroup of R.

I_2. $r \in R$ and $a \in \mathfrak{a}$ imply that $ra \in \mathfrak{a}$.

---

[3] See Sect. 2.6 in Chap. 2.

[4] It should be mentioned that in this chapter we only consider unitary commutative rings unless explicitly stated to the contrary.

Let R be a ring and S a subset of R. One can check that the intersection

$$\bigcap_{\substack{\mathfrak{a}\ \text{ideal} \\ S \subseteq \mathfrak{a}}} \mathfrak{a}$$

is still an ideal of R, called the *ideal generated by* S and denoted by (S). It is a classical exercise to verify that we have

$$(S) = \left\{ \sum_{i=1}^{n} a_i s_i : n \in \mathbb{N},\ s_i \in S,\ a_i \in R \right\}.$$

If $S = \{s_1, \ldots, s_r\}$ is finite, then one may deduce from above that

$$(S) = (s_1, \ldots, s_r) = \{a_1 s_1 + \cdots + a_r s_r : a_i \in R\}$$

which may be denoted by $(S) = Rs_1 + \cdots + Rs_r$. Since R is commutative, one usually writes $(S) = s_1 R + \cdots + s_r R$. Such ideals $\mathfrak{a} = (s_1, \ldots, s_r)$ are said to be *finitely generated*.

If $S = \{s\}$ has only one element, we shall say that $(S) = (s)$ is a *principal ideal*. Hence a principal ideal $\mathfrak{a}$ of R is of the form $\mathfrak{a} = (s) = Rs = sR = \{rs : r \in R\}$ for some $s \in R$. This leads to the following definition.

Definition 7.2 (PID) An integral domain in which all the ideals are principal is called a *Principal Ideal Domain*, abbreviated in PID for convenience.

The concept of *quotient ring* is also needed (see [Bou70]).

Definition 7.3 (Quotient ring) Let R be a commutative ring and $\mathfrak{a}$ be an ideal of R. The quotient ring $R/\mathfrak{a}$ is defined by

$$R/\mathfrak{a} = \{r + \mathfrak{a} : r \in R\}.$$

with addition $(r + \mathfrak{a}) + (s + \mathfrak{a}) = (r + s) + \mathfrak{a}$, multiplication $(r + \mathfrak{a})(s + \mathfrak{a}) = rs + \mathfrak{a}$ and identity element $1_R + \mathfrak{a}$. Furthermore, the ideals of $R/\mathfrak{a}$ are of the form $\mathfrak{b}/\mathfrak{a}$ where $\mathfrak{b}$ is an ideal of R such that $\mathfrak{b} \supset \mathfrak{a}$.

We have the useful *isomorphism theorems*.

Theorem 7.4 (Isomorphism theorems) *Let R, L be commutative rings.*

(i) *Let $f : R \longrightarrow L$ be a ring homomorphism. Then $R/\ker f \simeq \operatorname{Im} f$.*
(ii) *Let $\mathfrak{a}_1 \subset \mathfrak{a}_2$ be two ideals of R. Then $R/\mathfrak{a}_2 \simeq (R/\mathfrak{a}_1)/(\mathfrak{a}_2/\mathfrak{a}_1)$.*

Proof

(i) Set $K = \ker f$ which is an ideal in R, and define the (additive) group homomorphism $\widetilde{f} : R/K \longrightarrow \operatorname{Im} f$ by $\widetilde{f}(r + K) = f(r)$. One may check that $\widetilde{f}$ is well-defined and is a surjective ring homomorphism. Furthermore, if $\widetilde{f}(r + K) = 0$, then $r \in K$ so that $\widetilde{f}$ is also injective, and hence is a ring isomorphism.

(ii) Define the (additive) group homomorphism $f : R/\mathfrak{a}_1 \longrightarrow R/\mathfrak{a}_2$ by $f(r+\mathfrak{a}_1) = r + \mathfrak{a}_2$. Since $\mathfrak{a}_1 \subset \mathfrak{a}_2$, we infer that this map is well-defined and clearly surjective with kernel $\ker f = \mathfrak{a}_2/\mathfrak{a}_1$. One may check that $\mathfrak{a}_2/\mathfrak{a}_1$ is an ideal in $R/\mathfrak{a}_1$ and that f is a ring homomorphism. The result then follows by (i).

The proof is complete.   □

One can define two operations on the ideals of a ring R. First, the *sum* of two ideals $\mathfrak{a}$ and $\mathfrak{b}$ is by definition

$$\mathfrak{a} + \mathfrak{b} = \{a + b : a \in \mathfrak{a}, \ b \in \mathfrak{b}\}$$

so that $\mathfrak{a} + \mathfrak{b}$ is the ideal generated by $\mathfrak{a}$ and $\mathfrak{b}$. If $\mathfrak{a} + \mathfrak{b} = (1) = R$, then $\mathfrak{a}$ and $\mathfrak{b}$ are said to be *coprime*. The *product* of $\mathfrak{a}$ and $\mathfrak{b}$ is the ideal *generated* by all the products ab with $a \in \mathfrak{a}$ and $b \in \mathfrak{b}$ so that

$$\mathfrak{a}\mathfrak{b} = \left\{ \sum_i a_i b_i : a_i \in \mathfrak{a}, \ b_i \in \mathfrak{b} \right\}.$$

This product is commutative, associative and has the identity element $(1) = R$. Also, it is clear from the definition that, for all ideals $\mathfrak{a}$, $\mathfrak{b}$ of R, we have

$$\mathfrak{a}\mathfrak{b} \subseteq \mathfrak{a} \cap \mathfrak{b} \subseteq \mathfrak{a} + \mathfrak{b}.$$

Furthermore, if $\mathfrak{a}$ and $\mathfrak{b}$ are coprime, then

$$\mathfrak{a}\mathfrak{b} = \mathfrak{a} \cap \mathfrak{b}.$$

Indeed, since $\mathfrak{a}$ and $\mathfrak{b}$ are coprime, there exist $a \in \mathfrak{a}$ and $b \in \mathfrak{b}$ such that $a + b = 1$. Let $c \in \mathfrak{a} \cap \mathfrak{b}$. Then $c = ca + cb$ with $ca \in \mathfrak{a}\mathfrak{b}$ and $cb \in \mathfrak{a}\mathfrak{b}$, so that $\mathfrak{a} \cap \mathfrak{b} \subseteq \mathfrak{a}\mathfrak{b}$. One can easily extend this result by induction, showing that if $\mathfrak{a}_1, \ldots, \mathfrak{a}_n$ are pairwise coprime ideals of R, then

$$\bigcap_{i=1}^{n} \mathfrak{a}_i = \mathfrak{a}_1 \cdots \mathfrak{a}_n.$$

One may generalize the concept of congruence in the following way:

if $\mathfrak{a}$ is an ideal of R, we shall write $x \equiv y \pmod{\mathfrak{a}}$ to mean $x - y \in \mathfrak{a}$.

We are now in a position to prove the following version of the Chinese remainder theorem we shall need.

Proposition 7.5 (Chinese remainder theorem) *Let R be a ring, $n \in \mathbb{N}$ and $\mathfrak{a}_1, \ldots, \mathfrak{a}_n$ be pairwise coprime ideals of R. Then*

$$R/\mathfrak{a}_1 \cdots \mathfrak{a}_n \simeq R/\mathfrak{a}_1 \oplus \cdots \oplus R/\mathfrak{a}_n.$$

Therefore, given any $r_1, \ldots, r_n \in R$, there exists some $\alpha \in R$, unique modulo $\mathfrak{a}_1 \cdots \mathfrak{a}_n$, such that

$$
\begin{cases}
\alpha \equiv r_1 \ (\mathrm{mod}\ \mathfrak{a}_1), \\
\vdots \\
\alpha \equiv r_n \ (\mathrm{mod}\ \mathfrak{a}_n).
\end{cases}
$$

Proof Define the ring homomorphism

$$
\begin{array}{rcl}
F : & R & \longrightarrow & R/\mathfrak{a}_1 \oplus \cdots \oplus R/\mathfrak{a}_n, \\
& r & \longmapsto & (r + \mathfrak{a}_1, \ldots, r + \mathfrak{a}_n).
\end{array}
$$

▷ We have

$$
\ker F = \{r \in R : r \in \mathfrak{a}_i \text{ for } i = 1, \ldots, n\} = \bigcap_{i=1}^{n} \mathfrak{a}_i = \mathfrak{a}_1 \cdots \mathfrak{a}_n
$$

since the ideals $\mathfrak{a}_1, \ldots, \mathfrak{a}_n$ are pairwise coprime, and Theorem 7.4 (i) implies that

$$
R/\mathfrak{a}_1 \cdots \mathfrak{a}_n \simeq \operatorname{Im} F.
$$

▷ It remains to show that F is surjective. To this end, we first prove that, for $i \in \{1, \ldots, n\}$, the vector

$$
v_i = (\mathfrak{a}_1, \ldots, \mathfrak{a}_{i-1}, 1 + \mathfrak{a}_i, \mathfrak{a}_{i+1}, \ldots, \mathfrak{a}_n)
$$

is in $\operatorname{Im} F$. To see this, for $i \in \{1, \ldots, n\}$, put

$$
\mathfrak{b}_i = \prod_{j \neq i} \mathfrak{a}_j.
$$

Since $\mathfrak{a}_i$ and $\mathfrak{b}_i$ are coprime, there exist $x_i \in \mathfrak{a}_i$ and $y_i \in \mathfrak{b}_i$ such that $x_i + y_i = 1$ and hence

$$
F(y_i) = F(1 - x_i) = (\mathfrak{a}_1, \ldots, \mathfrak{a}_{i-1}, 1 + \mathfrak{a}_i, \mathfrak{a}_{i+1}, \ldots, \mathfrak{a}_n) = v_i
$$

implying the desired assertion. Now, for all $r_1, \ldots, r_n \in R$, we deduce that

$$
F\left(\sum_{i=1}^{n} r_i y_i\right) = \big(F_1(r_1), \ldots, F_n(r_n)\big)
$$

where the projections $F_i : R \longrightarrow R/\mathfrak{a}_i$ are surjective, concluding the proof.   □

One can define arithmetic tools in commutative integral rings in the following way. Let R be an integral domain and $a, b \in R$. We shall say that a divides b, written $a \mid b$, if there exists $c \in R$ such that $b = ca$. This is clearly equivalent to write $(b) \subseteq (a)$. Hence

$$a \mid b \quad \Longleftrightarrow \quad (b) \subseteq (a). \tag{7.1}$$

We also need the following concepts.

Definition 7.6 Let R be a ring and $a, b \in R$.

(i) a and b are *associate*, written $a \sim b$, if there exists a unit $u \in R^*$ such that $b = ua$. Hence we have

$$a \sim b \quad \Longleftrightarrow \quad a \mid b \text{ and } b \mid a \quad \Longleftrightarrow \quad (a) = (b).$$

Furthermore, u is a unit if and only if $(u) = (1) = R$.

(ii) a is *irreducible* if a is not a unit and the factorization $a = bc$ implies that b is a unit or c is a unit.

(iii) a is a *prime* if, for all $b, c \in R$, we have $a \mid bc \Longrightarrow a \mid b$ or $a \mid c$.

It is important to know whether the elements of an integral domain can be factorized into products of irreducible elements. A first tool is the following result introducing the *Norm map*.

Lemma 7.7 (Norm map) *Let R be an integral domain and suppose there exists a map $N : R \longrightarrow \mathbb{N}$, called a norm map, such that, for all $a, b \in R$, we have $N(ab) = N(a)N(b)$ and $N(a) = 1 \Longleftrightarrow a \in R^*$. Then every element of R can be written as a product of irreducible elements.*

Proof Let $b \in R$. We proceed by induction on $N(b)$. If b is irreducible, then we have nothing to prove, so assume that b is not irreducible. Then $b = ac$ with $a \notin R^*$ and $c \notin R^*$. Hence we have $N(b) = N(a)N(c)$ with $\max(N(a), N(c)) < N(b)$. If a and b are irreducible, then we are done. Otherwise, their norms are smaller than $N(b)$ and we conclude the proof by using the induction hypothesis. $\square$

Example 7.8 Let $\Delta \neq 0$ or 1 be a squarefree integer and consider $R = \mathbb{Z}[\sqrt{\Delta}]$. We define a map $N : R \longrightarrow \mathbb{N}$ such that $N(a + b\sqrt{\Delta}) = |a^2 - \Delta b^2|$ $(a, b \in \mathbb{Z})$. We have

$$N\big((a + b\sqrt{\Delta})(c + d\sqrt{\Delta})\big) = \big|(ac + bd\Delta)^2 - (ad + bc)^2\Delta\big|$$
$$= \big|(a^2 - \Delta b^2)(c^2 - \Delta d^2)\big|$$
$$= N(a + b\sqrt{\Delta})N(c + d\sqrt{\Delta})$$

and if $u = a + b\sqrt{\Delta}$ is a unit then there is $v \in R$ such that $uv = 1$. Therefore we have

$$1 = N(1) = N(uv) = N(u)N(v)$$

which implies that $N(u) = 1$ and $N(v) = 1$ since N is a positive integer-valued function. We deduce that N is a norm map, and hence every element of $\mathbb{Z}[\sqrt{\Delta}]$ can be written as a product of irreducible elements by Lemma 7.7.

A second tool is the concept of *nœtherian ring*.

Definition 7.9 An integral domain R is a *nœtherian ring* if every ideal of R is finitely generated.

One can characterize the nœtherian rings with the following three equivalent assertions (for a proof, see [AW04a, EM99, ST02] for instance).

Lemma 7.10 *Let R be an integral domain. The following conditions are equivalent.*

(i) *R is a nœtherian ring.*
(ii) *Every ascending chain of ideals of R stops, i.e. given an ascending chain of ideals*

$$\mathfrak{a}_1 \subseteq \mathfrak{a}_2 \subseteq \cdots \subseteq \mathfrak{a}_r \subseteq \cdots$$

there exists an integer n for which $\mathfrak{a}_n = \mathfrak{a}_{n+k}$ for all $k \geqslant 0$.
(iii) *Every non-empty set of ideals of R has a maximal element, i.e. an ideal which is not properly contained in every other ideal of the set.*

The next result shows that nœtherian rings make factorization into irreducibles always possible.

Proposition 7.11 *Let R be a nœtherian ring. Then every element of R can be written as a product of irreducible elements.*

Proof Suppose that there exists an element $a \in R \setminus R^*$ which cannot be written as a product of a finite number of irreducible elements of R. By assertion (iii) of Lemma 7.10, a can be chosen so that (a) is maximal. By definition, $a = bc$ where b and c are not units, so that $(a) \subseteq (b)$ by (7.1). If $(b) = (a)$, then a and b are associates and hence c is a unit by Definition 7.6 (i), which is not the case. We infer that $(a) \subsetneq (b)$, and similarly we have $(a) \subsetneq (c)$. Since (a) is maximal, b and c can necessarily be expressed as products of irreducible elements of R, which implies in turn that a can also be written as a product of irreducible elements of R, giving a contradiction. The proposition is proved.   □

The distinction between irreducibles and primes must be well understood. If a is a prime in R, then a is irreducible but the converse may not be true as we shall see in the following example.

In $R = \mathbb{Z}[\sqrt{-5}]$, 3 is irreducible (see Exercise 1) but it is not a prime since we have for instance $3 \mid (1 + 2\sqrt{-5})(1 - 2\sqrt{-5})$ but $3 \nmid 1 + 2\sqrt{-5}$ and $3 \nmid 1 - 2\sqrt{-5}$.

In fact, this really is the core of the problem of unique factorization, for which the failure of irreducibles to be primes is precisely the reason why unique factorization fails. This motivates the following definition.

Definition 7.12 (UFD) An integral domain R in which factorizations into irreducible elements are possible and all such factorizations are unique is called a *unique factorization domain*, abbreviated to UFD.

We have the important following result which gives a strong sufficient condition to solve the problem of unique factorization (for a proof, see [AW04a, EM99, GG04]).

Theorem 7.13 *Let R be a ring. If R is a PID, then R is a UFD.*

The Euclidean division plays an important role in the arithmetic of $\mathbb{Z}$ or $\mathbb{K}[X]$. One can mimic this tool in more general integral domains in the following way.

Definition 7.14 (ED) An integral domain R is a *Euclidean domain* if it has a *Euclidean function*, i.e. a map $\phi : R \setminus \{0\} \longrightarrow \mathbb{Z}_{\geqslant 0}$ such that, for all $a, b \in R \setminus \{0\}$, there exist $q, r \in R$ such that $a = bq + r$ and $r = 0$ or $\phi(r) < \phi(b)$.

$\mathbb{Z}$ is an ED with $\phi(n) = |n|$ and, if $\mathbb{K}$ is a commutative field, $\mathbb{K}[X]$ is also an ED with the Euclidean function $\phi(P) = \deg P$. Once again, one has a stronger sufficient condition to solve the problem of unique factorization.

Theorem 7.15 *Let R be a ring. If R is an ED, then R is a PID and hence R is a UFD.*

Proof Let $\mathfrak{a}$ be an ideal of R, which we may suppose to be non-zero. Take an element $a \in \mathfrak{a}$ such that $\phi(a)$ is minimal among all the elements of $\mathfrak{a}$ and let $b \in \mathfrak{a}$. Since R is an ED, one can find $q, r \in R$ such that $b = qa + r$ and $r = 0$ or $\phi(r) < \phi(a)$. Thus $r = b - qa$ so that $r \in \mathfrak{a}$ and therefore we cannot have $\phi(r) < \phi(a)$ since $\phi(a)$ is minimal. We infer that $r = 0$ and $b = qa$ for some $q \in R$, so that $\mathfrak{a}$ is principal.   □

7.2.2 Modules

The notion of vector space may be generalized in the concept of module, in which the set of scalars is only supposed to be a ring.

Let R be a commutative ring with an identity element. An abelian group $(M, +)$ with the operation

$$\begin{aligned} R \times M &\longrightarrow M, \\ (r, m) &\longmapsto r \cdot m \end{aligned}$$

is a *R-module* if, for all $a, b \in R$ and all $x, y \in M$, the following four rules hold.

$M_1.$ $a \cdot (b \cdot x) = (ab) \cdot x.$

$M_2.$ $(a + b) \cdot x = a \cdot x + b \cdot x.$

$M_3.$ $a \cdot (x + y) = a \cdot x + a \cdot y.$

$M_4.$ $1_R \cdot x = x.$

Example

1. If R is a field, then M is a vector space over R.
2. Let V be a vector space and R be the ring of all linear maps of V into itself. Then V is an R-module.
3. Any abelian group is a $\mathbb{Z}$-module.
4. An additive group consisting of 0 alone is a module over any ring.
5. Every ring R is an R-module.
6. Every ideal $\mathfrak{a}$ of a ring R is an R-module.

The notion of *sub-R-module* can be defined in a similar way, and one may check that, if M is an R-module and N is a non-empty subset of M, then N is a sub-R-module of M if, for all $a, b \in R$ and all $x, y \in M$, we have $a \cdot x + b \cdot y \in N$. The concept of quotient module is defined as

$$M/N = \{x + N : x \in M\}$$

where N is a sub-R-module of M.

Suppose here that R is an integral domain. The set denoted by M_{tors} of M and defined by

$$M_{\text{tors}} = \{x \in M : \exists r \in R \setminus \{0\}, \ r \cdot x = 0\}$$

is a sub-R-module of M and is called a *torsion sub-R-module* of M. We shall say that M is torsion free if $M_{\text{tors}} = \{0\}$ and M is a torsion module if $M_{\text{tors}} = M$.

The concept of *module homomorphism* is defined in the usual way and so are the concepts of the kernel and of the image of such a homomorphism, which are submodules. Replacing the ring by a module and the ideals by sub-R-modules, we see that Theorem 7.4 may be generalized and adapted to the modules. For instance, the proposition (ii) of Theorem 7.4 may be rewritten in the following form.

If $N \subseteq M \subseteq P$ are three sub-R-modules of a module, then we have

$$(P/N)/(M/N) \simeq P/M.$$

Similarly, one may define a *nœtherian module* in the same way as a nœtherian ring and the assertions (ii) and (iii) of Lemma 7.10 are still valid if we replace "ring" by "module" and "ideals" by "sub-R-modules".

As for the rings, if S is a subset of a module M, the set $(S)_R$, also sometimes denoted by RS, of the finite sums

$$(S)_R = RS = \left\{ \sum_{i=1}^n r_i s_i : n \in \mathbb{N}, \ r_i \in R, \ s_i \in S \right\}$$

is called a *sub-R-module of M generated by S*. If $S = \{s_1, \ldots, s_r\}$ is finite, we say that $(S)_R$ is *finitely generated*. In particular, a module M is said to be finitely generated if there exist a positive integer n and elements $x_1, \ldots, x_n \in M$ such that, for all $m \in M$, there exist $r_1, \ldots, r_n \in R$ satisfying

$$m = \sum_{i=1}^{n} r_i x_i.$$

A module M is said to be *free* if there exist a set I, finite or not, and elements $(x_i)_{i \in I}$ with $x_i \in M$ such that all $m \in M$ can be uniquely written in the form

$$m = \sum_{i \in I} r_i x_i$$

with $r_i \in R$. The set of elements $(x_i)_{i \in I}$ is called a *basis* of M.

We give some properties of these notions without proof. The reader interested in this subject could refer to [GG04, Sam71] for more information.

Proposition 7.16 *Let R be a commutative ring with identity element.*

 (i) *Let M be a finitely generated R-module. Then*

$$M \simeq R^n / N$$

where $n \in \mathbb{N}$ and N is a sub-R-module of the free R-module R^n.
 (ii) *An R-module M is a nœtherian module if and only if every sub-R-module of M is finitely generated.*
(iii) *Suppose that R is a PID. Then every sub-R-module of a finitely generated module is finitely generated. In particular, every sub-R-module of R^n is finitely generated and can be generated by at most n elements.*
(iv) *Suppose that R is a nœtherian ring. Then an R-module M is nœtherian if and only if M is finitely generated.*
 (v) *All bases of a free R-module M have the same number of elements, called the rank of M, written $\operatorname{rank} M$.*
(vi) *Suppose that R is a PID. Then every sub-R-module of a free R-module of rank n is free of rank $\leqslant n$.*

We now wish to focus on $R = \mathbb{Z}$ which is a PID. Recall that a $\mathbb{Z}$-module is an abelian group.

Definition 7.17

1. An abelian group G is said to be a *finitely generated abelian group* if G is finitely generated as a $\mathbb{Z}$-module, so that there exist $n \in \mathbb{N}$ and elements $g_1, \ldots, g_n \in G$ such that, for all $g \in G$, there exist $r_1, \ldots, r_n \in \mathbb{Z}$ such that

$$g = \sum_{i=1}^{n} r_i g_i.$$

2. We say that the elements $g_1, \ldots, g_n \in G$ are *linearly independent* over $\mathbb{Z}$ if

$$\sum_{i=1}^{n} r_i g_i = 0 \implies r_1 = \cdots = r_n = 0.$$

3. A linearly independent set which generates G is called a $\mathbb{Z}$-*basis* of G. If $\{g_1, \ldots, g_n\}$ is such a basis, then every $g \in G$ has a unique representation

$$g = \sum_{i=1}^{n} r_i g_i \quad (r_i \in \mathbb{Z}).$$

4. An abelian group G with a basis of n elements is called a *free abelian group* of rank[5] n.

Proposition 7.18 *Let G be a free abelian group such that* rank $G = n$.

(i) *Let $\{e_1, \ldots, e_n\}$ be a basis for G and set $f_i = \sum_{j=1}^{n} a_{ij} e_j$ with $a_{ij} \in \mathbb{Z}$. Then $\{f_1, \ldots, f_n\}$ is a basis for G if and only if $|\det(a_{ij})| = 1$.*

(ii) *Let H be a subgroup of G. The group G/H is finite if and only if* rank $G =$ rank H. *In this case, if $\{e_1, \ldots, e_n\}$ is a basis for G and $\{f_1, \ldots, f_n\}$ is a basis for H such that $f_i = \sum_{j=1}^{n} a_{ij} e_j$ with $a_{ij} \in \mathbb{Z}$, then*

$$(G : H) = \left| \det(a_{ij}) \right|.$$

Proof

(i) If $\{f_1, \ldots, f_n\}$ is a basis for G, then there exist $b_{ij} \in \mathbb{Z}$ such that $e_i = \sum_{j=1}^{n} b_{ij} f_j$. Set $A = (a_{ij})$, $B = (b_{ij}) \in \mathcal{M}_n(\mathbb{Z})$. Since $f_i = \sum_{j=1}^{n} a_{ij} e_j$, we deduce that $AB = I_n$, and hence $\det(AB) = 1$, so that $\det A \det B = 1$, and we conclude by noticing that $\det A$, $\det B \in \mathbb{Z}$.

Conversely, if $|\det A| = 1$, then the f_i are $\mathbb{Z}$-linearly independent and from the well-known formula $A^{-1} = (\det A)^{-1} \tilde{A}$ where $\tilde{A} \in \mathcal{M}_n(\mathbb{Z})$ is the adjoint matrix, we deduce that $A^{-1} = \pm \tilde{A}$, so that $e_i = \sum_{j=1}^{n} b_{ij} f_j$ with $A^{-1} = (b_{ij})$. Hence the f_i generate G, and thus $\{f_1, \ldots, f_n\}$ is a basis for G.

(ii) From the structure of subgroups of free abelian groups (see [Lan93] for instance), H is free of rank $m \leqslant n$. Furthermore, there exist a basis $\{g_1, \ldots, g_n\}$ for G and positive integers $r_1, \ldots, r_m$ such that $\{r_1 g_1, \ldots, r_m g_m\}$ is a basis for H. Hence G/H is the direct product of finite cyclic groups or orders $r_1, \ldots, r_m$ and of $\mathbb{Z}^{n-m}$. We infer that H is finite if and only if $n - m = 0$ and we then have in this case

$$(G : H) = |G/H| = r_1 \cdots r_m = r_1 \cdots r_n.$$

---

[5]By Proposition 7.16 (v), all bases of a finitely generated abelian group have the same number of elements.

Now write $g_i = \sum_{j=1}^{n} b_{ij}e_j$, $f_i = \sum_{j=1}^{n} c_{ij}r_j g_j$ and define $A = (a_{ij})$, $B = (b_{ij})$, $C = (c_{ij})$ and $D = \text{diag}(r_1, \ldots, r_n)$. Then we have

$$a_{ij} = \sum_{k=1}^{n} c_{ik} r_k b_{kj}$$

which may be written as $A = CDB$ so that

$$|\det A| = |\det C \det D \det B| = |\det D| = |r_1 \cdots r_n| = |G/H|.$$

The proof is complete.   □

7.2.3 Field Extensions

The polynomial $X^2 + 1$ has no root in $\mathbb{R}$ but has roots $\pm i$ in $\mathbb{C}$. Thus, working in a field $\mathbb{K}$, it may be interesting to work in a larger field $\mathbb{L}$ such that $\mathbb{K}$ is a subfield of $\mathbb{L}$. We then say that $\mathbb{L}$ is a *field extension*[6] of $\mathbb{K}$, written $\mathbb{L}/\mathbb{K}$.

A field extension $\mathbb{L}/\mathbb{K}$ has a natural structure of vector space over $\mathbb{K}$ with vector addition the addition in $\mathbb{L}$ and scalar multiplication the operation λv with $\lambda \in \mathbb{K}$ and $v \in \mathbb{L}$.

Definition 7.19 Let $\mathbb{L}/\mathbb{K}$ be a field extension.

 (i) A $\mathbb{K}$-*basis* of $\mathbb{L}$ is a basis of $\mathbb{L}$ as a vector space over $\mathbb{K}$.
 (ii) The dimension of the vector space $\mathbb{L}$ over $\mathbb{K}$ is called the *degree* of $\mathbb{L}/\mathbb{K}$ and is denoted by $[\mathbb{L} : \mathbb{K}]$.
(iii) If $[\mathbb{L} : \mathbb{K}] < \infty$, then $\mathbb{L}/\mathbb{K}$ is called a *finite extension*.

The degrees of field extensions have the property of being multiplicative as shown in the next result (for a proof, see [GG04, ST02]).

Lemma 7.20 *If* $\mathbb{k} \subseteq \mathbb{K} \subseteq \mathbb{L}$ *are fields such that* $\mathbb{L}/\mathbb{k}$ *is a finite extension, then*

$$[\mathbb{L} : \mathbb{k}] = [\mathbb{L} : \mathbb{K}] \times [\mathbb{K} : \mathbb{k}].$$

Furthermore, if $\mathbb{k}$ *and* $\mathbb{L}$ *are finite fields, then* $|\mathbb{L}| = |\mathbb{k}|^{[\mathbb{L}:\mathbb{k}]}$.

Let $\mathbb{L}/\mathbb{K}$ be a field extension and S be a subset of $\mathbb{L}$. The concept of *extension generated by S*, denoted by $\mathbb{K}(S)$, can be defined in the same way as that of ideal generated or module generated by a subset, i.e. the intersection of all sub-extensions of $\mathbb{L}/\mathbb{K}$ containing S. One can check that $\mathbb{K}(S)$ is the subfield of $\mathbb{L}$ generated by

---

[6]More generally, suppose we have a homomorphism $\varphi : \mathbb{K} \longrightarrow \mathbb{L}$. Since these sets are fields, φ is injective and one may identify $\mathbb{K}$ with its image $\varphi(\mathbb{K})$, which is a subfield of $\mathbb{L}$.

$\mathbb{K} \cup S$. If $S = \{\alpha_1, \ldots, \alpha_n\}$, then $\mathbb{K}(S)$ is written $\mathbb{K}(\alpha_1, \ldots, \alpha_n)$ and if $S = \{\alpha\}$, then $\mathbb{K}(S) = \mathbb{K}(\alpha)$ is called a *simple sub-extension* of $\mathbb{L}/\mathbb{K}$. One can prove (see [GG04] for instance) that the extensions $\mathbb{K}(\alpha_1, \ldots, \alpha_n)$ are characterized by

$$\mathbb{K}(\alpha_1, \ldots, \alpha_n) = \left\{ \frac{P(\alpha_1, \ldots, \alpha_n)}{Q(\alpha_1, \ldots, \alpha_n)} : P, Q \in \mathbb{K}[X_1, \ldots, X_n], \ Q(\alpha_1, \ldots, \alpha_n) \neq 0 \right\}.$$

7.2.4 Tools for Polynomials

Polynomials with coefficients in $\mathbb{Q}$ play an important role in this chapter. The main problem we shall have to deal with is to determine whether a given polynomial $P \in \mathbb{Q}[X]$ is irreducible over $\mathbb{Q}$ or not. After defining this concept, we will provide some useful irreducibility criteria.[7]

Definition 7.21

1. A polynomial $P \in \mathbb{Z}[X]$ is *irreducible*, or *irreducible over* $\mathbb{Z}$, provided that $P \neq \pm 1$ and whenever $P = QR$ with $Q, R \in \mathbb{Z}[X]$, either $Q = \pm 1$ or $R = \pm 1$.

   A polynomial not irreducible over $\mathbb{Z}$ and not 0, 1 or -1 is called *reducible over* $\mathbb{Z}$.

   The polynomials 0, 1 and -1 are considered neither irreducible nor reducible over $\mathbb{Z}$.

2. A polynomial $P \in \mathbb{Q}[X]$ is *irreducible*, or *irreducible over* $\mathbb{Q}$, provided that P is not constant and whenever $P = QR$ with $Q, R \in \mathbb{Q}[X]$, either $\deg Q = 0$ or $\deg R = 0$.

   A non-constant polynomial not irreducible over $\mathbb{Q}$ is called *reducible over* $\mathbb{Q}$.

   Constant polynomials are considered neither irreducible nor reducible over $\mathbb{Q}$.

For instance, the polynomial $X^2 + 1$ is irreducible over $\mathbb{Z}$ and $\mathbb{Q}$ and the polynomial $5X + 5$ is reducible over $\mathbb{Z}$ and irreducible over $\mathbb{Q}$. These examples suggest an important connection between irreducibilities over $\mathbb{Z}$ and $\mathbb{Q}$. Gauss's lemma implies the following answer.

Gauss's Lemma.   *If $P \in \mathbb{Z}[X]$ is irreducible over $\mathbb{Z}$, then it is irreducible over $\mathbb{Q}$. Furthermore, if $P \in \mathbb{Z}[X]$ is irreducible over $\mathbb{Q}$ and if* the gcd of its coefficients is equal to 1, *then P is irreducible over $\mathbb{Z}$.*

It follows that if $P \in \mathbb{Z}[X]$ is a reducible polynomial over $\mathbb{Q}$, then there exist $Q, R \in \mathbb{Z}[X]$ such that $P = QR$ with $\deg Q > 0$ and $\deg R > 0$. This is sometimes used to prove the irreducibility over $\mathbb{Q}$ of a polynomial $P \in \mathbb{Z}[X]$.

It should be pointed out that Definition 7.21 agrees with the algebraic concept of irreducible element of a ring seen in Definition 7.6 (ii). Furthermore, the following remarks, coming readily from the definition, can sometimes be of some help.

---

[7]We sometimes make use of the equality $P = P(X)$ where the right-hand side is the composition of the polynomial P with the polynomial X.

1. If $P \in \mathbb{Z}[X]$ is such that $P(0) \neq 0$ and if $n = \deg P$, then $P(X)$ is irreducible if and only if $X^n P(1/X)$ is irreducible.
2. Let $P \in \mathbb{Z}[X]$ and $a \in \mathbb{Z}$. If $P(X + a)$ is irreducible, then $P(X)$ is irreducible.
3. If $P \in \mathbb{Z}[X]$ is an irreducible polynomial over $\mathbb{Z}$, then P has no multiple roots.
   Indeed, if there exists $\alpha \in \mathbb{C}$ such that $(X - \alpha)^2$ divides P, then α is a common root of P and its formal derivative P', so that $\gcd(P, P')$ is a non-constant polynomial of degree $< \deg P$ dividing[8] P.

The first useful irreducibility criterion, often referred to as Eisenstein's criterion, was first proved by Schönemann and shortly afterwards by Eisenstein.

Proposition 7.22 (Schönemann–Eisenstein) *Let* $P = a_n X^n + \cdots + a_1 X + a_0 \in \mathbb{Z}[X]$ *with* $n \in \mathbb{N}$. *Suppose that there exists a prime number* p *such that* $p \nmid a_n$, $p \mid a_i$ *for all* $i < n$ *and* $p^2 \nmid a_0$. *Then* P *is irreducible over* $\mathbb{Q}$.

Proof Suppose that $P = QR$ with $Q, R \in \mathbb{Z}[X]$, $r = \deg Q > 0$ and $s = \deg R > 0$. Since $\mathbb{F}_p$ is a commutative field, the ring $\mathbb{F}_p[X]$ is a UFD by the examples given after Definition 7.14. Now we have

$$QR \equiv P \equiv a_n X^n \ (\mathrm{mod}\ p)$$

and since $p \nmid a_n$, the leading coefficients of Q and R are not multiples of p, so that there exists $b, c \in \mathbb{Z}$ such that

$$Q \equiv bX^r \ (\mathrm{mod}\ p) \quad \text{and} \quad R \equiv cX^s \ (\mathrm{mod}\ p).$$

Since $r, s > 0$, we get that p divides the constant terms of Q and R. This contradicts that $p^2 \nmid a_0$, which concludes the proof. □

Note that the condition "over $\mathbb{Q}$" of the proposition cannot be removed. Indeed, by using Eisenstein's criterion with $p = 5$, we see that the polynomial $P = 3X^7 + 15X^3 + 15$ is irreducible over $\mathbb{Q}$, but reducible over $\mathbb{Z}$.

A polynomial is said to be *monic* if its leading coefficient is equal to 1. With this definition, one may slightly simplify Proposition 7.22. Indeed, in this case, the condition $p \nmid a_n$ is trivially true and the gcd of all the coefficients of the polynomial is equal to 1. One may deduce the following consequence.

Corollary 7.23 *Let* $P = X^n + a_{n-1} X^{n-1} + \cdots + a_1 X + a_0 \in \mathbb{Z}[X]$ *be a monic polynomial with* $n \in \mathbb{N}$. *Suppose that there exists a prime number* p *such that* $p \mid a_i$ *for all* $0 \leqslant i \leqslant n - 1$ *and* $p^2 \nmid a_0$. *Then* P *is irreducible over* $\mathbb{Z}$.

Example 7.24

1. Let p be a prime and m be an integer such that $p \nmid m$. Then $X^n - mp$ is irreducible over $\mathbb{Z}$.

---

[8]More generally, one can prove that if $\mathbb{K}$ is a field such that $\mathrm{char}\,\mathbb{K} = 0$, then $P \in \mathbb{K}[X]$ can be written in the form $P = Q^2 R$ if and only if P and P' have a common factor of degree > 0.

2. Let p be a prime. Then the *cyclotomic polynomial* $P = X^{p-1} + X^{p-2} + \cdots + X + 1$ is irreducible over $\mathbb{Z}$.

    Indeed, it suffices to apply Corollary 7.23 to the polynomial

$$P(X+1) = X^{p-1} + \binom{p}{1}X^{p-2} + \binom{p}{2}X^{p-3} + \cdots + \binom{p}{p-1}.$$

The next criterion shows how the reduction modulo a prime of a polynomial may be helpful. If p is a prime number and $P \in \mathbb{Z}[X]$, we write $\overline{P} \in \mathbb{F}_p[X]$ its reduction modulo p.

Proposition 7.25 (Reduction modulo p) *Let* $P = a_n X^n + \cdots + a_1 X + a_0 \in \mathbb{Z}[X]$ *and* p *be a prime number such that* $p \nmid a_n$. *If* $\overline{P}$ *is irreducible over* $\mathbb{F}_p$, *then* P *is irreducible over* $\mathbb{Q}$.

Proof Suppose that $P = QR$ with $Q, R \in \mathbb{Z}[X]$, $r = \deg Q > 0$ and $s = \deg R > 0$ and write b_r and c_s the leading coefficients of Q and R. Since $\overline{P} = \overline{Q} \times \overline{R}$, we have $a_n \equiv b_r c_s \pmod{p}$. Since $\mathbb{F}_p$ is a field, the hypothesis $p \nmid a_n$ implies that $b_r \not\equiv 0$ $\pmod{p}$ and $c_s \not\equiv 0 \pmod{p}$. Since $\overline{P}$ is irreducible over $\mathbb{F}_p$, at least one of the two polynomials $\overline{Q}$ or $\overline{R}$ has degree 0. Suppose that $\deg \overline{Q} = 0$. Therefore $\deg Q = 0$, and P is irreducible over $\mathbb{Q}$.   □

Once again, if $P \in \mathbb{Z}[X]$ is monic, one can slightly simplify this result.

Corollary 7.26 *Let* $P = X^n + a_{n-1}X^{n-1} + \cdots + a_1 X + a_0 \in \mathbb{Z}[X]$ *be a monic polynomial and* p *be a prime number. If* $\overline{P}$ *is irreducible over* $\mathbb{F}_p$, *then* P *is irreducible over* $\mathbb{Z}$.

For instance, the reduction mod 2 of the polynomial $P = X^3 + 46246X^2 - 9987X + 258963$ is given by $P \equiv X^3 + X + 1 \pmod{2}$. Since $X^3 + X + \overline{1}$ has no roots in $\mathbb{F}_2$, it is irreducible over $\mathbb{F}_2$, and hence P is irreducible over $\mathbb{Z}$.

One may be careful that the converse of this result is generally untrue. There even exist irreducible polynomials over $\mathbb{Q}$ which are reducible over $\mathbb{F}_p$ for all primes p as can be shown in the next result.

Lemma 7.27 *Let* $a, b \in \mathbb{Z}$. *Then the polynomial* $P = X^4 + aX^2 + b^2$ *is reducible over* $\mathbb{F}_p$ *for all primes* p.

Proof If $p = 2$, there are only four polynomials of the form indicated, all reducible. Suppose $p > 2$ is a prime number. One can choose an integer c such that $a \equiv 2c \pmod{p}$ which readily gives

$$P \equiv \left(X^2 + c\right)^2 - \left(c^2 - b^2\right)$$

$$\equiv \left(X^2 + b\right)^2 - (2b - 2c)X^2$$

$$\equiv \left(X^2 - b\right)^2 - (-2b - 2c)X^2 \pmod{p}.$$

Hence the result follows from the fact that one of the numbers $c^2 - b^2$, $2b - 2c$ or $-2b - 2c$ is a quadratic residue modulo p. By Proposition 3.33, if x is not a quadratic residue modulo p, then $x^{(p-1)/2} \equiv -1 \pmod{p}$ and therefore if two integers are not quadratic residues modulo p, then their product is a quadratic residue modulo p. Now if $2b - 2c$ and $-2b - 2c$ are not quadratic residues modulo p, then $(2b - 2c)(-2b - 2c) = 4(c^2 - b^2)$ is a quadratic residue modulo p, and so is $c^2 - b^2$, which concludes the proof. $\qquad\square$

For instance, the polynomial $P = X^4 + 1$ is reducible over $\mathbb{F}_p$ for all primes p, but it is irreducible over $\mathbb{Z}$. Indeed, the only non-trivial real factors of P are the polynomials $X^2 \pm X\sqrt{2} + 1$ whose roots are $(1 \pm i)/2$ and $(-1 \pm i)/2$, but these polynomials do not belong to $\mathbb{Z}[X]$.

The next result relates the irreducibility of polynomials with the prime numbers, i.e. the irreducible elements of $\mathbb{Z}$ (see also Exercise 14 in Chap. 3).

Proposition 7.28 (Ore) *Let $P \in \mathbb{Z}[X]$ be of degree n. If there exist at least $n + 5$ integers m such that $|P(m)|$ is 1 or a prime number, then P is irreducible over $\mathbb{Z}$.*

Proof The starting point is the proof of the following assertion.

> *If $Q \in \mathbb{Z}[X]$ is of degree $d \geqslant 1$, then there exist at most $d + 2$*
>
> *integers m such that $Q(m) = \pm 1$.* $\qquad\qquad$ (7.2)

First note that, if $A, B \in \mathbb{Z}[X]$ such that $A(X) - B(X) = 2$ and if $a, b \in \mathbb{Z}$ such that $A(a) = B(b) = 0$, then $a - b \mid 2$. Now let $\alpha \in \mathbb{Z}$ be the greatest solution of the equation $(Q(x) + 1)(Q(x) - 1) = 0$. The factor vanished by α has at most d integer roots, and if β is an integer root of the other factor, then $\alpha \geqslant \beta$ and $\alpha - \beta \mid 2$ by the argument above applied with $A(X) = Q(X) + 1$ and $B(X) = Q(X) - 1$, so that $\alpha - \beta = 1$ or 2 which proves (7.2).

Now we may proceed as in Exercise 14 in Chap. 3 to show Proposition 7.28. Suppose that P is not irreducible. Since P is not identically 0, 1 or -1, we have $P = QR$ with $Q, R \in \mathbb{Z}[X]$, $Q \neq \pm 1$ and $R \neq \pm 1$. Let $r = \deg Q$ and $s = \deg R$. By (7.2), there are at most $r + 2$ integers m such that $Q(m) = \pm 1$ and there are at most $s + 2$ integers m such that $R(m) = \pm 1$. Now if $|P(m)| = |Q(m)||R(m)|$ is 1 or a prime, then either $Q(m) = \pm 1$ or $R(m) = \pm 1$, so that there are at most $r + s + 4 = n + 4$ integers m such that $|P(m)|$ is 1 or a prime, which proves the proposition by contraposition. $\qquad\square$

For instance, let $P = X^6 + 8X^5 + 22X^4 + 22X^3 + 5X^2 + 6X + 1$. It may be checked that $|P(m)|$ is 1 or a prime for $m \in \{-10, -6, -5, -3, -2, -1, 0, 2, 3, 4, 8\}$ so that P is irreducible[9] over $\mathbb{Z}$.

---

[9] The curve $y^2 = P(x)$ is an example of *hyperelliptic curve* of genus 2 whose jacobian was first treated with a 2-descent method in [FPS97]. Using a refinement of a profound result due to Chabauty and Coleman, it may be shown that this curve has only six rational points, i.e. the two points at infinity and the points $(0, \pm 1)$ and $(-3, \pm 1)$.

An irreducible polynomial P over $\mathbb{Q}$ such that $\deg P \geqslant 2$ has no roots in $\mathbb{Q}$, but the converse is untrue, since the polynomial $(X^2 + 1)(X^2 + 2)$ has no roots in $\mathbb{Q}$ but is reducible. On the other hand, if $\deg P = 2$ or 3, then P is irreducible over $\mathbb{Q}$ *if and only if* it has no roots in $\mathbb{Q}$, since any non-trivial factorization of P uses polynomials of degree 1 which have a root in $\mathbb{Q}$. These examples show that a link between roots and irreducibility must exist. The next result is another example that illustrates this subject.

Proposition 7.29 *Let $P \in \mathbb{Z}[X]$ such that $P \neq \pm 1$ and let $m \in \mathbb{Z}$ such that $|P(m)|$ is 1 or a prime number and such that P has no roots in the disc $\{z \in \mathbb{C} : |z - m| \leqslant 1\}$. Then P is irreducible over $\mathbb{Z}$.*

Proof Suppose that $P = QR$ with $Q, R \in \mathbb{Z}[X]$, $Q \neq \pm 1$ and $R \neq \pm 1$. By assumption, we have $|Q(m)| = 1$ or $|R(m)| = 1$. Without loss of generality, suppose that $|Q(m)| = 1$. Since $Q \neq \pm 1$, Q is not constant and we may write $Q = a \prod_{i=1}^{r} (X - \alpha_i)$ with $a \in \mathbb{Z} \setminus \{0\}$, $r \in \mathbb{N}$ and $\alpha_1, \dots, \alpha_r$ are the roots of Q counted to their multiplicity. Hence we have

$$1 = |Q(m)| = |a| \prod_{i=1}^{r} |m - \alpha_i|.$$

Since a is a non-zero integer, we have $|a| \geqslant 1$, so that there exists $j \in \{1, \dots, r\}$ such that $|m - \alpha_j| \leqslant 1$ and $P(\alpha_j) = Q(\alpha_j) R(\alpha_j) = 0$, which is impossible since P has no roots in the disc $\{z \in \mathbb{C} : |z - m| \leqslant 1\}$. The proof is complete. $\square$

This result supposes that we have at our disposal some tools to locate the roots of polynomials. The following lemma, due to Eneström and Kakeya (see [ASV79] for instance), is a useful tool to do the job.

Lemma 7.30 *Let $P = a_n X^n + \cdots + a_1 X + a_0 \in \mathbb{R}[X]$ such that $a_i > 0$ for all $i \in \{0, \dots, n\}$. Then the roots of P are contained in the annulus*

$$\min_{0 \leqslant i < n} \left(\frac{a_i}{a_{i+1}} \right) \leqslant |z| \leqslant \max_{0 \leqslant i < n} \left(\frac{a_i}{a_{i+1}} \right).$$

Example 7.31 Let $P = X^{p-1} + 2X^{p-2} + 3X^{p-3} + \cdots + (p-1)X + p$ where p is a prime number. By Lemma 7.30, all the roots of P are in the annulus $1 + 1/(p-1) \leqslant |z| \leqslant 2$. In particular, P has no roots in the disc $\{z \in \mathbb{C} : |z| \leqslant 1\}$ and since $|P(0)| = p$ is prime, we infer that P is irreducible over $\mathbb{Z}$ by Proposition 7.29 applied with $m = 0$.

The next result shows a more general context in which irreducibility and roots are once again related.

Proposition 7.32 *Let $P \in \mathbb{Z}[X]$ be a* monic *polynomial for which $P(0) \neq 0$. Suppose further that P has exactly one root α with multiplicity 1 such that $|\alpha| \geqslant 1$. Then P is irreducible over $\mathbb{Z}$.*

Proof Suppose that $P = QR$ with $Q, R \in \mathbb{Z}[X]$, $Q \neq \pm 1$ and $R \neq \pm 1$. Since P is monic, $\deg Q > 0$ and $\deg R > 0$ and one may suppose that they are also monic. Without loss of generality, suppose that $R(\alpha) = 0$. Since $P(0) \neq 0$, $Q(0)$ is a non-zero integer, so that $|Q(0)| \geqslant 1$ and one may write $Q = (X - \beta_1) \cdots (X - \beta_r)$ with $r \in \mathbb{N}$ and $\beta_1, \ldots, \beta_r$ are the roots of Q satisfying $|\beta_i| < 1$ for all $1 \leqslant i \leqslant r$. Therefore we get

$$1 \leqslant \left| Q(0) \right| = \prod_{i=1}^{r} |\beta_i| < 1$$

giving a contradiction. $\qquad\square$

The well-known *Rouché theorem* from complex analysis also provides some connections between irreducibility and roots. We state a particular version that will be needed.

Lemma 7.33 *Let $P, Q \in \mathbb{C}[X]$ such that, for all $z \in \mathbb{C}$ such that $|z| = 1$, the* strict *inequality $|P(z) + Q(z)| < |P(z)| + |Q(z)|$ holds. Then P and Q have the same total number of roots, counting multiplicity, in the open disc $\{z \in \mathbb{C} : |z| < 1\}$.*

Example 7.34 Let $P = X^{16} - 8X^{15} - 4X^{14} - 2X^{13} - \sum_{i=0}^{12} X^i$ and $Q = (2X - 1)P$. Thus

$$Q = 2X^{17} - 17X^{16} - \sum_{i=1}^{12} X^i + 1$$

and set $R = 17X^{16}$. Then for all $z \in \mathbb{C}$ such that $|z| = 1$, we have

$$\left| Q(z) + R(z) \right| \leqslant 2|z|^{17} + \sum_{i=1}^{12} |z|^i + 1 = 15 < 17 = |R(z)| \leqslant \left| Q(z) \right| + \left| R(z) \right|$$

so that by Lemma 7.33 we infer that Q and R have the same total number of roots, counting multiplicity, in the disc $\{z \in \mathbb{C} : |z| < 1\}$, and thus Q has exactly 16 roots in this disc. Since $\deg Q = 17$, this implies that Q has exactly one root α with multiplicity 1 such that $|\alpha| \geqslant 1$. We deduce that P has also exactly one root α with multiplicity 1 such that $|\alpha| \geqslant 1$. Furthermore, P is monic and $P(0) = -1$ so that P satisfies all the hypotheses of Proposition 7.32, and hence P is irreducible over $\mathbb{Z}$. Note also that the condition *monic* cannot be removed in Proposition 7.32, since the polynomial Q is indeed such that $Q(0) \neq 0$ and has exactly one root α such that $|\alpha| \geqslant 1$, but Q is not irreducible over $\mathbb{Z}$.

We end this section by providing some other irreducibility criteria and results about polynomials.

Proposition 7.35 *Let* $P = X^n + a_{n-1}X^{n-1} + \cdots + a_1 X + a_0 \in \mathbb{Z}[X]$ *be a monic polynomial such that* $n \geqslant 1$.

   (i) (Dumas). *Let* p *be a prime number such that* $p \mid a_0$, $(v_p(a_0), n) = 1$ *and satisfying* $n v_p(a_i) \geqslant (n - i) v_p(a_0)$ *for* $1 \leqslant i \leqslant n - 1$. *Then* P *is irreducible over* $\mathbb{Z}$.
  (ii) (Eisenstein generalized). *Let* p *be a prime number and* $k \in \{0, \ldots, n\}$ *such that* $p \mid a_i$ *for all* $0 \leqslant i < k$ *and* $p^2 \nmid a_0$. *Then* P *has an irreducible factor of degree* $\geqslant k$.
 (iii) (Perron). *Suppose that* $a_0 \neq 0$ *and* $|a_{n-1}| > 1 + |a_0| + \cdots + |a_{n-2}|$. *Then* P *is irreducible over* $\mathbb{Z}$.
 (iv) (Filaseta). *Suppose that* a_i *are such that* $0 \leqslant a_i \leqslant 10^{15}$. *If* $P(10)$ *is a prime number, then* P *is irreducible over* $\mathbb{Z}$.
  (v) *The* nth *cyclotomic polynomial* Φ_n *is irreducible over* $\mathbb{Z}$ *for all positive integers* n.
 (vi) *Almost all polynomials in* $\mathbb{Z}[X]$ *are irreducible over* $\mathbb{Q}$.

Definition 7.36 (Discriminant of a polynomial) Let R be an integral domain, $P \in R[X]$, $\Bbbk$ be the quotient field of R, $r \geqslant 2$ be an integer and let $P = a \prod_{i=1}^{r}(X - \alpha_i)$ be the factorization of P in an algebraic closure $\overline{\Bbbk}$ of $\Bbbk$. The *discriminant* of P is defined by[10]

$$\mathrm{disc}(P) = a^{r + \deg(P') - 1} \prod_{i=1}^{r} \prod_{j=i+1}^{r} (\alpha_i - \alpha_j)^2.$$

The following examples are important in practice.

Example 7.37 Let p be an odd prime number, $a, b, c \in \mathbb{Z}$ and $m \geqslant 2$ be an integer.

1. $\mathrm{disc}(aX^3 + bX^2 + cX + d) = -4b^3 d + (bc)^2 + 18abcd - 4ac^3 - 27(ad)^2$.
2. $\mathrm{disc}(X^4 - 2(a+b)X^2 + (a-b)^2) = \{64ab(a-b)\}^2$.
3. $\mathrm{disc}(X^4 - 2bX^2 + b^2 - ac^2) = (16ac^2)^2(b^2 - ac^2)$.
4. $\mathrm{disc}(X^m + aX + b) = (-1)^{m(m-1)/2}\{(-1)^{m-1}(m-1)^{m-1}a^m + m^m b^{m-1}\}$.
5. $\mathrm{disc}(X^{p-1} + \cdots + X + 1) = (-1)^{(p-1)/2}p^{p-2}$.

7.2.5 Algebraic Numbers

Let $\mathbb{K}/\Bbbk$ be a field extension and $\alpha \in \mathbb{K}$. Define the homomorphism

$$\begin{aligned} F_\alpha : \quad \Bbbk[X] &\longrightarrow \quad \mathbb{K}, \\ P &\longmapsto \quad P(\alpha). \end{aligned}$$

---

[10]This definition comes from the theory of *resultants*. The word "discriminant" indicates that $\mathrm{disc}(P)$ does not vanish if all the roots α_i are distinct so that $\mathrm{disc}(P)$ *discriminates* the roots of P.

Since $\mathbb{K}$ is a field, the image $\mathrm{Im}\, F_\alpha$, often denoted by $\mathbb{k}[\alpha]$, is a subring of $\mathbb{K}$ and is then an integral domain. By Theorem 7.4 (i), we have $\mathbb{k}[X]/\ker F_\alpha \simeq \mathbb{k}[\alpha]$ and since $\mathbb{k}[\alpha]$ is an integral domain, we infer that $\ker F_\alpha$ is a prime ideal by Lemma 7.80. There are two cases.

1. $\ker F_\alpha = (0)$.
2. $\ker F_\alpha \neq (0)$. In this case, since $\mathbb{k}$ is a field, the ring $\mathbb{k}[X]$ is a PID, so that we deduce that $\ker F_\alpha$ is a maximal ideal,[11] and is then of the shape $\ker F_\alpha = (P_\alpha)$ where P_α is an irreducible polynomial over $\mathbb{k}$.

Definition 7.38

1. The number α is said to be *algebraic* over $\mathbb{k}$ if $\ker F_\alpha \neq (0)$, otherwise α is called *transcendental* over $\mathbb{k}$. In other words, α is algebraic over $\mathbb{k}$ if there exists a non-zero irreducible polynomial P_α over $\mathbb{k}$ such that α is a root of P_α. One may choose this polynomial to be monic, in which case it is unique, and is called the *minimal polynomial* of α over $\mathbb{K}$, denoted by μ_α, and we have

$$\mathbb{k}(\alpha) \simeq \mathbb{k}[X]/(\mu_\alpha) \quad \text{and} \quad \mathbb{k}(\alpha) = \mathbb{k}[\alpha] = \big\{ P(\alpha) : P \in \mathbb{k}[X] \big\}.$$

2. The degree of μ_α is called the *degree* of α over $\mathbb{K}$, written $\deg \alpha$.
3. The extension $\mathbb{K}/\mathbb{k}$ is an *algebraic extension* if every element of $\mathbb{K}$ is algebraic over $\mathbb{k}$, otherwise it is called a *transcendental extension*.
4. The field $\mathbb{k}$ is said to be *algebraically closed* if every non-constant polynomial of $\mathbb{k}[X]$ has a root in $\mathbb{k}$. An *algebraic closure* of a field $\mathbb{k}$ is an algebraic extension $\overline{\mathbb{k}}/\mathbb{k}$ such that the field $\overline{\mathbb{k}}$ is algebraically closed.
5. $\theta \in \mathbb{C}$ is called an *algebraic number* if θ is algebraic over $\mathbb{Q}$. The set of algebraic numbers is $\overline{\mathbb{Q}}$. If θ is not an algebraic number, then it is called a *transcendental number*.

Note that $\overline{\mathbb{Q}}$ is *countable*. Indeed, it is the set of the roots of $P \in \mathbb{Z}[X]$. We define the following map

$$\Phi : \quad \begin{array}{ccc} \mathbb{Z}[X] & \longrightarrow & \mathbb{N}, \\ P = \sum_{i=0}^{n} a_i X^i & \longmapsto & N = \deg P + \sum_{i=0}^{n} |a_i|. \end{array}$$

Given $N \in \mathbb{N}$, there are only finitely many polynomials P such that $\Phi(P) = N$, so that $\mathbb{Z}[X]$ is countable as a countable union of finite sets, and hence $\overline{\mathbb{Q}}$ is countable. The set of transcendental numbers is then uncountable and has no structure since the sum or product of two transcendental numbers could be algebraic. For instance, e is shown to be transcendental[12] over $\mathbb{Q}$ but $e - e = 0$.

Let $\mathbb{K}/\mathbb{k}$ be an extension and $\alpha \in \mathbb{K}$. It can easily be seen that, if α is algebraic over $\mathbb{k}$ and if its minimal polynomial is of degree d, then $[\mathbb{k}(\alpha) : \mathbb{k}] = d$ and the set $\{1, \alpha, \ldots, \alpha^{d-1}\}$ is a $\mathbb{k}$-basis of $\mathbb{k}(\alpha)$.

---

[11] In a PID, every non-zero prime ideal is maximal. See [GG04, Théorème X.3.1] for instance.

[12] Hermite, 1873. In 1882, using essentially the same ideas, Lindemann proved that π is also transcendental over $\mathbb{Q}$ and hence showed that squaring the circle is impossible.

Example 7.39

1. The numbers i, $\sqrt{d}$ with $d \in \mathbb{Z} \setminus \{0, 1\}$ squarefree, $\sqrt[3]{2}$, $\rho = \zeta_3$, $\theta = \sqrt[4]{5} + \sqrt{5}$ are algebraic over $\mathbb{Q}$. Indeed, the minimal polynomials are respectively $X^2 + 1$, $X^2 - d$, $X^3 - 2$, $X^2 + X + 1$ since they are monic and have no roots in $\mathbb{Q}$ so that they are irreducible over $\mathbb{Z}$, and the minimal polynomial of θ is $X^4 - 10X^2 - 20X + 20$ since it is monic and irreducible over $\mathbb{Z}$ via Eisenstein's criterion applied with $p = 5$. We deduce that

$$[\mathbb{Q}(i) : \mathbb{Q}] = 2, \quad [\mathbb{Q}(\sqrt{d}) : \mathbb{Q}] = 2, \quad [\mathbb{Q}(\sqrt[3]{2}) : \mathbb{Q}] = 3,$$
$$[\mathbb{Q}(\rho) : \mathbb{Q}] = 2 \quad \text{and} \quad [\mathbb{Q}(\sqrt[4]{5} + \sqrt{5}) : \mathbb{Q}] = 4.$$

2. Let $\theta = \sqrt[3]{2} + \rho\sqrt[3]{4} = \sqrt[3]{2}(1 + \rho\sqrt[3]{2})$. We have

$$\theta^3 = 2(1 + 3\rho\sqrt[3]{2} + 3\rho^2\sqrt[3]{4} + 2) = 6(1 + \rho\sqrt[3]{2} + \rho^2\sqrt[3]{4}) = 6(1 + \rho\theta).$$

Hence $\theta^3 - 6\rho\theta - 6 = 0$. Since $2\rho = -1 + \sqrt{-3}$, we get $\theta^3 + 3\theta - 6 = 3\theta\sqrt{-3}$ from which we infer that $(\theta^3 + 3\theta - 6)^2 = -27\theta^2$ which means that θ is a root of the polynomial

$$P = X^6 + 6X^4 - 12X^3 + 36X^2 - 36X + 36.$$

It can be readily checked that $|P(m)|$ is prime whenever

$$m \in \{-43, -37, -23, -13, -11, -5, -1, 1, 11, 29, 41\}$$

so that P is irreducible over $\mathbb{Z}$ by Proposition 7.28. Since P is monic, it is the minimal polynomial of θ and we get

$$[\mathbb{Q}(\sqrt[3]{2} + \rho\sqrt[3]{4}) : \mathbb{Q}] = 6.$$

3. Similarly, the number $\theta = \sqrt{8 + 3\sqrt{7}}$ is a root of $P = X^4 - 16X^2 + 1$ and $|P(m)|$ is 1 or a prime number for $m \in \{-14, -12, -4, -2, 0, 2, 4, 12, 14\}$ so that P is irreducible over $\mathbb{Z}$ by Proposition 7.28. Since P is monic, it is the minimal polynomial of θ and we get

$$[\mathbb{Q}(\sqrt{8 + 3\sqrt{7}}) : \mathbb{Q}] = 4.$$

The following result is an easy consequence of Bézout's theorem in $\mathbb{Q}[X]$.

Lemma 7.40 *Let α be an algebraic number and μ_α be the minimal polynomial of α. If $P \in \mathbb{Q}[X]$ is such that $P(\alpha) = 0$, then μ_α divides P.*

Proof If μ_α does not divide P, then since μ_α is irreducible, we have $\gcd(\mu_\alpha, P) = 1$, so that there exist $U, V \in \mathbb{Q}[X]$ such that $U(X)\mu_\alpha(X) + V(X)P(X) = 1$. However, evaluating this identity at $X = \alpha$ gives a contradiction. $\qquad\square$

The next result relates finite and algebraic extensions.

Lemma 7.41 *Let* $\mathbb{K}/\Bbbk$ *be a field extension and* $\theta \in \mathbb{K}$. *Then* $\Bbbk(\theta)$ *is a finite-dimensional* $\Bbbk$-*vector space if and only if* θ *is algebraic over* $\Bbbk$. *In particular, every finite extension is algebraic.*

Proof $\Bbbk(\theta)$ is a sub-$\Bbbk$-vector space of $\mathbb{K}$ generated by $1, \theta, \theta^2, \ldots, \theta^n, \ldots$ If $\Bbbk(\theta)$ is a finite-dimensional $\Bbbk$-vector space, then $\dim \Bbbk(\theta) = \deg \mu_\theta$ by above. Conversely, if $\Bbbk(\theta)$ is a finite-dimensional $\Bbbk$-vector space with dimension n, then the vectors $1, \theta, \ldots, \theta^n$ are $\Bbbk$-linearly dependent so that there exist $(a_0, \ldots, a_n) \in \Bbbk^{n+1}$ such that $a_n \theta^n + a_{n-1} \theta^{n-1} + \cdots + a_0 = 0$, and hence θ is a root of the non-zero polynomial $a_n X^n + a_{n-1} X^{n-1} + \cdots + a_0$, and therefore θ is algebraic over $\Bbbk$. □

This result leads to the following definition.

Definition 7.42 An *algebraic number field* is a finite extension of $\mathbb{Q}$, written $\mathbb{K}/\mathbb{Q}$. The dimension of $\mathbb{K}$ as a $\mathbb{Q}$-vector space is called the *degree* of $\mathbb{K}/\mathbb{Q}$ and is denoted by $[\mathbb{K} : \mathbb{Q}]$.

Hence if $\mathbb{K}$ is an algebraic number field, then $\mathbb{K} = \mathbb{Q}(\alpha_1, \ldots, \alpha_n)$ for finitely many algebraic numbers $\alpha_1, \ldots, \alpha_n$, as for instance a $\mathbb{Q}$-basis of the $\mathbb{Q}$-vector space $\mathbb{K}$. The following result improves on this observation.

Lemma 7.43 *If* α, β *are algebraic numbers, then there exists an algebraic number* θ *such that*

$$\mathbb{Q}(\alpha, \beta) = \mathbb{Q}(\theta).$$

Proof Let μ_α and μ_β be the minimal polynomials of α and β. We want to show that we can find $q \in \mathbb{Q}$ such that $\theta = \alpha + q\beta$ and $\mathbb{Q}(\alpha, \beta) = \mathbb{Q}(\theta)$. If such a q exists, then clearly $\mathbb{Q}(\theta) \subseteq \mathbb{Q}(\alpha, \beta)$. Set $f(X) = \mu_\alpha(\theta - qX) \in \mathbb{Q}(\theta)[X]$. Since

$$f(\beta) = \mu_\alpha(\theta - q\beta) = \mu_\alpha(\alpha) = 0$$

we infer that β is a root of f. Now we choose q such that β is the only common root of f and μ_β. This can be done since only a finite number of choices of q are thus ruled out. Therefore $\gcd(f, \mu_\beta) = a(X - \beta)$ with $a \in \mathbb{C} \setminus \{0\}$. Then $a(X - \beta) \in \mathbb{Q}(\theta)[X]$ which implies that $a, a\beta \in \mathbb{Q}(\theta)$ and so $\beta \in \mathbb{Q}(\theta)$. Now $\theta = \alpha + q\beta \in \mathbb{Q}(\theta)$ which implies that $\alpha \in \mathbb{Q}(\theta)$ and hence $\mathbb{Q}(\alpha, \beta) \subseteq \mathbb{Q}(\theta)$, so that $\mathbb{Q}(\alpha, \beta) = \mathbb{Q}(\theta)$ as required. □

This result may be generalized quite easily by induction to show that for a set $\alpha_1, \ldots, \alpha_n$ of algebraic numbers, there exists an algebraic number θ such that

$$\mathbb{Q}(\alpha_1, \ldots, \alpha_n) = \mathbb{Q}(\theta).$$

We deduce that any algebraic number field $\mathbb{K}$ can be written as[13]

$$\mathbb{K} = \mathbb{Q}(\theta) \tag{7.3}$$

for some algebraic number θ.

Definition 7.44 Let $\mathbb{K} = \mathbb{Q}(\theta)$ be an algebraic number field and μ_θ be the minimal polynomial of the algebraic number θ, abbreviated to μ for convenience.

1. The polynomial μ is called a *defining polynomial* of $\mathbb{K}$.
2. The roots of μ, which are all distinct since μ is irreducible, are called the *conjugates* of θ. Hence if $\deg \mu = n$, then θ has n conjugates, including itself, sometimes denoted by $\theta = \theta_1, \theta_2, \ldots, \theta_n$.
3. For $i \in \{1, \ldots, n\}$, the field $\mathbb{K}_i = \mathbb{Q}(\theta_i)$ is called a *conjugate field* of $\mathbb{K}$.
4. The homomorphism $\sigma_i : \mathbb{K} \longrightarrow \mathbb{C}$ defined by $\sigma_i(\theta) = \theta_i$ is injective and called an *embedding* of $\mathbb{K}$ into $\mathbb{C}$.

   It has been seen above that the set $\{1, \theta, \ldots, \theta^{n-1}\}$ is a $\mathbb{Q}$-basis for $\mathbb{K} = \mathbb{Q}(\theta)$. Therefore if $\alpha = a_0 + a_1\theta + \cdots + a_{n-1}\theta^{n-1} \in \mathbb{K}$ with $a_i \in \mathbb{Q}$, then for all $i \in \{1, \ldots, n\}$, we get

   $$\sigma_i(\alpha) = a_0 + a_1\theta_i + \cdots + a_{n-1}\theta_i^{n-1}.$$

5. If $\sigma_i(\mathbb{K}) \subseteq \mathbb{R}$ which happens if and only if $\sigma_i(\theta) \in \mathbb{R}$, we say that σ_i is *real*, otherwise σ_i is *complex*. Since complex conjugation is an automorphism of $\mathbb{C}$, it follows that $\overline{\sigma_i}$ is an embedding σ_j of $\mathbb{K}$ for some j. Hence σ_i is real if and only if $\overline{\sigma_i} = \sigma_i$ and since $\overline{\overline{\sigma_i}} = \sigma_i$, the complex embeddings come in conjugate pairs. We may enumerate in such a way that the system of all embeddings is

   $$\underbrace{\sigma_1, \ldots, \sigma_{r_1}}_{\text{real}}, \underbrace{\sigma_{r_1+1}, \overline{\sigma_{r_1+1}}, \ldots, \sigma_{r_1+r_2}, \overline{\sigma_{r_1+r_2}}}_{\text{complex}}.$$

   Then $r_1 + 2r_2 = [\mathbb{K} : \mathbb{Q}]$ and the pair (r_1, r_2) is called the *signature* of $\mathbb{K}$. One may note that r_1 and $2r_2$ are also respectively the number of real and complex roots of μ.
6. The algebraic number field $\mathbb{K}$ is called a *Galois number field*, or *normal number field*, if, for all $\alpha \in \mathbb{K}$, the minimal polynomial μ_α has all its roots in $\mathbb{K}$. This is equivalent saying that all the conjugate fields of $\mathbb{K}$ are identical to $\mathbb{K}$.

   The set of all embeddings of a Galois number field $\mathbb{K}$ is a group called the *Galois group* of $\mathbb{K}$ and denoted by $\mathrm{Gal}(\mathbb{K}/\mathbb{Q})$.

   A Galois number field $\mathbb{K}$ is said to be *abelian* if $\mathrm{Gal}(\mathbb{K}/\mathbb{Q})$ is abelian and *cyclic* if $\mathrm{Gal}(\mathbb{K}/\mathbb{Q})$ is cyclic.

From Galois theory, we know that, if $\mathbb{K}/\mathbb{Q}$ is Galois, then $|\mathrm{Gal}(\mathbb{K}/\mathbb{Q})| = [\mathbb{K} : \mathbb{Q}] = n$ and the signature of $\mathbb{K}$ must be of the form $(n, 0)$ or $(0, n/2)$.

---

[13]This result is sometimes called the *theorem of the primitive element*.

For instance, if $d \in \mathbb{Z} \setminus \{0, 1\}$ is squarefree, then the number field $\mathbb{K} = \mathbb{Q}(\sqrt{d})$ is a cyclic number field with Galois group $\text{Gal}(\mathbb{K}/\mathbb{Q}) = \{\text{Id}, \sigma\}$ where Id is the identity and $\sigma(a + b\sqrt{d}) = a - b\sqrt{d}$. On the other hand, the algebraic number field $\mathbb{K} = \mathbb{Q}(\sqrt[3]{2})$ is not Galois since the two conjugate fields $\mathbb{Q}(\rho\sqrt[3]{2})$ and $\mathbb{Q}(\rho^2\sqrt[3]{2})$ are distinct from $\mathbb{K}$.

When $\mathbb{K} = \mathbb{Q}(\theta)$ is not Galois, one may define the *Galois closure* $\mathbb{K}^s$, or *normal closure*, of $\mathbb{K}$ to be the intersection of all subfields of $\overline{\mathbb{Q}}$ which are Galois and contain $\mathbb{K}$. This is also the splitting field of μ_θ, i.e. the field obtained by adjoining to $\mathbb{Q}$ all the roots of μ_θ. For instance, if $\mathbb{K} = \mathbb{Q}(\sqrt[3]{2})$, then $\mathbb{K}^s = \mathbb{Q}(\sqrt[3]{2}, \rho)$.

7.2.6   The Ring $\mathcal{O}_{\mathbb{K}}$

A number $\alpha \in \mathbb{C}$ is an *algebraic integer* if there is a *monic* polynomial $P \in \mathbb{Z}[X]$ such that $P(\alpha) = 0$.

For instance, $\sqrt{-2}$ is an algebraic integer since it is a root of $X^2 + 2$ but $1/3$ is not an algebraic integer. The number $\sqrt{2}/3$ is an algebraic number since it has a root of $9X^2 - 2$, but it is not an algebraic integer. Indeed, assume the contrary. Then there exists a monic polynomial $P = X^n + \sum_{i=0}^{n-1} a_i X^i$ with $a_i \in \mathbb{Z}$ such that $P(\sqrt{2}/3) = 0$. Clearing out the denominators we get

$$(\sqrt{2})^n + a_{n-1} \times 3 \times (\sqrt{2})^{n-1} + \cdots + a_0 \times 3^n = 0$$

which implies that $3 \mid 2^{n/2}$ if n is even and $3 \mid 2^{(n-1)/2}$ if n is odd, which is false in either case. Hence $\sqrt{2}/3$ is not an algebraic integer.

If α is an algebraic number of degree n, then the set $\{1, \alpha, \ldots, \alpha^{n-1}\}$ is a $\mathbb{Q}$-basis for $\mathbb{Q}(\alpha)$. Thus one may think of the structure of $\mathbb{Z}[\alpha] = \{P(\alpha) : P \in \mathbb{Z}[X]\}$ to try to determine whether α is an algebraic integer. This is the idea underlined in the following criterion.

Proposition 7.45 *Let $\alpha \in \mathbb{C}$. The following assertions are equivalent.*

- (i) α *is an algebraic integer.*
- (ii)   *The monic minimal polynomial μ_α lies in $\mathbb{Z}[X]$.*
- (iii) $\mathbb{Z}[\alpha]$ *is a finitely generated $\mathbb{Z}$-module.*
- (iv)   *There exists a non-zero finitely generated $\mathbb{Z}$-module M such that $\alpha M \subseteq M$.*

Proof

▷ (i) $\implies$ (ii). Let $P \in \mathbb{Z}[X]$ be a monic polynomial such that $P(\alpha) = 0$. By Lemma 7.40, we have $P = \mu_\alpha Q$ for some $Q \in \mathbb{Q}[X]$. We write $\mu_\alpha = (a/b)\mu^*$ with $a, b \in \mathbb{Z}$ and $\mu^* \in \mathbb{Z}[x]$ whose coefficients are coprime, and similarly $Q = (c/d)Q^*$ with $c, d \in \mathbb{Z}$ and $Q^* \in \mathbb{Z}[X]$ whose coefficients are also coprime. We infer that $bdP = ac\,\mu^*Q^*$ and Gauss's lemma implies that $bd = \pm ac$, so that $P = \pm\mu^*Q^*$. Therefore the leading coefficient of both μ^* and Q^* is ± 1 and since $\mu^*(\alpha) = 0$, we finally get $\mu_\alpha = \pm\mu^* \in \mathbb{Z}[X]$.

▷ (ii) $\implies$ (iii). Let $\mu_\alpha = X^n + a_{n-1}X^{n-1} + \cdots + a_0 \in \mathbb{Z}[X]$. It is sufficient to show that $\{1, \alpha, \ldots, \alpha^{n-1}\}$ generates $\mathbb{Z}[\alpha]$ as a $\mathbb{Z}$-module, i.e. for all $m \in \mathbb{N}$, α^m is a linear combination of $\{1, \alpha, \ldots, \alpha^{n-1}\}$. The result is clear for $m < n$, so assume $m \geqslant n$ and suppose this holds for α^j with $j < m$. Then we have

$$\alpha^m = \alpha^{m-n}\alpha^n = \alpha^{m-n}\left(-a_0 - a_1\alpha - \cdots - a_{n-1}\alpha^{n-1}\right) = \sum_{j=0}^{n-1}\alpha^j\left(-\alpha^{m-n}a_j\right)$$

and the result follows by induction hypothesis.

▷ (iii) $\implies$ (iv). Obvious by choosing $M = \mathbb{Z}[\alpha]$.

▷ (iv) $\implies$ (i). Let $m_1, \ldots, m_r$ be generators of M. Since $\alpha M \subseteq M$, we obtain the existence of $a_{ij} \in \mathbb{Z}$, with $(i, j) \in \{1, \ldots, r\}^2$, such that, for all $i \in \{1, \ldots, r\}$

$$\alpha m_i = \sum_{j=1}^{r} a_{ij} m_j$$

holds. Set $B \in \mathcal{M}_r(\mathbb{Z})$ the matrix with entries $b_{ij} = a_{ij} - \alpha\delta_{ij}$ where $\delta_{ij} = 1$ if $i = j$ and 0 otherwise. Since M is non-zero, not all m_i can vanish and we infer that $\det B = 0$. Expanding now this determinant gives an equation of the form $P(\alpha) = 0$ with $P \in \mathbb{Z}[X]$ monic.

The proof is complete. $\square$

Remark 7.46 One may prove in a similar way the following more general situation.

Let $\alpha \in B$ where B is an integral domain and let A be a subring of B. The following assertions are equivalent.

(i) *α is a root of a monic minimal polynomial $P \in A[X]$.*
(ii) *$A[\alpha]$ is a finitely generated A-module.*
(iii) *There exists a non-zero finitely generated A-module M such that $\alpha M \subseteq M$.*

Corollary 7.47 (The ring $\mathcal{O}_\mathbb{K}$) *Let $\mathbb{K}$ be an algebraic number field and define $\mathcal{O}_\mathbb{K}$ to be the set of all algebraic integers in $\mathbb{K}$. Then $\mathcal{O}_\mathbb{K}$ is a ring called the* ring of integers *of $\mathbb{K}$.*

Proof By Proposition 7.45 we know that for $\alpha, \beta \in \mathcal{O}_\mathbb{K}$, one can choose non-zero finitely generated $\mathbb{Z}$-modules M, N satisfying $\alpha M \subseteq M$ and $\beta N \subseteq N$. The $\mathbb{Z}$-module MN is finitely generated and non-zero, and we have

$$(\alpha \pm \beta)MN \subseteq MN \quad \text{and} \quad (\alpha\beta)MN \subseteq MN$$

so that $\alpha \pm \beta \in \mathcal{O}_\mathbb{K}$ and $\alpha\beta \in \mathcal{O}_\mathbb{K}$ by Proposition 7.45. $\square$

The next result is a refinement of (7.3).

Corollary 7.48 *Let* $\mathbb{K}$ *be an algebraic number field. Then there exists* $\theta \in \mathcal{O}_{\mathbb{K}}$ *such that*

$$\mathbb{K} = \mathbb{Q}(\theta).$$

Proof By (7.3) there exists an algebraic number α such that $\mathbb{K} = \mathbb{Q}(\alpha)$. Let

$$\mu_\alpha = X^n + a_{n-1}X^{n-1} + \cdots + a_0 \in \mathbb{Q}[X]$$

and choose $d \in \mathbb{Z} \setminus \{0\}$ such that $da_i \in \mathbb{Z}$ for all $i \in \{0, \ldots, n-1\}$. Then the number $d\alpha$ is a root of a monic polynomial of $\mathbb{Z}[X]$ and hence $d\alpha \in \mathcal{O}_{\mathbb{K}}$ by Proposition 7.45. Now it is clear that $\mathbb{Q}(d\alpha) = \mathbb{Q}(\alpha)$, which completes the proof by choosing $\theta = d\alpha$. $\square$

Remark 7.49 The ring $\mathcal{O}_{\mathbb{K}}$ is the most important object of the theory for it contains all the arithmetic information of $\mathbb{K}$. However, one must be careful with the following problem. Let $\mathbb{K} = \mathbb{Q}(\theta)$ with $\theta \in \mathcal{O}_{\mathbb{K}}$. Then $\mathbb{Z}[\theta] \subseteq \mathcal{O}_{\mathbb{K}}$ but it may be possible that $\mathbb{Z}[\theta] \neq \mathcal{O}_{\mathbb{K}}$. For instance, $\mathbb{K} = \mathbb{Q}(\sqrt{-3})$ is an algebraic number field and $\sqrt{-3}$ is an algebraic integer. But $\rho = \zeta_3 \in \mathbb{K}$ and since it is a root of $X^2 + X + 1$, we have $\rho \in \mathcal{O}_{\mathbb{K}}$, but $\rho \notin \mathbb{Z}[\sqrt{-3}]$.

This leads to the following definition.

Definition 7.50 (Index) Let $\mathbb{K} = \mathbb{Q}(\theta)$ be an algebraic number field with $\theta \in \mathcal{O}_{\mathbb{K}}$. The *index* of θ in $\mathcal{O}_{\mathbb{K}}$ is the number $f = [\mathcal{O}_{\mathbb{K}} : \mathbb{Z}[\theta]]$.

The arithmetic structure of f plays an important role in certain results of the theory. Dedekind provided the following criterion to determine whether a prime number p is not a divisor of f (see [Coh93]).

Proposition 7.51 (Dedekind) *Let* $\mathbb{K} = \mathbb{Q}(\theta)$ *be an algebraic number field with* $\theta \in \mathcal{O}_{\mathbb{K}}$, *$p$ a prime number and $f = [\mathcal{O}_{\mathbb{K}} : \mathbb{Z}[\theta]]$. Suppose that the decomposition of* $\mu_\theta \in \mathbb{Z}[X]$ *in* $\mathbb{F}_p[X]$ *is of the form*

$$\overline{\mu_\theta} = \prod_{i=1}^{g} \overline{P}_i^{e_i}$$

where $g, e_i \geqslant 1$ are integers and $\overline{P}_i$ is irreducible in $\mathbb{F}_p[X]$ for all $i \in \{1, \ldots, g\}$. Set

$$T(X) = \frac{1}{p}\left\{ \mu_\theta(X) - \prod_{i=1}^{g} P_i^{e_i}(X) \right\}.$$

Then we have

$$p \nmid f \quad \Longleftrightarrow \quad e_i = 1 \text{ or } \overline{P}_i \nmid \overline{T} \text{ in } \mathbb{F}_p[X] \quad (1 \leqslant i \leqslant g).$$

Diaz y Diaz restated Dedekind's criterion as follows (see [Coh00]).

Proposition 7.52 (Diaz y Diaz) *With the notation of Proposition 7.51, for all $i \in \{1, \ldots, g\}$, let R_i be the remainder of the Euclidean division of μ_θ by P_i. Then we have*

$$p \nmid f \quad \Longleftrightarrow \quad R_i \notin p^2 \mathbb{Z}[X] \quad \text{for all } i \in \{1, \ldots, g\} \text{ such that } e_i \geqslant 2.$$

We will prove in Exercise 6 the following third criterion when the minimal polynomial of θ is a p-Eisenstein, i.e. μ_θ satisfies the conditions of Corollary 7.23.

Proposition 7.53 *Let $\mathbb{K} = \mathbb{Q}(\theta)$ be an algebraic number field of degree n with $\theta \in \mathcal{O}_\mathbb{K}$ and write*

$$\mu_\theta = X^n + a_{n-1} X^{n-1} + \cdots + a_0 \in \mathbb{Z}[X].$$

Let p be a prime number such that $p \mid a_i$ for all $0 \leqslant i \leqslant n - 1$ and $p^2 \nmid a_0$. Then $p \nmid f$.

7.2.7 Integral Bases

Let $\mathbb{K} = \mathbb{Q}(\theta)$ be an algebraic number field of degree n with $\theta \in \mathcal{O}_\mathbb{K}$. We denote by $\theta_1, \ldots, \theta_n$ the conjugates of θ and $\sigma_1, \ldots, \sigma_n$ the embeddings of $\mathbb{K}$ in $\mathbb{C}$. It is a matter of fact that all the θ_i have the same minimal polynomial μ_θ. We first define the following two numbers.

Definition 7.54 (Norm and trace) Let $\alpha \in \mathbb{K}$. The *norm* and *trace* of α are defined by

$$N_{\mathbb{K}/\mathbb{Q}}(\alpha) = \prod_{i=1}^{n} \sigma_i(\alpha) \quad \text{and} \quad \mathrm{Tr}_{\mathbb{K}/\mathbb{Q}}(\alpha) = \sum_{i=1}^{n} \sigma_i(\alpha).$$

Since the σ_i are homomorphisms, we have clearly the following rules. Let $\alpha, \beta \in \mathbb{K}$.

1. $N_{\mathbb{K}/\mathbb{Q}}(\alpha\beta) = N_{\mathbb{K}/\mathbb{Q}}(\alpha) \, N_{\mathbb{K}/\mathbb{Q}}(\beta)$ and $\mathrm{Tr}_{\mathbb{K}/\mathbb{Q}}(\alpha + \beta) = \mathrm{Tr}_{\mathbb{K}/\mathbb{Q}}(\alpha) + \mathrm{Tr}_{\mathbb{K}/\mathbb{Q}}(\beta)$.
2. $N_{\mathbb{K}/\mathbb{Q}}(1) = 1$ and $\mathrm{Tr}_{\mathbb{K}/\mathbb{Q}}(1) = n$.
3. For all $q \in \mathbb{Q}$, we have $N_{\mathbb{K}/\mathbb{Q}}(q\alpha) = q^n N_{\mathbb{K}/\mathbb{Q}}(\alpha)$ and $\mathrm{Tr}_{\mathbb{K}/\mathbb{Q}}(q\alpha) = q \, \mathrm{Tr}_{\mathbb{K}/\mathbb{Q}}(\alpha)$.

The next result shows that the norm and trace are rational numbers.

Proposition 7.55 *Let $\alpha \in \mathbb{K}$. Then $N_{\mathbb{K}/\mathbb{Q}}(\alpha) \in \mathbb{Q}$ and $\mathrm{Tr}_{\mathbb{K}/\mathbb{Q}}(\alpha) \in \mathbb{Q}$. Furthermore, if $\alpha \in \mathcal{O}_\mathbb{K}$, then $N_{\mathbb{K}/\mathbb{Q}}(\alpha) \in \mathbb{Z}$ and $\mathrm{Tr}_{\mathbb{K}/\mathbb{Q}}(\alpha) \in \mathbb{Z}$.*

Proof We define the so-called *characteristic polynomial* of α

$$C_\alpha = \prod_{i=1}^{n} \big(X - \sigma_i(\alpha)\big) = X^n + b_{n-1} X^{n-1} + \cdots + b_0 \tag{7.4}$$

so that $N_{\mathbb{K}/\mathbb{Q}}(\alpha) = (-1)^n b_0$ and $\mathrm{Tr}_{\mathbb{K}/\mathbb{Q}}(\alpha) = -b_{n-1}$. Hence it suffices to show that $C_\alpha \in \mathbb{Q}[X]$. We first note that, if $\alpha = Q(\theta)$ for some $Q \in \mathbb{Q}[X]$, then $\sigma_i(\alpha) = Q(\theta_i)$ for all $i \in \{1, \dots, n\}$. Then we may write

$$C_\alpha = \prod_i (X - Q(\theta_i))$$

where the θ_i runs through all the roots of the minimal polynomial μ_θ of θ whose coefficients are in $\mathbb{Z}$. Expanding the product we see that the coefficients of C_α are of the form $R(\theta_1, \dots, \theta_n)$ where $R \in \mathbb{Q}[X_1, \dots, X_n]$ is a symmetric polynomial. Therefore we have $C_\alpha \in \mathbb{Q}[X]$.

The second part of the proposition follows from the following assertion

$$C_\alpha \text{ is a power of } \mu_\alpha. \tag{7.5}$$

Indeed, if (7.5) is true and if $\alpha \in \mathcal{O}_{\mathbb{K}}$, then $\mu_\alpha \in \mathbb{Z}[X]$ and $C_\alpha \in \mathbb{Z}[X]$ by Gauss's lemma. The rest of the text is devoted to the proof of (7.5). Since μ_α is irreducible, by factorizing C_α into irreducibles, we have $C_\alpha = \mu_\alpha^r P$ with μ_α and P are coprime and both monic. If P is not constant, then there exists $i \in \{1, \dots, n\}$ such that $\sigma_i(\alpha)$ is a root of P and hence θ_i is a root of the polynomial $P \circ Q$. By Lemma 7.40, we infer that $\mu_\theta \mid P \circ Q$, which implies in particular that $P \circ Q(\theta) = 0$ and thus

$$P(\alpha) = P \circ Q(\theta) = 0$$

so that $\mu_\alpha \mid P$ which is impossible since μ_α and P are coprime. We deduce that P is constant and monic, so that $P = 1$ and then $C_\alpha = \mu_\alpha^r$. $\qquad\square$

Example 7.56 The norm and trace can sometimes be useful to determine the ring of integers and the index of some algebraic number fields. For instance, let $\mathbb{K} = \mathbb{Q}(\sqrt{-5})$. This is an algebraic number field of degree 2 by Example 7.39 with embeddings $\{\mathrm{Id}, \sigma\}$ where $\sigma(a + b\sqrt{-5}) = a - b\sqrt{-5}$ for all $a, b \in \mathbb{Q}$. Hence if $\alpha = a + b\sqrt{-5} \in \mathbb{K}$, then

$$N_{\mathbb{K}/\mathbb{Q}}(\alpha) = a^2 + 5b^2 \quad \text{and} \quad \mathrm{Tr}_{\mathbb{K}/\mathbb{Q}}(\alpha) = 2a.$$

Assume now that $\alpha \in \mathcal{O}_{\mathbb{K}}$. By Proposition 7.55, we have $2a \in \mathbb{Z}$ and $a^2 + 5b^2 \in \mathbb{Z}$ which implies that the denominator of a, and hence also of b, is at most 2. Writing $a = c/2$ and $b = d/2$, we must have $(c^2 + 5d^2)/4 \in \mathbb{Z}$, or equivalently $c^2 + 5d^2 \equiv 0 \pmod 4$. Since all squares are $\equiv 0$ or $1 \pmod 4$, we infer that c and d are even, and hence $a, b \in \mathbb{Z}$. Therefore $\mathcal{O}_{\mathbb{K}} = \mathbb{Z}[\sqrt{-5}]$ and $f = 1$.

It has been seen above that we can choose a $\mathbb{Q}$-basis $\{\alpha_1, \dots, \alpha_n\}$ of $\mathbb{K}$ as a vector space over $\mathbb{Q}$. Now since $\mathcal{O}_{\mathbb{K}}$ is a $\mathbb{Z}$-module, we may ask for a $\mathbb{Z}$-basis for $\mathcal{O}_{\mathbb{K}}$.

Definition 7.57 (Integral bases) Let $\mathbb{K}$ be an algebraic number field of degree n and $\mathcal{O}_{\mathbb{K}}$ be its ring of integers. Then a $\mathbb{Z}$-basis of $\mathcal{O}_{\mathbb{K}}$ is called an *integral basis* for $\mathbb{K}$.

In other words, $\{\alpha_1, \ldots, \alpha_m\}$ is an integral basis if and only if $\alpha_i \in \mathcal{O}_{\mathbb{K}}$ and every element of $\mathcal{O}_{\mathbb{K}}$ can be uniquely expressed in the form

$$\sum_{i=1}^{m} a_i \alpha_i \quad \text{with} \quad a_i \in \mathbb{Z}.$$

By Corollary 7.48, it follows that an integral basis is a $\mathbb{Q}$-basis for $\mathbb{K}$, so that we have $m = n$.

Now one may wonder if such bases exist. If $\mathbb{K} = \mathbb{Q}(\theta)$, then a natural candidate is $\{1, \theta, \ldots, \theta^{n-1}\}$ which is a $\mathbb{Q}$-basis for $\mathbb{K}$ with $\theta \in \mathcal{O}_{\mathbb{K}}$ and we may indeed take this basis if $\mathcal{O}_{\mathbb{K}} = \mathbb{Z}[\theta]$, but we have seen in Remark 7.49 that this equality may be false. Thus we need more work to establish the existence of integral bases for *all* algebraic number fields. One possible proof[14] uses the following very important invariant of $\mathbb{K}$.

Definition 7.58 (Discriminants) Let $\mathbb{K}$ be an algebraic number field of degree n and $\alpha_1, \ldots, \alpha_n \in \mathbb{K}$.

1. The *discriminant* of $\alpha_1, \ldots, \alpha_n$ is the number defined by

$$\Delta_{\mathbb{K}/\mathbb{Q}}(\alpha_1, \ldots, \alpha_n) = \left(\det\left(\sigma_i(\alpha_j)\right)\right)^2$$

   where det denotes the determinant of the matrix with entries $\sigma_i(\alpha_j)$ in the ith row and jth column.
2. If $\{\alpha_1, \ldots, \alpha_n\}$ is an *integral basis* for $\mathbb{K}$, then $\Delta_{\mathbb{K}/\mathbb{Q}}(\alpha_1, \ldots, \alpha_n)$ is independent of the choice of that basis. It is called the *discriminant* of $\mathbb{K}$ and is denoted by $d_{\mathbb{K}}$.

When $(\alpha_1, \ldots, \alpha_n) = (1, \theta, \ldots, \theta^{n-1})$, one may check that we have

$$\Delta_{\mathbb{K}/\mathbb{Q}}\left(1, \theta, \ldots, \theta^{n-1}\right) = \prod_{i=1}^{n} \prod_{j=i+1}^{n} \left(\sigma_j(\theta) - \sigma_i(\theta)\right)^2 = \mathrm{disc}(\mu_\theta) \qquad (7.6)$$

by Definition 7.36, which explains the word "discriminant". The next result provides the first basic properties of the discriminants.

Proposition 7.59 Let $\mathbb{K} = \mathbb{Q}(\theta)$ be an algebraic number field of degree n with $\theta \in \mathcal{O}_{\mathbb{K}}$.

(i) Let $\alpha_1, \ldots, \alpha_n \in \mathbb{K}$. Then we have

$$\Delta_{\mathbb{K}/\mathbb{Q}}(\alpha_1, \ldots, \alpha_n) = \det\left(\mathrm{Tr}_{\mathbb{K}/\mathbb{Q}}(\alpha_i \alpha_j)\right).$$

---

[14]See also [EM99] for another proof using the properties of sub-$\mathbb{Z}$-modules of finitely generated $\mathbb{Z}$-modules.

(ii) *If $\{\alpha_1, \ldots, \alpha_n\}$ is a $\mathbb{Q}$-basis for $\mathbb{K}$, then $\Delta_{\mathbb{K}/\mathbb{Q}}(\alpha_1, \ldots, \alpha_n) \in \mathbb{Q}$. Furthermore, if $\alpha_i \in \mathcal{O}_{\mathbb{K}}$, then $\Delta_{\mathbb{K}/\mathbb{Q}}(\alpha_1, \ldots, \alpha_n) \in \mathbb{Z}$.*

In particular, we have $d_{\mathbb{K}} \in \mathbb{Z}$.

(iii) *Let $\alpha_1, \ldots, \alpha_n \in \mathbb{K}$. Then we have*

$$\Delta_{\mathbb{K}/\mathbb{Q}}(\alpha_1, \ldots, \alpha_n) = 0 \quad \Longleftrightarrow \quad \text{the } \alpha_i \text{ are } \mathbb{Q}\text{-linearly dependent.}$$

In particular, the discriminant of any $\mathbb{Q}$-basis for $\mathbb{K}$ is a non-zero rational number.

(iv) *Let $\{\alpha_1, \ldots, \alpha_n\}$ and $\{\beta_1, \ldots, \beta_n\}$ be two $\mathbb{Q}$-basis for $\mathbb{K}$ such that $\beta_i = \sum_{j=1}^{n} a_{ij}\alpha_j$ with $a_{ij} \in \mathbb{Q}$. Then we have*

$$\Delta_{\mathbb{K}/\mathbb{Q}}(\beta_1, \ldots, \beta_n) = \left(\det(a_{ij})\right)^2 \Delta_{\mathbb{K}/\mathbb{Q}}(\alpha_1, \ldots, \alpha_n).$$

Proof

(i) Consider the matrix $M = (\sigma_i(\alpha_j)) \in \mathcal{M}_n(\mathbb{C})$. Then we have $M^T M = (m_{ij})$ with

$$m_{ij} = \sum_{k=1}^{n} \sigma_k(\alpha_i)\sigma_k(\alpha_j) = \text{Tr}_{\mathbb{K}/\mathbb{Q}}(\alpha_i\alpha_j)$$

and we conclude by using the facts that $\det M^T = \det M$ and $\det(AB) = \det A \det B$.

(ii) This follows readily from (i) and Proposition 7.55.

(iii) If the α_i are $\mathbb{Q}$-linearly dependent, then so are the columns of the matrix M defined in (i) since $\mathbb{Q}$ is invariant by the σ_i. Conversely, assume that $\Delta_{\mathbb{K}/\mathbb{Q}}(\alpha_1, \ldots, \alpha_n) = 0$. This implies that $\ker M^T M \neq \{0\}$ and since $M^T M$ has entries in $\mathbb{Q}$, there exists $q_i \in \mathbb{Q}$ such that, for all j, $\text{Tr}_{\mathbb{K}/\mathbb{Q}}(x\alpha_j) = 0$ with $x = \sum_{i=1}^{n} q_i\alpha_i \neq 0$. If the α_i are $\mathbb{Q}$-linearly independent, they generate $\mathbb{K}$ as a $\mathbb{Q}$-vector space and we have $\text{Tr}_{\mathbb{K}/\mathbb{Q}}(xy) = 0$ for all $y \in \mathbb{K}$ with $x \neq 0$. But taking $y = x^{-1}$ gives $0 = \text{Tr}_{\mathbb{K}/\mathbb{Q}}(1) = n$ which is impossible. Then the α_i are $\mathbb{Q}$-linearly dependent.

(iv) Setting $M = (\sigma_i(\alpha_j))$, $N = (\sigma_i(\beta_j))$ and $A = (a_{ij})$, we infer that

$$\Delta_{\mathbb{K}/\mathbb{Q}}(\beta_1, \ldots, \beta_n) = (\det N)^2 = \left(\det M A^T\right)^2$$

$$= (\det A)^2 (\det M)^2$$

$$= \left(\det(a_{ij})\right)^2 \Delta_{\mathbb{K}/\mathbb{Q}}(\alpha_1, \ldots, \alpha_n)$$

as asserted.

The proof is complete. $\square$

Now we are in a position to answer the question of the existence of an integral basis in any algebraic number field.

Corollary 7.60 *Let $\mathbb{K} = \mathbb{Q}(\theta)$ be an algebraic number field of degree n and let $\mathcal{O}_{\mathbb{K}}$ be its ring of integers. Then $\mathcal{O}_{\mathbb{K}}$ has an integral basis. Furthermore, $\mathcal{O}_{\mathbb{K}}$ is a free $\mathbb{Z}$-module of rank n.*

Proof By Corollary 7.48, there exists a $\mathbb{Q}$-basis $\{\alpha_1, \ldots, \alpha_n\}$ for $\mathbb{K}$ with $\alpha_i \in \mathcal{O}_{\mathbb{K}}$. It remains to show that there exists such a basis which is a $\mathbb{Z}$-basis for $\mathcal{O}_{\mathbb{K}}$. By Proposition 7.59 (ii) and (iii), the discriminants of such bases are in $\mathbb{Z} \setminus \{0\}$. Thus we may choose a basis $\{\alpha_1, \ldots, \alpha_n\}$ with $\alpha_i \in \mathcal{O}_{\mathbb{K}}$ and discriminant, abbreviated here in $\Delta_{\mathbb{K}/\mathbb{Q}}(\alpha)$, such that $|\Delta_{\mathbb{K}/\mathbb{Q}}(\alpha)|$ is minimal. Suppose that this basis is not a $\mathbb{Z}$-basis for $\mathcal{O}_{\mathbb{K}}$. Then there exists $\beta \in \mathcal{O}_{\mathbb{K}}$ such that $\beta = \sum_{i=1}^{n} q_i \alpha_i$ with $q_i \in \mathbb{Q}$ and at least one of these q_i is not an integer. Without loss of generality, assume that $q_1 \notin \mathbb{Z}$ and write $q_1 = [q_1] + \{q_1\}$ with $0 < \{q_1\} < 1$. The matrix

$$\begin{pmatrix} \{q_1\} & q_2 & \cdots & q_n \\ 0 & 1 & \cdots & 0 \\ \vdots & \vdots & \vdots & \vdots \\ 0 & 0 & \cdots & 1 \end{pmatrix}$$

is non-singular since $\det A = \{q_1\} \neq 0$ so that $\{\gamma, \alpha_2, \ldots, \alpha_n\}$ is a $\mathbb{Q}$-basis for $\mathbb{K}$ by Proposition 7.59 (iii) where $\gamma = \beta - [q_1]\alpha_1$, and using (iv) we get

$$\left| \Delta_{\mathbb{K}/\mathbb{Q}}(\gamma, \alpha_2, \ldots, \alpha_n) \right| = \{q_1\}^2 \left| \Delta_{\mathbb{K}/\mathbb{Q}}(\alpha) \right| < \left| \Delta_{\mathbb{K}/\mathbb{Q}}(\alpha) \right|$$

which contradicts the minimality of $|\Delta_{\mathbb{K}/\mathbb{Q}}(\alpha)|$. Hence as a $\mathbb{Z}$-module, we get

$$\mathcal{O}_{\mathbb{K}} = \mathbb{Z}\alpha_1 \oplus \mathbb{Z}\alpha_2 \oplus \cdots \oplus \mathbb{Z}\alpha_n$$

which concludes the proof.   □

7.2.8 Tools for $\mathcal{O}_{\mathbb{K}}$

The determination of an explicit integral basis and of the discriminant of $\mathcal{O}_{\mathbb{K}}$ is not an easy task. Our aim here is to collect some tools which can sometimes be helpful.

In what follows, $\mathbb{K} = \mathbb{Q}(\theta)$ is an algebraic number field of degree n with $\theta \in \mathcal{O}_{\mathbb{K}}$, discriminant $d_{\mathbb{K}}$, signature (r_1, r_2). Let $f = [\mathcal{O}_{\mathbb{K}} : \mathbb{Z}[\theta]]$ be the index of θ in $\mathcal{O}_{\mathbb{K}}$ and $\mu_\theta \in \mathbb{Z}[X]$ be the minimal polynomial of $\mathbb{K}$. For convenience, we denote by $\Delta_{\mathbb{K}/\mathbb{Q}}(\theta)$ the discriminant of the $\mathbb{Q}$-basis $\{1, \theta, \ldots, \theta^{n-1}\}$.

Proposition 7.61

(i) *Let $\{\beta_1, \ldots, \beta_n\}$ be a $\mathbb{Q}$-basis for $\mathbb{K}$ such that $\beta_i \in \mathcal{O}_{\mathbb{K}}$. Then*

$$\Delta_{\mathbb{K}/\mathbb{Q}}(\beta_1, \ldots, \beta_n) = [\mathcal{O}_{\mathbb{K}} : N]^2 \times d_{\mathbb{K}}$$

where $N = \mathbb{Z}\beta_1 + \cdots + \mathbb{Z}\beta_n$. In particular, we have

$$\text{disc}(\mu_\theta) = f^2 \times d_{\mathbb{K}}.$$

(ii) (Stickelberger). *Let* $\alpha_1, \ldots, \alpha_n \in \mathcal{O}_{\mathbb{K}}$. *Then* $\Delta_{\mathbb{K}/\mathbb{Q}}(\alpha_1, \ldots, \alpha_n) \equiv 0 \ or \ 1 \ (\mathrm{mod}\, 4)$.

(iii) (Kronecker). *The sign of* $d_{\mathbb{K}}$ *is* $(-1)^{r_2}$.

(iv) *We have*

$$\Delta_{\mathbb{K}/\mathbb{Q}}(\theta) = (-1)^{n(n-1)/2} N_{\mathbb{K}/\mathbb{Q}}\big(\mu'_\theta(\theta)\big).$$

(v) *Let* $\{\beta_1, \ldots, \beta_n\}$ *be a* $\mathbb{Q}$-*basis for* $\mathbb{K}$ *such that* $\beta_i \in \mathcal{O}_{\mathbb{K}}$. *If* $\Delta_{\mathbb{K}/\mathbb{Q}}(\beta_1, \ldots, \beta_n)$ *is squarefree, then* $\{\beta_1, \ldots, \beta_n\}$ *is an integral basis for* $\mathbb{K}$.

(vi) *If* $\mathrm{disc}(\mu_\theta)$ *is squarefree or if* $\mathrm{disc}(\mu_\theta) = 4D$ *with* D *squarefree and* $D \not\equiv 1$ $(\mathrm{mod}\, 4)$, *then* $\{1, \theta, \ldots, \theta^{n-1}\}$ *is an integral basis for* $\mathbb{K}$ *and* $d_{\mathbb{K}} = \mathrm{disc}(\mu_\theta)$.

(vii) *Suppose that, for any prime number* p *such that* $p^2 \mid \Delta_{\mathbb{K}/\mathbb{Q}}(\theta)$, *the polynomial* μ_θ *is a* p-*Eisenstein (see Proposition 7.53). Then we have*

$$\mathcal{O}_{\mathbb{K}} = \mathbb{Z}[\theta].$$

(viii) *Suppose that* $f > 1$. *Then there exists* $\alpha \in \mathcal{O}_{\mathbb{K}}$ *of the form*

$$\alpha = p^{-1}\big(a_0 + a_1\theta + \cdots + a_{n-1}\theta^{n-1}\big)$$

where p *is a prime number such that* $p^2 \mid \Delta_{\mathbb{K}/\mathbb{Q}}(\theta)$ *and* $a_i \in \mathbb{Z}$ *such that* $0 \leqslant a_i < p$ *for all* $i \in \{0, \ldots, n-1\}$.

Proof

(i) Let $\{\alpha_1, \ldots, \alpha_n\}$ be an integral basis for $\mathbb{K}$. By Proposition 7.16 (vi), as a sub-$\mathbb{Z}$-module of the free $\mathbb{Z}$-module $\mathcal{O}_{\mathbb{K}}$, $\mathbb{Z}[\theta]$ is a free $\mathbb{Z}$-module and thus has a basis $\{\gamma_1, \ldots, \gamma_n\}$ such that $\gamma_i = \sum_{j=1}^{n} m_{ij}\alpha_j$ with $m_{ij} \in \mathbb{Z}$. By Propositions 7.59 (iv) and 7.18 applied with $G = \mathcal{O}_{\mathbb{K}}$ and $H = N$, we have

$$\Delta_{\mathbb{K}/\mathbb{Q}}(\gamma_1, \ldots, \gamma_n) = \big(\det(m_{ij})\big)^2 \times d_{\mathbb{K}} = [\mathcal{O}_{\mathbb{K}} : N]^2 \times d_{\mathbb{K}}.$$

Now by Proposition 7.59 (ii), the discriminants $\Delta_{\mathbb{K}/\mathbb{Q}}(\beta_1, \ldots, \beta_n)$ and $\Delta_{\mathbb{K}/\mathbb{Q}}(\gamma_1, \ldots, \gamma_n)$ are integers and, with Proposition 7.59 (iv) again, we infer that $\Delta_{\mathbb{K}/\mathbb{Q}}(\beta_1, \ldots, \beta_n)$ divides $\Delta_{\mathbb{K}/\mathbb{Q}}(\gamma_1, \ldots, \gamma_n)$ and $\Delta_{\mathbb{K}/\mathbb{Q}}(\gamma_1, \ldots, \gamma_n)$ divides $\Delta_{\mathbb{K}/\mathbb{Q}}(\beta_1, \ldots, \beta_n)$ and their signs are equal, so that

$$\Delta_{\mathbb{K}/\mathbb{Q}}(\gamma_1, \ldots, \gamma_n) = \Delta_{\mathbb{K}/\mathbb{Q}}(\beta_1, \ldots, \beta_n)$$

which gives the asserted result.

The assertion $\mathrm{disc}(\mu_\theta) = f^2 \times d_{\mathbb{K}}$ follows by applying this result to $\{1, \theta, \ldots, \theta^{n-1}\}$ and using (7.6).

(ii) Expanding the determinant $\det(\sigma_i(\alpha_j))$ and using the $n!$ terms allows us to write

$$\det\big(\sigma_i(\alpha_j)\big) = P - N$$

where P is the contribution of the terms corresponding to permutations of even signature and N is the contribution of the terms corresponding to odd permutations. Hence

$$\Delta_{\mathbb{K}/\mathbb{Q}}(\alpha_1, \ldots, \alpha_n) = (P - N)^2 = (P + N)^2 - 4PN.$$

Now since $\sigma_i(P + N) = P + N$ and $\sigma_i(PN) = PN$ we have $P + N, PN \in \mathbb{Q}$ by Galois theory, and we have in fact $P + N, PN \in \mathbb{Z}$ since $\alpha_i \in \mathcal{O}_\mathbb{K}$. The result follows from the fact that a square is always congruent to 0 or 1 modulo 4.

(iii) By (7.6) and (i) we have

$$d_\mathbb{K} = f^{-2} \prod_{i<j} \big(\sigma_j(\theta) - \sigma_i(\theta)\big)^2.$$

Now a case-by-case examination shows that when conjugate terms are paired, all the factors become positive except for

$$\prod_{r_1 < i \leqslant r_1 + r_2} \big(\sigma_{i+r_2}(\theta) - \sigma_i(\theta)\big)^2$$

whose sign is $(-1)^{r_2}$ since $\sigma_{i+r_2}(\theta) - \sigma_i(\theta)$ is purely imaginary.

(iv) Writing $\mu_\theta = \prod_{j=1}^n (X - \sigma_j(\theta))$ we get

$$\mu'_\theta(x) = \sum_{j=1}^n \frac{\mu_\theta(x)}{x - \sigma_j(\theta)}$$

and thus

$$\mu'_\theta\big(\sigma_i(\theta)\big) = \prod_{j \neq i} \big(\sigma_i(\theta) - \sigma_j(\theta)\big).$$

We deduce that

$$N_{\mathbb{K}/\mathbb{Q}}\big(\mu'_\theta(\theta)\big) = \prod_{i=1}^n \mu'_\theta\big(\sigma_i(\theta)\big) = \prod_{i=1}^n \prod_{j \neq i} \big(\sigma_i(\theta) - \sigma_j(\theta)\big)$$

$$= \prod_{i=1}^n \prod_{j=i+1}^n \Big(-\big(\sigma_i(\theta) - \sigma_j(\theta)\big)^2\Big)$$

$$= (-1)^{n(n-1)/2} \Delta_{\mathbb{K}/\mathbb{Q}}(\theta)$$

as asserted.

(v) Let $\{\alpha_1, \ldots, \alpha_n\}$ be an integral basis for $\mathbb{K}$ so that there exist $a_{ij} \in \mathbb{Z}$ satisfying $\beta_i = \sum_{j=1}^n a_{ij}\alpha_j$. By Proposition 7.59 (iv), we have

$$\Delta_{\mathbb{K}/\mathbb{Q}}(\beta_1, \ldots, \beta_n) = \big(\det(a_{ij})\big)^2 \Delta_{\mathbb{K}/\mathbb{Q}}(\alpha_1, \ldots, \alpha_n)$$

and we infer that $|\det(a_{ij})| = 1$ since $\Delta_{\mathbb{K}/\mathbb{Q}}(\beta_1, \ldots, \beta_n)$ is squarefree. We conclude by using Proposition 7.18 (i).

(vi) If $\mathrm{disc}(\mu_\theta)$ is squarefree, then we apply (v). If $\mathrm{disc}(\mu_\theta) = 4D$ with D as stated in the proposition, we first note that $f^2 \mid 4D$ by (i) and (7.6), and hence $f = 1$ or $f = 2$ since D squarefree. If $f = 2$, then we have $4D = 4d_\mathbb{K}$ so that $D =$

$d_{\mathbb{K}}$ and hence $d_{\mathbb{K}} \not\equiv 1 \pmod 4$, which contradicts (ii). Hence $f = 1$ and then $\mathrm{disc}(\mu_\theta) = d_{\mathbb{K}}$.

(vii) By Proposition 7.53, we infer that f is not divisible by p for any prime p such that $p^2 \mid \Delta_{\mathbb{K}/\mathbb{Q}}(\theta)$. Using (i) we get $f = 1$ as asserted.

(viii) Since $f > 1$, there exist $p \mid f$ and $\overline{\beta} \in \mathcal{O}_{\mathbb{K}}/\mathbb{Z}[\theta]$ of order p by Cauchy's theorem (Theorem 7.1 (i)), and thus $p\beta \in \mathbb{Z}[\theta]$. Since $\{1, \theta, \theta^2, \dots, \theta^{n-1}\}$ is a $\mathbb{Z}$-basis for $\mathbb{Z}[\theta]$, we get $p\beta = b_0 + b_1\theta + \cdots + b_{n-1}\theta^{n-1}$ with $b_i \in \mathbb{Z}$. Now the Euclidean division of b_i by p gives $b_i = pq_i + a_i$ with $0 \leqslant a_i < p$ and $q_i \in \mathbb{Z}$, so that the number $\alpha = \beta - \sum_{i=0}^{n-1} q_i \theta^i \in \mathcal{O}_{\mathbb{K}}$ satisfies the conditions of the proposition. Furthermore, we have $p \mid f$ so that $p^2 \mid \Delta_{\mathbb{K}/\mathbb{Q}}(\theta)$ by (i).

The proof is complete. $\qquad\qquad\qquad\qquad\qquad\qquad\qquad\qquad\qquad\qquad\qquad$ □

Example 7.62

1. Let $\mathbb{K} = \mathbb{Q}(6^{1/3})$. The polynomial $P = X^3 - 6$ satisfies Eisenstein's criterion with respect to 2 and 3 and hence is irreducible over $\mathbb{Z}$, so that $P = \mu_\theta$. We have $\mathrm{disc}(P) = -2^2 \times 3^5$ and the use of Proposition 7.61 (vii) implies that $\mathcal{O}_{\mathbb{K}} = \mathbb{Z}[\theta]$. Therefore $\mathbb{K}$ is an algebraic number field of degree 3, called a *pure cubic field*, signature $(1, 1)$ so that $\mathbb{K}$ is not Galois, discriminant $d_{\mathbb{K}} = -2^2 \times 3^5$ and $\{1, \theta, \theta^2\}$ is an integral basis for $\mathbb{K}$.

2. Let θ be a root of the polynomial $P = X^4 + X + 1$. We check that $|P(m)|$ is 1 or a prime for $m \in \{-6, -3, 0, 1, 2, 5, 6, 9, 11\}$, so that the polynomial P is irreducible over $\mathbb{Z}$ by Proposition 7.28. Let $\mathbb{K} = \mathbb{Q}(\theta)$ be the corresponding algebraic number field. Using Example 7.37, we obtain $\mathrm{disc}(P) = 229$. Since 229 is prime, we have $\mathcal{O}_{\mathbb{K}} = \mathbb{Z}[\theta]$ and $\{1, \theta, \theta^2, \theta^3\}$ is an integral basis for $\mathbb{K}$ by Proposition 7.61 (v). Hence $\mathbb{K}$ is an algebraic number field of degree 4, signature $(0, 2)$, discriminant $d_{\mathbb{K}} = 229$ and $\mathcal{O}_{\mathbb{K}} = \mathbb{Z}[\theta]$.

3. Let $\mathbb{K} = \mathbb{Q}(\sqrt{8 + 3\sqrt{7}})$ from Example 7.39. The minimal polynomial is

$$\mu_\theta = X^4 - 16X^2 + 1$$

whose discriminant is equal to

$$\mathrm{disc}(P) = \left(16 \times 7 \times 3^2\right)^2\left(8^2 - 7 \times 3^2\right) = 2^8 \times 3^4 \times 7^2$$

by Example 7.37. Hence the only prime factors of f are 2, 3 or 7. Proposition 7.52 gives $2 \nmid f$, $3 \mid f$ and $7 \nmid f$ so that $f = 3$ or $f = 9$. In particular $f > 1$, so that using Proposition 7.61 (viii) we infer that there exists $\alpha \in \mathcal{O}_{\mathbb{K}}$ of the form

$$\alpha = \frac{1}{3}\left(a + b\theta + c\theta^2 + d\theta^3\right)$$

with integers $0 \leqslant a, b, c, d \leqslant 2$. Since $\mathrm{Tr}_{\mathbb{K}/\mathbb{Q}}(\alpha) = 4(a + 8c)/3 \in \mathbb{Z}$, this implies that $3 \mid a + 8c$ so that $(a, c) = (0, 0)$, $(1, 1)$ or $(2, 2)$. In the first case we have

$$N_{\mathbb{K}/\mathbb{Q}}(\alpha) = \frac{1}{81}\left(b^4 + 32b^3d + 258(bd)^2 + 32bd^3 + d^4\right) \in \mathbb{Z}$$

which implies that $(b, d) = (1, 1)$. We verify that $(\theta + \theta^3)/3 \in \mathcal{O}_{\mathbb{K}}$. If $(a, c) = (1, 1)$, then

$$N_{\mathbb{K}/\mathbb{Q}}(\alpha) = \frac{1}{81}\left(b^4 + 32b^3 d + b^2\left(258d^2 - 36\right) + b\left(32d^3 - 576d\right) + d^4 \right.$$
$$\left. - 4572d^2 + 324\right)$$

and since $N_{\mathbb{K}/\mathbb{Q}}(\alpha) \in \mathbb{Z}$, we get $(b, d) = (0, 0)$, $(1, 1)$ or $(2, 2)$. We check that $\frac{1}{3}(1 + \theta^2) \in \mathcal{O}_{\mathbb{K}}$ and let Δ be the discriminant of $\{1, \theta, \frac{1}{3}(1 + \theta^2), \frac{1}{3}(\theta + \theta^3)\}$. Straightforward computations give $\Delta = 2^8 \times 7^2$. Hence $|\Delta|$ is minimal, so that $f = 9$ and $d_{\mathbb{K}} = \Delta = 2^8 \times 7^2$ by Proposition 7.61 (i). Furthermore

$$\left\{1, \theta, \frac{1 + \theta^2}{3}, \frac{\theta + \theta^3}{3}\right\}$$

is an integral basis for $\mathbb{K}$.

The algebraic number field $\mathbb{K}$ is said to be *monogenic*, or to have a *power basis*, if there exists $\alpha \in \mathcal{O}_{\mathbb{K}}$ such that $\mathcal{O}_{\mathbb{K}} = \mathbb{Z}[\alpha]$. The following example, due to Dedekind, shows that there exist some algebraic number fields which cannot be monogenic.

Example 7.63 (Dedekind) Let $\mathbb{K} = \mathbb{Q}(\theta)$ be an algebraic number field where θ is a root of the polynomial $P = X^3 - X^2 - 2X - 8$. One may check that $\mathrm{disc}(P) = -4 \times 503$ and that P is irreducible over $\mathbb{Q}$ by Exercise 12 in Chap. 3 for instance. Let $\beta = \frac{1}{2}(\theta^2 + \theta)$. A simple calculation shows that $\beta^3 - 3\beta^2 - 10\beta - 9 = 0$ so that $\beta \in \mathcal{O}_{\mathbb{K}}$ and, using Definition 7.58, we have

$$\Delta_{\mathbb{K}/\mathbb{Q}}\left(1, \theta, \theta^2\right) = -4 \times 503 \quad \text{and} \quad \Delta_{\mathbb{K}/\mathbb{Q}}(1, \theta, \beta) = \frac{1}{4}\Delta_{\mathbb{K}/\mathbb{Q}}\left(1, \theta, \theta^2\right) = -503.$$

Since $\Delta_{\mathbb{K}/\mathbb{Q}}(1, \theta, \beta)$ is squarefree, we deduce that $\{1, \theta, \beta\}$ is an integral basis for $\mathbb{K}$ by Proposition 7.61 (v). Now let $\alpha \in \mathcal{O}_{\mathbb{K}}$ and set $\alpha = a + b\theta + c\beta$ with $a, b, c \in \mathbb{Z}$. We get

$$\alpha^2 = \left(a^2 + 6c^2 + 8bc\right) + \left(2c^2 - b^2 + 2ab\right)\theta + \left(2b^2 + 3c^2 + 2ac + 4bc\right)\beta$$

so that

$$\Delta_{\mathbb{K}/\mathbb{Q}}\left(1, \alpha, \alpha^2\right) \equiv (bc)^2(3c + b)^2 \pmod{2}$$

which is an even number in all cases. Hence this discriminant cannot be equal to -503 which proves that $\mathcal{O}_{\mathbb{K}}$ has no integral basis of the form $\mathbb{Z}[\alpha]$.

7.2.9 Examples of Integral Bases

In this section, we intend to study some examples of algebraic number fields often used in the literature. The calculation of an integral basis for these fields requires both basic principles seen in the former sections and a touch of some arithmetic technicality.

Quadratic Fields

Proposition 7.64 *Let $d \in \mathbb{Z} \setminus \{0, 1\}$ be a squarefree integer and $\mathbb{K} = \mathbb{Q}(\sqrt{d})$ be a quadratic number field. We set*

$$\omega = \begin{cases} \frac{1}{2}(1 + \sqrt{d}), & \text{if } d \equiv 1 \pmod{4}, \\ \sqrt{d}, & \text{otherwise.} \end{cases}$$

Then

d	Integral basis	Discriminant	Polynomial
$d \equiv 1 \pmod{4}$	$(1, \omega)$	d	$X^2 - X + \frac{1-d}{4}$
$d \not\equiv 1 \pmod{4}$	$(1, \omega)$	$4d$	$X^2 - d$

Proof We first look at the algebraic integers in $\mathbb{K}$ and let $\alpha \in \mathbb{K}$. Then there exist $a, b, c \in \mathbb{Z}$ with $(a, b, c) = 1$ and $c > 0$ such that

$$\alpha = \frac{a + b\sqrt{d}}{c}.$$

If $\alpha \in \mathcal{O}_{\mathbb{K}}$, then $\text{Tr}_{\mathbb{K}/\mathbb{Q}}(\alpha) = 2a/c \in \mathbb{Z}$ and $N_{\mathbb{K}/\mathbb{Q}}(\alpha) = (a^2 - db^2)/c^2 \in \mathbb{Z}$. If there exists a prime p dividing both a and c, then from the norm we infer that $p \mid b$ since d is squarefree, which contradicts the fact that $(a, b, c) = 1$. Hence $(a, c) = 1$ and the condition on the trace implies that $c \mid 2$.

▷ Suppose that $d \not\equiv 1 \pmod{4}$. If $c = 2$, then from the norm we deduce that a and b have to be both odd and $a^2 - db^2 \equiv 0 \pmod{4}$. This implies that $d \equiv 1 \pmod{4}$, giving a contradiction. Hence $c = 1$ and since $\Delta_{\mathbb{K}/\mathbb{Q}}(1, \sqrt{d}) = 4d$ with d squarefree, we infer that $\mathcal{O}_{\mathbb{K}} = \mathbb{Z}[\sqrt{d}]$ in this case by Proposition 7.61 (vi).

▷ Suppose that $d \equiv 1 \pmod{4}$ and set $f = [\mathcal{O}_{\mathbb{K}} : \mathbb{Z}[\sqrt{d}]]$. If $P = X^2 - d$, then by assumption on d we get $P \equiv (X + 1)^2 \pmod{2}$, and hence $2 \mid f$ by Proposition 7.52. Since $f^2 \mid \text{disc}(P) = 4d$, we also have $f^2 \mid 4$ since d is squarefree. Thus $f = 2$. Note that $\omega = \frac{1}{2}(1 + \sqrt{d})$ is a root of the polynomial $X^2 - X - \frac{1}{4}(1 - d) \in \mathbb{Z}[X]$ so that $\omega \in \mathcal{O}_{\mathbb{K}}$. Now $\Delta_{\mathbb{K}/\mathbb{Q}}(1, \omega) = d$ so that $\mathcal{O}_{\mathbb{K}} = \mathbb{Z}[\frac{1}{2}(1 + \sqrt{d})]$ by Proposition 7.61 (v).

The proof is complete. $\qquad\qquad\qquad\qquad\qquad\qquad\qquad\qquad\qquad\qquad\qquad\quad\square$

Cyclotomic Fields

Let $\zeta_n = e_n(1) = e^{2i\pi/n}$. Then the n numbers $1, \zeta_n, \zeta_n^2, \ldots, \zeta_n^{n-1}$ are the nth roots of unity, forming a regular n-gon in the complex plane. If $(k, n) > 1$, then ζ_n^k is a root of unity or order $n/(n,k) < n$ whereas if $(n,k) = 1$, then ζ_n^k is not a root of lower order, and is called a *primitive* nth root of unity. The number of these roots is then $\varphi(n)$ where φ is Euler's totient function. We define the nth *cyclotomic polynomial* to be the monic polynomial Φ_n whose roots are the primitive nth root of unity so that

$$\Phi_n = \prod_{\substack{i=1 \\ (i,n)=1}}^{n} \left(X - \zeta_n^i\right).$$

We have

$$X^n - 1 = \prod_{k=1}^{n}\left(X - \zeta_n^k\right) = \prod_{d \mid n} \prod_{\substack{k=1 \\ (k,n)=n/d}}^{n} \left(X - \zeta_n^k\right) = \prod_{d \mid n} \prod_{\substack{i=1 \\ (i,d)=1}}^{d} \left(X - \zeta_n^{in/d}\right) = \prod_{d \mid n} \Phi_d$$

so that we get by the Möbius inversion formula

$$\Phi_n = \prod_{d \mid n}\left(X^d - 1\right)^{\mu(n/d)} \tag{7.7}$$

and hence $\Phi_n \in \mathbb{Z}[X]$ for all $n \geqslant 1$. In particular, if $n = p$ is a prime number, then

$$\Phi_p = X^{p-1} + X^{p-2} + \cdots + X + 1.$$

The irreducibility over $\mathbb{Z}$ of Φ_p has been proved in Example 7.24 and the irreducibility in the general case has been stated in Proposition 7.35. We now intend to show that Φ_n is irreducible over $\mathbb{Z}$ for all positive integers n. A possible proof uses the following lemma established by Schönemann.

Lemma 7.65 *Let $A = (X - a_1) \cdots (X - a_r) \in \mathbb{Z}[X]$ be a monic polynomial and let p be a prime number. Set $A_p = (X - a_1^p) \cdots (X - a_r^p)$. Then $\overline{A}_p = \overline{A}$ in $\mathbb{F}_p[X]$.*

We are now in a position to prove the irreducibility of Φ_n.

Lemma 7.66 *Let n be a positive integer. Then the polynomial Φ_n is irreducible over $\mathbb{Z}$.*

Proof We have clearly $\Phi_n \neq 0, \pm 1$ if $n \geqslant 1$. Suppose that $\Phi_n = PQ$ with $P, Q \in \mathbb{Q}[X]$ such that $\deg P > 0$. Gauss's lemma implies in fact that $P, Q \in \mathbb{Z}[X]$. Let p be a prime number such that $p \nmid n$ and we define the polynomial R by

$$R = \prod_{\zeta \in Z_P} \left(X - \zeta^p\right)$$

where Z_P is the set of the roots of P, so that $\overline{R} = \overline{P}$ in $\mathbb{F}_p[X]$ by Lemma 7.65. Note that if $F = X^n - 1$ and since $nF(x) - xF'(x) = -n$, we have $(\overline{F}, \overline{F'}) = 1$

if $p \nmid n$, and therefore $\overline{F}$ is squarefree. Hence $\overline{\Phi_n}$ is squarefree by (7.7). Now we have $(\overline{Q}, \overline{R}) \mid \overline{R} = \overline{P}$ and $(\overline{Q}, \overline{R}) \mid \overline{Q}$ so that $(\overline{Q}, \overline{R})^2 \mid \overline{PQ} = \overline{\Phi_n}$, and therefore $(\overline{Q}, \overline{R}) = 1$ since $\overline{\Phi_n}$ is squarefree. Hence $(Q, R) = 1$ and since $R \mid \Phi_n$, we infer that $R \mid P$ by Gauss's theorem. Since these polynomials are monic and have the same degree, we get $R = P$.

Now let ζ be a root of P. Then there exists a positive integer k such that $(k, n) = 1$ and ζ^k is a root of Φ_n. Write $k = p_1 \cdots p_r$ where all the non-necessarily distinct prime numbers p_i satisfy $p_i \nmid n$. By the argument above applied r times, we see that ζ^k is also a root of P, and applying this with all the roots of Φ_n, we deduce that $P = \phi_n$. Hence Φ_n is irreducible over $\mathbb{Z}$. $\qquad\square$

Lemma 7.66 allows the following definition. The algebraic number field $\mathbb{K} = \mathbb{Q}(\zeta_n)$ is called a *cyclotomic field*. Hence we have $[\mathbb{K} : \mathbb{Q}] = \varphi(n)$. It can be shown that $\mathbb{K}$ is monogenic (see [Was82]) so that $\mathcal{O}_{\mathbb{K}} = \mathbb{Z}[\zeta_n]$ for all $n \geqslant 1$. In what follows, the aim is to prove this result when $n = p$ is a prime number.

Proposition 7.67 *Let p be a prime number. Then $\mathbb{Q}(\zeta_p)$ is monogenic.*

Proof First note that since $\zeta_2 = -1$, we have $\mathbb{Q}(\zeta_2) = \mathbb{Q}$ and since $\mathbb{Q}(\zeta_3)$ is a quadratic field, we may suppose that $p \geqslant 5$. We write ζ instead of ζ_p for convenience. The number $\lambda = \zeta - 1$ plays an important part in the proof. Using Proposition 7.61 (iv) we get

$$\Delta_{\mathbb{K}/\mathbb{Q}}(\lambda) = \Delta_{\mathbb{K}/\mathbb{Q}}(\zeta) = (-1)^{(p-1)/2} p^{p-2}$$

as also stated in Example 7.37. Now μ_λ is a p-Eisenstein as seen in Example 7.24, so that $p \nmid f$ by Proposition 7.53. Since $f^2 \times d_{\mathbb{K}} = (-1)^{(p-1)/2} p^{p-2}$, we infer that $f = 1$ and hence $\mathbb{K}$ is monogenic. $\qquad\square$

The proof above shows that, if $\mathbb{K} = \mathbb{Q}(\zeta_p)$, then $d_{\mathbb{K}} = (-1)^{(p-1)/2} p^{p-2}$. More generally, one can prove (see [Was82]) that, if $\mathbb{K} = \mathbb{Q}(\zeta_n)$, then

$$d_{\mathbb{K}} = (-1)^{\varphi(n)/2} n^{\varphi(n)} \prod_{p \mid n} p^{-\varphi(n)/(p-1)}. \tag{7.8}$$

The next result shows that a cyclotomic field is an abelian algebraic number field.

Proposition 7.68 *The algebraic number field $\mathbb{K} = \mathbb{Q}(\zeta_n)$ is abelian with Galois group*

$$\mathrm{Gal}\big(\mathbb{Q}(\zeta_n)/\mathbb{Q}\big) \simeq (\mathbb{Z}/n\mathbb{Z})^*.$$

Proof Set $\mathbb{K} = \mathbb{Q}(\zeta_n)$. Since ζ_n is a primitive nth root of unity and since every nth root of unity is a power of ζ_n, we deduce that the extension $\mathbb{K}/\mathbb{Q}$ is normal. Since Φ_n is irreducible over $\mathbb{Q}$ and $\mathrm{char}\,\mathbb{Q} = 0$, we infer that the extension is separable, so that $\mathbb{K}/\mathbb{Q}$ is Galois. Now let $\sigma \in \mathrm{Gal}(\mathbb{Q}(\zeta_n)/\mathbb{Q})$. There exists an integer $a = a(\sigma)$

coprime to n such that $\sigma(\zeta_n) = \zeta_n^{a(\sigma)}$. Since the residue class $\bar{a}$ of $a(\sigma)$ modulo n is uniquely determined, one may define the map

$$\begin{array}{ccc} \mathrm{Gal}(\mathbb{Q}(\zeta_n)/\mathbb{Q}) & \longrightarrow & (\mathbb{Z}/n\mathbb{Z})^*, \\ \sigma & \longmapsto & \bar{a}. \end{array}$$

This map does not depend on the particular choice of the primitive nth root ζ_n. Furthermore, one may check that it is a group homomorphism. If $\bar{a} = \bar{1}$, then $\sigma(\zeta_n) = \zeta_n$, so that $\sigma = \mathrm{Id}$ and the map is injective. We infer that $\mathrm{Gal}(\mathbb{Q}(\zeta_n)/\mathbb{Q})$ is isomorphic to a subgroup of $(\mathbb{Z}/n\mathbb{Z})^*$ and hence is abelian. Since $\mathrm{Gal}(\mathbb{Q}(\zeta_n)/\mathbb{Q})$ and $(\mathbb{Z}/n\mathbb{Z})^*$ have the same order, the result follows.   □

Pure Cubic Fields

These are the fields of the form $\mathbb{K} = \mathbb{Q}(m^{1/3})$ where m is 3-free and written as $m = ab^2$ with $(a,b) = 1$, a, b 2-frees and we assume that if $3 \mid m$, then $3 \mid a$ and $3 \nmid b$.

Let $\theta = m^{1/3}$ which is a root of the polynomial $P = X^3 - m$. We have

$$\mathrm{disc}(P) = -3^3 \times m^2 = -3^3 \times a^2 \times b^4 = f^2 \times d_{\mathbb{K}}$$

so that writing $d_{\mathbb{K}} = -3^n \times a^\alpha \times b^\beta$, we get $f = 3^{(3-n)/2} \times a^{(2-\alpha)/2} \times b^{(4-\beta)/2}$. Hence $n = 1$ or 3, $\alpha = 0$ or 2 and $\beta = 0$, 2 or 4. Since P is a p-Eisenstein for any prime factor p of a, we have $p \mid a \Longrightarrow p \nmid f$ by Proposition 7.53. In particular, if $3 \mid m$, then $3 \nmid f$ and $27a^2 \mid d_{\mathbb{K}}$. If $3 \nmid a$, then $3a^2 \mid d_{\mathbb{K}}$ so that $\alpha \geqslant 2$ in all cases. Hence $\alpha = 2$.

Now let $\lambda = \tilde{m}^{1/3}$ with $\tilde{m} = a^2 b$, which is a root of the polynomial $Q = X^3 - a^2 b$. Since

$$\begin{pmatrix} 1 \\ \lambda \\ \lambda^2 \end{pmatrix} = A \begin{pmatrix} 1 \\ \theta \\ \theta^2 \end{pmatrix}$$

with $\det A = -a \neq 0$, we deduce that $\mathrm{disc}(Q) = f^2 \times d_{\mathbb{K}}$. Now

$$\mathrm{disc}(Q) = -3^3 \times a^4 \times b^2$$

which implies that $\beta \neq 4$. Since Q is a p-Eisenstein for any prime factor p of b, we have $p \mid b \Longrightarrow p \nmid f$ and hence $b^2 \mid d_{\mathbb{K}}$ so that $\beta \geqslant 2$. Therefore $\beta = 2$. We then get

$$d_{\mathbb{K}} = \begin{cases} -27(ab)^2, & \text{if } 3 \mid m, \\ -3(ab)^2 \text{ or } -27(ab)^2, & \text{otherwise.} \end{cases}$$

We consider the following three cases.

1. $m \not\equiv \pm 1 \pmod 9$. Thus $9 \nmid (m^3 - m)$ so that the polynomial $S = (X + m)^3 - m$ is a 3-Eisenstein. A root of S is $\theta - m$ and $\Delta_{\mathbb{K}/\mathbb{Q}}(\theta - m) = \Delta_{\mathbb{K}/\mathbb{Q}}(\theta) =$

$-3^3 \times m^2$. Then $3 \nmid f$ so that $f = b$. Since $1, \theta, \lambda \in \mathcal{O}_{\mathbb{K}}$ and $\Delta_{\mathbb{K}/\mathbb{Q}}(1, \theta, \lambda) = b^{-2}\Delta_{\mathbb{K}/\mathbb{Q}}(\theta) = d_{\mathbb{K}}$, we infer that

$$\left\{ 1, \theta, \frac{\theta^2}{b} \right\}$$

is an integral basis for $\mathbb{K}$.

2. $\widetilde{m} \equiv 1 \pmod 9$. Let $\nu = \frac{1}{3b}(b + ab\theta + \theta^2)$. One can observe that ν is a root of the polynomial

$$R = X^3 - X^2 - \frac{\widetilde{m} - 1}{3}X - \frac{(\widetilde{m} - 1)^2}{27} \in \mathbb{Z}[X]$$

so that $\nu \in \mathcal{O}_{\mathbb{K}}$. Since $3b\,\nu \in \mathbb{Z}[\theta]$, we infer that $\mathcal{O}_{\mathbb{K}}/\mathbb{Z}[\theta]$ has an element of order $3b$, so that $3b \mid f$ by Theorem 7.1 and then $f = 3b$ and $d_{\mathbb{K}} = -3(ab)^2$. Now $\Delta_{\mathbb{K}/\mathbb{Q}}(1, \theta, \nu) = (3b)^{-2}\Delta_{\mathbb{K}/\mathbb{Q}}(\theta) = d_{\mathbb{K}}$ so that

$$\left\{ 1, \theta, \frac{b + ab\theta + \theta^2}{3b} \right\}$$

is an integral basis for $\mathbb{K}$.

3. $\widetilde{m} \equiv -1 \pmod 9$. This case can be treated as above, except that we make use of $\varrho = \frac{1}{3b}(b + ab\theta - \theta^2)$ which is a root of the polynomial

$$S = X^3 - X^2 + \frac{\widetilde{m} + 1}{3}X - \frac{(\widetilde{m} + 1)^2}{27} \in \mathbb{Z}[X]$$

instead of ν.

Also note that $\widetilde{m} \equiv \pm 1 \pmod 9 \iff m \equiv \pm 1 \pmod 9$. We may sum up the discussion in the following result.

Proposition 7.69 *Let $\mathbb{K} = \mathbb{Q}(m^{1/3})$ be a pure cubic field where $m = ab^2$ is 3-free with $(a, b) = 1$ and a, b 2-frees. Set $\theta = m^{1/3} \in \mathcal{O}_{\mathbb{K}}$ and $\widetilde{m} = a^2 b$.*

▷ *If $m \not\equiv \pm 1 \pmod 9$, then $d_{\mathbb{K}} = -27(ab)^2$ and $\{1, \theta, \frac{\theta^2}{b}\}$ is an integral basis for $\mathbb{K}$.*
▷ *If $\widetilde{m} \equiv \pm 1 \pmod 9$, then $d_{\mathbb{K}} = -3(ab)^2$ and $\{1, \theta, \frac{b + ab\theta \pm \theta^2}{3b}\}$ is an integral basis for $\mathbb{K}$.*

Voronoï's Method for Cubic Fields

We investigate algebraic number fields $\mathbb{K} = \mathbb{Q}(\theta)$ where θ is a root of the polynomial $P = X^3 - aX + b \in \mathbb{Z}[X]$ with a 2-free or b 3-free. Using Example 7.37, we obtain $\mathrm{disc}(P) = 4a^3 - 27b^2$. In [Vor94], Voronoï devised a method to compute an integral basis for $\mathbb{K}$. A new proof of this method was given in [AW04b, AW04a] and rests on the evaluation of $d_{\mathbb{K}}$ which can be found in [LN83].

Proposition 7.70 (Voronoï) *Let* $\mathbb{K} = \mathbb{Q}(\theta)$ *be an algebraic number field where* θ *is a root of the polynomial* $P = X^3 - aX + b \in \mathbb{Z}[X]$ *with a 2-free or b 3-free.*

▷ *Suppose that* $a \not\equiv 3 \pmod 9$ *or* $b^2 \not\equiv a + 1 \pmod{27}$ *and let* n^2 *be the largest square dividing* $\mathrm{disc}(P)$ *for which the system of congruences*

$$\begin{cases} x^3 - ax + b \equiv 0 \pmod{n^2}, \\ 3x^2 - a \equiv 0 \pmod{n} \end{cases}$$

is solvable for x. *Then an integral basis for* $\mathbb{K}$ *is given by*

$$\left\{ 1, \theta, \frac{x^2 - a + x\theta + \theta^2}{n} \right\}.$$

▷ *Suppose that* $a \equiv 3 \pmod 9$ *and* $b^2 \equiv a + 1 \pmod{27}$ *and let* n^2 *be the largest square dividing* $\mathrm{disc}(P)/729$ *for which the system of congruences*

$$\begin{cases} x^3 - ax + b \equiv 0 \pmod{27n^2}, \\ 3x^2 - a \equiv 0 \pmod{9n} \end{cases}$$

is solvable for x. *Then an integral basis for* $\mathbb{K}$ *is given by*

$$\left\{ 1, \frac{\theta - x}{3}, \frac{x^2 - a + x\theta + \theta^2}{9n} \right\}.$$

Examples

1. Let $\mathbb{K} = \mathbb{Q}(\theta)$ where θ is a root of the polynomial $P = X^3 - 8X - 57$. Using Example 7.37, we get $\mathrm{disc}(P) = -5^2 \times 23 \times 149$ so that $n = 5$ and $x = 4$ and thus $\{1, \theta, \frac{8 + 4\theta + \theta^2}{5}\}$ is an integral basis for $\mathbb{K}$.

2. Let $\mathbb{K} = \mathbb{Q}(\theta)$ where θ is a root of the polynomial $P = X^3 - 12X + 65$. Similarly, we have $729^{-1} \mathrm{disc}(P) = -3 \times 7^2$ so that $n = 7$ and $x = -5$ and then $\{1, \frac{\theta + 5}{3}, \frac{13 - 5\theta + \theta^2}{63}\}$ is an integral basis for $\mathbb{K}$.

Some Monogenic Algebraic Number Fields

It is a long-standing problem in algebraic number theory to determine whether an algebraic number field $\mathbb{K}$ is monogenic. Indeed, the existence of an element $\theta \in \mathcal{O}_{\mathbb{K}}$ such that $\mathcal{O}_{\mathbb{K}} = \mathbb{Z}[\theta]$ makes the study of arithmetic in $\mathbb{K}$ considerably easier. From above we know that the quadratic fields and the cyclotomic fields are monogenic, but, in the general case, this property is relatively rare.

Let $\mathbb{K} = \mathbb{Q}(\theta)$ be an algebraic number field of degree n and $\{1, \alpha_2, \dots, \alpha_n\}$ be an integral basis for $\mathbb{K}$. There exists a form $I(X_2, \dots, X_n) \in \mathbb{Z}[X_2, \dots, X_n]$, called the *index form*, of degree $n(n-1)/2$ in $n-1$ variables $X_2, \dots, X_n$ such that

$$\Delta_{\mathbb{K}/\mathbb{Q}}(\alpha_2 X_2 + \cdots + \alpha_n X_n) = I(X_2, \dots, X_n)^2 d_{\mathbb{K}}.$$

Hence $\mathbb{K}$ is monogenic if and only if the *index form equation*

$$I(x_2, \ldots, x_n) = \pm 1$$

has a solution in $\mathbb{Z}^{n-1}$. The first effective upper bounds for the solutions of this equation were derived by Győry [Győ76] by using Baker's lower bounds for logarithmic forms. As a consequence, it follows that up to translation by elements of $\mathbb{Z}$, there exist only finitely many generators of power integral bases in an algebraic number field.

In what follows, we provide some examples of monogenic algebraic number fields of low degree.

▶ **Degree 3.** A *cyclic cubic field* is an algebraic number field $\mathbb{K}$ of degree 3 which is Galois, the Galois group being necessarily isomorphic to the cyclic group $C_3 \simeq \mathbb{Z}/3\mathbb{Z}$. A profound result[15] states that such a field is contained in a cyclotomic field $\mathbb{Q}(\zeta_{f_\mathbb{K}})$ with $f_\mathbb{K}$ minimal, called the *conductor* of $\mathbb{K}$. In the case of cyclic cubic fields, it can be shown that one can always write

$$f_\mathbb{K} = \frac{a^2 + 27b^2}{4}$$

with $b > 0$ and $a \equiv 1 \pmod 3$ if $f_\mathbb{K} \equiv 1 \pmod 3$, $a = 3a'$ with $a' \equiv 1 \pmod 3$ otherwise. In [Gra74], the following necessary and sufficient condition is proved.

Proposition 7.71 *With the notation above, $\mathbb{K}$ is monogenic if and only if the equation*

$$bu(u^2 - 9v^2) + av(u^2 - v^2) = 1$$

has solutions $u, v \in \mathbb{Z}$.

▶ **Degree 4.** A *biquadratic field* is an algebraic number field $\mathbb{K} = \mathbb{Q}(\sqrt{dm}, \sqrt{dn})$ with d, m, n squarefree, pairwise coprime such that $dm, dn, mn \neq 1$, $dm \equiv dn \pmod 4$, $d > 0$, $m > n$ and if $dm \equiv dn \equiv 1 \pmod 4$, then $d < \min(|m|, |n|)$. Such fields are Galois with Galois group $\mathrm{Gal}(\mathbb{K}/\mathbb{Q}) \simeq (\mathbb{Z}/2\mathbb{Z})^2$. Define $\delta \in \{0, 1\}$ such that $mn \equiv (-1)^\delta \pmod 4$. Then we have the following proposition [GT95].

Proposition 7.72 *Let $\mathbb{K} = \mathbb{Q}(\sqrt{dm}, \sqrt{dn})$ be a biquadratic field with d, m ,n as above.*

▷ *If $dm \equiv dn \equiv 1 \pmod 4$, then $\mathbb{K}$ is not monogenic.*
▷ *If $dm \equiv dn \not\equiv 1 \pmod 4$, then $\mathbb{K}$ is monogenic if and only if the following two conditions are fulfilled.*
   (i) $m - n = 2^{2-2\delta} d$.

---

[15] See Sect. 7.5 and the Kronecker–Weber theorem.

(ii) *The equation $2^8 m (u^2 - v^2)^2 - 2^8 n (u^2 + v^2)^2 = 4s$, where $s = \pm 1$, has solutions $u, v \in \mathbb{Z}$.*

▶ **Degrees 5 and 6.** In a series of papers [ESW07, LSWY05, SWW06], the authors consider some families of monic polynomials giving birth to infinitely many algebraic number fields of degrees 5 and 6 whose ring of integers are monogenic. The proofs rest on the following scheme.
  a. Prove that the polynomials are irreducible over $\mathbb{Z}$.
  b. Determine the Galois groups.
  c. Consider a subfamily of these polynomials, for instance those which have a squarefree discriminant, use results from elementary number theory to show that this subfamily contains infinitely many polynomials.
  d. Prove that the algebraic number fields defined by these polynomials are monogenic.

Proposition 7.73 *Let $m \in \mathbb{Z} \setminus \{1\}$.*

▷ *Let $P_m = X^5 - 2X^4 + (m+2)X^3 - (2m+1)X^2 + mX + 1$ such that the number $4m^3 + 28m^2 + 24m + 47$ is squarefree. Then P_m is irreducible over $\mathbb{Z}$, $\mathrm{Gal}(P_m/\mathbb{Q}) \simeq D_5$ and the algebraic number fields defined by P_m are distinct and monogenic.*

▷ *Let $P_m = X^6 + (2m+2)X^4 + (2m-1)X^2 - 1$ and assume that $4m^2 + 2m + 7$ is squarefree. Then P_m is irreducible over $\mathbb{Z}$, $\mathrm{Gal}(P_m/\mathbb{Q}) \simeq A_4$ and the algebraic number fields defined by P_m are distinct and monogenic.*

▷ *Let $P_m = X^6 - 4X^5 + 2X^4 - 3mX^3 + X^2 + 2X + 1$ and suppose that the number $729m^3 + 522m^2 + 1788m + 2648$ is squarefree. Then P_m is irreducible over $\mathbb{Z}$, $\mathrm{Gal}(P_m/\mathbb{Q}) \simeq \mathrm{PSL}(2,5) \simeq A_5$ and the algebraic number fields defined by P_m are distinct and monogenic.*

Proof (Sketch) We only provide a sketch of the proof of the third case.
    Suppose that P_m is not irreducible. Since $P_m \neq 0, \pm 1$ and

$$P_m \equiv (X - 1)\left(X^5 - X^3 - X^2 - 1\right) \pmod{3}$$

then we have $P_m = Q_m R_m$ where $Q_m, R_m \in \mathbb{Z}[X]$ are monic polynomials such that $\deg Q_m = 1$ and $\deg R_m = 5$. The condition $P_m(0) = 1$ implies that $Q_m = X \pm 1$. If $Q_m = X - 1$, then we have $0 = P_m(1) = 3 - 3m$ contradicting the fact that $m \neq 1$. If $Q_m = X + 1$, then similarly we have $0 = P_m(-1) = 3m + 7$ contradicting the fact that m is an integer. Hence P_m is irreducible over $\mathbb{Z}$.
    The factorization of P_m in $\mathbb{F}_3[X]$ above shows that $\mathrm{Gal}(P_m/\mathbb{Q})$ contains a 5-cycle so that 5 divides the order of $\mathrm{Gal}(P_m/\mathbb{Q})$. Since

$$\mathrm{disc}(P_m) = \left(729m^3 + 522m^2 + 1788m + 2648\right)^2$$

we have $\mathrm{Gal}(P_m/\mathbb{Q}) \subseteq A_6$ by Lemma 7.140. Among the sixteen transitive subgroups of S_6 which may be Galois groups of an irreducible polynomial of degree

6, the only groups having order divisible by 5 and contained in $\mathcal{A}_6$ are PSL$(2,5)$ and $\mathcal{A}_6$. The final step is given by using a result of [Hag00] where a factorization of a certain polynomial of degree 15 (see Theorem 7.148), along with the fact that disc(P_m) is a square, ensures that Gal$(P_m/\mathbb{Q}) \simeq$ PSL$(2,5)$.

A theorem by Erdős [Erd53] implies that there are infinitely many integers m, which are odd, such that $\delta(m) = 729m^3 + 522m^2 + 1788m + 2648$ is squarefree. Let p be an odd prime factor of disc$(P_m) = \delta(m)^2$. Using Corollary 7.103, we infer that $p \mid d_{\mathbb{K}}$. Since $\delta(m)$ is odd and squarefree, we get $p > 2$ and $v_p(\mathrm{disc}(P_m)) = 2$. If $p \mid f$, then the relation disc$(P_m) = f^2 \times d_{\mathbb{K}}$ and the fact that $p \mid d_{\mathbb{K}}$ give $p^3 \mid \mathrm{disc}(P_m)$, giving a contradiction with $v_p(\mathrm{disc}(P_m)) = 2$. Hence $f = 1$ and $\mathcal{O}_{\mathbb{K}} = \mathbb{Z}[\theta]$. $\square$

7.2.10 Units and Regulators

This section investigates the multiplicative group $\mathcal{O}_{\mathbb{K}}^*$ of the units in the ring $\mathcal{O}_{\mathbb{K}}$. The structure of this group has been entirely determined by Dirichlet who proved the following important theorem sometimes called *Dirichlet's unit theorem*.

Theorem 7.74 (Dirichlet) *Let* $\mathbb{K} = \mathbb{Q}(\theta)$ *be an algebraic number field with signature* (r_1, r_2) *and let* $\theta \in \mathcal{O}_{\mathbb{K}}$. *We denote by* $W_{\mathbb{K}}$ *the subgroup of* $\mathcal{O}_{\mathbb{K}}^*$ *consisting of roots of unity in* $\mathcal{O}_{\mathbb{K}}$. *Then* $\mathcal{O}_{\mathbb{K}}^*$ *is a finitely generated abelian group with*

$$\mathrm{rank}\, \mathcal{O}_{\mathbb{K}}^* = r_1 + r_2 - 1$$

and torsion subgroup equal to $W_{\mathbb{K}}$. *More precisely, setting* $r = r_1 + r_2 - 1$, *there exist units* $\varepsilon_1, \ldots, \varepsilon_r$, *called a* system of fundamental units, *such that every unit* $\varepsilon \in \mathcal{O}_{\mathbb{K}}$ *can be written uniquely in the form*

$$\varepsilon = \zeta \varepsilon_1^{n_1} \cdots \varepsilon_r^{n_r}$$

with $\zeta \in W_{\mathbb{K}}$ *and* $n_1, \ldots, n_r \in \mathbb{Z}$. *The number* r *is called the* Dirichlet rank *of* $\mathcal{O}_{\mathbb{K}}^*$. *In other words, we have*

$$\mathcal{O}_{\mathbb{K}}^* \simeq W_{\mathbb{K}} \times \mathbb{Z}^{r_1 + r_2 - 1}.$$

Example 7.75

1. **Imaginary quadratic fields.** Let $\mathbb{K} = \mathbb{Q}(\sqrt{-d})$ be an imaginary quadratic field with $d > 0$ squarefree. Hence $(r_1, r_2) = (0, 1)$ so that $\mathcal{O}_{\mathbb{K}}^* \simeq W_{\mathbb{K}}$. One may check that

$$\mathcal{O}_{\mathbb{K}}^* = \begin{cases} \{\pm 1, \pm i\}, & \text{if } d = 1, \\ \{\pm 1, \pm \rho, \pm \rho^2\}, & \text{if } d = 3, \\ \{\pm 1\}, & \text{otherwise.} \end{cases}$$

2. **Real quadratic fields.** Let $\mathbb{K} = \mathbb{Q}(\sqrt{d})$ be an imaginary quadratic field with $d > 0$ squarefree and $d \neq 1$. Let $\sigma : a + b\sqrt{d} \longmapsto a - b\sqrt{d}$ be the non-trivial

embedding of $\mathbb{K}$. Since $(r_1, r_2) = (2, 0)$, there exists a fundamental unit, denoted by γ_d, such that

$$\mathcal{O}_{\mathbb{K}}^* = \{\pm\gamma_d^k : k \in \mathbb{Z}\}.$$

The calculation of γ_d requires the theory of continued fractions in the following way (see [Sam71] for instance). Let ω be the number stated in Proposition 7.64 and assume that $-\sigma(\omega)$ has the continued fraction development $-\sigma(\omega) = [a_0, \overline{a_1, \ldots, a_t}]$. If p_{t-1}/q_{t-1} is the $(t-1)$th convergent, then

$$\gamma_d = p_{t-1} + \omega q_{t-1}$$

and we also have $N_{\mathbb{K}/\mathbb{Q}}(\gamma_d) = (-1)^t$.

We do not prove Dirichlet's unit theorem, but give the main ideas below. The reader interested in this subject may refer to [EM99, FT91, Jan96, Neu10, Sam71, ST02]. Our aim is to provide some useful facts about units and roots of unity of $\mathcal{O}_{\mathbb{K}}$ and to define the so-called *regulator* of $\mathbb{K}$.

A *root of unity* in $\mathbb{K}$ is a number $\zeta \in \mathbb{K}$ such that there exists a positive integer m such that $\zeta^m = 1$. Hence $\zeta \in \mathcal{O}_{\mathbb{K}}$. The following lemma will be useful.

Lemma 7.76 Let $A > 0$. The set $S = \{\alpha \in \mathcal{O}_{\mathbb{K}} : |\sigma_1(\alpha)| \leqslant A, \ldots, |\sigma_n(\alpha)| \leqslant A\}$ is finite.

Proof The characteristic polynomial (7.4) of $\alpha \in S$ belongs to $\mathbb{Z}[X]$ by (7.5) and its coefficients are all symmetric functions in the $\sigma_i(\alpha)$ and hence are bounded. We infer that there are only a finite number of possibilities for this polynomial, e.g. $(2A + 2)^{n^2}$ is such a bound. ☐

Proposition 7.77 *Let $\mathbb{K}$ be an algebraic number field of degree n, signature (r_1, r_2) and $\mathcal{O}_{\mathbb{K}}$ is the ring of integers of $\mathbb{K}$. We set $W_{\mathbb{K}}$ as in Theorem 7.74.*

(i) *ε is a unit in $\mathcal{O}_{\mathbb{K}}$ if and only if $|N_{\mathbb{K}/\mathbb{Q}}(\varepsilon)| = 1$.*
(ii) *There are only finitely many roots of unity in $\mathbb{K}$.*
(iii) *$W_{\mathbb{K}}$ is a cyclic group of even order. Furthermore if $r_1 > 0$, then $|W_{\mathbb{K}}| = 2$.*
(iv) *Let $\alpha \in \mathcal{O}_{\mathbb{K}}$ and μ_α be its minimal polynomial. If $\mu_\alpha(m) = \pm 1$ for some $m \in \mathbb{Z}$, then $\alpha - m$ is a unit in $\mathcal{O}_{\mathbb{K}}$.*

Proof

(i) If ε is a unit, then there exists $\lambda \in \mathcal{O}_{\mathbb{K}}$ such that $\lambda\varepsilon = 1$ so that $N_{\mathbb{K}/\mathbb{Q}}(\varepsilon) \times N_{\mathbb{K}/\mathbb{Q}}(\lambda) = 1$ and these norms are integers. Conversely, if $\varepsilon \in \mathcal{O}_{\mathbb{K}}$ satisfies $N_{\mathbb{K}/\mathbb{Q}}(\varepsilon) = \pm 1$, then one can write

$$\varepsilon \times \left(\pm \prod_{i=2}^{n} \sigma_i(\varepsilon) \right) = 1$$

so that ε is a unit in $\mathcal{O}_{\mathbb{K}}$ as all terms are algebraic integers.

(ii) Suppose that $\alpha^m = 1$. Then $|\alpha|^m = 1$ and hence $|\alpha| = 1$ and similarly $|\sigma_i(\alpha)| = 1$ for all $i \in \{1, \ldots, n\}$. The result follows from Lemma 7.76.

(iii) Let $\alpha_1, \ldots, \alpha_k$ be the roots of unity in $\mathbb{K}$. Since $\alpha_i^{m_i} = 1$ for some $m_i \geqslant 1$, we get $\alpha_i = e_{m_i}(n_i)$ for some $0 \leqslant n_i \leqslant m_i - 1$. If we set $m = m_1 \cdots m_k$, then each α_i belongs to the cyclic group generated by $e_m(1)$ and therefore $W_{\mathbb{K}}$ is a subgroup of this group, and hence is cyclic. Furthermore, since $\{-1, 1\} \subseteq W_{\mathbb{K}}$, then 2 divides $|W_{\mathbb{K}}|$ by Lagrange's theorem (see Theorem 7.1). Finally, if $r_1 > 0$, then $W_{\mathbb{K}} = \{\pm 1\}$ since all other roots of unity are non-real.

(iv) If $P(X) = \mu_\alpha(X + m)$, then $P \in \mathbb{Z}[X]$ is monic, satisfies $P(0) = \mu_\alpha(m) = \pm 1$ and $P(\alpha - m) = \mu_\alpha(\alpha) = 0$, so that the minimal polynomial of $\alpha - m \in \mathcal{O}_{\mathbb{K}}$ divides P by Lemma 7.40, and thus has constant term ± 1. We infer that $N_{\mathbb{K}/\mathbb{Q}}(\alpha - m) = \pm 1$ and then $\alpha - m \in \mathcal{O}_{\mathbb{K}}^*$ by (i).

The proof is complete.   □

Dirichlet's unit theorem can be proved by embedding the unit group in a *logarithmic space*. More precisely, if $\mathbb{K}$ is an algebraic number field of degree n, signature (r_1, r_2) and embeddings σ_i, we may define a map

$$\phi: \quad \mathbb{K} \quad \longrightarrow \quad \mathbb{R}^{r_1} \times \mathbb{C}^{r_2},$$
$$\alpha \quad \longmapsto \quad \phi(\alpha)$$

with $\phi(\alpha) = (\sigma_1(\alpha) \ldots, \sigma_{r_1}(\alpha), \sigma_{r_1+1}(\alpha), \ldots, \sigma_{r_1+r_2}(\alpha))$. The map ϕ is an injective ring homomorphism and one can prove that, if $\{\alpha_1, \ldots, \alpha_n\}$ is a $\mathbb{Q}$-basis for $\mathbb{K}$, then the vectors $\phi(\alpha_1), \ldots, \phi(\alpha_n)$ are $\mathbb{R}$-linearly independent. This map enables us to get a geometric representation of algebraic numbers.

Define the following map

$$L: \quad \mathbb{K} \setminus \{0\} \quad \longrightarrow \quad \mathbb{R}^{r_1+r_2},$$
$$\alpha \quad \longmapsto \quad (\delta_i \log|\sigma_i(\alpha)|)$$

where $\delta_i = 1$ if $i \in \{1, \ldots, r_1\}$ and $\delta_i = 2$ if $i \in \{r_1 + 1, \ldots, r_1 + r_2\}$. This map is called the *logarithmic representation* of $\mathbb{K} \setminus \{0\}$ and $\mathbb{R}^{r_1+r_2}$ is the *logarithmic space*. It is easy to check that $L(\alpha\beta) = L(\alpha) + L(\beta)$ for all $\alpha, \beta \in \mathbb{K} \setminus \{0\}$.

Now consider the restriction $L : \mathcal{O}_{\mathbb{K}}^* \longrightarrow \mathbb{R}^{r_1+r_2}$. This abelian group homomorphism is not injective, but one may determine its kernel and image. Whereas the first one is easily described as we will see below, the determination of $\text{Im}\, L$ uses the so-called *geometry of numbers* and more precisely a theorem by Minkowski applied to certain discrete additive subgroups of $\mathbb{R}^{r_1+r_2}$. A discrete[16] additive subgroup of $(\mathbb{R}^{r_1+r_2}, +)$ generated by the m linearly independent vectors $e_1, \ldots, e_m$ is called a *lattice* of dimension m. Hence we need to pass from the multiplicative group $\mathcal{O}_{\mathbb{K}}^*$ to an additive subgroup of $\mathbb{R}^{r_1+r_2}$, which is the reason why the logarithms are used.

---

[16] A subset of $\mathbb{R}^{r_1+r_2}$ is *discrete* if and only if it intersects every closed ball of center O in a finite set.

We may sketch the proof of Theorem 7.74. To this end, let $\varepsilon \in \mathcal{O}_{\mathbb{K}}^*$.

▷ From the definition of L above, we immediately see that $W_{\mathbb{K}} \subseteq \ker L$. Conversely, since $L(\varepsilon) = 0 \Longleftrightarrow |\sigma_i(\varepsilon)| = 1$ for all i, we infer that $\ker L$ is a finite subgroup of $\mathbb{K} \setminus \{0\}$ by Lemma 7.76. Hence $\ker L$ is cyclic and must consist entirely of roots of unity. Therefore $\ker L \simeq W_{\mathbb{K}}$, a result proved by Kronecker.

▷ Now set $L(\alpha) = (L_1(\alpha), \ldots, L_{r_1+r_2}(\alpha))$. Then

$$\sum_{i=1}^{r_1+r_2} L_i(\alpha) = \log \left| N_{\mathbb{K}/\mathbb{Q}}(\alpha) \right|$$

and Proposition 7.77 (i) implies

$$\sum_{i=1}^{r_1+r_2} L_i(\varepsilon) = 0$$

so that $\operatorname{Im} L \subseteq \mathcal{H}$ where $\mathcal{H}$ is the hyperplane of $\mathbb{R}^{r_1+r_2}$ of equation $x_1 + \cdots + x_{r_1+r_2} = 0$. We have $\dim \mathcal{H} = r_1 + r_2 - 1$. Now let $\eta > 0$ and let $\| \ \|_2$ be the Euclidean norm on $\mathbb{R}^{r_1+r_2}$. Suppose that $\|L(\varepsilon)\|_2 < \eta$. This readily implies that $|\sigma_i(\varepsilon)| < e^{\eta}$ for all $i \in \{1, \ldots, r_1\}$ and $|\sigma_i(\varepsilon)| < e^{\eta/2}$ for all $i \in \{r_1 + 1, \ldots, r_1 + r_2\}$, so that the set of points $\sigma(\varepsilon)$ such that $\|L(\varepsilon)\|_2 < \eta$ is finite by Lemma 7.76, and hence $\operatorname{Im} L$ is a lattice of dimension $\leqslant r_1 + r_2 - 1$.

Since $W_{\mathbb{K}}$ is finite and $\mathcal{O}_{\mathbb{K}}^*/W_{\mathbb{K}} \simeq \operatorname{Im} L$ by Theorem 7.4 and since $\operatorname{Im} L$ is a lattice, and then a free abelian group, we infer that $\mathcal{O}_{\mathbb{K}}^*$ is a finitely generated abelian group of rank $\leqslant r_1 + r_2 - 1$. Using topological tools (see [FT91, Lemma 4.7 page 171]), one may prove that the rank is actually equal to $r = r_1 + r_2 - 1$.

The volume of the lattice $\operatorname{Im} L$ plays an important role in the theory and is called the *regulator* of $\mathbb{K}$. More precisely, we have the following definition.

Definition 7.78 (Regulator) Let $\mathbb{K}/\mathbb{Q}$ be an algebraic number field of signature (r_1, r_2) and $r = r_1 + r_2 - 1$ be the Dirichlet rank of $\mathcal{O}_{\mathbb{K}}^*$. Let $\varepsilon_1, \ldots, \varepsilon_r$ be a system of fundamental units of $\mathcal{O}_{\mathbb{K}}^*$ and define the matrix $U \in \mathcal{M}_{r,r+1}(\mathbb{R})$ whose ith row is the row-vector $L(\varepsilon_i)$ for $1 \leqslant i \leqslant r$. Let U_r be any $r \times r$ submatrix extracted from U. The *regulator* of $\mathbb{K}$ is the number $\mathcal{R}_{\mathbb{K}}$ defined by

$$\mathcal{R}_{\mathbb{K}} = |\det U_r|.$$

Note that the regulator is well-defined since changing a $\mathbb{Z}$-basis of $\mathbb{Z}^r$ into another involves multiplication by a matrix of determinant ± 1 by Proposition 7.18 (i).

Example 7.79 Let θ be a root of the polynomial $P = X^3 - X^2 - 11X - 1$. If P has a rational root, then it must be ± 1, but $P(\pm 1) \neq 0$ and hence P is irreducible over $\mathbb{Z}$. Define $\mathbb{K} = \mathbb{Q}(\theta)$ and hence $\theta \in \mathcal{O}_{\mathbb{K}}$. The signature is $(r_1, r_2) = (3, 0)$ and therefore the Dirichlet rank of $\mathcal{O}_{\mathbb{K}}^*$ is $r = 2$. Since $P(0) = -1$, we have in fact $\theta \in \mathcal{O}_{\mathbb{K}}^*$. Now

$$(3 + \theta)^3 = \theta^3 + 9\theta^2 + 27\theta + 27 = 10\theta^2 + 38\theta + 28 = 2(\theta + 1)(5\theta + 14)$$

and we have $N_{\mathbb{K}/\mathbb{Q}}(1 + \theta) = -8$ and $N_{\mathbb{K}/\mathbb{Q}}(3 + \theta) = 4$ so that

$$64 = N_{\mathbb{K}/\mathbb{Q}}\big((3 + \theta)^3\big) = N_{\mathbb{K}/\mathbb{Q}}\big(2(\theta + 1)(5\theta + 14)\big) = -64 \times N_{\mathbb{K}/\mathbb{Q}}(5\theta + 14).$$

Therefore $N_{\mathbb{K}/\mathbb{Q}}(5\theta + 14) = -1$ and hence $5\theta + 14 \in \mathcal{O}_{\mathbb{K}}^*$ by Proposition 7.77 (i). If θ_2 and θ_3 are the conjugates of θ, we then have

$$U = \begin{pmatrix} \log|\theta| & \log|\theta_2| & \log|\theta_3| \\ \log|5\theta + 14| & \log|5\theta_2 + 14| & \log|5\theta_3 + 14| \end{pmatrix}$$

so that

$$\mathcal{R}_{\mathbb{K}} = \left| \det \begin{pmatrix} \log|\theta| & \log|\theta_2| \\ \log|5\theta + 14| & \log|5\theta_2 + 14| \end{pmatrix} \right| \approx 11.9265379\ldots$$

7.3 Ideal Theory

7.3.1 Arithmetic Properties of Ideals

Let R be a ring. A *maximal ideal* of R is an ideal $\mathfrak{a} \neq R$ such that there are no ideals of R strictly between $\mathfrak{a}$ and R, i.e. such that every ideal containing strictly $\mathfrak{a}$ is equal to R. A *prime ideal* of R is an ideal $\mathfrak{p} \neq R$ such that, for all $x, y \in R$ satisfying $xy \in \mathfrak{p}$, then either $x \in \mathfrak{p}$ or $y \in \mathfrak{p}$; in other words, $\mathfrak{p}$ is a prime ideal of R if, whenever $\mathfrak{a}$ and $\mathfrak{b}$ are ideals of R such that $\mathfrak{a}\mathfrak{b} \subseteq \mathfrak{p}$, then either $\mathfrak{a} \subseteq \mathfrak{p}$ or $\mathfrak{b} \subseteq \mathfrak{p}$.

The role of maximal ideals in a commutative ring is illustrated in *Krull's lemma* stating that, in a non-zero commutative ring R, every ideal $\mathfrak{a} \neq R$ is contained in a maximal ideal of R.

Since $R/(0) \simeq R$, we deduce that, if R is an integral domain, then the zero ideal is prime by Lemma 7.80, so that a principal ideal (p) is prime if and only if p is a prime number or zero. It may seem surprising that 0 is excluded from the list of prime numbers whereas (0) is a prime ideal.

The next result is a characterization of prime and maximal ideals.

Lemma 7.80 *Let R be a ring and $\mathfrak{a}$ be an ideal of R.*

(i) *$\mathfrak{a}$ is prime if and only if $R/\mathfrak{a}$ is an integral domain.*
(ii) *$\mathfrak{a}$ is maximal if and only if $R/\mathfrak{a}$ is a field.*

Proof

(i) Assume that $\mathfrak{a}$ is prime. Then $\mathfrak{a} \neq R$ and hence $R/\mathfrak{a} \neq \{\bar{0}\}$. Let $\bar{x}, \bar{y} \in R/\mathfrak{a}$ such that $\bar{x}\,\bar{y} = \bar{0}$. We get $\overline{xy} = \bar{0}$ so that $xy \in \mathfrak{a}$. By definition of a prime ideal, we have $x \in \mathfrak{a}$ or $y \in \mathfrak{a}$ so that $\overline{x} = \overline{0}$ or $\overline{y} = \overline{0}$ and hence $R/\mathfrak{a}$ is an integral domain.

Conversely, we have $\mathfrak{a} \neq R$ and if we suppose that $xy \in \mathfrak{a}$ with $x, y \in R$, then $\overline{xy} = \overline{0}$ and then $\overline{x}\,\overline{y} = \overline{0}$. Since $R/\mathfrak{a}$ is an integral domain, this implies that $\overline{x} = \overline{0}$ or $\overline{y} = \overline{0}$, and therefore $x \in \mathfrak{a}$ or $y \in \mathfrak{a}$ so that $\mathfrak{a}$ is prime.

(ii) Suppose that $\mathfrak{a}$ is maximal. Then $\mathfrak{a} \neq R$ and hence $R/\mathfrak{a} \neq \{\overline{0}\}$. Let $\overline{x} \neq \overline{0} \in R/\mathfrak{a}$. We have $x \notin \mathfrak{a}$ and the ideal $\mathfrak{a} + Rx$ is such that $\mathfrak{a} + Rx \supsetneq \mathfrak{a}$, and hence $\mathfrak{a} + Rx = R$ since $\mathfrak{a}$ is maximal. One may then write $1 = y + rx$ with $y \in \mathfrak{a}$ and $r \in R$, so that $\overline{1} = \overline{rx} = \overline{r}\,\overline{x}$ which means that $\overline{x}$ is invertible and hence $R/\mathfrak{a}$ is a field.

Conversely, we have $\mathfrak{a} \neq R$. Let $\mathfrak{b}$ be an ideal of R such that $\mathfrak{a} \subset \mathfrak{b} \subseteq R$ and let $b \in \mathfrak{b} \setminus \mathfrak{a}$. We then have $\overline{b} \neq \overline{0}$ and thus $\overline{b}$ is invertible, so that there exists $r \in R$ such that $\overline{r}\,\overline{b} = \overline{1}$ and hence $rb = 1 + c$ with $c \in \mathfrak{a}$, so that $1 = rb - c \in \mathfrak{b}$ which means that $\mathfrak{b} = R$, and therefore $\mathfrak{a}$ is maximal.

The proof is complete. $\square$

The following lemma is also needed in the sequel.

Lemma 7.81 *Let $\mathbb{K}/\mathbb{Q}$ be an algebraic number field of degree n. Let $\mathfrak{a}$ be a non-zero ideal in $\mathcal{O}_{\mathbb{K}}$. Then $\mathcal{O}_{\mathbb{K}}/\mathfrak{a}$ is finite.*

Proof By Corollary 7.60, $\mathcal{O}_{\mathbb{K}}$ is a free $\mathbb{Z}$-module of rank n. Since the ideal $\mathfrak{a}$ is non-zero and $\mathbb{Z}$ is a PID, $\mathfrak{a}$ is a free $\mathbb{Z}$-module of rank $r \leqslant n$ by Proposition 7.16 (vi). It remains to show that we actually have $r = n$. Let $\alpha \neq 0 \in \mathfrak{a}$. If $\mu_{\alpha} = X^r + a_{r-1}X^{r-1} + \cdots + a_0 \in \mathbb{Z}[X]$, then we have

$$a_0 = \alpha\left(-a_1 - \cdots - a_{r-1}\alpha^{r-2} - \alpha^{r-1}\right)$$

and hence $a_0 \in \mathfrak{a}$ since $-a_1 - \cdots - a_{r-1}\alpha^{r-2} - \alpha^{r-1} \in \mathcal{O}_{\mathbb{K}}$. Hence we get $a_0 \neq 0 \in \mathfrak{a}$ and $a_0 \in \mathbb{Z}$. Now let $\{\theta_1, \ldots, \theta_n\}$ be an integral basis for $\mathbb{K}$. Since $a_0\theta_i \in \mathfrak{a}$ for all $i = 1, \ldots, n$, we deduce that $\mathfrak{a}$ contains n $\mathbb{Q}$-linearly independent elements, so that $\mathfrak{a}$ is a free sub-$\mathbb{Z}$-module of rank n of $\mathcal{O}_{\mathbb{K}}$ and since $(a_0) \subseteq \mathfrak{a} \subseteq \mathcal{O}_{\mathbb{K}}$, the index $(\mathcal{O}_{\mathbb{K}} : \mathfrak{a})$ divides $(\mathcal{O}_{\mathbb{K}} : (a_0)) = |a_0|^n$ and hence $\mathcal{O}_{\mathbb{K}}/\mathfrak{a}$ is finite. $\square$

We deduce readily from Lemma 7.80 that every maximal ideal is a prime ideal, but the converse is untrue in general, as it may be shown with the ring $R = \mathbb{R}[x, y]$ and the prime ideal (x) which is not maximal since $R/(x) \simeq \mathbb{R}[y]$ is not a field. This leads to the following definition.

Definition 7.82 (Dedekind domain) Let R be an integral domain. R is a *Dedekind domain* if R is nœtherian, integrally closed[17] and such that every non-zero prime ideal is maximal.

---

[17]A domain R is *integrally closed* means that if α/β is a root of a monic polynomial lying in $R[X]$ with $\alpha, \beta \in R$, $\beta \neq 0$, then $\alpha/\beta \in R$.

The next result shows that the ring $\mathcal{O}_{\mathbb{K}}$ is the right tool to deal with.

Proposition 7.83 *Let* $\mathbb{K}/\mathbb{Q}$ *be an algebraic number field. Then* $\mathcal{O}_{\mathbb{K}}$ *is a Dedekind domain.*

Proof Let $\mathbb{A}$ be the set of all algebraic integers.

▷ The ring $\mathcal{O}_{\mathbb{K}}$ is an integral domain since $\mathbb{A}$ is an integral domain and $\mathcal{O}_{\mathbb{K}} = \mathbb{A} \cap \mathbb{K}$.
▷ The quotient field of $\mathcal{O}_{\mathbb{K}}$ is $\mathbb{K}$. Indeed, if $\mathbb{F}$ is this quotient field, then it is clear that $\mathbb{F} \subseteq \mathbb{K}$. Conversely, if $\alpha = a/b \in \mathbb{K}$ where a is an algebraic integer and $b \in \mathbb{Z} \setminus \{0\}$, then $a = \alpha b \in \mathbb{K}$ and hence $a \in \mathcal{O}_{\mathbb{K}}$, so that $\alpha = a/b \in \mathbb{F}$.
▷ $\mathcal{O}_{\mathbb{K}}$ is a finitely generated $\mathbb{Z}$-module and hence $\mathcal{O}_{\mathbb{K}}$ is nœtherian by Proposition 7.16 (iv).
▷ Let $\alpha \in \mathbb{K}$ be a root of a monic polynomial $P = X^n + a_{n-1}X^{n-1} + \cdots + a_0 \in \mathcal{O}_{\mathbb{K}}[X]$. As a $\mathbb{Z}$-module, $\mathbb{Z}[a_0, \ldots, a_{n-1}]$ is finitely generated since $a_i \in \mathcal{O}_{\mathbb{K}}$. Set $M = \mathbb{Z}[a_0, \ldots, a_{n-1}, \alpha]$. Since $P(\alpha) = 0$, we can write α^n as a $\mathbb{Z}[a_0, \ldots, a_{n-1}]$-linear combination of α^i for all $0 \leqslant i < n$, so that as a $\mathbb{Z}$-module, M is also finitely generated. Since $\alpha M \subseteq M$, we infer that α is integral over $\mathbb{Z}$ by Proposition 7.45, and hence $\alpha \in \mathcal{O}_{\mathbb{K}}$. Therefore $\mathcal{O}_{\mathbb{K}}$ is integrally closed.
▷ Let $\mathfrak{p}$ be a non-zero prime ideal of $\mathcal{O}_{\mathbb{K}}$. By Lemma 7.81, $\mathcal{O}_{\mathbb{K}}/\mathfrak{p}$ is finite. Since $\mathfrak{p}$ is prime, then $\mathcal{O}_{\mathbb{K}}/\mathfrak{p}$ is an integral domain by Lemma 7.80. Since a finite integral domain is a field, we infer that $\mathcal{O}_{\mathbb{K}}/\mathfrak{p}$ is a field, and therefore the ideal $\mathfrak{p}$ is maximal by Lemma 7.80.   □

It follows in particular that $\mathcal{O}_{\mathbb{K}}$ is a nœtherian ring so that every element of $\mathcal{O}_{\mathbb{K}}$ can be written as a product of irreducible elements by Proposition 7.11. We know that such decompositions may not be unique, but this problem may be overcome by considering ideals instead of elements of $\mathcal{O}_{\mathbb{K}}$. It remains to define more specific ideals which play the part of rational numbers.

7.3.2 Fractional Ideals

The study of the uniqueness of the factorization of ideals requires the behavior of these ideals under multiplication to be known. It has been seen that this operation has nearly all the properties needed, but inverses need not exist. One may overcome this difficulty if we remember that ideals in $\mathcal{O}_{\mathbb{K}}$ are also sub-$\mathcal{O}_{\mathbb{K}}$-modules of $\mathcal{O}_{\mathbb{K}}$. This gives birth to the concept of *fractional ideal*. Note that the following definition may be generalized in Dedekind domains.

Definition 7.84 Let $\mathbb{K}/\mathbb{Q}$ be an algebraic number field. A *fractional ideal* of $\mathcal{O}_{\mathbb{K}}$ is a non-zero sub-$\mathcal{O}_{\mathbb{K}}$-module $\mathfrak{a}$ of $\mathbb{K}$ such that there exists $\alpha \in \mathcal{O}_{\mathbb{K}}$, $\alpha \neq 0$, satisfying

$$\alpha \mathfrak{a} \subseteq \mathcal{O}_{\mathbb{K}}.$$

The collection of fractional ideals of $\mathcal{O}_{\mathbb{K}}$ is denoted by $\mathcal{I}(\mathbb{K})$. The collection of principal fractional ideals of $\mathcal{O}_{\mathbb{K}}$ is denoted[18] by $\mathcal{P}(\mathbb{K})$. Hence

$$\mathfrak{a} \in \mathcal{P}(\mathbb{K}) \quad \Longleftrightarrow \quad \mathfrak{a} = (a) = a\mathcal{O}_{\mathbb{K}} \quad (a \in \mathbb{K}).$$

Obviously, any non-zero ideal of $\mathcal{O}_{\mathbb{K}}$ is a fractional ideal by taking $\alpha = 1$. In order to underscore that the fractional ideal $\mathfrak{a}$ is actually contained in $\mathcal{O}_{\mathbb{K}}$, we will say that $\mathfrak{a}$ is an *integral ideal*.

We may define the multiplication of fractional ideals in the same way as we did for ideals. This product is associative and commutative with identity element $(1) = \mathcal{O}_{\mathbb{K}}$. It remains to define the inverse of a fractional ideal of $\mathbb{K}$ and to show that the set $\mathcal{I}(\mathbb{K})$ is a multiplicative abelian group.

We first generalize (7.1) by defining the notion of divisibility for integral ideals. Let $\mathfrak{a}$, $\mathfrak{b}$ be integral ideals. We shall say that $\mathfrak{a}$ divides $\mathfrak{b}$, written $\mathfrak{a} \mid \mathfrak{b}$ if and only if $\mathfrak{a} \supseteq \mathfrak{b}$ so that

$$\mathfrak{a} \mid \mathfrak{b} \quad \Longleftrightarrow \quad \mathfrak{b} \subseteq \mathfrak{a}.$$

Furthermore, the following observation, coming directly from the definition of a maximal ideal, will be often used.

Let $\mathfrak{p}$ be a maximal ideal of $\mathcal{O}_{\mathbb{K}}$ and $\mathfrak{a}$ be an integral ideal such that $(1) \mid \mathfrak{a} \mid \mathfrak{p}$. Then

$$\mathfrak{a} = \mathfrak{p} \quad or \quad \mathfrak{a} = (1). \tag{7.9}$$

The following first lemma will prove useful.

Lemma 7.85 *Let $\mathfrak{a}$ be an integral ideal. Then there exist prime ideals $\mathfrak{p}_1, \ldots, \mathfrak{p}_r$ such that*

$$\mathfrak{a} \mid \mathfrak{p}_1 \cdots \mathfrak{p}_r.$$

Proof Let S be the set of integral ideals that do not satisfy the result of the lemma. If $S \neq \varnothing$ and since $\mathcal{O}_{\mathbb{K}}$ is a nœtherian ring, there exists an ideal $\mathfrak{b} \in S$ which is maximal as an element of S. By assumption, $\mathfrak{b}$ is not prime so that there exist $\alpha_1, \alpha_2 \in \mathcal{O}_{\mathbb{K}}$ such that $\alpha_1 \alpha_2 \in \mathfrak{b}$ and $\alpha_1 \notin \mathfrak{b}$ and $\alpha_2 \notin \mathfrak{b}$. Now set $\mathfrak{d}_i = \mathfrak{b} + (\alpha_i)$ for $i \in \{1, 2\}$. Since $\mathfrak{b}$ is maximal as an element of S, we have $\mathfrak{d}_i \notin S$ and therefore both ideals contain a product of prime ideals. We deduce that

$$\mathfrak{d}_1 \mathfrak{d}_2 = \mathfrak{b}^2 + \mathfrak{b}(\alpha_1) + \mathfrak{b}(\alpha_2) + (\alpha_1 \alpha_2) \subseteq \mathfrak{b}$$

and hence $\mathfrak{b} \notin S$, giving a contradiction. $\qquad\qquad\square$

---

[18]In fact, we should rather denote these sets respectively by $\mathcal{I}(\mathcal{O}_{\mathbb{K}})$ and $\mathcal{P}(\mathcal{O}_{\mathbb{K}})$, but we have followed here the usual practice. Nevertheless, if R is any *number ring*, i.e. a domain for which the quotient field $\mathbb{K}$ is an algebraic number field, then the collection of all fractional ideals, resp. principal ideals, of R is denoted by $\mathcal{I}(R)$, resp. $\mathcal{P}(R)$. Similarly, the *class group* or *Picard group* of R, which is an invariant of R, is denoted by $\mathrm{Cl}(R)$ or $\mathrm{Pic}(R)$ and is defined as in Definition 7.110 by $\mathrm{Cl}(R) = \mathcal{I}(R)/\mathcal{P}(R)$. When R is the ring of integers of an algebraic number field $\mathbb{K}$, then $\mathrm{Cl}(\mathcal{O}_{\mathbb{K}})$ depends only on $\mathbb{K}$ and is usually denoted by $\mathrm{Cl}(\mathbb{K})$. Hence the class group in the usual sense may be viewed as an invariant of $\mathbb{K}$.

The next lemma is a technical tool.

Lemma 7.86 *Let $\mathfrak{a}$ be an integral ideal. If $\alpha \in \mathbb{K}$ satisfies $\mathfrak{a} \mid \alpha\mathfrak{a}$, then $\alpha \in \mathcal{O}_{\mathbb{K}}$.*

Proof As a finitely generated sub-$\mathbb{Z}$-module of $\mathcal{O}_{\mathbb{K}}$ with rank n, let $\beta_1, \ldots, \beta_n$ be a $\mathbb{Z}$-basis of $\mathfrak{a}$. Since $\alpha\mathfrak{a} \subseteq \mathfrak{a}$, there exists an integer matrix $A = (a_{ij}) \in M_n(\mathbb{Z})$ such that, for all $i \in \{1, \ldots, n\}$, we have

$$\alpha\beta_i = \sum_{j=1}^{n} a_{ij}\beta_j$$

which may be written as

$$(A - \alpha I_n) \begin{pmatrix} \beta_1 \\ \vdots \\ \beta_n \end{pmatrix} = 0$$

so that α is an eigenvalue of A and hence is an algebraic integer.   □

The natural candidate for the inverse of a fractional ideal $\mathfrak{a}$ will turn out to be

$$\mathfrak{a}^{-1} = \{\alpha \in \mathbb{K} : \alpha\mathfrak{a} \subseteq \mathcal{O}_{\mathbb{K}}\}.$$

We first check that $\mathfrak{a}^{-1} \in \mathcal{I}(\mathbb{K})$. It is obviously a non-zero sub-$\mathcal{O}_{\mathbb{K}}$-module and if $x \in \mathfrak{a}$ with $x \neq 0$, then $x\mathfrak{a}^{-1} \subseteq \mathcal{O}_{\mathbb{K}}$ as required. Also note that, if $\mathfrak{a} \mid \mathfrak{b}$, then $\mathfrak{b}^{-1} \mid \mathfrak{a}^{-1}$. In particular, if $\mathfrak{a} \neq 0$ is an integral ideal, then

$$\mathfrak{a}^{-1} \mid (1). \tag{7.10}$$

We are now in a position to show the main result of this section.

Proposition 7.87 $\mathcal{I}(\mathbb{K})$ *is an abelian multiplicative group, called the* ideal group.

Proof It remains to show that, for all $\mathfrak{a} \in \mathcal{I}(\mathbb{K})$, we have

$$\mathfrak{a}\mathfrak{a}^{-1} = (1). \tag{7.11}$$

1. Let $\mathfrak{p}$ be a prime ideal. By definition of $\mathfrak{p}^{-1}$, we have $(1) \mid \mathfrak{p}^{-1}\mathfrak{p}$ and, using (7.10), we get $\mathfrak{p}^{-1}\mathfrak{p} \mid \mathfrak{p}$ and therefore $(1) \mid \mathfrak{p}^{-1}\mathfrak{p} \mid \mathfrak{p}$, so that by (7.9) either $\mathfrak{p}^{-1}\mathfrak{p} = \mathfrak{p}$ or $\mathfrak{p}^{-1}\mathfrak{p} = (1)$. In the latter case, (7.11) holds for prime ideals, so suppose that $\mathfrak{p} = \mathfrak{p}^{-1}\mathfrak{p}$ and let $\alpha \in \mathfrak{p}$, $\alpha \neq 0$. By Lemma 7.85, there exist prime ideals $\mathfrak{p}_1, \ldots, \mathfrak{p}_r$ such that

$$\mathfrak{p} \mid (\alpha) \mid \mathfrak{p}_1 \cdots \mathfrak{p}_r$$

with r minimal. If $\mathfrak{p} \nmid \mathfrak{p}_i$ for all i, then we can choose for each i an element $c_i \in \mathfrak{p}_i$ such that $c_i \notin \mathfrak{p}$. But then the product of the c_i is in $\mathfrak{p}$ which is impossible since $\mathfrak{p}$ is a prime ideal. Therefore there exists a prime ideal $\mathfrak{p}_i$ such that $\mathfrak{p} \mid \mathfrak{p}_i$. Without loss of generality, assume that $\mathfrak{p} \mid \mathfrak{p}_1$. Since $\mathcal{O}_{\mathbb{K}}$ is a Dedekind domain, $\mathfrak{p}_1$ is maximal which implies that $\mathfrak{p} = \mathfrak{p}_1$. Because r is minimal, $(\alpha) \nmid \mathfrak{p}_2 \cdots \mathfrak{p}_r$ so that there exists $\beta \in \mathfrak{p}_2 \cdots \mathfrak{p}_r$ such that $\beta \notin (\alpha)$. Then

$$(\alpha) \mid \mathfrak{p}\mathfrak{p}_2 \cdots \mathfrak{p}_r \mid \mathfrak{p}(\beta)$$

which implies that $a = \beta/\alpha \in \mathfrak{p}^{-1}$. Since $\beta \notin (\alpha)$, we get $a \notin \mathcal{O}_{\mathbb{K}}$. Now using $\mathfrak{p} = \mathfrak{p}^{-1}\mathfrak{p}$, we get $\mathfrak{p} \mid a\mathfrak{p}$ and Lemma 7.86 implies that $a \in \mathcal{O}_{\mathbb{K}}$, giving a contradiction. Therefore (7.11) is true for prime ideals.

2. Now we prove that (7.11) also holds for integral ideals. Suppose the contrary. There exists a non-zero integral ideal $\mathfrak{a}$, taken maximal among all non-zero integral ideals, which does not have an inverse. Let $\mathfrak{p}$ be a prime ideal such that $\mathfrak{p} \mid \mathfrak{a}$. Multiplying both sides by $\mathfrak{p}^{-1}$ implies that $(1) \mid \mathfrak{p}^{-1}\mathfrak{a} \mid \mathfrak{a}$. Arguing as above, $\mathfrak{a} = \mathfrak{p}^{-1}\mathfrak{a}$ gives a contradiction. By the maximality of $\mathfrak{a}$, we infer that $\mathfrak{p}^{-1}\mathfrak{a}$ has an inverse $\mathfrak{b}$ and hence $\mathfrak{p}^{-1}\mathfrak{b}$ is an inverse of $\mathfrak{a}$, leading to a contradiction. We infer that (7.11) is also true for integral ideals.

3. Since every fractional ideal $\mathfrak{a}$ can be written as $\mathfrak{a} = d\mathfrak{b}$ for some $d \in \mathbb{K}$ and integral ideal $\mathfrak{b}$, we deduce from above that (7.11) is true for fractional ideals. $\square$

It is noteworthy that the divisibility relation between integral ideals may be now viewed as the divisibility relation in $\mathbb{Z}$.

> Let $\mathfrak{a}, \mathfrak{b}$ be two non-zero integral ideals. Then $\mathfrak{a} \mid \mathfrak{b}$ if and only if there exists a suitable non-zero integral ideal $\mathfrak{c}$ such that $\mathfrak{b} = \mathfrak{a}\mathfrak{c}$.

Indeed, if $\mathfrak{b} = \mathfrak{a}\mathfrak{c}$, then $\mathfrak{b} \subseteq \mathfrak{a}$. Conversely, if $\mathfrak{b} \subseteq \mathfrak{a}$, then we get $\mathfrak{b}\mathfrak{a}^{-1} \subseteq \mathfrak{a}\mathfrak{a}^{-1} = \mathcal{O}_{\mathbb{K}}$ so that $\mathfrak{c} = \mathfrak{b}\mathfrak{a}^{-1}$ is an integral ideal satisfying $\mathfrak{b} = \mathfrak{a}\mathfrak{c}$.

7.3.3 The Fundamental Theorem of Ideal Theory

We are now in a position to solve the problem of uniqueness in algebraic number fields. The following result shows that Kummer's favorite *ideal numbers* are the right tools to generalize the arithmetic in $\mathbb{Z}$.

Theorem 7.88 *Let $\mathfrak{a}$ be a non-zero integral ideal. Then $\mathfrak{a}$ can be written as a product*

$$\mathfrak{a} = \mathfrak{p}_1 \cdots \mathfrak{p}_r$$

of prime ideals of $\mathcal{O}_{\mathbb{K}}$ and this decomposition is unique up to order.[19]

---

[19]The zero ideal can also be written as a product of prime ideals, but the decomposition is not unique. Hence the case of the zero ideal is almost always excluded in this book.

Proof Let S be the set of non-zero integral ideals that cannot be written as a product of non-zero prime ideals in $\mathcal{O}_{\mathbb{K}}$. If $S \neq \varnothing$ and since $\mathcal{O}_{\mathbb{K}}$ is a nœtherian ring, there exists an ideal $\mathfrak{b} \in S$ which is maximal as an element of S. Now let $\mathfrak{p}$ be a non-zero prime ideal such that $\mathfrak{p} \mid \mathfrak{b}$. If $\mathfrak{b}\mathfrak{p}^{-1} = \mathfrak{b}$, then we get $\mathfrak{p}^{-1} = (1)$ by Proposition 7.87, which is false. We deduce that $\mathfrak{b}$ is strictly contained in $\mathfrak{b}\mathfrak{p}^{-1}$ and hence $\mathfrak{b}\mathfrak{p}^{-1} \notin S$. We infer that there exist non-zero prime ideals $\mathfrak{p}_1, \ldots, \mathfrak{p}_r$ such that $\mathfrak{b}\mathfrak{p}^{-1} = \mathfrak{p}_1 \cdots \mathfrak{p}_r$, and therefore we get $\mathfrak{b} = \mathfrak{p}_1 \cdots \mathfrak{p}_r \mathfrak{p}$ so that $\mathfrak{b} \notin S$, leading to a contradiction. Hence $S = \varnothing$ and every non-zero integral ideal can be written as a product of prime ideals.

Now let $\mathfrak{a}$ be a non-zero integral ideal and suppose that there exist non-zero prime ideals $\mathfrak{p}_1, \ldots, \mathfrak{p}_r$ and $\mathfrak{q}_1, \ldots, \mathfrak{q}_s$ such that

$$\mathfrak{a} = \mathfrak{p}_1 \cdots \mathfrak{p}_r = \mathfrak{q}_1 \cdots \mathfrak{q}_s.$$

If $\mathfrak{p}_1 \nmid \mathfrak{q}_i$ for all $i \in \{1, \ldots, s\}$, then there exist $c_i \in \mathfrak{q}_i$ such that $c_i \notin \mathfrak{p}_1$. By assumption, the product of the c_i is in $\mathfrak{p}_1 \cdots \mathfrak{p}_r$ which is a subset of $\mathfrak{p}_1$, and hence $\mathfrak{p}_1$ is not a prime ideal, giving a contradiction. Thus there exists $i \in \{1, \ldots, s\}$ such that $\mathfrak{p}_1 \mid \mathfrak{q}_i$, and without loss of generality, suppose that $\mathfrak{p}_1 \mid \mathfrak{q}_1$. Since $\mathfrak{q}_1$ is maximal, we get $\mathfrak{p}_1 = \mathfrak{q}_1$. By Proposition 7.87, we obtain $\mathfrak{p}_2 \cdots \mathfrak{p}_r = \mathfrak{q}_2 \cdots \mathfrak{q}_s$, and arguing as above we show that $\mathfrak{p}_2 = \mathfrak{q}_2$. Repeating the argument enables us to prove that $r = s$ and $\mathfrak{p}_i = \mathfrak{q}_i$ for all $i \in \{1, \ldots, r\}$ as required. $\square$

Remark 7.89

1. The reader may have seen the analogy between the proofs of Theorems 3.3 and 7.88.
2. As a consequence of Theorem 7.88, one can readily see that fractional ideals can also be written uniquely in the following form.

   Let $\mathfrak{a} \in \mathcal{I}(\mathbb{K})$. Then there exist prime ideals $\mathfrak{p}_1, \ldots, \mathfrak{p}_r$ and $\mathfrak{q}_1, \ldots, \mathfrak{q}_s$ such that

   $$\mathfrak{a} = (\mathfrak{p}_1 \cdots \mathfrak{p}_r)(\mathfrak{q}_1 \cdots \mathfrak{q}_s)^{-1}$$

   the decomposition being unique up to order.

3. As in the rational case, the fractional ideals can be uniquely written in the form

   $$\mathfrak{a} = \mathfrak{p}_1^{e_1} \cdots \mathfrak{p}_{\mathfrak{g}}^{e_{\mathfrak{g}}}$$

   where $\mathfrak{p}_i$ are *distinct* prime ideals and $e_i \in \mathbb{Z} \setminus \{0\}$ for all $i \in \{1, \ldots, g\}$. The ideal $(1) = \mathcal{O}_{\mathbb{K}}$ is regarded as the unique empty product of prime ideals. The integers e_i are the $\mathfrak{p}_i$-*adic valuations* of $\mathfrak{a}$, sometimes denoted by $v_{\mathfrak{p}_i}(\mathfrak{a})$. These valuations have the same properties as the ordinary valuations of rational numbers. In particular, if $\mathfrak{p}$ is a prime ideal and $\mathfrak{a}, \mathfrak{b}$ are non-zero fractional ideals of $\mathbb{K}$, we have

   $$v_{\mathfrak{p}}(\mathfrak{a}\mathfrak{b}) = v_{\mathfrak{p}}(\mathfrak{a}) + v_{\mathfrak{p}}(\mathfrak{b}),$$
   $$v_{\mathfrak{p}}(\mathfrak{a} + \mathfrak{b}) = \min(v_{\mathfrak{p}}(\mathfrak{a}), v_{\mathfrak{p}}(\mathfrak{b})),$$
   $$v_{\mathfrak{p}}(\mathfrak{a} \cap \mathfrak{b}) = \max(v_{\mathfrak{p}}(\mathfrak{a}), v_{\mathfrak{p}}(\mathfrak{b})).$$

Example 7.90 Let $\mathbb{K} = \mathbb{Q}(\sqrt{-5})$ which we know is not a UFD since

$$6 = 2 \times 3 = (1 + \sqrt{-5})(1 - \sqrt{-5}).$$

Theorem 7.88 shows how to restore unique factorization. Set

$$\mathfrak{p}_2 = (2, 1 + \sqrt{-5}) = 2\mathcal{O}_{\mathbb{K}} + (1 + \sqrt{-5})\mathcal{O}_{\mathbb{K}}$$

$\mathfrak{p}_3 = (3, 1 + \sqrt{-5})$ and $\mathfrak{p}'_3 = (3, 1 - \sqrt{-5})$. One can check that these ideals are prime and

$$(6) = \mathfrak{p}_3 \mathfrak{p}'_3 \mathfrak{p}_2^2.$$

7.3.4 Consequences of the Fundamental Theorem

Let $\mathbb{K}/\mathbb{Q}$ be an algebraic number field of degree n and $\mathcal{O}_{\mathbb{K}}$ be its ring of integers.

Generators of Fractional Ideals

As a $\mathbb{Z}$-module of rank n, any integral ideal can be generated by n elements of $\mathcal{O}_{\mathbb{K}}$. The purpose of this section is to prove that any fractional ideal can be generated as an $\mathcal{O}_{\mathbb{K}}$-module by at most two elements. We start with an application of the Chinese remainder theorem.

Lemma 7.91 *Let $\mathfrak{a}$ and $\mathfrak{b}$ be non-zero integral ideals. Then there exists $\alpha \in \mathfrak{a}$ such that the ideals $(\alpha)\mathfrak{a}^{-1}$ and $\mathfrak{b}$ are coprime.*

Proof Using Theorem 7.88, we can write $\mathfrak{b} = \mathfrak{p}_1^{f_1} \cdots \mathfrak{p}_g^{f_g}$ where $\mathfrak{p}_i$ are distinct prime ideals and f_i are non-negative integers. We set $\mathfrak{a} = \mathfrak{p}_1^{e_1} \cdots \mathfrak{p}_g^{e_g} \mathfrak{c}$ with $\mathfrak{b} + \mathfrak{c} = (1)$ and e_i are non-negative integers. For all $i \in \{1, \ldots, g\}$, we pick up an element $a_i \in \mathfrak{p}_i^{e_i} \setminus \mathfrak{p}_i^{e_i+1}$. Applying Proposition 7.5 with $n = g + 1$, $R = \mathcal{O}_{\mathbb{K}}$, $\mathfrak{a}_i = \mathfrak{p}_i^{e_i+1}$ for $i \in \{1, \ldots, g\}$ and $\mathfrak{a}_{g+1} = \mathfrak{c}$, we infer that there exists $\alpha \in \mathcal{O}_{\mathbb{K}}$ such that

$$\alpha \equiv a_i \ \left(\text{mod } \mathfrak{p}_i^{e_i+1}\right) \quad \text{and} \quad \alpha \equiv 0 \ (\text{mod } \mathfrak{c}).$$

Since $a_i \in \mathfrak{p}_i^{e_i} \setminus \mathfrak{p}_i^{e_i+1}$, we have $\alpha \equiv 0 \ (\text{mod } \mathfrak{p}_i^{e_i})$ and $\alpha \not\equiv 0 \ (\text{mod } \mathfrak{p}_i^{e_i+1})$ so that we get $(\alpha) = \mathfrak{a}\mathfrak{d}$ with $\mathfrak{b} + \mathfrak{d} = (1)$ which concludes the proof. $\square$

Proposition 7.92 *Let $\mathfrak{a} \in \mathcal{I}(\mathbb{K})$. Then there exists $\alpha, \beta \in \mathbb{K}$ such that*

$$\mathfrak{a} = (\alpha, \beta) = \alpha\mathcal{O}_{\mathbb{K}} + \beta\mathcal{O}_{\mathbb{K}}.$$

Proof The zero ideal is generated by one element and if $\mathfrak{a} \in \mathcal{I}(\mathbb{K})$ is not an integral ideal, then there exists $d \in \mathbb{K}$ such that $d\mathfrak{a} \in \mathcal{O}_{\mathbb{K}}$, and since $\mathfrak{a}$ and $d\mathfrak{a}$ have the same number of generators, we may suppose that $\mathfrak{a}$ is a non-zero integral ideal. Let $\beta \neq 0 \in \mathfrak{a}$ and set $\mathfrak{b} = (\beta)$. By Lemma 7.91, there exists $\alpha \in \mathfrak{a}$ such that the ideals $(\alpha)\mathfrak{a}^{-1}$ and $\mathfrak{b}$ are coprime. Since $\alpha, \beta \in \mathfrak{a}$, we have $\mathfrak{a} \mid (\alpha, \beta)$. Let $\mathfrak{p}$ be a prime divisor of (α, β) and e be its p-adic valuation. Since $(\alpha)\mathfrak{a}^{-1}$ and $\mathfrak{b}$ are coprime, we infer that $\mathfrak{p}^e \nmid (\alpha)\mathfrak{a}^{-1}$ but since $\mathfrak{p}^e \mid (\alpha) = (\alpha)\mathfrak{a}^{-1}\mathfrak{a}$, we deduce that $\mathfrak{p}^e \mid \mathfrak{a}$. Therefore we have $(\alpha, \beta) \mid \mathfrak{a}$ and hence $\mathfrak{a} = (\alpha, \beta)$ as asserted. $\qquad\square$

Unique Factorization

Theorem 7.13 asserts that if a ring R is a PID, then it is a UFD. We will prove that if R is a ring of integers of an algebraic number field, then the converse is true.

Theorem 7.93 *Let* $\mathbb{K}/\mathbb{Q}$ *be an algebraic number field of degree n. Then* $\mathcal{O}_{\mathbb{K}}$ *is a UFD if and only if it is a PID.*

Proof The factorization of elements of $\mathcal{O}_{\mathbb{K}}$ into irreducibles exists by Proposition 7.11. Suppose that this factorization is unique. By Theorem 7.88, it suffices to show that every prime ideal is principal. Let $\mathfrak{p}$ be a prime ideal and $\alpha \neq 0 \in \mathfrak{p}$. Let $\alpha = \pi_1 \cdots \pi_r$ be the decomposition of α into irreducibles in $\mathcal{O}_{\mathbb{K}}$. Then $(\alpha) = (\pi_1) \cdots (\pi_r)$. Now every (π) is a prime ideal for if $ab \in (\pi)$, then $\pi \mid ab$ and by assumption of unique factorization we get $\pi \mid a$ or $\pi \mid b$ and hence $a \in (\pi)$ or $b \in (\pi)$. Therefore $\mathfrak{p}$ divides a product of principal prime ideals and, by Theorem 7.88, we infer that $\mathfrak{p}$ is itself one of these principal prime ideals. $\qquad\square$

7.3.5 Norm of an Ideal

Let $\mathfrak{a}$ be a non-zero integral ideal. By Lemma 7.81, the order of $\mathcal{O}_{\mathbb{K}}/\mathfrak{a}$ is finite. This leads to the following definition.

Definition 7.94 Let $\mathbb{K}/\mathbb{Q}$ be an algebraic number field and let $\mathfrak{a}$ be a non-zero integral ideal. The *norm* of $\mathfrak{a}$ is the integer defined by

$$\mathcal{N}_{\mathbb{K}/\mathbb{Q}}(\mathfrak{a}) = |\mathcal{O}_{\mathbb{K}}/\mathfrak{a}|.$$

Example 7.95 Let $\mathbb{K} = \mathbb{Q}(\sqrt{-6})$. By Proposition 7.64, we get $\mathcal{O}_{\mathbb{K}} = \mathbb{Z}[\sqrt{-6}]$. Define the ideals $\mathfrak{a} = (1 + \sqrt{-6})$, $\mathfrak{b} = (2, \sqrt{-6})$ and $\mathfrak{c} = (3, \sqrt{-6})$.

1. Let $\alpha = a + b\sqrt{-6}$, $\beta = c + d\sqrt{-6} \in \mathbb{Z}[\sqrt{-6}]$. We have $\alpha \equiv \beta \pmod{\mathfrak{a}}$ if and only if

$$\frac{a - c + (b - d)\sqrt{-6}}{1 + \sqrt{-6}} \in \mathbb{Z}[\sqrt{-6}].$$

Since

$$\frac{a - c + (b - d)\sqrt{-6}}{1 + \sqrt{-6}} = \frac{a - c + 6(b - d)}{7} + \frac{b - d - a + c}{7}\sqrt{-6}$$

we infer that $\alpha \equiv \beta \pmod{\mathfrak{a}} \Longleftrightarrow a - c \equiv b - d \pmod{7}$ so that $\mathcal{N}_{\mathbb{K}/\mathbb{Q}}(\mathfrak{a}) = 7$.

2. One may check that $\mathfrak{b} = \{2a + b\sqrt{-6} : a, b \in \mathbb{Z}\}$ so that $a + b\sqrt{-6} \equiv c + d\sqrt{-6} \pmod{\mathfrak{b}}$ if and only if $a \equiv c \pmod{2}$ and hence $\mathcal{N}_{\mathbb{K}/\mathbb{Q}}(\mathfrak{b}) = 2$.

3. Similarly, we have $\mathfrak{c} = \{3a + b\sqrt{-6} : a, b \in \mathbb{Z}\}$ so that $a + b\sqrt{-6} \equiv c + d\sqrt{-6} \pmod{\mathfrak{c}}$ if and only if $a \equiv c \pmod{3}$ and hence $\mathcal{N}_{\mathbb{K}/\mathbb{Q}}(\mathfrak{c}) = 3$.

The norm of an integral ideal will in a way generalize the role of the integers. The next result summarizes the most important properties of this operator.

Theorem 7.96 *Let $\mathbb{K}/\mathbb{Q}$ be an algebraic number field of degree n with discriminant $d_{\mathbb{K}}$.*

(i) *Let $\mathfrak{a}$ be a non-zero integral ideal and $\{\alpha_1, \ldots, \alpha_n\}$ be a $\mathbb{Z}$-basis for $\mathfrak{a}$. Then*

$$\Delta_{\mathbb{K}/\mathbb{Q}}(\alpha_1, \ldots, \alpha_n) = \left(\mathcal{N}_{\mathbb{K}/\mathbb{Q}}(\mathfrak{a})\right)^2 d_{\mathbb{K}}.$$

(ii) *Let $\alpha \neq 0 \in \mathcal{O}_{\mathbb{K}}$. Then*

$$\mathcal{N}_{\mathbb{K}/\mathbb{Q}}((\alpha)) = \left|N_{\mathbb{K}/\mathbb{Q}}(\alpha)\right|.$$

(iii) *Let $\mathfrak{a}, \mathfrak{b}$ be non-zero integral ideals. Then*

$$\mathcal{N}_{\mathbb{K}/\mathbb{Q}}(\mathfrak{a}\mathfrak{b}) = \mathcal{N}_{\mathbb{K}/\mathbb{Q}}(\mathfrak{a})\,\mathcal{N}_{\mathbb{K}/\mathbb{Q}}(\mathfrak{b}).$$

Proof

(i) Let $\{\theta_1, \ldots, \theta_n\}$ be an integral basis for $\mathbb{K}$ and assume that

$$\alpha_i = \sum_{j=1}^{n} a_{ij}\theta_j$$

with $a_{ij} \in \mathbb{Z}$ for all $i, j \in \{1, \ldots, n\}$. By Proposition 7.18, we have

$$\mathcal{N}_{\mathbb{K}/\mathbb{Q}}(\mathfrak{a}) = |\mathcal{O}_{\mathbb{K}}/\mathfrak{a}| = \left|\det(a_{ij})\right|$$

and we conclude using Proposition 7.59 (iv).

(ii) Let $\{\theta_1, \ldots, \theta_n\}$ be an integral basis for $\mathbb{K}$. Then $\{\alpha\theta_1, \ldots, \alpha\theta_n\}$ is a $\mathbb{Z}$-basis for (α). Writing $\alpha\theta_i = \sum_{j=1}^{n} a_{ij}\theta_j$ with $a_{ij} \in \mathbb{Z}$, we have $N_{\mathbb{K}/\mathbb{Q}}(\alpha) = \det(a_{ij})$, so that by Proposition 7.59 (iv), we get

$$\Delta_{\mathbb{K}/\mathbb{Q}}(\alpha\theta_1, \ldots, \alpha\theta_n) = N_{\mathbb{K}/\mathbb{Q}}(\alpha)^2 d_{\mathbb{K}}.$$

We conclude by using (i) which implies that

$$\Delta_{\mathbb{K}/\mathbb{Q}}(\alpha\theta_1, \dots, \alpha\theta_n) = \mathcal{N}_{\mathbb{K}/\mathbb{Q}}((\alpha))^2 d_{\mathbb{K}}$$

and the fact that $\mathcal{N}_{\mathbb{K}/\mathbb{Q}}((\alpha))$ is a positive integer.

(iii) By Theorem 7.88 and induction on the number of factors, it is sufficient to prove

$$\mathcal{N}_{\mathbb{K}/\mathbb{Q}}(\mathfrak{ap}) = \mathcal{N}_{\mathbb{K}/\mathbb{Q}}(\mathfrak{a})\mathcal{N}_{\mathbb{K}/\mathbb{Q}}(\mathfrak{p})$$

where $\mathfrak{p}$ is a prime ideal. Since $\mathfrak{a} \mid \mathfrak{ap}$, using Theorem 7.4 (ii) applied with $R = \mathcal{O}_{\mathbb{K}}$, $\mathfrak{a}_1 = \mathfrak{ap}$ and $\mathfrak{a}_2 = \mathfrak{a}$, we get

$$\mathcal{N}_{\mathbb{K}/\mathbb{Q}}(\mathfrak{ap}) = \mathcal{N}_{\mathbb{K}/\mathbb{Q}}(\mathfrak{a}) \times |\mathfrak{a}/\mathfrak{ap}|.$$

It is therefore sufficient to prove that

$$|\mathfrak{a}/\mathfrak{ap}| = \mathcal{N}_{\mathbb{K}/\mathbb{Q}}(\mathfrak{p}). \tag{7.12}$$

First notice that from Theorem 7.88 we deduce that $\mathfrak{ap}$ is strictly contained in $\mathfrak{a}$. Now we will prove that there is no ideal $\mathfrak{b}$ strictly contained between $\mathfrak{a}$ and $\mathfrak{ap}$. Suppose the contrary. We then have $\mathfrak{a} \mid \mathfrak{b} \mid \mathfrak{ap}$ and multiplying by $\mathfrak{a}^{-1}$ gives $(1) \mid \mathfrak{a}^{-1}\mathfrak{b} \mid \mathfrak{p}$, and hence $\mathfrak{a}^{-1}\mathfrak{b} = (1)$ or $\mathfrak{a}^{-1}\mathfrak{b} = \mathfrak{p}$ by (7.9), so that $\mathfrak{b} = \mathfrak{a}$ or $\mathfrak{b} = \mathfrak{ap}$ as required.

Now choose $a \in \mathfrak{a} \setminus \mathfrak{ap}$ and define the map

$$F_a : \quad \mathcal{O}_{\mathbb{K}} \quad \longrightarrow \quad \mathfrak{a}/\mathfrak{ap},$$
$$\alpha \quad \longmapsto \quad a\alpha + \mathfrak{ap}$$

Since there are no ideals between $\mathfrak{a}$ and $\mathfrak{ap}$, this implies that $\mathfrak{a}$ is generated by a and $\mathfrak{ap}$, so that F_a is surjective. Furthermore, if $\alpha \in \mathfrak{p}$, then $a\alpha \in \mathfrak{ap}$ and hence $F_a(\alpha) = 0 + \mathfrak{ap}$ so that $\mathfrak{p} \subseteq \ker F_a$. Since $\mathfrak{p}$ is maximal, we get $\ker F_a$ is either (1) or $\mathfrak{p}$. But $F_a(1) = a + \mathfrak{ap} \neq 0 + \mathfrak{ap}$ since $a \notin \mathfrak{ap}$, so $\ker F_a = \mathfrak{p}$. By Theorem 7.4 (i), we deduce that $\mathcal{O}_{\mathbb{K}}/\mathfrak{p} \simeq \mathfrak{a}/\mathfrak{ap}$ proving (7.12).

The proof is complete. $\qquad\qquad\qquad\qquad\qquad\qquad\qquad\qquad\qquad\qquad\square$

Thus the norm map is completely multiplicative, which is certainly its most important property. Note also that, for each non-zero integral ideal $\mathfrak{a}$, we have

$$\mathcal{N}_{\mathbb{K}/\mathbb{Q}}(\mathfrak{a}) \in \mathfrak{a} \tag{7.13}$$

by applying Lagrange's theorem (Theorem 7.1 (i)) to the additive group $\mathcal{O}_{\mathbb{K}}/\mathfrak{a}$. Furthermore, if $\sigma : \mathbb{K} \longrightarrow \mathbb{K}$ is any field homomorphism, then

$$\mathcal{N}_{\mathbb{K}/\mathbb{Q}}(\sigma\mathfrak{a}) = \mathcal{N}_{\mathbb{K}/\mathbb{Q}}(\mathfrak{a}). \tag{7.14}$$

For instance, let $\mathbb{K} = \mathbb{Q}(\sqrt{-6})$ and take $\mathfrak{b} = (2, \sqrt{-6})$ of Example 7.95. If σ is the non-trivial embedding of $\mathbb{K}$, we have $\sigma \mathfrak{b} = (2, -\sqrt{-6})$ and hence by (7.14) and Theorem 7.96 (iii) and (ii), we get

$$\mathcal{N}_{\mathbb{K}/\mathbb{Q}}(\mathfrak{b})^2 = \mathcal{N}_{\mathbb{K}/\mathbb{Q}}(\mathfrak{b}\sigma\mathfrak{b}) = \mathcal{N}_{\mathbb{K}/\mathbb{Q}}(4, 2\sqrt{-6}, -2\sqrt{-6}, -6) = \mathcal{N}_{\mathbb{K}/\mathbb{Q}}((2))$$
$$= |N_{\mathbb{K}/\mathbb{Q}}(2)| = 4.$$

The next result provides some useful consequences of Theorem 7.96.

Corollary 7.97 *Let $\mathbb{K}/\mathbb{Q}$ be an algebraic number field of degree n.*

(i) *Let $\mathfrak{p}$ be a non-zero integral ideal. If $\mathcal{N}_{\mathbb{K}/\mathbb{Q}}(\mathfrak{p})$ is a prime number, then $\mathfrak{p}$ is a prime ideal.*

(ii) *There are only finitely many integral ideals with a given norm.*

Proof

(i) This follows from the complete multiplicativity of the norm.

(ii) If $\mathcal{N}_{\mathbb{K}/\mathbb{Q}}(\mathfrak{a}) = m$ for some positive integer m, then $\mathfrak{a} \mid (m)$ by (7.13). By uniqueness of factorization, m has only finitely many factors.

The proof is complete.   □

Example 7.98

1. Let $\mathbb{K} = \mathbb{Q}(\sqrt{-10})$. By using the PARI/GP system, we get

$$(14) = \mathfrak{p}_1 \mathfrak{p}_2 \mathfrak{p}_3^2$$

where $\mathfrak{p}_1 = (7, 2 + \sqrt{-10})$, $\mathfrak{p}_2 = \sigma \mathfrak{p}_1 = (7, 2 - \sqrt{-10})$ and $\mathfrak{p}_3 = (2, \sqrt{-10})$. Suppose that 14 lies in some integral ideal $\mathfrak{a}$. Then $\mathfrak{a} \mid (14)$ so that by Theorem 7.88 we get

$$\mathfrak{a} = \mathfrak{p}_1^{e_1} \mathfrak{p}_2^{e_2} \mathfrak{p}_3^{e_3}$$

with $e_1, e_2 \in \{0, 1\}$ and $e_3 \in \{0, 1, 2\}$ and hence 14 belongs only to a finite number of ideals. Also by Theorem 7.96 (ii) we get $\mathcal{N}_{\mathbb{K}/\mathbb{Q}}((14)) = |N_{\mathbb{K}/\mathbb{Q}}(14)| = 14^2$. Now how many ideals $\mathfrak{a}$ have norm 14 ? This can only happen when $\mathfrak{a} \mid (14)$ by (7.13), so that by above we get

$$14 = \mathcal{N}_{\mathbb{K}/\mathbb{Q}}(\mathfrak{a}) = 7^{e_1 + e_2} \times 2^{e_3}$$

which implies that $e_1 + e_2 = e_3 = 1$, giving two integral ideals. Note also that by a similar argument we have only one ideal with norm 2 and two ideals with norm 7. Let $\nu_{\mathbb{K}}(m)$ be the number of integral ideals with norm m. This example shows that $\nu_{\mathbb{K}}(2 \times 7) = \nu_{\mathbb{K}}(2) \nu_{\mathbb{K}}(7)$. We shall return to this function in Sect. 7.3.

2. Let $\mathbb{K} = \mathbb{Q}(\zeta_8)$ with defining polynomial $\Phi_8 = X^4 + 1$. We have $d_{\mathbb{K}} = 256$ and Theorem 7.105 below gives

$$(2) = \mathfrak{p}^4$$

with $\mathfrak{p} = (2, 1 + \zeta_8)$. A similar argument shows that $\mathfrak{p}$ is the only prime ideal with norm 2. Let us prove that $\mathfrak{p}$ is principal. It is first clear that $\mathfrak{p} \subseteq (1 + \zeta_8)$. Conversely, since

$$2 = (1 + \zeta_8)^2 (\zeta_8 - \zeta_8^2)$$

we infer that $(1 + \zeta_8) \subseteq \mathfrak{p}$. Therefore $\mathfrak{p} = (1 + \zeta_8)$ and then $\mathfrak{p}$ is principal. We will show later that $\mathcal{O}_{\mathbb{K}}$ is actually a PID.

7.3.6 Factorization of (p)

By definition, if $\mathfrak{p}$ is a prime ideal in $\mathcal{O}_{\mathbb{K}}$, then $\mathfrak{p} \cap \mathbb{Z}$ is a prime ideal in $\mathbb{Z}$, and is therefore of the form (p) for some prime number p. This leads to the following definition.

Definition 7.99 Let $\mathbb{K}/\mathbb{Q}$ be an algebraic number field, p be a prime number and $\mathfrak{p}$ be a prime ideal of $\mathcal{O}_{\mathbb{K}}$. We shall say that $\mathfrak{p}$ *lies above* p, written $\mathfrak{p} \mid p$, if $\mathfrak{p} \cap \mathbb{Z} = p\mathbb{Z}$.

When $\mathfrak{p}$ lies above p, then $p \in \mathfrak{p}$ so that $\mathfrak{p} \mid (p)$ as ideals of $\mathcal{O}_{\mathbb{K}}$, which explains the notation. We also notice that, for any prime ideal $\mathfrak{p}$ of $\mathcal{O}_{\mathbb{K}}$, there exists a unique prime number p such that $\mathfrak{p}$ lies above p. Indeed, suppose that there is another prime number $q \neq p$ such that $\mathfrak{p} \mid q$. We have $\mathfrak{p} \mid (p) + (q) = (1)$ which is impossible since $\mathfrak{p}$ is maximal.

To find prime ideals of $\mathcal{O}_{\mathbb{K}}$, we need to factorize ideals generated by prime numbers. Pay careful attention to the fact that the ideal (p) is not in general a prime ideal of $\mathcal{O}_{\mathbb{K}}$. Theorem 7.105 below, due to Kummer for a particular case and extended by Dedekind, relates the prime ideal factorization of (p) to the decomposition of $\overline{\mu}_\theta$ into irreducible polynomials in $\mathbb{F}_p[X]$. Kummer–Dedekind's theorem holds for any algebraic number field $\mathbb{K} = \mathbb{Q}(\theta)$ such that p does not divide the index f of θ. We do not have a similar result in the case where $p \mid f$, but algorithms do exist which perform the factorization of (p) in this case. However, recall that in practice $p^2 \nmid \Delta_{\mathbb{K}/\mathbb{Q}}(1, \theta, \ldots, \theta^{n-1})$ is sufficient to ensure that $p \nmid f$ and hence all but finitely many prime numbers p are covered by this result.

We start with the following general situation.

Lemma 7.100 *Let $\mathbb{K}/\mathbb{Q}$ be an algebraic number field of degree n, let p be a prime number and*

$$(p) = \prod_{i=1}^{g} \mathfrak{p}_i^{e_i} \tag{7.15}$$

be the factorization of (p) into prime ideals.

(i) *There exist positive integers f_i such that*

$$\mathcal{N}_{\mathbb{K}/\mathbb{Q}}(\mathfrak{p}_i) = p^{f_i} \quad and \quad \sum_{i=1}^{g} e_i f_i = n.$$

(ii) *The index $[\mathcal{O}_{\mathbb{K}}/\mathfrak{p}_i : \mathbb{Z}/p\mathbb{Z}]$ is finite and $f_i = [\mathcal{O}_{\mathbb{K}}/\mathfrak{p}_i : \mathbb{Z}/p\mathbb{Z}]$.*

Proof

(i) First, since $\mathfrak{p}_i \mid (p)$, we get $\mathcal{N}_{\mathbb{K}/\mathbb{Q}}(\mathfrak{p}_i) \mid \mathcal{N}_{\mathbb{K}/\mathbb{Q}}((p)) = p^n$, so that there exist integers $f_i \in \{1, \ldots, n\}$ such that $\mathcal{N}_{\mathbb{K}/\mathbb{Q}}(\mathfrak{p}_i) = p^{f_i}$. Furthermore, since $\mathfrak{p}_i^{e_i} + \mathfrak{p}_j^{e_j} = (1)$ for all $i \neq j$, the Chinese remainder theorem implies that

$$\mathcal{O}_{\mathbb{K}}/(p) \simeq \mathcal{O}_{\mathbb{K}}/\mathfrak{p}_1^{e_1} \oplus \cdots \oplus \mathcal{O}_{\mathbb{K}}/\mathfrak{p}_g^{e_g}$$

and hence, by Theorem 7.96, we get

$$p^n = \mathcal{N}_{\mathbb{K}/\mathbb{Q}}((p)) = \mathcal{N}_{\mathbb{K}/\mathbb{Q}}(\mathfrak{p}_1^{e_1}) \cdots \mathcal{N}_{\mathbb{K}/\mathbb{Q}}(\mathfrak{p}_g^{e_g}) = \mathcal{N}_{\mathbb{K}/\mathbb{Q}}(\mathfrak{p}_1)^{e_1} \cdots \mathcal{N}_{\mathbb{K}/\mathbb{Q}}(\mathfrak{p}_g)^{e_g}$$

which implies the asserted result.

(ii) The map

$$\begin{aligned} \mathbb{Z}/p\mathbb{Z} &\longrightarrow \mathcal{O}_{\mathbb{K}}/\mathfrak{p}_i, \\ \overline{a} &\longmapsto a + \mathfrak{p}_i \end{aligned}$$

is well-defined and one may check that it is a homomorphism. Now $\mathbb{Z}/p\mathbb{Z} \simeq \mathbb{F}_p$ is a finite field, and the same is true for $\mathcal{O}_{\mathbb{K}}/\mathfrak{p}_i$ by Lemmas 7.80 and 7.81 and the fact that $\mathfrak{p}_i$ is maximal. Hence the map is injective and $[\mathcal{O}_{\mathbb{K}}/\mathfrak{p}_i : \mathbb{Z}/p\mathbb{Z}]$ is the dimension of $\mathcal{O}_{\mathbb{K}}/\mathfrak{p}_i$ considered as a $\mathbb{Z}/p\mathbb{Z}$-vector space. Furthermore, we have

$$\mathcal{N}_{\mathbb{K}/\mathbb{Q}}(\mathfrak{p}_i) = |\mathcal{O}_{\mathbb{K}}/\mathfrak{p}_i| = p^{[\mathcal{O}_{\mathbb{K}}/\mathfrak{p}_i : \mathbb{Z}/p\mathbb{Z}]}$$

by Lemma 7.20, and we conclude by using (i).

The proof is complete.   □

Definition 7.101 Let $\mathbb{K}/\mathbb{Q}$ be an algebraic number field of degree n, p be a prime number and consider the factorization (7.15) of (p) into prime ideals.

1. (a) The integer e_i is called the *ramification index* of $\mathfrak{p}_i$ over $\mathbb{Z}$, denoted by $e(\mathfrak{p}_i \mid p)$.
   (b) The integer f_i is called the *inertial degree*, or *residue class degree*, of $\mathfrak{p}_i$ over $\mathbb{Z}$, denoted by $f(\mathfrak{p}_i \mid p)$.
   (c) The integer g is called the *decomposition number* of p over $\mathbb{Z}$, denoted by g_p.
2. (a) The prime number p is said to be *ramified* in $\mathcal{O}_{\mathbb{K}}$ if and only if there exists some $i \in \{1, \ldots, g\}$ such that $e_i \geqslant 2$. Similarly, the prime ideals $\mathfrak{p}_i$ such that $e_i \geqslant 2$ are called *ramified prime ideals*.

(b) The prime number p is said to be *unramified* in $\mathcal{O}_{\mathbb{K}}$ if $e_i = 1$ for all $i \in \{1, \ldots, g\}$. Two cases are then possible.

p *is inert*	$\begin{cases} g = 1, \\ e_1 = 1 \; f_1 = n \end{cases}$	$(p) = \mathfrak{p}$
p *splits completely*	$\begin{cases} g = n, \\ e_i = f_i = 1 \end{cases}$	$(p) = \mathfrak{p}_1 \cdots \mathfrak{p}_n$

Note that there are intermediate cases which do not deserve a special name. The ramified primes are characterized by the following result.

Theorem 7.102 *Let $\mathbb{K}/\mathbb{Q}$ be an algebraic number field with discriminant $d_{\mathbb{K}}$ and p be a prime number. Then p is ramified in $\mathbb{K}$ if and only if $p \mid d_{\mathbb{K}}$. In particular, there are $\omega(|d_{\mathbb{K}}|)$ ramified primes.*

We do not supply here a proof of this theorem but, as often in algebraic number theory, the result is more important than the proof itself. As a corollary, we provide the following criterion (see also [AW04a, SWW06]).

Corollary 7.103 *Let $\mathbb{K} = \mathbb{Q}(\theta)$ be an algebraic number field of degree n, discriminant $d_{\mathbb{K}}$, $\theta \in \mathcal{O}_{\mathbb{K}}$ and p be a prime number. Let $\mu_\theta = X^n + a_{n-1}X^{n-1} + \cdots + a_0 \in \mathbb{Z}[X]$ be the minimal polynomial of θ. Suppose that $p \parallel a_0$ and $p \mid a_1$. Then $p \mid d_{\mathbb{K}}$.*

Proof Suppose that p is not ramified. Then there exist distinct prime ideals $\mathfrak{p}_1, \ldots, \mathfrak{p}_g$ such that

$$(p) = \mathfrak{p}_1 \cdots \mathfrak{p}_g.$$

Since $p \parallel a_0$, we get $(a_0) = \mathfrak{p}_1 \cdots \mathfrak{p}_g(b)$ for some $b \in \mathbb{Z}$ with $p \nmid b$ and hence $\mathfrak{p}_i \nmid (b)$ for all $i \in \{1, \ldots, g\}$. Now $N_{\mathbb{K}/\mathbb{Q}}(\theta) = (-1)^n a_0 \equiv 0 \pmod{p}$ so that there exists a prime ideal $\mathfrak{p} \in \{\mathfrak{p}_1, \ldots, \mathfrak{p}_g\}$ such that $\mathfrak{p} \mid (\theta)$. As $\mathfrak{p} \mid (a_1)$, we get

$$a_0 = a_0 - \mu_\theta(\theta) = -a_1\theta - \cdots - a_{n-1}\theta^{n-1} - \theta^n$$

so that $\mathfrak{p}^2 \mid (a_0)$ contradicting the assumption $p \parallel a_0$. Hence p is ramified and we conclude by using Theorem 7.102. $\qquad\square$

In the case of Galois extensions, the result is simpler.

Proposition 7.104 *Let $\mathbb{K}/\mathbb{Q}$ be a Galois extension of degree n and p be a prime number. Then the ramification indexes are all equal, say to e_p, and the inertial degrees are all equal, say to f_p. Hence we have*

$$(p) = \left(\prod_{i=1}^{g_p} \mathfrak{p}_i \right)^{e_p}$$

with $e_p f_p g_p = n$. Furthermore, $\mathrm{Gal}(\mathbb{K}/\mathbb{Q})$ operates transitively on the prime ideal above p, i.e. for all prime ideals $\mathfrak{p}_i$, $\mathfrak{p}_j$ above p, there exists $\sigma \in \mathrm{Gal}(\mathbb{K}/\mathbb{Q})$ such that $\sigma(\mathfrak{p}_i) = \mathfrak{p}_j$.

Now we state the main result of this section. The idea is to prove and use the fact that, if $p \nmid f$, then

$$\mathcal{O}_{\mathbb{K}}/(p) \simeq \mathbb{F}_p[X]/(\overline{\mu_\theta}).$$

Theorem 7.105 (Kummer–Dedekind) *Let $\mathbb{K} = \mathbb{Q}(\theta)$ be an algebraic number field of degree n with $\theta \in \mathcal{O}_{\mathbb{K}}$, and p be a prime number such that $p \nmid f$. Let $\mu \in \mathbb{Z}[X]$ be the minimal polynomial of θ and suppose that μ factorizes over $\mathbb{F}_p[X]$ as*

$$\overline{\mu} = \prod_{i=1}^{g} \overline{P_i}^{e_i}$$

where the $\overline{P_i}$ are distinct monic irreducible polynomials in $\mathbb{F}_p[X]$. Then we have

$$(p) = \prod_{i=1}^{g} \mathfrak{p}_i^{e_i}$$

where the prime ideals $\mathfrak{p}_i$ are pairwise distinct and given by $\mathfrak{p}_i = (p, P_i(\theta)) = p\mathcal{O}_{\mathbb{K}} + P_i(\theta)\mathcal{O}_{\mathbb{K}}$ where P_i is any lifting of $\overline{P_i}$ in $\mathbb{Z}[X]$. We also have

$$\mathcal{N}_{\mathbb{K}/\mathbb{Q}}(\mathfrak{p}_i) = p^{\deg \overline{P_i}}.$$

Proof Let p be a prime number such that $p \nmid f$.

▷ **Step 1**. We will prove that

$$\mathcal{O}_{\mathbb{K}}/(p) \simeq \mathbb{F}_p[X]/(\overline{\mu}). \tag{7.16}$$

Let $\{\alpha_1, \ldots, \alpha_n\}$ be an integral basis for $\mathbb{K}$. Then $\{1, \theta, \ldots, \theta^{n-1}\}$ is a $\mathbb{Z}$-basis for $\mathbb{Z}[\theta]$ for which there exists a matrix $A = (a_{ij}) \in \mathcal{M}_n(\mathbb{Z})$ such that $\theta^{i-1} = \sum_{j=1}^{n} a_{ij}\alpha_j$ for all $i \in \{1, \ldots, n\}$. Since $\det(a_{ij}) = \pm f$ and $p \nmid f$, we infer that the reduction matrix $\overline{A} = (\overline{a_{ij}}) \in \mathcal{M}_n(\mathbb{F}_p)$ is invertible, and therefore induces an isomorphism

$$\mathcal{O}_{\mathbb{K}}/(p) \simeq \mathbb{Z}[\theta]/p\mathbb{Z}[\theta].$$

Now since μ is irreducible over $\mathbb{Z}$, we have $\mathbb{Z}[X]/(\mu) \simeq \mathbb{Z}[\theta]$ which implies that

$$\mathbb{Z}[\theta]/p\mathbb{Z}[\theta] \simeq \mathbb{Z}[X]/(\mu, p) \simeq \mathbb{F}_p[X]/(\overline{\mu})$$

and hence (7.16) is proved.

▷ **Step 2**. Let $(p) = \mathfrak{q}_1^{k_1} \cdots \mathfrak{q}_r^{k_r}$ given by Theorem 7.88 for some pairwise distinct prime ideals $\mathfrak{q}_i$ and positive integers r and k_i.

⋄ By Definition 7.3, the maximal ideals of $\mathcal{O}_{\mathbb{K}}/(p)$ are of the form $\mathfrak{q}/(p)$ where $\mathfrak{q}$ is a prime ideal dividing (p). We infer that $\mathfrak{q} \in \{\mathfrak{q}_1, \ldots, \mathfrak{q}_r\}$ and r is thus the number of maximal ideals in $\mathcal{O}_{\mathbb{K}}/(p)$, which are then of the form $\mathfrak{q}_i/(p)$ for all $i \in \{1, \ldots, r\}$.

⋄ Similarly, we infer that the maximal ideals of $\mathbb{F}_p[X]/(\overline{\mu})$ are the ideals $(\overline{P_i})/(\overline{\mu})$ and hence g counts the number of maximal ideals in $\mathbb{F}_p[X]/(\overline{\mu})$.

Hence we have proved that r is the number of maximal ideals in $\mathcal{O}_{\mathbb{K}}/(p)$ and g is that of maximal ideals in $\mathbb{F}_p[X]/(\overline{\mu})$. By (7.16), we get $r = g$.

▷ **Step 3**. For all $i \in \{1, \ldots, g\}$, let P_i be any lifting of $\overline{P_i}$ in $\mathbb{Z}[X]$ and set $\mathfrak{p}_i = (p, P_i(\theta))$. Similarly as for (7.16), one can prove the following isomorphisms for all $i \in \{1, \ldots, g\}$

$$\mathcal{O}_{\mathbb{K}}/\mathfrak{p}_i \simeq \mathbb{Z}[\theta]/\big(p, P_i(\theta)\big)\mathbb{Z}[\theta] \simeq \mathbb{Z}[X]/\big(p, P_i(X)\big)\mathbb{Z}[X] \simeq \mathbb{F}_p[X]/(\overline{P_i}).$$
(7.17)

Since the polynomials $\overline{P_i}$ are irreducible over $\mathbb{F}_p[X]$, the ideals $(\overline{P_i})$ are maximal. Therefore (7.17) implies that the ideals $\mathfrak{p}_i$ are maximal, and hence are prime ideals of $\mathcal{O}_{\mathbb{K}}$. We also infer that $\mathcal{N}_{\mathbb{K}/\mathbb{Q}}(\mathfrak{p}_i) = p^{\deg \overline{P_i}}$. Furthermore, we have

$$\mathfrak{p}_1^{e_1} \cdots \mathfrak{p}_g^{e_g} = \big(p, P_1(\theta)\big)^{e_1} \cdots \big(p, P_g(\theta)\big)^{e_g} \subseteq \big(p, P_1(\theta)^{e_1} \cdots P_g(\theta)^{e_g}\big)$$

and since $\mu \equiv P_1^{e_1} \cdots P_g^{e_g} \pmod{p}$ and $\mu(\theta) = 0$, we get $P_1(\theta)^{e_1} \cdots P_g(\theta)^{e_g} \in p\mathbb{Z}[\theta]$ and hence

$$\mathfrak{p}_1^{e_1} \cdots \mathfrak{p}_g^{e_g} \subseteq (p).$$

On the other hand, since $\mathfrak{p}_i = (p, P_i(\theta)) \supseteq (p)$, we get $\mathfrak{p}_i \mid (p)$ for all $i \in \{1, \ldots, g\}$. This implies that

$$(p) = \mathfrak{p}_1^{a_1} \cdots \mathfrak{p}_g^{a_g}$$

for some non-negative integers a_i satisfying $a_i \leqslant e_i$ for all $i \in \{1, \ldots, g\}$. Comparing the norms as in Lemma 7.100, we obtain

$$n = a_1 \deg \overline{P_1} + \cdots + a_g \deg \overline{P_g}$$

and by the Chinese remainder theorem we also have

$$\mathbb{F}_p[X]/(\overline{\mu}) \simeq \bigoplus_{i=1}^{g} \mathbb{F}_p[X]/(\overline{P_i})^{e_i}$$

implying that

$$n = e_1 \deg \overline{P_1} + \cdots + e_g \deg \overline{P_g}.$$

Setting $d_i = \deg \overline{P_i}$ and summarizing the above results, we have finally obtained $a_i \leqslant e_i$ and $a_1 d_1 + \cdots + a_g d_g = e_1 d_1 + \cdots + e_g d_g$, which implies that $a_i = e_i$ for all $i \in \{1, \ldots, g\}$. The proof is complete. □

Remark The fact that $\mathfrak{p}_i = (p, P_i(\theta))$ is not really useful in practice. What is more important yet is to know the *nature* of the prime ideals above p, e.g. whether they are ramified or not.

7.3.7 Prime Ideal Decomposition in Quadratic Fields

The Legendre–Jacobi–Kronecker Symbol

The Legendre symbol was seen in Exercise 17 in Chap. 3 where it was defined in the following way. Let p be an odd prime number and n be a positive integer. The Legendre symbol (n/p) is defined by $(n/p) = 0$ if $p \mid n$, and otherwise

$$\left(\frac{n}{p}\right) = \begin{cases} 1, & \text{if } n \text{ is residue quadratic mod } p \\ -1, & \text{if } n \text{ is not residue quadratic mod } p. \end{cases}$$

Among the usual properties already proved in this exercise, the main theorem is the *quadratic reciprocity law*. If p and q are distinct odd primes, then

$$\left(\frac{p}{q}\right)\left(\frac{q}{p}\right) = (-1)^{(p-1)(q-1)/4}.$$

We also have the two *complementary laws*

$$\left(\frac{-1}{p}\right) = (-1)^{(p-1)/2} \quad \text{and} \quad \left(\frac{2}{p}\right) = (-1)^{(p^2-1)/8}.$$

There are a lot of proofs in the literature. Nevertheless, it is noteworthy that the Legendre symbol (n/p) is a real primitive Dirichlet character modulo p. In fact, we have the following useful result (see [Coh07]).

Lemma 7.106 *Let $q \geqslant 2$ be an integer and χ be a real primitive Dirichlet character modulo q. If p is an odd prime, then*

$$\chi(p) = \left(\frac{\chi(-1)q}{p}\right).$$

As for all Dirichlet characters, the Legendre symbol may be generalized to all integers by complete multiplicativity. More precisely, if $m, n \in \mathbb{Z}$, the *Jacobi–Kronecker symbol*, still denoted by (n/m), is defined in the following way.

▷ $(n/1) = 1$, $(n/-1) = \operatorname{sgn}(n)$, $(-1/2) = 1$, $(-1/n) = (-1)^{(n-1)/2}$ if $n \geqslant 1$ is odd, and

$$\left(\frac{n}{2}\right) = \begin{cases} (-1)^{(n^2-1)/8}, & \text{if } 2 \nmid n, \\ 0, & \text{otherwise.} \end{cases}$$

▷ If $m = \pm p_1^{e_1} \cdots p_r^{e_r}$, then

$$\left(\frac{n}{m}\right) = \left(\frac{n}{\pm 1}\right) \prod_{i=1}^{r} \left(\frac{n}{p_i}\right)^{e_i}$$

where the (n/p_i) are Legendre symbols. A generalized quadratic reciprocity law still exists with this symbol. *Let $m, n \in \mathbb{Z} \setminus \{0\}$ and set $n = 2^e n_1$ and $m = 2^f m_1$ with n_1, m_1 odd. Then*

$$\left(\frac{n}{m}\right) = (-1)^{((m_1 - 1)(n_1 - 1) + (\text{sgn}(m) - 1)(\text{sgn}(n) - 1))/4} \left(\frac{m}{n}\right).$$

In particular, if D_1, D_2 are non-zero integers congruent to 0 or 1 modulo 4, then

$$\left(\frac{D_2}{D_1}\right) = (-1)^{((\text{sgn}(D_1) - 1)(\text{sgn}(D_2) - 1))/4} \left(\frac{D_1}{D_2}\right).$$

Lemma 7.106 was generalized by Dirichlet for Kronecker symbols (see [Coh07, Ros94]).

Lemma 7.107 *Let χ be a real primitive Dirichlet character modulo q. Then $D = \chi(-1)q$ is a discriminant of a quadratic field and*

$$\chi(n) = \left(\frac{D}{n}\right).$$

Conversely, if D is a discriminant of a quadratic field, then the Kronecker symbol defines a real primitive Dirichlet character modulo $q = |D|$.

Prime Ideal Decomposition of (p)

Let $d \in \mathbb{Z} \setminus \{0, 1\}$ squarefree and $\mathbb{K} = \mathbb{Q}(\sqrt{d})$ be a quadratic field with discriminant $d_{\mathbb{K}}$. By Proposition 7.64, $\{1, \omega\}$ is an integral basis for $\mathbb{K}$ where $\omega = \frac{1}{2}(d_{\mathbb{K}} + \sqrt{d_{\mathbb{K}}})$ in every case. Applying Corollary 7.103 and Theorem 7.105, we get the following result.

Proposition 7.108 *Let p be a prime number and $(d_{\mathbb{K}}/\cdot)$ be a Kronecker symbol. Then we have*

$(d_{\mathbb{K}}/p)$	Factorization of (p)	Nature	Remarks
-1	$(p) = \mathfrak{p}$	Inert	
1	$(p) = \mathfrak{p}_1 \mathfrak{p}_2$	Completely split	$\mathfrak{p}_i = (p, \omega - \frac{d_{\mathbb{K}} \pm b}{2})$ with b solution of $b^2 \equiv d_{\mathbb{K}} \pmod{4p}$
0	$(p) = \mathfrak{p}^2$	Totally ramified	$\mathfrak{p} = (p, \omega)$ except when $p = 2$ and $d_{\mathbb{K}} = 16$ where $\mathfrak{p} = (p, 1 + \omega)$

The Conductor of a Quadratic Field

Proposition 7.109 *Let* $d \in \mathbb{Z} \setminus \{0, 1\}$ *squarefree and* $\mathbb{K} = \mathbb{Q}(\sqrt{d})$ *be a quadratic field with discriminant* $d_{\mathbb{K}}$. *Then* $\mathbb{K}$ *is a subfield of* $\mathbb{Q}(\zeta_{|d_{\mathbb{K}}|})$.

Proof

▷ If $d_{\mathbb{K}}$ is odd, we have $d_{\mathbb{K}} = d \equiv 1 \pmod 4$. Write $d = \pm \prod_i p_i$ where p_i are pairwise distinct odd prime numbers. Define $p_i^* = (-1)^{(p_i-1)/2} p_i$ so that $p_i^* \equiv 1 \pmod 4$ and $d_{\mathbb{K}} = \prod_i p_i^*$. Using quadratic Gauss sums, one can show (see [Hin08, Proposition 4.1] for instance) that for all odd primes p

$$\sum_{j \pmod p} \left(\frac{j}{p} \right) \zeta_p^j = \sqrt{p^*}$$

which implies that $\mathbb{Q}(\sqrt{p_i^*}) \subset \mathbb{Q}(\zeta_{p_i})$. We infer that $\mathbb{Q}(\sqrt{p_i^*}) \subset \mathbb{Q}(\zeta_{|d_{\mathbb{K}}|})$ and hence each p_i^* is a square in $\mathbb{Q}(\zeta_{|d_{\mathbb{K}}|})$, so that $d_{\mathbb{K}}$ is also a square in $\mathbb{Q}(\zeta_{|d_{\mathbb{K}}|})$. Therefore $\mathbb{K}$ is a subfield of $\mathbb{Q}(\zeta_{|d_{\mathbb{K}}|})$ in this case.

▷ If $d_{\mathbb{K}}$ is even, we have $d_{\mathbb{K}} = 4d$ with $d \equiv 2, 3 \pmod 4$. We may suppose that $d \neq -1$ since $\mathbb{Q}(\sqrt{-1}) = \mathbb{Q}(\zeta_4)$. Set $d = \pm 2^e \prod_i p_i$ with $e \in \{0, 1\}$ and p_i are odd prime numbers and hence $d_{\mathbb{K}} = \pm 2^e \prod_i p_i^*$. As above, p_i^* is a square in $\mathbb{Q}(\zeta_{|d_{\mathbb{K}}|})$ and since $4 \mid d_{\mathbb{K}}$ and $\mathbb{Q}(\sqrt{-1}) = \mathbb{Q}(\zeta_4)$, -1 is also a square in $\mathbb{Q}(\zeta_{|d_{\mathbb{K}}|})$. Therefore $d_{\mathbb{K}}$ is a square in $\mathbb{Q}(\zeta_{|d_{\mathbb{K}}|})$ if $e = 0$. If $e = 1$, then $8 \mid d_{\mathbb{K}}$ and then $\mathbb{Q}(\zeta_8) \subset \mathbb{Q}(\zeta_{|d_{\mathbb{K}}|})$, so that 2 is a square in $\mathbb{Q}(\zeta_{|d_{\mathbb{K}}|})$, and hence $d_{\mathbb{K}}$ is a square in $\mathbb{Q}(\zeta_{|d_{\mathbb{K}}|})$ in this case, as required. $\square$

We say that $|d_{\mathbb{K}}|$ is the *conductor* of $\mathbb{Q}(\sqrt{d})$. Proposition 7.109 is a particular case of the Kronecker–Weber theorem (see Theorem 7.154) stating that every *abelian* extension $\mathbb{K}$ of $\mathbb{Q}$ is contained in a cyclotomic field $\mathbb{Q}(\zeta_f)$.

7.3.8 The Class Group

Let $\mathbb{K}/\mathbb{Q}$ be an algebraic number field of degree n, ring of integers $\mathcal{O}_{\mathbb{K}}$, signature (r_1, r_2) and embeddings $\sigma_1, \dots, \sigma_n$. Recall that $\mathcal{I}(\mathbb{K})$ is the abelian group of fractional ideals of $\mathcal{O}_{\mathbb{K}}$ and $\mathcal{P}(\mathbb{K})$ is the subset of $\mathcal{I}(\mathbb{K})$ of principal ideals. In view of $(a)(b)^{-1} = (ab^{-1})$, we see that $\mathcal{P}(\mathbb{K})$ is a subgroup of $\mathcal{I}(\mathbb{K})$, and since $\mathcal{I}(\mathbb{K})$ is abelian, we infer that $\mathcal{P}(\mathbb{K})$ is a normal subgroup of $\mathcal{I}(\mathbb{K})$, so that the quotient group $\mathcal{I}(\mathbb{K})/\mathcal{P}(\mathbb{K})$ is well-defined and abelian.

Definition 7.110 (Class group) The abelian quotient group $\mathcal{I}(\mathbb{K})/\mathcal{P}(\mathbb{K})$ is called the *ideal class group* of $\mathcal{O}_{\mathbb{K}}$ and is denoted[20] by $\mathrm{Cl}(\mathbb{K})$.

---

[20] See footnote 18 for the notation.

For $\mathfrak{a} \in \mathcal{I}(\mathbb{K})$, we write $\bar{\mathfrak{a}}$ the corresponding class in $\mathrm{Cl}(\mathbb{K})$. We have

$$\bar{\mathfrak{a}} = \overline{(1)} \quad \Longleftrightarrow \quad \mathfrak{a} = (a) = a\mathcal{O}_{\mathbb{K}}.$$

Example In $\mathbb{K} = \mathbb{Q}(\sqrt{6})$, set $\mathfrak{p} = (2, \sqrt{6})$. We have $\mathcal{N}_{\mathbb{K}/\mathbb{Q}}(\mathfrak{p}) = 2$ so that $\mathfrak{p}$ is not principal. Since $(2) = \mathfrak{p}^2$ by Theorem 7.105, we have $\bar{\mathfrak{p}}^2 = \overline{(1)}$ and hence $\bar{\mathfrak{p}}$ has order 2 in $\mathrm{Cl}(\mathbb{K})$.

The main result is the following fundamental finiteness theorem of $\mathrm{Cl}(\mathbb{K})$. The finiteness of $\mathrm{Cl}(\mathbb{K})$ is one of the three important finiteness theorems in algebraic number theory:

1. $\mathrm{Cl}(\mathbb{K})$ is finite.
2. $\mathcal{O}_{\mathbb{K}}^*$ is a finitely generated abelian group.
3. Given $D > 0$, the set of algebraic number fields $\mathbb{K}$ such that $|d_{\mathbb{K}}| \leqslant D$ is finite.

Theorem 7.111 *Let $\mathbb{K}/\mathbb{Q}$ be an algebraic number field of degree n. The ideal class group $\mathrm{Cl}(\mathbb{K})$ of $\mathbb{K}$ is a finite abelian group. Furthermore, $\mathrm{Cl}(\mathbb{K})$ is generated by the classes of prime ideals $\mathfrak{p}$ in $\mathcal{O}_{\mathbb{K}}$ such that $\mathcal{N}_{\mathbb{K}/\mathbb{Q}}(\mathfrak{p}) \leqslant c$ for some $c > 0$.*

This result leads to the following crucial definition.

Definition 7.112 (Class number) The order of $\mathrm{Cl}(\mathbb{K})$ is called the *class number* of $\mathbb{K}$ and is denoted by $h_{\mathbb{K}}$.

The proof of Theorem 7.111 requires the following lemma.

Lemma 7.113

(i) *There exists a constant $c_{\mathbb{K}} > 0$, called the* Hurwitz *constant and depending only on $\mathbb{K}$, such that in every non-zero integral ideal $\mathfrak{a}$ there is a non-zero β such that*

$$\left| \mathcal{N}_{\mathbb{K}/\mathbb{Q}}(\beta) \right| \leqslant c_{\mathbb{K}} \, \mathcal{N}_{\mathbb{K}/\mathbb{Q}}(\mathfrak{a}).$$

(ii) *Every ideal class of $\mathbb{K}$ contains an integral ideal $\mathfrak{a}$ such that $\mathcal{N}_{\mathbb{K}/\mathbb{Q}}(\mathfrak{a}) \leqslant c_{\mathbb{K}}$.*

Proof

(i) Let $\{\alpha_1, \ldots, \alpha_n\}$ be an integral basis for $\mathbb{K}$ and set

$$c_{\mathbb{K}} = \prod_{j=1}^{n} \left(\sum_{i=1}^{n} |\sigma_j(\alpha_i)| \right)$$

and $r = [\mathcal{N}_{\mathbb{K}/\mathbb{Q}}(\mathfrak{a})^{1/n}]$. The set

$$S = \{a_1\alpha_1 + \cdots + a_n\alpha_n : a_i \in \mathbb{Z}, \ 0 \leqslant a_i \leqslant r\}$$

has $(r+1)^n$ elements and so $|S| > \mathcal{N}_{\mathbb{K}/\mathbb{Q}}(\mathfrak{a}) = |\mathcal{O}_{\mathbb{K}}/\mathfrak{a}|$. The elements of S cannot lie in distinct cosets of $\mathfrak{a}$ in $\mathcal{O}_{\mathbb{K}}$ and therefore there are $\beta_1, \beta_2 \in S$ such that $\beta_1 \neq \beta_2$ and $\beta_1 \equiv \beta_2 \pmod{\mathfrak{a}}$ by the Dirichlet pigeon-hole principle. Set $\beta = \beta_1 - \beta_2$. Then $\beta \in \mathfrak{a} \setminus \{0\}$ and is such that $\beta = \sum_{i=1}^{n} b_i \alpha_i$ with $|b_i| \leqslant r$. Now we have

$$\left| \mathcal{N}_{\mathbb{K}/\mathbb{Q}}(\beta) \right| = \prod_{j=1}^{n} \left| \sigma_j(\beta) \right| \leqslant \prod_{j=1}^{n} \sum_{i=1}^{n} |b_i| \left| \sigma_j(\alpha_i) \right| \leqslant c_{\mathbb{K}} \, r^n \leqslant c_{\mathbb{K}} \, \mathcal{N}_{\mathbb{K}/\mathbb{Q}}(\mathfrak{a})$$

as asserted.

(ii) Let $\mathfrak{b}$ be a fractional ideal in the given ideal class. Without loss of generality, one may assume that $\mathfrak{b}^{-1}$ is an integral ideal. Choose an element $\beta \in \mathfrak{b}^{-1}$ such that $|\mathcal{N}_{\mathbb{K}/\mathbb{Q}}(\beta)| \leqslant c_{\mathbb{K}} \, \mathcal{N}_{\mathbb{K}/\mathbb{Q}}(\mathfrak{b}^{-1})$ and set $\mathfrak{a} = \beta \mathfrak{b}$.

▷ Since $\beta \in \mathfrak{b}^{-1}$ and $\mathfrak{b}\mathfrak{b}^{-1} = \mathcal{O}_{\mathbb{K}}$, we deduce that $\mathfrak{a}$ is an integral ideal.

▷ We have $\mathfrak{a}\mathfrak{b}^{-1} = (\beta)$ so that

$$\mathcal{N}_{\mathbb{K}/\mathbb{Q}}(\mathfrak{a}) \, \mathcal{N}_{\mathbb{K}/\mathbb{Q}}(\mathfrak{b}^{-1}) = \left| \mathcal{N}_{\mathbb{K}/\mathbb{Q}}(\beta) \right| \leqslant c_{\mathbb{K}} \, \mathcal{N}_{\mathbb{K}/\mathbb{Q}}(\mathfrak{b}^{-1})$$

which implies the desired result.

The proof is complete.   □

Now we are in a position to prove Theorem 7.111.

Proof of Theorem 7.111 Let $\overline{\mathfrak{b}}$ be an ideal class in $\mathrm{Cl}(\mathbb{K})$. By Lemma 7.113, $\overline{\mathfrak{b}}$ contains a non-zero integral ideal $\mathfrak{a}$ such that $\mathcal{N}_{\mathbb{K}/\mathbb{Q}}(\mathfrak{a}) \leqslant c_{\mathbb{K}}$. From Corollary 7.97, there are only finitely many integral ideals with a given norm so that there are only finitely many choices for $\mathfrak{a}$. Since $\overline{\mathfrak{b}} = \overline{\mathfrak{a}}$, we infer that there are only finitely many ideal classes $\overline{\mathfrak{b}}$.   □

Remark 7.114

1. The fractional ideals $\mathfrak{a}$ such that $\mathcal{O}_{\mathbb{K}} \subset \mathfrak{a}$ and $\mathcal{N}_{\mathbb{K}/\mathbb{Q}}(\mathfrak{a}) \leqslant c_{\mathbb{K}}$ represent all ideal classes, but that does not imply that the number of ideal classes, i.e. $h_{\mathbb{K}}$, is bounded above by $c_{\mathbb{K}}$ since several fractional ideals containing $\mathcal{O}_{\mathbb{K}}$ may have the same norm.

2. To get upper and lower bounds for $h_{\mathbb{K}}$ is one of the highlights in algebraic number theory, mainly because the class number is a sort of measure of the default of principality of an algebraic number field. Indeed, if $h_{\mathbb{K}} = 1$, then $\mathcal{I}(\mathbb{K}) = \mathcal{P}(\mathbb{K})$ and $\mathcal{O}_{\mathbb{K}}$ is a PID, and hence a UFD. Conversely, if $\mathcal{O}_{\mathbb{K}}$ is a UFD, it is also a PID by Theorem 7.93, so that $\mathcal{I}(\mathbb{K}) = \mathcal{P}(\mathbb{K})$ and thus $h_{\mathbb{K}} = 1$. We will provide some upper bounds using multiplicative methods in the next section.

3. The Hurwitz constant can be very large, and hence useless to help to compute $h_{\mathbb{K}}$. One can improve on $c_{\mathbb{K}}$ by using the following methods.

   ▷ The geometry of numbers enables us to obtain the *Minkowski bound*, i.e. *every ideal class of $\mathbb{K}$ contains an integral ideal $\mathfrak{a}$ such that*

   $$\mathcal{N}_{\mathbb{K}/\mathbb{Q}}(\mathfrak{a}) \leqslant M_{\mathbb{K}} |d_{\mathbb{K}}|^{1/2} \tag{7.18}$$

with

$$M_{\mathbb{K}} = \left(\frac{4}{\pi}\right)^{r_2} \frac{n!}{n^n}. \tag{7.19}$$

$M_{\mathbb{K}}$ is called the *Minkowski constant* attached to $\mathbb{K}$.
▷ Analytic methods may supersede the Minkowski constant (see Sect. 7.5).
4. The discussion above allows us to construct an algorithm to compute $h_{\mathbb{K}}$. Indeed, to determine the representatives of the ideal classes, we only need to look at the integral ideals with norms upper bounded by a fixed constant c. We infer that every prime ideal dividing these integral ideals has a norm bounded by this constant, and since the norm of a prime ideal is a power of a prime number, it is then sufficient to look at the prime numbers bounded by the constant. Indeed, forming all possible products of the prime ideals lying above $p \leqslant c$ will yield all ideals of norm $\leqslant c$. This gives the following algorithm.
   ▷ Given an algebraic number field $\mathbb{K} = \mathbb{Q}(\theta)$, determine its degree, signature, discriminant and compute $b_{\mathbb{K}} = M_{\mathbb{K}}|d_{\mathbb{K}}|^{1/2}$.
   ▷ Determine the prime ideal factorization of (p) for all primes $p \leqslant b_{\mathbb{K}}$.
   ▷ Find *all* dependence relations between the integral ideals having norm $\leqslant b_{\mathbb{K}}$. A useful method is to compute $N_{\mathbb{K}/\mathbb{Q}}(\theta + m)$ for some $m \in \mathbb{Z}$ and use those values that only involve the primes $p \leqslant b_{\mathbb{K}}$.
It should also be noticed that, if every prime number $p \leqslant b_{\mathbb{K}}$ factorizes into a product of prime ideals, each of which is principal, then $h_{\mathbb{K}} = 1$.

Example 7.115

1. Let $\mathbb{K} = \mathbb{Q}(\sqrt{229})$. A defining polynomial of $\mathbb{K}$ is $P = X^2 - X - 57$, and we have $n = 2$, $(r_1, r_2) = (2, 0)$, $\{1, \omega\}$ is an integral basis for $\mathbb{K}$ with $\omega = \frac{1}{2}(1 + \sqrt{229})$ and $d_{\mathbb{K}} = 229$. We also have $M_{\mathbb{K}}|d_{\mathbb{K}}|^{1/2} \approx 5.351$.
   We factorize (2), (3) and (5) using Theorem 7.105 or Proposition 7.108 which gives

$$(2) = \mathfrak{p}_2, \quad (3) = \mathfrak{p}_3\mathfrak{p}_3' \quad \text{and} \quad (5) = \mathfrak{p}_5\mathfrak{p}_5'.$$

Furthermore, we have

$$(7 - \omega) = \mathfrak{p}_3\mathfrak{p}_5 \quad \text{and} \quad (6 + \omega) = \mathfrak{p}_3'\mathfrak{p}_5'.$$

We infer that $\overline{\mathfrak{p}_2} = \overline{(1)}$, and

$$\left.\begin{array}{r}\overline{\mathfrak{p}_3} \cdot \overline{\mathfrak{p}_5} = \overline{(1)} \\ \overline{\mathfrak{p}_3} \cdot \overline{\mathfrak{p}_3'} = \overline{(1)}\end{array}\right\} \implies \overline{\mathfrak{p}_5} = \overline{\mathfrak{p}_3'} = (\overline{\mathfrak{p}_3})^{-1}$$

and

$$\left.\begin{array}{r}\overline{\mathfrak{p}_3'} \cdot \overline{\mathfrak{p}_5'} = \overline{(1)} \\ \overline{\mathfrak{p}_3} \cdot \overline{\mathfrak{p}_3'} = \overline{(1)}\end{array}\right\} \implies \overline{\mathfrak{p}_5'} = \overline{\mathfrak{p}_3}$$

and thus all the ideal classes lie among the following three distinct classes $\overline{(1)} = \overline{\mathfrak{p}_2}$, $\overline{\mathfrak{p}_3}$ and $(\overline{\mathfrak{p}_3})^{-1}$, so that $h_{\mathbb{K}} = 3$.

2. Let θ be a root of the polynomial $P = X^3 - X^2 - 11X - 1$. We have $n = 3$, $(r_1, r_2) = (3, 0)$, $d_{\mathbb{K}} = 1304$ and $M_{\mathbb{K}} |d_{\mathbb{K}}|^{1/2} \approx 8.025$.

It was seen in Example 7.79 that θ is a unit in $\mathbb{K}$ so that $(\theta) = (1)$. Furthermore, using Theorem 7.105, we get

$$(2) = \mathfrak{p}_2^2 \mathfrak{p}_2', \quad (3) = \mathfrak{p}_3 \mathfrak{p}_3', \quad (5) = \mathfrak{p}_5 \quad \text{and} \quad (7) = \mathfrak{p}_7.$$

We also have

$$(1 - \theta) = \mathfrak{p}_2 \mathfrak{p}_2' \mathfrak{p}_3, \quad (1 + \theta) = \mathfrak{p}_2 \mathfrak{p}_2'^2, \quad (2 + \theta) = \mathfrak{p}_3^2 \quad \text{and} \quad (3 + \theta) = \mathfrak{p}_2 \mathfrak{p}_2'.$$

Using $(1 - \theta)$, $(3 + \theta)$ and (3) we get

$$\left. \begin{array}{r} \overline{\mathfrak{p}_2} \cdot \overline{\mathfrak{p}_2'} \cdot \overline{\mathfrak{p}_3} = \overline{(1)} \\ \overline{\mathfrak{p}_2} \cdot \overline{\mathfrak{p}_2'} = \overline{(1)} \end{array} \right\} \quad \Longrightarrow \quad \overline{\mathfrak{p}_3} = \overline{\mathfrak{p}_3'} = \overline{(1)}$$

and using $(1 + \theta)$, $(3 + \theta)$ and (2) we get

$$\left. \begin{array}{r} \overline{\mathfrak{p}_2} \cdot \overline{\mathfrak{p}_2'}^2 = \overline{(1)} \\ \overline{\mathfrak{p}_2} \cdot \overline{\mathfrak{p}_2'} = \overline{(1)} \end{array} \right\} \quad \Longrightarrow \quad \overline{\mathfrak{p}_2} = \overline{\mathfrak{p}_2'} = \overline{(1)}$$

and since $\overline{\mathfrak{p}_5} = \overline{\mathfrak{p}_7} = \overline{(1)}$, we get $h_{\mathbb{K}} = 1$ and hence $\mathcal{O}_{\mathbb{K}}$ is a PID.

The following result is of great use in the theory of Diophantine equations.

Proposition 7.116 *Let $\mathbb{K}$ be an algebraic number field with class number $h_{\mathbb{K}}$ and $\mathfrak{a}$ be an integral ideal of $\mathbb{K}$. Let p be a prime number such that $p \nmid h_{\mathbb{K}}$. If $\mathfrak{a}^p$ is principal, then $\mathfrak{a}$ is principal.*

Proof By Bézout's theorem, we have $uh_{\mathbb{K}} + pv = 1$ for some $u, v \in \mathbb{Z}$. By assumption, we have $\overline{\mathfrak{a}}^p = \overline{(1)}$ and by Lagrange's theorem we infer that $\overline{\mathfrak{a}}^{h_{\mathbb{K}}} = \overline{(1)}$, so that

$$\overline{\mathfrak{a}} = \overline{\mathfrak{a}}^{uh_{\mathbb{K}} + pv} = \left(\overline{\mathfrak{a}}^{h_{\mathbb{K}}} \right)^u \left(\overline{\mathfrak{a}}^p \right)^v = \overline{(1)}$$

which implies the asserted result.   □

Example Let us have a look at Bachet's Diophantine equation $x^2 + 5 = y^3$. Let $(x, y) \in \mathbb{Z}^2$ be a solution of this equation. One may assume that y is odd, otherwise x is odd and we have $x^2 \equiv 3 \pmod 4$ which is impossible. One may also suppose that x and y are coprime. Consider the quadratic field $\mathbb{K} = \mathbb{Q}(\sqrt{-5})$. By Proposition 7.64, we have $\mathcal{O}_{\mathbb{K}} = \mathbb{Z}[\sqrt{-5}]$. In this ring, we have

$$(x + \sqrt{-5})(x - \sqrt{-5}) = y^3.$$

Since y is odd and x and y are coprime, we infer that the ideals $(x + \sqrt{-5})$ and $(x - \sqrt{-5})$ are coprime, and hence there exist two coprime integral ideals $\mathfrak{a}$ and $\mathfrak{b}$ such that $(x + \sqrt{-5}) = \mathfrak{a}^3$ and $(x - \sqrt{-5}) = \mathfrak{b}^3$. Therefore $\mathfrak{a}^3$ is principal. Using the method of Example 7.115, one has $h_{\mathbb{K}} = 2$ so that $\mathfrak{a}$ is principal by Proposition 7.116. By Example 7.75, the units in $\mathbb{K}$ are ± 1 which are cubes, hence we deduce that there exist $a, b \in \mathbb{Z}$ such that

$$x + \sqrt{-5} = (a + b\sqrt{-5})^3.$$

Equating imaginary parts implies that $3a^2 b - 5b^3 = 1$ and hence $b \mid 1$, so that $b = \pm 1$. This gives $3a^2 - 5 = \pm 1$ which is impossible in both cases. Hence Bachet's equation has no solution in $\mathbb{Z}^2$.

7.3.9 The PARI/GP System

The package PARI/GP is one of the main programs designed to deal with computational algebraic number theory problems. We introduce here some functions which enable us to compute the main invariants of algebraic number fields. In what follows, let $\mathbb{K} = \mathbb{Q}(\theta)$ be an algebraic number field with θ being a root of the monic polynomial $P \in \mathbb{Z}[X]$.

1. One may enter P in the classical way.

   Example $P = x^6 - x^5 + 8 * x^4 - 8 * x^3 + 22 * x^2 - 22 * x + 29$.

2. **poldisc**(P) and **nfdisc**(P) respectively give disc(P) and $d_{\mathbb{K}}$. One may use the function **factor** to provide a decomposition in prime numbers of these invariants.

   Example disc$(P) = -5^3 \times 7^5 \times 13^4$ and $d_{\mathbb{K}} = -5^3 \times 7^5$. We infer that $f = 13^2$.

3. **nfbasis**(P) gives an integral basis for $\mathbb{K}$.

   Example Set $\alpha_1 = 1, \alpha_2 = \theta, \alpha_3 = \theta^2, \alpha_4 = \theta^3$. Then $\{\alpha_1, \alpha_2, \alpha_3, \alpha_4, \alpha_5, \alpha_6\}$ is an integral basis for $\mathbb{K}$ with

   $$\alpha_5 = \frac{\theta^4 + 5\theta^3 + 4\theta^2 + 2\theta + 2}{13} \quad \text{and} \quad \alpha_6 = \frac{\theta^5 + 5\theta^3 - 5\theta^2 + 5\theta + 3}{13}.$$

4. **idealprimedec**$(\mathbf{nfinit}(P), p)$ and **idealfactor**$(\mathbf{nfinit}(P), a)$ respectively provide the prime ideal factorizations of the ideal (p) and the fractional ideal a.

   Example $(3) = \mathfrak{p}_3 \mathfrak{p}_3'$ with $\mathfrak{p}_3 = (3, -\alpha_1 + \alpha_2 + 2\alpha_4 - 4\alpha_5)$ and $\mathfrak{p}_3' = (3, -5\alpha_1 - \alpha_2 + 2\alpha_3 + 2\alpha_4 - 3\alpha_5 - \alpha_6)$. Furthermore, we also have $f(\mathfrak{p}_3 \mid 3) = f(\mathfrak{p}_3' \mid 3) = 3$.

5. **bnfclgp**(P) provides the class number and the class group. The given 3-component vector provides $h_\mathbb{K}$, the structure of $\mathrm{Cl}(\mathbb{K})$ as a product of cyclic groups of order m_i and the generators of $\mathrm{Cl}(\mathbb{K})$ of respective orders m_i.

   Example $h_\mathbb{K} = 2$.

6. **bnfreg**(P) provides the regulator of $\mathbb{K}$.

   Example $\mathcal{R}_\mathbb{K} \approx 2.101\,018\,72\ldots$

7. **bnfclassunit**(P) provides the class group, the regulator and a system of fundamental units. The result of this function is a 10-component vector v. In particular, v_2 is the signature, v_3 is $d_\mathbb{K}$ and f, v_4 is an integral basis, v_5 is $h_\mathbb{K}$ as in **bnfclgp**, v_6 is $\mathcal{R}_\mathbb{K}$ as in **bnfreg** and v_9 is a system of fundamental units expressed as polynomials.

8. **polgalois**(P) provides $\mathrm{Gal}(\mathbb{K}/\mathbb{Q})$.

   Example $\mathrm{Gal}(\mathbb{K}/\mathbb{Q}) \simeq C_6 \simeq \mathbb{Z}/6\mathbb{Z}$.

9. **rnfconductor**(**bnfinit**(y), P) provides the conductor $f_\mathbb{K}$ of the abelian extension $\mathbb{K}/\mathbb{Q}$.

   Example $f_\mathbb{K} = 35$ so that $\mathbb{K} \subseteq \mathbb{Q}(\zeta_{35})$ and $\mathbb{K} \nsubseteq \mathbb{Q}(\zeta_m)$ for $m < 35$. Furthermore, since $35 = 5 \times 7$, the only primes that ramify in $\mathbb{K}$ are 5 and 7.

10. **kronecker**(x, y) provides the Kronecker symbol (x/y) of $x, y \in \mathbb{Z}$.
11. **polcyclo**(n) gives Φ_n.
12. **polsubcyclo**(n, d) gives cyclic subfields of degree d of $\mathbb{Q}(\zeta_n)$ with $d \mid \varphi(n)$.

    Example $P = $ **polsubcyclo**$(35, 6)[2]$ gives $P = x^6 - x^5 + 8x^4 - 8x^3 + 22x^2 - 22x + 29$.

7.4 Multiplicative Aspects of the Ideal Theory

In what follows, $\mathbb{K} = \mathbb{Q}(\theta)$ is an algebraic number field of degree n and let $\mathcal{O}_\mathbb{K}$ be its ring of integers. The purpose of this section is to use tools from analytic number theory to get an answer to certain questions of algebraic number theory.

7.4.1 The Function $v_\mathbb{K}$

Definition 7.117 For all positive integers m, the function $v_\mathbb{K}$ is defined by

$$v_\mathbb{K}(m) = \sum_{\mathcal{N}_{\mathbb{K}/\mathbb{Q}}(\mathfrak{a})=m} 1.$$

Hence $v_\mathbb{K}(m)$ is the number of non-zero integral ideals with norm m.

The following result shows that $v_{\mathbb{K}}$ is a multiplicative function. Besides, we shall see later that this function is not really far from the usual Piltz–Dirichlet divisor function τ_n. Hence we will be able to apply to $v_{\mathbb{K}}$ the fundamental results from Chap. 4.

Proposition 7.118 *The function $v_{\mathbb{K}}$ is multiplicative. Furthermore, for all prime powers p^α, we have*

$$v_{\mathbb{K}}(p^\alpha) = \mathcal{D}_{(f_1,\ldots,f_g)}(\alpha)$$

where $\mathcal{D}_{(f_1,\ldots,f_g)}(\alpha)$ is the denumerant of α with respect to g and to the set $\{f_1,\ldots,f_g\}$ of all inertial degrees of the prime ideals above p, i.e. the number of non-negative integer solutions of the Diophantine linear equation $f_1 x_1 + \cdots + f_g x_g = \alpha$.

Proof

▷ Let $\mathfrak{a}$ be an integral ideal of norm $\mathcal{N}_{\mathbb{K}/\mathbb{Q}}(\mathfrak{a}) = bc$ with $(b,c) = 1$ and set $\mathfrak{a} = \prod_{\mathfrak{p}} \mathfrak{p}^{a_{\mathfrak{p}}}$ and

$$\mathfrak{b} = \prod_{\substack{\mathfrak{p} \\ (\mathcal{N}_{\mathbb{K}/\mathbb{Q}}(\mathfrak{p}),b)>1}} \mathfrak{p}^{a_{\mathfrak{p}}} \quad \text{and} \quad \mathfrak{c} = \prod_{\substack{\mathfrak{p} \\ (\mathcal{N}_{\mathbb{K}/\mathbb{Q}}(\mathfrak{p}),c)>1}} \mathfrak{p}^{a_{\mathfrak{p}}}.$$

Then $\mathfrak{a} = \mathfrak{b}\mathfrak{c}$ with $\mathcal{N}_{\mathbb{K}/\mathbb{Q}}(\mathfrak{b}) = b$ and $\mathcal{N}_{\mathbb{K}/\mathbb{Q}}(\mathfrak{c}) = c$. Hence, if $(b,c) = 1$, any ideal of norm bc can be uniquely written as the product of integral ideals of norm b and of norm c respectively. This establishes the multiplicativity of $v_{\mathbb{K}}$.

▷ Let p be a prime number. By Lemma 7.100, we have

$$(p) = \prod_{i=1}^{g} \mathfrak{p}_i^{e_i}$$

with $1 \leqslant g \leqslant n$ and $e_1 f_1 + \cdots + e_g f_g = n$.

If $\mathfrak{a}$ is a non-zero integral ideal satisfying $\mathcal{N}_{\mathbb{K}/\mathbb{Q}}(\mathfrak{a}) = p^\alpha$ for some $\alpha \in \mathbb{N}$, then $\mathfrak{a} \mid (p^\alpha)$ by (7.13) so that $\mathfrak{a} = \mathfrak{p}_1^{a_1} \cdots \mathfrak{p}_g^{a_g}$ for some $(a_1,\ldots,a_g) \in (\mathbb{Z}_{\geqslant 0})^g$, and comparing the norms we get $f_1 a_1 + \cdots + f_g a_g = \alpha$. Hence $\mathfrak{a}$ induces a solution $(a_1,\ldots,a_g) \in (\mathbb{Z}_{\geqslant 0})^g$ of the Diophantine equation $f_1 x_1 + \cdots + f_g x_g = \alpha$, and if $\mathfrak{b}$ is another non-zero integral ideal such that $\mathcal{N}_{\mathbb{K}/\mathbb{Q}}(\mathfrak{b}) = p^\alpha$ and inducing the same solution, then proceeding similarly we infer that $\mathfrak{b} = \mathfrak{p}_1^{b_1} \cdots \mathfrak{p}_g^{b_g}$ for some $(b_1,\ldots,b_g) \in (\mathbb{Z}_{\geqslant 0})^g$ satisfying $f_1 b_1 + \cdots + f_g b_g = \alpha$, and therefore $b_i = a_i$ for all $i \in \{1,\ldots,g\}$ so that $\mathfrak{b} = \mathfrak{a}$.

Conversely, if $(a_1,\ldots,a_g) \in (\mathbb{Z}_{\geqslant 0})^g$ is a solution of the equation $f_1 x_1 + \cdots + f_g x_g = \alpha$, then $\mathfrak{a} = \mathfrak{p}_1^{a_1} \cdots \mathfrak{p}_g^{a_g}$ is a non-zero integral ideal satisfying $\mathcal{N}_{\mathbb{K}/\mathbb{Q}}(\mathfrak{a}) = p^\alpha$ by Theorem 7.96 (iii).

We have thus defined a one-to-one correspondence between the set of non-zero integral ideals having a norm equal to p^α and the set of the solutions in $(\mathbb{Z}_{\geqslant 0})^g$

of the equation $f_1 x_1 + \cdots + f_g x_g = \alpha$, completing the proof since these two sets are finite. $\square$

The following particular cases are easy.

Corollary 7.119 *With the notation of Proposition 7.118, we have the following results.*

(i) *If $\mathbb{K}$ is Galois over $\mathbb{Q}$, then*

$$v_{\mathbb{K}}(p^\alpha) = \begin{cases} \tau_g(p^{\alpha/f_p}), & \text{if } f_p \mid \alpha, \\ 0, & \text{otherwise.} \end{cases}$$

(ii) *If all the prime ideals above p are of degree 1, then $v_{\mathbb{K}}(p^\alpha) = \tau_g(p^\alpha)$.*
(iii) *In every case, we have the inequality $v_{\mathbb{K}}(m) \leqslant \tau_n(m)$.*

Proof

(i) If $\mathbb{K}$ is Galois over $\mathbb{Q}$, then $f_1 = \cdots = f_g = f_p$ by Proposition 7.104, so that

$$v_{\mathbb{K}}(p^\alpha) = \mathcal{D}_{(1,\dots,1)}\left(\frac{\alpha}{f_p}\right) = \begin{cases} \tau_g(p^{\alpha/f_p}), & \text{if } f_p \mid \alpha, \\ 0, & \text{otherwise} \end{cases}$$

by (4.23).
(ii) If $f_i = 1$ for all $i \in \{1, \dots, g\}$, then $v_{\mathbb{K}}(p^\alpha) = \mathcal{D}_{(1,\dots,1)}(\alpha) = \tau_g(p^\alpha)$ by (4.23).
(iii) The two functions are multiplicative, so that it is sufficient to prove the inequality for prime powers. Now by (4.23) we get for all prime powers p^α

$$v_{\mathbb{K}}(p^\alpha) \leqslant \mathcal{D}_{(1,\dots,1)}(\alpha) = \tau_g(p^\alpha) \leqslant \tau_n(p^\alpha)$$

as required.

The proof is complete. $\square$

Remark 7.120

1. It should be noticed that from Corollary 7.119 (ii) we infer that $v_{\mathbb{K}}(p)$ counts the number of prime ideals above p with inertial degree 1.
2. The following special cases may be useful in practice. When $g = 1$, and then $ef = n$, we have

$$v_{\mathbb{K}}(p) = \begin{cases} 1, & \text{if } f = 1, \\ 0, & \text{otherwise} \end{cases} \quad \text{and} \quad v_{\mathbb{K}}(p^\alpha) = \begin{cases} 1, & \text{if } n \mid \alpha \times e, \\ 0, & \text{otherwise.} \end{cases}$$

Hence if p is inert, so that $e = g = 1$, we get

$$v_{\mathbb{K}}(p^\alpha) = \begin{cases} 1, & \text{if } n \mid \alpha, \\ 0, & \text{otherwise.} \end{cases}$$

7.4.2 The Dedekind Zeta-Function

As in the rational case, one may attach to an algebraic number field $\mathbb{K}$ a generating function which contains all the arithmetic information of $\mathbb{K}$.

Definition 7.121 Let $\mathbb{K}/\mathbb{Q}$ be an algebraic number field. The *Dedekind zeta-function* $\zeta_{\mathbb{K}}$ of $\mathbb{K}$ is the Dirichlet series of the multiplicative function $\nu_{\mathbb{K}}$, so that for all $s = \sigma + it \in \mathbb{C}$ such that $\sigma > 1$, we have

$$\zeta_{\mathbb{K}}(s) = \sum_{m=1}^{\infty} \frac{\nu_{\mathbb{K}}(m)}{m^s} = \sum_{\mathfrak{a}} \frac{1}{\mathcal{N}_{\mathbb{K}/\mathbb{Q}}(\mathfrak{a})^s} = \prod_{\mathfrak{p}} \left(1 - \frac{1}{\mathcal{N}_{\mathbb{K}/\mathbb{Q}}(\mathfrak{p})^s}\right)^{-1}$$

where the second sum is taken over non-zero integral ideals, the product runs through all prime ideals of $\mathcal{O}_{\mathbb{K}}$ and the last equality comes from Corollary 4.48. The absolute convergence in the half-plane $\sigma > 1$ follows from the estimate

$$\sum_{\mathcal{N}_{\mathbb{K}/\mathbb{Q}}(\mathfrak{p}) \leqslant x} \left| \frac{1}{\mathcal{N}_{\mathbb{K}/\mathbb{Q}}(\mathfrak{p})^s} \right| \leqslant n \sum_{p \leqslant x} \frac{1}{p^\sigma}$$

if $\sigma > 1$ and $x \geqslant 2$. It should be noticed that $\zeta_{\mathbb{K}}(s)$ has no zero in the region $\sigma > 1$ since none of the factors of the Euler product have any zeros therein.

Remark 7.122

1. We readily get $\zeta_{\mathbb{Q}}(s) = \zeta(s)$, so that the Dedekind zeta-function generalizes the ordinary Riemann zeta-function.
2. If $\mathbb{K}$ is Galois over $\mathbb{Q}$, then the Euler product simplifies

$$\zeta_{\mathbb{K}}(s) = \prod_{p} \left(1 - \frac{1}{p^{f_p s}}\right)^{-g_p}.$$

3. Using Corollary 7.119 (iii), we infer that for all real numbers $\sigma > 1$, we have

$$\zeta_{\mathbb{K}}(\sigma) \leqslant \zeta(\sigma)^n. \tag{7.20}$$

   We shall see below that the function $\zeta_{\mathbb{K}}(s)$ has an analytic continuation to a meromorphic function in the whole complex plane with a simple pole at $s = 1$. Therefore this inequality does not reflect the arithmetic nature of the Dedekind zeta-function, since $\zeta(s)^n$ has at $s = 1$ a multiple pole of order n.
4. It may be shown that if two algebraic number fields have the same Dedekind zeta-function, then they have the same degree, the same signature and the same discriminant (see [Coh93]).
5. One may define the *Möbius* function for $\mathbb{K}$ in the same way as in the rational case. Let $\mu_{\mathbb{K}}$ be the $\mathcal{O}_{\mathbb{K}}$-arithmetic function defined by $\mu_{\mathbb{K}}(\mathcal{O}_{\mathbb{K}}) = 1$ and, for all

non-zero integral ideals $\mathfrak{a}$

$$\mu_{\mathbb{K}}(\mathfrak{a}) = \begin{cases} (-1)^r, & \text{if } \mathfrak{a} = \mathfrak{p}_1 \cdots \mathfrak{p}_r, \\ 0, & \text{otherwise.} \end{cases}$$

One can prove that the Dirichlet series of $\mu_{\mathbb{K}}$ is $\zeta_{\mathbb{K}}(s)^{-1}$ and we also have for all $s \in \mathbb{C}$ such that $\sigma > 1$

$$\left| \zeta_{\mathbb{K}}(s)^{-1} \right| \leqslant \zeta_{\mathbb{K}}(\sigma).$$

In the sequel, the following specific notation is needed.

Let $\mathbb{K}/\mathbb{Q}$ be an algebraic number field of degree n, signature (r_1, r_2), discriminant $d_{\mathbb{K}}$, class number $h_{\mathbb{K}}$, regulator $\mathcal{R}_{\mathbb{K}}$ and let $w_{\mathbb{K}}$ be the number of roots of unity contained in $\mathbb{K}$ (see Proposition 7.77). According to the usual practice, we set

$$\Gamma_{\mathbb{K}}(s) = \Gamma(s/2)^{r_1} \Gamma(s)^{r_2} \quad \text{and} \quad A_{\mathbb{K}} = 2^{-r_2} \pi^{-n/2} |d_{\mathbb{K}}|^{1/2}$$

where $\Gamma(s)$ is the usual Gamma-function. The function $\Gamma_{\mathbb{K}}$ is sometimes called the *Gamma-function of*[21] $\mathbb{K}$.

The main properties of the Dedekind zeta-function generalize those of the Riemann zeta-function. In particular, $\zeta_{\mathbb{K}}(s)$ satisfies a functional equation and an approximate functional equation. The former was discovered by Hecke using n-dimensional analytic methods analogous to the case of ζ. Using adelic Chevalley's language, Tate gave another proof of the functional equation in his thesis.[22]

Theorem 7.123 (Functional equation)

(i) *The function $\zeta_{\mathbb{K}}(s)$ can be extended analytically in the whole complex plane to a meromorphic function having a simple pole at $s = 1$ with residue $\kappa_{\mathbb{K}}$ equal to*

$$\kappa_{\mathbb{K}} = \frac{2^{r_1} (2\pi)^{r_2} h_{\mathbb{K}} \mathcal{R}_{\mathbb{K}}}{w_{\mathbb{K}} |d_{\mathbb{K}}|^{1/2}}. \tag{7.21}$$

This identity is called the analytic class number formula.

(ii) *For all $s \in \mathbb{C} \setminus \{1\}$, the function $\xi_{\mathbb{K}}(s) = A_{\mathbb{K}}^s \Gamma_{\mathbb{K}}(s) \zeta_{\mathbb{K}}(s)$ satisfies the following functional equation*[23]

$$\xi_{\mathbb{K}}(s) = \xi_{\mathbb{K}}(1 - s).$$

Furthermore, for all real numbers $\sigma > 1$, we also have

$$\kappa_{\mathbb{K}} \leqslant (2\pi)^{r_2} \sigma (\sigma - 1) \xi_{\mathbb{K}}(\sigma) |d_{\mathbb{K}}|^{-1/2}. \tag{7.22}$$

---

[21] The factor $A_{\mathbb{K}}^s \Gamma_{\mathbb{K}}(s)$ is sometimes called the Euler factor of $\zeta_{\mathbb{K}}(s)$.

[22] See [Lan94, Nar04, Neu10] for an exhaustive account of Hecke's method and Tate's work.

[23] Some authors [Coh93, Neu10] consider the "weighted" function $2^{r_2} \xi_{\mathbb{K}}(s)$ instead.

(iii) *For all $t \in \mathbb{R}$, set $\tau = |t| + 3$. We have*

$$\zeta_\mathbb{K}(1 + it) \ll_\mathbb{K} \log \tau \quad \text{and} \quad \zeta_\mathbb{K}(it) \ll_\mathbb{K} \tau^{n/2} \log \tau.$$

By the Phragmén–Lindelöf principle, we infer that, for all $0 < \sigma < 1$, we have

$$\zeta_\mathbb{K}(\sigma + it) \ll_\mathbb{K} \tau^{n(1-\sigma)/2} \log \tau.$$

(iv) *In the region*

$$\sigma \geqslant 1 - \frac{1}{c_1 n^5} \left(\frac{\log \log t}{\log t} \right)^{2/3} \quad \text{and} \quad t \geqslant 1.1$$

we have

$$\left| \zeta_\mathbb{K}(s) \right| \leqslant e^{c_2 n^6 |d_\mathbb{K}|^2} \log t$$

for some absolute constants $c_1, c_2 > 0$.

For a proof of (iv), see [Bar78, Lemma 15].

The next step is to get a zero-free region for the Dedekind zeta-function. As for the Dirichlet L-functions, a Deuring–Heilbronn phenomenon appears for $\zeta_\mathbb{K}(s)$. We summarize in the following theorem the main results of this type and also provide the analogue of (3.35) for $\zeta_\mathbb{K}(s)$ which is used to get these zero-free regions. The proofs can be found in [Bar78, Sta79]. The result (ii) below is a refinement of classical theorems by Landau [Lan27] and Sokolovskiĭ [Sok68].

Theorem 7.124 (Zero-free regions) *Let $s = \sigma + it \in \mathbb{C}$ and the constants $c_1, c_3 > 0$ and $c_2 > 1$ be absolute.*

(i) *In the region $1 - 1/(n+1) \leqslant \sigma \leqslant 1$ and $t \geqslant e$, we have*

$$\left| \zeta_\mathbb{K}(s) \right| \leqslant e^{c_1 n^8 |d_\mathbb{K}|^2} t^{600 n^2 \{n(1-\sigma)\}^{3/2}} (\log t)^{2/3}.$$

(ii) *The function $\zeta_\mathbb{K}(s)$ has no zero in the region*

$$\sigma \geqslant 1 - \frac{1}{c_2 n^{11} |d_\mathbb{K}|^3 (\log t)^{2/3} (\log \log t)^{1/3}} \quad \text{and} \quad t \geqslant 4.$$

(iii) *The function $\zeta_\mathbb{K}(s)$ has no zero in the region*

$$\sigma > 1 - \frac{c_3}{\log(|d_\mathbb{K}|(|t| + 2)^n)} \quad \text{and} \quad t \in \mathbb{R}$$

except maybe for the exceptional zero *of $\zeta_\mathbb{K}(s)$. If the latter exists, then it is real and simple.*

Remark 7.125 As in the rational case (see Theorem 3.82), an approximate functional equation for $\zeta_{\mathbb{K}}(s)$ has been stated in [CN63] where it is proved that, for some $H > 0$, $x, y > H$ satisfying $xy = |d_{\mathbb{K}}|(\frac{|t|}{2\pi})^n$ and $c_1 < x/y < c_2$, we have

$$\zeta_{\mathbb{K}}(s) = \sum_{\mathcal{N}_{\mathbb{K}/\mathbb{Q}}(\mathfrak{a}) \leqslant x} \frac{1}{\mathcal{N}_{\mathbb{K}/\mathbb{Q}}(\mathfrak{a})^s} + A_{\mathbb{K}}^{1-2s} \frac{\Gamma_{\mathbb{K}}(1-s)}{\Gamma_{\mathbb{K}}(s)} \sum_{\mathcal{N}_{\mathbb{K}/\mathbb{Q}}(\mathfrak{a}) \leqslant y} \frac{1}{\mathcal{N}_{\mathbb{K}/\mathbb{Q}}(\mathfrak{a})^{1-s}}$$

$$+ O\left(x^{1-\sigma-1/n} \log x\right).$$

This implies that

$$\zeta_{\mathbb{K}}(s) \ll \tau^{n(1-\sigma)/2}$$

for all $0 < \sigma < 1$ where $\tau = |t| + 3$.

The functional equation shows that $\zeta_{\mathbb{K}}(s)$ may have trivial zeros at negative integers $-m$ with $m \in \mathbb{N}$. In fact, it can be shown that if m is even, then the order of the possible zero is equal to $r_1 + r_2$, while for m odd the order equals r_2. The Euler product implies that all other zeros $\rho = \beta + i\gamma$ satisfy $0 < \beta < 1$ and are called *non-trivial zeros*. Furthermore, we deduce that the only fields for which some of the values of $\zeta_{\mathbb{K}}(-m)$ can be non-zero are totally real fields. The next result shows that these values are nevertheless rational numbers.

Theorem 7.126 (Siegel–Klingen) *Let $\mathbb{K}/\mathbb{Q}$ be a totally real algebraic number field, i.e. $r_2 = 0$. For all positive integers m, we have $\zeta_{\mathbb{K}}(1 - 2m) \in \mathbb{Q}$.*

Combining this result with the functional equation and assuming $\mathbb{K}$ to be totally real of degree n, we get for all $m \in \mathbb{N}$

$$\zeta_{\mathbb{K}}(2m) = q_m \pi^{2mn} d_{\mathbb{K}}^{-1/2}$$

for some $q_m \in \mathbb{Q} \setminus \{0\}$.

7.4.3 Application to the Class Number

The problem of getting upper and lower bounds for the class number $h_{\mathbb{K}}$ of an algebraic number field $\mathbb{K}$ has a long history. In particular, the search for quadratic fields with class number one became in the early days one of the most important questions in algebraic number theory. For imaginary quadratic fields, the answer came in 1966. For real quadratic fields, Gauss conjectured that there are infinitely many such fields having class number one, but the question remains open (see Sect. 7.5).

One of the first ideas to get an upper bound for $h_{\mathbb{K}}$ is the use of the functional equation, in particular the inequality (7.22). Using (7.20) and choosing σ near 1 enables us to get very good upper estimates for $h_{\mathbb{K}} \mathcal{R}_{\mathbb{K}}$. In particular, this approach enables Louboutin [Lou98] to get the following estimate,[24] valid for all totally real algebraic number fields of degree $n \geqslant 2$

$$h_{\mathbb{K}} \mathcal{R}_{\mathbb{K}} \leqslant d_{\mathbb{K}}^{1/2} \left(\frac{e \log d_{\mathbb{K}}}{4n-4} \right)^{n-1}. \tag{7.23}$$

But precise lower bounds for the regulator are needed, and such results are very difficult to obtain in the general case. Nevertheless, it should be mentioned that non-trivial lower bounds have been discovered, notably by Cusick [Cus84], Silverman [Sil84], Friedman [Fri89] and Uchida [Uch94]. These estimates will be discussed in Sect. 7.5.

Another way to estimate $h_{\mathbb{K}}$ lies in Theorem 7.111 and the proof of Lemma 7.113. Let $b_{\mathbb{K}}$ be any positive real number such that every ideal class contains a non-zero integral ideal $\mathfrak{a}$ such that

$$\mathcal{N}_{\mathbb{K}/\mathbb{Q}}(\mathfrak{a}) \leqslant b_{\mathbb{K}}.$$

From Remark 7.114, we know that $b_{\mathbb{K}}$ could be the Minkowski bound, and we infer that

$$h_{\mathbb{K}} \leqslant \sum_{\mathcal{N}_{\mathbb{K}/\mathbb{Q}}(\mathfrak{a}) \leqslant b_{\mathbb{K}}} 1 = \sum_{m \leqslant b_{\mathbb{K}}} \nu_{\mathbb{K}}(m).$$

It is therefore important to have at our disposal upper bounds for the average order of $\nu_{\mathbb{K}}$ which *do not depend on* the invariants of $\mathbb{K}$. By Corollary 4.59, we get

$$\sum_{m \leqslant x} \nu_{\mathbb{K}}(m) = \kappa_{\mathbb{K}} x + o_{\mathbb{K}}(x)$$

but $\kappa_{\mathbb{K}}$ contains $h_{\mathbb{K}}$. Another interesting result is the estimate [CN63]

$$\sum_{m \leqslant x} \nu_{\mathbb{K}}(m)^2 \ll_{\mathbb{K}} x (\log x)^{n-1}$$

but the implied constant depends on the usual invariants of $\mathbb{K}$. Using Corollary 7.119 (iii) and Exercise 12 in Chap. 4, we get the following result.

Theorem 7.127 *Let $\mathbb{K}/\mathbb{Q}$ be an algebraic number field of degree n and class number $h_{\mathbb{K}}$. Let $b_{\mathbb{K}}$ such that every ideal class contains a non-zero integral ideal $\mathfrak{a}$ such that $\mathcal{N}_{\mathbb{K}/\mathbb{Q}}(\mathfrak{a}) \leqslant b_{\mathbb{K}}$. Then we have*

$$h_{\mathbb{K}} \leqslant b_{\mathbb{K}} \sum_{j=0}^{n-1} \binom{n-1}{j} \frac{(\log b_{\mathbb{K}})^j}{j!} \leqslant \frac{b_{\mathbb{K}}(\log b_{\mathbb{K}} + n - 1)^{n-1}}{(n-1)!}.$$

---

[24] A proof is supplied in Exercise 11.

In [Bor02], the function τ_n has been studied more carefully. This leads to the following improvement.

Theorem 7.128 *Let $\mathbb{K}/\mathbb{Q}$ be an algebraic number field of degree $n \geqslant 2$ and class number $h_{\mathbb{K}}$. Let $b_{\mathbb{K}} \geqslant 6$ such that every ideal class contains a non-zero integral ideal $\mathfrak{a}$ such that $\mathcal{N}_{\mathbb{K}/\mathbb{Q}}(\mathfrak{a}) \leqslant b_{\mathbb{K}}$. Then we have*

$$h_{\mathbb{K}} \leqslant 2\, b_{\mathbb{K}} (\log b_{\mathbb{K}})^{n-1}.$$

Applied with $b_{\mathbb{K}} = M_{\mathbb{K}} |d_{\mathbb{K}}|^{1/2}$ where $M_{\mathbb{K}}$ is the Minkowski constant (7.19), we get

$$h_{\mathbb{K}} \leqslant 2^{2-n} M_{\mathbb{K}} |d_{\mathbb{K}}|^{1/2} \left(\log M_{\mathbb{K}}^2 |d_{\mathbb{K}}| \right)^{n-1}$$

as soon as $|d_{\mathbb{K}}| \geqslant 36 M_{\mathbb{K}}^{-2}$.

7.4.4 Lower Bounds for $|d_{\mathbb{K}}|$

In 1881, Kronecker asked whether $|d_{\mathbb{K}}| > 1$ for all algebraic number fields $\mathbb{K}$ of degree $n \geqslant 2$. This question remained open until 1890 when Minkowski created the geometry of numbers and discovered the Minkowski bound (7.18). This lower bound follows readily by using the fact that $\mathcal{N}_{\mathbb{K}/\mathbb{Q}}(\mathfrak{a}) \geqslant 1$ and the easy bounds $r_2 \leqslant n/2$ and $n^n/n! \geqslant 2^{n-1}$. By Theorem 7.102, we infer that every algebraic number field $\mathbb{K} \neq \mathbb{Q}$ has at least a ramified prime number, which plays an important role in the proof of the Kronecker–Weber theorem. Furthermore, Minkowski observed that his method provides lower bounds tending to ∞ with the degree of $\mathbb{K}$. More precisely, using Stirling's bounds (1.2) we see that, for all $n \geqslant 3$, we have

$$\log |d_{\mathbb{K}}| \geqslant \left(2 + \log \frac{\pi}{4} \right) n + r_1 \log \frac{4}{\pi} - \log 2\pi n - \frac{1}{6n} > n - 1. \tag{7.24}$$

In the late 1960s and early 1970s, Stark [Sta75] used the functional equation and the Hadamard factorization theorem to get a lower bound which eventually supersedes the geometric methods. This is the purpose of the next result, in which the digamma function $\Psi(\sigma) = \Gamma'(\sigma)/\Gamma(\sigma)$ appears.

Proposition 7.129 *Let $\mathbb{K}/\mathbb{Q}$ be an algebraic number field of degree n, discriminant $d_{\mathbb{K}}$ and signature (r_1, r_2). For all real numbers $\sigma > 1$, we have*

$$\log |d_{\mathbb{K}}| \geqslant r_1 \left(\log \pi - \Psi \left(\frac{\sigma}{2} \right) \right) + 2 r_2 \left(\log 2\pi - \Psi(\sigma) \right) - \frac{2}{\sigma} - \frac{2}{\sigma - 1}.$$

Proof We proceed as in Proposition 3.89. Define

$$F_{\mathbb{K}}(s) = s(s-1)\xi_{\mathbb{K}}(s) = s(s-1) A_{\mathbb{K}}^s \Gamma_{\mathbb{K}}(s) \zeta_{\mathbb{K}}(s).$$

$F_{\mathbb{K}}(s)$ is an entire function of order 1, so that by the Hadamard factorization theorem, there exist suitable constants a, b such that

$$F_{\mathbb{K}}(s) = e^{a+bs} \prod_{\rho} \left(1 - \frac{s}{\rho}\right) e^{s/\rho}$$

where the product runs through all zeros $\rho = \beta + i\gamma$ of $F_{\mathbb{K}}(s)$ which are exactly the non-trivial zeros of $\zeta_{\mathbb{K}}(s)$. The logarithmic differentiation provides

$$\frac{F_{\mathbb{K}}'}{F_{\mathbb{K}}}(s) = b + \sum_{\rho} \left(\frac{1}{s-\rho} + \frac{1}{\rho}\right) \tag{7.25}$$

where the sum is absolutely convergent. Taking $s = 0$ gives $b = F'(0)/F(0)$, and using the functional equation of $\zeta_{\mathbb{K}}(s)$ gives $F_{\mathbb{K}}(s) = F_{\mathbb{K}}(1-s)$, so that

$$b = \frac{F_{\mathbb{K}}'(0)}{F_{\mathbb{K}}(0)} = -\frac{F_{\mathbb{K}}'(1)}{F_{\mathbb{K}}(1)} = -b - \sum_{\rho} \left(\frac{1}{\rho} + \frac{1}{1-\rho}\right).$$

Now if ρ is a zero of $F_{\mathbb{K}}(s)$, so is $\overline{\rho}$ and since $F_{\mathbb{K}}(s) = F_{\mathbb{K}}(1-s)$, we infer that $1 - \rho$ is also a zero of $F_{\mathbb{K}}$. This gives

$$b = -\frac{1}{2} \sum_{\rho} \left(\frac{1}{\rho} + \frac{1}{\overline{\rho}}\right)$$

and (7.25) becomes

$$\frac{F_{\mathbb{K}}'}{F_{\mathbb{K}}}(s) = \frac{1}{2} \sum_{\rho} \left(\frac{1}{s-\rho} + \frac{1}{s-\overline{\rho}}\right)$$

so that, by the definition of $F_{\mathbb{K}}(s)$, we get

$$\log|d_{\mathbb{K}}| = r_1 \left(\log \pi - \Psi\left(\frac{s}{2}\right)\right) + 2r_2 \left(\log 2\pi - \Psi(s)\right) - \frac{2}{s} - \frac{2}{s-1}$$
$$+ \sum_{\rho} \left(\frac{1}{s-\rho} + \frac{1}{s-\overline{\rho}}\right) - \frac{2\zeta_{\mathbb{K}}'}{\zeta_{\mathbb{K}}}(s).$$

Now we have

$$\sum_{\rho} \left(\frac{1}{\sigma-\rho} + \frac{1}{\sigma-\overline{\rho}}\right) = 2 \sum_{\rho} \frac{\sigma-\beta}{|\sigma-\rho|^2} > 0$$

and we have from the Euler product of Definition 7.121

$$-\frac{\zeta_{\mathbb{K}}'}{\zeta_{\mathbb{K}}}(\sigma) = \sum_{\mathfrak{p}} \frac{\log \mathcal{N}_{\mathbb{K}/\mathbb{Q}}(\mathfrak{p})}{\mathcal{N}_{\mathbb{K}/\mathbb{Q}}(\mathfrak{p})^\sigma - 1} > 0$$

which concludes the proof.   □

Using this result along with Gautschi's inequality, we get the following lower bounds for $|d_{\mathbb{K}}|$.

Corollary 7.130 *Let $\mathbb{K}/\mathbb{Q}$ be an algebraic number field of degree $n \geqslant 2$ and discriminant $d_{\mathbb{K}}$. Then we have*

$$|d_{\mathbb{K}}| > \max\left(e^{n-1}, e^{-8}(2\pi)^n\right).$$

Furthermore, if $n \geqslant 50$, then $|d_{\mathbb{K}}| > (2\pi)^n$.

Proof If $\mathbb{K}$ is a quadratic field, then $|d_{\mathbb{K}}| \geqslant 3 > e$, so that we may suppose that $n \geqslant 3$. The inequality $|d_{\mathbb{K}}| > e^{n-1}$ has been seen in (7.24). By Gautschi's inequality [EGP00, Corollary 3], we have for all $h > 0$

$$\Psi(1+h) < \log\left(h + e^{-\gamma}\right) \quad \text{and} \quad \Psi\left(\frac{1+h}{2}\right) < \log\left(\frac{h}{2} + e^{-\gamma - \log 4}\right).$$

Using these inequalities in Proposition 7.129 with $\sigma = 1 + h$ ($h > 0$), we get

$$
\begin{aligned}
\log|d_{\mathbb{K}}| &\geqslant r_1\left(\log\pi - \Psi\left(\frac{1+h}{2}\right)\right) + 2r_2\left(\log 2\pi - \Psi(1+h)\right) - \frac{2}{1+h} - \frac{2}{h} \\
&> \log\left\{\left(\frac{4\pi}{2h + e^{-\gamma}}\right)^{r_1}\left(\frac{2\pi}{h + e^{-\gamma}}\right)^{2r_2}\right\} - \frac{2}{1+h} - \frac{2}{h} \\
&= \log\left\{\left(\frac{2\pi}{h + e^{-\gamma}}\right)^{n}\left(\frac{2h + 2e^{-\gamma}}{2h + e^{-\gamma}}\right)^{r_1}\right\} - \frac{2}{1+h} - \frac{2}{h} \\
&\geqslant n\log\left(\frac{2\pi}{h + e^{-\gamma}}\right) - \frac{2}{1+h} - \frac{2}{h}.
\end{aligned}
$$

Choosing $h = \sqrt{2e^{-\gamma}/n}$ gives

$$\log|d_{\mathbb{K}}| > n\log 2\pi + f(n)$$

where

$$f(n) = \gamma n - 2\sqrt{2e^\gamma n}\left(1 + \frac{1}{\sqrt{2e^\gamma n} + 2}\right) > -8$$

for all $n \geqslant 2$ implying that $|d_{\mathbb{K}}| > e^{-8}(2\pi)^n$. If $n \geqslant 50$, the trivial bound

$$\frac{1}{\sqrt{2e^\gamma n} + 2} < \frac{1}{\sqrt{2e^\gamma n}}$$

implies

$$\log|d_{\mathbb{K}}| > n\log 2\pi + \gamma n - 2\sqrt{2e^\gamma n} - 2 > n\log 2\pi$$

as asserted. $\square$

Let us notice that the proof of Proposition 7.129 rests on the fact that, for all $\sigma > 1$, we have

$$-\frac{\zeta_{\mathbb{K}}'(\sigma)}{\zeta_{\mathbb{K}}(\sigma)} + \frac{1}{2}\sum_{\rho}\left(\frac{1}{\sigma - \rho} + \frac{1}{\sigma - \bar{\rho}}\right) > 0.$$

In [Odl76], Odlyzko showed that this quantity is in fact quite large and used this to get substantial improvements on the estimates of Corollary 7.130. Later Serre, Odlyzko and Poitou used Guinand's and Weil's explicit formulae to bound $|d_{\mathbb{K}}|$ in a somewhat much more elegant and efficient way than with Proposition 7.129, both under ERH[25] and unconditionally. For instance, one may prove [Poi77] without ERH that

$$\log|d_{\mathbb{K}}| \geqslant n(\gamma + \log 4\pi) + r_1 - 8.6n^{1/3}.$$

7.4.5 The Dedekind Zeta-Function of a Quadratic Field

Let $d \in \mathbb{Z} \setminus \{0, 1\}$ squarefree and $\mathbb{K} = \mathbb{Q}(\sqrt{d})$ be a quadratic field with discriminant $d_{\mathbb{K}}$. According to Lemma 7.107, the Kronecker symbol $(d_{\mathbb{K}}/\cdot)$ is a real primitive Dirichlet character. We define

$$L_{d_{\mathbb{K}}}(s) = \sum_{m=1}^{\infty} \frac{(d_{\mathbb{K}}/m)}{m^s}$$

its associated Dirichlet L-series, which is absolutely convergent in the half-plane $\sigma > 0$. Our aim is to prove the following factorization of $\zeta_{\mathbb{K}}$ which may be viewed as a sort of analytic translation of the quadratic reciprocity law.

Proposition 7.131 *For all real numbers $\sigma > 1$, we have $\zeta_{\mathbb{K}}(\sigma) = \zeta(\sigma) L_{d_{\mathbb{K}}}(\sigma)$.*

Proof First note that using the Euler product of L-functions, we get

$$L_{d_{\mathbb{K}}}(\sigma) = \prod_{p}\left(1 - \frac{(d_{\mathbb{K}}/p)}{p^{\sigma}}\right)^{-1} = \prod_{(d_{\mathbb{K}}/p)=1}\left(1 - \frac{1}{p^{\sigma}}\right)^{-1}\prod_{(d_{\mathbb{K}}/p)=-1}\left(1 + \frac{1}{p^{\sigma}}\right)^{-1}.$$

Now using Remark 7.122 and Proposition 7.108, we get for $\sigma > 1$

$$\zeta_{\mathbb{K}}(\sigma) = \prod_{(d_{\mathbb{K}}/p)=1}\left(1 - \frac{1}{p^{\sigma}}\right)^{-2}\prod_{(d_{\mathbb{K}}/p)=-1}\left(1 - \frac{1}{p^{2\sigma}}\right)^{-1}\prod_{p|d_{\mathbb{K}}}\left(1 - \frac{1}{p^{\sigma}}\right)^{-1}$$

---

[25] The *Extended Riemann Hypothesis* states that *all* non-trivial zeros of the Dedekind zeta-function lie on the line $\sigma = 1/2$. It should be mentioned that ERH is sometimes called GRH, for *Generalized Riemann Hypothesis*. However, GRH is referred to in this book as the conjecture asserting that the Dirichlet L-functions have *all* their zeros lying on the line $\sigma = 1/2$. See Footnote 26.

$$= L_{d_{\mathbb{K}}}(\sigma) \prod_{(d_{\mathbb{K}}/p)=1} \left(1 - \frac{1}{p^{\sigma}}\right)^{-1} \prod_{(d_{\mathbb{K}}/p)=-1} \left(1 - \frac{1}{p^{\sigma}}\right)^{-1} \prod_{p | d_{\mathbb{K}}} \left(1 - \frac{1}{p^{\sigma}}\right)^{-1}$$

$$= L_{d_{\mathbb{K}}}(\sigma) \prod_{p} \left(1 - \frac{1}{p^{\sigma}}\right)^{-1} = \zeta(\sigma) L_{d_{\mathbb{K}}}(\sigma)$$

as required.   □

Now letting $\sigma \longrightarrow 1$ in Proposition 7.131 and using the analytic class number formula gives $L_{d_{\mathbb{K}}}(1) = \kappa_{\mathbb{K}}$ which implies that

$$h_{\mathbb{K}} = \begin{cases} \dfrac{L_{d_{\mathbb{K}}}(1) d_{\mathbb{K}}^{1/2}}{2 \log \varepsilon_{\mathbb{K}}}, & \text{if } d > 0, \\[2ex] \dfrac{L_{d_{\mathbb{K}}}(1) |d_{\mathbb{K}}|^{1/2} w_{\mathbb{K}}}{2\pi}, & \text{if } d < 0 \end{cases} \qquad (7.26)$$

where $\varepsilon_{\mathbb{K}}$ is the fundamental unit of $\mathbb{K}$ in the case of $d > 0$ and $w_{\mathbb{K}} = 2, 4, 6$ according to whether $d_{\mathbb{K}} < -4$ or $d_{\mathbb{K}} = -4, -3$. This is the *Dirichlet class number formula for quadratic fields*.

7.5 Further Developments

7.5.1 Euler Polynomials and Gauss Class Number Problems

Let $P = X^2 + X + 41$. One may readily check that $P(n)$ is prime for $n \in \{0, \ldots, 39\}$ but $P(40) = 41^2$ is not a prime number. This polynomial, discovered by Euler in 1772, was one of the first polynomials which can provide a finite subset of prime numbers as long as n lies in a subset of non-negative integers. One may ask for a polynomial giving *all* the prime numbers, but we know by Example 3.12 that such a single-variable polynomial cannot exist. However, one may formulate the problem in the following way: are there polynomials of the form $P_q = X^2 + X + q$, with q prime, such that $P_q(n)$ is a prime number for all $n \in \{0, \ldots, q - 2\}$? The answer is given by the following result.

Theorem 7.132 *Let q be a prime number and $P_q = X^2 + X + q$. Then, $P_q(n)$ is a prime number for all $n \in \{0, \ldots, q - 2\}$ if and only if $q \in \{2, 3, 5, 11, 17, 41\}$.*

In fact, this result is the consequence of two profound theorems. The first one relates the values of q to the class number of the imaginary quadratic field $\mathbb{K} = \mathbb{Q}(\sqrt{1 - 4q})$. More precisely, we have the following theorem (see [Gol85, Rib88] and the references therein) which goes back to Rabinowitch.

Theorem 7.133 *Let q be a prime number, $P_q = X^2 + X + q$ and $\mathbb{K} = \mathbb{Q}(\sqrt{1-4q})$. Then $P_q(n)$ is a prime number for all $n \in \{0, \dots, q-2\}$ if and only if $h_{\mathbb{K}} = 1$.*

In 1966/7, with two different methods, Baker [Bak66] and Stark [Sta67] discovered *all* the imaginary quadratic fields with class number one, proving the following result which easily implies Theorem 7.132.

Theorem 7.134 *Let $d < 0$ squarefree and $\mathbb{K} = \mathbb{Q}(\sqrt{d})$. Then $h_{\mathbb{K}} = 1$ if and only if*

$$d \in \{-1, -2, -3, -7, -11, -19, -43, -67, -163\}.$$

In Articles 303 and 304 of his *Disquisitiones Arithmeticæ* [Gau86], Gauss used the language of binary quadratic forms to formulate several conjectures which remain open nowadays. Translated into the language of modern algebraic number theory, i.e. Dedekind's language, the two particular conjectures below can be stated as follows, where $\mathbb{K} = \mathbb{Q}(\sqrt{d})$ is a quadratic field with class number denoted here by $h(d)$.

▷ In Article 303, Gauss conjectured that $h(d) \longrightarrow \infty$ as $d \longrightarrow -\infty$. He also included the following table

$h(d)$	1	2	3	4	5
Number of fields	9	18	16	54	25
Largest $|d|$	163	427	907	1555	2683

translated into a table of quadratic fields with small class numbers and he surmised that this table is complete. Given any $h \in \mathbb{N}$, the problem of finding all imaginary quadratic fields of class number h is called *Gauss's class number h problem for imaginary quadratic fields*.

▷ For real quadratic fields, translated into modern language, Gauss surmised in Article 304 that there are infinitely many real quadratic fields with class number one. This conjecture still remains open today, and we do not know if there are infinitely many number fields *of arbitrary degree* having class number one, or even just bounded.

The conjecture $h(d) \longrightarrow \infty$ as $d \longrightarrow -\infty$ has a curious and interesting story (see [Gol85] for more information). In 1918, Landau published the following result which he attributed to a lecture given by Hecke.

Theorem 7.135 *Let $d < 0$ and χ be an odd, real and primitive Dirichlet character modulo $|d|$. If $L(\sigma, \chi) \neq 0$ for all real numbers $\sigma > 1 - c_1/\log|d|$, then we have*

$$h(d) > \frac{c_2|d|^{1/2}}{\log|d|}.$$

We infer from this result that the GRH[26] implies Gauss's conjecture.

But in the 1930s, Deuring, Mordell and Heilbronn proved that the *falsity* of GRH *also* implies Gauss's conjecture. Hence this conjecture is true and was the first result to be proved by assuming the truth and the falsity of GRH.

The flaw of this method is that the result is *not effective*, since if the GRH is false, all constants depend on an unknown zero of $L(s, \chi)$, namely the *Siegel's zero*, located off the line $\sigma = \frac{1}{2}$ (see Theorems 3.93 and 3.94). Refining this proof, in order to work with Gauss's class number one problem for imaginary quadratic fields, Heilbronn and Linfoot proved that there are at most ten imaginary quadratic fields with class number one, i.e. the nine fields of Theorem 7.134 plus possibly an unknown field. The existence of this tenth imaginary UFD quadratic field reflects the ineffectivity of the Deuring–Heilbronn phenomenon and would also contradict the truth of the GRH. Then one can imagine that this problem led to intense research.

The solution of Gauss's class number one problem was found by Baker [Bak66] and Stark [Sta67], with completely different methods. Baker used an idea of Gelfond and Linnik who proved that this problem could be solved if one had linear independence of three logarithms, whilst Stark showed that a tenth imaginary quadratic field cannot exist. At this point, it should be mentioned that fifteen years earlier, Heegner, a High School teacher, announced he had solved the class number one problem [Hee52]. Unfortunately, his work contained some "gaps" in his proof and his paper was dismissed at the time. As pointed out by Goldfeld [Gol85], Heegner died before anyone really understood his discoveries.

The general solution to Gauss's class number problem for imaginary quadratic fields comes from another area in number theory. Goldfeld showed that if an elliptic curve over $\mathbb{Q}$ having certain properties could exist, it would provide a lower bound of $h(d)$ sufficiently accurate to be effective, and then solve the conjecture. Gross and Zagier discovered such an elliptic curve, and the method, called today the *Goldfeld–Gross–Zagier* theorem, provides a solution to this long-standing problem.

Let us be more precise. To each elliptic curve E with minimal Weierstrass equation $y^2 = 4x^3 - g_2 x - g_3$ and discriminant $\operatorname{disc}(E) = g_2^3 - 27g_3^2 \neq 0$, one can define a positive integer N, called the *conductor* of E, having the same prime factors as $\operatorname{disc}(E)$ and dividing it, and also define an L-function, called the *Hasse–Weil L-function*, by the Euler product

$$L(E, s) = \prod_{p \nmid \operatorname{disc}(E)} \left(1 - \frac{a_p}{p^s} + \frac{1}{p^{2s-1}}\right)^{-1} \prod_{p \mid \operatorname{disc}(E)} \left(1 - \frac{a_p}{p^s}\right)^{-1}$$

where $a_p = p + 1 - |E(\mathbb{F}_p)|$ and $|E(\mathbb{F}_p)|$ is the number of points of E over $\mathbb{F}_p$, including the point at infinity. It has been shown by Hasse that $|a_p| \leqslant 2\sqrt{p}$, so that the Euler product is absolutely convergent in the half-plane $\sigma > \frac{3}{2}$. One of the most important results on elliptic curves over $\mathbb{Q}$ is a theorem proved by Wiles

---

[26]The *Generalized Riemann Hypothesis*, or GRH for short, asserts that the Dirichlet L-functions have *all* their zeros lying on the line $\sigma = 1/2$.

and other mathematicians[27] stating that the function $L(E, s)$ has an analytic continuation in the whole complex plane into a holomorphic function and satisfies a functional equation. Another fundamental result is Mordell's theorem stating that the group $E(\mathbb{Q})$ of rational points of E is a finitely generated abelian group. The rank of this group is called the *algebraic rank* of E, in comparison to the *analytic rank* of E which is defined as the order of vanishing of $L(E, s)$ at $s = 1$. These two ranks are related in one of the most famous conjectures in number theory.

Conjecture 7.136 (Birch & Swinnerton-Dyer) *For all elliptic curves over $\mathbb{Q}$, the algebraic rank and the analytic rank are equal.*

In recent decades, some progress has been made towards this conjecture. In particular, it can be proved that, *if the analytic rank is equal to 0, respectively 1, then the algebraic rank is equal to 0, respectively 1.*

The Goldfeld–Gross–Zagier method enables us to prove the following result [IK04, Theorem 23.2].

Theorem 7.137 *There exists an absolute, effectively computable constant $c > 0$ such that for all imaginary quadratic fields $\mathbb{K} = \mathbb{Q}(\sqrt{d})$ with $d < 0$, we have*

$$h(d) > c \prod_{p \mid d} \left(1 + \frac{1}{p}\right)^{-3} \left(1 + \frac{2p^{1/2}}{p+1}\right)^{-1} \log |d|.$$

7.5.2 The Brauer–Siegel Theorem

The functional equation of $\zeta_{\mathbb{K}}(s)$ and the inequality (7.22) imply that, for all $\sigma > 1$, we have

$$\kappa_{\mathbb{K}} \leqslant \sigma(\sigma - 1) \, 2^{r_2(1-\sigma)} \pi^{r_2 - n\sigma/2} |d_{\mathbb{K}}|^{(\sigma-1)/2} \Gamma_{\mathbb{K}}(\sigma) \zeta_{\mathbb{K}}(\sigma).$$

Alzer's inequality [Alz00] states that

$$\Gamma(x) < \begin{cases} x^{x-1-\gamma}, & \text{if } x > 1, \\ x^{\delta(x-1)-\gamma}, & \text{if } 0 < x < 1 \end{cases}$$

where $\gamma \approx 0.5772\ldots$ is the Euler constant and $\delta = \frac{1}{2}(\zeta(2) - \gamma) \approx 0.5338\ldots$, so that for all $1 < \sigma < 2$ we get

$$\Gamma_{\mathbb{K}}(\sigma) < (\sigma/2)^{r_1\{\delta(\sigma/2-1)-\gamma\}} \sigma^{r_2(\sigma-1-\gamma)}$$

$$= \sigma^{r_1\{\delta(\sigma/2-1)-\gamma\}+r_2(\sigma-1-\gamma)} \, 2^{-r_1\{\delta(\sigma/2-1)-\gamma\}}$$

$$= \sigma^{(\sigma-1)(r_1\delta+r_2)-\gamma(n-r_2)-r_1\delta\sigma/2} \, 2^{-r_1\{\delta(\sigma/2-1)-\gamma\}}$$

---

[27]Implying in particular, along with a result of Ribet, Fermat's last theorem.

and using inequality (7.20) along with $\zeta(\sigma) \leqslant \sigma/(\sigma - 1)$ and following the ideas of [Lou00], we may write

$$\kappa_{\mathbb{K}} \leqslant \frac{|d_{\mathbb{K}}|^{(\sigma-1)/2}}{(\sigma-1)^{n-1}} F(\sigma)$$

with

$$F(\sigma) = \sigma^{n+1+(\sigma-1)(r_1\delta+r_2)-\gamma(n-r_2)-r_1\delta\sigma/2} 2^{r_2(1-\sigma)-r_1\{\delta(\sigma/2-1)-\gamma\}} \pi^{r_2-n\sigma/2}.$$

Now we have for all $\sigma \geqslant 1$ and $n \geqslant 4$

$$2\sigma^2 \left(\frac{F'}{F}\right)'(\sigma) = 2n(\gamma - 1) + r_1\delta(\sigma + 2) + 2r_2(\sigma + 1 - \gamma) - 2$$

$$= n\{\delta\sigma + 2(\delta + \gamma - 1)\} + 2r_2\{\sigma(1 - \delta) + 1 - 2\delta - \gamma\} - 2$$

$$\geqslant n(3\delta + 2\gamma - 2) + 2r_2(2 - 3\delta - \gamma) - 2$$

$$\geqslant n\gamma - 2 > 0.$$

Since F is positive, this implies that F is convex on $[1, +\infty[$ so that we have for $\sigma \in [1, 2]$

$$F(\sigma) \leqslant \max\left(F(1), F(2)\right)$$

with

$$F(1) = 2^{(n-2r_2)(\gamma+\delta/2)} \pi^{r_2-n/2} \leqslant \left(\frac{2^{\gamma+\delta/2}}{\sqrt{\pi}}\right)^n < 2^{(n-1)/2}$$

$$F(2) = 2^{n+1-r_2\gamma} \pi^{r_2-n} \leqslant 2\left(\frac{2^{1-\gamma/2}}{\sqrt{\pi}}\right)^n < 2^{(n-1)/2}$$

for all $n \geqslant 3$. Now we choose

$$\sigma = 1 + \frac{2(n-1)}{\log |d_{\mathbb{K}}|}.$$

By Corollary 7.130, we have $1 < \sigma < 2$ as soon as $n \geqslant 3$, and hence using the inequality above, we get

$$\kappa_{\mathbb{K}} < \left(\frac{e}{\sqrt{2}} \frac{\log |d_{\mathbb{K}}|}{n-1}\right)^{n-1}$$

for $n \geqslant 4$. Using the class number formula (7.21), we infer the following result.

Proposition 7.138 Let $\mathbb{K}/\mathbb{Q}$ be an algebraic number field of degree $n \geqslant 4$, discriminant $d_{\mathbb{K}}$, class number $h_{\mathbb{K}}$, regulator $\mathcal{R}_{\mathbb{K}}$ and let $w_{\mathbb{K}}$ be the number of roots of unity *in* $\mathbb{K}$. *Then we have*

$$h_{\mathbb{K}}\mathcal{R}_{\mathbb{K}} < \frac{w_{\mathbb{K}}}{2}\left(\frac{2}{\pi}\right)^{r_2}\left(\frac{e}{2\sqrt{2}}\frac{\log|d_{\mathbb{K}}|}{n-1}\right)^{n-1}|d_{\mathbb{K}}|^{1/2}.$$

In particular, we have

$$h_{\mathbb{K}}\mathcal{R}_{\mathbb{K}} \ll |d_{\mathbb{K}}|^{1/2}(\log|d_{\mathbb{K}}|)^{n-1}.$$

It is natural to ask for lower bounds of $\kappa_{\mathbb{K}}$. Hecke's integral representation of $\zeta_{\mathbb{K}}(s)$ enables us to prove that, for $0 < \beta < 1$ satisfying $\zeta_{\mathbb{K}}(\beta) \leqslant 0$, then we have (see [Lan94, Nar04])

$$\kappa_{\mathbb{K}} \geqslant \beta(1-\beta)2^{-n}e^{-4\pi n}|d_{\mathbb{K}}|^{(\beta-1)/2}. \tag{7.27}$$

Using this inequality, one may show that, for all $\varepsilon > 0$, there exists $c_\varepsilon > 0$ such that, for all algebraic number fields $\mathbb{K}$ *Galois* over $\mathbb{Q}$, we have

$$\kappa_{\mathbb{K}} \geqslant c_\varepsilon|d_{\mathbb{K}}|^{-\varepsilon}. \tag{7.28}$$

This implies that there exists a constant $c_1 > 0$ such that, for all algebraic number fields $\mathbb{K}$ *Galois* over $\mathbb{Q}$, we have

$$|\log h_{\mathbb{K}}\mathcal{R}_{\mathbb{K}}| \geqslant c_1 \log|d_{\mathbb{K}}|^{1/2}.$$

Using this lower bound with Proposition 7.138, we get the *Brauer–Siegel theorem*.

Theorem 7.139 (Brauer–Siegel) *If* $\mathbb{K}$ *ranges over a sequence of algebraic number fields of degree n* Galois *over* $\mathbb{Q}$ *for which* $n/\log|d_{\mathbb{K}}|$ *tends to 0, then*

$$\log h_{\mathbb{K}}\mathcal{R}_{\mathbb{K}} \sim \log|d_{\mathbb{K}}|^{1/2}.$$

However, the lower bound (7.28) for all fields Galois over $\mathbb{Q}$ is *ineffective*. As in Chap. 3, any attempt at improving it effectively or at providing a value of c_ε for a sufficiently small $\varepsilon > 0$ has been unsuccessful. We have seen above that a possible exceptional zero of $\zeta_{\mathbb{K}}(s)$ causes trouble in the effectiveness of this constant. Stark [Sta74] also observed that algebraic number fields of small degrees, especially quadratic fields, are a real obstacle to any improvement.

We end this section by pointing out that one may improve on (7.27) in some special cases. For instance, if $\mathbb{K}/\mathbb{Q}$ is a totally imaginary algebraic number field of degree $n \geqslant 4$ and discriminant $d_{\mathbb{K}}$ satisfying $|d_{\mathbb{K}}| \geqslant 2683^n$, then Louboutin [Lou03] showed that

$$\kappa_{\mathbb{K}} \geqslant (1-\beta)|d_{\mathbb{K}}|^{(\beta-1)/2}$$

if $\zeta_{\mathbb{K}}(\beta) \leqslant 0$ for some $1 - 2/\log|d_{\mathbb{K}}| \leqslant \beta < 1$.

7.5.3 Computations of Galois Groups

The purpose of this section is to supply some usual tools from Galois theory to help compute some Galois groups of Galois extensions. We refer the reader to [Lan93, Pra04, Soi81] for more information and some proofs. We use group-theoretic notation for the usual transitive subgroups of $\mathcal{S}_n$, the *symmetric group* on $\{1, \ldots, n\}$. For instance $\mathcal{A}_n$ is the *alternating group* on $\{1, \ldots, n\}$, C_n is the *cyclic group* of order n, D_n is the *dihedral group* of order $2n$ which is the group of symmetries of a regular n-gon, and so on. For any positive integer n and prime number p, $\mathrm{PSL}(n, p)$ means $\mathrm{PSL}_n(\mathbb{F}_p)$.

Let $\mathbb{K}/\mathbb{Q}$ be a Galois extension of $\mathbb{Q}$ and $P \in \mathbb{Z}[X]$ be a defining polynomial of $\mathbb{K}$. Since P is irreducible, $\mathrm{Gal}(\mathbb{K}/\mathbb{Q}) \simeq \mathrm{Gal}(P/\mathbb{Q})$ is transitive considered as a subgroup of $\mathcal{S}_n$, i.e. for all $i, j \in \{1, \ldots, n\}$, there exists $\sigma \in \mathrm{Gal}(P/\mathbb{Q})$ such that $\sigma(i) = j$. The notation $a \in \mathbb{Q}^2$ means that a is a square in $\mathbb{Q}$.

The first criterion enables us to check whether $\mathrm{Gal}(P/\mathbb{Q}) \subseteq \mathcal{A}_n$ or not.

Lemma 7.140 $\mathrm{Gal}(P/\mathbb{Q}) \subseteq \mathcal{A}_n$ *if and only if* $\mathrm{disc}(P) \in \mathbb{Q}^2$.

Proof Since P is irreducible over $\mathbb{Q}$, the roots $\theta_1, \ldots, \theta_n$ of P are all distinct. By Definition 7.36, we have $\mathrm{disc}(P) = d^2$ with $d \neq 0$ and

$$d = \prod_{1 \leqslant i < j \leqslant n} (\theta_i - \theta_j).$$

If d is an algebraic integer, then, for all $\sigma \in \mathrm{Gal}(P/\mathbb{Q})$, we have $\sigma(d) = \epsilon(d)d$ where $\epsilon(d)$ is the signature of σ.

▷ If $\sigma \in \mathcal{A}_n$, then $\epsilon(d) = 1$ and therefore $\sigma(d) = d$ and we infer $d \in \mathbb{Z}$ from Galois theory.

▷ If $d \in \mathbb{Z}$, we have $\sigma(d) = d$ because σ fixes $\mathbb{Q}$. Since $d \neq 0$, we get $\epsilon(d) = 1$ and thus $\sigma \in \mathcal{A}_n$.

The proof is complete. □

The next tool, due to Dedekind, relies on the factorization of $\overline{P}$ in $\mathbb{F}_p[X]$ for some suitable prime number p.

Proposition 7.141 (Dedekind) *Let* $P \in \mathbb{Z}[X]$ *and* $p \nmid \mathrm{disc}(P)$ *be a prime number. Assume that in* $\mathbb{F}_p[X]$ *we have the factorization*

$$\overline{P} = \prod_{i=1}^{g} \overline{P_i}$$

where $\overline{P_i}$ *are irreducible polynomials over* $\mathbb{F}_p$. *Then* $\mathrm{Gal}(P/\mathbb{Q})$ *contains a permutation which is the product of distinct cycles* σ_i *of length* $\deg \overline{P_i}$.

For Galois groups over $\mathbb{F}_p$, the following proposition, which is a particular case of a result due to Frobenius, relies on the ramification of p in the corresponding Galois extension.

Proposition 7.142 (Frobenius) *Let $\mathbb{K}/\mathbb{Q}$ be a Galois extension of $\mathbb{Q}$ of degree n and $P \in \mathbb{Z}[X]$ be a defining monic irreducible polynomial of $\mathbb{K}$. Let $p \nmid \mathrm{disc}(P)$ be a prime number with inertial degree f_p and assume that $\overline{P}$ is squarefree in $\mathbb{F}_p[X]$. Then $\mathrm{Gal}(\overline{P}/\mathbb{F}_p)$ is the cyclic group of order f_p.*

The following tools are often useful in the determination of certain Galois groups.

Lemma 7.143

(i) *Let H be a transitive subgroup of $\mathcal{S}_n$. If H contains a transposition and a $(n-1)$-cycle, then $H = \mathcal{S}_n$.*

(ii) *Let p be a prime number and H be a subgroup of $\mathcal{S}_p$. If H contains a transposition and an element of order p, then $H = \mathcal{S}_p$.*

(iii) *Let $P \in \mathbb{Q}[X]$ such that $\deg P = p$ is a prime number. If P has exactly two non-real roots, then $\mathrm{Gal}(P/\mathbb{Q}) \simeq \mathcal{S}_p$.*

One may also notice the next lemma due to Jordan.

Lemma 7.144 (Jordan) *Let H be a transitive subgroup of $\mathcal{S}_n$. If H contains a p-cycle for some prime number p satisfying $n/2 < p < n-2$, then H contains $\mathcal{A}_n$.*

When the degree of P is small, it is easy to compute $\mathrm{Gal}(P/\mathbb{Q})$.

Lemma 7.145 *Let $P \in \mathbb{Q}[X]$ be irreducible.*

(i) *If $P = X^2 + pX + q$, then $\mathrm{Gal}(P/\mathbb{Q}) \simeq C_2$.*

(ii) *If $P = X^3 + pX + q$, then $\mathrm{Gal}(P/\mathbb{Q}) \simeq \begin{cases} \mathcal{A}_3 \simeq C_3, & \text{if } \mathrm{disc}(P) \in \mathbb{Q}^2 \\ \mathcal{S}_3, & \text{otherwise.} \end{cases}$*

(iii) *If $P = X^4 - 2bX^2 + b^2 - ac^2$ with $a, b, c \in \mathbb{Q}$ such that $a \notin \mathbb{Q}^2$ and $a(b^2 - ac^2) \in \mathbb{Q}^2$, then*

$$\mathrm{Gal}(P/\mathbb{Q}) \simeq C_4.$$

(iv) *If $P = X^4 - 2(a+b)X^2 + (a-b)^2$ with $a, b \in \mathbb{Q} \setminus \{0\}$ such that $a, b, ab^{-1} \notin \mathbb{Q}^2$, then*

$$\mathrm{Gal}(P/\mathbb{Q}) \simeq C_2 \times C_2 = V_4.$$

V_4 *is called the* **Klein 4-group.**

Resolvents

Let $F \in \mathbb{Z}[X_1, \ldots, X_n]$, G be a subgroup of $\mathcal{S}_n$ and $P \in \mathbb{Z}[X]$ with roots $\alpha_1, \ldots, \alpha_n$. The *stabilizer* H of F in G is the group

$$H = \{\sigma \in G : F(X_{\sigma(1)}, \ldots, X_{\sigma(n)}) = F(X_1, \ldots, X_n)\}.$$

The *resolvent* $\operatorname{Res}_G(F, P)$ associated with F and P is the polynomial defined by

$$\operatorname{Res}_G(F, P) = \prod_{\sigma \in G/H} \left(X - F(\alpha_{\sigma(1)}, \ldots, \alpha_{\sigma(n)})\right)$$

where the product ranges over $|G/H|$ cosets representatives of G/H. When $G = \mathcal{S}_n$, we omit the subscript in the notation $\operatorname{Res}_G(F, P)$. By the fundamental theorem on symmetric polynomials, the coefficients of $\operatorname{Res}_G(F, P)$ can be expressed as polynomials over $\mathbb{Q}$ in the coefficients of P. If P is monic, then these coefficients are algebraic integers, and hence rational integers. We also notice that $\operatorname{Res}_G(F, P)$ is independent of the ordering of the roots of P. An important special case is the *linear resolvent polynomial* defined with

$$F(X_1, \ldots, X_n) = \sum_{i=1}^{n} a_i X_i \quad (a_i \in \mathbb{Q}).$$

The resolvents are often used in computational algebra in order to construct algorithms to compute Galois groups of polynomials. Most of these algorithms rest on the following result (see [Coh93] for instance).

Proposition 7.146 *If* $\operatorname{Res}_G(F, P)$ *has a* simple *root in* $\mathbb{Z}$, *then* $\operatorname{Gal}(P/\mathbb{Q})$ *is conjugate under* G *to a subgroup of* H.

In practice, one does not need to compute explicitly the resolvent. It suffices to compute numerical approximations of the roots of P and determine numerically $F(\alpha_{\sigma(1)}, \ldots, \alpha_{\sigma(n)})$. These approximations are in general accurate enough to guarantee that we can correctly recognize when $\operatorname{Res}_G(F, P)$ has a simple root in $\mathbb{Z}$. Algorithms up to degree 7 are detailed in [Coh93] where the choices of polynomials F and corresponding systems of representatives of G/H are also given. It should be noticed that, for polynomials of degree 7, one can use the following simple resolvent

$$R = \prod_{1 \leqslant i < j < k \leqslant 7} (X - \alpha_i - \alpha_j - \alpha_k)$$

which is a polynomial of degree 35. It is an exercise in Galois theory to show that, if $R = R_1 R_2$ with R_i irreducibles such that $\deg R_1 = 7$ and $\deg R_2 = 28$, then

$$\operatorname{Gal}(P/\mathbb{Q}) \simeq \operatorname{PSL}(3, 2) \simeq \operatorname{PSL}(2, 7)$$

which is the unique simple group of order 168. For instance, this is the case for the polynomial $P = X^7 - 7X^3 + 14X^2 - 7X + 1$.

Examples

▶ deg $P = 4$. There are up to conjugacy five transitive subgroups of $\mathcal{S}_4$, i.e.

$$\mathcal{S}_4, \ \mathcal{A}_4, \ \mathcal{D}_4, \ \mathcal{V}_4, \ \mathcal{C}_4.$$

With $F = X_1 X_2 + X_3 X_4$ and $G = \mathcal{S}_4$, a system of representatives of G/H is given by

$$G/H = \{\mathrm{Id}, (12), (14)\}$$

and, if $P = X^4 + aX^3 + bX^2 + cX + d$, we have

$$\mathrm{Res}_G(F, P) = X^3 - bX^2 - (ac - 4d)X - (a^2 d + 4bd + c^2).$$

If $\mathrm{Res}_G(F, P)$ has no root in $\mathbb{Z}$, then $\mathrm{Gal}(P/\mathbb{Q}) \simeq \mathcal{S}_4$ or $\mathcal{A}_4$ by Proposition 7.146, and one can use Lemma 7.140 to determine precisely $\mathrm{Gal}(P/\mathbb{Q})$. If $\mathrm{Res}_G(F, P)$ has a root in $\mathbb{Z}$, then $\mathrm{Gal}(P/\mathbb{Q}) \simeq \mathcal{C}_4$, $\mathcal{V}_4$ or $\mathcal{D}_4$. We have $\mathcal{V}_4 \subset \mathcal{D}_4 \cap \mathcal{A}_4$ and hence if $\mathrm{disc}(P) \in \mathbb{Q}^2$, then $\mathrm{Gal}(P/\mathbb{Q}) \simeq \mathcal{V}_4$, otherwise $\mathrm{Gal}(P/\mathbb{Q}) \simeq \mathcal{C}_4$ or $\mathcal{D}_4$. One uses another resolvent to distinguish between the two.

When $P = X^4 + aX^2 + b$, we have the following more precise result.

Proposition 7.147 *Let $P = X^4 + aX^2 + b$ be irreducible over $\mathbb{Q}$ with roots $\pm\alpha$ and $\pm\beta$. We have*

$$\mathrm{Gal}(P/\mathbb{Q}) \simeq \begin{cases} \mathcal{C}_4, & \text{if } \alpha\beta^{-1} - \alpha^{-1}\beta \in \mathbb{Q} \\ \mathcal{V}_4, & \text{if } \alpha\beta \in \mathbb{Q} \text{ or } \alpha^2 - \beta^2 \in \mathbb{Q} \\ \mathcal{D}_4, & \text{otherwise.} \end{cases}$$

▶ deg $P = 5$. There are up to conjugacy five transitive subgroups of $\mathcal{S}_5$, i.e.

$$\mathcal{S}_5, \ \mathcal{A}_5, \ \mathcal{D}_5, \ M_{20} = \langle(12345), (2354)\rangle, \ \mathcal{C}_5.$$

We have the inclusions $\mathcal{C}_5 \subset \mathcal{D}_5 \subset \mathcal{A}_5 \cap M_{20}$. One uses a first resolvent with $G = \mathcal{S}_5$ and $H = M_{20}$ so that

$$G/H = \{\mathrm{Id}, (12), (13), (14), (15), (25)\}$$

and if $\mathrm{Res}_G(F, P)$ has no root in $\mathbb{Z}$, then $\mathrm{Gal}(P/\mathbb{Q}) \simeq \mathcal{S}_5$ or $\mathcal{A}_5$, and use again Lemma 7.140 to determine precisely $\mathrm{Gal}(P/\mathbb{Q})$. If $\mathrm{Res}_G(F, P)$ has a root in $\mathbb{Z}$ and $\mathrm{disc}(P) \notin \mathbb{Q}^2$, then $\mathrm{Gal}(P/\mathbb{Q}) \simeq M_{20}$, otherwise $\mathrm{Gal}(P/\mathbb{Q}) \simeq \mathcal{C}_5$ or $\mathcal{D}_5$. One uses again another resolvent to finalize the determination.

▶ deg $P = 6$. There are up to conjugacy sixteen transitive subgroups of $\mathcal{S}_6$. This case has been completely solved in [Hag00] where the author uses specializations of three resolvents of degrees 2, 10 and 15 denoted respectively by f_2, f_{10} and f_{15}. Now f_2 is just the polynomial $f_2 = X^2 - \mathrm{disc}(P)$ and the coefficients of f_{10} and f_{15} are given in [Hag00], and hence are known. Furthermore, there are twelve solvable subgroups of $\mathcal{S}_6$, i.e. the groups $\mathcal{C}_3^2 \rtimes \mathcal{D}_4$ and $\mathcal{S}_4 \times \mathcal{C}_2$ and their proper subgroups. As an example, let us give one of the results of this article.

Theorem 7.148 *Let $P \in \mathbb{Z}[X]$ be an irreducible polynomial of degree 6 with discriminant D and set $G = \mathrm{Gal}(P/\mathbb{Q})$.*

(i) *G is solvable if and only one of the following statements holds.*
   ▷ *f_{10} has a rational root and then $G \subseteq C_3^2 \rtimes D_4$.*
   ▷ *f_{15} has a rational root with multiplicity $\neq 5$.*
   ▷ *f_{15} has a rational root with multiplicity 5 and f_{10} is a product of quartic and sextic irreducible polynomials.*
(ii) *Assume that $G \simeq C_3^2 \rtimes D_4$, $C_3^2 \rtimes C_4$, $D_3 \times D_3$ or $C_3 \times D_3$. Then*

$$G \simeq C_3^2 \rtimes C_4 \quad \Longleftrightarrow \quad D \in \mathbb{Q}^2.$$

(iii) *If G is not solvable, then*
   ▷ *$G \simeq \mathcal{S}_6$ if and only if f_{15} is irreducible over $\mathbb{Q}$ and $D \notin \mathbb{Q}^2$.*
   ▷ *$G \simeq \mathcal{A}_6$ if and only if f_{15} is irreducible over $\mathbb{Q}$ and $D \in \mathbb{Q}^2$.*
   ▷ *$G \simeq \mathcal{S}_5 \simeq \mathrm{PGL}(2,5)$ if and only if f_{15} is reducible over $\mathbb{Q}$ and $D \notin \mathbb{Q}^2$.*
   ▷ *$G \simeq \mathcal{A}_5 \simeq \mathrm{PSL}(2,5)$ if and only if f_{15} is reducible over $\mathbb{Q}$ and $D \in \mathbb{Q}^2$.*

We end this section with a theorem due to Frobenius which is a particular case of Chebotarëv's density theorem (see [Nar04] for instance). We shall say that a subset S of prime numbers has a *Dirichlet density d*, or *analytic density d*, if

$$\frac{\sum_{p \in S} p^{-\sigma}}{\sum_p p^{-\sigma}} \longrightarrow d \quad \text{for} \quad \sigma \downarrow 1.$$

Proposition 7.149 *Let $P \in \mathbb{Z}[X]$ irreducible. Then the Dirichlet density of the set of prime numbers satisfying $p \nmid \mathrm{disc}(P)$ for which $\overline{P}$ splits completely in $\mathbb{F}_p[X]$ is equal to $|\mathrm{Gal}(P/\mathbb{Q})|^{-1}$.*

Example 7.150

1. $P = X^5 + 4X^3 + 7X^2 + 2X + 9$. Since $\overline{P} = X^5 + X^2 + \overline{1}$ is irreducible in $\mathbb{F}_2[X]$, we infer that P is irreducible over $\mathbb{Z}$ by Corollary 7.26. We also have $\mathrm{disc}(P) = 2503 \times 7759$ so that $\mathrm{Gal}(P/\mathbb{Q}) \not\subseteq \mathcal{A}_5$ by Lemma 7.140. Furthermore, we have

$$P \equiv X(X+1)(X^3 + 2X^2 + 2X + 2) \ (\mathrm{mod}\,3)$$

and

$$P \equiv (X+1)(X^4 + 2X^3 + 3X^2 + 3X + 3) \ (\mathrm{mod}\,5)$$

so that $\mathrm{Gal}(P/\mathbb{Q}) \simeq \mathcal{S}_5$ using Table 5C of [Soi81].
2. $P = X^6 + 2X^5 + 3X^4 + 4X^3 + 5X^2 + 6X + 7$. By Example 7.31, P is irreducible over $\mathbb{Z}$. We also have

$$P \equiv (X+1)(X^5 + 4X^4 + 11X^3 + 5X + 3) \ (\mathrm{mod}\,13)$$

so that $5 \mid \mathrm{Gal}(P/\mathbb{Q})$ by Proposition 7.141. Hence $\mathrm{Gal}(P/\mathbb{Q})$ lies among the following four groups

$$\mathrm{PSL}(2,5), \ \mathrm{PGL}(2,5), \ \mathcal{A}_6, \ \mathcal{S}_6$$

which are the only transitive subgroups of $\mathcal{S}_6$ having an order divisible by 5. Since

$$\mathrm{disc}(P) = -2^{16} \times 7^4$$

we have $\mathrm{Gal}(P/\mathbb{Q}) \not\subseteq \mathcal{A}_6$ by Lemma 7.140 and thus $\mathrm{Gal}(P/\mathbb{Q}) \not\simeq \mathrm{PSL}(2,5), \mathcal{A}_6$. Since $\mathrm{PGL}(2,5)$ and $\mathcal{S}_6$ are non-solvable groups, we infer that $\mathrm{Gal}(P/\mathbb{Q})$ is not solvable. One may compute the polynomial f_{15} of Theorem 7.148 and prove that it is reducible. Since $\mathrm{disc}(P)$ is not a square, we deduce from Theorem 7.148 that

$$\mathrm{Gal}(P/\mathbb{Q}) \simeq \mathrm{PGL}(2,5) \simeq \mathcal{S}_5.$$

Note that using the PARI/GP system, one may check that, of the first $10\,000$ prime numbers distinct from 2 and 7, P splits completely relative to 78 of them, which gives a proportion of $1/128$. Proposition 7.149 says that this proportion tends to $1/120$.

7.5.4 The Prime Ideal Theorem and the Ideal Theorem

Let $\mathbb{K}/\mathbb{Q}$ be an algebraic number field of degree n and $\pi_\mathbb{K}(x)$ be the number of prime ideals $\mathfrak{p}$ in $\mathcal{O}_\mathbb{K}$ such that $\mathcal{N}_{\mathbb{K}/\mathbb{Q}}(\mathfrak{p}) \leqslant x$, i.e.

$$\pi_\mathbb{K}(x) = \sum_{\mathcal{N}_{\mathbb{K}/\mathbb{Q}}(\mathfrak{p}) \leqslant x} 1.$$

Note that $\pi_\mathbb{Q}(x) = \pi(x)$ is the usual prime counting function.[28] The so-called *Ideal Theorem*, seen in Corollary 4.59, follows from classical contour integration methods. Similarly, this method may be also used to estimate $\pi_\mathbb{K}(x)$ in the same way as it was used in the rational case (see Theorem 3.85). The first version of the *Prime Ideal Theorem*, or PIT, was proved by Landau in [Lan03]. Subsequently, the error-term was gradually improved to an order of magnitude similar to that of (3.37) with a Vinogradov–Korobov type zero-free region as it was seen in Theorem 7.124 (ii). This gives the following result.

Theorem 7.151 (Prime Ideal Theorem) *For x sufficiently large, we have*

$$\pi_\mathbb{K}(x) = \mathrm{Li}(x) + O_\mathbb{K}\left\{ x \exp\left(-c_\mathbb{K}(\log x)^{3/5}(\log\log x)^{-1/5} \right) \right\}$$

where the implied constant and $c_\mathbb{K} > 0$ depend on the usual invariants of $\mathbb{K}$.

---

[28] See Definition 3.39.

Effective versions of the PIT have been stated by many authors. For instance, we quote the following estimate.

Theorem 7.152 (Effective Prime Ideal Theorem) *For all $x \geqslant e^{c_0 n \log^2 \sqrt{|d_{\mathbb{K}}|}}$, we have*

$$\left| \pi_{\mathbb{K}}(x) - \mathrm{Li}(x) \right| \leqslant \mathrm{Li}(x^{\beta_0}) + c_1 x \, \exp\left(-c_2 \sqrt{\frac{\log x}{n}} \right)$$

where the constants c_i are absolute and the term containing β_0 is present if and only if $\zeta_{\mathbb{K}}(s)$ has an exceptional zero β_0 in the region

$$1 - \frac{1}{4 \log |d_{\mathbb{K}}|} \leqslant \sigma < 1.$$

The analogue of Corollaries 3.50 and 3.51 in algebraic number fields does also exist [Leb07].

Theorem 7.153 (Mertens theorems in number fields) *There exist absolute constants $c_0, \ldots, c_4 > 0$ satisfying the following assertions.*

(i) *For all $x \geqslant \max(e^{c_0 n \log^2 \sqrt{|d_{\mathbb{K}}|}}, e^{1024 c_1^{-2} n \log^2(\sqrt{n}/c_2)})$, we have*

$$\sum_{\mathcal{N}_{\mathbb{K}/\mathbb{Q}}(\mathfrak{p}) \leqslant x} \frac{1}{\mathcal{N}_{\mathbb{K}/\mathbb{Q}}(\mathfrak{p})} = \log \log x + B_{\mathbb{K}} + R_{\mathbb{K}}(x)$$

with

$$B_{\mathbb{K}} = \log \kappa_{\mathbb{K}} + \gamma + \sum_{\mathfrak{p}} \left\{ \log\left(1 - \frac{1}{\mathcal{N}_{\mathbb{K}/\mathbb{Q}}(\mathfrak{p})} \right) + \frac{1}{\mathcal{N}_{\mathbb{K}/\mathbb{Q}}(\mathfrak{p})} \right\}$$

and

$$\left| R_{\mathbb{K}}(x) \right| \leqslant \frac{c_3}{\beta_0 \log x} \left(\frac{2 - \beta_0}{1 - \beta_0} \right) + \frac{2 c_4}{\log x}$$

and the term containing β_0 is present if and only if $\zeta_{\mathbb{K}}(s)$ has an exceptional zero β_0 in the region

$$1 - \frac{1}{4 \log |d_{\mathbb{K}}|} \leqslant \sigma < 1.$$

(ii) *For x sufficiently large, we have*

$$\prod_{\mathcal{N}_{\mathbb{K}/\mathbb{Q}}(\mathfrak{p}) \leqslant x} \left(1 - \frac{1}{\mathcal{N}_{\mathbb{K}/\mathbb{Q}}(\mathfrak{p})} \right)^{-1} = \kappa_{\mathbb{K}} e^{\gamma} \log x \left\{ 1 + O_{\mathbb{K}}\left(\frac{1}{\log x} \right) \right\}.$$

It should be noticed that there are few explicit upper bounds for the sum $\sum_{m \leqslant x} \nu_{\mathbb{K}}(m)$ in the literature. However, using elementary means based upon the geometry of numbers, the authors [MO07] derived the following result. Let r be the Dirichlet rank of $\mathcal{O}_{\mathbb{K}}^*$ and pick up fundamental units $\varepsilon_1, \ldots, \varepsilon_r$. Suppose that, for all $(i, j) \in \{1, \ldots, r\}^2$, we have $|\log|\sigma_i(\varepsilon_j)|| \leqslant R$. Then for all $x \geqslant 4$ we have

$$\sum_{m \leqslant x} v_{\mathbb{K}}(m) \leqslant C_{\mathbb{K}} \kappa_{\mathbb{K}} x \qquad (7.29)$$

where $C_{\mathbb{K}} = 2^{r_2 - r_1} n! (5\gamma e^{Rr})^n |d_{\mathbb{K}}| \mathcal{R}_{\mathbb{K}}^{-1}$.

7.5.5 Abelian Extensions and the Kronecker–Weber Theorem

The study of cyclotomic fields arose naturally in the early days of algebraic number theory owing to Fermat's equation. In order to generalize, one may ask which are the possible abelian extensions of $\mathbb{Q}$, i.e. Galois extensions $\mathbb{K}/\mathbb{Q}$ such that $\mathrm{Gal}(\mathbb{K}/\mathbb{Q})$ is abelian. Stated in this way, that question can certainly not be solved. But a great achievement of the 19th century is in the next important result which generalizes Proposition 7.109.

Theorem 7.154 (Kronecker–Weber) *If $\mathbb{K}/\mathbb{Q}$ is a finite abelian extension, then there exists a positive integer f such that $\mathbb{K} \subseteq \mathbb{Q}(\zeta_f)$.*

This result was first stated by Kronecker in 1853 who provided an incomplete proof which reveals some difficulties with extensions of degree 2^α for some α. In 1886, Weber gave the first proof but also had an error[29] at 2. Both authors used the theory of Lagrange resolvents. Later, Hilbert, and then Speiser, used ramification theory to give a proof which is now often considered as the classic one. It is noteworthy that Hilbert's strategy works partly because $\mathbb{Q}$ does not have any proper unramified abelian extension. Nowadays, many proofs do exist in the literature, mostly based upon Hilbert's method [Gre74], localization methods [Was82] or on the fact that they are a simple consequence of results belonging to class field theory.

We shall not prove this result and refer the reader to [Mol99, Rib01] for proofs using ramification theory. Nevertheless, it seems interesting to have a look at a particular case of Theorem 7.154 using the following lemma.

Lemma 7.155 (Reduction principle) *Assume that Theorem 7.154 is true for abelian number fields with prime power degrees. Then it is true for all abelian number fields.*

Proof Let $\mathbb{K}/\mathbb{Q}$ be an abelian number field with Galois group $G = \mathrm{Gal}(\mathbb{K}/\mathbb{Q})$. Since G is abelian and setting $|G| = p_1^{\alpha_1} \cdots p_r^{\alpha_r}$, we have by (3.1)

$$G \simeq H_1 \oplus \cdots \oplus H_r$$

---

[29]This error remained unnoticed for about 90 years.

where the H_i are the p_i-Sylow subgroups of G and hence $|H_i| = p_i^{\alpha_i}$. Set

$$E_i = \bigoplus_{j \neq i} H_j.$$

The E_i are subgroups of G and let $\mathbb{K}_i$ be the fixed field of E_i. From Galois theory, we have

$$[\mathbb{K}_i : \mathbb{Q}] = (G : E_i) = |H_i| = p_i^{\alpha_i}$$

and any automorphism that fixes the compositum $\mathbb{K}_1 \cdots \mathbb{K}_r$ fixes all of the $\mathbb{K}_i$ so that

$$\mathrm{Gal}(\mathbb{K}/\mathbb{K}_1 \cdots \mathbb{K}_r) \subseteq \bigcap_{i=1}^{r} E_i = \{0\}$$

which implies that $\mathbb{K} = \mathbb{K}_1 \cdots \mathbb{K}_r$. Now by assumption there exists a primitive root of unity ζ_i of order s_i such that $\mathbb{K}_i \subseteq \mathbb{Q}(\zeta_i)$ and hence

$$\mathbb{K} = \mathbb{K}_1 \cdots \mathbb{K}_r \subseteq \mathbb{Q}(\zeta_1, \ldots, \zeta_r) \subseteq \mathbb{Q}(\zeta_{[s_1, \ldots, s_r]})$$

which concludes the proof.   □

Now suppose that $\mathbb{K}/\mathbb{Q}$ is a finite abelian extension satisfying

$$[\mathbb{K} : \mathbb{Q}] = p^\alpha \quad \text{and} \quad d_\mathbb{K} = p^\beta$$

for some odd prime number p and positive integers α and β.

Define $\mathbb{L} = \mathbb{Q}(\zeta_{p^{\alpha+1}})$ with degree $[\mathbb{L} : \mathbb{Q}] = \varphi(p^{\alpha+1}) = p^\alpha(p - 1)$ and

$$d_\mathbb{L} = (-1)^{p^\alpha(p-1)/2} p^{p^\alpha(p(\alpha+1)-\alpha-2)}.$$

Let $\mathbb{F}$ be the fixed field of the subgroup H of $\mathrm{Gal}(\mathbb{L}/\mathbb{Q})$ of order $p - 1$ so that $[\mathbb{F} : \mathbb{Q}] = p^\alpha$. Since p is odd, $\mathrm{Gal}(\mathbb{L}/\mathbb{Q})$ is cyclic, then G/H is also cyclic and hence $\mathbb{F}/\mathbb{Q}$ is a cyclic number field. Furthermore, if q is a prime dividing $d_\mathbb{F}$, then q ramifies in $\mathbb{F}$ and then ramifies in $\mathbb{L}$, so that $q = p$ and $d_\mathbb{F}$ is a power of p.

We consider the compositum $\mathbb{K}\mathbb{F}$, see Fig. 7.1.

We have

$$[\mathbb{K}\mathbb{F} : \mathbb{Q}] = [\mathbb{K}\mathbb{F} : \mathbb{F}][\mathbb{F} : \mathbb{Q}] = [\mathbb{K} : \mathbb{K} \cap \mathbb{F}][\mathbb{F} : \mathbb{Q}]$$

so that $[\mathbb{K}\mathbb{F} : \mathbb{Q}]$ is a power of p. As above, let q be a prime dividing $d_{\mathbb{K}\mathbb{F}}$, and hence q is ramified in $\mathbb{K}\mathbb{F}$, which implies that q is ramified in $\mathbb{K}$ or q is ramified in $\mathbb{F}$, and therefore $q = p$ and $d_{\mathbb{K}\mathbb{F}}$ is a power of p. Next, it can be shown [Rib01] that $\mathbb{K}\mathbb{F}$ is a cyclic number field. We infer that the subgroup $\mathrm{Gal}(\mathbb{K}\mathbb{F}/\mathbb{K} \cap \mathbb{F})$ of $\mathrm{Gal}(\mathbb{K}\mathbb{F}/\mathbb{Q})$ is cyclic. Now we have

$$\mathrm{Gal}(\mathbb{K}\mathbb{F}/\mathbb{K} \cap \mathbb{F}) \simeq \mathrm{Gal}(\mathbb{K}/\mathbb{K} \cap \mathbb{F}) \oplus \mathrm{Gal}(\mathbb{F}/\mathbb{K} \cap \mathbb{F})$$

Fig. 7.1 The compositum $\mathbb{K}\mathbb{F}$

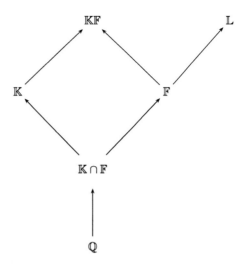

and using [Rib01, Lemma 2 p. 277] we infer that one of the groups of the right-hand side is trivial, so that either $\mathbb{K} = \mathbb{K} \cap \mathbb{F}$ or $\mathbb{F} = \mathbb{K} \cap \mathbb{F}$. In the first case, we deduce that $\mathbb{K} \subseteq \mathbb{F}$ whereas the second case gives $\mathbb{F} \subseteq \mathbb{K}$, but since they have the same degree, we get $\mathbb{F} = \mathbb{K}$ in both cases and hence

$$\mathbb{K} = \mathbb{F} \subseteq \mathbb{L} = \mathbb{Q}(\zeta_{p^{\alpha+1}}).$$

Using Proposition 7.109 and Lemma 7.155, we get the following result.

Lemma 7.156 *Let* $\mathbb{K} = \mathbb{K}_1 \cdots \mathbb{K}_r$ *be an abelian number field such that either* $[\mathbb{K}_i : \mathbb{Q}] = p_i^{\alpha_i}$ *and* $d_{\mathbb{K}_i} = p_i^{\beta_i}$ *for some odd primes* p_i *or* $[\mathbb{K}_i : \mathbb{Q}] = 2$. *Then* $\mathbb{K}$ *is contained in a cyclotomic field.*

7.5.6 Class Field Theory over $\mathbb{Q}$

From the discussion above, it turns out that describing all abelian extensions of an algebraic number field is a long-standing problem. In examining Abel's work, Kronecker observed that certain abelian extensions of imaginary quadratic fields may be generated by adjoining certain values of *automorphic functions* arising from elliptic curves. Roughly speaking, let $\mathcal{H} = \{s \in \mathbb{C} : \operatorname{Im} s > 0\}$ be the Poincaré half-plane and T be a subgroup of $\mathrm{SL}(2, \mathbb{Z})$. An automorphic function (for T) is a meromorphic function f defined on $\mathcal{H}$ having a Fourier series expansion of the form $f(z) = \sum_{n=N}^{\infty} a_n e(nz)$ for some $N \in \mathbb{Z}$, and satisfying

$$f\left(\frac{az+b}{cz+d}\right) = f(z)$$

for all $M = \left(\begin{smallmatrix} a & b \\ c & d \end{smallmatrix}\right) \in T$. For instance, the j-function defined on $\mathcal{H}$ by

$$j(z) = \frac{1728\,g_2(z)^3}{g_2(z)^3 - 27g_3(z)^2} \tag{7.30}$$

with

$$g_2(z) = \sum_{m,n} \frac{60}{(m + nz)^4} \quad \text{and} \quad g_3(z) = \sum_{m,n} \frac{140}{(m + nz)^6}$$

and where the summations are over all ordered pairs of integers $(m, n) \neq (0, 0)$, is an automorphic function. It can be shown that if $\mathbb{K} = \mathbb{Q}(\sqrt{-d})$ with $d > 0$ squarefree such that $d \not\equiv 3 \pmod 4$, then $\mathbb{K}(j(\sqrt{-d}))$ is an abelian extension[30] of $\mathbb{K}$.

Kronecker wondered whether *all* finite abelian extensions of imaginary quadratic fields could be obtained in this way.[31] The generalization of this problem was later addressed by Hilbert in 1900 when he presented a series of 23 problems for the new century at the International Congress of Mathematicians in Paris, and the generalization of Kronecker's conjecture is Hilbert's 12th problem.[32]

Historically, the class field theory began when Gauss tried to decide when the congruence equation $x^2 - a \equiv 0 \pmod p$ has a solution, where $p \nmid a$ is a prime. The answer is given by his quadratic reciprocity law stating that, *if p and q are odd primes not dividing a and $p \equiv q \pmod{4a}$, then $x^2 - a \equiv 0 \pmod p$ has a solution if and only if $x^2 - a \equiv 0 \pmod q$ has a solution.* Hence, whether or not the congruence $x^2 - a \equiv 0 \pmod p$ has a solution depends only on the arithmetic progression mod $4a$ to which p belongs.

Proposition 7.108 also shows that an odd prime p splits completely in a quadratic field $\mathbb{Q}(\sqrt{d})$ if and only if $p \nmid d$ and the congruence $x^2 - d \equiv 0 \pmod p$ has a solution.

Let $\mathbb{K} = \mathbb{Q}(\zeta_m)$ where $m \in \mathbb{N}$. It can be proved (see [Was82, Theorem 2.13]) that, if $p \neq m$ is a prime number and f is the smallest positive integer such that $p^f \equiv 1 \pmod m$, then $(p) = \mathfrak{p}_1 \cdots \mathfrak{p}_g$ with $g = \varphi(m)/f$. In particular, p splits completely in $\mathbb{Q}(\zeta_m)$ if and only if $p \equiv 1 \pmod m$.

All these examples above show that we have a decomposition theory in terms of congruence conditions. Therefore, the purposes of class field theory over $\mathbb{Q}$ are threefold:

1. Describe every finite abelian extension of $\mathbb{Q}$ in terms of the arithmetic of $\mathbb{Q}$.
2. Canonically realize the abelian group $\mathrm{Gal}(\mathbb{K}/\mathbb{Q})$ in terms of the arithmetic of $\mathbb{Q}$.
3. Describe the decomposition of a prime in terms of the arithmetic of $\mathbb{Q}$, i.e. by giving congruence conditions.

---

[30] This is in fact the *Hilbert class field* of $\mathbb{K}$. See Theorem 7.168.

[31] This is now referred to as *Kronecker Jurgendtraum* i.e. Kronecker's dream of his youth. This was completely proved by Takagi in 1920.

[32] Among the 23 problems introduced by Hilbert, some of them have been solved and others are still open. Note that Hilbert's 8th problem is the Riemann hypothesis.

In what follows, we intend to state the main results of class field theory over $\mathbb{Q}$ following [Gar81].

▶ Let p be a prime number and $\mathbb{F}_p$ be the field with p elements. By Fermat's little theorem (Theorem 3.15), the Frobenius map $x \in \mathbb{F}_p \longmapsto x^p$ is nothing but the identity. We generalize this concept to abelian number fields in the following way.

Proposition 7.157 *Let $\mathbb{K}/\mathbb{Q}$ be an abelian extension with a defining monic polynomial $P \in \mathbb{Z}[X]$ and let p be a prime number such that $p \nmid \mathrm{disc}(P)$. Then there exists an element $\phi \in \mathrm{Gal}(\mathbb{K}/\mathbb{Q})$ such that the Frobenius map in $\mathbb{F}_p$ is the reduction of ϕ modulo p. This element ϕ is called the* Artin symbol *and is denoted by* $\left(\frac{\mathbb{K}/\mathbb{Q}}{p}\right)$.

Example Let $\mathbb{K} = \mathbb{Q}(\theta)$ where θ is a root of the polynomial $P = X^3 - 3X - 1$. It can easily be checked that P is irreducible over $\mathbb{Q}$ since P has no integer root. Since $\mathrm{disc}(P) = 3^4 \in \mathbb{Q}^2$, we have $\mathrm{Gal}(\mathbb{K}/\mathbb{Q}) \simeq \mathcal{A}_3 \simeq C_3$ by Lemma 7.140, so that $\mathbb{K}/\mathbb{Q}$ is cyclic, and let σ be a generator of $\mathrm{Gal}(\mathbb{K}/\mathbb{Q})$. Since $-\theta^2 + 2$ and $\theta^2 - \theta - 2$ are the two other roots of P, set $\sigma(\theta) = -\theta^2 + 2$ so that $\sigma^2(\theta) = \theta^2 - \theta - 2$ and $\mathrm{Gal}(\mathbb{K}/\mathbb{Q}) = \{\mathrm{Id}, \sigma, \sigma^2\}$. Now we have

$$\theta^2 \equiv -\theta^2 + 2 \equiv \sigma(\theta) \ (\mathrm{mod}\, 2)$$

and hence

$$\left(\frac{\mathbb{K}/\mathbb{Q}}{2}\right) = \sigma.$$

Similarly, since

$$\theta^5 = \theta^3 \theta^2 = (1 + 3\theta)\theta^2 = \theta^2 + 3\theta^3 = \theta^2 + 9\theta + 3 \equiv \theta^2 - \theta - 2 \equiv \sigma^2(\theta) \ (\mathrm{mod}\, 5)$$

we have

$$\left(\frac{\mathbb{K}/\mathbb{Q}}{5}\right) = \sigma^2.$$

One may continue in this way to determine other values for primes not dividing 3. However, it seems more appropriate to look for a rule which can dictate the Artin symbol for all primes $p \neq 3$. This is the purpose of the *Artin reciprocity law*.

▶ Let $\mathbb{K}/\mathbb{Q}$ be an abelian number field. By Theorem 7.154, there exists a positive integer f such that $\mathbb{K} \subseteq \mathbb{Q}(\zeta_f)$. Such an integer is called an *admissible modulus* or *cycle* of $\mathbb{K}$.

Now let $a = \prod_{p \mid a} p^{v_p(a)} \in \mathbb{N}$ such that $(a, f) = 1$ where f is an admissible modulus of $\mathbb{K}$. The *Artin map* is defined by

$$\mathrm{Art}_{\mathbb{K}/\mathbb{Q}}: \quad (\mathbb{Z}/f\mathbb{Z})^* \quad \longrightarrow \quad \mathrm{Gal}(\mathbb{K}/\mathbb{Q})$$

$$\bar{a} \quad \longmapsto \quad \prod_{p\mid a}\left(\frac{\mathbb{K}/\mathbb{Q}}{p}\right)^{v_p(a)}.$$

Example Let m be a positive integer, $\mathbb{K} = \mathbb{Q}(\zeta_m)$ and let $p \nmid m$ be a prime number. By Proposition 7.157, we have $(\frac{\mathbb{Q}(\zeta_m)/\mathbb{Q}}{p}) = \sigma_p$ where, for all $a \in \mathbb{N}$ such that $(a, m) = 1$, σ_a is defined by $\sigma_a(\zeta_m) = \zeta_m^a$. Now if $a = \prod_p p^{v_p(a)} \in \mathbb{N}$ such that $(a, m) = 1$, then

$$\left(\frac{\mathbb{Q}(\zeta_m)/\mathbb{Q}}{a}\right) = \prod_p\left(\frac{\mathbb{Q}(\zeta_m)/\mathbb{Q}}{p}\right)^{v_p(a)} = \prod_p \sigma_p^{v_p(a)} = \sigma_a.$$

Furthermore, we also have

$$\left(\frac{\mathbb{Q}(\zeta_m)/\mathbb{Q}}{a}\right)(\zeta_m) = \zeta_m \iff \sigma_a(\zeta_m) = \zeta_m \iff \zeta_m^a = \zeta_m \iff a \equiv 1 \pmod{m}$$

which is called the *cyclotomic reciprocity law*.

It can be proved that the Artin map is *surjective*. One of the main purposes of the theory is then the study of the kernel denoted by $I_{\mathbb{K},f}$. Using $\mathrm{Gal}(\mathbb{Q}(\zeta_f)/\mathbb{Q}) \simeq (\mathbb{Z}/f\mathbb{Z})^*$, we have $I_{\mathbb{K},f} \simeq \mathrm{Gal}(\mathbb{Q}(\zeta_f)/\mathbb{K})$ and from Galois theory we have the correspondence

$$
\begin{array}{ccccc}
\mathbb{Q}(\zeta_f) & \supseteq & \mathbb{K} & \supseteq & \mathbb{Q} \\
\updownarrow & & \updownarrow & & \updownarrow \\
\mathrm{Gal}(\mathbb{Q}(\zeta_f)/\mathbb{Q}(\zeta_f)) & \subseteq & \mathrm{Gal}(\mathbb{Q}(\zeta_f)/\mathbb{K}) & \subseteq & \mathrm{Gal}(\mathbb{Q}(\zeta_f)/\mathbb{Q}) \\
\updownarrow & & \updownarrow & & \updownarrow \\
\{1\} & \subseteq & I_{\mathbb{K},f} & \subseteq & (\mathbb{Z}/f\mathbb{Z})^*
\end{array}
$$

so that $\mathbb{K}$ is the fixed field of $I_{\mathbb{K},f}$. This proves the first result of class field theory over $\mathbb{Q}$.

Theorem 7.158 (Artin reciprocity) *Let $\mathbb{K}/\mathbb{Q}$ be an abelian number field with an admissible modulus f. Then the following sequence is exact*

$$1 \longrightarrow I_{\mathbb{K},f} \hookrightarrow (\mathbb{Z}/f\mathbb{Z})^* \longrightarrow \mathrm{Gal}(\mathbb{K}/\mathbb{Q}) \longrightarrow 1$$

so that

$$\mathrm{Gal}(\mathbb{K}/\mathbb{Q}) \simeq (\mathbb{Z}/f\mathbb{Z})^*/I_{\mathbb{K},f}.$$

In other words, Theorem 7.158 realizes canonically $\mathrm{Gal}(\mathbb{K}/\mathbb{Q})$ in terms of arithmetic of $\mathbb{Q}$ and each abelian extension of $\mathbb{Q}$ is described in terms of arithmetic of $\mathbb{Q}$.

► The minimal admissible modulus of $\mathbb{K}$ is called the *conductor* of $\mathbb{K}$ and is denoted by $f_\mathbb{K}$. Since $\mathbb{K} \subseteq \mathbb{Q}(\zeta_f) \cap \mathbb{Q}(\zeta_{f_\mathbb{K}}) = \mathbb{Q}(\zeta_{(f,f_\mathbb{K})})$, we infer that $f_\mathbb{K} \mid f$. We have the following properties of the conductor [Gar81].

Lemma 7.159

(i) *Let $d \in \mathbb{Z}$ be a squarefree number such that $|d| > 1$. Then $|d_\mathbb{K}|$ is the conductor of $\mathbb{K} = \mathbb{Q}(\sqrt{d})$.*

(ii) *Let $\mathbb{K}/\mathbb{Q}$ be a cyclic extension with odd prime degree p. Then*

$$f_\mathbb{K} = p^\theta p_1 \cdots p_r$$

for some positive integer r, with $\theta \in \{0, 2\}$ and p_i are prime numbers satisfying $p_i \equiv 1 \pmod{p}$.

We have the following result on ramification.

Theorem 7.160 (Conductor-ramification theorem) *Let $\mathbb{K}/\mathbb{Q}$ be an abelian number field with conductor $f_\mathbb{K}$ and p be a prime number. Then p is ramified in $\mathbb{K}$ if and only if $p \mid f_\mathbb{K}$.*

This implies in particular that, if p is a prime number such that $p \nmid f$, then p is unramified in $\mathbb{Q}(\zeta_f)$. This may be generalized to any finite abelian extension of $\mathbb{Q}$.

Theorem 7.161 (Decomposition theorem) *Let $\mathbb{K}/\mathbb{Q}$ be an abelian number field with an admissible modulus f and let $p \nmid f$. Then the order of $\overline{p}$ in $(\mathbb{Z}/f\mathbb{Z})^*/I_{\mathbb{K},f}$ is its inertial degree f_p in $\mathbb{K}$.*

It is customary to denote by $\mathrm{Spl}(\mathbb{K}/\mathbb{Q})$ the set of prime numbers which split completely in $\mathbb{K}$. Using $e_p f_p g_p = [\mathbb{K} : \mathbb{Q}]$, we have

$$p \in \mathrm{Spl}(\mathbb{K}/\mathbb{Q}) \iff g_p = [\mathbb{K} : \mathbb{Q}] \iff e_p = f_p = 1$$

and applying Theorems 7.160 and 7.161 with $f = f_\mathbb{K}$, we get

$$p \in \mathrm{Spl}(\mathbb{K}/\mathbb{Q}) \iff p \nmid f_\mathbb{K} \text{ and } p \in I_{\mathbb{K},f_\mathbb{K}}. \tag{7.31}$$

We infer that

$$p \in \mathrm{Spl}(\mathbb{K}/\mathbb{Q}) \iff \exists \overline{a} \in I_{\mathbb{K},f_\mathbb{K}} : p \equiv a \pmod{f_\mathbb{K}}.$$

This clearly accomplishes the third goal of class field theory.

► Let $\mathbb{K}/\mathbb{Q}$ be an abelian extension with an admissible modulus f. We define the *character group* of $\mathbb{K}$ to be the group $X_f(\mathbb{K})$ of characters χ of $(\mathbb{Z}/f\mathbb{Z})^*$ such that $\chi(a) = 1$ for all $a \in I_{\mathbb{K},f}$. We first extend χ in a Dirichlet character modulo f, denoted again by χ, by setting $\chi(a) = 0$ for all a such that $(a, f) > 1$, and then to a primitive Dirichlet character. If $\chi \in X_f(\mathbb{K})$, we denote f_χ its conductor. The character group $X_f(\mathbb{K})$ satisfies the following properties (see [Nar04, Proposition 8.4]).

a. $X_f(\mathbb{K}) \simeq \mathrm{Gal}(\mathbb{K}/\mathbb{Q})$ and hence $|X_f(\mathbb{K})| = [\mathbb{K} : \mathbb{Q}]$.

b. $X_f(\mathbb{K}) \simeq (\mathbb{Z}/f\mathbb{Z})^* \Longleftrightarrow \mathbb{K} = \mathbb{Q}(\zeta_f)$ and $X_f(\mathbb{K}) = \{\chi_0\} \Longleftrightarrow \mathbb{K} = \mathbb{Q}$.

c. If $r_2 = 0$, then every character of $X_f(\mathbb{K})$ is even. If $r_1 = 0$, then $X_f(\mathbb{K})$ contains the same number of even and odd characters, and the set of even characters is the subgroup of $X_f(\mathbb{K})$ equal to $X_f(\mathbb{K}^+)$, where $\mathbb{K}^+$ is the maximal real subfield of $\mathbb{K}$.

The next result is another highlight of class field theory.

Theorem 7.162 (Conductor-discriminant formula) *Let $\mathbb{K}/\mathbb{Q}$ be an abelian extension with admissible modulus f and character group $X_f(\mathbb{K})$. Then $f_\mathbb{K} = \mathrm{lcm}\{f_\chi : \chi \in X_f(\mathbb{K})\}$ and*

$$|d_\mathbb{K}| = \prod_{\chi \in X_f(\mathbb{K})} f_\chi.$$

In particular, we have $f_\mathbb{K} \mid d_\mathbb{K}$ and hence we always have the following tower

$$\mathbb{Q} \subseteq \mathbb{K} \subseteq \mathbb{Q}(\zeta_{f_\mathbb{K}}) \subseteq \mathbb{Q}(\zeta_{|d_\mathbb{K}|}).$$

7.5.7 The Class Number Formula for Abelian Extensions

The Dirichlet class number formula (7.26) for quadratic fields may be generalized to any finite abelian extension of $\mathbb{Q}$ in the following way. Recall that, if p is a prime, then f_p and g_p are respectively its inertial degree and decomposition number in $\mathbb{K}$.

Theorem 7.163 (Class number formula) *Let $\mathbb{K}/\mathbb{Q}$ be an abelian number field with conductor $f_\mathbb{K}$ and character group $X_{f_\mathbb{K}}(\mathbb{K})$ denoted by X for convenience. Then*

$$h_\mathbb{K} \mathcal{R}_\mathbb{K} = \frac{\varphi(f_\mathbb{K})}{f_\mathbb{K}} \frac{|d_\mathbb{K}|^{1/2} w_\mathbb{K}}{2^{r_1}(2\pi)^{r_2}} \prod_{p \mid f_\mathbb{K}} \left(1 - \frac{1}{p^{f_p}}\right)^{-g_p} \prod_{\substack{\chi \in X \\ \chi \neq \chi_0}} L(1, \chi).$$

In particular, if $\mathbb{K}/\mathbb{Q}$ is cyclic with prime degree, then

$$h_\mathbb{K} \mathcal{R}_\mathbb{K} = \frac{|d_\mathbb{K}|^{1/2} w_\mathbb{K}}{2^{r_1}(2\pi)^{r_2}} \prod_{\substack{\chi \in X \\ \chi \neq \chi_0}} L(1, \chi).$$

Proof The starting point is the following identity sometimes called the *product formula for characters* (see [Hin08, Lemme IV.4.3] for a proof).

Let G be a finite abelian group and $\widehat{G}$ be the group of characters of G. If $a \in G$ is an element of order r, then we have for all z

$$\prod_{\chi \in \widehat{G}} \left(1 - \chi(a)z\right) = \left(1 - z^r\right)^{|G|/r}.$$

Let $p \nmid f_{\mathbb{K}}$ so that p is unramified in $\mathbb{Q}(\zeta_{f_{\mathbb{K}}})$. Applying the product formula with $z = p^{-\sigma}$ where $\sigma > 1$, $G = \mathrm{Gal}(\mathbb{K}/\mathbb{Q})$, $a = p$ so that $r = f_p$ by Theorems 7.158 and 7.161, we get

$$\prod_{\chi \in X} \left(1 - \frac{\chi(p)}{p^\sigma}\right)^{-1} = \left(1 - \frac{1}{p^{f_p \sigma}}\right)^{-g_p}$$

and multiplying out through all primes p such that $p \nmid f_{\mathbb{K}}$, we obtain

$$\prod_{\chi \in X} L(\sigma, \chi) = \prod_{p \nmid f_{\mathbb{K}}} \left(1 - \frac{1}{p^{f_p \sigma}}\right)^{-g_p}$$

so that

$$\zeta_{\mathbb{K}}(\sigma) = \prod_{p} \left(1 - \frac{1}{p^{f_p \sigma}}\right)^{-g_p} = \prod_{p \mid f_k} \left(1 - \frac{1}{p^{f_p \sigma}}\right)^{-g_p} \prod_{\chi \in X} L(\sigma, \chi).$$

Now splitting the second product into the cases $\chi = \chi_0$ and $\chi \neq \chi_0$ and using

$$L(\sigma, \chi_0) = \zeta(\sigma) \prod_{p \mid f_{\mathbb{K}}} \left(1 - \frac{1}{p^\sigma}\right)$$

we get

$$\frac{\zeta_{\mathbb{K}}(\sigma)}{\zeta(\sigma)} = \prod_{p \mid f_{\mathbb{K}}} \left(1 - \frac{1}{p^\sigma}\right)\left(1 - \frac{1}{p^{f_p \sigma}}\right)^{-g_p} \prod_{\substack{\chi \in X \\ \chi \neq \chi_0}} L(\sigma, \chi).$$

By Corollary 3.71, the function $L(s, \chi)$ is analytic in the half-plane $\sigma > 0$ as long as $\chi \neq \chi_0$. Hence letting $\sigma \longrightarrow 1^+$ in the previous identity and taking the residues of $\zeta(s)$ and $\zeta_{\mathbb{K}}(s)$ at $s = 1$ into account gives the first asserted result. The second proposition follows from the fact that if $\mathbb{K}/\mathbb{Q}$ is cyclic with prime degree, then $f_p = g_p = 1$ for all $p \mid f_{\mathbb{K}}$ and hence

$$\frac{\varphi(f_{\mathbb{K}})}{f_{\mathbb{K}}} \prod_{p \mid f_{\mathbb{K}}} \left(1 - \frac{1}{p^{f_p}}\right)^{-g_p} = \frac{\varphi(f_{\mathbb{K}})}{f_{\mathbb{K}}} \prod_{p \mid f_{\mathbb{K}}} \left(1 - \frac{1}{p}\right)^{-1} = 1$$

as required. $\square$

Remark 7.164 The class number formula may be expressed in a simpler form if we use the *primitive* characters instead. It is shown in [Nar04, page 416] that, if $p \mid f_{\mathbb{K}}$, then for all $\sigma > 1$, we still have

$$\prod_{\chi \in X} \left(1 - \frac{\chi^\star(p)}{p^\sigma} \right) = \left(1 - \frac{1}{p^{f_p \sigma}} \right)^{g_p}$$

where $\chi^\star$ is the primitive Dirichlet character that induces χ. This result along with (3.24) implies that, for all $\sigma > 1$, we have

$$\zeta_{\mathbb{K}}(\sigma) = \prod_{p \mid f_k} \left(1 - \frac{1}{p^{f_p \sigma}} \right)^{-g_p} \prod_{\chi \in X} L(\sigma, \chi)$$

$$= \prod_{\chi \in X} L\left(\sigma, \chi^\star\right) \prod_{p \mid f_k} \left(1 - \frac{1}{p^{f_p \sigma}} \right)^{-g_p} \prod_{\chi \in X} \left(1 - \frac{\chi^\star(p)}{p^\sigma} \right)$$

$$= \prod_{\chi \in X} L\left(\sigma, \chi^\star\right) = \zeta(\sigma) \prod_{\substack{\chi \in X \\ \chi \neq \chi_0}} L\left(\sigma, \chi^\star\right)$$

and arguing as above we obtain

$$h_{\mathbb{K}} \mathcal{R}_{\mathbb{K}} = \frac{|d_{\mathbb{K}}|^{1/2} w_{\mathbb{K}}}{2^{r_1} (2\pi)^{r_2}} \prod_{\substack{\chi \in X \\ \chi \neq \chi_0}} L\left(1, \chi^\star\right)$$

if $\mathbb{K}/\mathbb{Q}$ is abelian, and where $\chi^\star$ is the primitive Dirichlet character that induces χ.

7.5.8 *Primes of the Form $x^2 + ny^2$—Particular Cases*

The class field theory over $\mathbb{Q}$ may be generalized to any finite abelian extension $\mathbb{L}/\mathbb{K}$, but since the Kronecker–Weber theorem is only valid with the ground field $\mathbb{Q}$, the concept of admissible modulus must be rewritten.

The starting point is a generalization of the notion of prime ideals of an algebraic number field in the following way. Let $\mathbb{K}$ be an algebraic number field of degree n. An *absolute value* of $\mathbb{K}$ is a map $|\cdot| : \mathbb{K} \longrightarrow \mathbb{R}$ satisfying

(i) $|x| \geqslant 0$ and $|x| = 0 \Longleftrightarrow x = 0$.
(ii) $|xy| = |x||y|$.
(iii) $|x + y| \leqslant |x| + |y|$.

If we replace the condition (iii) by the stronger condition

(iv) $|x + y| \leqslant \max(|x|, |y|)$

then the absolute value is said to be *non-archimedean*, and *archimedean* otherwise. If $|\cdot|_1$ and $|\cdot|_2$ are two absolute values such that there exist constants $0 < c_0 \leqslant c_1$ such that $c_0 |x|_1 \leqslant |x|_2 \leqslant c_1 |x|_1$ for all $x \in \mathbb{K}$, then the absolute values are equivalent, and the set of equivalence classes is called the *places* of $\mathbb{K}$. There are two types of places.

▷ **The finite places.** If $\mathfrak{p}$ is a prime ideal of $\mathcal{O}_{\mathbb{K}}$ and p is the prime below $\mathfrak{p}$, then we define for all $x \in \mathbb{K}$

$$|x|_{\mathfrak{p}} = p^{-v_{\mathfrak{p}}(x)/e_{\mathfrak{p}}}$$

where $e_{\mathfrak{p}}$ is the ramification index of $\mathfrak{p}$ in $\mathbb{K}$ and $v_{\mathfrak{p}}(x) = v_{\mathfrak{p}}(x\mathcal{O}_{\mathbb{K}})$. This absolute value is non-archimedean and if $\mathfrak{p}_1 \neq \mathfrak{p}_2$, then $|\cdot|_{\mathfrak{p}_1}$ and $|\cdot|_{\mathfrak{p}_2}$ are not equivalent. Furthermore, any non-archimedean absolute value is equivalent to one of these.
▷ **The infinite places.** Let σ be one of the embeddings of $\mathbb{K}$. Then we define for all $x \in \mathbb{K}$

$$|x|_{\sigma} = |\sigma(x)|.$$

If $\sigma(\mathbb{K}) \subseteq \mathbb{R}$, then the place is called real, complex otherwise. For a complex place σ, the conjugate $\overline{\sigma}$ defines the same place. We infer that the number of infinite places of $\mathbb{K}$ is given by $r_1 + r_2$. For instance, if $\mathbb{K} = \mathbb{Q}$, there is only one infinite place, often denoted by ∞, given by $|x|_{\mathrm{Id}} = |x|$ where the right-hand side is the ordinary absolute value on $\mathbb{Q}$.

It can be proved that there is no other place for a number field $\mathbb{K}$, and therefore there are infinitely many finite places and finitely many infinite places in $\mathbb{K}$.

Let $\mathbb{L}/\mathbb{K}$ be a finite extension *unramified at all places*, which is then equivalent to the two following assertions. If $P = X^n + a_{n-1}X^{n-1} + \cdots + a_0 \in \mathcal{O}_{\mathbb{K}}[X]$ is a defining polynomial of $\mathbb{L}/\mathbb{K}$, then

(i) Every prime ideal of $\mathbb{K}$ is unramified in $\mathbb{L}$, or equivalently the relative discriminant $\mathfrak{D}_{\mathbb{L}/\mathbb{K}}$ is equal to $\mathcal{O}_{\mathbb{K}}$.
(ii) Let σ_i be the embeddings of $\mathbb{K}$ in $\mathbb{L}$. Then either $\sigma_i(\mathbb{K}) \not\subseteq \mathbb{R}$ or $\sigma_i(\mathbb{K}) \subseteq \mathbb{R}$ and P^{σ_i} has no non-real roots, where $P^{\sigma_i} = X^n + \sigma_i(a_{n-1})X^{n-1} + \cdots + \sigma_i(a_0)$.

Note that if $r_1 = 0$, then no infinite place ramifies in $\mathbb{L}$, so that $\mathbb{L}/\mathbb{K}$ unramified at all places is *equivalent* to the point (i) above. On the other hand, in the case $r_1 \geqslant 1$, we say that $\mathbb{L}/\mathbb{K}$ is unramified *outside* ∞ if and only if the sole point (i) above is satisfied.

Example The extension $\mathbb{Q}(\zeta_{12})/\mathbb{Q}(\sqrt{3})$ is unramified outside ∞ since

$$\mathfrak{D}_{\mathbb{Q}(\zeta_{12})/\mathbb{Q}(\sqrt{3})} = \mathbb{Z}[\sqrt{3}]$$

but is not unramified since the two infinite places ramify in $\mathbb{Q}(\zeta_{12})$. Indeed, we have

$$P = X^2 - X\sqrt{3} + 1$$

and the two infinite places of $\mathbb{Q}(\sqrt{3})$ are $\{\mathrm{Id}, \sigma\}$ where $\sigma(a + b\sqrt{3}) = a - b\sqrt{3}$. We have $\sigma_i(\mathbb{Q}(\sqrt{3})) \subseteq \mathbb{R}$ and the polynomials P^{σ_i} have non-real roots.

We have previously seen that Hilbert's proof of the Kronecker–Weber theorem works in part thanks to the fact that $\mathbb{Q}$ has no unramified abelian extension larger than $\mathbb{Q}$. It is certainly the reason why Hilbert focused on unramified abelian extensions. However, it should be mentioned that, at that time, the word "unramified" meant at finite places only, so that Hilbert's class fields were studied in the *narrow sense*, i.e. the Galois group is isomorphic to the narrow ideal class group, where fractional ideals are identified if and only if their ratio is a principal ideal having a totally positive generator.[33]

The next result, conjectured by Hilbert, is an important tool in class field theory.

Theorem 7.165 (Hilbert class field) *Let $\mathbb{K}$ be a number field. Then there exists a unique maximal unramified abelian extension of $\mathbb{K}$ denoted by $\mathbb{K}(1)/\mathbb{K}$, in the sense that each unramified abelian extension of $\mathbb{K}$ is isomorphic to a sub-extension of $\mathbb{K}(1)$. The extension $\mathbb{K}(1)/\mathbb{K}$ is called the* Hilbert class field *of $\mathbb{K}$ and satisfies the following properties.*

(i) $\mathrm{Gal}(\mathbb{K}(1)/\mathbb{K}) \simeq \mathrm{Cl}(\mathbb{K})$ *and therefore $h_{\mathbb{K}} = [\mathbb{K}(1) : \mathbb{K}]$.*

(ii) *Let $\mathfrak{p}$ be a prime ideal in $\mathbb{K}$. Then $\mathfrak{p}$ splits completely in $\mathbb{K}(1)$ if and only if $\mathfrak{p}$ is principal in $\mathbb{K}$.*

(iii) *Every integral ideal of $\mathbb{K}$ becomes principal as an integral ideal of $\mathbb{K}(1)$.*

Example The following extensions are examples of Hilbert's class fields over their ground fields.

$$\mathbb{Q}(\sqrt{2}, \sqrt{5}) \qquad \mathbb{Q}(\sqrt{-23}, \alpha) \qquad \mathbb{Q}(\sqrt{-14}, \sqrt{2\sqrt{2}-1}) \qquad \mathbb{Q}(\theta, \sqrt{u(5-\theta^2)})$$

$$\Big|\, 2 \qquad\qquad \Big|\, 3 \qquad\qquad\qquad \Big|\, 4 \qquad\qquad\qquad\qquad \Big|\, 2$$

$$\mathbb{Q}(\sqrt{10}) \qquad\quad \mathbb{Q}(\sqrt{-23}) \qquad\qquad \mathbb{Q}(\sqrt{-14}) \qquad\qquad\qquad \mathbb{Q}(\theta)$$

Here we have $\alpha^3 = \alpha - 1$, $\theta^3 = 11$ and $u = 89 + 40\theta + 18\theta^2$. The last example is picked up from [Jan96].

Hilbert proved the existence of the Hilbert class field of $\mathbb{K}$ when $\mathbb{K}$ is a quadratic field with class number 2. Hilbert's student Furtwängler proved the point (i) in 1907 and the point (iii), called the *capitulation property*, in 1930.

The point (ii) may be generalized as follows.

Proposition 7.166 *If $\mathfrak{p}$ is a prime ideal of $\mathbb{K}$ and if f is the smallest power of $\mathfrak{p}$ such that $\mathfrak{p}^f$ is a principal ideal of $\mathbb{K}$, then $\mathfrak{p}$ splits into $h_{\mathbb{K}}/f$ distinct prime ideals of $\mathbb{K}(1)$ with degree f.*

The next result studies the normality of the extension $\mathbb{K}(1)/\mathbb{Q}$.

---

[33] For a definition of totally positive numbers in algebraic number fields, see [Nar04, page 44].

Lemma 7.167 *If $\mathbb{K}/\mathbb{Q}$ is Galois, then $\mathbb{K}(1)/\mathbb{Q}$ is Galois.*

Proof Considering $\mathbb{K}$ and $\mathbb{K}(1)$ in a fixed algebraic closure $\overline{\mathbb{Q}}$ of $\mathbb{Q}$, let $\sigma \in \mathrm{Gal}(\overline{\mathbb{Q}}/\mathbb{Q})$. Since $\mathbb{K}$ is Galois over $\mathbb{Q}$, we have $\sigma(\mathbb{K}) = \mathbb{K}$. But $\sigma(\mathbb{K}(1))$ is the Hilbert class field of $\sigma(\mathbb{K}) = \mathbb{K}$, so that using the maximality of $\mathbb{K}(1)$, we infer that $\sigma(\mathbb{K}(1)) = \mathbb{K}(1)$ and hence $\mathbb{K}(1)$ is Galois over $\mathbb{Q}$. $\qquad\square$

One of the most beautiful applications of the Hilbert class field lies in the solution to the following problem. Let $n \in \mathbb{N}$ be a squarefree integer satisfying

$$n \not\equiv 3 \ (\mathrm{mod}\,4). \tag{7.32}$$

We ask for necessary and sufficient conditions for a prime number p to be expressed in the form $p = x^2 + ny^2$ for some integers x, y. We use the imaginary quadratic field $\mathbb{K} = \mathbb{Q}(\sqrt{-n})$ and notice that (7.32) and Proposition 7.64 imply that $\mathcal{O}_{\mathbb{K}} = \mathbb{Z}[\sqrt{-n}]$. The theoretic solution to this problem is given by the following result.

Theorem 7.168 *Let $n \in \mathbb{N}$ be a squarefree number satisfying (7.32). Let $\mathbb{K} = \mathbb{Q}(\sqrt{-n})$ and $\mathbb{K}(1)$ be the Hilbert class field of $\mathbb{K}$. If p is an odd prime such that $p \nmid n$, then*

$$p = x^2 + ny^2 \quad \Longleftrightarrow \quad p \in \mathrm{Spl}\big(\mathbb{K}(1)/\mathbb{Q}\big).$$

Proof First note that $d_{\mathbb{K}} = -4n$ by (7.32) and Proposition 7.64, and hence since $p \nmid n$ is odd, we infer that $p \nmid d_{\mathbb{K}}$ and therefore p is unramified in $\mathbb{K}$.

▷ Assume that $p = x^2 + ny^2$. Then $(p) = (x + \sqrt{-n}\,y)(x - \sqrt{-n}\,y) = \mathfrak{p}\overline{\mathfrak{p}}$, say. Since p is unramified, we get $\mathfrak{p} \neq \overline{\mathfrak{p}}$. Conversely, if $(p) = \mathfrak{p}\overline{\mathfrak{p}}$ with $\mathfrak{p}$ principal, then $\mathfrak{p} = (x + \sqrt{-n}\,y)$ for some integers x, y since $\mathcal{O}_{\mathbb{K}} = \mathbb{Z}[\sqrt{-n}]$. This implies that $(p) = (x^2 + ny^2)$, and then $p = x^2 + ny^2$. At this step, we have then proved

$$p = x^2 + ny^2 \quad \Longleftrightarrow \quad (p) = \mathfrak{p}\overline{\mathfrak{p}},\ \mathfrak{p} \neq \overline{\mathfrak{p}} \text{ and } \mathfrak{p} \text{ principal}.$$

▷ By Theorem 7.165 (ii), we get

$$p = x^2 + ny^2 \quad \Longleftrightarrow \quad (p) = \mathfrak{p}\overline{\mathfrak{p}},\ \mathfrak{p} \neq \overline{\mathfrak{p}} \text{ and } \mathfrak{p} \in \mathrm{Spl}\big(\mathbb{K}(1)/\mathbb{K}\big)$$

which is in turn equivalent to the fact that p splits completely in $\mathbb{K}$ and that some prime ideal of $\mathbb{K}$ dividing p splits completely in $\mathbb{K}(1)$. By Lemma 7.167, we deduce that $\mathbb{K}(1)$ is Galois over $\mathbb{Q}$ which implies that the previous assertion is equivalent to p splitting completely in $\mathbb{K}(1)$. $\qquad\square$

Therefore, the problem is twofold. We first have to determine the Hilbert class field of an imaginary quadratic field. The next elegant theorem gives an answer by using the elliptic modular function j defined in (7.30).

Theorem 7.169 *Let $n \in \mathbb{N}$ be a squarefree number satisfying (7.32). Let $\mathbb{K} = \mathbb{Q}(\sqrt{-n})$ and $\mathbb{K}(1)$ be the Hilbert class field of $\mathbb{K}$. Then*

$$\mathbb{K}(1) = \mathbb{K}\big(j(\sqrt{-n})\big).$$

However, it should be mentioned that in practice it is a very difficult matter to compute these j-invariants, so that many authors provided results in certain particular cases. For instance, Herz [Her66] proved a quite general theorem giving any unramified cyclic cubic extension of an imaginary quadratic field. Also for real quadratic fields $\mathbb{K}$ with class number 2, it is known [CR00] that there exists a proper divisor δ of $d_{\mathbb{K}}$ satisfying $\delta \equiv 0, 1 \pmod 4$ such that $\mathbb{K}(1) = \mathbb{K}(\sqrt{\delta})$.

The second point to solve is to determine which primes split completely in $\mathbb{K}(1)/\mathbb{Q}$. We might use (7.31), but the computation of $I_{\mathbb{K}(1),f}$ is very often uneasy. The following alternative tool provides a useful criterion in this direction (see [Cox89, Proposition 5.29]).

Proposition 7.170 *Let $\mathbb{K}$ be an imaginary quadratic field and $\mathbb{L}$ be a finite extension of $\mathbb{K}$ which is Galois over $\mathbb{Q}$. Then there exists an algebraic integer θ such that $\mathbb{L} = \mathbb{K}(\theta)$ and if $P \in \mathbb{Z}[X]$ is the monic minimal polynomial of θ and p is a prime such that $p \nmid \operatorname{disc}(P)$, then*

$$p \in \operatorname{Spl}(\mathbb{L}/\mathbb{Q}) \quad \Longleftrightarrow \quad \left(\frac{d_{\mathbb{K}}}{p}\right) = 1 \text{ and } P(x) \equiv 0 \pmod p \text{ has an integer solution.}$$

Example 7.171 Let $p \neq 5$ be an odd prime number. We will prove that

$$p = x^2 + 5y^2 \quad \Longleftrightarrow \quad p \equiv 1 \text{ or } 9 \pmod{20}.$$

Proof Let $\mathbb{K} = \mathbb{Q}(\sqrt{-5})$. It is known that $j(\sqrt{-5}) = 282\,880\,\sqrt{5} + 632\,000$ so that $\mathbb{K}(1) = \mathbb{K}(\sqrt{5})$ by Theorem 7.168. Using Theorem 7.169 and Proposition 7.170, we get

$$p = x^2 + 5y^2 \quad \Longleftrightarrow \quad \left(\frac{-20}{p}\right) = 1 \text{ and } x^2 - 5 \equiv 0 \pmod p \text{ has an integer solution}$$

$$\Longleftrightarrow \quad \left(\frac{5}{p}\right) = \left(\frac{-5}{p}\right) = 1$$

where we used $(-20/p) = (4/p) \times (-5/p) = (-5/p)$. Now by the quadratic reciprocity law and since p is odd, we infer that

$$\left(\frac{5}{p}\right) = 1 \quad \Longleftrightarrow \quad \left(\frac{p}{5}\right) = 1 \quad \Longleftrightarrow \quad p \equiv \pm 1 \pmod 5.$$

Using $(-5/p) = (-1/p) \times (5/p) = (-1)^{(p-1)/2}(5/p)$, we get

$$\left(\frac{-5}{p}\right) = 1 \quad \Longleftrightarrow \quad p \equiv 1, 3, 7, 9 \pmod{20}.$$

Hence

$$p = x^2 + 5y^2 \quad \Longleftrightarrow \quad p \equiv \pm 1 \ (\mathrm{mod}\, 5) \text{ and } p \equiv 1,\ 3,\ 7,\ 9 \ (\mathrm{mod}\, 20)$$

implying the stated result. $\qquad\qquad\qquad\qquad\qquad\qquad\qquad\qquad\qquad\qquad\qquad\qquad$ $\square$

7.5.9 Primes of the Form $x^2 + ny^2$—General Case

If the integer n is not squarefree or does not satisfy (7.32), then $\mathbb{Z}[\sqrt{-n}]$ need not be the maximal order of $\mathbb{Q}(\sqrt{-n})$. We then look for a generalization of the Hilbert class field, which will be given by class field theory. Let $\mathcal{O}$ be any order of an imaginary quadratic field of conductor f and class group $\mathrm{Cl}(\mathcal{O})$. Then the existence theorem from class field theory implies that there exists an abelian extension $\mathbb{L}/\mathbb{K}$, called a *ring class field* of $\mathcal{O}$, such that all primes of $\mathbb{K}$ ramified in $\mathbb{L}$ divide $f\mathcal{O}_{\mathbb{K}}$ and the Artin map induces the following isomorphism

$$\mathrm{Gal}(\mathbb{L}/\mathbb{K}) \simeq \mathrm{Cl}(\mathcal{O}).$$

We first provide some information about orders in quadratic fields. In what follows, $\mathbb{K}$ is a quadratic field.

1. An *order* in $\mathbb{K}$ is a free $\mathbb{Z}$-module of rank 2 of $\mathbb{K}$ containing 1.
2. An order $\mathcal{O}$ has always a finite index f in $\mathcal{O}_{\mathbb{K}}$, called the *conductor* of $\mathcal{O}$. Furthermore, if $d_{\mathcal{O}}$ is the discriminant of $\mathcal{O}$, then

$$d_{\mathcal{O}} = f^2 d_{\mathbb{K}}$$

   and if $\mathcal{O} = \mathbb{Z}[\sqrt{-n}]$, then $d_{\mathcal{O}} = -4n$.
3. If $I(\mathcal{O})$ is the set of invertible fractional ideals of the order $\mathcal{O}$ and $P(\mathcal{O})$ is the set of principal ideals of $\mathcal{O}$, then $\mathrm{Cl}(\mathcal{O}) = I(\mathcal{O})/P(\mathcal{O})$ is the *ideal class group*, or *Picard group*, of $\mathcal{O}$ and $h_{\mathcal{O}} = |\mathrm{Cl}(\mathcal{O})|$ is the *class number* of $\mathcal{O}$.
4. If $\mathcal{O} \subsetneq \mathcal{O}_{\mathbb{K}}$, then $\mathcal{O}$ is *not* in general a Dedekind ring.

We summarize the main properties of orders of an imaginary quadratic field in the next lemma.

Lemma 7.172 *Let $n \in \mathbb{N}$ squarefree and $\mathbb{K} = \mathbb{Q}(\sqrt{-n})$ with discriminant $d_{\mathbb{K}}$ and class number $h_{\mathbb{K}}$. Let $\mathcal{O}$ be any order of $\mathbb{K}$ with discriminant $d_{\mathcal{O}}$, class number $h_{\mathcal{O}}$ and conductor f.*

(i) *We have*

$$h_{\mathcal{O}} = \frac{h_{\mathbb{K}} \times f}{[(\mathcal{O}_{\mathbb{K}})^* : \mathcal{O}^*]} \prod_{p \mid f}\left(1 - \left(\frac{d_{\mathbb{K}}}{p}\right)\frac{1}{p}\right)$$

where the index $[(\mathcal{O}_{\mathbb{K}})^* : \mathcal{O}^*]$ *is given by*

$$[(\mathcal{O}_{\mathbb{K}})^* : \mathcal{O}^*] = \begin{cases} 2, & \text{if } n = 1 \\ 3, & \text{if } n = 3 \\ 1, & \text{otherwise.} \end{cases}$$

(ii) *A representative of a class of invertible fractional ideals of* $\mathcal{O}$ *is given by*

$$\left(a, \ \frac{-b + \sqrt{d_\mathcal{O}}}{2} \right)$$

with $0 < a \leqslant \sqrt{-d_\mathcal{O}/3}$, $|b| \leqslant a \leqslant c$ *where* $c = \frac{b^2 - d_\mathcal{O}}{4a} \in \mathbb{N}$ *is such that* $(a, b, c) = 1$. *Furthermore, if* $b \geqslant 0$, *then either* $|b| = a$ *or* $a = c$.

The elliptic modular invariant j seen in (7.30) may be defined on orders of $\mathbb{K}$ in the following way. For all $\tau \in \mathcal{H}$ and orders $(1, \tau)$, we set

$$j\big((1, \tau)\big) = j(\tau).$$

It can be shown that, if $\mathcal{O}$ is any order in $\mathbb{K}$, then $j(\mathcal{O}) \in \mathbb{R}$. Furthermore, if the class of $\mathfrak{a}$ has an order $\leqslant 2$ in $\mathrm{Cl}(\mathcal{O})$, then $j(\mathfrak{a}) \in \mathbb{R}$ (see [Cox89, Exercise 11.1]).

The next result generalizes Theorems 7.168 and 7.169.

Theorem 7.173 *Let n be a positive integer, $\mathbb{K} = \mathbb{Q}(\sqrt{-n})$ and $\mathcal{O} = \mathbb{Z}[\sqrt{-n}]$ with class number $h_\mathcal{O}$. Let $\mathfrak{a}$ be any representative of a class of invertible fractional ideals of $\mathcal{O}$. Then the extension*

$$\mathbb{L} = \mathbb{K}\big(j(\mathfrak{a})\big)$$

is the ring class field of $\mathcal{O}$. Furthermore, if $\mathfrak{a}_1, \ldots, \mathfrak{a}_{h_\mathcal{O}}$ are representatives of the classes of invertible fractional ideals of $\mathcal{O}$, then the polynomial

$$H_\mathcal{O} = \prod_{k=1}^{h_\mathcal{O}} \big(X - j(\mathfrak{a}_k)\big) \in \mathbb{Z}[X]$$

is a defining polynomial of $\mathbb{L}$.

That result contains Theorem 7.169 since, if n is squarefree and satisfies (7.32), then $\mathbb{Z}[\sqrt{-n}] = \mathcal{O}_{\mathbb{K}}$ so that $f = 1$, $d_\mathcal{O} = d_{\mathbb{K}}$, $h_\mathcal{O} = h_{\mathbb{K}}$ and

$$\mathbb{K}\big(j(\mathfrak{a})\big) = \mathbb{K}\big(j(\mathcal{O}_{\mathbb{K}})\big) = \mathbb{K}\big(j(\sqrt{-n})\big) = \mathbb{K}(1).$$

Let $n \in \mathbb{N}$. Proceeding similarly as in Theorem 7.168 (see [Cox89, Theorem 9.4]), one may prove that, if $p \nmid n$ is an odd prime, then

$$p = x^2 + ny^2 \quad \Longleftrightarrow \quad p \in \mathrm{Spl}(\mathbb{L}/\mathbb{Q})$$

where $\mathbb{L}$ is the ring class field of the order $\mathbb{Z}[\sqrt{-n}]$ of $\mathbb{Q}(\sqrt{-n})$. Now applying Proposition 7.170 and Theorem 7.173, we obtain the general solution to the problem of detecting primes of the form $x^2 + ny^2$.

Corollary 7.174 *Let n be a positive integer and $p \nmid n$ be an odd prime. With the notation of Theorem 7.173, we have*

$$p = x^2 + ny^2 \iff \left(\frac{-n}{p}\right) = 1 \text{ and } H_{\mathcal{O}}(x) \equiv 0 \pmod{p} \text{ has an integer solution.}$$

Example 7.175 We wish to determine which primes p can be expressed in the form $p = x^2 + 11y^2$. We work in $\mathbb{K} = \mathbb{Q}(\sqrt{-11})$ with $d_{\mathbb{K}} = -11$ and $h_{\mathbb{K}} = 1$. The order $\mathcal{O} = \mathbb{Z}[\sqrt{-11}]$ has discriminant $d_{\mathcal{O}} = -44$ and conductor $f = \sqrt{44/11} = 2$. Using Lemma 7.172 (i), we get

$$h_{\mathcal{O}} = \frac{1 \times 2}{1} \prod_{p|2} \left(1 - \left(\frac{-11}{p}\right) \frac{1}{p} \right) = 3.$$

By Lemma 7.172 (ii), representatives of the three classes of invertible fractional ideals of $\mathcal{O}$ are given by

$$\mathfrak{a}_1 = (1, \sqrt{-11}), \quad \mathfrak{a}_2 = (3, -1 + \sqrt{-11}) \quad \text{and} \quad \mathfrak{a}_3 = (3, 1 + \sqrt{-11}).$$

Using PARI we infer that

$$H_{\mathcal{O}} = X^3 - aX^2 + bX - \left(2^4 \times 11 \times 17 \times 29 \right)^3 \tag{7.33}$$

where $a = 2^4 \times 1709 \times 41\,057$ and $b = 2^8 \times 3 \times 11^4 \times 24\,049$. As in Example 7.171, we have

$$\left(\frac{-11}{p}\right) = \left(\frac{-1}{p}\right)\left(\frac{11}{p}\right) = (-1)^{(p-1)/2}\left(\frac{11}{p}\right) = \begin{cases} (11/p), & \text{if } p \equiv 1 \pmod 4, \\ -(11/p), & \text{if } p \equiv 3 \pmod 4. \end{cases}$$

Now using the quadratic reciprocity law, if p and q are primes such that $q \equiv 3 \pmod 4$, then

$$\left(\frac{q}{p}\right) = 1 \iff p \equiv \pm \alpha^2 \pmod{4q}$$

for some odd integer α such that $q \nmid \alpha$. This implies that

$$\left(\frac{11}{p}\right) = 1 \iff p \equiv \pm 1, \pm 5, \pm 9, \pm 25, \pm 37 \pmod{44}$$

so that

$$\left(\frac{-11}{p}\right) = 1 \iff p \equiv 1, 3, 5, 9, 15, 23, 25, 27, 31, 37 \pmod{44}.$$

This gives the following result.

Corollary 7.176 *Let $p \neq 11$ be an odd prime number. Then $p = x^2 + 11y^2$ if and only if*

▷ $p \equiv 1, 3, 5, 9, 15, 23, 25, 27, 31, 37 \pmod{44}$
▷ *the equation $x^3 - ax^2 + bx - (2^4 \times 11 \times 17 \times 29)^3 \equiv 0 \pmod{p}$ has an integer solution, where $a = 2^4 \times 1709 \times 41\,057$ and $b = 2^8 \times 3 \times 11^4 \times 24\,049$.*

Remark 7.177 The theory predicts that the constant term of $\mathcal{H}_{\mathcal{O}}$ is indeed a perfect cube, which happens if and only if $3 \nmid d_{\mathcal{O}}$. In this case, Gross and Zagier, improving on an earlier work of Deuring, showed that, if p is a prime factor of the constant term of $\mathcal{H}_{\mathcal{O}}$, then $(d_{\mathcal{O}}/p) \neq 1$ and either $p = 3$ or $p \equiv 2 \pmod 3$. Furthermore, we also have

$$p \leqslant \frac{3|d_{\mathcal{O}}|}{4}$$

which explains why the prime factors of the constant term of $\mathcal{H}_{\mathcal{O}}$ are so small in (7.33).

7.5.10 Analytic Methods for Ideal Classes

Let $\mathbb{K}/\mathbb{Q}$ be an algebraic number field of degree n, signature (r_1, r_2), discriminant $d_{\mathbb{K}}$ and let $\zeta_{\mathbb{K}}(s)$ be its Dedekind zeta-function. Recall that

$$\Gamma_{\mathbb{K}}(s) = \Gamma(s/2)^{r_1} \Gamma(s)^{r_2} \quad \text{and} \quad A_{\mathbb{K}} = 2^{-r_2} \pi^{-n/2} |d_{\mathbb{K}}|^{1/2}$$

where $\Gamma(s)$ is the usual Gamma-function. We have previously seen that bounds for class numbers may require the computation of a minimal positive real number $b_{\mathbb{K}}$ such that every ideal class contains a non-zero integral ideal $\mathfrak{a}$ satisfying $\mathcal{N}_{\mathbb{K}/\mathbb{Q}}(\mathfrak{a}) \leqslant b_{\mathbb{K}}$. From the work of Minkowski who created and used deep results from the geometry of numbers, we know that $b_{\mathbb{K}} = M_{\mathbb{K}} |d_{\mathbb{K}}|^{1/2}$ is admissible, where the Minkowski constant $M_{\mathbb{K}}$ is defined in (7.19). In the 1950s, Rogers, and later Mulholland, improved on Minkowski's bound using essentially the same ideas.

Stark's work as previously seen in Proposition 7.129 was the starting point of the resurgence of analytic methods in algebraic number theory. In the late 1970s, using the functional equation, given below, of the partial Dedekind zeta-function associated to a given ideal class and its conjugate class, Zimmert [Zim81] succeeded in establishing an inequality which supersedes the previous results. The *conjugate class* $\mathcal{C}^\star$ of an ideal class $\mathcal{C}$ is defined by $\mathcal{C}\mathcal{C}^\star = \mathcal{D}$ where $\mathcal{D}$ is the class of the different[34] of $\mathcal{O}_{\mathbb{K}}$.

---

[34]The *different* of $\mathcal{O}_{\mathbb{K}}$ is the integral ideal $\mathcal{D}_{\mathbb{K}/\mathbb{Q}}$ defined as the inverse of the fractional ideal

$$\{y \in \mathbb{K} : \mathrm{Tr}_{\mathbb{K}/\mathbb{Q}}(xy) \in \mathbb{Z} \text{ for all } x \in \mathcal{O}_{\mathbb{K}}\}$$

Let $\zeta(s, \mathcal{C})$ be the *partial Dedekind zeta-function* associated to the ideal class $\mathcal{C}$ defined by

$$\zeta(s, \mathcal{C}) = \sum_{\mathfrak{a} \in \mathcal{C}} \frac{1}{\mathcal{N}_{K/\mathbb{Q}}(\mathfrak{a})^s}.$$

As for $\zeta_K(s)$, this function has a functional equation of a similar form (see [Lan94, Neu10]).

Define $\xi_K(s, \mathcal{C}) = A_K^s \Gamma_K(s) \zeta(s, \mathcal{C})$. Then this function is analytic except for simple poles at $s = 0$, $s = 1$, and

$$\xi_K(s, \mathcal{C}) = \xi_K(1 - s, \mathcal{C}^\star).$$

Now assume that the two Dirichlet series $F(s) = \sum_{n=1}^{\infty} f(n) n^{-s}$ and $G(s) = \sum_{n=1}^{\infty} g(n) n^{-s}$ are such that $f(n), g(n) \geqslant 0$, have finite abscissas of convergence and satisfy the following hypotheses.

1. F and G can be extended analytically in the whole complex plane to meromorphic functions having a simple pole at $s = 1$.
2. They satisfy the functional equation $\xi_F(s) = \xi_G(1 - s)$ where

$$\xi_F(s) = \left(2^{-b} A\right)^s \Gamma\left(\frac{s}{2}\right)^{a-b} \Gamma(s)^b F(s)$$

   and similarly for $\xi_G(s)$, for some positive real number A and integers $a \geqslant b \geqslant 0$.
3. The function $s(s - 1)\xi_F(s)$ is entire of order 1.
4. The residues of $F(s)$ and $G(s)$ at $s = 1$ are equal to κ.

Using tools borrowed from complex analysis and contour integration methods, Zimmert proved the following general inequality.

For all real numbers $x > 0$, $\sigma > 1$ and $0 \leqslant \alpha < \sigma - 1$, we have

$$G(\sigma)\left\{\log x - 2\frac{\xi_G'}{\xi_G}(\sigma) + \frac{G'}{G}(\sigma) - \frac{2}{\sigma - \alpha - 1}\right\}$$

$$\leqslant 2^{1-b} \pi^{(a-b)/2} A \kappa \left\{x^\sigma \frac{G(2\sigma)}{\xi_G(2\sigma)} t(\sigma, \alpha) - x^{\sigma-1} \frac{G(2\sigma - 1)}{\xi_G(2\sigma - 1)} t(\sigma - 1, \alpha - 1)\right\}$$

(7.34)

where

$$t(\sigma, \alpha) = \frac{\alpha + 1}{\sigma^2(2\sigma - \alpha - 1)}.$$

---

called the *codifferent*. It can be proved that the ramified prime ideals of $\mathcal{O}_K$ are the prime ideals dividing $\mathcal{D}_{K/\mathbb{Q}}$ and $\mathcal{N}_{K/\mathbb{Q}}(\mathcal{D}_{K/\mathbb{Q}}) = |d_K|$. In fact, much more is true. Indeed, let p be a prime number and $\mathfrak{p}$ be a prime ideal above p. If e is the ramification index $e(\mathfrak{p} \mid p)$, then $\mathfrak{p}^{e-1} \mid \mathcal{D}_{K/\mathbb{Q}}$.

The starting point of the proof is the inequality

$$\frac{1}{2\pi i}\int_{2-i\infty}^{2+i\infty}\frac{\xi_F(z)F(z+2\sigma-1)R_{\alpha,\sigma}(z)}{\xi_F(z+2\sigma-1)}x^z dz \geqslant 0 \qquad (7.35)$$

valid for all $x > 0$, $\sigma > 1$ and where $R_{\alpha,\sigma}(z)$ is a certain rational function of z depending on the parameters α and σ (and possibly on other parameters) and satisfying $R_{\alpha,\sigma}(x) > 0$ for all $x > 0$. By the hypotheses above, one may shift the line of integration of (7.35) from $\mathrm{Re}\, z = 2$ to $\mathrm{Re}\, z = 1 - \sigma$, and picking up the residues at $s = 0$, $s = 1$ and at the poles of $R_{\alpha,\sigma}(z)$, and using the functional equation above, Theorem 3.54 and estimates for $\zeta_{\mathbb{K}}(s,\mathcal{C})$ in the spirit of Theorem 7.123, the inequality (7.34) follows after some tedious, but straightforward, computations.

We may apply (7.34) with

$$F(s) = \zeta(s,\mathcal{C}^\star) \quad \text{and} \quad G(s) = \zeta(s,\mathcal{C})$$

so that $a = r_1 + r_2$, $b = r_2$ and $A = 2^b A_{\mathbb{K}}$, and choosing

$$x = \frac{G(2\sigma-1)\xi_G(2\sigma)}{\xi_G(2\sigma-1)G(2\sigma)}\times\frac{t(\sigma-1,\alpha-1)}{t(\sigma,\alpha)} = A_{\mathbb{K}}\times\frac{\Gamma_{\mathbb{K}}(2\sigma)}{\Gamma_{\mathbb{K}}(2\sigma-1)}\times\frac{t(\sigma-1,\alpha-1)}{t(\sigma,\alpha)}$$

we get

$$\log A_{\mathbb{K}} + \frac{G'}{G}(\sigma) \leqslant 2\frac{\xi_G'}{\xi_G}(\sigma) - \log\frac{\Gamma_{\mathbb{K}}(2\sigma)}{\Gamma_{\mathbb{K}}(2\sigma-1)} - \log\frac{t(\sigma-1,\alpha-1)}{t(\sigma,\alpha)} + \frac{2}{\sigma-\alpha-1}.$$

Note that

$$\frac{\xi_G'}{\xi_G}(\sigma) = \log A_{\mathbb{K}} + \frac{\Gamma_{\mathbb{K}}'}{\Gamma_{\mathbb{K}}}(\sigma) + \frac{G'}{G}(\sigma)$$

and hence we obtain

$$-2\frac{\Gamma_{\mathbb{K}}'}{\Gamma_{\mathbb{K}}}(\sigma) + \log\frac{\Gamma_{\mathbb{K}}(2\sigma)}{\Gamma_{\mathbb{K}}(2\sigma-1)} + \log\frac{t(\sigma-1,\alpha-1)}{t(\sigma,\alpha)} - \frac{2}{\sigma-\alpha-1} \leqslant \log A_{\mathbb{K}} + \frac{G'}{G}(\sigma).$$

Now let $\mathfrak{a}_0$ be an ideal in $\mathcal{C}$ with smallest norm. Then for all $\sigma > 1$ we have

$$G'(\sigma) = -\sum_{\mathfrak{a}\in\mathcal{C}}\frac{\log\mathcal{N}_{\mathbb{K}/\mathbb{Q}}(\mathfrak{a})}{\mathcal{N}_{\mathbb{K}/\mathbb{Q}}(\mathfrak{a})^\sigma} \leqslant -G(\sigma)\log\mathcal{N}_{\mathbb{K}/\mathbb{Q}}(\mathfrak{a}_0)$$

and therefore

$$\log\mathcal{N}_{\mathbb{K}/\mathbb{Q}}(\mathfrak{a}_0) \leqslant \log A_{\mathbb{K}} + 2\frac{\Gamma_{\mathbb{K}}'}{\Gamma_{\mathbb{K}}}(\sigma) - \log\frac{\Gamma_{\mathbb{K}}(2\sigma)}{\Gamma_{\mathbb{K}}(2\sigma-1)} - \log\frac{t(\sigma-1,\alpha-1)}{t(\sigma,\alpha)}$$

$$+ \frac{2}{\sigma-\alpha-1}.$$

Choosing α optimally implies Zimmert's theorem in the following shape.

Theorem 7.178 (Zimmert) *Let $\sigma > 1$ be a real number. Then every ideal class contains a non-zero integral ideal $\mathfrak{a}$ such that*

$$\mathcal{N}_{\mathbb{K}/\mathbb{Q}}(\mathfrak{a}) \leqslant |d_{\mathbb{K}}|^{1/2} \exp\left\{ 2 \frac{\Gamma'_{\mathbb{K}}}{\Gamma_{\mathbb{K}}}(\sigma) - \log \frac{\Gamma_{\mathbb{K}}(2\sigma)}{\Gamma_{\mathbb{K}}(2\sigma - 1)} - \log\left(2^{r_2} \pi^{n/2}\right) + r(\sigma) \right\}$$

where

$$r(\sigma) = \frac{2s(\sigma)}{\sigma(\sigma - 1)} + 2\log\left(1 - \frac{1}{\sigma}\right) + \log\left(\frac{(s(\sigma) + \sigma)(s(\sigma) - \sigma + 1)}{(s(\sigma) - \sigma)(s(\sigma) + \sigma - 1)}\right)$$

and $s(\sigma) = \sqrt{3\sigma^2 - 3\sigma + 1}$.

Example 7.179 The following table supplies numbers $C_{\mathbb{K}} > 1$ such that every ideal class contains a non-zero integral ideal $\mathfrak{a}$ such that

$$\mathcal{N}_{\mathbb{K}/\mathbb{Q}}(\mathfrak{a}) \leqslant C_{\mathbb{K}}^{-1} |d_{\mathbb{K}}|^{1/2}.$$

The signatures $(2, 0)$ and $(0, 1)$ are due to Gauss and one has to suppose $d_{\mathbb{K}} \geqslant 8$ for the real quadratic field case.

(r_1, r_2)	$(2, 0)$	$(0, 1)$	$(3, 0)$	$(1, 1)$	$(4, 0)$	$(2, 1)$	$(0, 2)$	$(5, 0)$
$C_{\mathbb{K}}$	$\sqrt{8}$	$\sqrt{3}$	4.636	3.355	14.45	9.749	6.792	50.21

(r_1, r_2)	$(3, 1)$	$(1, 2)$	$(6, 0)$	$(0, 3)$	$(8, 0)$	$(0, 4)$	$(10, 0)$	$(0, 5)$
$C_{\mathbb{K}}$	32.12	21.11	188.1	46.74	3088	385.5	58 540	3560

Choosing $\sigma = 1 + 2n^{-1/2}$, Theorem 7.178 implies that every ideal class contains a non-zero integral ideal $\mathfrak{a}$ such that

$$\mathcal{N}_{\mathbb{K}/\mathbb{Q}}(\mathfrak{a}) \leqslant |d_{\mathbb{K}}|^{1/2} e^{-r_1 (\gamma + \log 4) - 2 r_2 (\gamma + \log \sqrt{2\pi}) + O(\sqrt{n})}$$

as $n \longrightarrow \infty$.

It can be checked (see [Fri89, (4.1)] for instance) that the inequality (7.35) still holds if $R_{\alpha, \sigma}(z)$ belongs to the class of rational functions defined by

$$\mathcal{R}(z) = \left(\frac{z}{z + 2\sigma - 1}\right)^a \left(\frac{z + 1}{z + 2\sigma}\right)^b \sum_{k=0}^{m} a_k \prod_{j=0}^{r_k} \frac{1}{z + b_{jk}}$$

with $\sigma > 1$, $m \in \mathbb{Z}_{\geqslant 0}$, $a_k, b_{jk} \geqslant 0$ and $r_k \in \mathbb{N}$. In [dlM01], the author refined Zimmert's results for small degrees by taking another rational function in this class.

7.5.11 Lower Bounds for the Regulator

As can be seen in (7.23) or Proposition 7.138, a lower bound for $\mathcal{R}_{\mathbb{K}}$ is needed to get a good estimate for the class number. In the 1930s, Remak [Rem32] provided such lower bounds depending on the degree n of $\mathbb{K}$ when this field is totally real, i.e. $r_1 = n$. Using the geometry of numbers, he proved that there exist constants $c_0 > 0$ and $c_1 > 1$ such that $\mathcal{R}_\mathbb{K} \geqslant c_0 c_1^n$. This was later improved by Pohst [Poh77] who showed that, if $\mathbb{K}$ is totally real, then

$$\mathcal{R}_\mathbb{K} > n^{-1/2} \Gamma\left(\frac{n+3}{2}\right)^{-1} \left(\sqrt{\frac{n\pi}{2}} \log \Phi\right)^{n-1}$$

where $\Phi = \frac{1+\sqrt{5}}{2}$ is the golden ratio.

In view of (7.23), one may ask for some bounds depending on the discriminant. This was first achieved in the early 1950s by Remak [Rem52] who proved that if $\mathbb{K}$ *is not* a totally complex quadratic extension of a totally real field,[35] then there exists a constant $c_n > 0$ such that

$$\mathcal{R}_\mathbb{K} \geqslant c_n \log \frac{|d_\mathbb{K}|}{n^n}$$

and for algebraic number fields without proper subfields, he showed that there exists a constant $C_n > 0$ such that

$$\mathcal{R}_\mathbb{K} \geqslant C_n \left(\log \frac{|d_\mathbb{K}|}{n^n}\right)^r$$

where $r = r_1 + r_2 - 1$ is the Dirichlet rank of $\mathcal{O}_\mathbb{K}^*$. This result was generalized independently by Silverman [Sil84], Friedman [Fri89] and Uchida [Uch94], still using the geometry of numbers. Their main theorems are summarized in the next result.

Theorem 7.180 *Let $\mathbb{K}$ be an algebraic number field of degree $n \geqslant 2$, signature (r_1, r_2), discriminant $d_\mathbb{K}$, regulator $\mathcal{R}_\mathbb{K}$ and let $r = r_1 + r_2 - 1$ be the Dirichlet rank of $\mathcal{O}_\mathbb{K}^*$.*

(i) *(Silverman–Friedman). Let ρ be the maximal Dirichlet rank of the unit groups of all proper subfields of $\mathbb{K}$. Then there exists an effectively computable absolute constant $c > 0$ such that*

$$\mathcal{R}_\mathbb{K} > \frac{c}{n^{2n}} \left(\log \frac{|d_\mathbb{K}|}{n^n}\right)^{r-\rho}.$$

(ii) *(Uchida). Let $\alpha > 1$ and $\mathbb{F}$ be a maximal subfield of $\mathbb{K}$ such that $|d_\mathbb{F}| < |d_\mathbb{K}|^{m^{-\alpha}}$ where $m = [\mathbb{K} : \mathbb{F}]$. If λ is the Dirichlet rank of $\mathcal{O}_\mathbb{F}^*$, then there exist constants $c_n, d_n > 0$ depending only on n such that*

$$\mathcal{R}_\mathbb{K} > c_n (\log d_n |d_\mathbb{K}|)^{r-\lambda}.$$

---

[35] Such fields are usually called CM-fields.

Sharpening the constants in the above results is a very difficult problem as n grows. However, for small degrees, there exist some optimal results. For $n = 2$, we have the following easy lemma.

Lemma 7.181 Let $d \geq 5$ be a squarefree number and $\mathbb{K} = \mathbb{Q}(\sqrt{d})$ be a real quadratic field. Then

$$\mathcal{R}_{\mathbb{K}} \geq \log\left(\frac{\sqrt{d} + \sqrt{d-4}}{2}\right).$$

The equality occurs when $d = 5$.

Proof If $\varepsilon_d = \frac{1}{2}(u_1 + v_1\sqrt{d})$ is the fundamental unit of $\mathbb{K}$, then ε_d is equal to the fundamental solution of the equation $u^2 - dv^2 = \pm 4$, and hence

$$\varepsilon_d = \frac{u_1 + v_1\sqrt{d}}{2} \geq \frac{\sqrt{dv_1^2 - 4} + v_1\sqrt{d}}{2}$$

implying the asserted result since $\mathcal{R}_{\mathbb{K}} = \log \varepsilon_d$. □

For cubic and quartic fields, Cusick [Cus84] showed the following lower bounds.

1. If $\mathbb{K}$ is a cubic field with signature $(3, 0)$, then

$$\mathcal{R}_{\mathbb{K}} \geq \frac{1}{16}\left(\log\frac{d_{\mathbb{K}}}{4}\right)^2$$

   and the constant $\frac{1}{16}$ is sharp.
2. If $\mathbb{K}$ is a cubic field with signature $(1, 1)$, then

$$\mathcal{R}_{\mathbb{K}} \geq \frac{1}{3}\log\frac{|d_{\mathbb{K}}|}{27}.$$

3. If $\mathbb{K}$ is a quartic field with signature $(4, 0)$, then

$$\mathcal{R}_{\mathbb{K}} \geq \frac{1}{80\sqrt{10}}\left(\log\frac{d_{\mathbb{K}}}{16}\right)^3.$$

4. If $\mathbb{K}$ is a quartic field with signature $(0, 2)$, then

$$\mathcal{R}_{\mathbb{K}} \geq \frac{1}{4}\log\frac{d_{\mathbb{K}}}{256}.$$

Finally, for a quintic cyclic number field $\mathbb{K}$, Schoof and Washington [SW88] proved that

$$\mathcal{R}_{\mathbb{K}} \geq \frac{1}{6400}\left(\log\frac{d_{\mathbb{K}}}{16}\right)^4.$$

The analytic tools can also be used to derive lower bounds for the regulator. Using (7.34) applied with $\alpha = 0$ and G the partial Dedekind zeta function associated to the trivial class, Zimmert [Zim81] was able to prove that, for each algebraic number field $\mathbb{K}$ with signature (r_1, r_2) and number of roots of unity $w_{\mathbb{K}}$, we have for all $\sigma > 1$

$$\frac{\mathcal{R}_{\mathbb{K}}}{w_{\mathbb{K}}} \geq 2^{-(r_1+1)} \sigma (2\sigma - 1) \Gamma_{\mathbb{K}}(2\sigma) \exp\left(-2\sigma \frac{\Gamma'_{\mathbb{K}}}{\Gamma_{\mathbb{K}}}(\sigma) - \frac{3\sigma - 1}{\sigma - 1}\right)$$

and he chose $\sigma = 2$ to get

$$\frac{\mathcal{R}_{\mathbb{K}}}{w_{\mathbb{K}}} \geq 0.02 \, e^{0.46 r_1 + 0.2 r_2}.$$

This implies in particular that the smallest regulator of an algebraic number field is ≥ 0.056 and that the smallest regulator of a totally real algebraic number field is $\log \Phi$, i.e. the regulator of $\mathbb{Q}(\sqrt{5})$.

Zimmert's ideas were later extended by Friedman and Skoruppa. The former showed in [Fri89] that the ratio $\mathcal{R}_{\mathbb{K}}/w_{\mathbb{K}}$ can be expressed as a sum of rapidly convergent series

$$\frac{\mathcal{R}_{\mathbb{K}}}{w_{\mathbb{K}}} = \sum_{\mathfrak{a}} f\left(\frac{\mathcal{N}_{\mathbb{K}/\mathbb{Q}}(\mathfrak{a})^2}{|d_{\mathbb{K}}|}\right) + \sum_{\mathfrak{b}} f\left(\frac{\mathcal{N}_{\mathbb{K}/\mathbb{Q}}(\mathfrak{b})^2}{|d_{\mathbb{K}}|}\right) \tag{7.36}$$

where $\mathfrak{a}$ runs through the principal ideals of $\mathcal{O}_{\mathbb{K}}$, $\mathfrak{b}$ runs through the integral ideals in the ideal class of the different of $\mathcal{O}_{\mathbb{K}}$ and $f :]0, \infty[\longrightarrow \mathbb{R}$ is defined by

$$f(x) = \frac{2^{-(r_1+1)}}{2\pi i} \int_{2-i\infty}^{2+i\infty} \xi_{\mathbb{K}}(s) \zeta_{\mathbb{K}}(s)^{-1} (x |d_{\mathbb{K}}|)^{-s/2} (2s - 1) \, ds.$$

This function is C^{∞}, takes both positive and negative values, and, for all $r_1 + r_2 \geq 2$, we have

$$\lim_{x \to 0^+} f(x) = -\infty.$$

However, studying more carefully the behavior of f, Friedman [Fri07] proved that $f(x)$ has a single simple zero and that this conclusion still holds for all derivatives $f^{(k)}$ of f. This implies that

$$\frac{\mathcal{R}_{\mathbb{K}}}{w_{\mathbb{K}}} \geq f\left(\frac{1}{|d_{\mathbb{K}}|}\right).$$

Indeed, assume that $f(|d_{\mathbb{K}}|^{-1}) > 0$ otherwise the inequality is trivial, then by the previous observations we infer that $f(x) > 0$ as long as $x \geq |d_{\mathbb{K}}|^{-1}$, and the result follows by dropping out all terms in (7.36) except the one corresponding to the ideal $(1) = \mathcal{O}_{\mathbb{K}}$.

Friedman pointed out that this bound is useful only when the discriminant lies in a certain range, otherwise Zimmert's bound gives better results. For instance, let $\mathbb{K}$

be the totally complex algebraic number field of degree 36 with discriminant $d_{\mathbb{K}} = 3^{18} \times 4057^9$ taken from [Coh00, § 12.2.2]. Friedman's result provides $\mathcal{R}_{\mathbb{K}} w_{\mathbb{K}}^{-1} > 27\,839$ whereas Zimmert's bound only gives[36] $\mathcal{R}_{\mathbb{K}} w_{\mathbb{K}}^{-1} > 121$, but we also have the useless bound $f(10^{-17} d_{\mathbb{K}}^{-1}) \approx -3.6 \times 10^7$.

7.6 Exercises

1 Show that 3, 7, $1 \pm 2\sqrt{-5}$ are irreducibles in $\mathbb{Z}[\sqrt{-5}]$.

2 Show that $\theta = \sqrt[3]{2} + \sqrt[5]{2}$ is algebraic over $\mathbb{Q}$.

3 Let $P = X^3 - X + 1$ and α be a root of P. Set $\beta = \frac{1}{2\alpha^2 - 3\alpha + 2}$. Determine the minimal polynomial of β.

4 Show that $\mathbb{Q}(\sqrt{5}, \sqrt[4]{2}) = \mathbb{Q}(\sqrt{5} + \sqrt[4]{2})$. Deduce $[\mathbb{Q}(\sqrt{5}, \sqrt[4]{2}) : \mathbb{Q}]$.

5 (Schur) Let $n \in \mathbb{N}$ and set

$$P_n = \frac{X^n}{n!} + \frac{X^{n-1}}{(n-1)!} + \cdots + X + 1.$$

The aim of this exercise is to show the following result due to Schur.

Theorem 7.182 (Schur) *P_n is irreducible over $\mathbb{Q}$.*

The proof will make use of the following generalization of Bertrand's postulate seen in Corollary 3.44 (see [Erd34] for instance).

Lemma 7.183 *Let $k \leqslant l$ be positive integers. Then at least one of the numbers $l + 1, l + 2, \ldots, l + k$ is divisible by a prime number $p > k$.*

Suppose that $F_n = n! P_n$ is reducible over $\mathbb{Q}$. Since F_n is monic, it must have an irreducible monic factor $A_m \in \mathbb{Z}[X]$ of degree $m \leqslant n/2$ defined by

$$A_m = X^m + a_{m-1} X^{m-1} + \cdots + a_0.$$

1. Prove that each prime divisor of $n(n-1) \cdots (n - m + 1)$ divides a_0.
2. Let p be a prime factor of a_0 and θ be a root of A_m. Define $\mathbb{K} = \mathbb{Q}(\theta)$ so that $[\mathbb{K} : \mathbb{Q}] = m$.
   (a) Prove that there is some prime ideal $\mathfrak{p}$ lying above p and dividing (θ).

---

[36] PARI/GP provides $\mathcal{R}_{\mathbb{K}} w_{\mathbb{K}}^{-1} \approx 172\,495$.

(b) Write $(\theta) = \mathfrak{p}^\alpha \mathfrak{a}$ and $(p) = \mathfrak{p}^e \mathfrak{b}$ for some $\alpha, e \in \mathbb{N}$ and integral ideals $\mathfrak{a}$ and $\mathfrak{b}$ not divisible by $\mathfrak{p}$. Using the fact that $F_n(\theta) = 0$, prove that there exists an integer $k \in \{1, \ldots, n\}$ such that

$$v_\mathfrak{p}\left(\frac{n!\theta^k}{k!}\right) \leqslant v_\mathfrak{p}(n!)$$

and using the identity $v_\mathfrak{p}(r!) = ev_p(r!)$, deduce that $k\alpha \leqslant ev_p(k!)$.
(c) Using Exercise 12 in Chap. 3, deduce that $p \leqslant m$.
3. Using Lemma 7.183, conclude the proof of Schur's theorem.

6 This exercise provides a proof of Proposition 7.53 from which we take the notation. We set $M = \mathbb{Z} + \mathbb{Z}\theta + \cdots + \mathbb{Z}\theta^{n-1}$.

(a) Show that $\theta^n/p \in \mathcal{O}_\mathbb{K}$ and that $N_{\mathbb{K}/\mathbb{Q}}(\theta) \not\equiv 0 \pmod{p^2}$.
    In what follows, we suppose that $p \mid f$.
(b) Explain why there exist $\alpha \in \mathcal{O}_\mathbb{K} \setminus M$ and $b_0, \ldots, b_{n-1} \in \mathbb{Z}$ not all divisible by p and such that

$$p\alpha = b_0 + b_1\theta + \cdots + b_{n-1}\theta^{n-1}.$$

(c) Let j be the smallest index such that $p \nmid b_j$ and set

$$\beta = \alpha - \frac{1}{p}\left(b_0 + b_1\theta + \cdots + b_{j-1}\theta^{j-1}\right).$$

Show that $\beta \in \mathcal{O}_\mathbb{K}$ and deduce that $(b_j\theta^{n-1})/p \in \mathcal{O}_\mathbb{K}$. Prove the contradiction by taking the norm of this number.

7 Let $P = X^3 - 189X + 756$ and θ be a root of P.

(a) Show that P is irreducible over $\mathbb{Z}$ and let $\mathbb{K} = \mathbb{Q}(\theta)$ be the corresponding algebraic number field.
(b) Determine an integral basis for $\mathbb{K}$.
(c) Compute the Galois group $\mathrm{Gal}(\mathbb{K}/\mathbb{Q})$ and the regulator $\mathcal{R}_\mathbb{K}$ of $\mathbb{K}$.

8

(a) Factorize (3) in $\mathbb{K} = \mathbb{Q}(\sqrt{-2})$.
    Deduce the factorization of the principal ideal $\mathfrak{a} = (1+2\sqrt{-2})$ and that there exists no non-negative integer n such that $(1 + 2\sqrt{-2})^n = 3^n$.
(b) Show that $\arccos(1/3)/\pi \notin \mathbb{Q}$.

9 Let a, b be two squarefree numbers such that $a \equiv 1 \pmod 3$ and $b \equiv 1 \pmod 3$. Prove that the biquadratic field $\mathbb{K} = \mathbb{Q}(\sqrt{a}, \sqrt{b})$ is not monogenic.

10 Let $\mathbb{K}$ be a cubic field such that $d_{\mathbb{K}} < 0$ and p be any prime number. It can be shown [Has30] that the factorization of (p) in $\mathcal{O}_{\mathbb{K}}$ can be divided into the following five different types.

Type	I	II	III	IV	V
(p)	$\mathfrak{p}\mathfrak{p}'\mathfrak{p}''$	$\mathfrak{p}$	$\mathfrak{p}\mathfrak{q}$	$\mathfrak{p}^2\mathfrak{q}$	$\mathfrak{p}^3$
Degree	1	3	1 and 2	1	1

Let $\alpha \in \mathbb{N}$. Compute $v_{\mathbb{K}}(p^\alpha)$ in each case.

11 The purpose of this exercise is to provide a proof of the inequality (7.23) following Louboutin's ideas [Lou98]. Let $\mathbb{K}/\mathbb{Q}$ be a totally real algebraic number field of degree $n \geqslant 2$, discriminant $d_{\mathbb{K}} > 1$ and class number $h_{\mathbb{K}}$.

(a) Set $\sigma_0 = 1 + \frac{2n-2}{\log d_{\mathbb{K}}}$. Prove that $1 < \sigma_0 < 2$.
(b) Let $\sigma > 1$. Using the bounds $\zeta_{\mathbb{K}}(\sigma) \leqslant \zeta(\sigma)^n \leqslant (\sigma/(\sigma - 1))^n$ and writing the inequality (7.22) in the form

$$\kappa_{\mathbb{K}} \leqslant \frac{d_{\mathbb{K}}^{(\sigma-1)/2}}{(\sigma - 1)^{n-1}} f_n(\sigma)$$

show that the function $g_n : \sigma \longmapsto \frac{2\sigma f_n'(\sigma)}{n f_n(\sigma)}$ is convex on $]0, +\infty[$.
(c) Deduce that, for all $\sigma \in [1, 2]$, we have $g_n(\sigma) \leqslant 0$.
(d) Finalize the proof of (7.23).

12 In [Ram01], Ramaré proved that, if χ is a primitive even Dirichlet character of conductor f_χ, then

$$|L(1, \chi)| \leqslant \frac{1}{2} \log f_\chi. \tag{7.37}$$

Let $\mathbb{K}/\mathbb{Q}$ be an abelian real number field of degree n, discriminant $d_{\mathbb{K}}$, class number $h_{\mathbb{K}}$ and regulator $\mathcal{R}_{\mathbb{K}}$. With the help of (7.37) and the class number formula, show that

$$h_{\mathbb{K}} \mathcal{R}_{\mathbb{K}} \leqslant d_{\mathbb{K}}^{1/2} \left(\frac{\log d_{\mathbb{K}}}{4n - 4} \right)^{n-1}$$

thus improving the bound (7.23) by a factor e^{n-1}.

13 Determine the prime numbers $p \geqslant 7$ which can be expressed in the form $p = x^2 + 15y^2$.

References

[Alz00]   Alzer H (2000) Inequalities for the Gamma function. Proc Am Math Soc 128:141–147

[ASV79]   Anderson N, Saff EB, Varga RS (1979) On the Eneström–Kakeya theorem and its sharpness. Linear Algebra Appl 28:5–16

[AW04a]   Alaca S, Williams KS (2004) Introductory algebraic number theory. Cambridge University Press, Cambridge

[AW04b]   Alaca S, Williams KS (2004) On Voronoï's method for finding an integral basis of a cubic field. Util Math 65:163–166

[Bak66]   Baker A (1966) Linear forms in the logarithms of algebraic numbers. Mathematika 13:204–216

[Bar78]   Bartz KM (1978) On a theorem of Sokolovskii. Acta Arith 34:113–126

[Bor02]   Bordellès O (2002) Explicit upper bounds for the average order of $d_n(m)$ and application to class number. JIPAM J Inequal Pure Appl Math 3, Art. 38

[Bou70]   Bourbaki N (1970) Algèbre I, chapitres 1 à 3. Hermann, Paris

[CN63]   Chadrasekharan K, Narasimhan R (1963) The approximate functional equation for a class of zeta-functions. Math Ann 152:30–64

[Coh93]   Cohen H (1993) A course in computational algebraic number theory. GTM, vol 138. Springer, Berlin

[Coh00]   Cohen H (2000) Advanced topics in computational algebraic number theory. GTM, vol 193. Springer, Berlin

[Coh07]   Cohen H (2007) Number theory Volume I: Tools and Diophantine equations. GTM, vol 239. Springer, Berlin

[Cox89]   Cox DA (1989) Primes of the form $x^2 + ny^2$. Wiley, New York

[CR00]   Cohen H, Roblot X-R (2000) Computing the Hilbert class field of real quadratic fields. Math Comput 69:1229–1244

[Cus84]   Cusick TW (1984) Lower bounds for regulators. In: Number theory. Lect. notes math., vol 1068, pp 63–73. Proc J Arith, Noordwijkerhout/Neth 1983

[dlM01]   de la Maza A-C (2001) Bounds for the smallest norm in an ideal class. Math Comput 71:1745–1758

[EGP00]   Elezović N, Giordano C, Pečarić J (2000) The best bounds in Gautschi's inequality. Math Inequal Appl 3:239–252

[EM99]   Esmonde J, Murty MR (1999) Problems in algebraic number theory. GTM, vol 190. Springer, Berlin

[Erd34]   Erdős P (1934) A theorem of Sylvester and Schur. J Lond Math Soc 9:282–288

[Erd53]   Erdős P (1953) Arithmetic properties of polynomials. J Lond Math Soc 28:416–425

[ESW07]   Eloff D, Spearman BK, Williams KS (2007) A_4-sextic fields with a power basis. Missouri J Math Sci 19:188–194

[FPS97]   Flynn EV, Poonen B, Schaefer EF (1997) Cycles of quadratic polynomials and rational points on a genus-2 curve. Duke Math J 90:435–463

[Fri89]   Friedman E (1989) Analytic formulas for the regulator of a number field. Invent Math 98:599–622

[Fri07]   Friedman E (2007) Regulators and total positivity. Publ Mat 51:119–130

[FT91]   Fröhlich A, Taylor MJ (1991) Algebraic number theory. Cambridge studies in advanced mathematics, vol 27. Cambridge University Press, Cambridge

[Gar81]   Garbanati D (1981) Class field theory summarized. Rocky Mt J Math 11:195–225

[Gau86]   Gauss CF (1986) Disquisitiones Arithmeticæ. Springer-Verlag, Berlin. Reprint of the Yale University Press, New Haven, 1966

[GG04]   Gras G, Gras M-N (2004) Algèbre Fondamentale, Arithmétique. Ellipses, Paris

[Gol85]   Goldfeld D (1985) Gauss' class number problem for imaginary quadratic fields. Bull Am Math Soc 13:23–37

[Gra74]   Gras M-N (1974) Nombre de classes, unités et bases d'entiers des extensions cubiques cycliques de $\mathbb{Q}$. Mem SMF 37:101–106

[Gre74]   Greenberg M (1974) An elementary proof of the Kronecker–Weber theorem. Am Math Mon 81:601–607. Correction (1975), 81:803

[GT95]   Gras M-N, Tanoé F (1995) Corps biquadratiques monogènes. Manuscr Math 86:63–79

[Győ76]   Győry K (1976) Sur les polynômes à coefficients entiers et de discriminant donné, III. Publ Math (Debr) 23:141–165

[Hag00]   Hagedorn TR (2000) General formulas for solving solvable sextic equations. J Algebra 233:704–757

[Has30]   Hasse H (1930) Arithmetischen Theorie der kubischen Zahlkörper auf klassenkörpertheoretischer Grundlage. Math Z 31:565–582

[Hee52]   Heegner K (1952) Diophantische Analysis und Modulfunktionen. Math Z 56:227–253

[Her66]   Herz CS (1966) Construction of class fields. In: Seminar on complex multiplication. Lecture notes in math., vol 21. Springer, Berlin, pp VII-1–VII-21

[Hin08]   Hindry M (2008) Arithmétique. Calvage & Mounet, Paris

[IK04]   Iwaniec H, Kowalski E (2004) Analytic number theory. Colloquium publications, vol 53. Am. Math. Soc., Providence

[Jan96]   Janusz G (1996) Algebraic number fields. Graduate studies in mathematics, vol 7, 2nd edn. Amer. Math. Soc., Providence

[Lan03]   Landau E (1903) Neuer Beweis des Primzahlsatzes und Beweis des Primidealsatzes. Math Ann 56:645–670

[Lan27]   Landau E (1927) Einführung in die elementare und analytische Theorie der algebraischen Zahlen und der Ideale. Teubner, Leipzig. 2nd edition: Chelsea, 1949

[Lan93]   Lang S (1993) Algebra, 3rd edn. Addison-Wesley, Reading

[Lan94]   Lang S (1994) Algebraic number theory. GTM, vol 110, 2nd edn. Springer, Berlin

[Leb07]   Lebacque F (2007) Generalized Mertens and Brauer–Siegel theorems. Acta Arith 130:333–350

[LN83]   Llorente P, Nart E (1983) Effective determination of the decomposition of the rational primes in a cubic field. Proc Am Math Soc 87:579–582

[Lou98]   Louboutin S (1998) Majorations explicites du résidu au point 1 des fonctions zêta de certains corps de nombres. J Math Soc Jpn 50:57–69

[Lou00]   Louboutin S (2000) Explicit bounds for residues of Dedekind zeta functions, values of L-functions at $s = 1$, and relative class number. J Number Theory 85:263–282

[Lou03]   Louboutin S (2003) Explicit lower bounds for residues of Dedekind zeta functions at $s = 1$ and relative class number of CM-fields. Trans Am Math Soc 355:3079–3098

[LSWY05]   Lavallee MJ, Spearman BK, Williams KS, Yang Q (2005) Dihedral quintic fields with a power basis. Math J Okayama Univ 47:75–79

[MO07]   Murty MR, Order JV (2007) Counting integral ideals in a number fields. Expo Math 25:53–66

[Mol99]   Mollin RA (1999) Algebraic number theory. Chapman and Hall/CRC, London

[Nar04]   Narkiewicz W (2004) Elementary and analytic theory of algebraic numbers. SMM, 3rd edn. Springer, Berlin

[Neu10]   Neukirch J (2010) Algebraic number theory. A series of comprehensive studies in mathematics, vol 322. Springer, Berlin

[Odl76]   Odlyzko AM (1976) Lower bounds for discriminants of number fields. Acta Arith 29:275–297

[Poh77]   Pohst M (1977) Regulatorabschätzungen für total reelle algebraische Zahlkörper. J Number Theory 9:459–492

[Poi77]   Poitou G (1977) Minorations de discriminants (d'après AM Odlyzko). Séminaire Bourbaki, vol 1975/76, 28ème année. Springer, Berlin, pp 136–153

[Pra04]   Prasolov VV (2004) Polynomials. ACM, vol 11. Springer, Berlin

[Ram01]   Ramaré O (2001) Approximate formulæ for $L(1, \chi)$. Acta Arith 100:245–266

[Rem32]   Remak R (1932) Über die Abchätzung des absoluten Betrages des Regulator eines algebraisches Zahlkörpers nach unten. J Reine Angew Math 167:360–378

[Rem52]   Remak R (1952) Über Grössenbeziehungen zwischen Diskriminante und Regulator eines algebraischen Zahlkörpers. Compos Math 10:245–285

[Rib88]   Ribenboim P (1988) Euler's famous prime generating polynomial and the class number of imaginary quadratic fields. Enseign Math 34:23–42

[Rib01]   Ribenboim P (2001) Classical theory of algebraic numbers. Universitext. Springer,
           Berlin
[Ros94]   Rose HE (1994) A course in number theory. Oxford Science Publications, London
[Sam71]   Samuel P (1971) Théorie Algébrique des Nombres. Collection Méthodes. Hermann,
           Paris
[Sil84]   Silverman J (1984) An inequality relating the regulator and the discriminant of a num-
           ber field. J Number Theory 19:437–442
[Soi81]   Soicher L (1981) The computation of Galois groups. PhD thesis, Concordia Univ.,
           Montréal
[Sok68]   Sokolovskiĭ AV (1968) A theorem on the zeros of the Dedekind zeta function and the
           distance between neighbouring prime ideals (Russian). Acta Arith 13:321–334
[ST02]   Stewart I, Tall D (2002) Algebraic number theory and Fermat's last theorem, 3rd edn.
           AK Peters, Wellesley
[Sta67]   Stark HM (1967) A complete determination of the complex quadratic fields of class-
           number one. Mich Math J 14:1–27
[Sta74]   Stark HM (1974) Some effective cases of the Brauer–Siegel theorem. Invent Math
           23:123–152
[Sta75]   Stark HM (1975) The analytic theory of algebraic numbers. Bull Am Math Soc
           81:961–972
[Sta79]   Stas W (1979) On the order of the Dedekind zeta-function near the line $\sigma = 1$. Acta
           Arith 35:195–202
[SW88]   Schoof R, Washington LC (1988) Quintic polynomials and real cyclotomic fields with
           large class number. Math Comput 50:543–556
[SWW06]   Spearman BK, Watanabe A, Williams KS (2006) PSL$(2, 5)$ sextic fields with a power
           basis. Kodai Math J 29:5–12
[Uch94]   Uchida K (1994) On Silverman's estimate of regulators. Tohoku Math J 46:141–145
[Vor94]   Voronoï G (1894) Concerning algebraic integers derivable from a root of an equation
           of the third degree. Master's thesis, St. Petersburg
[Was82]   Washington LC (1982) Introduction to cyclotomic fields. GTM, vol 83. Springer, New
           York. 2nd edition: 1997
[Zim81]   Zimmert R (1981) Ideale kleiner Norm in Idealklassen und eine Regulatorab-
           schätzung. Invent Math 62:367–380

Appendix
Hints and Answers to Exercises

A.1 Chapter 1

1. By assumption we have $a = bq + r$ with $q \leqslant r < b$ so that $q(b+1) \leqslant a < b(q+1)$. Now if $a = q_1(b+1) + r_1$ with $0 \leqslant r_1 < b+1$, we deduce that

$$q(b+1) \leqslant q_1(b+1) + r_1 < b(q+1)$$

and hence

$$q - \frac{r_1}{b+1} \leqslant q_1 < \frac{b}{b+1}(q+1) + \frac{r_1}{b+1}$$

implying $q - 1 < q_1 < q + 1$ and therefore $q_1 = q$.

2. If $1 \leqslant q < a$, let b_q be the quotient of the Euclidean division of a by q. We have

$$b \in \mathcal{S}_q \iff \left[\frac{a}{b} \right] = q \iff \frac{a}{b} - 1 < q \leqslant \frac{a}{b} \iff \frac{a}{q+1} < b \leqslant \frac{a}{q}$$

and since $b \in \mathbb{N}$, this is equivalent to $b_{q+1} < b \leqslant b_q$. Hence we have

$$\sum_{q=1}^{a} |\mathcal{S}_q| = \sum_{q=1}^{a-1} |\mathcal{S}_q| + 1 = \sum_{q=1}^{a-1} (b_q - b_{q+1}) + 1 = b_1 - b_a + 1 = a - 1 + 1 = a.$$

3. For (i), the Euclidean division of n by m gives $n = qm + r$ with $0 \leqslant r \leqslant m - 1$ so that

$$\frac{n+1}{m} - 1 = q + \frac{r+1}{m} - 1 \leqslant q + \frac{m-1+1}{m} - 1 = q = \left[\frac{n}{m} \right].$$

O. Bordellès, *Arithmetic Tales*, Universitext,
DOI 10.1007/978-1-4471-4096-2, © Springer-Verlag London 2012

For (ii), the proof is similar except that we use $m \nmid n \implies 1 \leqslant r \leqslant m - 1$. The identity (iii) is obvious if $m \mid n$. Otherwise, we have by (i) and (ii)

$$0 \leqslant \left[\frac{n}{m}\right] - \left[\frac{n-1}{m}\right] \leqslant 1 - \frac{1}{m} < 1$$

and we conclude the proof by noticing that the difference above is an integer.

4. The first identity follows from Theorem 1.14 (i) and letting $x \longrightarrow \infty$. The second identity follows from

$$\sum_{n>x} f(n)g(n) = \sum_{n=1}^{\infty} f(n)g(n) - \sum_{n\leqslant x} f(n)g(n).$$

5. Using Theorem 1.14 (i) we get for all integers $N > M$

$$\sum_{n=1}^{N} \frac{a_n}{n} = \frac{1}{N}\sum_{n=1}^{N} a_n + \int_1^M \frac{1}{t^2}\left(\sum_{n\leqslant t} a_n\right) dt + \int_M^N \frac{1}{t^2}\left(\sum_{n\leqslant t} a_n\right) dt.$$

By assumption, the first term on the right-hand side tends to 0 as $N \longrightarrow \infty$ and the second integral converges by Rule 1.20 since $|\sum_{n\leqslant t} a_n| \leqslant M$. Hence we obtain

$$\left|\sum_{n=1}^{\infty} \frac{a_n}{n}\right| \leqslant \int_1^M \frac{1}{t^2}\left|\sum_{n\leqslant t} a_n\right| dt + \int_M^{\infty} \frac{1}{t^2}\left|\sum_{n\leqslant t} a_n\right| dt$$

$$\leqslant \int_1^M \frac{dt}{t} + M\int_M^{\infty} \frac{dt}{t^2} = \log M + 1$$

as asserted.

6. One may assume that $N \geqslant 2$. By Abel's summation as stated in Remark 1.15, we get

$$\sum_{k=1}^{n-1} \frac{k}{N}(a_{k+1} - a_k) + \sum_{k=n}^{N-1} \frac{k-N}{N}(a_{k+1} - a_k)$$

$$= \sum_{k=1}^{N-1} \frac{k}{N}(a_{k+1} - a_k) - \sum_{k=n}^{N-1}(a_{k+1} - a_k)$$

$$= \frac{1}{N}\left\{(N-1)\sum_{k=1}^{N-1}(a_{k+1} - a_k) - \sum_{k=1}^{N-2}\sum_{j=1}^{k}(a_{j+1} - a_j)\right\} - a_N + a_n$$

$$= \left(1 - \frac{1}{N}\right)(a_N - a_1) - \frac{1}{N}\sum_{k=1}^{N-2}(a_{k+1} - a_1) - a_N + a_n$$

$$= a_n - \frac{a_1 + a_N}{N} - \frac{1}{N} \sum_{k=2}^{N-1} a_k = a_n - \frac{1}{N} \sum_{k=1}^{N} a_k$$

so that

$$a_n = \frac{1}{N} \sum_{k=1}^{N} a_k + \sum_{k=1}^{N} \Delta_N(n,k)(a_{k+1} - a_k)$$

where

$$\Delta_N(n,k) = \frac{1}{N} \times \begin{cases} k, & \text{if } 1 \leqslant k \leqslant n-1, \\ k-N, & \text{if } n \leqslant k \leqslant N \end{cases}$$

and the result follows from the trivial estimate $|\Delta_N(n,k)| \leqslant 1$.

7.   This is an immediate consequence of Theorem 1.14 (ii) using

$$\sum_{p \leqslant x} f(p) = \sum_{p \leqslant x} \left(\frac{f(p)}{\log p} \times \log p \right).$$

8.

(a) By integration by parts

$$\mathrm{Li}(x) = \frac{x}{\log x} - \frac{2}{\log 2} + \int_2^x \frac{dt}{(\log t)^2}$$

as required.

(b) Using Exercise 7 with $f(x) = 1$ and the previous question, we get

$$\begin{aligned}
\pi(x) &= \frac{\theta(x)}{\log x} + \int_2^x \frac{\theta(t)}{t(\log t)^2} \, dt \\
&= \frac{x}{\log x} + \int_2^x \frac{dt}{(\log t)^2} + \frac{\theta(x) - x}{\log x} + \int_2^x \frac{\theta(t) - t}{t(\log t)^2} \, dt \\
&= \mathrm{Li}(x) + \frac{2}{\log 2} + \frac{\theta(x) - x}{\log x} + \int_2^x \frac{\theta(t) - t}{t(\log t)^2} \, dt
\end{aligned}$$

and hence

$$\left| \pi(x) - \mathrm{Li}(x) \right| \leqslant \frac{2}{\log 2} + \frac{R(x)}{\log x} + \int_2^x \frac{R(t)}{t(\log t)^2} \, dt$$

with

$$\int_2^x \frac{R(t)}{t(\log t)^2} \, dt = \left(\int_2^{\sqrt{x}} + \int_{\sqrt{x}}^x \right) \frac{R(t)}{t(\log t)^2} \, dt$$

$$< \frac{\sqrt{x}}{(\log 2)^2} + R(x) \int_{\sqrt{x}}^{x} \frac{dt}{t(\log t)^2}$$

$$= \frac{\sqrt{x}}{(\log 2)^2} + \frac{R(x)}{\log x}$$

$$\leqslant \frac{R(x)}{\log x}\left(1 + \frac{1}{(\log 2)^2}\right)$$

and therefore

$$\left|\pi(x) - \mathrm{Li}(x)\right| < \frac{R(x)}{\log x}\left(2 + \frac{1}{(\log 2)^2}\right) + \frac{2}{\log 2} < \frac{5R(x)}{\log x}.$$

A.2 Chapter 2

1.

(a) If $d = (a, b)$, then d divides $2a$ and $2b$, so that d divides $2(a, b) = 2$.

(b) If $d = (a, b)$, then d divides $a(a + b) - ab = a^2$ and $b(a + b) - ab = b^2$, so that d divides $(a^2, b^2) = 1$.

(c) If $b = ka$ for some integer k, then $b^n = k^n a^n$ and thus $a^n \mid b^n$. Conversely, assume that $a^n \mid b^n$ and set $d = (a, b)$ and write $a = da'$ and $b = db'$ so that $(a', b') = 1$. We have $a'^n \mid b'^n$ and since $(a'^n, b'^n) = (a', b')^n = 1$, we infer that $a'^n = 1$ and then $a' = 1$. Thus $a = d$ and therefore $b = ab'$, so that $a \mid b$.

(d) Set $d = (a, b)$ and $D = (ax + by, az + bt)$ with $|xt - yz| = 1$. We have clearly $d \mid D$. Conversely, assuming $ax + by \geqslant 0$ and $az + bt \geqslant 0$, we have

$$\left.\begin{array}{r} D \mid ax + by \\ D \mid az + bt \end{array}\right\} \quad \Longrightarrow \quad D \mid \left\{X(ax + by) + Y(az + bt)\right\}$$

for all $(X, Y) \in \mathbb{Z}^2$. Taking $X = t$ and $Y = -y$ we obtain $D \mid \pm a$ and taking $X = z$ and $Y = -x$ gives $D \mid \pm b$, and hence $D \mid d$.

2.

(a) We have $|\mathcal{S}^2| = ([\sqrt{p}] + 1)^2 > p = |\{0, \dots, p - 1\}|$ so that f is not injective by the Dirichlet pigeon-hole principle.

(b) We have $f(u_1, v_1) = f(u_2, v_2) \iff au_1 - v_1 \equiv au_2 - v_2 \pmod{p} \iff au \equiv v \pmod{p}$.

Furthermore, $|u| = |u_1 - u_2| \leqslant [\sqrt{p}] < \sqrt{p}$ and similarly $|v| < \sqrt{p}$.

If $u = 0$, then we have $v \equiv 0 \pmod{p}$ and hence $v = 0$ since $|v| < \sqrt{p}$. This is impossible in view of the condition $(u_1, v_1) \neq (u_2, v_2)$.

If $v = 0$, then we have $au \equiv 0 \pmod{p}$ and since $p \nmid a$ and p is prime, then we get $u = 0$ by Lemma 3.4, which is also impossible.

3. Let us first notice that $q_k \geq 1$ for all $1 \leq k < n$ and $q_n \geq 2$.

(a) The identity is true when $k = 0$ and $k = 1$. Assume it is true for some $1 \leq k \leq n$. Then

$$r_{k+1} = r_{k-1} - r_k q_k$$
$$= s_{k-1}a + t_{k-1}b - q_k(s_k a + t_k b)$$
$$= a(s_{k-1} - q_k s_k) + b(t_{k-1} - q_k t_k)$$
$$= s_{k+1}a + t_{k+1}b$$

proving the asserted result by induction.

(b) We show the identity $\operatorname{sgn}(s_k) = (-1)^k$ by induction, where we set $\operatorname{sgn}(0) = \pm 1$ by convention. The identity is true when $k = 0$ and $k = 1$. Assume it is true for some $1 \leq k \leq n$. Then $\operatorname{sgn}(s_{k+1}) = \operatorname{sgn}(-q_k s_k + s_{k-1})$ and by induction hypothesis we have

$$\operatorname{sgn}(-q_k s_k) = -\operatorname{sgn}(s_k) = (-1)^{k+1} = \operatorname{sgn}(s_{k-1})$$

and hence $\operatorname{sgn}(s_{k+1}) = (-1)^{k+1}$ as required. Similarly, we have $\operatorname{sgn}(t_k) = (-1)^{k+1}$. We conclude the proof using $x = \operatorname{sgn}(x)|x|$.

For all $k \in \{1, \ldots, n\}$, we then have

$$s_{k+1} = -q_k s_k + s_{k-1} \iff (-1)^{k+1}|s_{k+1}| = (-1)^{k+1}q_k|s_k| + (-1)^{k-1}|s_{k-1}|$$

and simplifying by $(-1)^{k+1}$ gives the desired result. The proof is analogous for the identity

$$|t_{k+1}| = q_k|t_k| + |t_{k-1}|.$$

(c) Define the sequence (u_k) by $u_k = |t_k|r_{k-1} + |t_{k-1}|r_k$ for all $k \in \{1, \ldots, n+1\}$. Using the previous questions, we get

$$u_{k+1} = |t_{k+1}|r_k + |t_k|r_{k+1}$$
$$= \left(q_k|t_k| + |t_{k-1}|\right)r_k + |t_k|(r_{k-1} - r_k q_k)$$
$$= |t_k|r_{k-1} + |t_{k-1}|r_k = u_k$$

and hence $u_k = u_1 = |t_1|r_0 + |t_0|r_1 = a$ for all $k \in \{1, \ldots, n+1\}$, as asserted.

4.

(a) By induction.

(b) Set $d = (u_n, u_{n-1})$ so that d divides $u_{n-1}^2 - u_n = 2$, and since u_n and u_{n-1} are odd, we get $d = 1$.

(c) By induction, the result being clearly true when $n = 3$ since

$$u_3 - 2 = u_2^2 - 4 = (u_2 + 2)(u_2 - 2) = u_1^2(u_2 - 2).$$

Assume it is true with n replaced by $n-1$. We have

$$u_n - 2 = u_{n-1}^2 - 4$$
$$= (u_{n-1} + 2)(u_{n-1} - 2)$$
$$= u_{n-2}^2 (u_{n-1} - 2)$$

and using induction hypothesis we get

$$u_n - 2 = u_{n-2}^2 u_{n-3}^2 \cdots u_1^2 (u_2 - 2)$$

concluding the proof.

(d) For all $r \in \{2, \ldots, n-1\}$, define $d_r = (u_n, u_{n-r})$. We have $d_r \mid u_n$ and $d_r \mid u_{n-r}$ so that d_r divides

$$u_n - u_{n-r}^2 u_{n-2}^2 \cdots u_{n-r+1}^2 u_{n-r-1}^2 \cdots u_1^2 (u_2 - 2) = 2$$

and we conclude the proof by using the fact that both u_n and u_{n-r} are odd.

5.

▷ The first equation is equivalent to $19x + 14y = 3$ and has solutions

$$(9 + 14k, -12 - 19k) \quad \text{with } k \in \mathbb{Z}$$

by Proposition 2.15.

▷ Let $(x, y) \in \mathbb{N}^2$ be a solution of the second equation and set $d = (x, y)$ and $x = dx'$ and $y = dy'$ so that $(x', y') = 1$. The equation is equivalent to $5d(x' + y')^2 = 147x'y'$ and hence $(x' + y')^2$ divides $147x'y'$. By Exercise 1(b), we have $((x' + y')^2, x'y') = 1$ and Theorem 2.12 implies that

$$(x' + y')^2 \mid 147 = 3 \times 7^2$$

and therefore $x' + y' \mid 7$. Since $x' + y' > 1$, we get $x' + y' = 7$, implying that $5d = 3x'y'$ and hence $5 \mid x'y'$ by Theorem 2.12 since $(3, 5) = 1$. This implies that $(x', y') \in \{(2, 5), (5, 2)\}$ and then $d = 6$, and

$$(x, y) \in \{(12, 30), (30, 12)\}.$$

Conversely, one can check that these pairs are solutions of the equation.

▷ The system is equivalent to

$$\begin{cases} x \equiv 2 \ (\text{mod } 5), \\ x \equiv 6 \ (\text{mod } 7) \end{cases}$$

and by Theorem 2.27, we infer that the solution of this system is $x \equiv 27 \ (\text{mod } 35)$.

6.

(a) Write $a = da'$ and $b = db'$ and thus $(a', b') = 1$. The line (OA) has equation $y = (b'/a')x$ so that a point $N\langle x, y\rangle$ is an integer point of the segment $]OA]$ if and only if $x, y \geqslant 1$, $x \leqslant a$ and $a'y = b'x$. Then $a' \mid b'x$ and Theorem 2.12 implies that $a' \mid x$. Hence the number of integer points lying on the segment $]OA]$ is equal to the number of non-zero multiples of a' which are $\leqslant a$, and this number is in turn equal to $[a/a'] = d$ by Proposition 1.11 (v).

(b) The result follows at once using Pick's formula applied to the triangle OAB with area$(OAB) = ab/2$ and $\mathcal{N}_{\partial\mathcal{P}} = a + b + d$ by the previous question.

7.   The answer is "no" as can be seen with the solution $(x, y, z) = (2, 2, 2)$.

8.

(a) We have $n \equiv k \pmod 4$ with $k \in \{0, \pm 1, 2\}$ which implies that $n^2 \equiv k^2 \pmod 8$ with $k^2 \in \{0, 1, 4\}$. We infer that the sum of three squares can only be congruent to $0, 1, 2, 4, 5, 6$ modulo 8.

(b) Similarly, we have $n \equiv k \pmod 3$ with $k \in \{0, \pm 1\}$ so that $n^3 \equiv k^3 \pmod 9$ with $k^3 \in \{0, \pm 1\}$. Thus the sum of three cubes can only be congruent to $0, 1, 2, 3, 6, 7, 8$ modulo 9.

(c) Let $(x, y, z) \in \mathbb{N}^3$ be a solution. We have $x^3 + y^3 + z^3 = 2005^2 \equiv 4 \pmod 9$ contradicting the previous question. Thus the equation has no solution in $\mathbb{N}^3$.

9.

▷ FIRST METHOD. We use $641 = 2^4 + 5^4 = 5 \times 2^7 + 1$ giving

$$2^{32} = 2^4 \times 2^{28} = \left(641 - 5^4\right) \times 2^{28}$$

$$= 641 \times 2^{28} - \left(5 \times 2^7\right)^4$$

$$= 641 \times 2^{28} - (641 - 1)^4$$

$$= 641 \times \left(2^{28} - 641^3 + 4 \times 641^2 - 6 \times 641 + 4\right) - 1$$

and then

$$2^{32} + 1 = 641 \times 6\,700\,417.$$

▷ SECOND METHOD. We use $2^{32} + 1 = 16 \times 2^{28} + 1 = (1 + 3 \times 5) \times (2^7)^4 + 1$ and notice that $3 = 128 - 125 = 2^7 - 5^3$ so that

$$2^{32} + 1 = \left\{1 + 5 \times \left(2^7 - 5^3\right)\right\} \times \left(2^7\right)^4 + 1$$

$$= \left(1 + 5 \times 2^7 - 5^4\right) \times \left(2^7\right)^4 + 1$$

$$= \left(1 + 5 \times 2^7\right) \times 2^{28} + 1 - \left(5 \times 2^7\right)^4$$

$$= \left(1 + 5 \times 2^7\right) \times 2^{28} + \left\{1 - \left(5 \times 2^7\right)^2\right\}\left\{1 + \left(5 \times 2^7\right)^2\right\}$$

$$= \left(1 + 5 \times 2^7\right) \times 2^{28} + \left(1 + 5 \times 2^7\right)\left(1 - 5 \times 2^7\right)\left(1 + \left(5 \times 2^7\right)^2\right)$$
$$= \left(1 + 5 \times 2^7\right) \times \left\{2^{28} + \left(1 - 5 \times 2^7\right)\left(1 + \left(5 \times 2^7\right)^2\right)\right\}$$
$$= 641 \times 6\,700\,417.$$

Remark Euler was the first to obtain this result, disproving the old conjecture stating that the *Fermat numbers* $F_n = 2^{2^n} + 1$ are all primes. Euler proved that $641 \mid F_5$ although F_n was already known to be prime for $n \in \{0, \dots, 4\}$. Currently, it is known that F_n is composite for $n \in \{5, \dots, 19\}$. The largest known prime Fermat number is $F_4 = 65\,537$ and the largest known composite Fermat number is $F_{23\,471}$. The Fermat number whose complete prime factorization is known are F_5, F_6, F_7, F_8, F_9 and F_{11}. The smallest Fermat number for which no prime factor is known is F_{14}. Finally, the following questions are still open:

1. Do there exist infinitely many prime Fermat numbers?
2. Do there exist infinitely many composite Fermat numbers?
3. Is every Fermat number squarefree?

10.   Let us first show that $x > y$. Indeed, if $x = y$, then $a = b$ which is impossible by assumption. If $x < y$, then we have

$$0 < by^2 - ax^2 = x - y < 0$$

giving a contradiction.

(a) Let us notice that

$$ax^2 + x = by^2 + y \quad \Longleftrightarrow \quad x - y = by^2 - ax^2$$

and hence

$$(x - y)\left\{1 + b(x + y)\right\} = x - y + b\left(x^2 - y^2\right) = by^2 - ax^2 + bx^2 - by^2$$
$$= (b - a)x^2 = (mx)^2.$$

The second identity is similar.

(b) Define $d = (x, y)$ and $D = (b - a, x - y)$.
  ▷ Set $x - y = DA$ and $m^2 = DB$ with $(A, B) = 1$. By the previous question, we have

$$A\left(1 + b(x + y)\right) = Bx^2 \quad \text{and} \quad A\left(1 + a(x + y)\right) = By^2$$

so that $A \mid (Bx^2, By^2) = Bd^2$. Since $(A, B) = 1$, Theorem 2.12 implies that $A \mid d^2$.
  ▷ On the other hand, since $d^2 \mid x^2$ and $x^2 \mid A(1 + b(x + y))$, we deduce that

$$d^2 \mid A\left(1 + b(x + y)\right).$$

Suppose that $(d^2, 1 + b(x + y)) > 1$ and let p be a prime factor of this gcd. Hence p divides x, y and $1 + b(x + y)$, so that p must divide $1 + b(x + y) - b(x + y) = 1$ which is impossible. Hence $(d^2, 1 + b(x + y)) = 1$ and using Theorem 2.12 we get $d^2 \mid A$.

▷ We thus have $d^2 = A$ and then

$$x - y = DA = Dd^2 = (b - a, x - y) \times (x, y)^2$$

as required.

(c) $x - y$ is not always a square as can be seen by taking $(a, b, x, y) = (109, 334, 7, 4)$ or $(a, b, x, y) = (3, 199, 8, 1)$.

▷ Suppose that $b = a + 1$. Hence $(b - a, x - y) = 1$ and then $x - y = (x, y)^2$.

▷ Now suppose that there exist integers $m \geqslant 1$ and $n \geqslant 2$ such that $a = m^2(n - 1)$ and $b = m^2 n$. Then

$$x - y = by^2 - ax^2 = m^2 (ny^2 - (n - 1)x^2)$$

and then $m^2 = b - a$ divides $x - y$ so that $(b - a, x - y) = b - a = m^2$ and hence $x - y = (md)^2$.

11.

(a) Since $(a, b) = 1$, there exist $U, V \in \mathbb{Z}$ such that $aU + bV = 1$ by Corollary 2.6. Since $a, b \in \mathbb{N}$, we have $UV \leqslant 0$, and without loss of generality, one may assume that $U \leqslant 0$ and $V \geqslant 0$. Setting $u = -U$ and $v = V$, we get two non-negative integers u, v such that $-au + bv = 1$.

By Proposition 2.15, the solutions of the equation $ax + by = n$ are given by the pairs

$$(-un + bk, vn - ak) \quad \text{with } k \in \mathbb{Z}$$

and the non-negative solutions are obtained by solving the inequalities $-un + bk \geqslant 0$ and $vn - ak \geqslant 0$ implying

$$\frac{un}{b} \leqslant k \leqslant \frac{vn}{a}.$$

Hence $\mathcal{D}_2(n)$ is equal to the number of integers in the interval $[un/b, vn/a]$ so that

$$\mathcal{D}_2(n) = \left[\frac{vn}{a} - \frac{un}{b} \right] + r$$

and we conclude the proof with

$$\frac{vn}{a} - \frac{un}{b} = \frac{n(-au + bv)}{ab} = \frac{n}{ab}.$$

(b) Using Theorem 2.31 we get

$$r = 0 \quad \Longleftrightarrow \quad \left[\frac{n}{ab} \right] = \frac{n}{ab} - 1 + \frac{aa' + bb'}{ab} \quad \Longleftrightarrow \quad \left\{ \frac{n}{ab} \right\} = 1 - \frac{aa' + bb'}{ab}$$

which is equivalent to

$$0 \leqslant 1 - \frac{aa' + bb'}{ab} < 1$$

giving the asserted result. The case $r = 1$ is similar.

12.

(a) The first identity is trivial if $a = 1$. If $a \geqslant 2$, it follows from the logarithmic derivative of (2.9) and taking $x = 1$. For the second identity, we have

$$\prod_{\substack{j=1 \\ j \neq k}}^{a} \frac{1}{e_a(k) - e_a(j)} = e_a(-k(a-1)) \prod_{\substack{j=1 \\ j \neq k}}^{a} \frac{1}{1 - e_a(j-k)}$$

$$= e_a(k) \prod_{h=1}^{a-1} \frac{1}{1 - e_a(h)} = \frac{e_a(k)}{a}$$

where we used (2.9) with $x = 1$.

(b) By (2.8), we infer that $F(z) = z^{n+1} f(z)$ is the generating function of $\mathcal{D}_2(n)$. Therefore

$$f(z) = \frac{1}{z^{n+1}} \sum_{k=0}^{\infty} \mathcal{D}_2(k) z^k = \frac{\mathcal{D}_2(n)}{z} + \sum_{\substack{k=0 \\ k \neq n}}^{\infty} \mathcal{D}_2(k) z^{k-n-1}$$

so that

$$\operatorname*{Res}_{z=0} f(z) = \mathcal{D}_2(n).$$

The non-zero poles of f are respectively 1 (order 2), $e_a(k)$ (order 1) for all $1 \leqslant k \leqslant a - 1$ and $e_b(k)$ (order 1) for all $1 \leqslant k \leqslant b - 1$. Since

$$z^a - 1 = \prod_{j=1}^{a} (z - e_a(j)) \quad \text{and} \quad z^b - 1 = \prod_{j=1}^{b} (z - e_b(j))$$

we get

$$\operatorname*{Res}_{z=1} f(z) = G'(1)$$

where

$$G(z) = (z-1)^2 f(z) = z^{-n-1} \prod_{k=1}^{a-1} (z - e_a(k))^{-1} \prod_{k=1}^{b-1} (z - e_b(k))^{-1}$$

so that

$$\frac{G'}{G}(z) = -\frac{n+1}{z} - \sum_{k=1}^{a-1} \frac{1}{z - e_a(k)} - \sum_{k=1}^{b-1} \frac{1}{z - e_b(k)}.$$

Now (2.9) with $x = 1$ implies that $G(1) = (ab)^{-1}$ and then we get using the previous question

$$\operatorname*{Res}_{z=1} f(z) = \frac{1}{ab} \left\{ -n - 1 - \sum_{k=1}^{a-1} \frac{1}{1 - e_a(k)} - \sum_{k=1}^{b-1} \frac{1}{1 - e_b(k)} \right\}$$

$$= \frac{1}{ab} \left\{ -n - 1 - \frac{a-1}{2} - \frac{b-1}{2} \right\}$$

$$= -\frac{a+b+2n}{2ab}.$$

Finally, for all $k \in \{1, \ldots, a-1\}$, we have using the previous question

$$\operatorname*{Res}_{z=e_a(k)} f(z) = \frac{1}{e_a(kb) - 1} \times \frac{1}{e_a(k(n+1))} \times \prod_{\substack{j=1 \\ j \neq k}}^{a} \frac{1}{e_a(k) - e_a(j)}$$

$$= \frac{1}{a\, e_a(kn)(e_a(kb) - 1)}$$

and similarly

$$\operatorname*{Res}_{z=e_b(k)} f(z) = \frac{1}{b\, e_b(kn)(e_b(ka) - 1)}$$

for all $k \in \{1, \ldots, b-1\}$.

(c) Since $\lim_{|z| \to \infty} z f(z) = 0$, Jordan's (first) lemma implies that

$$\lim_{R \to \infty} \frac{1}{2\pi i} \int_{|z|=R} f(z)\, dz = 0$$

and Cauchy's residue theorem then gives

$$\mathcal{D}_2(n) = \frac{a+b+2n}{2ab} + \frac{1}{a} \sum_{k=1}^{a-1} \frac{1}{e_a(kn)(1 - e_a(kb))} + \frac{1}{b} \sum_{k=1}^{b-1} \frac{1}{e_b(kn)(1 - e_b(ka))}$$

$$\tag{A.1}$$

which is similar to (2.4).

(d) If $b = 1$, then the equation is $ax + y = n$ so that $n/a \geqslant x$ and then

$$\mathcal{D}_2(n) = \left| \left[0, \frac{n}{a} \right] \cap \mathbb{Z} \right| = \left[\frac{n}{a} \right] + 1 = \frac{n}{a} - \left\{ \frac{n}{a} \right\} + 1.$$

Replacing in (A.1) gives

$$\frac{n}{a} - \left\{\frac{n}{a}\right\} + 1 = \frac{a+1+2n}{2a} + \frac{1}{a}\sum_{k=1}^{a-1}\frac{1}{e_a(kn)(1-e_a(k))}$$

so that

$$\frac{1}{a}\sum_{k=1}^{a-1}\frac{1}{e_a(kn)(1-e_a(k))} = \frac{a-1}{2a} - \left\{\frac{n}{a}\right\}.$$

(e) By above we get

$$\frac{1}{a}\sum_{k=1}^{a-1}\frac{1}{e_a(kn)(1-e_a(kb))} = \frac{1}{a}\sum_{k=1}^{a-1}\frac{1}{e_a(k\overline{b}n)(1-e_a(k))} = \frac{a-1}{2a} - \left\{\frac{n\overline{b}}{a}\right\}$$

and similarly

$$\frac{1}{b}\sum_{k=1}^{b-1}\frac{1}{e_b(kn)(1-e_b(ka))} = \frac{b-1}{2b} - \left\{\frac{n\overline{a}}{b}\right\}$$

and therefore

$$\mathcal{D}_2(n) = \frac{a+b+2n}{2ab} + \frac{a-1}{2a} - \left\{\frac{n\overline{b}}{a}\right\} + \frac{b-1}{2b} - \left\{\frac{n\overline{a}}{b}\right\}$$

$$= \frac{n}{ab} + 1 - \left\{\frac{n\overline{b}}{a}\right\} - \left\{\frac{n\overline{a}}{b}\right\}.$$

A.3 Chapter 3

1. Let $A = 3^{4^5} + 4^{5^6}$. Using Sophie Germain's identity

$$m^4 + 4n^4 = (m^2 + 2mn + 2n^2)(m^2 - 2mn + 2n^2)$$

with $m = 3^{4^4}$ and $n = 4^{3906}$, we get $A = BC$ with $B > 1$ and $C > 1$, implying that A is composite.

2. The inequality may be numerically checked for all $n \in \{33, \dots, 65\}$ so that we suppose that $n \geqslant 66$. Among the integers $\{2, \dots, n\}$, we remove the $[n/2] - 1$ even integers $\neq 2$ and the $[n/3] - 1$ integers $\neq 3$ multiples of 3. However, the $[n/6]$

integers multiples of 6 have been removed twice, and removing the numbers 25, 35, 55 and 65, we get

$$\pi(n) \leqslant n - \left(\left[\frac{n}{2}\right] - 1\right) - \left(\left[\frac{n}{3}\right] - 1\right) + \left[\frac{n}{6}\right] - 4$$

and the inequalities $x - 1 < [x] \leqslant x$ imply the asserted result.

3.   This sequence has $6k - 5$ integers. Furthermore, if $n \geqslant 6$ is even, then $n^2 + 2$ is also even and then is composite. Similarly, if $n \equiv \pm 1 \pmod{6}$, then $n^2 + 2 \equiv 3 \pmod{6}$ and $n^2 + 2$ is odd and composite. We infer that the number of primes in the sequence is

$$\leqslant 6k - 5 - (3k - 2) - 2(k - 1) = k - 1 < k.$$

4.

(a) Let $N \geqslant 2$ be an integer. Using Theorem 1.14 (ii) and a weak version of Corollary 3.50, we get

$$\sum_{p \leqslant N} \frac{1}{p \log p} = \frac{1}{\log N} \sum_{p \leqslant N} \frac{1}{p} + \int_2^N \frac{1}{t(\log t)^2} \left(\sum_{p \leqslant t} \frac{1}{p}\right) dt$$

$$= \frac{\log \log N + O(1)}{\log N} + \int_2^N \frac{\log \log t + O(1)}{t(\log t)^2} dt = O(1)$$

implying the asserted result.

(b) The sum is clearly convergent. By Exercise 7 in Chap. 1 and Corollary 3.98, we have[1]

$$\sum_p \frac{\log p}{p^2} = \left(\sum_{p \leqslant 100} + \sum_{p > 100}\right) \frac{\log p}{p^2}$$

$$= \sum_{p \leqslant 100} \frac{\log p}{p^2} - \frac{\theta(100)}{10^4} + 2 \int_{100}^\infty \frac{\theta(t)}{t^3} dt$$

$$< 0.484 - 0.0075 + 2.000162 \int_{100}^\infty \frac{dt}{t^2} < \frac{1}{2}$$

as required.

---

[1]Exercise 7 in Chap. 1 is used in the following form. If $f \in C^1[2, +\infty[$ such that $f(x)\theta(x)/\log x \longrightarrow 0$ as $x \longrightarrow \infty$, then

$$\sum_{p > x} f(p) = -\frac{f(x)\theta(x)}{\log x} - \int_x^\infty \theta(t) \frac{d}{dt}\left(\frac{f(t)}{\log t}\right) dt.$$

5.   By Theorem 3.37 we get

$$v_p\{(pa)!\} = \sum_{k=1}^{\infty}\left[\frac{a}{p^{k-1}}\right] = a + v_p(a!)$$

so that

$$v_p\left\{\binom{pa}{pb}\right\} = v_p\{(pa)!\} - v_p\{(pb)!\} - v_p\{(p(a-b))!\}$$

$$= v_p(a!) - v_p(b!) - v_p((a-b)!) = v_p\left\{\binom{a}{b}\right\}$$

as required.

6.   If (x, y) is a solution, set $d = (x, y)$ and write $x = dx'$ and $y = dy'$ so that $(x', y') = 1$. The equation is equivalent to $dx'y' = p(x' + y')$ and hence $x'y'$ divides $p(x' + y')$. Since $(x' + y', x'y') = 1$ by Exercise 1 in Chap. 2, we infer that $x'y' \mid p$ by Theorem 2.12, and thus

$$x'y' = 1 \quad \text{or} \quad x'y' = p$$

and hence $(x', y') \in \{(1, 1), (1, p), (p, 1)\}$. This implies that $d = 2p$ if $(x', y') = (1, 1)$ and $d = p + 1$ otherwise, so that we get

$$(x, y) \in \{(2p, 2p), (p + 1, p(p + 1)), (p(p + 1), p + 1)\}.$$

Conversely, one easily checks that these pairs are solutions.

7.   2 and 3 are not solutions, but 5 is a solution. Suppose now that $p \geqslant 7$ is prime. We then have $(p - 1)(p + 1) = 2^3 \times q$ with $q \geqslant 7$ prime. This implies that either q divides $p - 1$ or q divides $p + 1$ by Lemma 3.4.

▷ If $p - 1 = hq$ for some $h \in \mathbb{N}$ then $8 = h(hq + 2)$, so that $h \mid 8$. We then get

h	1	2	4	8
q	6	1	0	$\notin \mathbb{N}$

and since q is prime, we see that this case does not provide any solution.
▷ Similarly, if $p + 1 = kq$ then $8 = k(kq - 2)$, so that $k \mid 8$. We then get

k	1	2	4	8
q	10	3	1	$\notin \mathbb{N}$

and since q is prime, $q = 3$ is the only admissible value, giving $p = 5$.

8.   Let (x, y, z) be a solution. Note that

$$x + y + z + xy + yz + zx + xyz + 1 = (x + y + xy + 1) + z(x + y + xy + 1)$$
$$= (x + y + xy + 1)(z + 1)$$
$$= (x + 1)(y + 1)(z + 1)$$

so that the equation is equivalent to $(x + 1)(y + 1)(z + 1) = 2010 = 2 \times 3 \times 5 \times 67$. Furthermore, since $x < y < z$, this implies that $(x + 1)^3 < 2010$ so that $x \leqslant 11$. We deduce that $x \in \{2, 4, 5, 9\}$ and we obtain the triples

$$(2, 4, 133), \quad (2, 9, 66) \text{ and } (4, 5, 66).$$

Conversely, one easily checks that these triples are solutions.

9.   Observe first that $2 \nmid a$ and $5 \nmid a$ since $(a, 10) = 1$.

(a) Since a is odd, we have $a^8 \equiv 1 \pmod{2}$. Furthermore, since $5 \nmid a$, Fermat's little theorem implies that $a^4 \equiv 1 \pmod{5}$ and hence $a^8 \equiv 1 \pmod{5}$. Finally, since $(2, 5) = 1$, one may apply Proposition 2.13 (vi) which gives $a^8 \equiv 1 \pmod{10}$.
   The congruence $a^{8 \times 10^k} \equiv 1 \pmod{10^{k+1}}$ can be proved by induction as in Exercise 10.
   Taking $k = 8$ and multiplying by a gives $a^{800\,000\,001} \equiv a \pmod{10^9}$.

(b) Since $(123\,456\,789, 10) = 1$, we get by the previous question

$$123\,456\,789^{800\,000\,001} \equiv 123\,456\,789 \ \left(\mathrm{mod}\ 10^9\right)$$

so that if $x = 123\,456\,789^{266\,666\,667}$, then $x^3 \equiv 123\,456\,789 \pmod{10^9}$ as required.

10.   We prove the congruence by induction, the case $k = 0$ being clear using Fermat's little theorem. Assume that the result is true for some $k \geqslant 0$. By Lemma 1.6, we have

$$a^{p^{k+2}} - a^{p^{k+1}} = \left(a^{p^{k+1}} - a^{p^k}\right) \sum_{j=1}^{p} a^{j \times p^k (p-1)}.$$

By induction hypothesis, we have $a^{p^{k+1}} - a^{p^k} \equiv 0 \pmod{p^{k+1}}$ and since $a^{p-1} \equiv 1 \pmod{p}$ by Fermat's little theorem, we infer that

$$\sum_{j=1}^{p} a^{j \times p^k (p-1)} \equiv p \equiv 0 \pmod{p}$$

so that $p \times p^{k+1} = p^{k+2}$ divides $a^{p^{k+2}} - a^{p^{k+1}}$, completing the proof.

11.

(a) If $n \equiv 1 \pmod{p}$, then $0 \equiv n^2 + n + 1 \equiv 3 \pmod{p}$ which is impossible since $p \geqslant 5$.

If $n^2 \equiv 1 \pmod{p}$, then $n \equiv \pm 1 \pmod{p}$ by Lemma 3.4. Since $n \not\equiv 1 \pmod{p}$, we obtain $n \equiv -1 \pmod{p}$ and then $0 \equiv n^2 + n + 1 \equiv 1 \pmod{p}$ giving a contradiction again.

Since $n^3 - 1 = (n-1)(n^2 + n + 1)$, we deduce that $n^3 \equiv 1 \pmod{p}$ and hence $\operatorname{ord}_p(n) = 3$.

We infer that $3 \mid (p-1)$ by (3.4). If $p = 1 + 3k$ for some even integer k, then $p \equiv 1 \pmod{6}$. If $p = 1 + 3(1 + 2h) = 2(2 + 3h)$ for some $h \in \mathbb{N}$, then p is composite, giving a contradiction. Hence $p \equiv 1 \pmod{6}$.

(b) Assume that the set of primes of the form $p \equiv 1 \pmod{6}$ is finite and write all these primes as

$$p_1 = 7 < p_2 < \cdots < p_m.$$

Set $M = (p_1 \cdots p_m)^2 + p_1 \cdots p_m + 1$ supposed to be composite without loss of generality. Using the previous question, the prime factors of M are all congruent to 1 modulo 6 and then there exists an index $i \in \{1, \ldots, m\}$ such that $p_i \mid M$, giving a contradiction since we also have $p_i \mid (M-1)$.

12. Let $p \leqslant n$ be a prime number and set $N = [\log n / \log p]$. By Theorem 3.37 and the inequalities $x - 1 < [x] \leqslant x$, we get

$$n \sum_{k=1}^{N} \frac{1}{p^k} - N < v_p(n!) \leqslant n \sum_{k=1}^{N} \frac{1}{p^k}$$

so that

$$\frac{n}{p-1}\left(1 - \frac{1}{p^N}\right) - N < v_p(n!) \leqslant \frac{n}{p-1}\left(1 - \frac{1}{p^N}\right)$$

and using $\log n / \log p - 1 < N \leqslant \log n / \log p$ gives

$$\frac{n}{p-1}\left(1 - \frac{p}{n}\right) - \frac{\log n}{\log p} < v_p(n!) \leqslant \frac{n}{p-1}\left(1 - \frac{1}{n}\right)$$

implying the asserted inequalities. This also may be written in the form

$$\frac{1}{(p-1)\log n} \leqslant \frac{\frac{n}{p-1} - v_p(n!)}{\log n} < \frac{1}{\log p} + \frac{p}{(p-1)\log n}$$

and since $p \geqslant 2$, we get

$$0 < \frac{\frac{n}{p-1} - v_p(n!)}{\log n} < \frac{1}{\log 2} + \frac{2}{\log n} = O(1)$$

as required.

Remark Used with $p = 5$, this asymptotic formula shows that, if n is sufficiently large, the decimal expansion of $n!$ ends up with approximately $n/4$ zeros.

13. Since a square is congruent to 0 or 1 modulo 4, we see that $a^2 - b^2$ cannot be congruent to 2 modulo 4. Conversely, let $n \not\equiv 2 \pmod 4$. Then either n is odd or it is a multiple of 4. If n is odd, then

$$n = \left(\frac{n+1}{2}\right)^2 - \left(\frac{n-1}{2}\right)^2$$

is a difference of two squares. If $4 \mid n$, then

$$n = \left(\frac{n}{4} + 1\right)^2 - \left(\frac{n}{4} - 1\right)^2$$

is also a difference of two squares. Finally, if $n \geqslant 4$, then $n! \equiv 0 \pmod 4$ and hence $n!$ can be expressed as a difference of two squares in this case. For instance, we have $13! = 78\,912^2 - 288^2 = 112\,296^2 - 79\,896^2$. Furthermore, $2! \equiv 2 \pmod 4$ and $3! \equiv 2 \pmod 4$ so that neither $2!$ nor $3!$ can be expressed as a difference of two squares.

14. The proof is the same as in Proposition 7.28. Suppose that P is not irreducible over $\mathbb{Z}$. Then $P = QR$ for some $Q, R \in \mathbb{Z}[X]$ such that $Q, R \neq \pm 1$. Set $d = \deg Q$ and $\delta = \deg R$ so that $n = d + \delta$. Since $Q \neq \pm 1$, each polynomial $Q \pm 1$ has at most d roots. Therefore, there are at most d integers m such that $Q(m) = 1$ and at most d integers m such that $Q(m) = -1$, so that there are at most $2d$ integers m such that $Q(m) = \pm 1$. Similarly, there are at most 2δ integers m such that $R(m) = \pm 1$. Now if $|P(m)| = |Q(m)| \times |R(m)|$ is prime, then either $Q(m) = \pm 1$ or $R(m) = \pm 1$. We infer that there are at most $2d + 2\delta = 2n$ integers m such that $|P(m)|$ is prime, as required.

The polynomials P_1 and P_2 are both irreducible over $\mathbb{Z}$ by applying this criterion respectively with $m \in \{2, 3, 4, 5, 6, 9, 11, 12, 15\}$ and $m \in \{3, 5, 8, 9, 12, 14, 15, 17, 21\}$.

15. Since $(7, 15) = 1$, the sequence $7 - 15k$ contains infinitely many primes by Theorem 3.63. Now $7 - 15k + 2 = 3(3 + 5k)$ and $7 - 15k - 2 = 5(1 + 3k)$ and hence these two numbers are composite. We deduce that the primes contained in the sequence $7 - 15k$ cannot lie in a pair of twin primes.

16. The sequence (d_j) of the positive divisors of n is strictly increasing so that $d_j \geqslant j$ for all j. Note also that $d_j d_{k+1-j} = n$ for all $j \in \{1, \ldots, k\}$ so that

$$d_j = \frac{n}{d_{k+1-j}} \leqslant \frac{n}{k+1-j}$$

and hence

$$\sum_{j=2}^{k} d_{j-1} d_j \leqslant \sum_{j=2}^{k} \frac{n^2}{(k+2-j)(k+1-j)}$$

$$= n^2 \sum_{j=2}^{k} \left(\frac{1}{k+1-j} - \frac{1}{k+2-j} \right)$$

$$= n^2 \left(1 - \frac{1}{k} \right) < n^2.$$

17.

(a) Since $P(a/b) = 0$ we get

$$c_n \left(\frac{a}{b} \right)^n + c_{n-1} \left(\frac{a}{b} \right)^{n-1} + \cdots + c_1 \left(\frac{a}{b} \right) + c_0 = 0$$

so that

$$c_n a^n + c_{n-1} a^{n-1} b + \cdots + c_1 a b^{n-1} + c_0 b^n = 0. \tag{A.2}$$

This may be written as $c_0 b^n = ha$ with

$$h = -c_n a^{n-1} - c_{n-1} a^{n-2} b - \cdots - c_1 b^{n-1} \in \mathbb{Z}$$

so that $a \mid c_0 b^n$ and since $(a, b) = 1$, Theorem 2.12 implies that $a \mid c_0$. Similarly, (A.2) may be written as $c_n a^n = kb$ with

$$k = -c_{n-1} a^{n-1} - \cdots - c_1 a b^{n-2} - c_0 b^{n-1} \in \mathbb{Z}$$

and we conclude as above.

(b) By the previous question, if a/b is a root of a monic polynomial, then $b = \pm 1$.

(c) The roots of $X^2 - p$ are $\pm\sqrt{p}$. By above, these roots are either integer or irrational, and since p is prime, we have $\sqrt{p} \notin \mathbb{Z}$.

18.

Part A.

(a) This follows easily from the fact that a square is congruent to 0 or 1 modulo 4 and that p is odd.

(b) Using Theorem 3.19, we have

$$-1 \equiv (p-1)! \equiv 1 \times 2 \times \cdots \times \frac{p-1}{2} \times \frac{p+1}{2} \times \cdots \times (p-1)$$

$$\equiv 1 \times 2 \times \cdots \times \frac{p-1}{2} \times \left(-\frac{p-1}{2} \right) \times \cdots (-2) \times (-1)$$

$$\equiv (-1)^{(p-1)/2} \left\{ \left(\frac{p-1}{2} \right)! \right\}^2$$

$$\equiv \left\{ \left(\frac{p-1}{2} \right)! \right\}^2 \equiv x^2 \pmod{p}.$$

(c) Using Thue's lemma, there exist two integers u, v such that

$$\begin{cases} xu \equiv v \pmod{p}, \\ 1 \leqslant |u| < \sqrt{p}, \\ 1 \leqslant |v| < \sqrt{p}. \end{cases}$$

Hence

$$u^2 + v^2 \equiv u^2 + x^2 u^2 \equiv u^2 (x^2 + 1) \equiv 0 \pmod{p}$$

so that $p \mid (u^2 + v^2)$, and we also have

$$2 \leqslant u^2 + v^2 < 2p$$

implying that $p = u^2 + v^2$.

Remark We have then proved that, if $p \equiv 1 \pmod 4$, then -1 is quadratic residue modulo p. This is also true for $p = 2$ since $-1 \equiv 1^2 \pmod 2$. By Example 3.35, the converse is also true so that we may state the following result.

-1 *is quadratic residue modulo p if and only if either $p = 2$ or $p \equiv 1 \pmod 4$.*

Part B.

(a) The sequence (r_n) of the remainders is a strictly decreasing sequence of non-negative integers and since $p > \sqrt{p} > 1$, we infer that there exists an index k such that $r_{k-1} > \sqrt{p} > r_k$.

(b) By Exercise 3 in Chap. 2, we have $p = |t_k| r_{k-1} + |t_{k-1}| r_k$ and hence

$$p \geqslant |t_k| r_{k-1} > |t_k| \sqrt{p}$$

so that $|t_k| < \sqrt{p}$. Since $r_k = s_k p + t_k x$, we get $r_k \equiv t_k x \pmod p$ and hence the pair (t_k, r_k) satisfies the conditions of Thue's lemma.

(c) 9733 is prime and satisfies $9733 \equiv 1 \pmod 4$. By above, we have $9733 = r_k^2 + t_k^2$ where the (r_n) are the successive remainders in the Euclidean division of 9733 by $x = 7024$, and the index k is given by $r_{k-1} > \sqrt{9733} \approx 98.7 > r_k$. Using Exercise 3 in Chap. 2, we get

k	1	2	3	4	5	6
r_k	7024	2709	1606	1103	503	97
q_k	1	2	1	1	2	5
t_k	1	-1	3	-4	7	-18

so that $9733 = r_6^2 + t_6^2 = 97^2 + 18^2$.

19.   Using Lemma 3.42 and Theorem 3.49, we get

$$\prod_p p^{\lfloor n/p \rfloor} = \prod_{p \leqslant n} p^{\lfloor n/p \rfloor} = \exp\left\{ \sum_{p \leqslant n} \left[\frac{n}{p} \right] \log p \right\}$$

$$= \exp\left\{ n \sum_{p \leqslant n} \frac{\log p}{p} + O\big(\theta(n)\big) \right\}$$

$$= \exp\{ n \log n + O(n) \}.$$

20.   Let $\sigma > 1$. We have

$$N^{\sigma-1} \sum_{n=N}^{\infty} \frac{1}{n^{\sigma}} = N^{\sigma-1} \sum_{j=1}^{\infty} \sum_{n=jN}^{(j+1)N-1} \frac{1}{n^{\sigma}}$$

$$= \frac{1}{N} \sum_{j=1}^{\infty} \sum_{n=jN}^{(j+1)N-1} \left(\frac{N}{n} \right)^{\sigma}$$

$$\leqslant \frac{1}{N} \sum_{j=1}^{\infty} \left(\frac{N}{jN} \right)^{\sigma} \sum_{n=jN}^{(j+1)N-1} 1 = \zeta(\sigma).$$

The second inequality is similar.

21.

(a) Set $d = (m,n)$ and $d^* = (m,n)^*$. Since $d^* \mid m$ and $d^* \mid n$, we have $d^* \mid d$. Now set

$$m = dm' = d^*m'',$$
$$n = dn' = d^*n''$$

with $(m',n') = (d^*,m'') = (d^*,n'') = 1$. Thus $m'n'' = m''n'$ and then $m' \mid n'm''$ hence $m' \mid m''$ using Theorem 2.12. Write $m'' = m_1 m'$. Since $(d^*,m'') = 1$, we have $(d^*,m_1) = 1$. Thus, we have $dm' = d^*m'' = d^*m_1 m'$, and then $d = d^*m_1$ with $(d^*,m_1) = 1$, showing that d^* is a unitary divisor of d.

(b) Follows at once from the fact that the unitary divisors of p^e are 1 and p^e.

(c) If either $f_i = 0$ or $f_i = e_i$ then $\min(f_i, e_i - f_i) = 0$ and then

$$(d, n/d) = p_1^{\min(f_1, e_1 - f_1)} \cdots p_r^{\min(f_r, e_r - f_r)} = 1$$

so that d is a unitary divisor of n. Conversely, if d is a unitary divisor of n, then $(d, n/d) = 1$ so that $\min(f_i, e_i - f_i) = 0$ for all $i = 1, \ldots, r$ which implies that either $f_i = 0$ or $f_i = e_i$.

Let $n = p_1^{e_1} \cdots p_r^{e_r}$. By above, there are exactly two possible choices for the valuation of a prime p_i. Thus, the number of unitary divisors of n is equal to $2^{\omega(n)}$.

Example

n	Unitary divisors
$6615 = 2^4 \times 3^3 \times 7$	$1, 5, 27, 49, 135, 245, 1323, 6615$
$3024 = 3^3 \times 5 \times 7^2$	$1, 7, 16, 27, 112, 189, 432, 3024$

(d) Let $n = p_1^{e_1} \cdots p_r^{e_r}$ and $m = p_1^{f_1} \cdots p_r^{f_r}$, $d^* = (m, n)^*$ and set $\delta = p_1^{g_1} \cdots p_r^{g_r}$ where g_i are given in the exercise. By the previous question, δ is a unitary common divisor of m and n, and then $\delta \leqslant d^*$. Conversely, we have by above $d^* = p_1^{h_1} \cdots p_r^{h_r}$ where either $h_i = 0$ or $h_i = \min(e_i, f_i)$, and then $d^* \leqslant \delta$.

Example $(6615, 3024)^* = 2^0 \times 3^3 \times 5^0 \times 7^0 = 27$.

A.4 Chapter 4

1.

(a) To each divisor d of n corresponds a unique divisor d' such that $dd' = n$. Hence either d or d' must be $\leqslant \sqrt{n}$ so that

$$\tau(n) \leqslant 2 \sum_{\substack{d \mid n \\ d \leqslant \sqrt{n}}} 1 \leqslant 2\sqrt{n}.$$

(b) By Example 4.8, we have $\sigma = \mathbf{1} \star \mathrm{Id}$ and hence

$$\sigma(n) = (\mathbf{1} \star \mathrm{Id})(n) = \sum_{d \mid n} \frac{n}{d}.$$

Now let $t \in [1, n]$ be a parameter at our disposal and write

$$\sum_{d \mid n} \frac{1}{d} = \sum_{\substack{d \mid n \\ d \leqslant t}} \frac{1}{d} + \sum_{\substack{d \mid n \\ d > t}} \frac{1}{d} \leqslant \sum_{d \leqslant t} \frac{1}{d} + \frac{\tau(n)}{t}.$$

Now using the previous question we get

$$\sum_{d \mid n} \frac{1}{d} \leqslant \log t + 1 + \frac{2\sqrt{n}}{t}.$$

and choosing $t = 2\sqrt{n}$ implies the asserted result.

(c) Since n is composite, it has a prime factor q such that $q \leqslant \sqrt{n}$ by Proposition 3.1 and hence

$$\varphi(n) = n \prod_{p \mid n} \left(1 - \frac{1}{p}\right) \leqslant n \left(1 - \frac{1}{q}\right) \leqslant n \left(1 - \frac{1}{\sqrt{n}}\right)$$

as required.

2. The following ideas are due to Pólya (see [HW38] for instance). Fix a small real number $\varepsilon > 0$. By assumption, the product

$$\prod_{\substack{p^\alpha \\ |f(p^\alpha)| \geqslant 1}} f(p^\alpha)$$

is finite and set M its value. Furthermore, except for finitely many integers, each integer n has at least a prime power p^A such that

$$\left| f(p^A) \right| < \frac{\varepsilon}{|M|}.$$

Thus

$$
\begin{aligned}
|f(n)| &= \left| f\left(p_1^{\alpha_1} \cdots p^A \cdots\right) \right| \\
&= \left| f(p_1^{\alpha_1}) \right| \left| f(p_2^{\alpha_2}) \right| \cdots \left| f(p^A) \right| \cdots \\
&< |M| \times \frac{\varepsilon}{|M|} = \varepsilon.
\end{aligned}
$$

This implies the asserted result since ε may be as small as we want. We apply now this result to the positive multiplicative function

$$f(n) = \frac{\tau(n)}{n^\varepsilon}.$$

In view of the inequality

$$\frac{\alpha + 1}{p^{\varepsilon\alpha}} \leqslant \frac{2(1 + \log p^\alpha)}{p^{\varepsilon\alpha}} \xrightarrow[p^\alpha \to \infty]{} 0$$

we get

$$\lim_{p^\alpha \to \infty} f(p^\alpha) = \lim_{p^\alpha \to \infty} \frac{\alpha + 1}{p^{\varepsilon\alpha}} = 0$$

and hence $\tau(n) = O(n^\varepsilon)$.

3. The method is similar for the four identities, this is why we only give the details for the first one.

We have to show that $\tau^3 \star 1 = (\tau \star 1)^2$. As in Example 4.11, we need to verify this identity only for prime powers. Using the well-known identity

$$\sum_{j=0}^{N} j^3 = \left(\sum_{j=0}^{N} j \right)^2$$

we get

$$\left(\tau^3 \star 1\right)\left(p^\alpha\right) = \sum_{j=0}^{\alpha}(j+1)^3 = \left(\sum_{j=0}^{\alpha}(j+1)\right)^2 = (\tau \star 1)^2\left(p^\alpha\right).$$

4.

(a) We make use of the convolution identity $\Lambda_j = \log^j \star \mu$ and, by the Möbius inversion formula, we also have $\log^j = \Lambda_j \star 1$ so that

$$\Lambda_k(n) = \sum_{d|n} \mu(d) \log^k(n/d) = \sum_{d|n} \mu(d) \log^{k-1}(n/d) \log(n/d)$$

$$= \log n \sum_{d|n} \mu(d) \log^{k-1}(n/d) - \sum_{d|n} \mu(d) \log d \log^{k-1}(n/d)$$

$$= \Lambda_{k-1}(n) \log n - \left(\log^{k-1} \star \mu \log\right)(n)$$

and the identity $\mu \star 1 = e_1$ together with the use of Lemma 4.9 implies that

$$\log^{k-1} \star \mu \log = \left(\log^{k-1} \star \mu\right) \star (1 \star \mu \log) = -(\Lambda_{k-1} \star \Lambda)$$

giving the asserted identity.

(b) We proceed as in Theorem 4.10. Since $(m, n) = 1$ and using Newton's formula, we get

$$\Lambda_k(mn) = \sum_{a|m} \sum_{b|n} \mu(ab) \log^k\left(\frac{mn}{ab}\right)$$

$$= \sum_{a|m} \mu(a) \sum_{b|n} \mu(b) \left(\log(m/a) + \log(n/b)\right)^k$$

$$= \sum_{a|m} \mu(a) \sum_{b|n} \mu(b) \sum_{j=0}^{k} \binom{k}{j} \log^j(m/a) \log^{k-j}(n/b)$$

$$= \sum_{j=0}^{k} \binom{k}{j} \left(\sum_{a|m} \mu(a) \log^j(m/a)\right) \left(\sum_{b|n} \mu(b) \log^{k-j}(n/b)\right)$$

$$= \sum_{j=0}^{k} \binom{k}{j} \Lambda_j(m) \Lambda_{k-j}(n).$$

5. Define the multiplicative function f by $f(1) = 1$ and, for all prime powers

$$f\left(p^\alpha\right) = -\frac{\binom{2\alpha}{\alpha}}{4^\alpha(2\alpha - 1)}.$$

Then $f \star f$ is multiplicative by Theorem 4.10 and hence $(f \star f)(1) = 1 = \mu(1)$. Furthermore, for all primes p, we have

$$(f \star f)(p) = 2p = -\frac{2}{4} \times \binom{2}{1} = -1 = \mu(p)$$

and for all prime powers p^α such that $\alpha \geqslant 2$, we have

$$(f \star f)(p^\alpha) = \sum_{j=0}^{\alpha} f(p^j) f(p^{\alpha-j})$$

$$= \sum_{j=0}^{\alpha} \left(-\frac{\binom{2j}{j}}{4^j(2j-1)} \right) \left(-\frac{\binom{2\alpha-2j}{\alpha-j}}{4^{\alpha-j}(2\alpha-2j-1)} \right)$$

$$= 4^{-\alpha} \sum_{j=0}^{\alpha} \frac{\binom{2j}{j}\binom{2\alpha-2j}{\alpha-j}}{(2j-1)(2\alpha-2j-1)} = 0 = \mu(p^\alpha)$$

where we used [Gou72, identity 3.93], which concludes the proof.

6. We proceed as in Theorem 4.10 or in Exercise 4 above.

$$(f \star g)(mn) = \sum_{a|m} \sum_{b|n} f(ab) g\left(\frac{mn}{ab}\right)$$

$$= \sum_{a|m} \sum_{b|n} (f(a) + f(b)) g\left(\frac{m}{a}\right) g\left(\frac{n}{b}\right)$$

$$= \sum_{a|m} f(a) g\left(\frac{m}{a}\right) \sum_{b|n} g\left(\frac{n}{b}\right) + \sum_{a|m} g\left(\frac{m}{a}\right) \sum_{b|n} f(b) g\left(\frac{n}{b}\right)$$

$$= (f \star g)(m)(g \star 1)(n) + (g \star 1)(m)(f \star g)(n)$$

as required. We apply this identity with $g = \mu$ and f any additive function. First, $(f \star \mu)(1) = f(1) = 0$. If $n = p_1^{\alpha_1} p_2^{\alpha_2}$, we get using (4.6)

$$(f \star \mu)(n) = (f \star \mu)(p_1^{\alpha_1}) e_1(p_2^{\alpha_2}) + (f \star \mu)(p_2^{\alpha_2}) e_1(p_1^{\alpha_1}) = 0$$

and by induction this result is still true for all $n = p_1^{\alpha_1} \cdots p_r^{\alpha_r}$ and $r \geqslant 2$. Note that if $f(p^\alpha) \neq f(p^{\alpha-1})$, then $(f \star \mu)(p^\alpha) \neq 0$ by Example 4.11.

7.

(a) One may assume that $n \geqslant 2$ is expressed in the form $n = p_1^{\alpha_1} \cdots p_r^{\alpha_r}$ with $\alpha_j \geqslant 1$. Then $n^n = p_1^{n\alpha_1} \cdots p_r^{n\alpha_r}$ and hence

$$\varphi(n)\sigma\left(n^n\right) = n\prod_{j=1}^{r}\left(1 - \frac{1}{p_j}\right)\prod_{j=1}^{r}\left(\frac{p_j^{n\alpha_j+1} - 1}{p_j - 1}\right)$$

$$= n\prod_{j=1}^{r} p^{n\alpha_j}\left(1 - \frac{1}{p_j^{n\alpha_j+1}}\right)$$

$$= n^{n+1}\prod_{j=1}^{r}\left(1 - \frac{1}{p_j^{n\alpha_j+1}}\right)$$

which implies the asserted upper bound by noticing that the product is $\leqslant 1$. For the lower bound, we use $\alpha_j \geqslant 1$ which provides

$$\varphi(n)\sigma\left(n^n\right) \geqslant n^{n+1}\prod_{j=1}^{r}\left(1 - \frac{1}{p_j^{n+1}}\right) \geqslant n^{n+1}\prod_{p}\left(1 - \frac{1}{p^{n+1}}\right) = \frac{n^{n+1}}{\zeta(n+1)}$$

as asserted.

(b) By the previous question, we first have $f(n) \geqslant 0$ and

$$f(n) \leqslant \frac{n}{\varphi(n)} - \frac{n^{n+1}}{n^n\varphi(n)\zeta(n+1)} = \frac{n(\zeta(n+1) - 1)}{\varphi(n)\zeta(n+1)}.$$

Now by Exercise 20 in Chap. 3, we get

$$\zeta(n+1) - 1 = \sum_{k=2}^{\infty}\frac{1}{k^{n+1}} \leqslant \frac{\zeta(n+1)}{2^n}$$

and the estimate

$$\frac{n}{\varphi(n)} < e^\gamma \log\log n + \frac{2.51}{\log\log n}$$

valid for all $n \geqslant 3$ and which may be found in [Rn62], implies that

$$0 \leqslant f(n) < 2^{-n}\left(e^\gamma \log\log n + \frac{2.51}{\log\log n}\right)$$

for all $n \geqslant 3$, and hence the series $\sum_{n\geqslant 1} f(n)$ converges. Using PARI/GP we obtain

$$\sum_{n=1}^{\infty}\left(\frac{n}{\varphi(n)} - \frac{\sigma\left(n^n\right)}{n^n}\right) \approx 0.298\,603\ldots$$

8.   This is [HT88, Lemma 61.1].

9.

1. (a) Let m be the smallest period of x. The Euclidean division of n by m gives $n = mq + r$ with $0 \leqslant r < m$ and then

$$x = F^n(x) = F^{mq+r}(x) = F^r\left((F^m)^q(x)\right) = F^r(x).$$

If $r \geqslant 1$, then $r \geqslant m$ since m is the smallest positive integer k such that $F^k(x) = x$, contradicting the inequality $0 \leqslant r < m$. Hence $r = 0$ and thus $m \mid n$.

(b) It is sufficient to show that

$$\mathrm{Per}_n(F) = \bigcup_{d \mid n} \mathrm{Per}_d^*(F). \tag{A.3}$$

Note first that, since each point of E has at most a smallest period, the union is disjoint. Furthermore, if $d \mid n$ and $F^d(x) = x$ for some $x \in E$, then $F^n(x) = (F^d)^{n/d}(x) = x$ so that

$$\bigcup_{d \mid n} \mathrm{Per}_d^*(F) \subseteq \mathrm{Per}_n(F).$$

Conversely, if $F^n(x) = x$ for some $x \in E$, then using the previous question we infer that x has a smallest period d dividing n and therefore

$$\mathrm{Per}_n(F) \subseteq \bigcup_{d \mid n} \mathrm{Per}_d^*(F).$$

The proof of (A.3) is thus complete and implies at once the first identity. The second one follows by using the Möbius inversion formula.

(c) Let x have smallest period n. By definition of $\mathcal{O}_x$, we have

$$\left\{x, F(x), F^2(x), \ldots, F^{n-1}(x)\right\} \subseteq \mathcal{O}_x.$$

Conversely, let $F^k(x) \in \mathcal{O}_x$ for some $k \in \mathbb{Z}_{\geqslant 0}$. The Euclidean division of k by n gives $k = nq + r$ with $0 \leqslant r < n$. Thus we have

$$F^k(x) = F^r\left((F^n)^q(x)\right) = F^r(x) \in \left\{x, F(x), F^2(x), \ldots, F^{n-1}(x)\right\}$$

and then

$$\mathcal{O}_x = \left\{x, F(x), F^2(x), \ldots, F^{n-1}(x)\right\}.$$

Finally, suppose there exist two integers $0 \leqslant i < j < n$ such that $F^i(x) = F^j(x)$. Then we have

$$x = F^n(x) = F^{n-i}\left(F^i(x)\right) = F^{n-i}\left(F^j(x)\right) = F^{j-i}\left(F^n(x)\right)$$

and hence $F^{j-i}(x) = x$, so that $j - i$ is a period of x, so $j - i \geq n$, contradicting the inequalities $0 \leq i < j < n$. We infer that the elements of $\{x, F(x), F^2(x), \ldots, F^{n-1}(x)\}$ are pairwise distinct and thus

$$|\mathcal{O}_x| = \left|\{x, F(x), F^2(x), \ldots, F^{n-1}(x)\}\right| = n.$$

(d) Let x have smallest period n. By the previous question, it suffices to show that $F^k(x)$ has smallest period n for some $k \in \{0, \ldots, n-1\}$.

Since n is a period for x, we have $F^n(F^k(x)) = F^k(F^n(x)) = F^k(x)$ and thus n is a period for $F^k(x)$. Assume that $m \leq n$ is the smallest period of $F^k(x)$. Then $F^m(F^k(x)) = F^k(x)$ and

$$F^{n-k}\left\{F^m\left(F^k(x)\right)\right\} = F^{n-k}\left(F^k(x)\right)$$

and then $F^m(F^n(x)) = F^n(x) = x$ so that m is a period for x. This implies that $m \geq n$ and thus $m = n$, as required.

(e) We will only prove the symmetry, leaving the reflexivity and the transitivity to the reader.

Let $x \in \mathcal{O}_y$. By above, there exists an integer $0 \leq r < n$ such that $x = F^r(y)$ and hence

$$y = F^n(y) = F^{n-r}\left(F^r(y)\right) = F^{n-r}(x)$$

showing that $y \in \mathcal{O}_x$. Therefore the relation $\sim$ is symmetric.

The equivalence class of x is the set $\{y \in \mathrm{Per}_n^*(F) : y \in \mathcal{O}_x\} = \mathcal{O}_x$ since $\mathcal{O}_x \subseteq \mathrm{Per}_n^*(F)$. Since $|\mathcal{O}_x| = n$, we deduce that, if $\mathrm{Per}_n^*(F)$ is a finite set, then d divides $|\mathrm{Per}_n^*(F)|$.

(f) Let $u = (u_n)$ be a realizable sequence. By definition, there exist a set E and a map $F : E \longrightarrow E$ such that $u_n = |\mathrm{Per}_n^*(F)|$. By the previous questions, we have

$$\sum_{d \mid n} \left|\mathrm{Per}_d^*(F)\right| \mu(n/d) \equiv 0 \ (\mathrm{mod}\, n). \tag{A.4}$$

2. We apply (A.4) to the sequences (i) and (ii). For Fermat's little theorem for integer matrices, we first restrict ourselves to matrices with non-negative integer entries since each matrix has such a representative modulo p. The use of (A.4) with $n = p$ gives the desired result.

3. (a) Let $P = X^3 - X - 1$. Since A is the companion matrix of P, A is diagonalizable in $\mathcal{M}_3(\mathbb{C})$ and P is the minimal polynomial of A. Note that P is also the characteristic polynomial of the sequence (u_n). Thus, if λ_1, λ_2 and $\overline{\lambda_2}$ are the eigenvalues of A, then there exist three constants $a, b, c \in \mathbb{C}$ such that, for all $n \in \mathbb{Z}_{\geq 0}$, we have

$$u_n = a\lambda_1^n + b\lambda_2^n + c\overline{\lambda_2}^n.$$

The initial values of the sequence, together with the easy identities

$$\mathrm{Tr}(A) = 0 = \lambda_1 + \lambda_2 + \overline{\lambda_2} \quad \text{and} \quad \mathrm{Tr}(A^2) = 2 = \lambda_1^2 + \lambda_2^2 + \overline{\lambda_2}^2$$

provide the following Vandermonde system of equations

$$\begin{cases} a' + b' + c' = 0, \\ a'\lambda_1 + b'\lambda_2 + c'\overline{\lambda_2} = 0, \\ a'\lambda_1^2 + b'\lambda_2^2 + c'\overline{\lambda_2}^2 = 0 \end{cases}$$

where $a' = a - 1$, $b' = b - 1$ and $c' = c - 1$. Since the eigenvalues are distinct, this system has the unique solution $(a', b', c') = (0, 0, 0)$ and thus $a = b = c = 1$. We deduce that, for all $n \in \mathbb{Z}_+$, we have

$$u_n = \lambda_1^n + \lambda_2^n + \overline{\lambda_2}^n = \mathrm{Tr}(A^n).$$

(b) By Fermat's little theorem for integer matrices, we get for all primes p

$$u_p = \mathrm{Tr}(A^p) \equiv \mathrm{Tr}(A) \equiv 0 \pmod{p}.$$

Remark The converse is untrue: Adam and Shanks [AS82] discovered that, if $n = 521^2$, then $n \mid u_n$.

10. By the convolution identity $\varphi = \mu \star \mathrm{Id}$ and Proposition 4.17, we get

$$\sum_{n \leqslant x} \varphi(n) = \sum_{d \leqslant x} \mu(d) \sum_{k \leqslant x/d} k = \frac{1}{2} \sum_{d \leqslant x} \mu(d) \left[\frac{x}{d} \right] \left(\left[\frac{x}{d} \right] + 1 \right)$$

$$= \frac{1}{2} \sum_{d \leqslant x} \mu(d) \left\{ \frac{x^2}{d^2} + O\left(\frac{x}{d} \right) \right\}$$

$$= \frac{x^2}{2} \sum_{d=1}^{\infty} \frac{\mu(d)}{d^2} - \frac{x^2}{2} \sum_{d > x} \frac{\mu(d)}{d^2} + O\left(x \sum_{d \leqslant x} \frac{1}{d} \right)$$

$$= \frac{x^2}{2\zeta(2)} + O(x) + O(x \log x) = \frac{x^2}{2\zeta(2)} + O(x \log x)$$

where we used the bound of Exercise 20 in Chap. 3.

11. We have

$$\sum_{i=1}^{n} f((i, n)) = \sum_{d \mid n} f(d) \sum_{\substack{k \leqslant n/d \\ (k, n/d) = 1}} 1 = \sum_{d \mid n} f(d) \varphi\left(\frac{n}{d} \right) = (f \star \varphi)(n).$$

12.

(a) By partial summation, we obtain

$$\sum_{n\leqslant z} n\tau(n) = z\sum_{n\leqslant z}\tau(n) - \int_1^z\left(\sum_{n\leqslant t}\tau(n)\right)dt$$

$$= z\{z(\log z + 2\gamma - 1) + O(z^{\theta+\varepsilon})\}$$

$$- \int_1^z\{t(\log t + 2\gamma - 1) + O(t^{\theta+\varepsilon})\}\,dt$$

$$= \frac{z^2\log z}{2} + z^2\left(\gamma - \frac{1}{4}\right) + O(z^{1+\theta+\varepsilon})$$

as asserted.

(b) Exercise 11 with $f = \mathrm{Id}$ implies that $S = \varphi \star \mathrm{Id}$ and (4.7) gives

$$S = \mu \star \mathrm{Id} \star \mathrm{Id} = \mu \star \mathrm{Id} \times (\mathbf{1} \star \mathbf{1}) = \mu \star (\mathrm{Id} \times \tau)$$

where we used the complete multiplicativity of the function Id.

(c) By above and Proposition 4.17, we have

$$\sum_{n\leqslant x} S(n) = \sum_{d\leqslant x}\mu(d)\sum_{k\leqslant x/d} k\tau(k)$$

$$= \sum_{d\leqslant x}\mu(d)\left\{\frac{x^2}{d^2}\left(\frac{1}{2}\log\frac{x}{d} + \gamma - \frac{1}{4}\right) + O\left(\left(\frac{x}{d}\right)^{1+\theta+\varepsilon}\right)\right\}$$

$$= x^2\left\{\left(\frac{\log x}{2} + \gamma - \frac{1}{4}\right)\sum_{d\leqslant x}\frac{\mu(d)}{d^2} - \sum_{d\leqslant x}\frac{\mu(d)\log d}{2d^2}\right\}$$

$$+ O\left(x^{1+\theta+\varepsilon}\sum_{d\leqslant x}\frac{1}{d^{1+\theta+\varepsilon}}\right).$$

Now using Exercise 20 in Chap. 3 we get as usual

$$\sum_{d\leqslant x}\frac{\mu(d)}{d^2} = \sum_{d=1}^{\infty}\frac{\mu(d)}{d^2} + O\left(\sum_{d>x}\frac{1}{d^2}\right) = \frac{1}{\zeta(2)} + O\left(\frac{1}{x}\right)$$

and the use of Theorem 4.41 implies that

$$\sum_{d\leqslant x}\frac{\mu(d)\log d}{d^2} = \sum_{d=1}^{\infty}\frac{\mu(d)\log d}{d^2} + O\left(\frac{\log x}{x}\right) = \frac{\zeta'(2)}{\zeta(2)^2} + O\left(\frac{\log x}{x}\right)$$

so that

$$\sum_{n\leqslant x} S(n) = \frac{x^2}{\zeta(2)}\left(\frac{\log x}{2} + \gamma - \frac{1}{4}\right) - \frac{x^2\zeta'(2)}{2\zeta(2)^2} + O\left(x^{1+\theta+\varepsilon}\right)$$

$$= \frac{x^2}{2\zeta(2)}\left(\log x + 2\gamma - \frac{1}{2} - \frac{\zeta'(2)}{\zeta(2)}\right) + O\left(x^{1+\theta+\varepsilon}\right).$$

The best value for θ to date is given by Huxley in Theorem 6.40. We get for all $\varepsilon > 0$

$$\sum_{n\leqslant x} S(n) = \frac{x^2}{2\zeta(2)}\left(\log x + 2\gamma - \frac{1}{2} - \frac{\zeta'(2)}{\zeta(2)}\right) + O\left(x^{547/416+\varepsilon}\right).$$

13.

(a) Using $\tau_{k+1} = \tau_k \star \mathbf{1}$, we get with Proposition 4.17

$$S_{k+1}(x) = \sum_{n\leqslant x}(\tau_k \star \mathbf{1})(n) = \sum_{d\leqslant x} \tau_k(d)\left[\frac{x}{d}\right] \qquad (A.5)$$

and the inequalities $x - 1 < [x] \leqslant x$ then give

$$x\sum_{d\leqslant x}\frac{\tau_k(d)}{d} - S_k(x) < S_{k+1}(x) \leqslant x\sum_{d\leqslant x}\frac{\tau_k(d)}{d}$$

and using Theorem 1.14 we get

$$\sum_{d\leqslant x}\frac{\tau_k(d)}{d} = \frac{S_k(x)}{x} + \int_1^x \frac{S_k(t)}{t^2}\,dt \qquad (A.6)$$

which concludes the proof.

(b) The inequalities are true for $k = 1$. Assume they are true for some positive integer k. By the previous question and the induction hypothesis, we have

$$S_{k+1}(x) \leqslant x\sum_{j=0}^{k-1}\binom{k-1}{j}\frac{(\log x)^j}{j!} + x\int_1^x \frac{1}{t}\left(\sum_{j=0}^{k-1}\binom{k-1}{j}\frac{(\log t)^j}{j!}\right)dt$$

$$= x\sum_{j=0}^{k-1}\binom{k-1}{j}\frac{1}{j!}\left\{(\log x)^j + \int_1^x \frac{(\log t)^j}{t}\,dt\right\}$$

$$= x\sum_{j=0}^{k-1}\binom{k-1}{j}\frac{1}{j!}\left\{(\log x)^j + \frac{(\log x)^{j+1}}{j+1}\right\}$$

$$= x\left(1+\frac{(\log x)^k}{k!}\right) + x\sum_{j=1}^{k-1}\left\{\binom{k-1}{j}+\binom{k-1}{j-1}\right\}\frac{(\log x)^j}{j!}$$

$$= x\left(1+\frac{(\log x)^k}{k!}\right) + x\sum_{j=1}^{k-1}\binom{k}{j}\frac{(\log x)^j}{j!} = x\sum_{j=0}^{k}\binom{k}{j}\frac{(\log x)^j}{j!}$$

and

$$S_{k+1}(x) > x\sum_{j=0}^{k-1}\frac{(-1)^{k+j+1}}{j!}\int_1^x\frac{(\log t)^j}{t}\,dt + (-1)^k x\int_1^x\frac{dt}{t^2}$$

$$= x\sum_{j=0}^{k-1}(-1)^{k+j+1}\frac{(\log x)^{j+1}}{(j+1)!} + (-1)^k(x-1)$$

$$= x\sum_{j=1}^{k}(-1)^{k+j+2}\frac{(\log x)^j}{j!} + (-1)^{k+2}x + (-1)^{k+1}$$

$$= x\sum_{j=0}^{k}(-1)^{k+j+2}\frac{(\log x)^j}{j!} + (-1)^{k+1}$$

completing the proof.

(c) The result follows from the upper bound of the previous question and the inequality

$$\frac{1}{j!} = \frac{1}{(k-1)!}\times\prod_{i=0}^{k-j-2}(i+j+1) \leqslant \frac{(k-j-2+j+1)^{k-j-1}}{(k-1)!} = \frac{(k-1)^{k-j-1}}{(k-1)!}$$

so that

$$S_k(x) \leqslant \frac{x}{(k-1)!}\sum_{j=0}^{k-1}\binom{k-1}{j}(\log x)^j(k-1)^{k-1-j} = \frac{x(\log x+k-1)^{k-1}}{(k-1)!}$$

by Newton's formula.

Remark One may proceed slightly differently by using the arithmetic-geometric mean inequality which implies that, for all $k \geqslant 3$ and $0 \leqslant j \leqslant k-3$, we have

$$\prod_{i=0}^{k-j-2}(i+j+1) = (k-1)\prod_{i=0}^{k-j-3}(i+j+1)$$

$$\leqslant (k-1)\left(\frac{1}{k-j-2}\sum_{i=0}^{k-j-3}(i+j+1)\right)^{k-j-2}$$

$$= (k-1)\left(\frac{k+j-1}{2}\right)^{k-j-2} \leqslant (k-1)(k-2)^{k-j-2}$$

$$= \frac{k-1}{k-2}(k-2)^{k-j-1}$$

since $k \geqslant 3$, and this inequality remains clearly true if $j \in \{k-2, k-1\}$, so that for all $k \geqslant 3$, we get

$$S_k(x) \leqslant \frac{x(\log x + k - 2)^{k-1}}{(k-2)(k-2)!}.$$

It was proved in [Bor06] that the denominator may be replaced by $(k-1)!$.

(d) The identity is true for $k = 2$ by Corollary 4.20. Suppose it is true for some integer $k \geqslant 2$. By (A.5), (A.6) and induction hypothesis, we get

$$S_{k+1}(x) = x \sum_{d \leqslant x} \frac{\tau_k(d)}{d} + O\big(S_k(x)\big)$$

$$= x \int_1^x \frac{S_k(t)}{t^2}\, dt + O\big(x(\log x)^{k-1}\big)$$

$$= x \int_1^x \left\{ \frac{(\log t)^{k-1}}{(k-1)!} + O\big((\log t)^{k-2}\big) \right\} \frac{dt}{t} + O\big(x(\log x)^{k-1}\big)$$

$$= \frac{x(\log x)^k}{k!} + O\big(x(\log x)^{k-1}\big)$$

completing the proof.

(e) Define

$$S_k^\star(x) = \sum_{n \leqslant x} \tau_k^\star(n)$$

and we prove the inequality by induction on k, the case $k = 1$ being clearly true via

$$S_1^\star(x) = [x] - 1 < x.$$

Assume the inequality is true for some $k \in \mathbb{N}$. By induction hypothesis we have

$$S_{k+1}^\star(x) = \sum_{2 \leqslant n \leqslant x2^{-k}} S_k^\star\left(\frac{x}{n}\right) \leqslant \frac{x}{(k-1)!} \sum_{2 \leqslant n \leqslant x2^{-k}} \frac{1}{n}\left(\log \frac{x}{n}\right)^{k-1}.$$

Now when $x < 2^{k+1}$, we have $S_{k+1}^\star(x) = 0$ and otherwise

$$S_{k+1}^\star(x) \leqslant \frac{x}{(k-1)!} \int_1^{x2^{-k}} \left(\log \frac{x}{t}\right)^{k-1} \frac{dt}{t} = \frac{x(\log x)^k}{k!} - \frac{x(k \log 2)^k}{k!} < \frac{x(\log x)^k}{k!}$$

as required.

14.   Using Corollary 3.7 (v) we obtain

$$\sum_{n\leqslant x} s_2(n) = \sum_{\substack{a^2b^3\leqslant x \\ \mu_2(b)=1}} 1 = \sum_{b\leqslant x^{1/3}} \mu_2(b) \sum_{a\leqslant (x/b^3)^{1/2}} 1$$

$$\leqslant x^{1/2} \sum_{b\leqslant x^{1/3}} \frac{\mu_2(b)}{b^{3/2}} \leqslant \frac{\zeta(3/2)\,x^{1/2}}{\zeta(3)} < 3x^{1/2}$$

where we used Lemma 3.58 in the last inequality.

The second estimate follows from partial summation in the form of Exercise 4 in Chap. 1. Indeed, using this and the above estimate, we get

$$\sum_{n>x} \frac{s_2(n)}{n} = -\frac{1}{x}\sum_{n\leqslant x} s_2(n) + \int_x^\infty \left(\sum_{n\leqslant t} s_2(n)\right)\frac{dt}{t^2} < 3\int_x^\infty \frac{dt}{t^{3/2}} = \frac{6}{x^{1/2}}$$

as asserted.

15.   Define $g = f \star \mu$. Then g is multiplicative by Theorem 4.10, $|g(p)| = |f(p) - 1| \leqslant p^{-1}$ and, for all prime powers p^α with $\alpha \geqslant 2$, we have $|g(p^\alpha)| \leqslant 1$, so that $|g(n)| \leqslant 1$ for all $n \in \mathbb{N}$. Let $x \geqslant 2$ be a large real number. We have

$$\sum_{p\leqslant x}\sum_{\alpha=1}^\infty \frac{|g(p^\alpha)|}{p^\alpha} = \sum_{p\leqslant x} \frac{|f(p)-1|}{p} + \sum_{p\leqslant x}\sum_{\alpha=2}^\infty \frac{|g(p^\alpha)|}{p^\alpha}$$

$$\leqslant \sum_{p\leqslant x} \frac{1}{p^2} + \sum_{p\leqslant x}\sum_{\alpha=2}^\infty \frac{1}{p^\alpha}$$

$$\leqslant 3\sum_{p\leqslant x} \frac{1}{p^2} < \frac{3}{2}$$

by (4.20), so that the series $\sum_{d\geqslant 1} g(d)/d$ converges absolutely by Theorem 4.47 and we have

$$1 + \sum_{\alpha=1}^\infty \frac{g(p^\alpha)}{p^\alpha} = 1 + \sum_{\alpha=1}^\infty \frac{f(p^\alpha) - f(p^{\alpha-1})}{p^\alpha} = \left(1 - \frac{1}{p}\right)\left(1 + \sum_{\alpha=1}^\infty \frac{f(p^\alpha)}{p^\alpha}\right).$$

Using Theorem 4.13 and Proposition 4.17, we get

$$\sum_{n\leqslant x} f(n) = \sum_{n\leqslant x}(g\star\mathbf{1})(n) = \sum_{d\leqslant x} g(d)\left[\frac{x}{d}\right]$$

$$= x\sum_{d\leqslant x} \frac{g(d)}{d} + O\left(\sum_{d\leqslant x} |g(d)|\right)$$

$$= x \sum_{d=1}^{\infty} \frac{g(d)}{d} + O\left(\sum_{d \leqslant x} |g(d)| + x \sum_{d > x} \frac{|g(d)|}{d} \right)$$

and by partial summation in the form of Exercise 4 in Chap. 1, we obtain

$$\sum_{d > x} \frac{|g(d)|}{d} = -\frac{1}{x} \sum_{d \leqslant x} |g(d)| + \int_{x}^{\infty} \left(\sum_{d \leqslant t} |g(d)| \right) \frac{dt}{t^2}$$

and

$$\sum_{d=1}^{\infty} \frac{g(d)}{d} = \prod_{p} \left(1 + \sum_{\alpha=1}^{\infty} \frac{g(p^{\alpha})}{p^{\alpha}} \right) = \prod_{p} \left(1 - \frac{1}{p} \right) \left(1 + \sum_{\alpha=1}^{\infty} \frac{f(p^{\alpha})}{p^{\alpha}} \right)$$

so that

$$\sum_{n \leqslant x} f(n) = x \prod_{p} \left(1 - \frac{1}{p} \right) \left(1 + \sum_{\alpha=1}^{\infty} \frac{f(p^{\alpha})}{p^{\alpha}} \right) + R(x)$$

with

$$|R(x)| \ll \sum_{d \leqslant x} |g(d)| + x \int_{x}^{\infty} \left(\sum_{d \leqslant t} |g(d)| \right) \frac{dt}{t^2}.$$

It remains to estimate the sum $\sum_{d \leqslant x} |g(d)|$. To do this we use the unique decomposition of each positive integer d in the form $d = ab$ with $\mu_2(a) = s_2(b) = 1$ and $(a, b) = 1$. Also note that, for all squarefree numbers a, we have $|g(a)| \leqslant a^{-1}$. Using Exercise 14, we obtain

$$\sum_{d \leqslant x} |g(d)| = \sum_{a \leqslant x} \mu_2(a) |g(a)| \sum_{b \leqslant x/a} s_2(b) |g(b)|$$

$$\leqslant \sum_{a \leqslant x} \frac{\mu_2(a)}{a} \sum_{b \leqslant x/a} s_2(b)$$

$$< 3x^{1/2} \sum_{a \leqslant x} \frac{\mu_2(a)}{a^{3/2}} \leqslant \frac{3 \zeta(3/2) x^{1/2}}{\zeta(3)}.$$

Hence

$$|R(x)| \ll x^{1/2} + x \int_{x}^{\infty} \frac{dt}{t^{3/2}} \ll x^{1/2}$$

completing the proof.

16.   This is a direct application of Exercise 15 since $\beta(p) = 1$ and $\beta \star \mu = s_2$.

17.   Again a direct application of Exercise 15 with $f(n) = \varphi(n)\gamma_2(n)/n^2$ since, for all prime powers p^α, we have

$$f(p) = 1 - \frac{1}{p} \quad \text{and} \quad (f \star \mu)(p^\alpha) = \begin{cases} -1/p, & \text{if } \alpha = 1, \\ -p^{-\alpha}(p-1)^2, & \text{if } \alpha \geqslant 2. \end{cases}$$

18.

(a) Let $g(n) = \mu(n)/\tau(n)$ and $G = g \star 1$. By Theorem 4.10, G is multiplicative and, for all prime powers p^α, we have

$$G(p^\alpha) = 1 + \sum_{j=1}^{\alpha} g(p^j) = 1 - \frac{1}{\tau(p)} = \frac{1}{2}$$

so that, for all $n \in \mathbb{N}$, we get

$$G(n) = 2^{-\omega(n)}$$

and hence using Proposition 4.17 we obtain

$$\sum_{n \leqslant x} G(n) = \sum_{n \leqslant x} (g \star 1)(n) = \sum_{d \leqslant x} g(d) \left[\frac{x}{d} \right]$$

$$= x \sum_{d \leqslant x} \frac{g(d)}{d} - \sum_{d \leqslant x} g(d) \left\{ \frac{x}{d} \right\}$$

$$= x F(x) - \sum_{d \leqslant x} g(d) \left\{ \frac{x}{d} \right\}$$

as required.

(b) Using the inequalities $0 \leqslant \{x\} < 1$ we get

$$\left| \sum_{n \leqslant x} \frac{\mu(n)}{\tau(n)} \left\{ \frac{x}{n} \right\} \right| < \sum_{n \leqslant x} \frac{\mu^2(n)}{\tau(n)} = \sum_{n \leqslant x} \frac{\mu^2(n)}{2^{\omega(n)}} \leqslant \sum_{n \leqslant x} \frac{1}{2^{\omega(n)}}$$

so that

$$\sum_{n \leqslant x} \frac{1}{2^{\omega(n)}} + \sum_{n \leqslant x} \frac{\mu(n)}{\tau(n)} \left\{ \frac{x}{n} \right\} > 0.$$

Furthermore, we also have

$$F(x) = |F(x)| \leqslant \frac{1}{x} \left(\sum_{n \leqslant x} \frac{1}{2^{\omega(n)}} + \left| \sum_{n \leqslant x} \frac{\mu(n)}{\tau(n)} \left\{ \frac{x}{n} \right\} \right| \right) < \frac{2}{x} \sum_{n \leqslant x} \frac{1}{2^{\omega(n)}}.$$

The function $2^{-\omega}$ satisfies the Wirsing conditions (4.21) with $\lambda_1 = 1/2$ and $\lambda_2 = 1$ so that we may apply Theorem 4.22 with $(a, b) = (\log 2, 3/2)$ by Lemma 4.23. This gives

$$\sum_{n \leqslant x} \frac{1}{2^{\omega(n)}} \leqslant e^{3/2} \left(\frac{5}{2} + \log 2 \right) \frac{x}{\log ex} \exp\left(\frac{1}{2} \sum_{p \leqslant x} \frac{1}{p} \right).$$

By Corollary 3.99, we have

$$\sum_{p \leqslant x} \frac{1}{p} < \log\log x + \frac{1}{2}$$

as soon as $x \geqslant 8$ implying that

$$\sum_{n \leqslant x} \frac{1}{2^{\omega(n)}} < \frac{19x}{(\log x)^{1/2}}$$

and hence we finally get for all $x \geqslant 8$

$$0 < F(x) < 38 \, (\log x)^{-1/2}.$$

19.

(a) We have $f(p^\alpha) = t^\alpha - t^{\alpha-1}$ and using Theorem 4.13 and Proposition 4.17, we get

$$\sum_{n \leqslant x} t^{\Omega(n)} = \sum_{n \leqslant x} (f \star \mathbf{1})(n) = \sum_{d \leqslant x} f(d) \left[\frac{x}{d} \right]$$

$$\leqslant x \sum_{d \leqslant x} \frac{f(d)}{d} \leqslant x \prod_{p \leqslant x} \left(1 + \sum_{\alpha=1}^{\infty} \frac{t^\alpha - t^{\alpha-1}}{p^\alpha} \right)$$

$$= x \prod_{p \leqslant x} \left(1 + \frac{t-1}{p-t} \right).$$

We treat the cases $p = 2$ and $p \geqslant 3$ separately which gives

$$\sum_{n \leqslant x} t^{\Omega(n)} \leqslant \frac{x}{2-t} \prod_{3 \leqslant p \leqslant x} \left(1 + \frac{t-1}{p-t} \right)$$

$$\leqslant \frac{x}{2-t} \prod_{3 \leqslant p \leqslant x} \left(1 + \frac{1}{p-2} \right)$$

$$\leqslant \frac{x}{2-t} \exp\left(\sum_{3 \leqslant p \leqslant x} \frac{1}{p-2} \right) \ll \frac{x \log x}{2-t}$$

as required.

(b) We have

$$N_k(x) = \sum_{\substack{n \leqslant x \\ \Omega(n)=k}} t^{-\Omega(n)} t^{\Omega(n)} \leqslant t^{-k} \sum_{n \leqslant x} t^{\Omega(n)} \ll \frac{t^{-k} x \log x}{2 - t}$$

and the choice of

$$t = \frac{2k}{k + 1}$$

gives the asserted estimate.

20.

1. (a) Since n is squarefree, a positive integer d is a divisor of n if and only if either
   d divides n/p or $d \mid n$ is a multiple of p, so that

$$\sum_{d \mid n} f(d) = \sum_{d \mid (n/p)} f(d) + \sum_{\substack{d \mid n \\ p \mid d}} f(d) = \sum_{d \mid (n/p)} f(d) + \sum_{k \mid (n/p)} f(kp).$$

Since f is multiplicative and $k \mid (n/p)$ implies that $(k, p) = 1$, we infer that

$$\sum_{d \mid n} f(d) = \sum_{d \mid (n/p)} f(d) + f(p) \sum_{k \mid (n/p)} f(k) = \big(1 + f(p)\big) \sum_{k \mid (n/p)} f(k)$$

giving (4.49).

   (b) First note that $f(d) \geqslant 0$ for all divisors d of n necessarily squarefree. Thus,
   using (4.49), we get

$$\sum_{d \mid n} f(d) \log d = \sum_{d \mid n} f(d) \sum_{p \mid d} \log p = \sum_{p \mid n} \log p \sum_{k \mid (n/p)} f(kp)$$

$$= \sum_{p \mid n} f(p) \log p \sum_{k \mid (n/p)} f(k)$$

$$= \sum_{p \mid n} \left(\frac{f(p) \log p}{1 + f(p)} \sum_{d \mid n} f(d) \right)$$

$$= \left(\sum_{d \mid n} f(d) \right) \left(\sum_{p \mid n} \frac{f(p) \log p}{1 + f(p)} \right)$$

as required. Now the function $t \longmapsto t/(1 + t)$ is increasing on $[0, \lambda]$ and
since $f(d) \geqslant 0$ for all divisors d of n, we deduce that

$$\sum_{d \mid n} f(d) \log d \leqslant \frac{\lambda}{1 + \lambda} \left(\sum_{d \mid n} f(d) \right) \left(\sum_{p \mid n} \log p \right) = \frac{\lambda \log n}{1 + \lambda} \left(\sum_{d \mid n} f(d) \right).$$

2. First note that $\log(n^{1/a}/d) \leqslant \log(n^{1/a})$ and since $f(d) \geqslant 0$, we get

$$\sum_{\substack{d|n \\ d \leqslant n^{1/a}}} f(d) \frac{\log(n^{1/a}/d)}{\log(n^{1/a})} \leqslant \sum_{\substack{d|n \\ d \leqslant n^{1/a}}} f(d).$$

Note also that, if $d > n^{1/a}$, then $\log(n^{1/a}/d) < 0$ and hence

$$\sum_{d|n} f(d) \frac{\log(n^{1/a}/d)}{\log(n^{1/a})} \leqslant \sum_{\substack{d|n \\ d \leqslant n^{1/a}}} f(d) \frac{\log(n^{1/a}/d)}{\log(n^{1/a})}.$$

We infer that

$$\sum_{\substack{d|n \\ d \leqslant n^{1/a}}} f(d) \geqslant \sum_{d|n} f(d) \frac{\log(n^{1/a}/d)}{\log(n^{1/a})} = \sum_{d|n} f(d) - \frac{a}{\log n} \sum_{d|n} f(d) \log d$$

and using (4.50) gives

$$\sum_{\substack{d|n \\ d \leqslant n^{1/a}}} f(d) \geqslant \sum_{d|n} f(d) \left(1 - \frac{\lambda a}{1 + \lambda} \right)$$

implying the desired estimate since $\lambda < (a - 1)^{-1}$.

21. Let $S_n = (s_{ij})$ and $T_n = (t_{ij})$ be the matrices defined by

$$s_{ij} = \begin{cases} 1, & \text{if } i \mid j, \\ 0, & \text{otherwise} \end{cases} \quad \text{and} \quad t_{ij} = \begin{cases} M(n/i), & \text{if } j = 1, \\ 1, & \text{if } i = j \geqslant 2, \\ 0, & \text{otherwise.} \end{cases}$$

We will prove the following result.

Lemma We have $R_n = S_n T_n$. In particular we have $\det R_n = M(n)$.

Proof Set $S_n T_n = (x_{ij})$. If $j = 1$ we have

$$x_{i1} = \sum_{k=1}^{n} s_{ik} t_{k1} = \sum_{\substack{k \leqslant n \\ i|k}} M\left(\frac{n}{k}\right) = \sum_{d \leqslant n/i} M\left(\frac{n/i}{d}\right) = 1 = r_{i1}$$

by (4.15). If $j \geqslant 2$, then $t_{1j} = 0$ and thus

$$x_{ij} = \sum_{k=2}^{n} s_{ik} t_{kj} = s_{ij} = \begin{cases} 1, & \text{if } i \mid j, \\ 0, & \text{otherwise} \end{cases} = r_{ij}$$

which is the desired result. The second assertion follows at once from

$$\det R_n = \det S_n \det T_n = \det T_n = M(n).$$

The proof is complete.   □

22.

1. We may suppose $n > 1$. Let p^α be a prime power. Using Bernoulli's inequality, we get

$$\left(t^\omega \star 1\right)\left(p^\alpha\right) = 1 + \alpha t \leqslant (1+t)^\alpha = (1+t)^{\Omega(p^\alpha)}$$

implying the first inequality by multiplicativity.

2. Write $n = p_1^{\alpha_1} \cdots p_r^{\alpha_r}$ and, for each $j \in \{1, \ldots, k\}$, let d_j be a divisor of n with $\omega(d_j) = j \leqslant k$. The number of such divisors is at most equal to the number of integers which are the products of j prime powers from the list

$$p_1, p_1^2, \ldots, p_1^{\alpha_1}, p_2, p_2^2, \ldots, p_2^{\alpha_2}, \ldots, p_r, p_r^2, \ldots, p_r^{\alpha_r}.$$

Since this list contains $\Omega(n)$ elements, we infer that the number of divisors d_j is at most $\binom{\Omega(n)}{j}$ and hence

$$\sum_{\substack{d\mid n \\ \omega(d)\leqslant k}} t^{\omega(d)} \leqslant 1 + \sum_{j=1}^{k} \binom{\Omega(n)}{j} t^j = \sum_{j=0}^{k} \binom{\Omega(n)}{j} t^j$$

as asserted. It is easy to see that this inequality generalizes the previous one and, with a little more work, one can prove that

$$\sum_{\substack{d\mid n \\ \omega(d)\leqslant k}} t^{\omega(d)} = \sum_{j=0}^{k} t^j \sum_{1\leqslant i_1 < i_2 < \cdots < i_j \leqslant \omega(n)} \alpha_{i_1} \cdots \alpha_{i_j}.$$

23.   This exercise follows readily from the convolution identities

$$\tau = 1 \star 1 \quad \text{and} \quad 1 \star \Lambda = \log$$

since then

$$\tau \star \mu \star \Lambda = 1 \star 1 \star \mu \star \Lambda = 1 \star \Lambda = \log$$

and

$$\sum_{n\leqslant N} (\tau \star \mu \star \Lambda)(n) = \sum_{n\leqslant N} \log n = \log(N!).$$

A.5 Chapter 5

1.   We have $0 \leqslant \|x\| \leqslant 1/2$, and the function ψ is odd and 1-periodic, implying that the function $\|\cdot\|$ is even and 1-periodic. For the first inequality, it suffices to suppose that $|x|, |y| \leqslant \frac{1}{2}$ giving in this case

$$|\|x\| - \|y\|| = ||x| - |y|| \leqslant |x - y|.$$

The second inequality may be proved similarly, noticing that for all $x \in \mathbb{R}$, there exists a unique $\theta_x \in \ ] - \frac{1}{2}, \frac{1}{2}]$ such that $x = \lfloor x \rfloor + \theta_x$, so that by periodicity we get $\|x + y\| = \|\theta_x + \theta_y\|$ showing that we may suppose that $|x|, |y| \leqslant \frac{1}{2}$, and we also may restrict ourselves to $0 \leqslant x, y \leqslant \frac{1}{2}$ since the function $\|\cdot\|$ is even. In this case, we finally have

$$\|x + y\| = \begin{cases} x + y = \|x\| + \|y\|, & \text{if } 0 \leqslant x + y \leqslant \frac{1}{2}, \\ 1 - (x+y) \leqslant x + y = \|x\| + \|y\|, & \text{if } \frac{1}{2} \leqslant x + y \leqslant 1 \end{cases}$$

as required.

2.   The functions $x \longmapsto |\sin(\pi x)|$ and $x \longmapsto \|x\|$ are both even and 1-periodic, so that it suffices to prove the asserted inequality for all $x \in [0, \frac{1}{2}]$. In this interval, the inequality takes the shape

$$2x \leqslant \sin(\pi x) \leqslant \pi x$$

which is well-known. For instance, if f is the function $x \longmapsto \sin(\pi x) - 2x$, then $f''(x) = -\pi^2 \sin(\pi x) \leqslant 0$ for all $x \in [0, \frac{1}{2}]$, so that f is concave on this interval and therefore

$$f(x) \geqslant \min\big(f(0), f(1/2)\big) = 0$$

as required.

3.

(a) If $N < x^{1/5}$, one may take $T = \mathcal{S}(f, N, \delta)$ since then $x^{-1/6} N^{5/6} < 1$. Now suppose that $N \geqslant x^{1/5}$ and let $a, b \in \mathbb{N}$ and $n, n+a$ and $n+a+b$ be three consecutive elements of $\mathcal{S}(f, N, \delta)$ such that

$$1 \leqslant a, b \leqslant 2^{2/3} x^{-1/6} N^{5/6}.$$

As in Lemma 5.32, we will show that there are only two possibilities for the choice of b. The result will then follow by taking each 4th element of $\mathcal{S}(f, N, \delta)$. There exist non-zero integers m_i and real numbers δ_i such that

$$f(n) = m_1 + \delta_1,$$
$$f(n+a) = m_2 + \delta_2,$$
$$f(n+a+b) = m_3 + \delta_3$$

with $|\delta_i| < \delta$ for $i \in \{1, 2, 3\}$. In fact, each integer m_i is positive since, for all $u \in [N, 2N]$, we have $f(u) \geqslant \sqrt{x}/\sqrt{2N} \geqslant 1/\sqrt{2}$ and $\delta \leqslant c_0 N^{-1}$. Using the given polynomials P and Q, we get

$$f(n)P(n, a) - f(n+a)Q(n, a) = \left(\frac{x}{n}\right)^{1/2}(4n + a) - \left(\frac{x}{n+a}\right)^{1/2}(4n + 3a)$$

$$= \left(\frac{x}{n}\right)^{1/2}\frac{(n+a)^{1/2}(4n+a) - n^{1/2}(4n+3a)}{(n+a)^{1/2}}$$

$$= \frac{x^{1/2}a^3}{n^{1/2}(n+a)^{1/2}D(n, a)}$$

where

$$D(n, a) = (n+a)^{1/2}(4n+a) + n^{1/2}(4n+3a) \geqslant 8N^{3/2}$$

so that

$$0 < f(n)P(n, a) - f(n+a)Q(n, a) \leqslant \frac{x^{1/2}a^3}{8N^{5/2}} < \frac{1}{2}.$$

On the other hand, we have

$$f(n)P(n, a) - f(n+a)Q(n, a) = m_1 P(n, a) - m_2 Q(n, a) + \varepsilon$$

with

$$|\varepsilon| \leqslant 7\delta(n+a) \leqslant 14N\delta < \frac{1}{2}.$$

Hence by Lemma 5.31, we obtain

$$m_1 P(n, a) - m_2 Q(n, a) = 0. \tag{A.7}$$

Similarly we have

$$m_2 P(n+a, b) - m_3 Q(n+a, b) = 0,$$

$$m_1 P(n, a+b) - m_3 Q(n, a+b) = 0$$

and eliminating m_3 we obtain

$$m_2 P(n+a, b)Q(n, a+b) - m_1 P(n, a+b)Q(n+a, b) = 0$$

implying that

$$3b^2(m_1 - m_2) + \kappa_1 b + 2\kappa_2 = 0 \tag{A.8}$$

where

$$\kappa_1 = a(7m_1 - 15m_2) - 16n(m_1 - m_2),$$

$$\kappa_2 = a^2(m_1 - 3m_2) - an(-20m_1 + 28m_2) - 16n^2(m_1 - m_2).$$

If $m_1 = m_2$, then by (A.7) we have $P(n, a) = Q(n, a)$ and then $4n + a = 4n + 3a$, so that $a = 0$ which is impossible since $a \geqslant 1$. Therefore $m_1 \neq m_2$ and (A.8) is a quadratic equation in b, concluding the proof.

(b) Hence we deduce that

$$\mathcal{R}\left(\sqrt{\frac{x}{n}}, N, \delta\right) \ll |T| + 1 \ll \frac{N}{x^{-1/6} N^{5/6}} + 1 \ll (Nx)^{1/6}.$$

4.   We follow the proof of Corollary 5.35 from which we borrow the notation. If $16 \leqslant y \leqslant x^{1/5}$, then obviously

$$\left| \sum_{x < n \leqslant x+y} \mu_2(n) - \frac{y}{\zeta(2)} \right| < 3y \leqslant 3 x^{1/15} y^{2/3}.$$

Suppose that $x^{1/5} < y < x^{1/2}/4$. We may write

$$\sum_{2\sqrt{y} < n \leqslant \sqrt{x}} \left(\left[\frac{x+y}{n^2} \right] - \left[\frac{x}{n^2} \right] \right) = \left(\sum_{2\sqrt{y} < n \leqslant c_0^{-1} y} + \sum_{c_0^{-1} y < n \leqslant \sqrt{x}} \right) \left(\left[\frac{x+y}{n^2} \right] - \left[\frac{x}{n^2} \right] \right)$$

$$= \Sigma_1 + \Sigma_2$$

where c_0 is the constant appearing in (5.33). We use Theorem 5.23 (i) with $k = 4$ for Σ_1 and Theorem 5.30 for Σ_2 which gives

$$\Sigma_1 \ll \left(x^{1/10} y^{2/5} + y^{2/3} + (xy)^{1/7} \right) \log x \ll x^{1/15} y^{2/3} \log x$$

and

$$\Sigma_2 \ll \max_{c_0^{-1} y < n \leqslant \sqrt{x}} \left(x^{1/5} + x^{1/15} y N^{-1/3} \right) \log x \ll \left(x^{1/5} + x^{1/15} y^{2/3} \right) \log x$$

$$\ll x^{1/15} y^{2/3} \log x$$

since $y > x^{1/5}$. This completes the proof since clearly $y^{1/2} \leqslant x^{1/15} y^{2/3}$.

5.   The proof is exactly the same as that of Corollary 5.35 except that Theorem 5.30 is replaced by Theorem 5.36 and Theorem 5.23 (i) is used with $k = 2r$ instead of $k = 4$. We omit the details.

6.

(a) Let $L < T \leqslant (x + y)^{1/3}$ be any parameter at our disposal. We have

$$\sum_{L < b \leqslant (x+y)^{1/3}} \left(\left[\sqrt{\frac{x+y}{b^3}} \right] - \left[\sqrt{\frac{x}{b^3}} \right] \right)$$

$$= \left(\sum_{L < b \leqslant T} + \sum_{T < b \leqslant (x+y)^{1/3}} \right) \left(\left[\sqrt{\frac{x+y}{b^3}} \right] - \left[\sqrt{\frac{x}{b^3}} \right] \right)$$

$$= \sum_{L < b \leqslant T} \left(\left[\sqrt{\frac{x+y}{b^3}} \right] - \left[\sqrt{\frac{x}{b^3}} \right] \right) + \sum_{T < b \leqslant (x+y)^{1/3}} \sum_{x < a^2 b^3 \leqslant x+y} 1$$

$$= \sum_{L < b \leqslant T} \left(\left[\sqrt{\frac{x+y}{b^3}} \right] - \left[\sqrt{\frac{x}{b^3}} \right] \right) + \sum_{a \leqslant \sqrt{\frac{x+y}{T^3}}} \sum_{(\frac{x}{a^2})^{1/3} < b \leqslant (\frac{x+y}{a^2})^{1/3}} 1$$

$$= \sum_{L < b \leqslant T} \left(\left[\sqrt{\frac{x+y}{b^3}} \right] - \left[\sqrt{\frac{x}{b^3}} \right] \right) + \sum_{a \leqslant \sqrt{\frac{x+y}{T^3}}} \left(\left[\left(\frac{x+y}{a^2} \right)^{1/3} \right] - \left[\left(\frac{x}{a^2} \right)^{1/3} \right] \right)$$

$$= \sum_{L < b \leqslant T} \left(\left[\sqrt{\frac{x+y}{b^3}} \right] - \left[\sqrt{\frac{x}{b^3}} \right] \right)$$

$$+ \left(\sum_{a \leqslant L} + \sum_{L < a \leqslant \sqrt{\frac{x+y}{T^3}}} \right) \left(\left[\left(\frac{x+y}{a^2} \right)^{1/3} \right] - \left[\left(\frac{x}{a^2} \right)^{1/3} \right] \right)$$

$$\ll \sum_{L < b \leqslant T} \left(\left[\sqrt{\frac{x+y}{b^3}} \right] - \left[\sqrt{\frac{x}{b^3}} \right] \right) + \sum_{a \leqslant L} \left(\frac{y}{(ax)^{2/3}} + 1 \right)$$

$$+ \sum_{L < a \leqslant \sqrt{\frac{x+y}{T^3}}} \left(\left[\left(\frac{x+y}{a^2} \right)^{1/3} \right] - \left[\left(\frac{x}{a^2} \right)^{1/3} \right] \right)$$

$$\ll \sum_{L < b \leqslant T} \left(\left[\sqrt{\frac{x+y}{b^3}} \right] - \left[\sqrt{\frac{x}{b^3}} \right] \right) + y x^{-2/3} L^{1/3} + L$$

$$+ \sum_{L < a \leqslant \sqrt{\frac{2x}{T^3}}} \left(\left[\left(\frac{x+y}{a^2} \right)^{1/3} \right] - \left[\left(\frac{x}{a^2} \right)^{1/3} \right] \right).$$

Using (5.45), we get $y x^{-2/3} L^{1/3} < L$ and, for all $A, B > L$, we infer

$$\frac{y}{\sqrt{x B^3}} < \frac{y}{\sqrt{x L^3}} = x^{1/4} y^{-1/2} (\log x)^{3/4} \leqslant \frac{1}{4},$$

$$\frac{y}{(Ax)^{2/3}} < \frac{y}{(xL)^{2/3}} = \left(x^{-1} y \log x \right)^{1/3} \leqslant \frac{1}{4}$$

as soon as $x \geqslant 3$. Now the choice of $T = (2x)^{1/5}$ and the usual splitting argument provide the final result.

(b) Using Corollary 3.7 (v) and the previous question, we have

$$\sum_{x<n\leqslant x+y} s_2(n) = \sum_{b\leqslant (x+y)^{1/3}} \mu_2(b) \sum_{\sqrt{\frac{x}{b^3}}<a\leqslant\sqrt{\frac{x+y}{b^3}}} 1$$

$$= \left(\sum_{b\leqslant L} + \sum_{L<b\leqslant(x+y)^{1/3}}\right)\mu_2(b)\left(\left[\sqrt{\frac{x+y}{b^3}}\right]-\left[\sqrt{\frac{x}{b^3}}\right]\right)$$

$$= \sum_{b\leqslant L}\mu_2(b)\left(\left[\sqrt{\frac{x+y}{b^3}}\right]-\left[\sqrt{\frac{x}{b^3}}\right]\right)+O\{(R_1+R_2)\log x+L\}$$

and since

$$\sqrt{\frac{x+y}{b^3}}-\sqrt{\frac{x}{b^3}} = \frac{y}{2\sqrt{xb^3}}+O\left(\frac{y^2}{(bx)^{3/2}}\right)$$

we infer that

$$\sum_{b\leqslant L}\mu_2(b)\left(\left[\sqrt{\frac{x+y}{b^3}}\right]-\left[\sqrt{\frac{x}{b^3}}\right]\right)$$

$$= \frac{y}{2x^{1/2}}\sum_{b\leqslant L}\frac{\mu_2(b)}{b^{3/2}}+O\left(\frac{y^2}{x^{3/2}}+L\right)$$

$$= \frac{y}{2x^{1/2}}\sum_{b=1}^{\infty}\frac{\mu_2(b)}{b^{3/2}}+O\left(L+\frac{y}{x^{1/2}}\sum_{b>L}\frac{1}{b^{3/2}}\right)$$

$$= \frac{\zeta(3/2)}{2\zeta(3)}\frac{y}{x^{1/2}}+O\left(L+\left(y^2x^{-1}\log x\right)^{1/4}\right)$$

$$= \frac{\zeta(3/2)}{2\zeta(3)}\frac{y}{x^{1/2}}+O(L)$$

where we used (5.45) which implies that L dominates all the other terms.

(c) To bound R_2, we split the sum into two subsums estimating trivially in the interval $]L, x^{2/15}]$ and using Theorem 5.22 in the interval $]x^{2/15}, (2x)^{1/5}]$ giving

$$\mathcal{R}\left(\left(\frac{x}{a^2}\right)^{1/3}, A, \frac{y}{(Ax)^{2/3}}\right) \ll (Ax)^{1/9}+y\left(Ax^{-2}\right)^{1/3}+\left(x^{-1}y^3A^{-4}\right)^{1/6}+Ayx^{-1}$$

so that

$$\max_{x^{2/15}<A\leqslant(2x)^{1/5}}\mathcal{R}\left(\left(\frac{x}{a^2}\right)^{1/3}, A, \frac{y}{(Ax)^{2/3}}\right) \ll x^{2/15}+yx^{-3/5}+y^{1/2}x^{-23/90}.$$

Now (5.45) gives

$$R_2 \ll x^{2/15} + \frac{L}{\log x}.$$

We treat R_1 by using Theorem 5.23 (i) with $k = 3$ giving

$$\mathcal{R}\left(\sqrt{\frac{x}{b^3}}, B, \frac{y}{\sqrt{xB^3}}\right) \ll \left(B^3 x\right)^{1/12} + \left(yB^{-1}\right)^{1/4} + y(Bx)^{-1/2} + B\left(yx^{-1}\right)^{1/2}$$

so that

$$\max_{L < B \leqslant (2x)^{1/5}} \mathcal{R}\left(\sqrt{\frac{x}{b^3}}, B, \frac{y}{\sqrt{xB^3}}\right) \ll x^{2/15} + y^{1/2}x^{-1/4}(\log x)^{1/4} + y^{1/2}x^{-3/10}$$

and using (5.45) implies also that

$$R_1 \ll x^{2/15} + \frac{L}{\log x}$$

concluding the proof.

(d) ▷ *Bounds for* R_2. In the range $]L, (64x)^{1/8}]$, we use Theorem 5.23 (i) with $k = 3$ giving

$$\mathcal{R}\left(\left(\frac{x}{a^2}\right)^{1/3}, A, \frac{y}{(Ax)^{2/3}}\right)$$
$$\ll \left(A^7 x\right)^{1/18} + \left(A^2 x^{-1} y^3\right)^{1/12} + y\left(x^{-2}A\right)^{1/3} + A\left(yx^{-1}\right)^{1/2}$$

so that

$$\max_{L < A \leqslant (64x)^{1/8}} \mathcal{R}\left(\left(\frac{x}{a^2}\right)^{1/3}, A, \frac{y}{(Ax)^{2/3}}\right)$$
$$\ll x^{5/48} + y^{1/4}x^{-1/16} + yx^{-5/8} + y^{1/2}x^{-3/8}.$$

In the range $](64x)^{1/8}, (2x)^{1/5}]$, we use Theorem 5.26 implying that

$$\mathcal{R}\left(\left(\frac{x}{a^2}\right)^{1/3}, A, \frac{y}{(Ax)^{2/3}}\right)$$
$$\ll \left\{(Ax)^{1/10} + \left(A^4 x^3\right)^{1/15} + y\left(Ax^{-2}\right)^{1/3} + \left(A^{-2}xy^3\right)^{1/24}\right.$$
$$\left. + \left(A^8 x^{-1} y^3\right)^{1/21} + \left(Ax^{-1}y^2\right)^{1/5} + A\left(yx^{-1}\right)^{1/2}\right\}(\log A)^{2/5}$$

so that

$$\max_{(64x)^{1/8} < A \leqslant (2x)^{1/5}} \mathcal{R}\left(\left(\frac{x}{a^2}\right)^{1/3}, A, \frac{y}{(Ax)^{2/3}}\right)$$

$$\ll \left\{ x^{3/25} + yx^{-3/5} + x^{1/32}y^{1/8} + x^{1/35}y^{1/7} \right.$$
$$\left. + y^{2/5}x^{-4/25} + y^{1/2}x^{-3/10} \right\} (\log x)^{2/5}$$

and the lower bound of (5.46) implies that

$$R_2 \ll x^{3/25}(\log x)^{2/5} + \frac{L}{\log x}.$$

▷ *Bounds for R_1.* In the range $]L, (16x)^{1/6}]$, we use Theorem 5.23 (i) with $k = 3$ giving

$$\mathcal{R}\left(\sqrt{\frac{x}{b^3}}, B, \frac{y}{\sqrt{xB^3}} \right) \ll \left(B^3 x \right)^{1/12} + \left(yB^{-1} \right)^{1/4} + y(Bx)^{-1/2} + B\left(yx^{-1} \right)^{1/2}$$

so that

$$\max_{L < B \leqslant (16x)^{1/6}} \mathcal{R}\left(\sqrt{\frac{x}{b^3}}, B, \frac{y}{\sqrt{xB^3}} \right)$$
$$\ll x^{1/8}(\log x)^{1/8} + y^{1/2}x^{-1/4}(\log x)^{1/4} + y^{1/2}x^{-3/8}.$$

In the range $](16x)^{1/6}, (2x)^{1/5}]$, we use Theorem 5.26 implying that

$$\mathcal{R}\left(\sqrt{\frac{x}{b^3}}, B, \frac{y}{\sqrt{xB^3}} \right)$$
$$\ll \left\{ \left(xB^{-1} \right)^{3/20} + (Bx)^{1/10} + y(Bx)^{-1/2} + \left(xy^2 B^{-3} \right)^{1/16} \right.$$
$$\left. + (By)^{1/7} + \left(x^{-1}y^4 B^{-3} \right)^{1/10} + B\left(yx^{-1} \right)^{1/2} \right\} (\log B)^{2/5}$$

so that

$$\max_{(16x)^{1/6} < B \leqslant (2x)^{1/5}} \mathcal{R}\left(\sqrt{\frac{x}{b^3}}, B, \frac{y}{\sqrt{xB^3}} \right)$$
$$\ll \left\{ x^{1/8} + yx^{-7/12} + x^{1/32}y^{1/8} + x^{1/35}y^{1/7} \right.$$
$$\left. + y^{2/5}x^{-3/20} + y^{1/2}x^{-3/10} \right\} (\log x)^{2/5}$$

and the lower bound of (5.46) implies that

$$R_1 \ll x^{1/8}(\log x)^{2/5} + \frac{L}{\log x}.$$

The proof is complete.

Remark Splitting R_1 into more parts and using Theorem 5.28 in the "critical" part, the author [Tri02] proved that, under a more restricted range than (5.46), the exponent $1/8$ may be reduced to $19/154 \approx 0.12333\ldots$ Note that the result obtained for

R_2 above shows that we might expect to get the bound $3/25 = 0.12$. However, this estimate for R_1 still remains open.

7.   Let $r \geqslant 2$ be an integer and define the multiplicative function $\tau^{(r)}$ by

$$\tau^{(r)}(n) = \sum_{d^r \mid n} 1.$$

Clearly, we have $\tau^{(r)}(p^\alpha) = 1 + [\alpha/r]$ so that $\tau^{(r)}$ satisfies the hypotheses of Theorem 4.62. Furthermore, we have using Lemma 5.2

$$\mathcal{R}\left(\frac{x}{n^r}, N\delta\right) \leqslant \sum_{N \leqslant d \leqslant 2N} \left(\left[\frac{x}{d^r} + \delta\right] - \left[\frac{x}{d^r} - \delta\right]\right) = \sum_{N \leqslant d \leqslant 2N} \sum_{x - d^r\delta < md^r \leqslant x + d^r\delta} 1$$

$$\leqslant \sum_{N \leqslant d \leqslant 2N} \sum_{x - (2N)^r\delta \leqslant md^r \leqslant x + (2N)^r\delta} 1 = \sum_{x - (2N)^r\delta \leqslant n \leqslant x + (2N)^r\delta} \sum_{\substack{d^r \mid n \\ N \leqslant d \leqslant 2N}} 1$$

$$\leqslant \sum_{x - (2N)^r\delta \leqslant n \leqslant x + (2N)^r\delta} \tau^{(r)}(n)$$

$$\ll \frac{2^{r+1} N^r \delta}{\log x} \exp\left(\sum_{p \leqslant x} \frac{\tau^{(r)}(p)}{p}\right) + x^\varepsilon \ll_{r,\varepsilon} N^r\delta$$

where we used Theorem 4.62 with the fact that $r \geqslant 2$, and Corollary 3.50.

8.

(a) Using respectively the inequalities $\delta^{-1} \geqslant c_0^{-1} N$ and $N^{-1} > (c_0^{-1}x)^{-1/3}$, we get

$$\frac{A}{R} = N^{-2/3} x^{1/6} \delta^{-1/6} \geqslant c_0^{-1/6} N^{-1/2} > c_0^{-1/6} \left(c_0^{-1}x\right)^{-1/6} x^{1/6} = 1$$

and similarly using $N \geqslant 2(x\delta)^{-1/2}$, we obtain

$$\frac{N}{2A} = 2^{-2/3} N^{1/3} (x\delta)^{1/6} \geqslant (x\delta)^{1/6} (x\delta)^{-1/6} = 1$$

so that

$$R < A \leqslant \frac{N}{2}.$$

(b) Inserting this value of A in the proof of Theorem 5.30 we get

$$|T| \ll \left(N^2 \delta x\right)^{1/6} + \left(x\delta^{-2} N^{-4}\right)^{1/3} + \left(x\delta^7 N^{11}\right)^{1/9}$$

implying the asserted result.

(c) For all $c_0^{-1} y \leqslant N \leqslant (c_0^{-1} x)^{1/3}$, we infer that

$$\mathcal{R}\left(\frac{x}{n^2}, N, \frac{y}{N^2}\right) \ll x^{1/5} + (xy)^{1/6} + \left(xy^{-2}\right)^{1/3} + \left(xy^7 N^{-3}\right)^{1/9}$$

so that

$$\max_{c_0^{-1} y \leqslant N \leqslant (c_0^{-1} x)^{1/3}} \mathcal{R}\left(\frac{x}{n^2}, N, \frac{y}{N^2}\right) \ll x^{1/5} + (xy)^{1/6} + \left(xy^{-2}\right)^{1/3} + \left(xy^4\right)^{1/9}$$

$$\ll \left(xy^4\right)^{1/9}$$

since $x^{1/5} \leqslant y \leqslant x^{1/3}$. Furthermore, it has been proved in Corollary 5.35 that

$$\max_{2\sqrt{y} \leqslant N \leqslant c_0^{-1} y} \mathcal{R}\left(\frac{x}{n^2}, N, \frac{y}{N^2}\right) \ll x^{1/10} y^{2/5} + (xy)^{1/7} + y^{2/3} + \left(x^{-1} y^4\right)^{1/3}$$

$$\ll \left(xy^4\right)^{1/9}$$

since $x^{1/5} \leqslant y \leqslant x^{1/3}$. Finally, by (5.29), we have

$$\max_{N \leqslant 2c_0^{2/3} x^{1/3}} \mathcal{R}\left(\sqrt{\frac{x}{n}}, N, \frac{y}{\sqrt{Nx}}\right) \ll x^{1/5} (\log x)^{2/5}$$

if c_0 is sufficiently small. Clearly $y^{1/2} \leqslant (xy^4)^{1/9}$, so that Lemma 5.3 with $A = 2\sqrt{y}$ and $B = (c_0^{-1} x)^{1/3}$ implies the asserted result. Note also that, since $y \geqslant x^{1/5}$, then

$$x^{1/15} y^{2/3} \geqslant \left(xy^4\right)^{1/9}.$$

9.   Using Theorem 5.22 we get for all $4y < N \leqslant x$

$$\mathcal{R}\left(\frac{x}{n}, N, \frac{y}{N}\right) \ll x^{1/3} + y + (xy)^{1/2} N^{-1} + N\left(yx^{-1}\right)$$

so that

$$\max_{4y < N \leqslant x} \mathcal{R}\left(\frac{x}{n}, N, \frac{y}{N}\right) \ll x^{1/3} + y + \left(xy^{-1}\right)^{1/2}$$

and the second term dominates the others in view of $y \geqslant x^{1/3}$.

10.   We use induction on k, the case $k = 2a$ coming from (5.26) and the fact that $k = 2a \geqslant 4$. Assume that the estimate is true for some $k \geqslant 2a$. By induction hypothesis and (5.26) used with $k + 1$ instead of k, we get $\mathcal{R}(f, N, \delta) \ll \min(E, F)$ where

$$E = \max\left(T^{\frac{2}{(k+1)(k+2)}} N^{\frac{k}{k+2}}, N\delta^{\frac{2}{k(k+1)}}, N\left(\delta T^{-1}\right)^{\frac{1}{k+1}}\right) = \max(e_1, e_2, e_3)$$

and

$$F = \max\left(T^{\frac{2}{k(k+1)}} N^{\frac{k-1}{k+1}}, N\delta^{\frac{1}{a(2a-1)}}\right) = \max(f_1, f_2)$$

say. The result is proved except in the cases $\min(e_2, f_1)$ and $\min(e_3, f_1)$. As in Proposition 5.24, the following inequality

$$\min(x, y) \leqslant x^a y^{1-a}$$

with $0 \leqslant a \leqslant 1$, is used.

▷ *Case* $\min(e_2, f_1)$. We choose $a = \frac{k-2a}{(k-a)(k+2)} \in [0, 1]$ which gives

$$\mathcal{R}(f, N, \delta) \ll T^{\frac{2}{(k+1)(k+2)}} N^{\frac{k}{k+2}} \left(TN^{-a}\delta^{1-2a/k}\right)^{\frac{2}{(k-a)(k+1)(k+2)}} \ll T^{\frac{2}{(k+1)(k+2)}} N^{\frac{k}{k+2}}$$

by (5.50) and the fact that $k \geqslant 2a$ and $\delta < \frac{1}{4}$.

▷ *Case* $\min(e_3, f_1)$. We choose $a = \frac{2}{k+2}$ which gives

$$\mathcal{R}(f, N, \delta) \ll T^{\frac{2}{(k+1)(k+2)}} N^{\frac{k}{k+2}} \left(N\delta T^{-1}\right)^{\frac{2}{(k+1)(k+2)}} \ll T^{\frac{2}{(k+1)(k+2)}} N^{\frac{k}{k+2}}$$

by (5.50) again. This completes the proof.

Using this result with $a = 2$ we get

$$\mathcal{R}(f, N, \delta) \ll T^{\frac{2}{k(k+1)}} N^{\frac{k-1}{k+1}} + N\delta^{1/6}$$

if $N\delta \leqslant T \leqslant N^2$. This result is useful since the condition $T \leqslant N^2$ (i.e. $\lambda_2 \leqslant 1$) is often satisfied in the usual applications.

A.6 Chapter 6

1.

(a) We have

$$\left|e(x) - e(y)\right| = \left|e(y)\right| \times \left|e(x - y) - 1\right| = \left|e(x - y) - 1\right| = 2\left|\sin\pi(x - y)\right|$$

and we conclude using Exercise 2 in Chap. 5.

(b) If $\alpha \in \mathbb{Z}$, then $e(\alpha n + \beta) = e(\beta)$ so that

$$\left|\sum_{n=M+1}^{N} e(\alpha n + \beta)\right| = \left|\sum_{n=M+1}^{N} 1\right| = N - M.$$

Assume that $\alpha \notin \mathbb{Z}$. Using the previous inequality we get

$$\left|\sum_{n=M+1}^{N} e(\alpha n + \beta)\right| = \left|\sum_{n=M+1}^{N} e(\alpha n)\right| = \frac{|e(N\alpha) - e(M\alpha)|}{|e(\alpha) - 1|} \leqslant \frac{2}{4\|\alpha\|} = \frac{1}{2\|\alpha\|}$$

as asserted.

2.

(a) We have

$$\left| \sum_{N < n \leqslant N_1} e\big(\pm f(n)\big) \right| \leqslant \left| \sum_{\substack{N < n \leqslant N_1 \\ \|f'(n)\| < \delta}} e\big(\pm f(n)\big) \right| + \left| \sum_{\substack{N < n \leqslant N_1 \\ \|f'(n)\| \geqslant \delta}} e\big(\pm f(n)\big) \right|$$

$$\leqslant \mathcal{R}\big(f', N, \delta\big) + \left| \sum_{\substack{N < n \leqslant N_1 \\ \|f'(n)\| \geqslant \delta}} e\big(\pm f(n)\big) \right|.$$

Since f' is non-decreasing and $f'(N_1) - f'(N) \ll N\lambda_2$ by the mean-value theorem, the interval $f'([N, N_1])$ has at most $\ll N\lambda_2 + 1$ integers by Proposition 1.11 (vi). It follows that the set $\{x \in [N, N_1] : \|f'(x)\| \geqslant \delta\}$ can be partitioned in at most $\ll N\lambda_2 + 1$ subintervals, and the Kusmin–Landau inequality (Corollary 6.7) applied on each of these intervals implies the asserted estimate.

(b) Applying Theorem 5.6 we get

$$\sum_{N < n \leqslant N_1} e\big(f(n)\big) \ll N\lambda_2 + N\delta + \delta\lambda_2^{-1} + N\lambda_2\delta^{-1} + \delta^{-1} + 1$$

$$\ll N\lambda_2\delta^{-1} + N\delta + \delta\lambda_2^{-1} + \delta^{-1}$$

and choosing $\delta = \lambda_2^{1/2}$ gives

$$\sum_{N < n \leqslant N_1} e\big(f(n)\big) \ll N\lambda_2^{1/2} + \lambda_2^{-1/2}$$

as required.

3.   Squaring out we get

$$\left| \sum_{h=0}^{H-1} e(ha) \right|^2 = \sum_{h_1=0}^{H-1} e(h_1 a) \sum_{h_2=0}^{H-1} e(-h_2 a) = \sum_{h_1=0}^{H-1} \sum_{h_2=0}^{H-1} e\big((h_1 - h_2)a\big).$$

Now set $h_1 = h + k$ and $h_2 = k$ so that

$$\begin{cases} 0 \leqslant h_1 \leqslant H - 1, \\ 0 \leqslant h_2 \leqslant H - 1 \end{cases} \quad \Longleftrightarrow \quad \begin{cases} 0 \leqslant k \leqslant H - 1, \\ -h \leqslant k \leqslant H - 1 - h \end{cases}$$

$$\Longleftrightarrow \quad \begin{cases} |h| \leqslant H - 1, \\ 0 \leqslant k \leqslant H - 1 - |h| \end{cases}$$

and hence

$$\left| \sum_{h=0}^{H-1} e(ha) \right|^2 = \sum_{|h| \leqslant H-1} \sum_{k=0}^{H-1-|h|} e(ha) = \sum_{|h| \leqslant H-1} (H - |h|) e(ha)$$

as required.

4. We may obviously assume that $\mathcal{R}(f, N, \delta) \neq 0$, otherwise the inequality (6.29) is trivial.

(a) There exist $m \in \mathbb{Z}$ and $\delta_0 \in \mathbb{R}$ such that $f(n) = m + \delta_0$ with $|\delta_0| < \delta$, so that

$$hf(n) = hm + h\delta_0 \quad \text{with} \quad |h\delta_0| < (H - 1)\delta \leqslant (K - 1)\delta \leqslant \frac{1}{8}$$

and thus

$$\text{Re}\{e(hf(n))\} = \cos(2\pi hf(n)) = \cos(2\pi hm + 2\pi h\delta_0) = \cos(2\pi h\delta_0) > \frac{\sqrt{2}}{2}$$

since $2\pi |h\delta_0| < \pi/4$.

(b) Summing the previous inequality over n and h running respectively through the whole set $\mathcal{S}(f, N, \delta)$ and the integers $\{0, \ldots, H - 1\}$, we obtain

$$H\mathcal{R}(f, N, \delta) \leqslant \sqrt{2} \sum_{n \in \mathcal{S}(f,N,\delta)} \text{Re}\left(\sum_{h=0}^{H-1} e(hf(n)) \right)$$

$$\leqslant \sqrt{2} \sum_{n \in \mathcal{S}(f,N,\delta)} \left| \sum_{h=0}^{H-1} e(hf(n)) \right|.$$

Applying the Cauchy–Schwarz inequality gives

$$\mathcal{R}(f, N, \delta) \leqslant \frac{\sqrt{2}}{H} \left(\sum_{n \in \mathcal{S}(f,N,\delta)} 1 \right)^{1/2} \left(\sum_{n \in \mathcal{S}(f,N,\delta)} \left| \sum_{h=0}^{H-1} e(hf(n)) \right|^2 \right)^{1/2}$$

$$= \frac{\sqrt{2}}{H} \mathcal{R}(f, N, \delta)^{1/2} \left(\sum_{n \in \mathcal{S}(f,N,\delta)} \left| \sum_{h=0}^{H-1} e(hf(n)) \right|^2 \right)^{1/2}$$

so that squaring out we get

$$\mathcal{R}(f, N, \delta) \leqslant \frac{2}{H^2} \sum_{n \in \mathcal{S}(f,N,\delta)} \left| \sum_{h=0}^{H-1} e(hf(n)) \right|^2 \leqslant \frac{2}{H^2} \sum_{N \leqslant n \leqslant 2N} \left| \sum_{h=0}^{H-1} e(hf(n)) \right|^2$$

as asserted.

Now using Exercise 3 we obtain

$$\mathcal{R}(f, N, \delta) \leqslant \frac{2}{H^2} \sum_{N \leqslant n \leqslant 2N} \sum_{|h| \leqslant H-1} (H - |h|)e\big(hf(n)\big)$$

and treating the cases $h = 0$ and $h \neq 0$ separately we get

$$\mathcal{R}(f, N, \delta) \leqslant \frac{2(N+1)}{H} + \frac{2}{H^2} \sum_{N \leqslant n \leqslant 2N} \sum_{\substack{|h| \leqslant H-1 \\ h \neq 0}} (H - |h|)e(hf(n))$$

$$\leqslant \frac{4N}{H} + \frac{2}{H} \sum_{N \leqslant n \leqslant 2N} \sum_{\substack{|h| \leqslant H-1 \\ h \neq 0}} \left(1 - \frac{|h|}{H}\right)e(hf(n))$$

$$= \frac{4N}{H} + \frac{2}{H} \sum_{h=1}^{H-1} \left(1 - \frac{h}{H}\right) \sum_{N \leqslant n \leqslant 2N} \{e(hf(n)) + e(-hf(n))\}$$

$$\leqslant \frac{4N}{H} + \frac{4}{H} \sum_{h=1}^{H-1} \mathrm{Re}\left(\sum_{N \leqslant n \leqslant 2N} e(hf(n))\right)$$

$$\leqslant \frac{4N}{H} + \frac{4}{H} \sum_{h=1}^{H-1} \left| \sum_{N \leqslant n \leqslant 2N} e(hf(n)) \right|$$

completing the proof of (6.29).

5. By Definition 6.34 and the inequality (6.29), we get for all integers $1 \leqslant H \ll \delta^{-1}$

$$\mathcal{R}(f, N, \delta) \ll NH^{-1} + H^{-1} \sum_{h \leqslant H} \big((hT)^k N^{l-k} + N(hT)^{-1}\big)$$

$$\ll NH^{-1} + (HT)^k N^{l-k} + NT^{-1} \ll NH^{-1} + (HT)^k N^{l-k}$$

since $N \leqslant 8T$. We conclude the proof by using Lemma 5.5.

6.

▷ *Squarefree problem.* Let x, y be real numbers satisfying (5.5). Using the exponent pair (6.19), we get for all $N \leqslant 2x^{1/3}$

$$\mathcal{R}\left(\sqrt{\frac{x}{n}}, N, \frac{y}{\sqrt{Nx}}\right) \ll (x^{97} N^{167})^{1/696} + (x^{97} N^{-27})^{1/502} + y(Nx^{-1})^{1/2}$$

so that

$$\max_{N \leqslant 2x^{1/3}} \mathcal{R}\left(\sqrt{\frac{x}{n}}, N, \frac{y}{\sqrt{Nx}}\right) \ll x^{229/1044} + yx^{-1/3}.$$

Next we apply the transformation BA to Huxley's exponent pair $(\frac{32}{205} + \varepsilon, \frac{1}{2} + \frac{32}{205} + \varepsilon)$ giving the exponent pair

$$\left(\frac{269}{948} + \varepsilon, \frac{269}{474} + \varepsilon \right)$$

where the small real number $\varepsilon > 0$ need not have the same occurrence at each computation. This implies for all $2\sqrt{y} < N \leqslant 2x^{1/3}$ and all $\varepsilon > 0$

$$\mathcal{R}\left(\frac{x}{n^2}, N, \frac{y}{N^2} \right) \ll x^{269/1217+\varepsilon} + \left(xN^{-1} \right)^{269/948+\varepsilon} + yN^{-1}$$

so that

$$\max_{x^{1/4} < N \leqslant 2x^{1/3}} \mathcal{R}\left(\frac{x}{n^2}, N, \frac{y}{N^2} \right) \ll x^{269/1217+\varepsilon} + y^{1/2}.$$

The exponent pair $(\frac{1}{6}, \frac{2}{3})$ provides the bound

$$\mathcal{R}\left(\frac{x}{n^2}, N, \frac{y}{N^2} \right) \ll (N^2 x)^{1/7} + (Nx)^{1/6} + yN^{-1}$$

for all $2\sqrt{y} < N \leqslant 2x^{1/3}$, so that

$$\max_{2\sqrt{y} < N \leqslant x^{1/4}} \mathcal{R}\left(\frac{x}{n^2}, N, \frac{y}{N^2} \right) \ll x^{3/14} + y^{1/2}.$$

Hence using Lemma 5.3 with $A = 2\sqrt{y}$ and $B = x^{1/3}$ we get for all x, y satisfying (5.5) and $\varepsilon > 0$

$$\sum_{x < n \leqslant x+y} \mu_2(n) = \frac{y}{\zeta(2)} + O_\varepsilon\left(x^{269/1217+\varepsilon} + y^{1/2} \right).$$

▷ *Square-full problem.* We take up the notation of Exercise 6 in Chap. 5 and let x, y be real numbers satisfying (5.45). Using the exponent pair (6.19), we get

$$\mathcal{R}\left(\left(\frac{x}{a^2} \right)^{1/3}, A, \frac{y}{(Ax)^{2/3}} \right) \ll (x^{97} A^{202})^{1/1044} + (x^{97} A^{-89})^{1/753} + y(Ax^{-2})^{1/3}$$

so that

$$\max_{L < A \leqslant (2x)^{1/5}} \mathcal{R}\left(\left(\frac{x}{a^2} \right)^{1/3}, A, \frac{y}{(Ax)^{2/3}} \right) \ll x^{229/1740} + yx^{-3/5}.$$

Using (6.19) again, we get

$$\mathcal{R}\left(\sqrt{\frac{x}{b^3}}, B, \frac{y}{\sqrt{xB^3}} \right) \ll (x^{97} B^{-27})^{1/696} + (x^{97} B^{-221})^{1/502} + \frac{y}{\sqrt{Bx}}$$

so that

$$\max_{x^{59/313} < B \leqslant (2x)^{1/5}} \mathcal{R}\left(\sqrt{\frac{x}{b^3}}, B, \frac{y}{\sqrt{xB^3}}\right) \ll x^{124/939} + yx^{-186/313}.$$

The exponent pair $BA^2B(0,1) = (\frac{2}{7}, \frac{4}{7})$ provides the estimate

$$\mathcal{R}\left(\sqrt{\frac{x}{b^3}}, B, \frac{y}{\sqrt{xB^3}}\right) \ll (Bx)^{1/9} + \left(xB^{-1}\right)^{1/7} + \frac{y}{\sqrt{Bx}}$$

valid for all $L < B \leqslant (2x)^{1/5}$, so that

$$\max_{x^{1/8} < B \leqslant x^{59/313}} \mathcal{R}\left(\sqrt{\frac{x}{b^3}}, B, \frac{y}{\sqrt{xB^3}}\right) \ll x^{124/939} + yx^{-9/16}.$$

Completing the proof with the trivial bound for R_1 in the range $]L, x^{1/8}]$ we finally get for all x, y satisfying (5.45)

$$\sum_{x < n \leqslant x+y} s_2(n) = \frac{\zeta(3/2)}{2\zeta(3)} \frac{y}{x^{1/2}} + O\left(x^{124/939} \log x + L\right).$$

7.

(a) This is [GK91, Lemma 2.11].

(b) Let $1 \leqslant T < t$ and let (k, l) be an exponent pair. We have

$$\left|\sum_{n \leqslant t} n^{-\sigma-it}\right| \leqslant \left|\sum_{n \leqslant T} n^{-\sigma-it}\right| + \left|\sum_{T < n \leqslant t} n^{-\sigma-it}\right|$$

$$\ll \sum_{n \leqslant T} n^{-\sigma} + \max_{T < N \leqslant t}\left|\sum_{N < n \leqslant 2N} n^{-\sigma-it}\right| \log t$$

$$\ll \sum_{n \leqslant T} n^{-\sigma} + \max_{T < N \leqslant t} N^{-\sigma} \max_{N \leqslant N_1 \leqslant 2N}\left|\sum_{N \leqslant n \leqslant N_1} n^{-it}\right| \log t$$

$$\ll T^{1-\sigma} + \max_{T < N \leqslant t} N^{-\sigma} \max_{N \leqslant N_1 \leqslant 2N}\left(t^k N^{l-k} + Nt^{-1}\right) \log t$$

$$\ll T^{1-\sigma} + \max_{T < N \leqslant t}\left(t^k N^{l-k-\sigma} + t^{-1} N^{1-\sigma}\right) \log t.$$

Now since $l - k \leqslant \frac{1}{2}$ and $\frac{1}{2} \leqslant \sigma \leqslant 1$, we deduce that $k + \sigma \geqslant l$. This implies that

$$\sum_{n \leqslant t} n^{-\sigma-it} \ll T^{1-\sigma} + t^k T^{l-k-\sigma} \log t$$

and the choice of $T = t^{\frac{k}{1+k-l}}$ gives

$$\sum_{n \leqslant t} n^{-\sigma - it} \ll t^{\frac{k(1-\sigma)}{1+k-l}} \log t.$$

Note that $0 \leqslant k \leqslant \frac{1}{2} \leqslant l \leqslant 1$ and $l - k \leqslant \frac{1}{2}$ imply that $\frac{1}{2} \leqslant 1 + k - l \leqslant 1$. Using the previous question, we get

$$\zeta(\sigma + it) \ll \left(t^{\frac{k(1-\sigma)}{1+k-l}} + t^{1-2\sigma}\right) \log t$$

and the second term is clearly absorbed by the first one. With $\sigma = \frac{1}{2}$ this gives for all $t \geqslant 3$

$$\zeta\left(\frac{1}{2} + it\right) \ll t^{\frac{k}{2(1+k-l)}} \log t$$

and Huxley's exponent pair $(\frac{32}{205} + \varepsilon, \frac{1}{2} + \frac{32}{205} + \varepsilon)$ provides the bound

$$\zeta\left(\frac{1}{2} + it\right) \ll t^{32/205 + \varepsilon}$$

for all $t \geqslant 3$ and $\varepsilon > 0$, which is the best result up to now.

A.7 Chapter 7

1. Let $\mathbb{K} = \mathbb{Q}(\sqrt{-5})$. Suppose that 3 is not an irreducible so that $3 = rs$ with $N_{\mathbb{K}/\mathbb{Q}}(r) \neq 1$ and $N_{\mathbb{K}/\mathbb{Q}}(s) \neq 1$. Since $9 = N_{\mathbb{K}/\mathbb{Q}}(3) = N_{\mathbb{K}/\mathbb{Q}}(r) N_{\mathbb{K}/\mathbb{Q}}(s)$, we must then have $N_{\mathbb{K}/\mathbb{Q}}(r) = N_{\mathbb{K}/\mathbb{Q}}(s) = 3$ and hence $a^2 + 5b^2 = 3$ for some $a, b \in \mathbb{Z}$. This implies that $b = 0$ and thus $a^2 = 3$ which is impossible.

Similarly, if $7 = rs$ where neither r nor s is a unit, then we must have $a^2 + 5b^2 = 7$, implying that either $b = 0$ and $a^2 = 7$ or $b = \pm 1$ and $a^2 = 2$, both cases being impossible.

If $1 \pm 2\sqrt{-5} = rs$ where neither r nor s is a unit, then

$$21 = N_{\mathbb{K}/\mathbb{Q}}(1 \pm 2\sqrt{-5}) = N_{\mathbb{K}/\mathbb{Q}}(r) N_{\mathbb{K}/\mathbb{Q}}(s)$$

and hence either $N_{\mathbb{K}/\mathbb{Q}}(r) = 3$ or $N_{\mathbb{K}/\mathbb{Q}}(s) = 3$, which is impossible as was seen above.

2. θ is algebraic over $\mathbb{Q}$ as sum of two algebraic numbers over $\mathbb{Q}$. This gives the answer to the exercise, but does not provide the minimal polynomial of θ.

To do this, one may use the following lemma, useful for small degrees (see [Coh00, Proposition 2.1.7]).

Lemma *Let α, β be algebraic over $\mathbb{Q}$ and P, $Q \in \mathbb{Q}[X]$ such that $P(\alpha) = Q(\beta) = 0$. Then the resultant*

$$R = \mathrm{Res}_Y\big(P(X), Q(Y - X)\big)$$

satisfies $R \in \mathbb{Q}[Y]$ and $R(\alpha + \beta) = 0$.

Proof We have clearly $R \in \mathbb{Q}[Y]$. Furthermore, R is equal to zero if and only if P and Q have a common root. But $R(\alpha + \beta)$ is the resultant of $P(X)$ and $Q(\alpha + \beta - X)$ which have α as a common root, and hence $R(\alpha + \beta) = 0$. $\qquad\square$

Applying this result with $P = X^5 - 2$ and $Q = X^3 - 2$ we get

$$R = \begin{vmatrix} 1 & 0 & 0 & 0 & 0 & -2 & 0 & 0 \\ 0 & 1 & 0 & 0 & 0 & 0 & -2 & 0 \\ 0 & 0 & 1 & 0 & 0 & 0 & 0 & -2 \\ -1 & 3Y & -3Y^2 & Y^3 - 2 & 0 & 0 & 0 & 0 \\ 0 & -1 & 3Y & -3Y^2 & Y^3 - 2 & 0 & 0 & 0 \\ 0 & 0 & -1 & 3Y & -3Y^2 & Y^3 - 2 & 0 & 0 \\ 0 & 0 & 0 & -1 & 3Y & -3Y^2 & Y^3 - 2 & 0 \\ 0 & 0 & 0 & 0 & -1 & 3Y & -3Y^2 & Y^3 - 2 \end{vmatrix}$$

$$= Y^{15} - 10Y^{12} - 6Y^{10} + 40Y^9 - 360Y^7 - 80Y^6 + 12Y^5$$
$$- 1080Y^4 + 80Y^3 - 240Y^2 - 240Y - 40.$$

Furthermore, this polynomial is irreducible over $\mathbb{Z}$ by applying Ore's criterion since $|P(m)|$ is prime for

$$m \in \{-653, -579, -532, 459, -447, -429, -427, -367, -337, -271, -81, -43$$
$$51, 209, 213, 339, 423, 509, 521, 581\}.$$

Hence $\deg(2^{1/3} + 2^{1/5}) = 15$.

3. P is irreducible over $\mathbb{Z}$ since $\deg P = 3$ and P has no rational root. Indeed, if P has such a root, then it must be ± 1 by using Exercise 17 in Chap. 3, and $P(\pm 1) = 1$. This implies that $2\alpha^2 - 3\alpha + 2 \neq 0$ and then β is well-defined. Furthermore, $[\mathbb{Q}(\alpha) : \mathbb{Q}] = 3$ and $\{1, \alpha, \alpha^2\}$ is a $\mathbb{Q}$-base of $\mathbb{Q}(\alpha)/\mathbb{Q}$. Therefore there exist x, y, $z \in \mathbb{Q}$ such that

$$\frac{1}{2\alpha^2 - 3\alpha + 2} = x + y\alpha + z\alpha^2.$$

This may be written as $(2\alpha^2 - 3\alpha + 2)(x + y\alpha + z\alpha^2) = 1$ and expanding the product and using the relations $\alpha^3 = \alpha - 1$ and $\alpha^4 = \alpha^2 - \alpha$ we get

$$\alpha^2(2x - 3y + 4z) + \alpha(-3x + 4y - 5z) + 2x - 2y + 3z - 1 = 0$$

implying that $x = z = 1$ and $y = 2$, so that

$$\frac{1}{2\alpha^2 - 3\alpha + 2} = 1 + 2\alpha + \alpha^2.$$

This gives $\beta^2 = 7\alpha^2 + 7\alpha - 3$ and $\beta^3 = 25\alpha^2 + 15\alpha - 24$, so that

$$\beta^3 - 5\beta^2 + 10\beta - 1 = 0.$$

One easily checks that the polynomial $Q = X^3 - 5X^2 + 10X - 1$ is irreducible over $\mathbb{Z}$, and hence Q is the minimal polynomial of β.

4. Set $\theta = \sqrt{5} + \sqrt[4]{2}$. We have obviously $\mathbb{Q}(\theta) \subseteq \mathbb{Q}(\sqrt{5}, \sqrt[4]{2})$. Conversely, since $(\theta - \sqrt{5})^4 = 2$, expanding the product we get

$$\theta^4 + 30\theta^2 + 23 = \sqrt{5}(4\theta^3 + 20\theta)$$

so that

$$\sqrt{5} = \frac{\theta^4 + 30\theta^2 + 23}{4\theta^3 + 20\theta}$$

and hence $\sqrt{5} \in \mathbb{Q}(\theta)$. Thus

$$\sqrt[4]{2} = \theta - \sqrt{5} \in \mathbb{Q}(\theta, \sqrt{5}) \subseteq \mathbb{Q}(\theta).$$

Therefore we get

$$\mathbb{Q}(\sqrt{5}, \sqrt[4]{2}) \subseteq \mathbb{Q}(\sqrt{5} + \sqrt[4]{2})$$

as required. Now using the lemma of Exercise 2 we infer that θ is a root of the polynomial

$$P = X^8 - 20X^6 + 146X^4 - 620X^2 + 529$$

and $|P(m)|$ is prime for $\pm m \in \{6, 12, 18, 60, 66, 120, 132\}$ so that P is irreducible over $\mathbb{Z}$ by Proposition 7.28. Hence

$$[\mathbb{Q}(\sqrt{5} + \sqrt[4]{2}) : \mathbb{Q}] = 8.$$

5. We first have

$$F_n = n! \, P_n = \sum_{k=0}^{n} \frac{n!}{k!} X^k.$$

1. Let p be a prime factor of $n(n - 1) \cdots (n - m + 1) = \frac{n!}{(n-m)!}$. Therefore for all $0 \leqslant k \leqslant n - m$, p divides $(n!/k!)$ which is the coefficient of X^k in $F_n(X)$, so that $F_n(X) \bmod p$ is divisible by X^{n-m+1}.

If $F_n = \overline{A_m} \, B$ with $B \in \mathbb{Z}[X]$ is monic such that $\deg B = n - m$, then X^{n-m+1} divides $\overline{A_m} \times \overline{B}$ in $\mathbb{F}_p[X]$. Since $\deg \overline{B} = n - m$, we get $X \mid \overline{A_m}$. This implies that $\overline{A_m}(0) = \overline{0}$ as required.

2. a. First note that $\theta \in \mathcal{O}_{\mathbb{K}}$ since A_m is monic. By the previous question, we get

$$N_{\mathbb{K}/\mathbb{Q}}(\theta) = \pm a_0 \equiv 0 \pmod{p}$$

so that $p \mid \mathcal{N}_{\mathbb{K}/\mathbb{Q}}((\theta))$.

b. Since $F_n(\theta) = 0$, we have

$$-n! = \sum_{k=1}^{n} \frac{n! \theta^k}{k!}$$

hence there exists an index $k \in \{1, \ldots, n\}$ such that

$$v_{\mathfrak{p}}\left(\frac{n! \theta^k}{k!}\right) \leqslant v_{\mathfrak{p}}(n!).$$

Since

$$v_{\mathfrak{p}}\left(\frac{n! \theta^k}{k!}\right) = v_{\mathfrak{p}}(n!) + k\alpha - v_{\mathfrak{p}}(k!) = v_{\mathfrak{p}}(n!) + k\alpha - e v_p(k!)$$

we get

$$k\alpha - e v_p(k!) \leqslant 0.$$

c. By Exercise 12 in Chap. 3, we get

$$k\alpha \leqslant \frac{e(k-1)}{p-1}$$

and hence

$$(p-1)\alpha \leqslant \frac{e(k-1)}{k} < e \leqslant m$$

so that

$$p < \frac{m}{\alpha} + 1 \leqslant m + 1$$

implying that $p \leqslant m$.

3. By the first question, all the prime factors of $n, n-1, \ldots, n-m+1$ divide a_0 and the previous question shows that each of these prime factors is $\leqslant m$. Thus the numbers $n, n-1, \ldots, n-m+1$ form a sequence of m consecutive integers all greater than m which have no prime factor greater than m, contradicting Lemma 7.183.

Remark In [Col87], the author provided an elegant proof of Schur's result based upon the theory of Newton polygons for polynomials belonging to $\mathbb{Q}_p[X]$. Let us compute the discriminant of P_n. If $\alpha_1, \ldots, \alpha_n$ are the roots of P_n in an algebraic closure of $\mathbb{Q}$, we have by Definition 7.36

$$\mathrm{disc}(P_n) = (n!)^{2-2n} \prod_{1 \leqslant i < j \leqslant n} (\alpha_i - \alpha_j)^2.$$

We proceed as in the proof of Proposition 7.61 (iv). Writing $P_n = (n!)^{-1} \prod_{i=1}^n (X - \alpha_i)$, we infer that

$$\frac{P_n'}{P_n} = \sum_{i=1}^n \frac{1}{X - \alpha_i}$$

so that

$$P_n' = \frac{1}{n!} \sum_{i=1}^n \prod_{j \neq i} (X - \alpha_j)$$

and thus for all $i \in \{1, \ldots, n\}$, we get

$$n! \, P_n'(\alpha_i) = \prod_{j \neq i} (\alpha_i - \alpha_j).$$

This implies that

$$\prod_{i=1}^n n! \, P_n'(\alpha_i) = \prod_{i=1}^n \prod_{j \neq i} (\alpha_i - \alpha_j)$$

$$= \prod_{i=1}^n \prod_{i < j} (-(\alpha_i - \alpha_j)^2)$$

$$= (-1)^{n(n-1)/2} (n!)^{2n-2} \, \mathrm{disc}(P_n).$$

Note that $P_n' = P_{n-1}$ and $P_n(X) = P_{n-1}(X) + x^n/n!$ so that

$$P_n'(\alpha_i) = P_n(\alpha_i) - \frac{\alpha_i^n}{n!} = -\frac{\alpha_i^n}{n!}$$

and hence

$$\mathrm{disc}(P_n) = (-1)^{n(n-1)/2} (n!)^{2-2n} \prod_{i=1}^n (-\alpha_i^n)$$

$$= (-1)^{n(n-1)/2+n} (n!)^{2-2n} \left(\prod_{i=1}^n \alpha_i \right)^n$$

$$= (-1)^{n(n-1)/2+n} (n!)^{2-2n} \left((-1)^n n! \right)^n$$

$$= (-1)^{n(n-1)/2} (n!)^{2-n}.$$

We deduce that if $n \equiv 2$ or $3 \pmod 4$, then $\mathrm{disc}(P_n) < 0$ and hence $\mathrm{disc}(P_n)$ is not a square in $\mathbb{Q}$. If $n \equiv 0 \pmod 4$, then $n-2$ is even and thus $\mathrm{disc}(P_n)$ is a square in $\mathbb{Q}$. Now assume that $n \equiv 1 \pmod 4$. By Corollary 3.44, for all $n \geqslant 2$, there exists a prime number p such that $n/2 < p \leqslant n$, so that $v_p(n!) = 1$. This implies that $v_p((n!)^n) = n$ is odd, and thus $(n!)^n$ is not a square in $\mathbb{Q}$ if $n \equiv 1 \pmod 4$ and $n \geqslant 2$. Therefore $\mathrm{disc}(P_n)$ cannot be a square in $\mathbb{Q}$ in this case. Furthermore, it can be proved that $\mathrm{Gal}(P_n/\mathbb{Q})$ contains a p-cycle for some prime number satisfying $n/2 < p < n-2$ (see [Col87] for instance). Using Lemma 7.144, we get the following result due to Schur too.

Proposition (Schur) *Let $n \in \mathbb{Z}_{\geqslant 2}$. Then*

$$\mathrm{Gal}(P_n/\mathbb{Q}) \simeq \begin{cases} \mathcal{A}_n, & \text{if } n \equiv 0 \pmod 4, \\ \mathcal{S}_n, & \text{otherwise.} \end{cases}$$

6.

(a) Since $\theta^n = -a_{n-1}\theta^{n-1} - \cdots - a_1\theta - a_0$ and $p \mid a_i$, we infer that $\theta^n/p \in M \subseteq \mathcal{O}_{\mathbb{K}}$ and that $N_{\mathbb{K}/\mathbb{Q}}(\theta) = a_0 \not\equiv 0 \pmod{p^2}$ by assumption.

(b) Since $p \mid f$, we deduce that there is an element of order p in $\mathcal{O}_{\mathbb{K}}/M$ by Theorem 7.1 (ii), so that there exists $\alpha \in \mathcal{O}_{\mathbb{K}}$ such that $\alpha \notin M$ and $p\alpha \in M$. Hence

$$p\alpha = b_0 + b_1\theta + \cdots + b_{n-1}\theta^{n-1}$$

where not all the b_i are divisible by p, otherwise $\alpha \in M$.

(c) $\beta \in \mathcal{O}_{\mathbb{K}}$ since both α and $b_0 p^{-1} + \cdots + b^{j-1}\theta^{j-1}p^{-1}$ are in $\mathcal{O}_{\mathbb{K}}$. This implies that

$$\beta\theta^{n-j-1} = \frac{b_j\theta^{n-1}}{p} + \frac{\theta^n}{p}\left(b_{j+1} + b_{j+2}\theta + \cdots + b_n\theta^{n-j-2}\right)$$

is also in $\mathcal{O}_{\mathbb{K}}$. Now by the first question $\theta^n/p \in \mathcal{O}_{\mathbb{K}}$ and also $b_{j+1} + b_{j+2}\theta + \cdots + b_n\theta^{n-j-2} \in \mathcal{O}_{\mathbb{K}}$, so that

$$\frac{b_j\theta^{n-1}}{p} \in \mathcal{O}_{\mathbb{K}}.$$

By Proposition 7.55, we infer that the norm of this element must be an integer. But

$$N_{\mathbb{K}/\mathbb{Q}}\left(\frac{b_j\theta^{n-1}}{p}\right) = \left(\frac{b_j}{p}\right)^n N_{\mathbb{K}/\mathbb{Q}}(\theta)^{n-1} = \frac{b_j^n a_0^{n-1}}{p^n}$$

cannot be an integer since $p \nmid b_j$ and $p^2 \nmid a_0$.

7.

(a) P is irreducible over $\mathbb{Z}$ by Eisenstein's criterion with $p = 7$.

(b) Since $189 = 3^3 \times 7$ and $756 = 2^2 \times 3^3 \times 7$, we have $\mathbb{K} = \mathbb{Q}(\alpha)$ where $\alpha = \theta/3$ is a root of $Q = X^3 - 21X + 28$. We have $\mathrm{disc}(Q) = 2^2 \times 3^4 \times 7^2$ and use Proposition 7.70. The largest square n^2 dividing $\mathrm{disc}(Q)$ for which the system of congruences

$$\begin{cases} x^3 - 21x + 28 \equiv 0 \ (\mathrm{mod}\, n^2), \\ 3x^2 - 21 \equiv 0 \ (\mathrm{mod}\, n) \end{cases}$$

is solvable for x is given by $n = 2$ and we get $x = 1$, so that

$$\left\{ 1, \alpha, \frac{-1 + \alpha + \alpha^2}{2} \right\} = \left\{ 1, \frac{\theta}{3}, -\frac{1}{2} + \frac{\theta}{6} + \frac{\theta^2}{18} \right\}$$

is an integral basis for $\mathbb{K}$.

(c) Since $\mathrm{disc}(P)$ is a square in $\mathbb{Q}$, we get $\mathrm{Gal}(\mathbb{K}/\mathbb{Q}) \simeq \mathcal{A}_3 \simeq C_3$ by Lemma 7.145 (ii) or Lemma 7.140.

Since $(r_1, r_2) = (3, 0)$, we have $\mathcal{O}_{\mathbb{K}}^* \simeq W_{\mathbb{K}} \times \mathbb{Z}^2$ by Dirichlet's unit theorem (Theorem 7.74). Using Theorem 7.105 we get

$$(3) = \mathfrak{p}_3^3$$

with $\mathfrak{p}_3 = (3, \theta/3 + 1)$ and using the PARI/GP system we obtain

$$(6 - \theta) = \mathfrak{p}_2 \mathfrak{p}_3^4 \quad \text{and} \quad (12 - \theta) = \mathfrak{p}_2^3 \mathfrak{p}_3^3$$

with $\mathfrak{p}_2 = (2, \theta)$. This implies that

$$(6 - \theta)^3 = (3)^3 (12 - \theta)$$

and hence there exists a unit u such that $(6 - \theta)^3 = 27u(12 - \theta)$. Now expanding $(6 - \theta)^3$ and using $\theta^3 = 189\theta - 756$, we deduce that

$$18\theta^2 - 297\theta + 972 = 27u(12 - \theta)$$

so that

$$9(\theta - 12)(2\theta - 9) = 27u(12 - \theta)$$

and then $u = 3 - 2\theta/3$. Using PARI, the second unit is $u' = \theta^2/9 - 5\theta/3 + 5$ so that

$$\mathcal{R}_{\mathbb{K}} = \left| \det \begin{pmatrix} \log|3 - 2\theta/3| & \log|\theta^2/9 - 5\theta/3 + 5| \\ \log|3 - 2\theta'/3| & \log|\theta'^2/9 - 5\theta'/3 + 5| \end{pmatrix} \right| \approx 12.594\,188\,956\ldots$$

8.   Since $-2 \not\equiv 1 \ (\mathrm{mod}\, 4)$, we have $d_{\mathbb{K}} = -8$.

(a) ▷ Since $-8 \equiv 1 \ (\mathrm{mod}\, 3)$, we get $(-8/3) = (1/3) = 1$ so that 3 splits completely in $\mathbb{K}$ by Proposition 7.108 and then

$$(3) = \mathfrak{p}_3 \overline{\mathfrak{p}_3}$$

where $\mathfrak{p}_3 = (1 + \sqrt{-2})$ and $\overline{\mathfrak{p}_3} = (1 - \sqrt{-2})$.

▷ Since $(1 - \sqrt{-2})^2 = -1 - 2\sqrt{-2}$, we infer that $\mathfrak{a} = \overline{\mathfrak{p}_3}^2$, so that the equality $(1 + 2\sqrt{-2})^n = 3^n$ contradicts Theorem 7.88.

(b) Suppose that $\arccos(1/3)/\pi \in \mathbb{Q}$. There exists $(p, q) \in \mathbb{Z} \times \mathbb{Z}_{\geqslant 1}$ such that $(p, q) = 1$ and

$$\arccos\left(\frac{1}{3}\right) = \frac{p\pi}{q}.$$

Since

$$1 + 2\sqrt{-2} = 3e^{i \arccos(1/3)} = 3e^{ip\pi/q}$$

we have

$$(1 + 2\sqrt{-2})^{2q} = 3^{2q}$$

contradicting the previous question. Therefore, $\arccos(1/3)/\pi \notin \mathbb{Q}$.

9. First note that, if $\Bbbk_1 = \mathbb{Q}(\sqrt{a})$ and $\Bbbk_2 = \mathbb{Q}(\sqrt{b})$, then by assumption on a and b we get $d_{\Bbbk_i} \equiv 1 \pmod 3$, so that 3 splits completely in $\Bbbk_1$ and in $\Bbbk_2$ by Proposition 7.108. This implies that 3 splits completely in $\mathbb{K}$. Now assume that $\mathcal{O}_{\mathbb{K}} = \mathbb{Z}[\theta]$ for some $\theta \in \mathbb{K}$. Then $\mathbb{K} = \mathbb{Q}(\theta)$ and the minimal polynomial μ of θ is of degree 4. By the previous observation, we infer that the reduction $\overline{\mu}$ in $\mathbb{F}_3[X]$ can be expressed as a product of four distinct monic linear polynomials, which is impossible since $\mathbb{F}_3$ has only three distinct elements.

10. We use Proposition 7.118.

▷ **Type I.** We have $\nu_{\mathbb{K}}(p^\alpha) = \mathcal{D}_{(1,1,1)}(\alpha) = \binom{\alpha+2}{2}$.
▷ **Type II.** In this case, p is inert so that by Remark 7.120 we get

$$\nu_{\mathbb{K}}(p^\alpha) = \begin{cases} 1, & \text{if } 3 \mid \alpha, \\ 0, & \text{otherwise.} \end{cases}$$

▷ **Type III.** We have $\nu_{\mathbb{K}}(p^\alpha) = \mathcal{D}_{(1,2)}(\alpha)$ which can be computed by using Theorem 2.32 or Popoviciu's result of Exercise 12 in Chap. 2. For instance, applying Popoviciu's theorem with $a = 1$, $b = 2$ and $n = \alpha$, we get $\overline{a} = 1$ and thus

$$\nu_{\mathbb{K}}(p^\alpha) = \frac{\alpha}{2} + 1 - \left\{\frac{\alpha}{2}\right\} = \begin{cases} (\alpha + 2)/2 & \text{if } \alpha \equiv 0 \pmod 2, \\ (\alpha + 1)/2, & \text{if } \alpha \equiv 1 \pmod 2. \end{cases}$$

▷ **Type IV.** We have $\nu_{\mathbb{K}}(p^\alpha) = \mathcal{D}_{(1,1)}(\alpha) = \binom{\alpha+1}{1} = \alpha + 1$.
▷ **Type V.** We have $g = 1$ and thus we may use Remark 7.120, and since $e = 3$, we get

$$\nu_{\mathbb{K}}(p^\alpha) = 1.$$

11.

(a) This is done in the proof of Proposition 7.138 using Corollary 7.130.

(b) We have $f_n(\sigma) = \sigma^{n+1}\pi^{-n\sigma/2}\Gamma(\sigma/2)^n$ and hence

$$g_n(\sigma) = \frac{2}{n} - (\log\pi + \gamma)\sigma + \sum_{k=1}^{\infty}\left(\frac{\sigma}{k} - \frac{\sigma}{k+\sigma/2}\right)$$

and thus

$$g_n''(\sigma) = 8\sum_{k=1}^{\infty}\frac{k}{(\sigma+2k)^3} > 0$$

so that g_n is convex on $]0, +\infty[$.

(c) For all $\sigma \in [1,2]$, we deduce that

$$g_n(\sigma) \leqslant \max\big(g_n(1),\, g_n(2)\big)$$

$$= \max\left(2 - \gamma - \log(4\pi) + \frac{2}{n},\, 2\left(1 - \gamma - \log\pi + \frac{1}{n}\right)\right)$$

and since $n \geqslant 2$ we obtain

$$g_n(\sigma) \leqslant \max\big(3 - \gamma - \log(4\pi),\, 3 - 2(\gamma + \log\pi)\big) < 0.$$

(d) The previous question implies that f_n is decreasing on $[1,2]$ so that

$$f_n(\sigma_0) \leqslant f_n(1) = 1.$$

Therefore

$$\kappa_{\mathbb{K}} \leqslant \frac{d_{\mathbb{K}}^{(\sigma_0-1)/2}}{(\sigma_0-1)^{n-1}} = \left(\frac{e\log d_{\mathbb{K}}}{2n-2}\right)^{n-1}$$

and (7.21) gives then (7.23).

12.   Since $\mathbb{K}$ is real, we have $(r_1, r_2) = (n, 0)$, $w_{\mathbb{K}} = 2$ and every character of $X(\mathbb{K}) = X$ is even, so that the class number formula seen in Remark 7.164 can be written in this case as

$$h_{\mathbb{K}}\mathcal{R}_{\mathbb{K}} = \frac{d_{\mathbb{K}}^{1/2}}{2^{n-1}}\prod_{\substack{\chi\in X \\ \chi\neq\chi_0}} L(1, \chi^\star)$$

where $\chi^\star$ is the primitive even Dirichlet character that induces χ. Now using (7.37) and the arithmetic-geometric mean inequality, we get

$$\prod_{\substack{\chi\in X \\ \chi\neq\chi_0}} |L(1, \chi^\star)| \leqslant \left(\frac{1}{2n-2}\sum_{\substack{\chi\in X \\ \chi\neq\chi_0}}\log f_{\chi^\star}\right)^{n-1}$$

and the conductor–discriminant formula (Theorem 7.162) implies that

$$\prod_{\substack{\chi \in X \\ \chi \neq \chi_0}} |L(1, \chi^\star)| \leq \left(\frac{\log d_{\mathbb{K}}}{2n - 2} \right)^{n-1}$$

as required.

13.   We proceed as in Example 7.175. We have $15 \equiv 3 \pmod 4$, the order $\mathcal{O} = \mathbb{Z}[\sqrt{-15}]$ has discriminant -60 and conductor $f = \sqrt{60/15} = 2$. By Lemma 7.172 (i), the class number of $\mathcal{O}$ is given by

$$h_{\mathcal{O}} = 2 \times 2 \times \left(1 - \frac{1}{2} \left(\frac{-15}{2} \right) \right) = 2$$

and, by Lemma 7.172 (ii), representatives of the two classes of invertible fractional ideals of $\mathcal{O}$ are

$$(1, \sqrt{-15}) \quad \text{and} \quad (3, \sqrt{-15})$$

and hence using PARI/GP we obtain

$$H_{\mathcal{O}} = X^2 - \left(3^3 \times 5^3 \times 10\,968\,319 \right) X + \left(3^2 \times 5 \times 29 \times 41 \right)^3.$$

By Corollary 7.174, we infer that a prime $p \geq 7$ can be expressed in the form $p = x^2 + 15y^2$ if and only if $(-15/p) = 1$ and the equation

$$x^2 - \left(3^3 \times 5^3 \times 10\,968\,319 \right) x + \left(3^2 \times 5 \times 29 \times 41 \right)^3 \equiv 0 \pmod p$$

has a solution in $\mathbb{Z}$.

References

[AS82]   Adams W, Shanks D (1982) Strong primality tests that are not sufficient. Math Comput 39:255–300

[Bor06]   Bordellès O (2006) An inequality for the class number. JIPAM J Inequal Pure Appl Math 7:87

[Coh00]   Cohen H (2000) Advanced topics in computational algebraic number theory. GTM, vol 193. Springer

[Col87]   Coleman R (1987) On the Galois groups of the exponential Taylor polynomials. Enseign Math 33:183–189

[GK91]   Graham SW, Kolesnik G (1991) Van der Corput's method of exponential sums. London math. soc. lect. note series, vol 126. Cambridge University Press, Cambridge

[Gou72]   Gould HW (1972) Combinatorial identities. A standardized set of tables listing 500 binomial coefficients summations. Morgantown, West Virginia

[HT88]   Hall RR, Tenenbaum G (1988) Divisors. Cambridge University Press, Cambridge

[HW38]   Hardy GH, Wright EM (1938) An introduction to the theory of numbers. Oxford, London

[Rn62] Rosser JB, Schœnfeld L (1962) Approximate formulas for some functions of prime numbers. Ill J Math 6:64–94

[Tri02] Trifonov O (2002) Lattice points close to a smooth curve and squarefull numbers in short intervals. J Lond Math Soc 65:309–319

Index

A

Abel summation formula, 9, 16
Abscissa of absolute convergence, 198, 201, 210
Abscissa of convergence, 201, 205, 207
Absolute convergence, 198–203, 210, 431
Absolute value, 204, 462, 463
Additive characters, 155, 156
Additive function, vi, 226, 241, 506
Admissible modulus, 457–460, 462
Algebraic closure, 374, 375, 465, 541
Algebraic integers, 355, 379–381, 391, 400, 405, 407, 446, 448, 466
Algebraic number fields, v, 167, 355, 377–381, 383–385, 390, 393, 395–399, 402, 405, 406, 408, 411, 415, 417, 423–427, 431, 434–436, 445, 452, 455, 462, 464, 474, 476
Algebraic numbers, 356, 374–379, 381, 401, 436
Algebraic rank of an elliptic curve, 443
Alladi, 246
Alternating group, 446
Analytic class number formula, 432, 440
Analytic rank of an elliptic curve, 443
Approximate functional equation, 98, 124, 125, 432, 434
Archimedean, 463
Arithmetic large sieve, 238
Artin, 73, 75
Artin map, 457, 458, 467
Artin reciprocity law, 457, 458
Artin symbol, 457
Associate, 361, 362
Automorphic functions, 455, 456
Average order, vi, 177, 186, 192, 193, 213, 216, 226, 297, 340, 435

B

Bachet–Bézout's theorem, 27, 28, 31, 42, 54, 426
Bachet's Diophantine equation, 426
Baker, A, 397, 441, 442, 480
Baker, R. C, 351
Barban, 221
Basis, 365–367, 383–391, 395, 396, 404, 412, 418, 421, 423, 425, 427, 428, 478, 543
Berkane, 349, 351
Bernoulli functions, 20
Bernoulli numbers, 19, 20
Bernoulli polynomial, 19, 21
Bertrand's postulate, 83, 84, 477
Bézout's coefficients, 28, 31
Biquadratic fields, 397, 478
Birch, 443
Birch & Swinnerton-Dyer conjecture, 443
Bonferroni's inequalities, 122, 227
Bordellès, 161, 162, 247, 349, 351, 480, 546
Branton, 275, 295
Brauer–Siegel theorem, 443, 445
Brun, 66, 121–123, 161, 227
Brun–Titchmarsh inequality, 141, 233, 239
Brun's pure sieve, 122
Burgess, 157, 162

C

Capitulation property, 464
Carmichael number, 67
Cauchy, 22, 48, 128, 137, 202, 208, 356, 389, 493
Cauchy–Binet's identity, 309
Cauchy–Schwarz's inequality, 237, 308–310, 348, 533
Characteristic, xiii, 107, 108, 120, 155, 166, 357

Characteristic polynomial, 382, 400, 509
Chebotarëv's density theorem, 75, 450
Chebyshev's estimates, xiv, 81, 84, 343, 344
Chen, 234, 351
Chevalley, 432
Chinese remainder theorem, 39, 122, 359, 410,
    416, 419
Class field theory, v, 453, 455–460, 462, 464,
    467
Class group, 406, 422, 423, 428, 464, 467
Class number, v, 117, 423, 424, 426, 428,
    432, 434–436, 440–442, 444, 464,
    466–468, 470, 473, 479, 545,
    546
Class number formula, 432, 440, 444, 460,
    462, 479, 545
CM-fields, 474
Cohen, 480, 546
Complementary laws, 420
Completely additive function, 168, 506
Completely multiplicative function, vi, xviii,
    xix, 122, 167–169, 172, 173, 177, 186,
    187, 190, 193, 209, 210, 212–222,
    224–227, 230, 232, 240, 241, 245, 246,
    248, 347, 413, 429, 431, 504, 505,
    529
Completely split, 421
Complex character, 114
Composite number, 67
Conditional convergence, 201, 202, 204
Conductor, 116, 219, 397, 422, 428, 442, 459,
    460, 467, 469, 479, 546
Conductor-Discriminant formula, 460, 546
Conductor-Ramification theorem, 459
Congruences, 35, 36, 39, 67, 68, 71, 72, 122,
    176, 177, 224, 359, 396, 456, 497, 543
Conjugate class, 470
Conjugate field, 378, 379
Conjugates, 378, 382, 403, 448, 463
Conrey, 149, 161
Coprime, 27, 28, 39–41, 44, 52, 68, 82, 105,
    106, 108, 109, 138, 155, 160, 168, 169,
    222, 224, 233, 241, 359, 360, 379, 383,
    394, 397, 410, 411, 426, 427
Core, 166, 363
Critical strip, 98, 102, 103, 321
Cusick, 435, 475, 480
Cycle, 398, 446, 447, 457, 542
Cyclic cubic field, 397
Cyclic group, 42, 366, 397, 400, 401, 428, 446,
    447
Cyclic number field, 379, 454, 475
Cyclotomic fields, 355, 391, 393, 396, 397,
    422, 453, 455

Cyclotomic polynomial, 106, 370, 374, 392
Cyclotomic reciprocity law, 458

D
Davenport, 234
De La Vallée Poussin, 85, 102, 103, 111, 126,
    142
Decomposition number, 416, 460
Decomposition theorem, 459
Dedekind, vii, 381, 390, 415, 441, 446
Dedekind domain, 404, 405, 408
Dedekind function, 166
Dedekind zeta-function, vii, 219, 431–433,
    439, 470, 471
Defining polynomial, 378, 414, 425, 446, 463,
    468
Degree of an algebraic number field, 219, 220,
    382–384, 386, 389, 396, 398, 400, 401,
    404, 410–412, 414–418, 422, 423, 428,
    432, 435, 436, 438, 444, 445, 451, 462,
    470, 474, 477, 479
Degree of an element, 367, 375, 377
Density hypothesis, 132, 133
Denumerant, 43, 49, 189, 429
Deuring, 442, 470
Deuring-Heilbronn phenomenon, 433, 442
Diaz y Diaz, 381, 382
Different, 57, 112, 206, 252, 268, 330, 441,
    442, 470, 476, 479
Digamma function, 436
Dihedral group, 446
Dirichlet character, 108–112, 114–116, 125,
    133, 138–141, 155–157, 167, 169, 175,
    178, 179, 185, 186, 203, 420, 421, 439,
    441, 459, 460, 462, 479, 545
Dirichlet class number formula, 440, 460
Dirichlet convolution product, 171, 176, 195,
    197
Dirichlet divisor problem, vi, 151, 184, 185,
    243, 298, 304, 307, 314, 325, 327, 328,
    334
Dirichlet hyperbola principle, 183, 185
Dirichlet L-function, 141, 433, 439, 442
Dirichlet pigeon-hole principle, 136, 278, 424,
    486
Dirichlet series, vi, 95, 102, 111, 125–127,
    138, 140, 150, 182, 196–201, 203–207,
    209–212, 217, 219, 345, 431, 432, 471
Dirichlet–Piltz divisor function, 165
Dirichlet's approximation theorem, 330
Dirichlet's theorem, vi, xviii, xix, 66, 74, 92,
    93, 95, 102, 105, 107–112, 114–116,
    118, 125–127, 133, 136, 138–141, 150,
    151, 155-157, 160, 162, 165, 167, 169,

171, 173, 175, 176, 178, 179, 18–186,
195–201, 203–207, 209–212, 217, 219,
243, 247, 248, 278, 295, 298, 299, 303,
304, 307, 308, 314, 324, 325, 327, 328,
330, 334, 345, 349, 351, 352, 399–403,
420, 421, 424, 429, 431–433, 439–442,
450, 453, 459, 460, 462, 471, 474, 479,
486, 543, 545
Dirichlet's unit theorem, 399–401, 543
Discrete Hardy–Littlewood method, v, 298,
328
Discriminant of a polynomial, 374
Discriminant of an algebraic number field,
220, 386, 438, 442, 445
Divided differences, 6, 262, 264, 275, 281,
283, 289
Double large sieve inequality, 332
Dumas, 374
Duplication formula, 94, 219
Dusart, 144, 145, 162

E
ED, 363
Eigenvalue, 152, 153, 407, 509, 510
Eisenstein's criterion, 369
Elliptic curve, 442, 443, 455
Embedding, 378, 400, 401, 414
Erdős, 64, 66, 75, 85, 162, 215, 216, 224, 246,
247, 399, 480
Euclid, vi, 29, 57, 58, 64–66, 105, 107
Euclidean algorithm, 29–31, 51, 160
Euclidean division, 1, 2, 22, 23, 28, 29, 36, 37,
62, 71, 152, 363, 382, 389, 483, 501,
508
Euler, 12, 20, 64, 65, 92, 94, 96, 105, 107, 197,
440, 490
Euler product, 93, 117, 197, 431, 434, 437,
439, 442
Euler summation formula, 19, 20
Euler–MacLaurin summation formula, 19, 21
Euler–Mascheroni constant, 12, 135, 144
Euler's totient function, 40, 42, 173, 340, 392
Explicit formula, 41, 133, 136, 139, 142, 208,
439
Exponent pair conjecture, 324
Exponent pairs, 321, 323–325, 328, 333, 335,
351, 535, 536
Exponential sums, vi, xx, 124, 129, 146, 154,
156, 162, 163, 248, 295, 297, 298,
300–302, 304, 306, 308, 310, 312, 314,
316, 318, 320, 322, 324–330, 332, 334,
336, 338, 340, 342, 344, 346, 348,
350–353, 546
Extended Riemann hypothesis, 75, 439

F
Fermat, 67
Fermat equation, v, 355, 356
Fermat numbers, 490
Fermat's last theorem, v, 53, 356, 443
Fermat's little theorem, 37, 67, 69–72, 77, 457,
497
Fermat's little theorem for integer matrices,
242, 509, 510
Field, v, vi, xi, 75, 117, 355, 357, 363, 364,
367, 369, 370, 374, 375, 378, 379, 389,
391, 393, 395, 397, 403–406, 413, 416,
421, 422, 434, 442, 454, 457, 458,
462–464, 474–476, 479
Field extensions, 367, 374, 377
Filaseta, vi, 250, 275, 281–283,286, 288, 289,
295, 374
Finite places, 463, 464
Finitely generated, 358, 362, 365, 366, 379,
380, 384, 399, 402, 405, 407, 423,
443
First Bernoulli's function, xii
First Chebyshev function, 79
First derivative test, 256, 257
First derivative test for integrals, 317
First Mertens's theorem, 87
Fogels, 250, 295
Ford, 141, 143, 162, 351
Fractional ideal, 405–410, 424, 427, 464,
467–470,l 546
Fractional part, xii, 4, 54
Free abelian group, 366, 402
Friedman, 435, 474, 476, 477, 480
Frobenius, 43, 447, 450, 457
Fundamental units, 399, 400, 402, 428, 440,
453, 475
Furtwängler, 464

G
Gallagher, 24, 236, 247
Galois, 378, 379, 389, 393, 397, 430, 431, 445,
465, 466
Galois group, 378, 379, 393, 397, 398,
446–448, 453, 464, 478
Galois number field, 378
Gamma function, 94, 95
Gauss, v, 27, 35, 434, 441, 456, 473, 480
Gauss circle problem, 328
Gauss sums, 155–157, 330, 331, 422
Gauss's class number one problem, 442
Gauss's lemma, 368, 379, 383, 392
Gauss's theorem, 31–37, 59, 63, 67, 76, 77, 82,
224, 393
Gautschi's inequality, 438

Gelfond, 442
General divisor problem, 299
Generalized Riemann hypothesis, 439,
    442
Generating functions, 43, 47, 48
Geometry of numbers, 401, 424, 436, 453,
    470, 474
Golden ratio, 29, 30, 474
Goldfeld, 442, 443, 480
Gorny's inequality, 264
Graham, 250, 295, 325, 352, 546
Grekos, 325, 352
Gross, 442, 443, 470
Guinand, 439

H

H-functions, 48, 49
Hadamard, 85, 102, 103, 111, 126, 264, 266
Hadamard factorization theorem, 134, 436,
    437
Halberstam, vi, 225, 234, 248, 351, 352
Hall, 225, 245, 248, 546
Hardy, 125, 147, 148, 154, 162, 298, 299, 352,
    546
Hardy function, 147
Hardy–Littlewood's circle method, 328
Harman, 347, 351, 352
Harmonic number, 11, 14, 18, 151
Hasse, 299, 352, 442, 481
Hasse–Weil L-function, 442
Heath-Brown, 74, 154, 162, 346, 347, 352
Hecke, 432, 441, 445
Heegner, 442, 481
Heilbronn, 442
Hilbert, vii, 146, 453, 456, 464
Homomorphism theorems, 358
Hooley, 75, 158, 162, 224, 352
Hooley divisor function, 166, 169, 342
Hurwitz constant, 423, 424
Huxley, v, vi, 102, 133, 154, 162, 264, 277,
    278, 291, 295, 298, 326, 333, 334,
    352, 512

I

Ideal, 357–359, 362–364, 367, 403–415,
    423–427, 429, 435, 436, 464, 470, 472,
    473, 476
Ideal group, 407
Ideal theorem, 220, 451, 452
Inclusion-exclusion principle, 118, 176
Index, xiii, 259, 356, 381, 383, 386, 404,
    415, 416, 467, 468, 478, 498, 501,
    540
Index form, 396, 397

Inert, 417, 421, 430, 544
Inertial degree, 416, 417, 429, 430, 447, 459,
    460
Infinite places, 463
Integer points, 52, 53, 184, 249, 251, 256, 278,
    281, 290, 291, 334, 489
Integral basis, 382–391, 395–397, 404, 412,
    418, 421, 423, 425, 427, 428, 478, 543
Integral part, xi, 78, 297
Integrally closed, 404, 405
Irreducible, 159, 178, 224, 355, 361, 362,
    368–374, 376, 378, 381, 383, 389, 390,
    392, 393, 398, 402, 418, 419, 446, 447,
    449, 450, 457, 477, 478, 499, 537–539,
    542
Irreducible polynomial, 177, 178, 369, 370,
    372, 375, 398, 415, 418, 446, 447, 450
Ivić, 125, 126, 132, 147, 149, 154, 162, 166,
    248, 299, 352
Iwaniec, v, 123, 124, 162, 213, 217, 219, 220,
    247, 248, 328, 351, 352, 481

J

j-function, 456
Jarnik, 290
Jordan, 447
Jordan totient function, 166
Jordan–Hölder theorem, 57
Jordan's lemma, 493

K

$\mathbb{K}$-basis, 367
k-free number, 59, 60, 62, 166, 170
k-full number, 59, 60, 63, 166
Karacuba, 339, 352
Klein 4-group, 447
Kloosterman sums, 157
Kolesnik, 250, 295, 325, 328, 352, 546
Korobov, 129
Kronecker, vii, 203, 387, 402, 436, 453, 455,
    456
Kronecker Jurgendtraum, 456
Kronecker symbol, 175, 421, 428, 439
Kronecker–Weber theorem, 397, 422, 436,
    453, 462, 464
kth derivative test, 262
Kummer, v, vii, 356, 408, 415
Kusmin–Landau's inequality, 301, 302, 304,
    316, 323, 532

L

Lagarias, 14, 151
Lagrange polynomial, 7, 263, 267, 273
Lagrange's theorem, 67, 69, 71, 356, 413
Lambert series, 117

Lamé, 29, 356
Landau, xiv, 129, 137, 140, 142, 206, 220, 248, 266, 301, 433, 441, 451, 481
Landau–Hadamard–Kolmogorov inequalities, 264, 266
Large sieve, vi, 24, 234, 236, 237, 332
lattice, 251, 262, 401, 402
Legendre, 78
Legendre symbol, 420, 421
Legendre–Jacobi–Kronecker symbol, 420
Lindelöf hypothesis, 102, 133, 151, 298, 299
Linfoot, 442
Linnik, 234, 238, 248, 442
Liouville function, 165
Littlewood, 125, 150, 154, 299
Logarithmic integral, 24
Logarithmic representation, 401
Logarithmic space, 401
Long sum, 222
Louboutin, 435, 445, 479, 481

M

Majors arcs, 267, 268, 271–274, 329, 332
Maximal ideal, 375, 403, 404, 406, 419
Mean-value theorem, 6, 8, 256-258, 260-262, 305, 306, 532
Mertens conjecture, 150, 151
Mertens constant, 89, 144, 227
Mertens function, 150, 152, 247
Minimal polynomial, 375–380, 382, 383, 386, 389, 400, 401, 417, 418, 466, 477, 509, 537, 539, 544
Minkowski, vii, 401, 436, 470
Minkowski bound, 424, 435, 436
Minkowski constant, 425, 436, 470
Minors arcs, 329, 332, 333
Möbius function, 74, 120, 121, 150, 165, 169, 176, 182, 200, 214, 343, 431
Möbius inversion formula, vi, 73, 167, 176–178, 215, 228, 229, 392, 508
Module, 363–365, 367
Monogenic, 390, 393, 396–398, 478
Monomial function, 322
Montgomery, 25, 163, 239, 248, 352
Mordell, 352, 442, 443
Mozzochi, v, 328, 352
Mulholland, 470
Multiplicative characters, 155, 156, 330
Multiplicative function, vi, 167–169, 172, 173, 177, 186, 187, 190, 193, 209, 210, 212–222, 224, 226, 227, 230, 232, 240, 241, 245, 246, 429, 431, 504, 505, 529

Multiplicative order, 70

N

Nair, 86, 163, 224, 248
Newton's formula, 46, 87, 505
Nœtherian module, 364, 365
Nœtherian ring, 362, 364, 365, 405, 406, 409
Non-archimedean, 463
Non-trivial zeros, 103, 132, 134, 142, 143, 146, 147, 210, 434, 437, 439
Norm map, 361, 362, 413
Norm of an element, 542
Norm of an ideal, 411
Normal closure, 379
Normal number field, 378
nth power residue, 76, 77
nth root of unity, 355, 392, 393
Number ring, 406

O

Odlyzko, 55, 151, 163, 439, 481
Order, xiv, 19, 57, 72, 73, 93, 102, 111, 112, 115, 134, 138, 157, 167, 170, 171, 176, 177, 180, 297, 312, 323, 324, 329, 336, 338, 389, 392, 395, 398–400, 408, 409, 411, 423, 428, 431, 434, 437, 443, 446–448, 451, 454, 459, 461, 467–469, 471, 542
Ore's criterion, 538
Ostrowski, 317, 322, 323

P

p_i-adic valuation, 409
Padé approximants, 289
Page, 140
PARI, 158, 414, 427, 451, 469, 507, 543, 546
Parseval's identity, 235, 237
Partial summation, 8–10, 13, 23, 46, 89, 113, 114, 123, 124, 132, 143, 150, 199, 203, 215, 225, 245, 330, 511, 515, 516
Perrin sequences, 241, 243
Perron, 374
Perron summation formula, 126
Pétermann, 339, 352
Phillips, vi, 321
Phragmén–Lindelöf principle, 101, 206, 433
Piatetski-Shapiro, 145, 146
PID, 358, 363, 365, 375, 404, 411, 415, 424, 426
Pila, 290, 291
Pillai function, 218, 243
Places, 463, 464
Pochhammer's symbol, 322
Pohst, 474, 481

Poincaré half-plane, 455
Poisson summation formula, vi, 318, 325, 331
Poitou, 439, 481
Pólya–Vinogradov inequality, 156, 157, 185, 186
Popoviciu's theorem, 544
Power basis, 390
Prime, xiii, 27, 42, 43, 57, 58, 64, 66–71, 73, 74, 76, 79, 82, 105, 106, 118, 122, 157, 159, 344, 356, 361, 362, 369–372, 376, 389, 391, 403–406, 410, 420, 440, 454, 456, 459–461, 463, 465, 466, 468–470, 477, 486, 490, 496, 499–502, 538, 539, 546
Prime counting function, 79, 451
Prime ideal, 375, 403–405, 407–409, 411, 413–415, 417–421, 425, 427, 463–465, 471, 477
Prime Ideal Theorem, 451, 452
Prime number, vi, xi, 9, 24, 42, 51, 57–59, 62, 64, 65, 67–72, 75, 76, 78, 84, 85, 87, 105, 106, 118, 122, 144, 146, 158–161, 177, 212, 243, 343, 357, 369–372, 374, 376, 381, 382, 387, 392, 393, 403, 414–418, 420, 421, 425, 426, 429, 440, 441, 446, 447, 456–459, 465, 466, 470, 471, 477, 479, 498
Prime Number Theorem, xiv, 78, 81, 102, 104, 111, 126, 129, 138, 145, 149
Prime Number Theorem for arithmetic Progressions, 111, 138
Primitive character, 116, 117, 141, 156, 462
Primitive nth root of unity, 355, 392, 393
Primitive roots, 72, 73, 76, 77, 177, 454
Principal ideal, 358, 403, 406, 422, 464, 467, 476, 478
Product formula for characters, 460
Proper major arc, 268, 272
Pure cubic fields, 394, 395

Q
Q-basis, 377–379, 383–387, 401
Quadratic character, 111, 117, 139–141, 175, 186
Quadratic fields, 112, 116, 175, 186, 211, 391, 393, 396, 420–422, 426, 434, 438–441, 445, 456, 460, 464, 467, 475
Quadratic reciprocity law, 420, 421, 439, 456, 469
Quadratic residue, 76, 77, 106, 371, 501
Quotient module, 364
Quotient ring, 358

R
Ram Murty, 74, 106
Ramachandra, 343, 352
Ramaré, 135, 144, 163, 349, 351, 352, 479, 481
Ramification index, 416, 417, 463, 471
Ramified, 416, 417, 420, 421, 436, 454, 459, 467, 471
Rankin's trick, 191
Real character, 111, 115
Real quadratic fields, 399, 434, 441, 466
Realizable sequences, 241, 242, 509
Redheffer, 151, 163, 247
Reduction modulo p, 370
Reduction principle, 259, 275, 453
Regulator, v, 399, 400, 402, 428, 432, 435, 444, 473, 474, 476, 478, 479
Remak, 473, 474, 481
Renyi, 234
Residue class degree, 416
Resolvent, 448, 449, 453
Richert, vi, 225, 248, 250, 295
Riemann, 92, 93, 96, 132, 146, 147
Riemann hypothesis, 14, 75, 133, 146, 147, 149–153, 157, 186, 210, 214, 298, 343, 439, 442, 456
Riemann zeta-function, vi, 92, 93, 96, 98, 103, 104, 132, 142, 147, 153, 207, 345, 351, 431, 432
Riemann–Siegel formula, 147
Riemann–Siegel function, 147
Riemann–Stieltjes integral, 10, 14–16
Riemann–von Mangoldt formula, 132
Ring, 41, 42, 108, 195, 196, 355–361, 363–365, 368, 369, 375, 379–381, 383, 386, 398–401, 403–406, 410, 411, 422, 426, 428
Rivat, 146, 163, 353
Robin, 151
Rogers, 470
Rolle, 6
Rosser, 143, 144, 163, 547
Rosser–Iwaniec sieve, 345
Roth, vi, 234, 250, 295
Rouché's theorem, 373

S
Saddle point method, 48
Sargos, vi, ix, 146, 163, 248, 264, 275–277, 295, 353
Schmidt, 290, 351
Schœnfeld, 143, 144, 163, 547
Schoof, 475, 482
Schur, 106, 477, 478, 540, 542

Second Chebyshev function, 79
Second derivative test, 259, 260
Second derivative test for integrals, 305
Second Mertens's theorem, 194, 225
Selberg, 95, 149, 227, 228, 231, 235, 245
Selberg's sieve, vi, 227, 233
Serre, v, 439
Serret's algorithm, 160
Shiu, 55, 222, 248, 295
Shiu's theorem, 222–224, 249, 342
Short sums, 222, 224
Siegel, 140, 147
Siegel's zero, 239, 442
Sieve of Eratosthenes, 118, 121, 123, 238, 347
Sieves methods, 120, 224, 227, 234, 237
Silverman, 435, 474, 482
Simple groups, 57, 448
Singular value, 153
Singularity, 206, 207
Skoruppa, 476
Sokolovskiĭ, 433, 480, 482
Sophie Germain's identity, 494
Soundararajan, 150, 163
Speiser, 453
Square-full number, 60, 62, 255, 292
Square-full number problem, 255, 292, 351
Squarefree kernel, 166
Squarefree number, 74, 213, 220, 249, 250,
        255, 262, 275, 279, 281, 351, 459, 465,
        466, 475, 478, 516
Squarefree number problem, 253, 255, 264,
        274, 277, 280
Srinivasan's optimization lemma, 256
Stabilizer, 448
Stark, 436, 441, 442, 445, 470, 482
Stationary phase, vi, 220, 318, 319
Stirling's estimate, 3
Stirling's formula, 10, 49, 94
Strongly additive function, 168
Strongly multiplicative function, 122
Sub-additive function, 168
Sub-multiplicative function, 168, 225, 226
Sub-R-module, 364, 365
Super-additive function, 168
Super-multiplicative function, 168
Swinnerton-Dyer, vi, 278, 290, 443
Symmetric group, 446

T
Taylor–Lagrange formula, 262, 270
Te Riele, 151
Tenenbaum, 163, 224, 225, 245, 248, 546
Theorem of the primitive element, 378
Theta function, 94

Thue's lemma, 51, 160, 501
Titchmarsh, xiv, 163, 248
Tong, 298, 353
Trace, 382, 383, 391
Trifonov, vi, 250, 275, 277, 279, 281–283,
        286, 288, 289, 295, 547
Trivial zeros, 98, 133, 136, 434
Truncated Poisson summation formula, 318,
        331
Twin primes, 66, 122, 123, 238, 499

U
Uchida, 435, 474, 482
UFD, 196, 356, 363, 369, 410, 411, 424, 442
Unitary commutative ring, 195, 357
Unitary convolution product, 196
Unitary divisors, 161, 502, 503
Units, 41, 42, 108, 355, 357, 361, 362, 399,
        400, 426–428, 537, 543
Unramified extension at all places, 463
Unramified extension outside ∞, 463
Unrestricted partitions, 171, 195

V
Vaaler, 246, 299, 353
Vaaler's theorem, 299, 301
Van der Corput, vi, 301, 304, 305, 308, 310,
        317, 318, 321
Van der Corput's A-process, vi, 310, 312, 321,
        324, 325, 333, 348
Van der Corput's B-process, vi, 319–321, 324,
        325, 333
Van der Corput's inequality, 304, 315
Vaughan, 25, 152, 163, 239, 248, 351, 352
Vinogradov, A. I, 129
Vinogradov, I. M, 129, 336, 353
Vinogradov integral, 337
Vinogradov's mean-value theorem, 332, 338
Vinogradov's method, 129, 334, 337, 338
Von Mangoldt function, 79–81, 167, 169, 212,
        213
Voronoï, 163, 220, 248, 308, 327, 349, 353,
        395, 396
Voronoï summation formula, 126

W
Walfisz, 339, 340, 353
Washington, 163, 475, 482
Watt, v, 333, 352
Weber, 453
Weierstrass equation, 442
Weierstrass's double series theorem, 204
Weil, v, 142, 157, 163
Weyl, vi

Weyl's shift, 308–311, 336
Wiles, v, 53, 356, 442
Wilson's theorem, 68
Wirsing conditions, 187, 188, 193, 216, 222, 226, 518
Wu, 248, 339, 352, 353

Z
$\mathbb{Z}$-basis, 366, 383, 386, 389, 402, 407, 412, 418
Zagier, 442, 470
Zero-free region, vii, 102, 104, 129, 139, 142, 143, 433, 451
Zimmert, 470, 471, 473, 476, 482